NATURAL PHILOSOPHY

III

Astronomy. History of Astronomy.
Mathematical Geography.
Physical Geography, and Navigation

With
an Explanation of Scientific Terms,
and
an Index

Elibron Classics
www.elibron.com

Elibron Classics series.

ISBN 0-543-94987-7 (paperback)
ISBN 0-543-94986-9 (hardcover)

This Elibron Classics Replica Edition is an unabridged facsimile of the edition published in 1834 by Baldwin and Cradock, London.

LIBRARY

OF

USEFUL KNOWLEDGE.

NATURAL PHILOSOPHY.

III.

ASTRONOMY.
HISTORY OF ASTRONOMY.
MATHEMATICAL GEOGRAPHY.

PHYSICAL GEOGRAPHY,
AND
NAVIGATION.

WITH

AN EXPLANATION OF SCIENTIFIC TERMS,

AND

AN INDEX.

LONDON:
BALDWIN AND CRADOCK, PATERNOSTER-ROW.

MDCCCXXXIV.

LONDON:
Printed by WILLIAM CLOWES,
Duke-street, Lambeth.

COMMITTEE.

LOCAL COMMITTEES.

ADVERTISEMENT.

In presenting to the public this third and concluding volume of *Natural Philosophy*, it seems requisite to offer a few remarks concerning the plan adopted for the arrangement of the originally disjointed portions. To those who would have wished that it had been more systematic, it may be observed that the several Treatises were composed so that the subjects might, as much as possible, be individually perfect, and independent of one another; and this necessarily tended, not only to occasion repetitions when collected into volumes, but, in some degree, to break the thread which would have united them in one general science.

The object of *Natural Philosophy* is the investigation of those principles which are to be considered as inherent in *matter*, and by the agency of which the changes in the relative positions, modifications, and internal combinations of the several masses (or separate bodies) are produced. These are,—

I. The Effects of Force or Impulse, and its modes of Propagation.
II. The Effects of the Pressure and Motion of non-elastic Fluids.
III. The Effects of Air, and similar elastic Fluids.
IV. The Effects of Caloric, or the principle of Heat.
V. The Phenomena of Light and Colours:
VI. The Phenomena and Effects of the Electric and Magnetic Fluids.
VII. The Effects of the principle of Gravitation.

The First of the preceding divisions is investigated in the three *Treatises* on *Mechanics*, in Vol. I.; and in the *Introduction to Mechanics*, Vol. II.

The Second is treated of under the titles *Hydrostatics* and *Hydraulics*, Vol. I., and the *Introduction to Hydrostatics*, Vol. II.

The Third is treated of under the titles *Pneumatics*, Vol. I., and *Introduction to Pneumatics*, Vol. II.

The Fourth is treated of under the titles *Heat*, Vol. I., and *Thermometer* and *Pyrometer*, Vol. II.

The Fifth is copiously treated under the several heads of *Optics*, *Double Refraction*, and *Polarization of Light*, Vol. I.; and of *Introduction to Optics*, *Sir Isaac Newton's Optics*, and *Optical Instruments*, Vol. II.

The Sixth, with an account of the latest discoveries, will be found under *Electricity*, *Galvanism*, *Magnetism* and *Electro-magnetism*, Vol. II. And

The Seventh is the *Astronomy* of the present volume, an *Introduction* to which is prefixed to Vol. II.

To the *Treatises on Astronomy*, and the *History* of that science, it was judged proper to append *Mathematical Geography*. The Treatise on *Physical Geography* is less connected with the principal portion of the volume, but naturally follows that division of Geography which is termed *Mathematical*. *Navigation*, as far as it is a science, is wholly dependent on *Astronomy*, and as such is not considered to be out of place.

The discoveries of modern *Chemistry* have raised it from an art to the rank of a science; and there would have been no impropriety in including it in the general system of which we now speak. In such an arrangement it would have stood thus:—

VIII. The Effects of Corpuscular Attraction on the Combination and Decomposition of Bodies;

and the Treatises on the subject would have formed a Fourth Volume of Natural Philosophy. It has, however, been judged advisable to follow the usual practice; and, therefore, the several Treatises (of which *four* have already appeared) will, when completed, be collected and published in a separate volume, under the title of Chemistry.

October, 1834.

CONTENTS.

ERRATA.

ASTRONOMY.

Page 5, column 2, line 17 from the bottom, *dele* 'each of.'
10, 2, 8, *for* 'of motion,' *read* 'of his motion.'
19, 1, 16 from the bottom, *for* ES, Es, *read* ES_1, Es_1.
13 from the bottom, *for* Es_2, *read* Es_1.
2, 14, for $s_1 s_2 s^3$, read $s_1 s_2 s_3$.
20, 2, 33, for t_1, read t.
34, for $s_1 t_1$, read $s_1 t$.
26, 1, 41 and 44, for $t' e' r'$, read $t_1' e' r_1'$.
28, 2, 4 in the note, *for* S, *read* S_1.
12 in the note, *for* ES^1, *read* ES_1.
29, 2, 45, for t_1 e' r_1', read t_1' e' r_1'.
32, 2, last, *for* S, *read* S_1.
33, 2, add to the first note, 'At the points Z and Y therefore these circles are for a short space parallel to each other, or the sun has no perceptible motion Northward or Southward: at the point ♈ his course makes an angle = ZY with the equator, which is necessarily its greatest inclination to it: or his motion Northward is then most rapid: the conclusions referred to in the note in page 23, col. i.'
34, 2, 13, in the note, *for* Ps_1 Pt, *read* Ps, Pt.
39, 1, 31 and 32, *for* 'a motion from left to right in those already referred to,' *read* 'from left to right where the North pole is above the horizon.'
47, 1, 35, *for* 'on its' *read* 'on in its.'
61, 2, 10, *for* '110^8' *read* '110^3'; and *for* '133100' *read* '1331000.'
64, 1, 35, *for* 'from' *read* 'for.'
81, 1, top line to be transferred to the foot of column 2.
111, figure 31 reversed.
231, 1, 12, prefix VI. to mark the section.
253, 1, *for* 'CHAPTER XIII,' *read* 'CHAPTER XII.'

HISTORY OF ASTRONOMY.

96, 1, 10, *for* 'being carried' *read* 'carrying.'

ASTRONOMY.

Introductory Observations.

In treating of any science which is grounded upon physical facts and appearances, two courses are generally open. We may begin with a statement of the results observed, and by gradual investigation extricate from them the principles on which they depend: or else, if these principles have been ascertained, we may begin by stating them, and may deduce from them the consequences which would follow on the supposition of their truth; and finally, by comparing these consequences with the appearances presented by Nature, and finding them to correspond, we may satisfy ourselves of the truth of those principles which we originally assumed. The former is necessarily the course of discovery; the latter is often the most concise and convenient method of instruction after the discovery has been made. In some cases there is little practical distinction between the two methods; for instance, the fundamental principle of Hydrostatics is the equal pressure of fluids in all directions; and the fact that they do press so is one of the first and most obvious results of observation and experiment; and from the time that it is ascertained, the experimental and hypothetical mode of discussing the subject may very nearly coincide. In other sciences, on the contrary, the first effect of observation is to lead us to conclusions very distant from the fundamental principles which we finally adopt, or even at variance with them. In these cases the simpler and shorter mode of instruction will generally be the second which we have mentioned. Among these sciences Astronomy is eminently distinguished; for almost all the immediate results of observation are contrary to its true principles, and it is not without much labour and reasoning that the truth can be extricated from the mass of error in which it is involved. Astronomy has been a favourite study from the earliest periods: it is only within the last two hundred years that its true principles have been at all generally received; it is only from the time of Newton that they have been adequately explained.

In the present treatise, nevertheless, the results of observation will not be explained from principles assumed in the first instance, but the principles of astronomy will be deduced, as far as they can be so without complicated mathematical investigation, from observation. There are several reasons which seem to render this the most desirable course of proceeding, although adopted at the sacrifice of much conciseness, and of any very logical precision of arrangement.

The present treatise is principally addressed to a class of readers not habitually accustomed to severe reasoning, and is intended for those who know nothing of astronomy when they enter upon its perusal; and to them the course which we have preferred will probably be at once more interesting and more intelligible than the other. The general appearances of the heavens, the succession of day and night, the apparent courses of the heavenly bodies, are objects of interest and curiosity to all, however ignorant of the laws which regulate them, or the consequences which may be deduced from them. And they are not only interesting, but to a certain extent familiar; sufficiently so to perplex the reader of statements apparently at variance with them, and to deprive him of the greatest satisfaction that the student of a new science can feel, the power of at once comparing his deductions with facts, and convincing himself experimentally of the soundness of his reasonings by the accuracy of the results to which they lead him. A mind habituated to close reasoning upon merely hypothetical truth may be satisfied without such confirmation; yet it is agreeable to all, and to those which have not been thus exercised it is almost necessary; but, on the hypothetical system, it is the last point they arrive at. We have therefore preferred that course, which begins by stating and classifying appearances of general interest, and by explaining them in a manner consistent with popular observation; especially as we thus very early arrive at conclusions of high importance, which depend on those appearances, and follow from them as necessary consequences, in whatever manner they are themselves finally explained.

This however is not the only object sought in the adoption of the proposed course. One of the most important and laborious exercises of the mind consists in extracting from complicated results the simple principles on which they depend, in developing truth from the mass of confusion which often conceals it. There is no department of science which furnishes so long and so curiously connected a train of this kind of investigation as Astronomy: none in which the process is more curious, or the results more satisfactory. It has also this additional advantage, that it is capable of being made intelligible in its general outlines, almost without reference to mathematical investigation; and with none which requires more than a very limited portion of mathematical acquirement. It consequently seems to furnish to those, whose attention has not been directed to the severer studies, some opportunity of applying that peculiar discipline to the mind which those studies best furnish, and which moral and metaphysical inquiries, from the more vague and qualified character of the elements on which they depend, almost entirely fail in affording.

In speaking of the truths of Astronomy as admitting of explanation, almost without recourse to mathematical investigation, we only mean that the fundamental observations and principles admit of being stated intelligibly, and that much reasoning may be grounded upon them without introducing it. Of course however a science which is conversant entirely with mathematical relations can only be established on mathematical principles. Those therefore who are unacquainted with the elements of mathematics, cannot follow the whole process of the deduction; but they may trace at least its general course, assuming the correctness of those investigations which they are unable to prosecute, but taking their results as the elements of a new train of reasoning, which will, in many cases, be perfectly intelligible to every one who is acquainted with the common relations of number and quantity. With a view to facilitate such a course, all the more complicated mathematical investigations (of which, however, none run into difficulty, for the fuller development of the subject is reserved for a treatise of a purely scientific character) have, as far as possible, been removed from the text into notes *, and certain portions of the text itself, of too much importance to the general progress of the deduction to be thus removed, are included between brackets, in cases where they have appeared likely to offer difficulty to any class of readers. In these instances care has been taken, wherever the result deduced is of importance to the future course of the reasoning, to state it, if possible, in terms intelligible to those who are unable to pursue its investigation, and thus to enable them to resume the thread of the general argument. With all these assistances, however, it is impossible to remove some degree of difficulty and abstruseness even from the most popular parts of the treatise: a long process of investigation cannot be followed without thought and diligence, and the results of reasoning from their very nature can only be comprehended by some effort of reason and attention.

It may be necessary to remark, that while we profess to deduce our principles from the first and fundamental observations, we do not pretend to follow the historical course of discovery. For our purpose the class of objects which present the most simple phenomena requires the first attention: in the historical course of investigation, the most remarkable appearances were probably those which first attracted attention, and thus furnished the groundwork of theory. All that we profess to do, is to point out those observations from which we may most readily and conclusively deduce the principles of which we are in search: and in so doing we shall often call attention at an early period to facts and appearances, easily ascertained when the attention of an observer is directed to them, but which probably excited little attention till the progress of science showed their importance.

* There are some notes also, which have been inserted in that shape, not on account of their difficulty, but as bearing only an incidental relation to the text. That there may be no confusion between the two classes of notes, the notes which present any mathematical difficulty will, as well as the similar passages of the text, be included in brackets.

Chapter I.

Section 1.—*First Observations on the Stars—Division into Constellations—Daily rotation of the Heavens.*

To an observer situated anywhere upon the earth, the heavens offer the appearance of a vast concave vault, in the centre of which he is himself placed, and which is bounded by a plane extending to the sky; this plane is called *the horizon*, and itself appears circular. In this vault we perceive several different sorts of bodies: the sun visible by day only (for its sinking below the horizon terminates the day, when that word is used in opposition to night); the moon visible either by day or by night; and the stars (among which, for the present, we include the planets and comets) visible generally by night only. All these bodies appear to the eye to be situated in the vault itself: for we have no means, by mere ocular observation at a single place, of estimating their respective distances, and we consequently refer all alike to one imaginary surface in which they appear to be placed, and which, as we find no apparent difference in their distances, we imagine to be spherical. A slight degree of observation will shew that all these objects partake generally of one common motion. The sun and moon, from their great apparent size, and the remarkable changes of their situation and appearance, more forcibly attract our attention than any others; but the stars, on account of their number and the opportunities which they in consequence afford for comparative observations, as well as from other circumstances which will hereafter be explained, are the preferable objects for the first and fundamental observations.

A very moderate degree of attention is enough to assure us that the stars, with the exception of a few called *planets**, from a Greek word signifying *wanderers*, always hold very nearly the same positions with respect to each other. If we consider a few stars to form a group, we may observe this group night after night, and its shape and appearance will be always the same. There are not anywhere in the heavens different groups of stars of considerable extent, so resembling each other that an observer can be in any danger of mistaking one group for the other. And as the groups cannot be mistaken, the individual stars composing them may thus be certainly recognized, however any single stars in each may resemble each other in magnitude, colour, and brightness. Being thus able to recognize a star which we have once observed, we may prosecute our observations upon it night after night, and year after year. For the immediate observations of an individual no more than this is requisite; but when he wants to register their results, or to inform others of their nature, it is evident that he can no longer be satisfied with this mere power of identifying to his own satisfaction the particular star which he observes at different times, but he must have some means of distinguishing between the different stars which he has himself observed, and of announcing to others which it is, among all the heavenly bodies, to which he has especially applied his attention. For this purpose we again have recourse to those groups of stars, by which we originally distinguished each particular star from every other. These groups, when divided for the convenience of reference, are called *constellations*, (*i. e. collections of stars*,) a name which is also applied to those portions of the heavens which they respectively occupy; and the whole surface of the heavens has been long divided in this manner. The divisions are arbitrary in themselves, and often, perhaps, ill chosen; but as the only real use of them is for the convenience of reference, the one important object is to have a single received standard; and it would consequently be very undesirable to alter them even for the purpose of making what would originally have been a simpler and more distinct division. The surface of the heavens being thus divided into constellations, consisting each of a moderate number of stars, the stars in each are catalogued, and are arranged nearly in the order of their apparent brightness; the brightest stars being designated by the earlier letters of the Greek alphabet, the less bright by the later letters of the same; and if there be more stars in the constellation than there are letters in that alphabet, other alphabets, generally the small Italic and large Roman alphabets, are used to denote the stars next in importance, and if all the stars in the constellations are not thus distinguished, the least considerable of all are catalogued by num-

* The comets may be considered as a class of planets.

bers. These stars being registered on maps of stars*, or on celestial globes, or their places being defined in the manner we shall hereafter explain, become known bodies; and any astronomer making observations, on any particular star, may communicate them to any other, who will at once know the star in question, and be able to compare those results with his own observations and conclusions. Besides this mode of distinction, some of the brighter and more remarkable stars have been distinguished by particular names: and a vague classification has been made of stars, according to their lustre, into what are called different *magnitudes;* the brightest which appear the largest to the naked eye, although none have any ascertainable diameter when viewed through a telescope, being called of the first magnitude, the next class of the second, and so on. The stars of the fifth magnitude are barely discernible by the unassisted eye; and beyond these there are stars of the sixth, seventh, eighth, and even lower magnitudes, which we discover by means of telescopes. The number of stars of each magnitude increases as their brilliancy diminishes: thus in the Astronomical Society's catalogue of 2881 stars, there are only 21 stars above the second magnitude, (18 of which are called of the first magnitude, and 3 between the first and second magnitude), about 50 of the second magnitude and between the second and third, about 180 of the third magnitude, and between the third and fourth; and so on. As an instance of these different modes of characterizing stars, we may take the principal stars of the constellation Orion. The brightest star, α of Orion, is also of the first magnitude, and is sometimes called Betelgeuse, though this name is nearly obsolete; the next, β of Orion, is also of the first magnitude, and is called Rigel†; then there are four of nearly equal brilliancy, γ, δ, ε, ζ, of Orion, all of the second magnitude, and none having any received name; then follow stars of the third magnitude, then of the fourth, and so in succession‡.

Let us now examine what are the appearances presented to an observer in the course of a single bright night. He sees a vast variety of stars distributed about the heavens, and all, at first sight, seeming to be at rest. A very short course of observation, however, proves that this appearance of rest is fallacious. The stars all continue at the same apparent distances from each other, but their positions vary with respect to the horizon and the observer. On one side of the heavens they are seen gradually to rise higher above the horizon, and some stars appear above it, which at first were not seen at all; on the other, they sink towards the horizon, and some which were originally visible, sink below it and disappear. Confining our observations to a particular star, we shall see it rise above the horizon at a particular spot, gradually increase in elevation till it reaches its highest point, when it is said to *culminate*, and then sink by like degrees, and finally fall below the horizon at another spot which may also be ascertained. The same observation may be made with other stars, and extended to any number. It is not even necessary that the observations of different stars should be made at the same time, or that the observations of the rising and setting of any star be made on the same night; for the same star is invariably found to rise and set in the same spots, although at different periods of the night. Now each arch of the horizon intercepted between the points of rising and setting of any star may be *bisected*, or divided into two equal parts; and the points of bisection will themselves be at the extremities of a diameter of the circle. It is found from observation, that this diameter is the same for every star; and the points of the horizon through which it passes are called *the North and South points*. The side of the heavens where the stars rise is the *East side*, that where they set, *the West;* the East and West points being those equi-distant on each side from the North and South points. If through the North and South points a plane pass perpendicular to the horizon, the intersection of this plane with the celestial hemisphere will be a semi-circle; and this semi-circle is called the *meridian of the observer*, or

* A map of the stars (comprised in six sheets) is published under the Superintendence of the Society for the Diffusion of Useful Knowledge, in which these constellations are represented.

† These, and many of the names of the stars, were given by the Arabians, who, during the eighth century, and for a long period after, cultivated astronomy more diligently than any other people.

‡ In the map referred to, Orion will be found nearly in the centre of Plate 2. This constellation may be best seen in the heavens for the purpose of comparing the original with the representation, about twelve o'clock at night in the middle of December, ten in the middle of January, and eight in the middle of February.

of the place of observation. A figure will explain these positions, although it is difficult sufficiently to represent the various dimensions of the sphere on a plane surface; and the reader should therefore, if possible, examine a globe with reference to these points.

Let H R N T (*fig.* 1.) represent the horizon of an observer; O, the centre of that circle and of the hemisphere

Fig. 1.

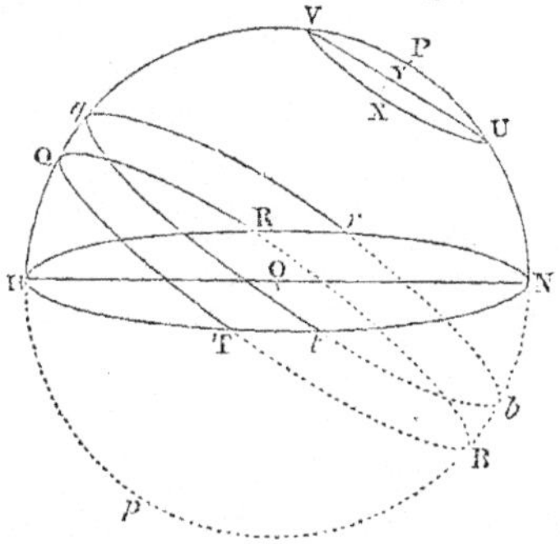

H Q *q* N, being his position. Let T be the spot where a star first rises, T Q R its path above the horizon, and R the point where it sets; and let the arch T H R of the horizon, intercepted in one direction between the points T, R, be bisected in the point H, and the arch T N R intercepted in the other direction, be bisected in the point N. It is clear that H and N are at the two extremities of a diameter, for HR = HT, and RN = NT, and therefore, H R N (the sum of H R and R N) = H T N (the sum of H T and T N); or H R N, H T N are each of them semi-circles. Now, suppose *t q r* to be the path of another star above the horizon, *t* its point of rising, *r* its point of setting; it will be found that the arch *t* H = the arch H *r*, and *t* N = N *r*; that is to say, the points N and H in this case also bisect the arches of the horizon intercepted between the points of rising and setting. The same will be found to be true of every star (with some apparent exceptions to be presently noticed), and the points N and H are consequently fixed points, and independent of the particular stars by observation of which they were ascertained. Let H Q *q* N be the intersection with the celestial sphere of the plane passing through N H and perpendicular to the horizon; this is the meridian of the observer at O, and as the points N and H are fixed, the vertical plane drawn through them must be fixed also, and this meridian therefore is a fixed line. It will also be found by observation, that the points Q, *q*, where the paths of the different stars meet this meridian, are the most elevated points of those paths respectively, and consequently that a star which rises and sets attains its greatest height above the horizon, or *culminates*, when it is upon the meridian; and we may add, that the paths, T Q R, *t q r*, of the stars observed, appear to be parallel to each other, and that each path is divided by the meridian into parts T Q, Q R, or *t q*, *q r*, apparently equal and similar; that is to say, T Q equal and similar to Q R, and *t q* to *q r*.

In one part of the heavens the appearances are a little different. If we look to the Northern part we shall see many stars which never sink below the horizon. These stars pass the meridian twice, once at the lowest, and once at the highest point of their course. In moving from the lowest to the highest point, their course is entirely to the East of the meridian; in passing from the highest to the lowest point, it is entirely to the West of that line. These parts also in this case, as well as in that already mentioned, are similar and equal; and if the course of one of those stars be represented in the figure by the line U X V Y, every point in that line will be found to be equidistant from a certain point P, situated in the sphere, and U X V Y will be a circle, and the point P its pole. Besides this, it will be found that the path of every one of these stars which never set is a circle described at a given distance from the same point P, which itself appears stationary; and the position of this point being ascertained, it will further be found, that each of the paths T Q R, *t q r*, described by stars which alternately rise above the horizon and fall below it, and which appear, as we have already mentioned, to be parallel to each other, are themselves also portions of circles, every point of which is equidistant from the same point P, and which therefore are parallel to each other, and to the path U X V Y, described by a star which never sinks below the horizon. The point P therefore is the pole of all the parallel circles described by the stars; and if we suppose the sphere to be completed, as it is by the dotted lines in the figure, and *p* to be the point directly opposite to P, *p* will be the other pole of the same circles; or there are two fixed points, one P found by observation, the other *p* deduced from it, to which the motion of the stars may be equally referred. O being the centre of the sphere, will be a point in the line which joins P, *p*. The stars which are seen to describe their whole circle round the pole are called *circumpolar* stars. We shall also find, that in the time in which any particular star describes a third or fourth part, or any given portion of its circle, all others describe the same portion of theirs; and consequently they all continue in the same positions with respect to each other, though their places vary with respect to the horizon and the observer.

Having thus far ascertained the appearances which the stars present, let us see if we can thence deduce any conclusions respecting the occasion of them. For this purpose let us suppose the hemisphere, H Q P N, to be made a complete sphere, as is done by the dotted lines in the figure. Let us also suppose that the whole sphere has a motion of rotation round a line joining P, *p*, which is called *the axis;* but that H R N T, the horizon, or the line in which the plane bounding the visible hemisphere meets the heavens, continues fixed: and let us see what would be the appearances presented in such a case. Let us take the case of a star upon the horizon at T. It is clear that, as by the rotation of the sphere it was transferred from T, it would appear to move in a line of which every point was equidistant from P, for every point in that line would be determined by the actual distance of the star from P. Its apparent path therefore would be a portion of a circle, every point of which is equidistant from P; and, in point of fact, we have already seen that it is so. In the same manner, if there be a star which never falls below the horizon, and whose distance from P is P V, its apparent path would be a circle, of which every point is at the same distance P V from P, or it would be represented by a circle, U X V Y, which we have already seen to represent the apparent path of a star which never sets *. Each of these circles, thus described by the motion of different stars, having every point in it equidistant from the same point P, they all would, as before, be parallel circles. Again, as each of them is described in consequence of the same general motion of rotation of the whole sphere, each would be described in the same time; namely, the time of that rotation; and in the same manner, in any portion of that time, each star would describe the same portion of its own circle; namely, the same portion of that circle which the sphere describes of a complete revolution. All these are the appearances which we have already seen that the heavens, in fact, present.

The appearances presented by the motions of the stars may then be accounted for on the supposition that the sphere of the heavens revolves round an axis joining P, *p*. They cannot be explained however on this supposition, except by supposing that the sphere goes through a complete revolution. The motion of a star which is seen to rise and set, as that whose path is T Q R, might be explained by imagining the sphere to make only a part of a revolution; and the magnitude of that part would depend on the proportion which the visible path T Q R bore to the whole circle of which it formed a part; but the stars which never set, as that whose path is U X V Y, are seen to describe the whole circle, and their motion therefore can only be thus explained on the supposition of a complete revolution of the heavens on the axis P *p*. If however this be the case, the motions of the stars which sink below the horizon must also be continued below it, or they will describe below it the remaining parts (those represented by the dotted lines) of the circles T Q R B, *t q r b*. Let us see if we have any means of discovering, by observation or reasoning, whether they do so.

The first remark that occurs on this question, is that the supposition that they describe below the horizon the remainder of the circle, of which they are seen to describe part above it, at once accounts for one circumstance that seems

* These positions may be thus illustrated :—If you take a top, or any body which you can spin with great steadiness and accuracy, and place a spot of ink upon it, and then spin it with great velocity, so that the spot returns to the same place in less time than is necessary for the eye to lose the impression last made by it (See Treatise on Optics, p. 44), the effect produced will be that of a line encompassing the top, and forming a circle upon it, of which every point will be equidistant from the extremity of the axis upon which the top spins. The apparent track of the point is of course the same, whether its motion be quick or slow; but by the rapidity of its motion we gain the advantage of actually seeing the whole track at once.

to admit of no other explanation. We trace the path of a star from its rising at T, to its setting at R, and then lose sight of it; but on the next night we again see it appear at the same point, T. We know therefore that the star is in some way transferred from the point R, where it sets, to the point T, where it rises; and the most probable way in which we can suppose this transference effected, is the continuation below the horizon of the same motion which it had when above it, or the description of that circle, R B T, which it would describe on the supposition that the whole heavens revolve round the observer.

If however we take the case of a star rising just at the time when the stars begin to appear in the evening, and setting as day breaks on the following morning, it is evident that its path below the horizon, if it be described at all, must be described by day; or that the same motion of revolution continues by day, which we seem to have ascertained to exist by night. Does observation then confirm, or disprove this conclusion? The sun and moon are visible by day, but their motions, although they generally confirm it, are of a more complicated nature, and we therefore do not wish to draw our inferences on this point from them; and the stars are not visible to the unassisted eye when the sun is above the horizon. The telescope however, in the hands of a skilful observer, for only such a one can make the observations necessary for this purpose, removes this difficulty; with it he can, even when the day is brightest, ascertain the positions from time to time, and consequently the motions, of many of the brighter stars; and the result of these observations is, that the stars are ascertained to describe in the day-time the same courses which they are easily seen to trace in the night: and we consequently come to the conclusion that their motions may be accurately comprehended and explained, on the supposition that the whole heavens revolve about an axis, passing through the position of the observer, and carry the particular stars with them in their revolution.

If this be so, and the meridian of the place, H Q P N, be continued, as by the dotted line N B *p* H, below the horizon, so as to complete the circle, this lower part of the circle will again intersect the circles T Q R B, *t q r b*, in the opposite points, B, *b*, to those, Q, *q*, where the upper part of it met them above the horizon; and as Q, *q*, were the points most elevated above the horizon, B, *b*, will be those most depressed below it; or in other words, every heavenly body, which sinks below the horizon of a particular place, will be most depressed below it, when it passes the meridian of that place below the horizon, and of course below the pole. We have already seen a corresponding result with respect to circumpolar stars, when they cross the meridian below the pole though above the horizon.

As yet we have only considered the conclusions which an observer, confined to a single point on the earth's surface, would arrive at on this subject. We will now proceed to examine how they will be affected by a comparison with the results of other observations, made at a different place. The account which we have given of the observations made at one place, applies with equal correctness to all; that is to say, an observer situated anywhere upon the earth, finds that the apparent paths of the stars are circles, or portions of circles, each having every point in it equidistant from two fixed points, one in the observed heavens, and one in the other part of the sphere, supposed to be completed, and each bisected by a line passing through the visible fixed point, and dividing the visible heavens into two equal portions. In each case therefore, this line is what we have termed the meridian of the place of observation; and every place therefore has a meridian, passing through a fixed and immoveable point in the heavens. The position of this point may be ascertained by observation at each particular place, and it is found to be the same at all; the other extremity of the axis also is the same in each place.

We come therefore to this conclusion, that the axis P *p* round which the revolution of the heavens takes place is a fixed and determined line, not depending on the situation of the observer: and this is one circumstance necessary to the establishment of our theory, that the apparent motions of the stars may be attributed to the revolution of the heavens round a fixed axis; for if observations made at each place gave a different axis, they would be inconsistent with such a supposition. The points, P, *p*, are only imaginary points, being those where the axis Pp meets the imaginary sphere of the heavens; they are however important to be known, and go by the name of the *poles of the heavens*. They are points, as we have already seen, in the meridian of every place, and therefore they no where appear either in the East or West side of the heavens; if however we conceive the heavens divided by a vertical plane, passing through the East and West point at any place*, the points P and p will always be on opposite sides of this plane; that is to say, the one on the North side of it, the other on the South, and the same point P is always on the same side of the plane. If therefore (in *fig.* 1.) P represent the pole, which to an observer at O is on the Northern side of the heavens, P is always on the Northern side, and is called on this account the *North Pole* of the heavens, and in like manner p is the *South Pole*†. There are however two points on the earth, (the poles of the earth) where the points P, p, are the one directly over the head, the other directly under the feet of the observer; here therefore there is no North or South point, and we shall hereafter see that the phenomena from which we deduced our definition of these points,—namely, the rising and setting of stars, do not take place at these situations. We have already seen that P, p, are points in the meridian of every place; all these meridians therefore intersect each other at the two poles. If p, the South Pole, be above the horizon, P, the North Pole, will of course be below it.

One circumstance may here require explanation before we proceed farther. We have already seen that the centre of the heavenly sphere is a point in the axis Pp, and that this centre appears to be the situation of the observer; and we have also said that the results of observation are the same, wherever on the earth's surface he be placed. If two observers be at situations, the one one thousand miles East of the other, the situation of both cannot be in the line Pp; but if the one is in it, the other must be nearly one thousand miles out of it: yet they both appear to be in it. We know from very simple reasoning, or we may easily satisfy ourselves by trial, that a small change of position in the observer does not affect the apparent position of a very distant object. Thus, if there be two trees, or two spires distant ten miles from each other, and two men stand half-way between them, the one precisely in the line joining them, and the other a yard on one side of it, each will alike feel that, to all common observation, he is exactly in the line which unites them. The angle between the two directions in this case, would be considerably less than half a minute*, and would not be observable except by instruments of some delicacy. In the same manner, if the distance to the points P, p, be excessively great in proportion to the distance between the situations of different observers, each observer will seem to be in the same position with respect to the points P, p, and the line joining them. There is therefore nothing absurd or contradictory in the apparent coincidence of each situation with the line Pp, if we only suppose the points P, p, so remote from the earth, that any line drawn on its surface is too small to be estimated in comparison with that distance; and we get therefore a notion of the vast distance of those points, instead of a difficulty affecting the notion of such a revolution, as we have supposed to take place. If however every point on the earth's sur-

* The circle made by the intersection of such a plane with the sphere of the heavens is called the *prime vertical*.

† A well-known star, α of the Little Bear, called also *Polaris* or the Pole Star, is at the distance of only one degree and fifty minutes (see the next note) from the North Pole P. Its motion, like that of all other stars, is in a circle, every point of which is equidistant from P. As P therefore is a fixed point, and the Polestar always very near it, the observation of the Polestar furnishes a very easy method of finding very nearly the fixed North Pole of the heavens. The meridian being a vertical circle, and passing through the pole to the horizon, the points where it intersects the horizon are those points of the horizon respectively nearest and most distant from the pole, and thus the North point of the horizon is that nearest to, and the South point of the horizon that most distant from, the North Pole.

* Every circle is conceived to be divided into 360 parts called degrees, each degree into 60 minutes, each minute into 60 seconds, each second into 60 thirds, and so on. One degree is written 1°; one minute $1'$, one second $1''$; and so on; though the divisions beyond seconds are quite as frequently expressed in decimal parts of a second. Thus nine degrees, fifteen minutes, twelve seconds and twenty-four thirds, are written $9^\circ\ 15'\ 12''\ 24'''$, or, as 24 thirds are $\frac{24}{60}$, or $\frac{4}{10}$ of a second, they are also written $9^\circ\ 15'\ 12''.4$. A degree therefore is the 360th part of a circle, a minute the 21600th part of it; half a minute, the quantity mentioned in the text, the 43200th part of a circle. The degrees, minutes, &c., of one circle, will of course occupy more space than those of another, exactly in the same proportion as the one circle is longer than the other, for they are in each the same proportional part of the whole circle. They correspond to the same angle in every case, but differ in linear magnitude, as the circles on which they are measured differ.

face be apparently in the line P p, so must its centre also, which lies in the midst between these points. The axis P p therefore may be considered to pass apparently through the centre of the earth; and we shall hereafter see the strongest physical reasons for believing that it actually does so.

There are several reasons which convince us, on very slight examination, that the earth itself is, speaking loosely, of a globular figure. They are collected and explained in the first pages of the "Treatise on Mathematical Geography;" and it is therefore unnecessary here to go into any account of them *. But it is important here to point out how we are enabled by this globular figure of the earth to verify the conclusions we have already formed respecting the motions of the stars. The earth being convex at all points, the horizons of different places, which are always planes touching the earth at those points, will be inclined to each other at all different angles; and the height of the pole above the horizon will vary in consequence. The height of the pole above the horizon measures what is called *the latitude* of the place. We shall hereafter explain why: it is sufficient now to state, that whenever we speak of the latitude of a place, as having a particular value, for instance 50°, we mean that the pole is there at the height of 50° above the horizon. We have already seen that there are certain stars which never sink below the horizon at a particular place, and which at that place are called *circumpolar* stars. If we take another situation, so that the height of the pole above the horizon be greater there than at the former place, some of those stars which before rose and set, will now (if the supposition, that they continue to describe below the horizon the remainder of the circle which they were seen to describe in part above it, is a true one) have their whole course above the horizon: in other words, they also will be circumpolar stars at the second place of observation; and we may there ascertain, by complete observation, whether they do actually describe the whole circle, as we have supposed they would: and in every case it is found that they do so. Again, the earth being globular, there are points on its surface from which an observer will actually see the opposite part of the heavens from that presented at any given places; and the question, whether the same motions be continued below the horizon which we observe above it, may in this way also be brought to the test of complete trial, and the truth of the doctrine will be established, if we find that in every part of the earth the same appearances of circular motion are observed in the paths of the stars above the several horizons of each observer. And this we find to be universally the case. In every way therefore the original conclusions, which we draw from observations at a single place, are confirmed by a comparison of those made at several.

We shall hereafter see that there is another and simpler mode of accounting for these appearances, by ascribing a motion of rotation to the earth itself. The appearances themselves however are exactly those which would result from the rotation of the heavens; and our only proof that the other theory will account for them, will be by shewing that it must necessarily produce the same appearances as the actual rotation of the heavens would do. We may therefore properly, for the present, treat the rotation of the heavens, not as an established fact, but as a supposition enabling us to account for, and represent all these appearances.

Section 2.—*First Observations on the apparent Motions of the Sun—Diurnal Motion—Annual Motion.*

Having thus ascertained the apparent rotation of the heavens and the stars round an axis, we may proceed to consider the more complicated appearances presented by the sun. The first notion which we gain from observation is, that he also follows the same course of revolution as the stars; for he is seen to rise in the Eastern, and set in the Western, part of the heavens, culminating in the meridian, and rising and setting at points apparently equidistant, or very nearly so, from the North point of the horizon. So far the appearances of a single day seem to correspond with those already noticed in the stars. But when we register the observations of a long period, a year for example, we find a striking difference between the two cases. The star always

* The argument in that treatise deduced from the appearances of *heavenly* bodies cannot properly be applied to the purposes of this treatise, as the reader must be supposed hitherto ignorant of the nature of these appearances. Omitting these, we have sufficient evidence, in p. 3 of that Treatise, of the globular figure of the earth in a very general sense, and that is all which we want for the present purpose.

appears in the meridian at the same spot, it rises always in one point of the East, and sets in one point of the West; but all these things are different in the case of the sun. If we observe his height when he crosses the meridian, we find that it is less on the 21st day of December, than at any other period during the year; that it is greatest on the following 21st day of June; and that in the interval it is on each succeeding day greater than on the preceding one: and again, after the 21st of June the sun passes the meridian at points successively lower on each day, until, on the 21st of December, he again returns to his least elevation. The variation of this elevation is not uniform; it is most rapid about the 21st of March and the 21st of September; and for a few days before and after the 21st of June and the 21st of December it is hardly perceptible. At these periods therefore the sun has loosely been said to stand still; and they have in consequence, and from the seasons at which in Europe they happen, gained the name of the summer and winter *solstices*. The same observation may be made also with respect to the points where the sun rises and sets; indeed it is a necessary consequence of the sun rising from day to day higher on the meridian, that his points of rising and setting should also approach the pole which is above the horizon; and, on the other hand, when he is falling on the meridian, that the points of rising and setting should in like manner recede from it. It is also found by nice observation, that during the period in which his meridian heights are continually increasing, he always, when it is the North Pole which is above the horizon, sets nearer to the Northern point than he rises, and rises on the following day still nearer the North; and conversely, that when his meridian heights are continually diminishing, his point of setting on any day is more Southward than that of his rising that day, but less so than that of his rising on the following day. All these observations correspond exactly to the supposition, that for the six months extending from the winter to the summer solstice, the sun, besides partaking of the general rotation of the heavens, has a proper motion of his own in the heavens which continually carries him Northward; and again, that in the interval between the summer and the winter solstice, he has also a proper motion, but that its direction is opposite to the former, that is to say, from North to South. The same observations may be made where the South Pole is above the horizon, with similar results; the only difference being, that in that case his motion Southwards produces effects corresponding to those of motion Northwards in the other.

This however is not the only motion which we can discover in the sun. Every one who has observed the nightly appearances of the heavens with any attention, is aware that they continually differ. On each succeeding night, or it may be more convenient, for the sake of marking the changes more strongly, to take nights at a considerable interval, as for instance a month, from each other, some stars become visible which had not been so at the last time of observation, and others cease to appear which had then been seen. The new stars which from time to time make their first appearance, do so invariably in the Eastern portion of the heavens a little before sun-rise; those which have ceased to appear always were last seen in the West a little after sun-set. It is found also, that stars which rose a little before the sun at the former observations, rise longer before him at the latter; and, in the same manner, that those which set a certain time after him at the former, set a less time after him at the latter observations. It is easily seen that these changes can only be occasioned in one way. We have already seen that the stars themselves keep the same positions with respect to each other; but the sun evidently approaches the regions of the heavens which are above, but near his point of setting in the Western part of the horizon, at the time of his setting; for having previously set before stars there situated, he now sets at the same time with them. In the same manner, he recedes from the regions of the heavens which are near, but above his point of rising in the eastern part of the horizon when he rises; for having previously risen at the same time with stars there situated, he now rises after them. He evidently therefore has a motion from West to East. These observations hold alike in every period of the year, and at the close of the year the sun is in the same position relatively to the stars, as at its beginning. We have therefore ascertained that he has a proper motion from North to South, and back from South to North, which restores him at the end of the year to the same altitude when on the meridian which he had at its beginning; and that he has also a motion from West to East, which in the same period brings him back to the same place, but which, being continually in one direction, can only do so by making him describe in that direction the complete circuit of the heavens.

SECTION 3.—*Mode of ascertaining the motions and places of heavenly Bodies —Measure of time—Pendulum—Solar and sidereal day—Measure of place—Right ascension—Declination —Sun's path, the ecliptic —Equinoctial points—Longitude–Latitude.*

We are still however far from a sufficient knowledge of the motions either of the sun or stars. We see that they revolve round the earth, but that they do not do so in the same time, and we cannot yet tell whether either revolve uniformly, or with a variable motion. To know any thing certain concerning motion we must know the space passed over and the velocity with which it is so. But the velocity can only be measured by the space passed over in a given portion of time, and the first great object therefore is to find out some method of computing and measuring time. If we knew that the motion either of the sun or of the stars was equable, we might adopt that motion as a standard; but it is evident that we have as yet no reason for so supposing. An hour appears a period of a very different length according to the mode in which we pass it: almost imperceptible sometimes in sleep, short in agreeable mental or bodily exertion or amusement, long in irksome employment or tedious idleness; and we have no power, by mere sensation, of accurately discriminating or comparing the lengths of several unconnected periods. But the periods of the successive returns of a heavenly body, whether it be sun or star, to the meridian are unconnected; and we consequently have no means by mere mental perception of comparing their durations, and saying whether they are equal or unequal. And this being the case, we can have no right to assume either as a standard of time; for time, like everything else, can only be measured by reference to something of a fixed and determinate value. Our notions of time indeed are so complicated with the words describing the periods into which we find it convenient to divide it, that it requires some attention to feel the full force of the difficulty. A day is the period to which we commonly refer everything, and which we consider as of a fixed and uniform duration; a day is also the period between the successive appearances of the sun on the meridian; and in common language it is also the period between his successive risings or settings. Being thus in the habit of using the word both as an expression for a certain fixed and uniform period of time, and for the interval between certain events which do in fact happen very nearly at that distance of time from each other, we cannot readily separate the two meanings of the word, or perceive how difficult it would be to discover the equality of the intervals between these events unless we had some certain and definite standard, independent of themselves, whereby to measure them. The difficulty however certainly exists; and we may in some degree be convinced of its force, even by an argument drawn from the difference between the two latter uses of the word *day* which we have mentioned. During the period in which the sun's motion is from South to North, we have already seen that he rises farther North on each successive morning, and it necessarily follows (a consequence which will be fully proved hereafter), that on each successive day a larger portion of his circle of rotation is above the horizon. If his motion be such that each successive interval between his appearances on the meridian is equal, or nearly so, his successive appearances on the horizon will be rapidly accelerated for a considerable period, then slowly so, and finally retarded. This will be more fully explained hereafter; but we all know the fact from the common tables of the time of sun rise. Thus, in our climate, in March, sun-rise precedes the arrival of the sun on the meridian by a period increasing from day to day by about two minutes; in June the alteration is hardly perceptible; and in September the interval diminishes as rapidly as it increased in March. In March then, the day, understood as the interval from sun-rise to sun-rise, is about two minutes less than the day, understood as the interval from noon to noon; while in June they are equal, and in September again, the day from sun-rise to sun-rise is as much longer than that from noon to noon, as it was shorter in March. No man could pretend to be conscious of these minute differences of duration; and if therefore we were to depend on sensation merely for our measure of time, we might just as well take the interval from sun-rise to sun-rise for our standard, as that from noon to noon; and either as that from the successive appearances of a given star on the meridian. The differences between the actual lengths of these different periods may seem trifling; but such is the nicety required in astronomical observations and computations, that no inaccuracy can safely be overlooked; nor, if we once force ourselves to give up the appearance of security given by the popular use of the words "day," "year," &c., are we justified in believing that our errors will be confined within any narrow limits.

We have however one standard of time independent of any assumption whatever, except that of the truth of those general principles, called the laws of motion, which are involved in all the deductions of mechanical science, and which, although inferred from observation and reasoning, do not admit of strict and absolute demonstration. All bodies are found, by experiments made anywhere on the earth's surface, to fall there towards the earth in a straight line, called a vertical line, in consequence of the action of some force acting on them in that direction, which we call the force of gravitation. If at that point the force act at every instant of time with the same intensity, a pendulum properly adjusted will perform its successive oscillations in equal times; that is to say, it will always take the same time in swinging backwards and forwards. This depends on no assumption whatever; it is a necessary mathematical consequence of the action of a constant force, tending vertically downwards, on such a body so suspended. If therefore we can ascertain that the force by which a body falls to the ground at a particular place is always equal, we know that the oscillation of a pendulum there will give a fixed standard of time. We might perhaps venture at once to assume that this force is invariable, on the principle that, all the circumstances under which it acts being, as far as we can perceive, the same at all times, there is no reason whatever for supposing it to vary. It is, however, satisfactory, if possible, to ascertain by actual experiment, whether the force be variable or not. Of course, when the object is to ascertain the mode of measuring time, no measure of time must be used in the process. Nothing, therefore, which depends upon the measurement of velocity can be admitted, for velocity only means space described *in a certain time*. We have however sufficient, though not very ready means of ascertaining the fact. Space is easily measured; and the momentum of a body can be, although not very accurately, yet sufficiently estimated by the effects it produces. If the quantity of matter in a body be represented by A, and the force acting on each particle of matter in it by F, and the space through which it falls, by S, it may be proved, as a necessary consequence of the meaning of these terms, that the momentum varies as $A.\,F.^{\frac{3}{2}},\ S.^{\frac{1}{2}}$* If therefore we try the experiment with the same body falling through the same space at different times, the momentum will vary as $F^{\frac{3}{2}}$, and will increase and diminish with the increase and diminution of the force. If then we find that the body always sinks to the same depth in similar bodies placed to receive it, that it produces the same contraction of a spring attached to the point on which it falls, and generally that it produces effects always equal in manners which may be varied according to the pleasure of the experimenter, we are led to the conclusion that the force always acts alike at the same point. Now this is the conclusion found by experiment. We have therefore in the pendulum a standard of duration; not a convenient one as yet, because we have nothing whereby to fix on any particular length of pendulum as the standard; nor a general one, for we have as yet no reason to form an opinion whether the force of gravity at different points on the earth's surface is the same or different, and we therefore cannot tell whether our pendulum of one place would swing at the same rate at another. We know only that at the place where the experiments were tried its successive oscillations are always of equal lengths, and consequently, that we may there ascertain by it whether the revolutions of the sun, or stars, or either of them, be of uniform length; for whatever be the duration of the uniform oscillation

---

* [Momentum $\propto$ W. V.

$\propto$ A. F. V. (but $V. \propto \sqrt{F.\,S.}$)

$\propto A.\,F.\sqrt{F.\,S.}$

$\propto A.\,F.^{\frac{3}{2}}\,S^{\frac{1}{2}}.$

of our pendulum, if we observe that either of these periods of revolution always includes the same number of these oscillations, that is an uniform period, and may itself be adopted as a standard of duration. If the number of oscillations be different in different revolutions, the period is not uniform or fit for a standard*.

Now the period of revolution of the stars is found by this sort of comparison to be always accurately the same. Each star is found to have precisely the same interval between its successive appearances on, or as they are also termed, *appulses* to, the meridian, and this interval is called a *sidereal* day. It is also found that the different portions of each revolution are described in proportional periods: thus, if two different stars are in the same circle of rotation, but the one distant 180° from the other, the half of a sidereal day elapses between their appulses to the meridian; if the distance of two such stars be 90°, the interval between their appulses is a quarter of a sidereal day; and so in like manner for every proportion of distance.† We find therefore, not only that the duration of a sidereal day is constant, but that during every part of it the rotation goes on uniformly; or in other words, *that the heavens revolve round their axis continually with an uniform velocity.* The discoveries of an advanced state of science furnish the most complete confirmation of this result, by shewing that none of the causes which produce variations and disturbance in other motions, produce any in that from which the apparent motion of the heavens round an axis proceeds.

The interval between the successive appearances of the sun upon the meridian, or from noon to noon, is necessarily longer than that between those of a star; for as the motion of the heavens is from East to West, and the proper motion of the sun from West to East, the sun on each successive day, when the point of the heavens where he was at noon on the day before returns to the meridian, is to the Eastward of that point, and, consequently, to the Eastward of the meridian; and he therefore only returns to the meridian after the rotation of the heavens has continued for some additional period, long enough to bring his new place to the meridian. The interval between two successive appulses of the sun to the meridian, is called a *solar day*. We find by observations with our pendulum, that the length of the solar day is continually varying, but still that its variations succeed each other in a regular succession, and go through all their changes in a certain period of time, a year. Although therefore the solar day is of variable length, we can, as we know all its variations and their period, ascertain its *mean* or average length; and this quantity is called a *mean solar day*.

The pendulum, although it furnishes us with the means of ascertaining that the motions of the stars are uniform, and those of the sun variable, is not, for the reasons we have given, adapted in the first instance for furnishing the common standard of time. The solar day, again, does not seem well suited to the purpose, on account of its variable length. On the other hand, as the continual appearances and re-appearances of the sun above the horizon furnish the most remarkable distinctions between different portions of time, and do practically regulate the occupations of men, and determine the periods of labour and those of rest, it would be inconvenient so to fix the standard of time used respecting the common occurrences of life, that the periods by which we measure it should continually have

* It must not be understood that the considerations stated in the text were those from which the expediency of using the pendulum as a standard of time was in fact deduced, though they seem to furnish the mode of deducing it, which involves the fewest assumptions. It is, indeed, probable that the experiments mentioned in the text have never been made with any degree of minuteness. Huygens, who first demonstrated the mathematical principles of the motion of a pendulum, though Galileo had accidentally observed that its oscillations appeared to be all of the same length, seems to have proceeded on the principle mentioned in the text, of assuming that the force of gravitation was constant, from the absence of any apparent reason for its varying. From this assumption he deduced the laws of the motion of a pendulum; and it has ever since been adopted as the standard of time.

† These observations of distance cannot be made with any great degree of accuracy, unless by means, hereafter pointed out, which proceed on the supposition that the heavens do revolve continually with an uniform velocity. They may, however, be made with certainty enough to convince us that the limits within which any variation must be confined must be exceedingly small: for, by multiplying observations, and taking *a mean*, or average, between their results, we may be sure, where there is no cause affecting all the observations in the same way, to obtain a value very near the truth. This is an obvious and necessary consequence from the supposition that there is no constant cause of error affecting all the observations *alike*: any accidental cause is likely to affect them in different manners, sometimes increasing and sometimes diminishing the apparent distance, and the errors thus compensating each other, the mean result, even where the individual errors are considerable, will not differ much from the truth.

their commencement at different parts of this, which we may call *the working day*. This inconvenience however would clearly occur if we took the sidereal day for our ordinary standard; that is, if we fixed the commencement of the day, at the instant when a particular star is on the meridian: for as the sun, moving continually Eastward, is successively at all distances East of the star, he must come to the meridian at all intervals of time after it; and consequently the star will be on the meridian at one period of the year when the sun is so also, or as we say at noon; at others when he is just rising, just setting, or midway below the horizon, or as we say at midnight. The *mean* solar day, however, is liable to no such objection: being a period deduced from computation of the *average* of the actual solar days, it is a fixed and invariable period; and as no solar day differs much from it, and the differences are some in excess, and others in defect, the period of the observed commencement of the real solar day never differs so far from the computed or registered commencement of the mean solar day as to occasion the inconvenience which must result when these standards materially differ. In fact, the difference between the commencement of the real and the mean solar day never much exceeds sixteen minutes*.

Taking then the *mean* solar day as our standard of time, it is divided into twenty-four hours, each hour into sixty minutes, and each minute into sixty seconds; and these are each of fixed and determinate length. It is established by the principles of mechanics, that we can, by varying the length of a pendulum, make its oscillations of any exact length that we please: and as the second is the smallest division of time in common use, it is usual to make the pendulum of a clock of such a length that its oscillations are of a second each. If the force of gravity is different at different places on the earth (as we shall hereafter see that it is), the lengths of the pendulums vibrating seconds at these different places will differ; but they may be so adjusted that the period of the oscillation shall be accurately the same in all. The length of a pendulum vibrating seconds in the latitude of London, 51° 31.′1, in a vacuum at the level of the sea, is 39,13929 inches.

Taking these common divisions of time into hours, minutes, and seconds, the length of the sidereal day is found to be uniformly 23 hours, 56 minutes, or more accurately, $23^h\ 56^m\ 4^s.092$. This period may be divided into 24 equal portions, and each of those subdivided into 60, and those again into 60 other equal portions; and these divisions will be respectively *sidereal hours, minutes, and seconds*, bearing the same relation to the sidereal day that the common hour, minute, and second do to the mean solar day. Of course a pendulum may be made of such a length that its oscillations will be of a sidereal second each; and in fact astronomical clocks are usually made so. Computations and observations made by sidereal time are of course easily transferred to common (or mean solar) time, and the reverse.

Having now got our measures of time, we proceed to see how, by means of them, we can ascertain the actual position of the sun in the heavens on each particular day. His light so far overpowers that of every other body, that there is some difficulty in observing his situation accurately, by immediate comparison with that of others; for, although we can see the stars through telescopes in the day time, they are not very readily found, and cannot easily be observed unless we have some previous knowledge of their positions, and adjust the telescope accordingly; they consequently are ill suited for the earliest observations. The moon indeed is often visible at the same time with the sun, and may be compared in her position with him by day, and afterwards with known stars by night; but she has, as we shall see, a very complicated proper motion of her own, in the interval between the two observations; and consequently furnishes only an inconvenient method of determining the position of another body. We must therefore have recourse to other means, and for this purpose we must explain a few of the simpler properties of the sphere.

In *fig.* 2. let P*p* represent the axis of the heavens, P, *p*, being the poles, and let PE*p*Q be the meridian of the observer. Suppose also that a plane is

* Although the commencement of the mean solar day seems the most convenient division of time, for the reasons given in the text, it is by no means universally adopted. The French use the *true* solar day; that is to say, they fix the commencement of each day at the instant when the sun is actually on the meridian. The length, therefore, of the day varies, and their clocks require to be regulated accordingly.

Fig. 2.

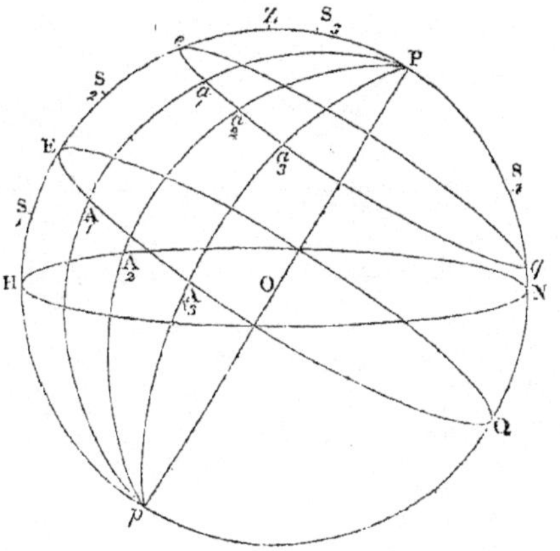

drawn through O, the centre of the sphere, so that Pp shall be perpendicular to it; the intersection of this plane with the sphere will be a great circle of the sphere *, and this great circle will bisect the meridian $PEpQ$, and every other circle drawn through the two points P, p, as PA_1p, PA_2p, PA_3p. This circle EA_1A_2Q is called the *equator.* The meridian of every place is a great circle passing through the two poles, and every circle so passing must be the meridian of some particular places, according to our original definition of a meridian. Generally therefore they are called *meridians*, or *meridional lines*, and the equator bisects all the meridians. In this sense of the word the meridians are lines in the sphere partaking of its general revolution. The meridian of a particular place, on the contrary, is a fixed line, with which each of the others in the course of its revolution coincides. Thus, referring to the common celestial globe for illustration, lines drawn from pole to pole on the surface of the globe would be *meridians*, using that term generally: the brass meridian corresponds to the *meridian of the place*, for which the globe is adjusted.

The equator EA_1Q being a circle bisecting every meridian, has every point equidistant from P, or p: it is therefore a circle of rotation, or any point situate in the equator describes a course round the earth coincident with the equator itself. Let ea_1a_2q be any small circle parallel to the equator. It will, by the properties of the sphere, have every point equidistant from P or p, and will consequently be a circle of rotation. If e, a_1, a_2, a_3, be the several points where the different meridians $PeEp$, Pa_1A_1p, Pa_2A_2p, Pa_3A_3p, cut this small circle, and E, A, A_2, A_3, the points where the same meridians cut the equator, then it is a property of the sphere, that whatever be the proportions that the several arcs EA_1, EA_2, EA_3, bear to the whole circle EA_1Q, the arcs ea_1, ea_2, ea_3, bear the same proportions respectively to the whole circle ea_1q. But the periods of rotation of a star in the small circle ea_1q, or in the great circle EA_1Q, are equal; for they are each a sidereal day; and the motions of rotation in each are uniform: the periods of rotation therefore through the arcs a_1e, a_2e, a_3e, are to the whole sidereal day, in the proportions of the respective arcs themselves to the

* Any circle formed by the intersection with the sphere of a plane passing through the centre of the sphere, is called a *great circle* of the sphere. If a diameter of the sphere be drawn perpendicular to this plane, the extremities of this diameter are called the *poles* of the great circle; and if any number of great circles be drawn through these poles, they will each be in planes perpendicular to the first circle, and are called *secondaries* to that great circle. Being in planes perpendicular to it, they are so, of course, to all other circles parallel to it. Thus, taking the circles whose names are given in the text, the north and south poles of the heavens are the *poles* of the equator, and every meridian is a *secondary* to the equator, and perpendicular to it, and to every circle parallel to it: or every meridian is perpendicular to every circle of daily rotation, for all such circles are parallel, and the equator is one of them.

It may also be convenient here to mention that all great circles in the same sphere are equal, and that they all bisect each other.

On all matters relating to the properties of the sphere, see Treatise on Geom. Book VI.

whole circle $e a_1 q$; and the periods of rotation through the arcs A_1E, A_2E, A_3E, are to the same whole sidereal day in the proportions also of those respective arcs to the whole circle $E A_1 Q$, the same proportion as the preceding ones. The periods of rotation therefore through the arcs $a_1 e$, A_1E, are equal; and so are those through $a_2 e$, and A_2E, and again through $a_3 e$, and A_3E. But when any point of one of these meridians coincides with the meridian of the place, the whole meridian coincides with it; or every point of a meridian is on the meridian of the place at the same time; and the periods of rotation through corresponding arcs being equal, every point of a meridian, when not on the meridian of the place, requires the same time to arrive at it; and that time bears the same proportion to a sidereal day that the arc of the equator, intercepted between the meridian of the place and the meridian in question, bears to the whole circumference of the equator. And conversely, if we observe the time when one star is on the meridian of the place, and that when another body appears there on the same side of the pole (for this qualification is necessary in the case of circumpolar stars which appear on the meridian of the place both above and below the pole), and note their interval, we know the interval of the equator intercepted between the two meridians; for it is the same proportion of the whole equator which the observed interval of time is of the whole sidereal day. It is of course convenient to have some fixed point of the equator as a standard of reference, and that point, which we shall presently see the reason of choosing, is called the first point of Aries, and is often marked thus ♈. The distance, measured Eastward, from ♈ to the intersection with the equator of the meridian passing through any heavenly body, is called *the right ascension* of the body, and is obviously the same for every body upon the same meridian, and upon the same side of the pole, and for none off it. The right ascension may evidently be estimated in time, as it is ascertained by it. The period of a whole revolution, or of passing through 360°, is twenty-four hours of sidereal time: in one such hour therefore a body passes through 15° of right ascension; in four minutes through 1°; and these right ascensions may be equally well called either 1° and 15°, or four minutes, and one hour. In other words, an astronomical clock marks $0^h\ 0^m$ when the first point of Aries is on the meridian; when it marks 4^m the bodies then on the meridian are said to have 1° or 4^m of right ascension: when it marks 1^h, the bodies on the meridian are said to have 15° or 1^h of right ascension.

By the observation of time therefore we can ascertain the right ascension of a heavenly body, and consequently the meridian on which it is. If we can also observe its distance from the pole, or its distance from the equator, which is called *its declination*, and is called North or South declination as the body is North or South of the equator, we ascertain its position upon that meridian. The declination being measured along a meridian, the meridians have also the name of *circles of declination.* Every point in a small circle parallel to the equator, is of course at the same distance from the equator, or, in other words, has the same declination: such a small circle is therefore called a *parallel of declination.*

The observation of declination is easily made. The horizon of any given place is a fixed circle: the point therefore where a perpendicular to the horizon meets the heavens above the spectator's head (which is called the *zenith**) is a fixed point also. This point may always be ascertained, for a plumb line hangs directly in a line from it. The zenith is obviously a point in the meridian; for the plane of the meridian passes through the place of observation perpendicular to the horizon, and the perpendicular to the horizon at the place must be a line in that plane.

The zenith is the pole of the horizon, and if any great circles be drawn through the zenith, they will be secondaries to the horizon. These are called *vertical circles:* the meridian of the place is of course one of them. The arch of a vertical circle intercepted between a heavenly body and the horizon, is called the *altitude* of the body, and can always be measured by proper instruments†.

* The opposite point, where the perpendicular produced below the horizon meets the opposite hemisphere of the heavens, is called the *nadir.* These, like many others of the terms of astronomy, are derived from the Arabian observers.

† The observer being placed apparently in the centre of the sphere, the angle which two objects in the heavens subtend at his eye will accurately measure the arc of the great circle of the heavens which lies between them. This arc therefore, or their distance, is accurately measured by observation of the angle between them; or rather, as we know nothing of their actual distances, we can only observe the angles, which give us the relative *directions* of the bodies; and then we suppose

The distance from the body to the zenith, or its *zenith distance*, together with the altitude, make up the whole distance of the zenith from the horizon, or 90°. The zenith distance, therefore, is the *complement* of the altitude.

The pole is a fixed point in the meridian; it is therefore at a fixed distance from the zenith of any given place, and from the North and South points of the horizon there. If therefore, when any heavenly body is on the meridian, its distance from either the zenith or the North or South point of the horizon can be observed, its distance from the pole, or its *North Polar distance*, can be ascertained from it. And it is obvious that if an instrument be adjusted so as to move in the plane of the meridian, these distances may be ascertained by mere observations of altitude made with it: for the altitude of a body when on the meridian, is its distance either from the North or South point of the horizon; and the complement of the altitude is the zenith distance. The adjustment of the instrument to move in the plane of the meridian, is more complicated than that required merely to make it move in a vertical circle: it is however generally made with the best instruments, which are fixed so as to move in the plane of the meridian, and in that only. Having an instrument so adjusted, the north polar distances of all the stars may be ascertained.

Let S_1, S_2, S_3, S_4, in *fig.* 2, represent any heavenly bodies on the meridian: $S_1 E$, $S_2 E$, $S_3 E$, $S_4 Q$, are obviously their declinations. Now, observing that PE, or $PQ = 90°$, it is obvious that whenever the North polar distance, as PS_2, PS_3, PS_4, is less than 90°, the declination $S_2 E, S_3 E, S_4 Q = 90° -$ North polar distance; and in these cases the declination is North. When the North polar distance, as PS_1, is greater than 90°, the declination $S_1 E =$ North polar distance $- 90°$. In each case therefore the declination is the difference between the North polar distance and 90°, being North when the North polar distance is less, and South, when it is greater, than 90°.

We can then always ascertain the declination of a body by observing its altitude when on the meridian of the place. Its right ascension may be ascertained in the manner already pointed out: or, if the right ascension of any particular star has been previously accurately ascertained by a sufficient number of observations, then the right ascension of any other star may be found by observing the time which elapses between its appearance on the meridian and that of the star, whose place we suppose to be already determined. The observation of right ascension gives us the meridian on which the body is situated in the heavens, and that of declination the precise point of that meridian, and the two combined give us the exact place of the body: for whenever we can ascertain the distance of a body in a known surface from a given point, measured in two directions perpendicular to each other (as in this case along the equator, and along a circle of declination), or, indeed, inclined at any given angle to each other, there is only one point which can fulfil both conditions, and that point therefore is completely ascertained.

In this manner we can determine the place of the sun on every day in the year; and the observation of all these points will give us so many of his successive positions in the heavens, and his path among the stars must, therefore, be a line passing through all these points. This line is found, when the observations are carefully made and registered, to be a great circle of the sphere, which intersects the equator at an angle of 23° 28′ nearly, and is called *the ecliptic**. All great circles of the sphere

them ranged in the surface of a sphere, for the convenience of computation, by applying to them the principles of spherical geometry.

The altitude of a body is easily ascertained. The instrument with which the observation is made can be mechanically adjusted, so as to move in a vertical plane, and, consequently, in the plane of a vertical circle. The place of the horizon may be ascertained, though some nicety is required in the ascertainment, into the details of which it is not necessary here to enter. All instruments intended for observation have a graduated circular limb attached to them, and when the plane of the horizon is ascertained, the points on the limb which are in that plane may be found; and the angle between the diameter joining them and the direction of the instrument, which obviously measures the altitude of the body, observed by means of the graduation on the limb. There are additional and more delicate adjustments, which need not be detailed here.

* The ecliptic being a great circle of the sphere, has, of course, two poles, which are distant from the poles of the equator by an arc of 23° 28′, equal to the inclination of the one circle to the other. If from the pole of the ecliptic a secondary be drawn through any heavenly body to the ecliptic, the arc of this circle between the body and the ecliptic is called the *latitude* of the body, and the arc of the ecliptic intercepted between the first point of Aries, and the intersection of the secondary with the ecliptic, is called the *longitude* of the body. The place of a body is of course as

intersect and bisect each other; and one of the points where the ecliptic intersects the equator is called *the first point of Aries* (the point from which we have said that the right ascension is measured); the other, *the first point of Libra.* For reasons which will afterwards be seen, the former of these is also called the *vernal equinox*, the latter the *autumnal equinox.* The first point of Libra is often marked thus, ♎.

SECTION 4.—*Variation of Sun's apparent diameter — Sun's orbit an Ellipse, the earth being in the focus —Variation of angular velocity — Equable description of areas by Sun's radius vector.*

WE have already stated, that if any plane pass through the centre of the sphere, its intersection with the sphere is a great circle; and that the ecliptic, or sun's apparent path, is such a circle. As yet we know nothing of the distance of the sun from the earth; but it is clear that we have no reason to suppose his distance equal to that of the stars, or always the same: if we have no means of comparing his distance with that of other objects, we can only ascertain the direction in which he appears. Thus the sun, the moon, the stars all appear alike situated in the concave surface of the heavens, although we shall hereafter see that their distances are almost immeasurably different; and any one who has seen a bright meteor at night, such as those commonly spoken of as shooting, or falling stars, knows that they also appear to the eye to be similarly distant. The very name indeed of shooting *stars* is evidently deduced from this circumstance. In their case indeed we know them really to be meteors engendered in the atmosphere, and therefore know their distance to be comparatively small, or we may sometimes ascertain it to be so by comparing observations made at different places; and in the case of the moon, we may perhaps imagine that she looks less distant than the sun or stars: but this notion, when it obtains, is probably really owing to our habitual knowledge that she is so, or to our being able to distinguish different parts of her surface from each other, by the different degrees of their brightness. If however we abstract ourselves from such considerations, we shall find ourselves alike unable to estimate by the eye the distances of any of these objects, and consequently unable to say which are furthest from us. All therefore that we can conclude, from finding the apparent path of the sun to be a great circle of the heavens, is, that his motion is in a plane passing through the earth; but what the shape of his circuit is, whether his distance be always the same, or be variable, and whether, if it vary, it vary gradually and regularly, or suddenly and without any fixed rule, we as yet are quite ignorant. We know only that we continually see him more Eastward at every successive observation, and consequently conclude that his motion is always in that direction, and not occasionally East, and occasionally West.

To determine the form of his orbit, we need only know the proportions of his distances from the earth at different times. For this purpose the actual distances are unimportant; for instance, if his distance is always the same, his orbit is a circle, whatever be that distance, which is the radius of the circle. Now the apparent diameter of an object, or the angle which it subtends at the eye of the observer, increases in the same proportion as the distance diminishes, so long as the real diameter continues unchanged, and the apparent diameter is very small: * or in such cases, *the apparent diameter varies inversely as the distance.* Now the apparent diameter of the sun may be very accurately measured, and it is found to be continually varying: his distance therefore is so also, for we have no reason whatever to suppose that his actual magnitude changes; and the proportions of his apparent diameters at different times being ascertained by observation, the proportions of his cor-

well ascertained by knowing its longitude and latitude, as by knowing its right ascension and declination; but the latter are the best adapted for observation, and the former may always be deduced from them. The terms *longitude* and *latitude*, as applied to a heavenly body, are established; but it is to be regretted that they have been chosen, as their sense in that application differs widely from their meaning with respect to places on the earth, being, indeed, analogous measures in the two cases, but referred to different circles and planes.

* [In strictness, the *tangent* of the apparent radius (i. e., of the angle subtended by the radius) = real radius divided by the distance: and the real radius being always the same in the same body, the *tangent* of the apparent radius varies inversely as the distance. Whenever, therefore, the apparent radius is so small, that the arc and its tangent may be treated as equal, the apparent radius varies inversely as the distance: and so of course does the apparent diameter, which is double the apparent radius.]

responding distances may be so also. If then we conceive lines drawn from the earth to every point of his orbit, the observation of his right ascension and declination will always determine on which of these lines he is situated; and if we take portions of these lines always in the inverse proportion to each other of his apparent diameters, the curve joining the extremities of these portions will be similar to his orbit. This perhaps may be rendered clearer by a figure, and an instance. Thus let us suppose the sun to be observed on any day, and that his apparent diameter is $31\frac{1}{2}'$; again, and that his place in the ecliptic is then 70° from that of the first observation, and his apparent diameter then of $32'$; and a third time, and that his place in the ecliptic is then 180° from that of the first observation (or exactly opposite to it), and his apparent diameter then of $32\frac{1}{2}'$. Take

Fig. 3.

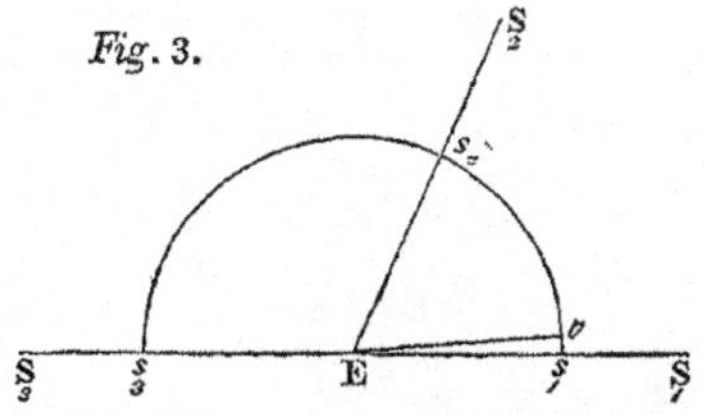

any line $E S_1$ to represent the direction of the sun at the first observation; and produce it, in the opposite direction, to S_3: then $E S_3$ will represent the direction of the sun at the third observation; and if $E S_2$ be drawn making an angle of 70° with $E S_1$ it will represent the direction of the sun at the second observation. Now the apparent diameter of the sun at the first observation is $31\frac{1}{2}'$, at the second it is $32'$; his corresponding distances, therefore, are in the proportion of 32 to $31\frac{1}{2}$, or of 64 to 63. If, therefore, we take in the line E S, a portion $E s$, of any magnitude we choose, and measure upon $E S_2$ a portion $E s_2$ which shall be to $E s_2$ in the proportion of 63 to 64, the distances $E s_1$, $E s_2$ will be proportionate to the distances of the sun at those two observations. In the same manner, the apparent diameter of the sun at the third observation being of $32\frac{1}{2}'$, his distance at that observation is to his distance at the first observation in the proportion of $31\frac{1}{2}$ to $32\frac{1}{2}$, or of 63 to 65, and it is to his distance at the second observation in the proportion of 32 to $32\frac{1}{2}$, or of 64 to 65. If, therefore, we take in the line $E S_3$, a portion $E s_3$, which shall be to $E s$ in the proportion of 63 to 65, or to $E s_2$ in the proportion of 64 to 65, the distance $E s_3$ will be proportionate to the distance of the sun at the third observation. All the distances, therefore, $E s_1$, $E s_2$, $E s_3$, will be proportionate to the distances of the sun at the different observations; and the points s_1, s_2, s_3, will accurately represent his situations at those times. In the same manner his situations at other and intermediate times may be represented, and the line joining all these points (the curve $s_1\ s_2\ s^3$) will represent his orbit. It is found, by careful observation, that this line is an ellipse, of which the earth is in the focus*.

The ellipse is an oval curve, divided into two equal and similar parts by a line drawn through its foci, which is called its *major axis*, or *transverse axis*. If from one of the foci lines be drawn to every point in the curve, the greatest of these lines is that portion of the major axis which is drawn through the other focus to meet the curve, and the least is the other portion of the major axis; and the lines continually diminish from the greatest to the least, and continually increase from the least to the greatest. It follows therefore, if the sun moves in an ellipse round the earth in the focus, that his apparent diameter should continually increase from its least amount, which is when the sun is at his greatest distance, or in his *apogee* (from two Greek words signifying *away from the earth*) to its greatest amount, which is when the sun is at his least distance, or in his *perigee* (from two Greek words signifying *about* or *near the earth*); and as the two portions of the ellipse are equal and similar, the diminution of his distance in the one case will correspond with the increase of it in the other, and at the end of a complete revolution he will again be at his original distance. And so we actually find it. The least apparent diameter of the sun is of $31\frac{1}{2}'$ nearly, and is at present observed about the 30th of June; he is therefore then in his

* The ellipse is one of the curves called *conic sections*. It is the curve made by cutting a cone by a plane, which passes through it without intersecting the base. Its fundamental property is this: there are two points within it, called the *foci*, such that the sum of the distances of any point in the curve from the two foci is always the same. This property furnishes an easy mode of drawing the curve. If a thread be fixed at the two foci, and a pencil carried round within the thread, so as always to keep it stretched, the pencil will describe an ellipse; for the whole length of the thread is always the same, and it constitutes the two distances from the foci.

apogee: his apparent diameter then continually increases until (about the 30th of December) he arrives at his perigee, when his apparent diameter is of 32′ 35″ nearly: and then again it continually diminishes until (about the 30th of June) it is again of 31½′, from which value it again begins to increase. The variations of the apparent diameter being so small, those of the distance are so also, or the ellipse differs but little from a circle.

Having thus ascertained that the sun appears to move in an ellipse round the earth, we next inquire what is the rule of his motion, whether it be regular or irregular, uniform or variable. We have already seen that we can, by ascertaining the right ascension and declination of a body, determine its place in the heavens, or the direction in which it is seen from the earth: we may therefore ascertain at each particular instant the point of the ecliptic in which the sun is, and, consequently, the arc through which he has moved in each particular interval, or the angles which the directions in which he is successively seen make with each other. We have thus the means of ascertaining his angular velocity, as referred to the earth. Now this angular velocity is found to be variable, and it is also found to be greatest when the apparent diameter is greatest or the distance least, and least when the apparent diameter is least or the distance greatest; and generally, the greater the distance the less we find the angular velocity. Its inequalities however are greater than those of the apparent diameter or distance. We have already seen that the least apparent diameter is to the greatest nearly in the proportion of 30 to 31; but the proportion of the least angular velocity to the greatest is nearly that of 30 to 32, and hardly so great. On accurate comparison of the different angular velocities with the distances, it is found that they vary *inversely as the squares of the distances*, that is, that they diminish in the same proportion as the squares of the distances increase, and increase in the same proportion as the squares of the distances diminish. For instance, if the distance is doubled, the angular velocity is reduced to a quarter of its former amount; if the distance is diminished by a third, the angular velocity is increased in the proportion of 1 to the square of two-thirds, or of 9 to 4.

These velocities however, and the distances themselves, may be considered for very short periods of time as constant, for the changes of distance in such periods are so small that they may be neglected. The whole variation of distance is only about a thirtieth part of the least distance; the greatest difference between the angular velocities only about a fifteenth part of the least angular velocity; and these differences are the accumulated differences of months: for an hour therefore, or a day, there will be no perceptible difference, and none which can at all affect the results which we shall deduce from the supposition that, during such a period, they are uniform.

In *fig.* 3, let t be the place of the sun in its orbit, when one of these very short periods has elapsed since it was at s_1. If we conceive a straight line drawn from the earth to the sun, and moving round the earth with the sun, as for instance, if they were joined by a wire, which we must suppose to be lengthened and shortened as the sun recedes from and approaches the earth, so as always to extend from the centre of the one to the centre of the other, and no farther, this line will originally have coincided with E s_1, and its position, when the sun is at t, will be E t. While the sun has moved from s_1 to t_1, therefore, or described the arc s_1 t_1 this line or wire will have described the small area E s_1 t. This line is called the *radius vector*, and it is of great importance to ascertain the areas which it describes, and these we shall find to be always equal in equal portions of time.

[This area may be considered as a triangle, for the arc s_1 t, being very small, differs insensibly from the straight line joining the points s_1 t; and the area of a triangle is equal to half the product of its base, and the perpendicular from its vertex. Again, the angle t E s_1 being very small, the perpendicular from the vertex t is indistinguishable from a small circular arc, whose centre is E, and radius E t, or E s_1; and these also may be considered as equal. The magnitude of such an arc varies as the product of the angle t E s_1, and the radius E t, or E s_1. The area t E s_1 therefore (which is equal to half the product of its base and the perpendicular from its vertex) varies as the distance E s_1 multiplied by the product of the angle t E s_1 and the radius E s_1 (for its base is the distance E s_1, and the perpendicular we have already seen to vary as the product of the two latter quantities); or it varies as the square of the distance E s_1, multiplied into the angle t E s_1. If, however, the area E s_1 t be supposed to be that described in some given portion of time, as an hour or a minute, and the angular velocity be uniform during the time of describing it, the angle t E s_1, varies as the angular velocity; for it will increase or diminish exactly in the same proportion as that velocity increases or diminishes. The area t E s_1 therefore, described in such a period of time, varies as the square of the distance E s_1, multiplied by the angular velocity. But we have seen that the angular velocity varies inversely as the square of the distance: the area t E s_1 therefore is not affected by the distance at all, or it is constant; for in whatever proportion it is increased by reason of the increase of the distance E s_1, in the same proportion it is diminished by the diminution of the angle t E s_1.]

If however the areas described by the radius vector in small equal portions of time be equal in every part of the orbit, the whole areas described by the radius vector in equal portions of time of any magnitude will also be equal; for they will be the sums of equal numbers of these small equal portions: and generally, *the areas described by the radius vector in any times whatever will be proportional to those times.* This therefore is ascertained to be a law of the motion of the sun round the earth, and it is one which we shall hereafter find of the very greatest importance.

We have now ascertained these facts with respect to the sun: that its annual motion round the earth takes place in an ellipse, of which the earth is in one of the foci; that the plane in which he moves is inclined at an angle of 23° 28′ to the plane of the equator; and that his radius vector describes areas proportional to the time. These are the principal phenomena of his motions: they require however some qualifications, which we shall hereafter mention; and we shall also see that we may explain his apparent motions on the supposition that he is at rest, and the earth moves round him, just as we have said that we shall be able to explain the apparent diurnal motions of the stars, by supposing them at rest, and the earth spinning on an axis.

We may however at once proceed to explain how these facts produce some of the most important phenomena which we observe, especially the difference of the seasons in the same country, the different lengths of the day (as distinguished from the night) at different periods of the year, and the different climates of different countries; and also the equation of time, or the manner and degree in which the real differs from the mean solar time.

SECTION 5.—*Comparative length of time during which bodies at different Declinations continue above the horizon—Variation and length of Day at the same place—Tropics—Equinoxes—Seasons produced by the Sun.*

WE have already seen that the apparent diurnal motions of the heavenly bodies are performed in circles, every point of which is equidistant from the pole: of these circles the equator is one, and is a great circle of the sphere; the others are all small circles parallel to the equator. The horizon, being a great circle as well as the equator, bisects that circle. It will immediately appear, also, from inspection of any of the figures, as *fig.* 1, or *fig.* 2, that of the parallel circles a larger portion is continually above the horizon, and a smaller continually below, as they approach nearer that pole which is above the horizon. Thus, in *fig.* 1, P q being less than P Q, $t\,q\,r$ is a larger portion of the circle $t\,q\,r\,b$ than T Q R is of T Q R B: and the portions continually go on increasing, until, finally, the whole circle is above the horizon, as in the case of $e\,a_1\,a_2\,a_3\,q$ in *fig.* 2, or of X V Y U in *fig.* 1, when the distance, P q or P U, of the body from the pole is less than the elevation of the pole above the horizon. In the same manner, if a body move in a circle distant from the other pole, p, less than its depression below the horizon, or than H p, it will never appear above the horizon at all. We have also seen that the diurnal motions of the heavenly bodies are equable; that is, that the times of passing through every successive portion of their circles of motion bear the same proportion to the whole time of rotation, that the portion described bears to the whole circle; and that the times of rotation in every circle are the same. The larger, therefore, the portion of each circle of rotation which is above the horizon, or the nearer that circle approaches to the pole which is above the horizon, the longer the body is above it. If then the North Pole be above the horizon, as it is here, those bodies which have the least South, or the greatest North declination, are longest above the horizon: and in like manner, where the South Pole is above the horizon, the most Southerly bodies are longest visible. The equator is in all cases bisected by the horizon; in every place therefore a body in the equator is for an equal length of time above and below the horizon; and it is from this property that the circle itself derives its name.

These positions, which are deduced from the supposition that the diurnal motions take place in parallel circles, and uniformly, are not strictly and accurately true for the sun, whose place on each succeeding day is either North or South of that of the day before, and whose diurnal motion, being compounded of the diurnal rotation of the whole heavens and his own motion in his orbit, cannot be strictly uniform, because his motion in his orbit is not so. Neither of these inequalities however need here be taken into consideration, so small is the proportion which they bear to the general diurnal motion which they affect. The heavens complete their rotation, or move through 360°, in a day. In that time the sun's motion Northwards or Southwards, or his motion in declination, is never so much as 24′, and seldom near it: a quantity so small that we cannot err in considering, for purposes of general explanation, the diurnal motion of the sun to take place in a circle parallel to the equator, and distant from it by the sun's distance from it when he is on the meridian. The other inequality arising from the inequality of the sun's motion is still less perceptible: the whole motion of the sun in his orbit is less than 1° in a day on an average, and very little more when greatest; the whole difference between the greatest and least motions is only about a fifteenth part of this, or about 4′, a quantity which, for all purposes of explanation, though not of computation, may safely be neglected in comparison with 360°. Even this, which occasions some difference of length in days taken at periods thus distant from each other, is the accumulated difference of half a year; the difference in a single day will be quite imperceptible, and the motion during one day may therefore be considered as perfectly uniform. Thus we may with safety apply to the sun our conclusions with respect to the length of time for which a body is above the horizon*, and say that, where the North Pole is above the horizon, he is longer above the horizon as he comes farther North, and less so as he returns Southward; or generally, as his being above or below the horizon constitutes the distinction between day and night, that the days are longer as he is more to the North, and shorter as he is more to the South. Where the South Pole is above the horizon, the days are of course longer when he is more to the South, and shorter as he moves Northward. Whatever conclusions, indeed, are true with respect to the parts of the earth where the North Pole is above the horizon, or of which the latitude is North, are true also for those where the South Pole is so, or of which the latitude is South, if we change the positions of the words, and attribute the same results to South declination, &c., which, in the other case, we ascribe to North; and we shall not in future express both, but suppose the North Pole above the horizon, and deduce our results on that supposition. The reader will find no difficulty in transferring them to the other case.

We have then the sun moving in an orbit, of which the most Southern point is 23° 28′ South of the equator, and the most Northern 23° 28′ North of it. Taking his motion from the first of these points, where he is about the 21st of December, he continually travels Northward till about the 20th of March, when he is in the equator, and still Northward till about the 21st of June, when he attains his greatest Northern declination; he then returns towards the South, passes the equator again about the 22nd of September, and returns to his greatest South declination about the 21st of the following December, when he again returns Northward. His successive places, from December the 21st to June the 21st, being continually more and more Northward, his time of continuing above the horizon, or the day, conti-

* The sun also being thus considered to move on any given day in a circle parallel to the equator, will have the part of that circle which is above the horizon bisected, like any other heavenly body, by the meridian; or he will be an equal time above the horizon on each side of the meridian. He will, therefore, be on the meridian at mid-day; and this is the origin of the word meridian, which is derived from *meridies*, a Latin word signifying mid-day. The primary application of the word therefore was to the meridian of the place, the line on which the sun appears at noon; the general application to all great circles passing through the two poles is only secondary, and derived from the circumstance, that each of these must correspond at some time with the meridian of the place.

nually lengthens; and in the same manner it continually shortens from that time till the 21st of December. The longest day is the 21st of June, the summer solstice, or the day on which he has his greatest Northern declination; the shortest, the 21st of December, the winter solstice, or the day on which he has the greatest Southern declination; and the circles which on those days he describes are called *the tropics*, from a Greek word signifying *to turn*, because, although still proceeding in the same direction in his orbit, he has reached the extreme North or South point of it, and, therefore, as it were, turns back, and travels in an opposite direction (as to North and South) to his former course. At the two periods when the sun is in the equator, or when he is at the first point of Aries, and the first point of Libra, the parts of his daily course above and below the horizon are equal, or the day and night are equal; these therefore are called *equinoxes*, and taking place in spring and autumn, they are called the *vernal* and the *autumnal* equinox, and the equator itself is sometimes called *the equinoctial* on the same account. The rate of the sun's motion Northward or Southward is most rapid at the time that his orbit intersects the equator, and is hardly perceptible just when he is at his greatest distance from that circle*: the variation of the length of the day is therefore most rapid about the equinoxes, and least so about the solstices. The rate of variation is also affected by the rate of his motion in his orbit; for as his motion in declination is produced by that, the greater his motion in his orbit is, the greater, all other things continuing the same, is his motion in declination. But his motion in his orbit is greatest when he is in and near his perigee, and least when in his apogee; and as he is in his apogee nearly at the summer solstice, and in his perigee nearly in the winter solstice, his motion is slower in summer than in winter, and the days vary still more slowly in length at the summer than at the winter solstice. For the same reason, as the two portions of the ellipse from perigee to apogee, and from apogee to perigee are equal and similar, the sun will take longer in describing those 180°, or that half of his orbit in the course of which he is in apogee, than that half during which he is in perigee: and as he is in apogee between the vernal and autumnal equinox, and in perigee between the autumnal and the vernal, the former interval must be longer than the latter, or the summer in these parts of the world longer than the winter. In fact, the difference at present is of 7 days, 16 hours, and 50 minutes.

* We have already stated this as a fact ascertained by observation: it will be hereafter (in the section on the equation of time) shewn to be a necessary consequence of the inclination of the ecliptic to the equator.

Having thus seen how the motion of the sun produces the various lengths of the day, it is easy to perceive how it also produces the different seasons of the year. Light and heat being essential to all the great operations of nature, and the sun being the great source of light and heat, his position and the length of time he is above the horizon must be the great causes of the variations of season. The variation of his distance would of course tend to make his influence greatest when he is nearest, and this would seem contrary to observation, as he is nearest in the winter. In fact however other circumstances are much more powerful in increasing and modifying his effect than this, which could not in its extreme variations produce a difference of more than about one-fifteenth*. The main causes of the different warmth of the seasons are to be found in the different lengths of the time during which he is above the horizon, and the different elevation he attains above it. The difference of the length of the day evidently tends to increase the temperature by giving the sun a longer time to act; and as with us the longest day is more than twice the length of the shortest, this cause by itself is far more than enough to counterbalance the effect of the increased distance of the sun. The other cause, the difference of elevation, requires a little more explanation. In the first place, the greater the elevation of the sun, the more directly do his rays strike upon the earth, and the greater proportion of them does a given surface receive. Thus, in *fig.* 4 let the parallel

Fig. 4.

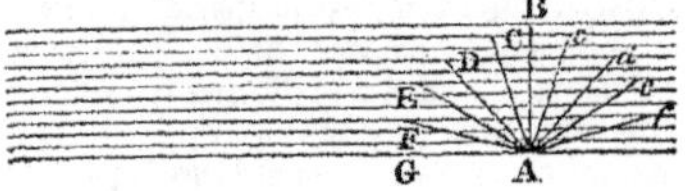

* All other things continuing the same, the intensity of light and heat diminishes in the same proportion that the square of the distance from their source increases. In the case of the sun and earth, the extreme difference is about one-fifteenth.

and equidistant lines represent rays of light proceeding from the sun, and A B an object of any given magnitude exposed to them in a direction perpendicular to them, so as to intercept them all. Now, if A B be gradually turned round the point A, so as to form an angle continually more and more acute with the direction of the rays, as at A C, A D, A E, A F, or continually more obtuse as at A *c*, A *d*, A *e*, A *f*, it is evident, from inspection of the figure, that it will in each successive position receive fewer of them; and this will be the case however close the rays be to each other, if they are uniformly so; for, however we may multiply the number of rays which pass between A F and A G, or which fall on the object in the position A F, the number which it ceases to receive by passing from the position A B to A F, or the number of those above F in the figure, is increased in the same proportion. Besides this, we know that both light and heat are absorbed in passing through the atmosphere; and as the height of the atmosphere above the surface of the earth is the same, or nearly so, in all places, the course of a ray through the atmosphere is shortest when it falls most directly upon it, and longer as its deviation from the perpendicular increases*. On this account also, the greater the elevation of the sun, the greater is the light and heat which he affords: and we accordingly observe, that the mid-day sun does actually give more light and heat than that of the morning or afternoon, and the summer sun than that of winter.

It is true that the greatest heat of the day is generally after noon, and the greatest heat of the year after Midsummer; but these facts are not inconsistent with the principles we have stated. There is a continual tendency in nature to equalize the heat of different bodies; those more heated parting with their heat to the cooler. This process is continually going on; but, as long as the sun is above the horizon, and furnishing new supplies of heat, the heat will increase if he supplies it faster than by the ordinary processes of cooling it is diminished; and it will diminish as soon as his supplies are less than the ordinary consumption. For a considerable period after noon he generally supplies heat faster than it is distributed, and consequently, although the actual supplies diminish, the whole stock increases: just as if you pour water rapidly into a vessel with a hole at the bottom, the water will rise in the vessel as long as you pour it in more rapidly than it runs out, although you may supply it less copiously after a time than you did at first; or, as the mercury in a thermometer brought from a frosty air, and placed near a fire, will continue to rise in the room after it is removed from the fire, if it has not been kept there so long as to render its temperature equal to that of the atmosphere of the room. The same reason also explains the greater heat of July and August than of June. As long as the surface of the earth and the atmosphere in the course of twenty-four hours receive more heat than they part with, the general heat of the weather increases; and this is the case long after the summer solstice, though the actual quantity of heat then received is greatest. The heat of particular days is principally influenced by other local and accidental causes; but the general difference of the seasons depends on these principles, and on the different degrees of moisture and cloud in the air, and the different quantity of evaporation and radiation which consequently takes place at different times. Any theory of the seasons resting merely on the sun's position, without reference to these other causes, would be imperfect and fallacious; but, as far as the sun's position affects them, they are thus to be explained.

Section 6. — *Differences as to the length of Day and the Seasons at different places—Days at the Equinoxes everywhere of the same length—Longest Days greater, and shortest Days less, as the latitude increases—If latitude above 66° 32′, there are some periods of twenty-four hours during which the Sun never sets, others, during which he never rises; and these are longer as the latitude is greater—At the latitude 90°, there is only one Day, and one Night—At the latitude 0°, the Days and Nights are all equal—Length of Day increased by the Sun's magnitude, and by refraction—diminished by parallax—Effect of the Sun on climate at different places.*

We have thus far spoken of the difference of days and seasons at a particular place; we have still to shew how

* See this more fully explained in a note in the section on Refraction.

different places are affected, to explain the diversity of climates, and the exceptions which particular parts of the earth (the circumpolar and intertropical regions) offer to the general account which we have given.

For this purpose, let us suppose several places, at each of which the sun, or any other heavenly body, or any point of the heavens, comes upon the meridian at the same time, or where the meridian of the place is the same great circle of the heavens. This supposition is only made for the sake of simplifying a necessarily complicated figure; for every conclusion which is true for a place having this meridian, will be equally so for a place having any other, if the elevation of the sun when on the meridian is the same at each. We have already seen, that at every place half the equator is always above, and half below the horizon; the equator and horizon being both of them great circles. Now as half the equator is above the horizon at each place, and the point of the equator on the meridian is the middle point of that half, and that point, on the supposition we have made, is the same at each place, the points of the equator which are upon the horizon must be the same at each place; or, in other words, the horizons of different places under the same meridian always intersect the equator at the same points. If, therefore, in *fig.* 5, we take the circle

Fig. 5.

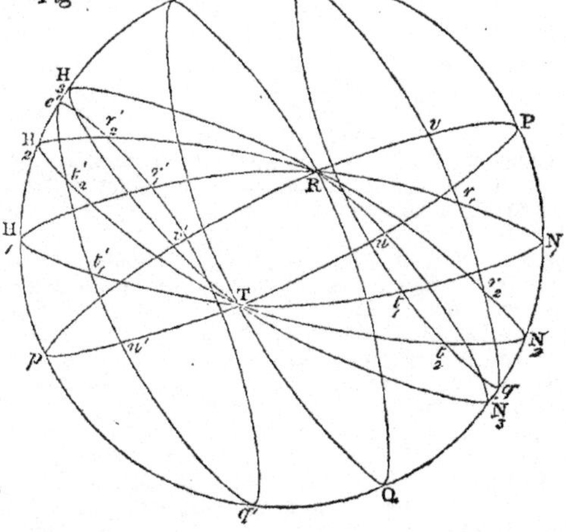

E R Q T to represent the equator, and $H_1 R N_1 T$ to represent the horizon of a particular place, R and T, being the points where the equator and this horizon intersect each other, will also be the points where the equator and the horizon of any other place which has the same meridian intersect each other; and consequently, if through the points R and T, we draw other great circles $H_2 R N_2 T$, $H_3 R N_3 T$, these will represent the horizons of other places under the same meridian. If P be the pole, the elevation of the pole at these different situations will be PN_1, PN_2, PN_3. Now, let $e\,q$, $e'\,q'$ be two small circles parallel to the equator E Q, the one North of it, the other South of it; they therefore are circles of diurnal rotation: and let the points where they respectively cut the different horizons, be r_1, t_1, r_2, t_2, r_1', t_1', r_2', t_2'; and we shall immediately see, by inspection of the figure, how the elevation of the pole, or the latitude of the place, affects the time during which any heavenly body is above the horizon, and consequently the length of the day at different periods of the year.

The equator is bisected by all the horizons; and of course a body moving in it is everywhere half its time above, and half below the horizon; when the sun therefore is in the equator, day and night are everywhere equal, so that those periods are called the equinoxes, not from any accidental equality affecting a particular place of observation only, but from an universal fact.

Let us next take the case of a circle

of rotation North of the equator, as $e\,q$, and let us conceive a great circle described through the points R and T, and passing also through the poles P and p. By a known property of the sphere, this secondary to the equator E Q bisects it, and every circle parallel to it. If, therefore, u and v be the points where it cuts the circle $e\,q$, $u\,e\,v$ is half that circle. Now the part of the circle $e\,q$ which is above the horizon $H_1\,N_1$, is $t_1\,e\,r_1$, which is evidently greater than $u\,e\,v$, or more than half the circle of rotation $e\,q$ is above the horizon; and the part of the circle $e\,q$ which is above the horizon $H_2\,N_2$ is $t_2\,e\,r_2$, which is evidently greater not only than $u\,e\,v$, but also than $t_1\,e\,r_1$. But the elevation of the pole above the horizon $H_1\,N_1$, is $P\,N_1$, its elevation above the horizon $H_2\,N_2$ is $P N_2$; the greater portion, therefore, of the circle $e\,q$ is above that horizon above which the pole is most elevated, or the horizon of that place which has the greater latitude. Now $e\,q$ being a circle of rotation North of the equator, may represent the sun's line of diurnal motion when his declination is North; and we consequently deduce the following general rule, that *wherever the North Pole is above the horizon, and the sun's declination is North, the day is longer than the night; and that in different places, the day is longer where the North Pole is more elevated above the horizon, and shorter where it is less so.*

In the same manner, taking the case of a circle of rotation South of the equator, as $e'\,q'$, the circle P R p T divides it also into equal parts $u'\,e'\,v'$, $u'\,q'\,v'$, and $t'\,e'\,r'$, the part above the horizon $H_1\,N_1$ is evidently less than $u'\,e'\,v'$; and $t_2'\,e'\,r_2'$, the part above the horizon $H_2\,N_2$, is evidently still less than $t'\,e'\,r'$. *Whenever the North Pole, therefore, is above the horizon, and the sun's declination is South, the day is shorter than the night; and in different places, the day is shorter where the North Pole is more elevated above the horizon, and longer when it is less so.*

As the days both lengthen and shorten more and more as the elevation of the pole increases, of course the inequality of their lengths increases. Thus at London, where the elevation of the pole is of 51° 31′, the longest day is of $16^h\,34^m$; the shortest of $7^h\,44^m$, and the difference of $8^h\,50^m$. At Paris, where the pole is only elevated 48° 50′, the longest day is of $16^h\,7^m$; the shortest of $8^h\,11^m$, and the difference of $7^h\,56^m$. And at Edinburgh, where the elevation is of 55° 57′, the longest day is of $17^h\,25^m$; the shortest of $6^h\,53^m$, and the difference of $10^h\,32^m$.

We have, in our figure, a third horizon, $H_3\,N_3$, of which we have yet made no use, but which will furnish us with some very important observations. It will be seen from the figure, that neither the circle $e\,q$, nor the circle $e'\,q'$, ever meet this horizon at all; the circle $e\,q$ being entirely above it, the circle $e'\,q'$ entirely below it. A body moving in the former circle is then always above, in the latter always below, this horizon. The elevation of the pole P is greater above the horizon $H_3 N_3$, than above either $H_1\,N_1$ or $H_2\,N_2$: as this elevation increases therefore, some circles of rotation become entirely above the horizon, which were not so before; and others disappear entirely below it. It is easy to ascertain when this takes place for any particular circle. Taking the case of the circle $e\,q$, the point most elevated above the horizon is e, where it cuts the meridian on one side of the pole; the point least elevated above, or most depressed below, the horizon is q, where it cuts the meridian on the other side of the pole. Whenever therefore q is above the horizon, the whole circle is so. Now, the zenith is 90° from the horizon, and the pole 90° from the equator; and these equal arcs of 90° are made up in one case of the distance of the pole from the zenith, and the latitude of the place, or elevation of the pole above the horizon, in the other of the same elevation of the pole above the horizon, and the depression of the point Q of the equator below it. The zenith distance of the pole, therefore, and the depression of the equator at the point Q below the horizon are equal. Now, it is clear that whenever the North declination of a body, as Q q, is less than the depression of the equator below the horizon, as it is in the cases of the horizons $H_1\,N_1$, $H_2\,N_2$, the body when it comes to the meridian at q is below the horizon; but that when the declination Q q is greater than that depression, as it is in the case of the horizon $H_3\,N_3$, the body at q is above the horizon; and then its whole circle of rotation is so. When the declination of the body, and the depression of the point Q are exactly equal, the body just touches the horizon at q, and all the rest of its course is completely above it. Generally therefore, *wherever the North declination of a body is not less than the depression of the point Q of the equator, or than the zenith distance of the pole, no part of its whole course of diurnal rotation is below the horizon.*

In the same manner, taking the case of the circle $e' q'$, we shall find that the most elevated point e' never rises above the horizon $H_3 N_3$, when the South declination E e' exceeds the elevation of the point E of the equator above the horizon; and as the 90° from the horizon to the zenith are made up of this elevation and the zenith distance of the point E, and the 90° from the point E to the pole are made up of the same zenith distance of the point E and the zenith distance of the pole, the elevation of the point E, and the zenith distance of the pole are equal. Generally therefore, *wherever the South declination of a body is not less than the zenith distance of the pole, no part of its whole course of diurnal rotation is above the horizon.*

The greatest North and South declinations of the sun are of 23° 28′ each: for, although there is some little difference in the observations of the two, they may without sensible error, for this purpose, be treated as equal. These are his declinations at the solstices. Where therefore the zenith distance of the pole is of 23° 28′, or the latitude of 66° 32′, the sun at the summer solstices, where he is at his greatest North declination, will, at his lowest point, only just touch the horizon, and all the rest of his course for that day will be above it; or there will be one day of 24 hours, with no night: and, in the same manner, there will be one period of 24 hours, when the sun is at the winter solstice, during which he will, at his highest point, only just touch the horizon, and the rest of his course will be below it; or there will be a night of 24 hours, and no day. If the zenith distance of the pole be less than 23° 28′, or the latitude greater than 66° 32′, (as, for instance, if the zenith distance be 15°, or the latitude 75°,) then, as soon as the sun attains a North declination equal to that zenith distance, (in this instance 15°,) his daily course will only just touch the horizon; and from that time forward till he attains his greatest North declination, and again till he returns to the same North declination (15°), his whole course will be above the horizon; or, for a considerable period (in the instance put, from the 30th of April to the 12th of August), there will be uninterrupted daylight. In the same manner, as soon as the South declination is equal to this zenith distance (here 15°), the sun's course will only just touch the horizon; and from that time till he attains his greatest South declination, and again till he returns to the same South declination (15°), his whole course will be below the horizon; so that for a considerable period (in this instance from the 2d of November to the 8th of February), there will be uninterrupted night. These intervals of uninterrupted day and night are obviously longer as the latitude increases; for then the zenith distance of the pole, or the declination at which the sun begins to be continually above or continually below the horizon, diminishes, and the sun in consequence attains that declination sooner after one equinox, and does not return to it till a shorter time before the other; or the interval during which he has that or greater declination is longer.

When the pole coincides with the zenith, or the latitude is 90°, the equator coincides with the horizon. In this case therefore, every circle parallel to the equator, and North of it, is entirely above the horizon, and every point of it at the same elevation; and all circles parallel to the equator, and South of it, are entirely below the horizon. Here then all bodies which have North declinations are always above the horizon, and describe circles in their daily rotation parallel to it; and all bodies which have South declination are always below the horizon. At the vernal equinox therefore, the sun, which then comes upon the equator, just coincides with the horizon; he is then continually above it while his declination is North, or until he returns to the equator at the autumnal equinox, and from that time continually below it, while his declination is South, or until he returns again to the equator at the succeeding vernal equinox. In this case, therefore, there is uninterrupted daylight for half the year, and then uninterrupted night for the remainder.

The pole may also coincide with the horizon, which, in this case, will be represented by the circle P p, which, as we have already observed, bisects, and is perpendicular to, every circle of rotation, as E Q, $e\ q$, and $e'\ q'$. In this case then every heavenly body is an equal time above and below the horizon; and the sun therefore is so whatever be his declination, or at every period of the year.

We have now generally ascertained the manner in which the length of the day differs at different places. There are however two causes which make the length of the day greater and that of the night less than we have stated it. In speaking of the sun, as in the equator or horizon, we mean that his centre is so. At these instants therefore, his Northern limb, or the Northern part of his visible circumference, is North of the equator, his upper limb above the horizon; and as the mean apparent semidiameter of the sun is about 16′, the sun begins to appear on the horizon, or the equator or any circle parallel to the equator, sooner than we have stated, by the time corresponding to that difference of elevation in the one case, and of declination in the other; that is to say, his upper limb appears on the horizon when his centre is 16′ below it, and thus the length of each day is increased; and the greatest declination of the Northern limb being 16′ greater than the declination of the centre, or being 23° 44′, some part of the sun is always above the horizon at the summer solstice, where the zenith distance of the pole is 23° 44′, instead of 23° 28′. So also, where the pole and zenith coincide, as the North limb of the sun is on the equator, when the declination of the centre is 16′ South, some part of the sun is continually seen on the horizon, not merely from the time of his coming to the equinox, but from the time that his South declination is less than 16′. There is another cause also, which produces a similar effect, and to a greater degree. In speaking of the place of the sun, we have hitherto given the results of observation, as they are obtained after allowing for the operation of certain causes which materially complicate them in the first instance, and which we shall hereafter explain under the heads of parallax and refraction. Parallax tends to make the apparent place of a body lower than that which we consider it really to occupy, and which we call its real place; refraction to make it higher; and as the effect of refraction is much greater than that of parallax upon the sun, the joint effect of the two is to render the apparent higher than the real place. When the sun therefore is really upon the horizon he appears above it; and when he appears on it, he is really about 33′ below it. The length of daylight is increased by the time corresponding to this difference of elevation, and the zenith distance of the pole at the points where some part of the sun is first seen never to set at the solstice, and the South declination of the sun when first seen on the horizon where the pole and zenith coincide, are increased in exactly the same manner by this cause, as we have already seen that they are by the apparent magnitude of the sun himself.

A short notice of the manner in which the climates of different places differ will be sufficient. We have already seen that the sun's influence depends on the length of time during which he continues above the horizon, and the elevation he attains above it. The greatest elevation of the sun, like that of all other heavenly bodies in their daily rotation, is always when he is on the meridian; and his distance at that time from the Southern point of the horizon is always equal to the distance of the intersection of the meridian and equator from that point, increased by the declination, when North, and diminished by it when South *; or, as the distance of the intersection of the meridian and equator from the South point of the horizon is equal to the zenith distance of the pole, the sun's distance from that point when he is on the meridian is equal to the zenith distance of the pole increased by his declination when North, or diminished by it when South. When the declination is South, and greater than the zenith distance of the pole, this expression becomes negative; and the sun never rises. When it has a positive value, it continually increases as the South declination diminishes, or the North declination increases. Now, the whole arch of the meridian from South to North is 180°; and consequently when the meridian distance of the sun from the South point exceeds 90°, he is nearer the North point than the South point: his meridian elevation, therefore, in that case is his distance from the North point, and dimi-

* These conclusions will, perhaps, appear more plainly by a reference to a figure. In *fig.* 2, H being the South point of the horizon, and E the intersection of the equator and meridian, S, and S_2 may represent two situations of the sun in the meridian, S_1 being a situation where his declination is South, S_2 where it is North. S_1 H and S2 H are evidently his distances at those times from H, the South point of horizon; and S_1 H is the distance, E H, of the intersection of the meridian and equator from that point, diminished by E S1, the sun's South declination; S_2 H is the same distance E H, increased by E S_2, the sun's North declination.

nishes as his declination increases, after it has once attained that value which makes it equal to 90°.

The greatest value of the declination is 23° 28′: whenever, therefore, the zenith distance of the pole is less than 66° 32′, the sun's meridian distance from the South point can never exceed 90°: this distance therefore is the sun's altitude at the time, and increases for all places as the sun moves Northward; and it is greatest on the same day, for different places, as the zenith distance of the pole increases, or the elevation of the pole diminishes. The influence of the sun then, as far as it is determined by his elevation, increases as the elevation of the pole diminishes, within these limits. We have already seen that the length of the day increases, when the sun's declination is North, as the elevation of the pole increases. These two causes of heat therefore are opposed to each other, and we cannot easily tell in what degree they may counteract each other; and in fact we know that in high latitudes, or places where the elevation of the pole is great, the summers are often very hot, although the elevation of the sun is small.

Where the zenith distance of the pole is greater than 66° 32′, we are still less able to arrive at any satisfactory conclusion. In this case we have already seen that the meridian elevation of the sun may have to be measured Northward, and that it will then decrease as the declination increases, or as the length of the day increases. Thus the two causes which affect the solar power are here opposed, so as to prevent us from even saying when it is greatest at the place itself; and of course we have another difficulty, added to that already mentioned, in comparing it with the corresponding power in other situations. When the pole is in the horizon, the length of the day, as we have already seen, is always equal. In this case the power of the sun to communicate heat depends only on his elevation. Now here the equator passes through the zenith, and the meridian elevation of the sun is therefore greatest when he is in the equator, and continually less as he recedes from it, either towards the North or the South. Here therefore the equinoxes are the periods of the greatest, the solstices of the least solar power.

The periods of the least solar power, which correspond to the winter of different places, are more easily ascertained. The least distance from the South point is always the difference between the elevation of the equator where it crosses the meridian, and the greatest South declination: and this is necessarily the least meridian elevation, except where the poles are in the horizon, and there are equal elevations on the North and South side of the zenith. The least elevation therefore increases, as the elevation of the point of intersection of the equator and meridian increases, or as the zenith distance of the pole increases; and the greater this zenith distance the greater is the least meridian elevation of the sun. But we have already seen that, as the zenith distance of the pole increases, or its elevation diminishes, the length of the shortest days increases; and consequently on both accounts the influence of the sun, during the period of his South declination, is increased as the elevation of the North Pole is diminished. We should expect, therefore, to find the winters diminish in severity, as the elevation of the North Pole diminishes.

It may seem also, that, notwithstanding the difficulty of comparing the sun's extreme power at different places, his average power may be accurately compared. The whole time for which the sun in the course of a year is above the horizon, is everywhere the same or nearly so. If, in *fig.* 5, $e\,q$ be a circle of rotation North of the equator, and $e'\,q'$ be one South of it, and at the same distance from it, the portion $t_1\,e\,r_1$ of the first circle which is above the horizon $H_1\,N_1$, and the part $t_1'\,q'\,r_1'$ of the second circle which is below it, are in all cases equal; and so, of course, are the parts remaining. The parts above the horizon therefore, $t_1\,e\,r_1$ and $t_1\,e'\,r_1'$ are together equal to the whole of one of these circles; or the time for which the sun is above the horizon in the two days during which he describes the two circles $e\,q$, $e'\,q'$, is equal to the time of describing one of the circles, or to a day, and the time during which he is below the horizon is so also; and, in this manner, as the sun in his course has equal North and South declination, two days may continually be found, in which, taken together, the whole time of the sun's being above the horizon is equal to that of his being below it; and the whole year may be divided into these pairs of days, or only not so, because the sun's motion being rather more rapid in the Southern than the Northern part of his orbit, the days during which he is longer below than above the horizon, are not, in places North of the equator, quite so many as those when he is longer above it. In the same manner, if the circle eq is entirely above the horizon H_3N_3, the circle $e'q'$ is entirely below it; and thus the period, during which the sun never sinks below the horizon, is counterbalanced by a period during which he never rises above it.

Taking the extreme case, when the pole is in the zenith, these periods are each of them from equinox to equinox, or they comprehend each half the sun's course, and they only differ therefore, like the others, by the small inequality occasioned by his variable rate of motion. When the pole is in the horizon we have already seen that the day and night are always equal. Neglecting therefore the slight inequality we have mentioned, (which makes the whole period during which, in the course of a year, the sun is above the horizon, somewhat longer when the North Pole is so, and shorter when the South Pole is, and thus tends to render the heat of the former greater than that of the latter climates) we may consider that everywhere the sun is half the year above, and half below the horizon; and his influence to produce heat throughout the year, and consequently the average heat, as far as he occasions it, will depend upon his average elevation above the horizon, and be greatest where that is greatest, and least where that is least. It would however lead to very complicated investigation, if we were to endeavour to determine the manner in which this average elevation differs at different places. It is sufficient to state generally that it is greatest when the pole is in the horizon, and continually diminishes as the elevation of the pole increases; or the influence of the sun for the whole year is greatest when the latitude is nothing, it gradually diminishes with the increase either of North or South latitude, and is least where the latitude is of 90°. And it is familiarly known to every one that these deductions actually correspond with the general results of observation; that, generally, the countries near *the equinoctial line* (for so the line where the latitude is nothing is called, the day and night being there always equal) are the hottest, those in high latitudes the coldest. The ancients indeed put so much faith in these considerations, that they divided the whole earth into climates accordingly; and having no actual knowledge of the high Northern regions, nor of those near the equinoctial line, or South of it, they conceived the former to be uninhabitable from extreme cold, and the countries near the equinoctial to be equally so from extreme heat. The experience of modern times has proved that both suppositions were extravagant: and it has farther shewn that the heat of different places where the elevation of the pole is the same is very different; that it is indeed so much affected by local causes, as the elevation of the country above the level of the sea, its degree of cultivation, its geological constitution, the average moisture of the soil and atmosphere, the prevalent winds, the extent of continent with which it is connected, and other similar circumstances, that the mere consideration of the average influence of the sun, as deduced from his average elevation, is quite inadequate to give any information as to the comparative temperature of different places. Thus Edinburgh, Moscow, and Copenhagen, have all nearly the same latitude; but the Baltic is frozen up every year, while the sea near Edinburgh is unencumbered with ice, and the severities of a Russian winter, and the early period at which they commence, have too lately been the subjects of history, for us to want any other proof of the comparative mildness of the climate of Scotland*. It would therefore be useless to enter into any laborious research to discover the exact laws according to which a power, which we find not to have the importance once attributed to it, varies; nor should we have given to it even so much consideration as we have, unless the investigation had involved in it some facts and principles which are really of value; the explanations, we mean, of the manner in which the greatest and least solar power vary at different places, and the important general fact that the whole period of daylight during the year is equal, or very nearly so, at every place on the earth's surface.

* Our intercourse with our North American possessions furnishes a still more striking example. The mouth of the river St. Lawrence is in the latitude of about 49° N., more than 2° south of London; and this is the most Northerly point of the river; yet all access to Canada is stopped by the frost which closes the river at an early period of the winter, and the ice does not break up, and navigation recommence, until the month of May.

SECTION 7.—*On the equation of Time—Inequality arising from unequal motion in his Orbit—that from the inclination of the Ecliptic—Deduction of the nature of the equation throughout the Year.*

A GREAT number of the most important elements involved in astronomical researches are variable in their amount. Their variations, however, generally succeed each other in a certain order, and are confined within certain limits; and when these limits, and all the varying values are ascertained, it is of course possible to take an average among them, and this average value is termed *a mean value.*

It is indeed always possible to take an average between any number of observations, or of ascertained values of a particular element; but unless the observations are so taken that the whole course or cycle of the variation is included, it is not usual to call the average a mean value; or rather it is not the absolute mean value of the thing itself, it is only its average or mean value for a certain time. For instance, we have already seen that the length of the day, considered as the interval from sunrise to sunset is continually varying, but that it goes through all its changes in the interval from one solstice to another. Its average duration for this whole time, then, is its mean value: it is a little more than twelve hours, and it is very nearly the same everywhere; it would be everywhere exactly twelve hours, if the sun always moved at the same rate, and there were no parallax or refraction. An average, however, might be taken of the lengths of this day for a portion only of this interval, for instance, from the vernal equinox to the summer solstice; in this case, the shortest length of the day would be a little more than twelve hours, and the average length at London about fourteen hours and three quarters. This would be a correct average of the lengths observed; but as the time of observation would not comprehend all the variations of the element in question, it would not be the mean length of the day absolutely, though it might be called the mean length for the period of observation.

In the same manner as we have taken an instance of mean duration, we might have an instance of mean motion; that is to say, if a body moves with a variable motion; but if the whole course of its variation is ascertained, its average rate of motion during this whole course may be found, and this will be called *its mean motion.* A body moving with this mean motion, and of course moving uniformly, for the whole time occupied by the whole series of the real motions, would move through the same space as the real body, but its place at many, or all intermediate periods, would be different from the place of the real body, on account of the difference between the real and mean motions. The place of a body so moving, or the place which the real body would occupy on the supposition that it moved uniformly, and described in the time occupied by the whole series of its real motions the same spaces which it actually does, is called *the mean place of the body.* In the same manner an event which happens at various intervals which succeed each other in a certain and recurring order, will have *a mean time* of occurrence.

Now it very generally happens in astronomy that it is less inconvenient first to compute the mean place of a body, or the mean time of an event, and then to ascertain the difference between the mean and the true, than to go through the computations necessary to find the true time and place in the first instance.

When once the mean values have been ascertained, the mean motion of a body during a known period, its mean place at a known time, the mean time of the occurrence of a given event, are easily found; for the intervals of the mean time, and the rate of the mean motion being always the same, we only want to know how often the event has occurred, or how long the motion has been continued. If, from consideration of the manner in which the difference between the true and mean values arises, we can ascertain the amount of that difference in each particular instance, we can find what is to be added to or subtracted from the mean value to arrive at the true; and the quantity so added or subtracted is called *an equation.* The mean value thus leads to the true value, and of course it furnishes an approximation to it; and as the subjects of astronomical inquiry generally have their variations confined within narrow limits, so that the difference between the true and mean motion's times and places is not very great, the approximation is not very distant.

We shall find several instances of the application of the terms above explained, and of the use made of these mean values and results in treating of the equation of time, of which we have still to speak, and then the more obvious appearances of the sun, and their principal effects, will be for the present sufficiently explained.

The solar day is longer than the sidereal day in consequence of the motion of the sun Eastward in his orbit. It is evident that the degree of its excess above the sidereal day must be affected by the quantity of that motion, and must, when other circumstances are the same, be greatest when that motion is greatest, or when the sun is in his perigee, and least when that motion is least, or the sun is in his apogee. The motion of the sun goes through all its variations in the course of one revolution of the sun in his orbit; it admits, therefore, on the principles already explained, of a mean value. Let us call the sun S, and let us suppose a fictitious body, which we call S_1, to move uniformly in the ecliptic, and to perform a complete revolution in the same time as S: the motion of S_1, therefore, will be the *mean motion*, and its place, the *mean place*, of S. The difference between the places of S and S_1 will be *an equation;* it is called *the equation of the centre.* Let us suppose the two bodies, S, and S_1, to be together when the sun is in apogee. As the revolution of the supposed body S_1, is completed at the same time as that of S, they will be again together at the end of the year, or when the sun returns to his apogee. Besides this, the times of the sun's motion from apogee to perigee, and from perigee back to apogee, are equal; they are therefore each equal to half the time of his whole revolution, or to half the time of the revolution of S, or to the time taken by S_1, which moves uniformly, to pass through half the ecliptic. But the sun's apogee and perigee are at the distance of half the ecliptic from each other; consequently, as S and S_1, were together at the apogee, they are so also at the perigee, each taking the same time (half of the year) to pass through half of their orbits. We find therefore that the real and mean places of the sun coincide at the apogee and perigee.

It is also plain that they correspond nowhere else. The sun's distance from the earth continually decreases from apogee to perigee, and his angular velocity continually increases during the same period. It is evident then that as his whole real angular motion for that period is equal to his mean angular motion for the same time, the real motion will at first be less, and afterwards greater than the mean motion. His real place therefore will at first fall behind his mean place, and the distance between them will increase day by day until his real motion becomes equal to his mean motion; the distance will then diminish as the real motion becomes greater than the mean motion, and this excess will finally bring them together again. It will not however do this until they reach the perigee, for we have already seen that they are then together, and this could not be the case if they had been so before; for as they are only brought together by the real motion exceeding the mean motion, and as the real motion continually increases from apogee to perigee, if at any period before the perigee S had come up with S_1, at the following instant S would have passed S_1 by the excess of its motion, and would have continued from day to day to increase the distance between them by the continuing and growing excess of the real above the mean motion; and the consequence would be, contrary to the fact, that S would arrive at the perigee before S_1. We arrive therefore at this conclusion, that the real place of the sun is *behind* his mean place, as he passes from apogee to perigee, the distance between them continually increasing for a certain time, then continually diminishing till the two places again coincide at the perigee. Exactly in the same manner we find that the real and mean place never coincide from perigee to apogee, only with this difference, that as in this half of the orbit the real motion is at first greater, and afterwards less, than the mean motion, the real place is always *before* the mean place, until, on the return to the apogee, they both again correspond. The greatest difference in the time of the approach of S and S_1 to the meridian cannot exceed $8^m 24^s$.

It is not however only by the variation in the sun's rate of motion, that the length of the solar day, or the interval between his successive appearances on the meridian, is affected; it is also influenced by the inclination of the ecliptic to the equator. The motion of rotation of the heavens is uniform; and consequently, if we suppose another fictitious body, S_2, to be at the point ♈ at the same time with the supposed body S, to move uniformly Eastward in the equator, and to pass completely round the heavens in that circle in the course of a solar year, as this body will every day have moved Eastward by an equal portion, it will always come to the meridian later by an equal period than the point which it occupied on the preceding day; and as this point returns to the meridian in a sidereal day, which we have seen to be always of the same length, and the excess of time above this sidereal day before S_2 returns to the meridian is also equal, the intervals between the successive arrivals of S_2 on the meridian are themselves equal; and each of them, as S_2 performs its whole revolution in a solar year, must be equal to a *mean solar day*. We now then have two fictitious bodies, S_1 and S_2, each moving uniformly, the first in the ecliptic, the second in the equator; and, as these circles are equal, the actual motions of each are equal. Still we have to inquire whether their periods of arriving at the meridian are the same. For this purpose let us suppose (in *fig.* 6) that P represents the pole of the heavens, ♈ Y ♎ half the equator, ♈ Z ♎ half the ecliptic, (for these circles bisect each other in the points ♈ and ♎,) and let Y and Z be the points of bisection of the arcs ♈ Y ♎, ♈ Z ♎, respectively. Each of the arcs, ♈ Y, ♈ Z, therefore is 90°, or they are equal; and, consequently, the fictitious bodies, S_1 and S_2, which were together at ♈, and whose

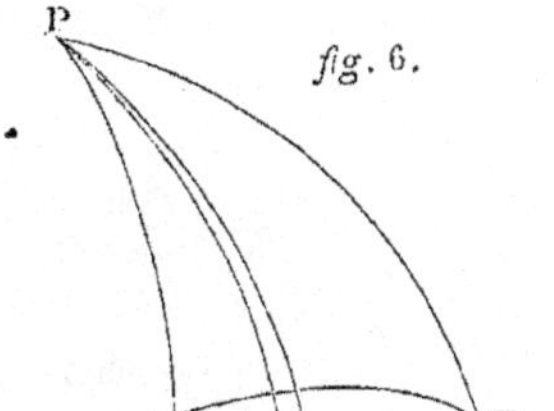

rates of motion are equal, would arrive at the same time at Z and Y. Besides this, each of the arcs, ♈ Y, ♈ Z, being 90°, the angles ♈ Z Y, ♈ Y Z (by a well-known property of the sphere) are each 90° also*, or Z Y, the arc of a great circle which joins the points Z Y, is perpendicular to the equator ♈ Y at Y, and therefore is part of a meridian, P Z Y, passing through those points. Z therefore, being a point upon the same meridian as Y, comes upon the meridian of the place at the same time with Y, and, as the bodies S_1 and S_2 are at Z and Y at the same time, they will there come on the meridian of the place together. But Z, being the middle point of the arc ♈ Z ♎, is the sun's place when in the tropic.

From the time when the body S_1 leaves the point ♈, until its arrival at Z, we shall find that it will always come upon the meridian of the place before S_2. To show this, let S_1 represent any intermediate position of that body, and let P S_1 *s* be a meridian drawn through it; of course therefore the points S_1, *s*, come upon the meridian of the place together. If therefore *s* be the place of the supposed body S_2 at that time, the bodies S_1, S_2, will then come on the meridian of the place together; but if the body S_2 have advanced beyond the point *s*, then S_2 will come on the meridian of the place later than *s*, and consequently than S_1, for *s* and S_1 arrive there together. But the actual motion of S_2 is equal to that of S_1, or to ♈ S_1. Whenever therefore ♈ S_1 is greater than ♈ *s*, the body S_2 is more advanced than the point *s*, or comes on the meridian later than S_1. Now it is a general property of all spherical triangles, as well as plane ones, that the greater side is opposite to the greater angle; and in the spherical triangle S_1 ♈ *s*, as long as the side ♈ *s* is less than 90° or than ♈ Y, the right angle S_1 *s* ♈ is greater than the angle ♈ S_1 *s*.† Until S_1 and S_2 therefore arrive at the points Z, Y, ♈ S_1, which is opposite to the right angle ♈ *s* S_1, is always greater than ♈ *s*, and conse-

* [The arcs ♈ Z, ♈ Y, being each 90°, the point ♈ is the pole of the circle Z Y, and consequently the arcs ♈ Z, ♈ Y, are secondaries to Z Y, or perpendicular to it.]

† [Cos. ♈ S_1 *s* = sin. S_1 ♈ *s*, cos. ♈ *s*, by Napier's rules. Whenever, therefore, ♈ *s* is less than 90°, or at all points between ♈ and Z or Y, cos. ♈ *s* is positive, and consequently cos. ♈ S_1 *s* is so likewise, or ♈ S_1 *s* is less than 90°; for the other factor, sin. S_1 ♈ *s*, is positive, since S_1 ♈ *s* is 23° 28′. When ♈ *s* is 90°, or = ♈ Y, cos. ♈ *s* = 0, consequently cos. ♈ S_1 *s* = 0, or ♈ S_1 *s* (in this case ♈ Z Y) = 90°, as we have already seen. When ♈ *s* is greater than 90°, cos. ♈ *s* is negative; therefore cos. ♈ S_1 *s* is so, or ♈ S_1 *s* is greater than 90° or than ♈ *s* S_1, and consequently ♈ S_1 is less than ♈ *s*, or the body S_1 comes later on the meridian than S_2, as is afterwards independently shown in the text.]

quently S_2 always comes to the meridian after S_1 in the course of their passage from ♈ to the points Y and Z, or during the sun's passage from the equinox to the tropic.

In passing from Z and Y to the other equinox ♎ the result will be exactly contrary: for in this case, exactly in the same manner as in the other, if a meridian be drawn through the sun's place on any day, the arc of the equator intercepted between it and the equinox will be less than that of the ecliptic so intercepted; and consequently, the point of the equator which comes on the meridian of the place at the same time with S_1 will be nearer than S_1, and consequently than S_2, to the equinox, *towards* which the bodies are moving. The interval of time therefore between the arrival of S_2 and the point ♎ on the meridian of the place, will be greater than that between the arrival of S_1 and the point ♎ there; and as ♎ comes on the meridian of the place *after* them, of course S_2 which precedes it by the longer interval, will arrive there before S_1. We see then that in proceeding from the tropic to the equinox, S_2 always comes on the meridian of the place before S_1. The same conclusions may be deduced in exactly the same way for the remaining two quadrants of the ecliptic and equator: and we draw the general conclusion, that when the sun is either at an equinox or a tropic, S_1 and S_2 come on the meridian of the place together: but that while he is moving from equinox to tropic, S_1 is always earlier on the meridian of the place than S_2, and always later while he is moving from tropic to equinox.

It is further found, by a process of computation sufficiently easy, and the details of which will be found in the note*, that the difference between the times of the appearance of S_1 and S_2 on the meridian of the place continually increases from the equinox or tropic till

---

* [It is of some importance to investigate the ratio which the motion in right ascension bears to the motion in longitude at different periods.

For this purpose, let $S_1 T_1$ represent the motion in longitude for a very short period, and let $P T_1 V t$ be a meridian passing through T_1 and $S_1 V$, an arc of a small circle parallel to the equator. Of course $s t$ is the difference of right ascension corresponding to the difference of longitude $S_1 T_1$. Now the motion of S_2 is entirely in right ascension and uniform; and whenever the effect of the motion $S_1 T_1$ on the right ascension is equal to the uniform motion in right ascension of S_2, then, at whatever distance the point s was from the corresponding situation of S_2, the point t will be at the same distance from the situation of S_2 corresponding to it; and they will gain or lose on S_2 as the effect of the motion $S_1 T_1$ estimated in right ascension, is greater or less than the uniform motion of S_2, or of S_1.

Taking then $S_1 T_1$ to represent this uniform motion—

$s t$ (the effect in right ascension) $= \frac{S_1 V}{\cos. S_1 s}$ (for they are similar arcs of parallel circles, of which the radii are to each other as $1 : \cos. S_1 s$.)

$= \frac{S_1 T_1 . \sin. S_1 T_1 V}{\cos. S_1 s}$ (for the very small triangle $S_1 T_1 V$, may be considered as a plane triangle, and the angle $S_1 V T_1$ is a right angle.)

$= \frac{S_1 T_1 \sin. \Upsilon S_1 s}{\cos. S_1 s}$ (for the angles $\Upsilon S_1 s$, $\Upsilon T_1 V$ as the meridians, $P s$, $P t$ are very close to each other, may be considered as equal.)

But $\cos. S_1 \Upsilon s = \sin. \Upsilon S_1 s \cos. S_1 s$, or $\sin. \Upsilon S_1 s = \frac{\cos. S_1 \Upsilon s}{\cos. S_1 s}$; and, consequently, $s t = S_1 T_1 \frac{\cos. S_1 \Upsilon s}{\cos^2. S_1 s}$

$= S_1 T_1 \frac{\cos. \text{obliquity of ecliptic}}{\cos^2. \text{declination}}$, a quantity which continually increases, as the declination does so.

At the equinox, cos. declination $= 1$, and $s t = S_1 T_1$ cos. obliquity of the ecliptic.

At the solstice, the declination $=$ the obliquity of the ecliptic: and then, therefore,

$$\therefore s t = S_1 T_1 \frac{\cos. \text{obliquity of the ecliptic}}{\cos^2. \text{obliquity of the ecliptic}} = \frac{S_1 T_1}{\cos. \text{obliquity of the ecliptic.}}$$

To find the point where the effect of the motion of S_1 in right ascension is equal to the motion of S_1 itself, or $s t = S_1 T_1$ we have this equation—

$$S_1 T_1 = S_1 T_1 . \frac{\cos. \text{obliquity of ecliptic}}{\cos^2. \text{declination}}$$

or $\cos^2.$ declination $=$ cos. obliquity of ecliptic $= \cos. 23° 28'$.

$$\text{Hence log. cos. declination} = \frac{1}{2} (10 + \log. \cos. 23° 28')$$

$$= \frac{19.9625076}{2} = 9.9812538$$

or declination $= 16° 42' 49''$.

Again, $\sin. \Upsilon s = \cot. S_1 \Upsilon s \tan. S_1 s = \frac{\tan. S_1 s}{\tan. S_1 \Upsilon s}$

$$= \frac{\tan. \text{declination}}{\tan. \text{obliquity}}$$

$\therefore \log. \sin \Upsilon s = 10 + \log. \tan. \text{decl.} - \log. \tan. \text{obl.}$
$10 + \log. \tan. 16° 42' 49'' - \log. \tan. 23° 28'$.
$= 19.4775198 - 9.6376106 = 9.8399092$
and $\Upsilon s = 43° 45' 49\frac{1}{2}''$.

Again, $\cos. \Upsilon S_1 = \cos. \Upsilon s \cos. S_1 s$
or $\log. \cos. \Upsilon S_1 = \log. \cos. \Upsilon s + \log. \cos. \text{declination} - 10$
$= \log. \cos. 43° 45' 49\frac{1}{2}'' + \log. \cos. 16° 42'' 49 - 10$
$= 9.8586563 + 9.9812538 - 10 = 9.8399101$
or $\Upsilon S_1 = 46° 14' 10''$.

The difference, therefore, between ΥS_1 and Υs, at this time $= 46° 14' 10'' - 43° 45' 49\frac{1}{2}''$
$= 2° 28' 20\frac{1}{2}''$ in space,
or 9m. $53\frac{1}{2}$s. in time.

This is evidently the greatest difference which can be thus occasioned; for from the equinox until this point, the effect of the motion in right ascension is always less than the actual motion, and the difference is therefore continually accumulating up to this period: from this period onwards to the solstice, the effect of the motion in right ascension is continually greater than the actual motion, and the difference therefore continually diminishing.]

S_1 has moved through about half the quadrant (more correctly till S_1 is 46° 14′ 10″ from the equinox), and then diminishes till S_1 arrives at the tropic or equinox towards which it is moving, and that the greatest difference in time thus occasioned is about $9^m\ 53\frac{1}{2}^s$.

We have thus two sources of inequality; one arising from the unequal motion of the sun in his orbit, the other from the inclination of that orbit to the equator. To ascertain how the time at which S comes to the meridian of the place, or the true solar time, differs from that at which S_2 comes to the same meridian, or the mean solar time, we must combine the two. To see how this combination is to be effected, let us begin from the winter solstice, and examine what will be the relative positions of S and S_2 for the year: the difference in time between the instants of their being on the meridian is called *the equation of time* *, being the difference between the actual time of the sun's being on the meridian and the beginning of the mean solar day †. The sun is in perigee about the 30th of December. We have already seen, that in passing from apogee to perigee, S_1 is always more advanced, or comes later on the meridian, than S: and consequently, at the winter solstice, which is a little before the sun comes to perigee, S_1 is more advanced than S. But at the solstices S_1 and S_2 come on the meridian together, and consequently, S_2 is then also more advanced than S, or the mean solar time is later than the apparent, or the sun is on the meridian before noon by the clock, or the clock is after the sun. From the solstice to the equinox S_2 continually is less advanced than S_1, but at perigee, S_1 and S coincide, therefore at perigee S_2 is less advanced than S, or the mean solar time is earlier than the apparent, or the clock before the sun: and as a few days before it was later, and all the changes are gradual and continuous, there must have been some intermediate instant when S and S_2 were on the meridian together, or the mean and apparent time coincided, or the equation of time was nothing. This would be on December 24th, 1829; and from that time till the perigee, the mean would be earlier than the apparent time, or the clock before the sun. From the perigee onwards, S_1 is continually less advanced than S; and from the solstice to the equinox S_2 is continually less advanced than S_1. On both accounts therefore, during the whole period from the perigee to the vernal equinox S_2 is less advanced than S, or the mean solar time is earlier than the apparent, or the clock is before the sun. After the vernal equinox until the apogee, S continues more advanced than S_1; but from the equinox to the solstice S_2 is more advanced also than S_1: when therefore these differences are equal, the equation of time becomes nothing. Now, the greatest difference in time between the approach of S and S_1 to the meridian never amounts to more than $8^m\ 24^s$; the greatest difference between the approach of S_1 and S_2 to the meridian is of about $9^m\ 53^s$, and takes place a little more than half way between the equinox and the solstice, or about May 8th. By May 8th therefore, the difference between S_1 and S_2 has become greater than that between S and S_1, having been less at the equinox, and between these times the differences must have been equal, or S and S_2 must have been on the meridian together, or the equation of time must have been nothing. In point of fact, it is so on April 15th, 1830. After this time, S_2 and S are both more advanced than S_1, but S_2 more so than S; or S_2 is more advanced than S, or the mean time later than the apparent, or the clock after the sun. At the summer solstice however, S and S_1 are again together: but the apogee does not take place till after the summer solstice (on June 30th), therefore S continues more advanced than S_1, or, at the summer solstice, S, which on April 16th was less advanced than S_2, has become more so, and there will therefore have been an instant when they came on the meridian together, and the equation of time was nothing. This is on June 15th, 1830. From that time to the solstice, it is clear that S is more ad-

* The equation of time is only registered from noon to noon, when S and S_2 are on the meridian; but the term applies with equal correctness to the difference of time between the instants of their arrival at any given distance from the meridian. It is plain that this is correctly *an equation*, being the difference of the real and mean times of the occurrence of a particular event.

† The mean solar day is considered, for astronomical purposes, to begin when S_2 is on the meridian, or at noon. The period of its commencement may of course be arbitrarily fixed, and for astronomical purposes this is the most convenient; for civil purposes that is the most convenient which includes all the active period of day-light within one day, and the civil day, therefore, begins at midnight, when the sun is on the meridian below the horizon, or midway between his setting and rising. Each astronomical day, therefore, contains the last twelve hours of one, and the first twelve of the next, civil day.

vanced than S_2, or the apparent later than the mean time, or the clock before the sun. From the solstice to the autumnal equinox, S_1 is continually more advanced than S_2, but until the apogee S is more advanced than S_1, and of course, than S_2: for this period therefore, the apparent time is later than the mean time, or the clock before the sun. After the apogee however, S is continually less advanced than S_1, and as at the autumnal equinox S_1 and S_2 are together, S, which at the apogee was more advanced than S_2, has then become less so, and there has been some intervening instant at which S and S_2 have come on the meridian together, or the equation of time has been nothing. This is on September 1st, 1830; and from that time to the equinox S is less advanced than S_2, and the apparent is earlier than the mean time, or the clock is after the sun. After the equinox, until the winter solstice, the sun is still moving from apogee towards perigee, and S is consequently less advanced than S_1, and in the interval between equinox and solstice, S_1 is also less advanced than S_2: on both accounts therefore S is less advanced, during this whole period, than S_2, or the apparent time is earlier than the mean time, or the clock after the sun.

We have now gone through the year, and we may collect our results thus: that in the course of the year there are four days, and only four, namely, December 24th, April 15th, June 15th, and Sept. 1st, when the apparent and mean time are the same, or the equation of time is nothing: and that in the interval between the first and second of these, and again in that between the third and fourth, the apparent is always later than the mean time, or the clock before the sun: and that between the second and third, and again between the fourth and first, the apparent is always earlier than the mean time, or the clock after the sun. These results correspond with those in the common tables of the equation of time.

It is also evident, from the manner in which these results have been deduced, that they depend entirely on the relative positions of the apogee, and of the equinoxes. If these are fixed points, or hold always the same relative position, the results we have obtained will serve alike for every year: if they vary, the equation of time will vary also; and this consideration leads us to inquire whether there be any motion of the equinoxes, and whether the apogee and perigee be or be not fixed points. As far also as the magnitude of the equation is concerned, it is evident that any variation in the inclination of the ecliptic to the equator would affect it; for the angle ♈ S_1 *s* (in *fig.* 6), and the declination of the sun at any point of the ecliptic, would both be affected by this change; and both these quantities are involved in the solution of that part of the question which arises from the motion of the sun in a plane inclined to the equator.

In point of fact, it is found that the inclination of the ecliptic and equator does undergo some slight variation. This is not sufficient to produce any material alteration in the results, or to call for more extended notice here; but it furnishes one reason why the results obtained for the equation of time cannot, as far as their numerical values are concerned, apply accurately, except to the particular periods for which they are computed. The other considerations are more important in themselves, and will deserve separate consideration. They are also of practical importance, as connected with the division of time into longer periods than we have yet used, except in a loose and popular way of speaking. Our next section therefore will treat of *the precession of the equinoxes*, and the *progression of the apogee*; the following one of the length of the year, and the consequent corrections and adjustment of the calendar.

Section 8.—*Precession of the Equinoxes produced by the retrograde motion of the Equator on the Ecliptic—Effects on the longitude, latitude, declination, and right ascension of the Heavenly Bodies—Progressive motion of the Sun's apogee and perigee.*

In speaking of the mode of ascertaining the declination of a heavenly body, we have only referred to observations made on the meridian. And they are the best adapted to that purpose generally; it may, however, also be computed from the altitude and azimuth* of the object when observed out of the meridian. But when we take the case

* The *azimuth* of a body is the arc of the horizon intercepted between the meridian and a vertical circle passing through the body. The situation of a body with respect to a particular place on the earth, is determined by its altitude and azimuth, just as its situation, with respect to the heavens in general, is determined by its declination and right ascension or by its latitude and longitude.

of a body whose declination, like that of the sun, continually varies, it is clear that we cannot be sure of ascertaining the time at which his declination is of any given value by observations on the meridian; for he is upon the meridian and above the horizon, only for one instant (or in some cases two), in twenty-four hours; and there is no greater likelihood that his declination will then be that concerning which we inquire, than at any other instant in that time. Nor can we even be sure of ascertaining the required time by the means mentioned at the beginning of this section, for he may attain the required declination while he is below the horizon, and then no direct observation can be made. We are not however without the means of ascertaining the required period. The sun's change of declination, although very different in amount at different periods of his course, may with little error be considered as uniform during the small space of twenty-four hours; and we consequently have the means, by observations made on the meridian on two successive days, of computing his declination at particular times in the interval between them. Thus, if at noon on September 20th the sun's declination were 11′ North, and at noon on the following day 13′ South, we might safely estimate his change of declination at 1′ hourly (the whole difference in the twenty-four hours being of 24′); and we should therefore say, that at six o'clock in the evening of September 20th, his declination was 5′ North; and that it was nothing, or that the equinox took place, at eleven o'clock on the same evening. For purposes of greater accuracy, there are easy means of making the same computations, allowing for the variation of the rate of the sun's change of declination.

Having thus the means of ascertaining the time at which the sun is at any particular declination, and consequently, among others, the time of the equinox, we find, especially by comparing together the observations made at distant times, that when the sun now comes to the meridian at the equinoxes, the same stars are not on the meridian which formerly were so when he came thither at that part of his annual course. The position of the equinox being ascertained at a particular time, a catalogue of stars can be formed, and their right ascensions and declinations registered. Their longitudes and latitudes may also be registered, or obtained by computation from the corresponding right ascensions and declinations. The same process may be gone through at a subsequent time. If this be done, we shall find the right ascensions and declinations of all stars altered, but in various manners and degrees, the latitude of all remaining very nearly the same, and the longitude of all increased by very nearly the same quantity. We shall also find that the different stars keep the same position with respect to each other; they continue at the same distance from, and make the same angles with, each other. The alteration therefore in their right ascensions, &c. of which we have spoken, does not proceed from any motion among themselves; it must proceed from some alteration in some of the arbitrarily assumed points or lines from which these elements are measured, and by which they are estimated. And these are the equinoxes, and the equator and ecliptic.

Let us first take the simpler class of phenomena. We have already stated that the longitude of every heavenly body is increased by the same quantity, and that their latitude is not affected. This would evidently be the case if the position of the ecliptic itself continued unaltered, but the point from which arcs are measured along it were removed backward; for the latitude, being the perpendicular distance between the star and the fixed circle, remains the very same arc, the longitude, being the distance from the same point (the intersection of the secondary passing through the star with the ecliptic) to the arbitrary standard whence the measurement is taken, is increased exactly as much as that standard is removed. The longitude and latitude therefore are affected as they would be if the ecliptic remained unmoved itself, but the first point of Aries receded upon it.

The first point of Aries however is the intersection of the ecliptic with the equator: and if this is moved, while the ecliptic remains stationary, it can only be so by moving the whole circle of the equator. The *inclination* of one of these circles to the other is always found to be very nearly the same: the distance of their poles is always equal to the inclination, and it therefore is always the same. If then the equinox moves backwards gradually, the equator is continually changing its position; its pole herefore is continually moving, but still always at the same distance from that of the ecliptic; and consequently the pole of the equator would describe a circle or part of a circle about that of the ecliptic. The position of any given star continuing fixed, the line of the secondary to the equator which passes through it would vary, for it passes through the star and the pole of the equator, and this latter point has changed its position; the position of the equator itself would vary also, and so also would that of the first point of Aries. The declination therefore would vary; for the pole of the equator, moving while the star continued fixed, would generally approach towards or recede from it; and the North polar distance being thus altered, and the North polar distance and the declination together making 90°, the declination would vary also. This variation also would be different for different stars; for the same motion which brings the pole nearer to some carries it further from others; and this at a different rate, as its motion is directly towards or from the star, or oblique with respect to it. In the same manner the right ascension of different stars would be differently affected. All would have a certain effect produced on them by the alteration of the point from which the measurement is taken: but this would not be the only cause of alteration. The right ascension is the distance from the first point of Aries to the intersection with the equator of a secondary to that circle passing through the star. As these secondaries all pass through the pole of the equator, the old and new secondaries would intersect each other at the star, and make different angles with each other in the case of different stars, as those stars are situated nearer to or farther from the poles, and in one direction or another with respect to them. Intersecting each other at the star, they would diverge after they passed it, and, consequently, even if the angles made were the same in two different instances, the space by which the secondaries would have separated before meeting the equator, would be different as the distance of the stars from that circle, or the declination, differed. The variation therefore of right ascension would be different on all these accounts in different stars. These different results and variations may be with advantage illustrated by a figure, as far as relates to the right ascensions and declinations. The positions with respect to the longitude and latitude are too simple to require such illustration, and the figure would be inconveniently complicated by the introduction of the circles relating to them.

Let ♈ t T (in *fig.* 7) represent one position of the equator, ♈′ t' T′ another at some considerable distance of time, and let P & P′ be the corresponding positions

Fig. 7.

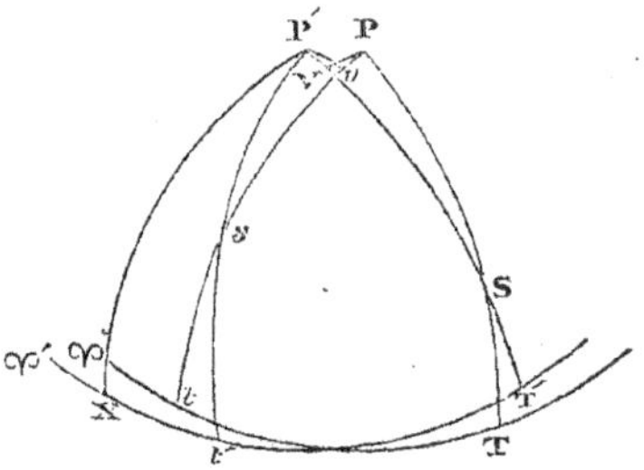

of the pole. Let S and s be any two stars, and through S draw P S T, P′ S T′ secondaries to the first and second positions of the equator respectively; and in like manner P s t, P′ s t', through s. It is evident that S T, ♈ T, represent the declination and right ascension of the star S, when the equator is in the position ♈ t T, and that ST′, ♈′ T′, represent them when the equator is in the position ♈′ t' T′. In the same manner, s t, ♈ t, are the declination and right ascension of the star s in the first position of the equator; s t', ♈′ t', in the second. If S V be taken equal to S P, and s v equal to s P′, it is evident from inspection of the figure that S V is less than S P′, and s v than s P. In the former case therefore the new North polar distance is greater than the old one, or the declination has diminished: in the latter the new North polar distance is less than the old one, or the declination has increased.

We next proceed to examine the variation in right ascension. For this purpose let us draw through ♈, the former position of the equinox, P′ ♈ X, a secondary to the new position of the equator. On the new equator therefore the right ascension of ♈ is ♈′ X, and the right ascensions of all the heavenly bodies are increased by this quantity. This however is not the only variation. The remaining portion of the new right ascension, X T′, or X t', is obviously not necessarily equal to ♈ T, or ♈ t. In the figure as drawn it is greater in each case; it may in other positions of the star be less. But in every case it is evident that the difference between them will depend upon the magnitude of the angle T S T′ or $t s t'$, or, which is the same thing, on that of the angles P S P′, P s P′, and also on the distance of S and s from the circles ♈ t T, ♈′ t' T′. It will therefore vary for every star: and this is all which we are here desirous of illustrating.

We have thus seen generally what would be the nature of the effect produced by the retrocession of the equator upon the ecliptic: that the latitudes of stars would remain the same; that their longitudes would be increased uniformly; that their declinations would be differently affected, some being increased, others diminished, and this in unequal amounts and proportions; and that the right ascensions also, although with many exceptions (in cases where the quantity X T′ is less than ♈ T, and their difference exceeds ♈′X) would generally increase, but at different rates in different cases*. We have seen also that these phenomena are actually observed to take place; and we therefore lay it down as an established fact, that while the ecliptic continues immoveable, the equator has a retrograde motion upon it, or a motion from left to right in those already referred to. The amount of this, subject to some small inequalities, is 50″.1 in a year; that is to say, the first point of Aries recedes annually 50″.1 upon the ecliptic: the retrocession, therefore, is 1° 23′ 30″ in a century, or a degree in about 71½ years; and the first point of Aries will have receded through the whole circle, and consequently will return to its present position in about 25868 years. This retrograde motion is called, on account of its effects on the time of the occurrence of the equinox, which it accelerates, *the precession of the equinoxes*, i. e., their going forward.

So slow an alteration may seem of little importance, except in a very long series of years. In a science however, where none except the most accurate results are of practical value and importance, no cause of error is to be neglected; and especially where, as in the present case, the error is of a nature continually accumulating. Thus the difference occasioned in the equation of time by this alteration of the position of the equinox, would at present be but slight, and generally only affect the actual numerical value of that correction for a considerable period; but we have already seen that the nature of the correction is mainly dependent on the relative positions of the equinoxes or solstices and the sun's apogee or perigee. The perigee is now nearly 10° more advanced, or has 10° greater longitude, than the winter solstice: in the year 1250, it coincided with the winter solstice, and before that time it preceded it; the combined operation of the retrocession of the equinox on the ecliptic, and a progressive motion of the perigee itself, having since then brought them into their present relative positions. At those times therefore the equation of time would not only differ in amount from its present values, but the considerations used in deducing the periods at which the apparent is before or after the mean time, would themselves differ.

The motion which we have just mentioned to exist in the perigee, or apogee, (for as the two are always 180° distant from each other, they must move alike) is also deduced from observation. If the very instant of the sun's being in perigee or apogee could be readily determined, this motion would easily be ascertained; for his place at the time would be determined, and the alteration of that place, when he was next in the like situation, would be the motion required. The variations however of his apparent diameter, or of his angular motion, by which alone we can immediately estimate those of his distance, are too slow to admit of any very accurate estimation of very small

---

* It is not desirable here to introduce the calculations on which the results depend; but the results themselves may be given with advantage. The precession in declination is found to be positive, that is to say, the declination is increased by the effect of precession, wherever the right ascension of the star is less than 90°, or greater than 270°; the precession is negative, or the declination is diminished when the right ascension is between 90° and 270°: it is nothing, or the declination is not affected when the right ascension is 90° or 270°. The angle formed by a secondary to the equator and a secondary to the ecliptic, each passing through the star, is called the *angle of position* of the star. The precession in right ascension is positive, or the right ascension is increased wherever the angle of position is less than 90°; it is nothing, when that angle is 90°; it is negative, or the right ascension is diminished, when that angle is greater than 90°.

These are the effects of a small variation in the position of the equinoctial points, or of the precession for a short period. The effects of precession for a long time, when they become considerable, must be deduced from computation of the accumulated effects of these minute variations; for the right ascensions being continually changed by the effect of precession, its effect on declination, which depends on them, will continually change also; and the pole of the equator changing its place, the angle of position, which is determined by it, will vary also, and thus the variation of right ascension will also itself be changeable.

differences. The time required may however be ascertained within certain limits; and if such observations be taken at great intervals of time, any inaccuracy in the estimate will be of less importance. There are however better methods of making the computation. The principle on which they depend is very simple. We have seen that the radius vector of the sun describes equal areas in equal times. Now the only straight line which can be drawn through the focus of an ellipse, so as to divide the ellipse into two equal parts, is the transverse axis. The sun's position at one extremity of any line passing through the earth is distant by 180° from its position at the other extremity; if, therefore, he be observed at any two points 180° distant from each other, he is then at the two extremities of a line passing through the focus of the ellipse, and the portions of the ellipse on each side of that line must be unequal, unless the line be the transverse axis which passes through the apogee and perigee. If the portions are unequal, his time of passing through them is unequal also: if the times, therefore, are found to be equal, the instant of observation is the instant of his being in perigee or apogee: if unequal, it is not so, but the instant of his being in perigee or apogee may be ascertained from the observations made, by calculations of which it is not necessary here to enter into any detail. The general result is that the perigee or apogee (or the *apsides* of the sun's orbit, as they are also termed) have a progressive motion on the ecliptic of about 11″. 8 annually. The longitude of the perigee and apogee therefore increases at the rate of about 62″ annually, for it increases by the actual motion forwards of those points themselves, and also by the whole amount of the retrocession of the equinox, from which it is measured; or by the sum of the two quantities 11″. 8 and 50″. 1.

The principle above referred to, that any inaccuracy of observation may be rendered less material by comparing observations made at distant periods, is one of very great and general importance. If we suppose ourselves unable to discover the exact situation of the equinox or the perigee within a minute, it is plain, assuming the amount we have already assigned to their motions to be correct, that we cannot, by observing two successive positions of the sun in either of these points, form any estimate of the precession of the equinox or the progression of the apogee on which we can at all rely; for the motion of the point in the interval between the two observations will be less than the probable error of the observations themselves, and we shall be unable to tell whether the difference between the observed positions is the effect of a motion in the object observed, or merely the result of the inaccuracy of observation. If however there be any continuing motion, the distance between the places of the point observed will be increased if the distance between the times of observing be so; and it may therefore become greater than the amount of any probable inaccuracy of observation. For instance, taking the case of precession, in twelve years the equinox would have receded about 10′, a quantity much greater than that which we have supposed to be the limit of the errors of observation; and therefore, from observations made at this interval of time, we should be able to pronounce with certainty that a retrograde motion existed. Still we should be unable to determine its amount with any great accuracy. Having supposed the probable error of observation to be 1′, we could not say whether the real retrogression might not be either 9′ or 11′, quantities respectively 1′ less, and greater than the observed retrogression of 10′; and as this would be the retrogression of twelve years, the annual retrogression might be as little as 45″, or as great as 55″. This would be a great and important uncertainty; but it may be very greatly diminished by a further extension of the same principle, that is to say, by taking still longer intervals of observation. For instance, let us suppose the interval 400 years, and the observed precession 5° 34′, or 20,040″. If we suppose, as before, that the amount of probable error is 1′, the least possible amount of precession will be 5° 33′, or 19,980″; and the greatest 5° 35′, or 20,100″; and the least possible annual amount 49″. 95, and the greatest 50″. 25; two quantities approaching very near to each other, and the mean between them, or 50″ 1, would probably differ very little indeed from the true value.

The principle on which this power of approximating to a correct value depends is obvious. The error of observation, whatever its amount may be, occurs only at the times of observation.

If the element whose value is sought occurs only once during the interval, it is affected by the whole error; if it occurs oftener, the whole result is affected by the whole error, but the whole result being the sum of so many repetitions of the element, the value of the element itself is only affected by a proportional part of the error, the whole error being divided among all the repetitions. If therefore the element be sufficiently often repeated, the proportion of the error involved in its value will become exceedingly small; and this may be the case, even when the actual amount of the probable error of observation is increased, if the number of repetitions among which the error is to be distributed is increased in a greater proportion.

SECTION 9.—*Of the different years—Equinoctial (or tropical) year—Sidereal year—Anomalistic year—Construction of the calendar—Julian correction — Julian year—Leap year—Gregorian correction—Persian correction.*

WE now see that there are three different periods at which the sun may, in different senses, be said to return to the same position: when he returns to the same equinox at which he was before; when he returns to the same spot in his orbit; and when, having been in perigee or apogee, he returns to it again; or, which is the same thing, when having been at a given distance from any of these points, he returns to the same point with respect to them. Each of these may be said to be the completion of a revolution of the sun; and a revolution of the sun is called *a year*. The year from equinox to equinox is called the *equinoctial year*, or sometimes *the tropical year;* for his time of returning from tropic to tropic, they being situations always holding the same relation to the equinox for the time being, is obviously the same as that from equinox to equinox. The year from any point in the ecliptic to the same point again is the *sidereal year*, for the sun is then in the same position as before, with relation to the stars. The sun's angular distance from the apogee is called *the true anomaly*, and the period between his leaving and returning to a given situation with respect to the apogee is therefore called the *anomalistic year*.

It is evident that the equinoctial is the shortest, the anomalistic the longest of these years. When the sun starts from the equinox, it is a given point of his orbit; before he returns to it, the equinox has receded on the ecliptic, and he therefore meets it again sooner than he returns to the same spot in his orbit. The effect therefore of the retrograde motion of the equinoctial point on the ecliptic is to *bring forward* the time of the equinox (or the instant at which the sun is upon the equator); and hence, as we have already mentioned, the phenomenon is known by the name of *the precession* of the equinoxes. In the mean time however, the apogee has moved forward on the ecliptic; and the sun therefore, after returning to the same spot in his orbit where he was at the former equinox, has still a further arc to describe before he arrives at his original position with respect to the apogee, and the time of his doing so is of course later.

The *mean length of the equinoctial year* is $365^d\ 5^h\ 48^m\ 51^s.6$, (or, expressing it decimally, $365^d.242264$) of mean solar time. After this, the sun has to describe $50''.1$ to return to the same point of his orbit at which he was at the commencement of the year, or to complete the sidereal year; and the *mean length of the sidereal year* is thus made $365^d\ 6^h\ 9^m\ 11^s.5$, or $365^d\ 256383$. He then has to describe a further arc of $11''.8$ to arrive at his original position with respect to the apogee, and the *length of the anomalistic year* is thus made $365^d\ 6^h\ 13^m\ 58^s.8$, or $365^d\ 259708$. In ascertaining all these lengths, it is of course important to resort to the principle explained at the end of the last section, and to deduce their value from observations made at long intervals, so that any error of observation may be distributed among many periods of the length required.

The lengths assigned to the equinoctial and sidereal years are only *mean* lengths; that given to the anomalistic year is a true one. We shall hereafter shew, from other considerations, that the length of the anomalistic year does not vary. For the present, we will assume that fact, and then it is obvious that the length of the equinoctial and sidereal years must continually vary; for each of these years is shorter than the anomalistic year by the time which the sun takes to describe a given angle of his orbit; in one case $62''$, in the other $11''.8$. Now the rate of the sun's motion is different in different parts of his orbit, faster as he is further from the apogee, slower as he approaches it; and, consequently, his times of describing these spaces of 62″ and 11″.8 continually vary, as they are differently situated with respect to the apogee. The times therefore which are to be subtracted from the uniform length of the anomalistic year, to ascertain those of the equinoctial and sidereal years respectively, are themselves of variable duration; and the lengths of the equinoctial and sidereal year are necessarily so too. The variation however is very small, and the mean differs from the true length at any period by a very inconsiderable quantity.

It is obviously necessary, for many purposes, not only of chronology and history, but even of personal and domestic convenience, that we should have the means of dividing time into definite periods of considerable length; and the most obvious and natural period to adopt, is that which includes all the various operations and appearances which succeed each other in regular order, which comprehends seed-time and harvest, summer and winter. All these are included in the space of *an equinoctial year*. The position of the apogee, as we have already seen, has some effect on the *length* of the seasons; but it is the position of the sun with respect to the equinox, which determines what the season is, and in his passage through his whole round from one equinox till he returns to the same again, he occasions the whole variety and succession of spring, summer, autumn, and winter. The length of this revolution therefore has been adopted as the unit of long duration; and a period, assigned with more or less accuracy, but intended to represent this duration, has been uniformly adopted by all civilised nations, and called by the name of *year*, or the *civil year*.

It would be productive of great inconvenience, if the beginning of the year did not correspond with the beginning of a day; and this would be the case, if we took the exact period of $365^d\ 5^h\ 48^m\ 51^s.6$ for the length of the *civil year*. If, for instance, at any given time, the beginning of the year exactly corresponded with the beginning of a day, the following year would begin $5^h\ 48^m\ 51^s.6$ after the beginning of the 366th day; the year after that would begin $11^h\ 37^m\ 43^s\ 2$. after the beginning of the 731st day, and so on. We should therefore continually have days belonging in part to two different years; and our chronology would become confused and inaccurate. It is obviously better to fix some length for the civil year which shall be free from these inconveniences.

On the other hand, it is evident that if the length assigned to the civil year were to differ materially from the true length of the astronomical year, great confusion would before long be produced. Thus if it were shorter by a whole day, in 100 years its commencement would precede that of the astronomical year by 100 days, more than a quarter of the whole year; or if it were shorter by a quarter of a day, the same error would be produced in 400 years. Thus the commencement of the civil year would, at these intervals of time, correspond to completely different stages of the seasons; if, at the one period, it were at mid-winter, at the other it would be early in the autumn; if at one it were seed-time, at the other it would be the depth of winter. This would obviously be inconvenient. We want to know, when we hear that an event took place at a particular period of the year, whether it took place in spring, summer, autumn, or winter; if a battle, whether at the opening or the close of a campaign; if the discovery of a country, whether during the season of abundance or scarcity.

It is therefore desirable so to fix the length of the civil year, that its commencement shall always be at the beginning of a day; and at the same time to adopt some mode which shall prevent it from ever being far distant from the commencement of the astronomical year. In this consists the adjustment of the calendar; and many attempts have been made, at different times, and in different places, to establish an accurate and complete one. We need not here enter into the history of all these attempts; but there is one so famous, both from the celebrity of the man under whose auspices it was made, and as the ground-work of the adjustment now used, that it may be well to mention it, especially as the simplicity of the numbers involved in it makes it the easiest example which can be given of the principle of such a correction.

The Roman Calendar had fallen into great confusion from the causes already explained, when it was determined to remedy the inconvenience which resulted from its condition, and to rectify it for the future. This was done under the auspices of Julius Cæsar, and it is in consequence called the Julian correction; Sosigenes of Alexandria was the astronomer who made the calculations. The length of the equinoctial year was not then accurately known; it was however supposed to consist of $365^d\ 6^h$, (an amount, as we have seen, not very far from the truth), and that time was assumed as its accurate length. Now four years, each of $365^d\ 6^h$, together make up a period of 1461 days; and these 1461 days may be divided into three periods of 365 days, and one of 366. If therefore these successive civil years were made to consist each of 365 days, and then a fourth were made to consist of 366, the length of the four would be equal to that of four astronomical years; and if at the beginning of the first year the commencement of the civil and astronomical years coincided, at the end of the fourth they would do so again. It is true that the civil years would be of unequal length, and that none of them would accurately correspond in duration with the astronomical year, three of them being each six hours shorter, the fourth eighteen hours longer than it; and it is true also that the commencement of each, except the first of the four years, would not coincide with that of the astronomical year, that of the second being six, of the third twelve, and of the fourth eighteen hours before it. None of these inconveniences however were material; the first would not be felt, as soon as the people became familiar with the order in which the longer and shorter years succeeded each other; the second would only be felt by astronomers, and they would have abundant means of removing any difficulty which it occasioned; and the third, although of serious importance if the error accumulated so as to produce a great interval between the commencement of the different years, was of no moment, when, as it seemed, the error would never exceed eighteen hours. On these grounds the ordinary length of the civil year was fixed at 365 days, but it was ordained that every fourth year should consist of 366. The fourth year was called *bissextile*, from the manner in which the additional, or *intercalary* day was inserted*. It corresponds to our *leap* year.

If the length of the astronomical year had been accurately $365^d\ 6$, this adjustment of the calendar would have been sufficient. But the true length of the equinoctial year is only $365^d\ 5^h\ 48^m\ 51^s.6$; or it falls short of $365^d\ 6^h$ by $11^m\ 8.^s4$. Four equinoctial years therefore would fall short of four years of $365^d\ 6^h$ each, or of four Julian years, three of 365 and one of 366 days, by the space of $44^m\ 33^s.6$; and 100 equinoctial years would fall short of 100 Julian years by 25 times that space of time, or by $18^h\ 34^m$; and 400 equinoctial years would fall short of 400 Julian years by $74^h\ 16^m$, or by a little more than three days. In consequence of this inaccuracy a further correction was required; and this was carried into effect by Pope Gregory XIII. in the year 1582, at which time the vernal equinox fell ten days earlier in the civil year than it had done in the year 325, at the Council of Nice, the period which he chose for the correct standard of the commencement of the civil year. At the Council of Nice, the vernal equinox had been on the 21st March; at the time of Pope Gregory, it was on the 11th; but he omitted ten days in the current year, and made the 15th of October in that year immediately succeed the 4th, so that in the next year the vernal equinox again took place, as it had done in 325, on the 21st of March. In this manner the error which had already taken place was rectified. To prevent it from again occurring, the following additional adjustment was devised.

We have seen that the Julian correction made the length of 400 years too great by a little more than three days. This inaccuracy then would very nearly be removed, by omitting the additional day in three years, taken in the course of the 400. It is of great practical convenience to have these corrections easily remembered, and the three days were in consequence omitted in three of the four years which completed centuries. It was determined that, dividing time into portions of 400 years each, every fourth year, except those which terminated the first three centuries of such a period, should be of 366 days, but that those three, like the common years, should each be of 365 days only. Thus the years 1600,

* The Romans called the first day of each month the Calends (*Calendæ*) of that month. Hence the word, Calendar. They then reckoned the latter days of the preceding month, by their distance from the Calends of the following one. Thus the 1st of March being the Calends of March, the 28th of February was the day before the Calends of March, the 27th, the third day of the Calends of March, and so on. The 24th of February, therefore was the sixth (*sextus*) day of the Calends of March. The fourth year was lengthened by adding an additional day; at this time there were two days reckoned as the sixth day of the Calends of March, and hence the year was called *bissextile*, as having a *double sixth day* of these Calends.

2000, 2400, would be *leap years*, or have 366 days; the years 1700, 1800, 1900, 2100, 2200, 2300, would be common years of 365 days each.

This mode of correcting the calendar has been adopted at different times in almost all civilized nations. As it was introduced by a Pope, for the express purpose of regulating the ecclesiastical year, and ascertaining the time for keeping the festivals celebrated in the Roman Church, it was immediately adopted by all states professing that form of Christianity. The Russians, who are members of the Greek Church, have not yet adopted it. In England it was not introduced till the year 1752; and as one of the years concluding a century, in which the additional or *intercalary* day was to be omitted (the year 1700) had passed since the correction by Pope Gregory, it was necessary to omit eleven instead of ten days in the current year. The alteration was proposed in Parliament by the Earl of Chesterfield, and he, at that time, received most of the praise due to so important a reform; but the Earl of Macclesfield, then president of the Royal Society, a nobleman of much talent and science, was the real author of the measure, and bore a great part in the details of its execution.

Even this correction does not arrive at complete accuracy. The excess of 400 Julian over 400 astronomical years is more than three days, by the space of $2^h\ 16^m$, and the calendar consequently, even after the Gregorian adjustment which omits the three days, still makes the 400 years too long by $2^h\ 16^m$. In ten times 400 years, or 4000 years, this error would be increased ten fold, or would become of $22^h\ 40^m$, or nearly of a day. It would therefore be a further, though still not quite an accurate correction, to make also every 4000th year (which on the Gregorian adjustment would be a leap year) a common year. But this extreme accuracy is hardly necessary in a matter which, after all, is more of popular than scientific importance. An error of less than a day in 4000 years cannot derange the ordinary calculations of time; and astronomers, in their computations, refer to standards of time much more accurately fixed than the civil year.

There is no occasion therefore for seeking for any more accurate adjustment of the calendar, and the convenience of having the correction occurring at a period so easily remembered as that of four years, and the further adjustments all happening at times so remarkable as the years which complete centuries, render it the most eligible that has been devised. There is one however which may be here mentioned, because it is curious for its great accuracy, especially when we consider the early period at which it was discovered. It was adopted by the Persians in the course of the eleventh century, and consists in intercalating (or adding to the 365 days which constitute the ordinary length of the year) 8 days in 33 years, instead of in 32 as according to the Julian adjustment. On this system, there are seven leap years, each the fourth after the preceding one, and then one which is the fifth; or, if this period be supposed to begin at the year 1801, the years 1804, 1808, 1812, 1816, 1820, 1824, 1828, 1833, 1837, &c. would be leap years. It is evident that this, however simple in theory, is less easily remembered and applied to practice than the Gregorian adjustment. It is however remarkable, considering the time when it was made and the people who discovered it, that it exceeds the Gregorian adjustment in accuracy; and if it be further corrected by the omission of the intercalary day once in 4000 years, the accuracy of the approximation becomes truly remarkable, being within $\frac{1}{70000}$ part of a day, or the remaining error would not amount to a day in 70,000 years.*

The length of the year being determined, the exact time of its commencement, and its division into minor portions (which we call months, as their duration has been fixed with some reference to the changes of the moon) may be arbitrarily assumed. It might seem natural to make the beginning correspond with some remarkable period of the astronomical year, as a solstice or an equinox. But this is not material: the length being accurately fixed, the beginning of the year will always be at very nearly the same distance from the solstices and equinoxes; and this is just as convenient as its coinciding with either of them. It takes place about eleven days after the winter solstice.

* In the Treatise on Algebra, p. 111, $365\frac{7}{29}$ is deduced as an approximation to the length of the year, by the method of continued fractions. If the process had been continued one step further, the result would have been $365\frac{8}{33}$, the exact approximation adopted in this Persian adjustment.

Chapter II.

Section 1.—*Necessity of Corrections —Enumeration of them.*

We must now revert to the consideration of the fixed stars. We have hitherto considered them as always occupying the same place. It is however evident, from what we have already said concerning precession, that, supposing them really to do so, that place will at least be differently described at different times, that its right ascension, longitude, and declination will all vary. We must therefore, in comparing the observations made on a star at different times, make allowances for this alteration; we must ascertain what the right ascension and declination, or else what the longitude and latitude of the point where the star was at the time of the earlier, would have become, in consequence of the precession of the equinoxes, at the time of the later observation; and if these computed quantities correspond with those then observed, we conclude that the star occupies really the same place in the heavens, and is what we call a fixed star, although that place is differently described, in consequence of the alteration of the points and circles to which its situation is referred. By making these observations we find, subject to certain apparent, and some real, exceptions, which will hereafter be noticed, that the places of the stars in general do continue the same.

The importance of this computation will appear very familiarly, by an example which shews the great amount of the inaccuracy produced by neglecting it. A space of eight degrees on each side of the ecliptic was early distinguished from the rest of the heavens, because the planets (whose nature and motions will be the subject of future consideration) were never found out of its limits. This space is called *the zodiac.* There were found within, or principally within it, twelve constellations, situated not accurately at equal distances from each other, but so nearly so, that when it was thought convenient to divide the ecliptic into twelve equal parts of 30° each, it was found that each of these would be in the immediate neighbourhood of one of these constellations. This division was accordingly adopted; the portions into which the ecliptic was divided were each named after the constellation which was near it; and the divisions themselves, and also the constellations which gave name to them, were called *the signs of the zodiac,* and characters were invented to express them. The names of the constellations, or signs, and the characters used to express them, are as follows:—Aries, or the Ram, ♈; Taurus, or the Bull, ♉; Gemini, or the Twins, ♊; Cancer, or the Crab, ♋; Leo, or the Lion, ♌; Virgo, or the Virgin, ♍; Libra, or the Balance, ♎; Scorpio, or the Scorpion, ♏; Sagittarius, or the Archer, ♐; Capricornus, or the Wild Goat, ♑; Aquarius, or the Water-carrier, ♒; and Pisces, or the Fishes, ♓.

At the time that these constellations were fixed on to determine the names of the subdivisions of the ecliptic, the vernal equinox was a point very near the constellation of the Ram, the summer solstice was near the Crab, the autumnal equinox near the Balance, and the winter solstice near the Wild Goat. The vernal equinox therefore, or the point from which right ascension and longitude are measured, got the name of *the first point of Aries,* which we have seen to be still used; the autumnal equinox was the first point of Libra; and *the tropics,* the circles parallel to the equinoctial which pass through the solstices (pp. 10, 23), were called the tropics of Cancer and of Capricorn. These names also continue in use. But though the names continue, the circumstances from which they took their origin have ceased to exist. The precession of the equinoxes has carried the vernal equinox backwards, away from the constellation of the Ram; and in the same manner all the other constellations from which the signs are named have changed their situations with respect, not to the circle of the ecliptic, but to the point with reference to which their position is stated. Not only however do the points of the equinoxes and solstices retain their name, but the whole ecliptic is still divided into twelve portions, which are called signs, and retain the names of the constellations after which they were originally called. Thus the longitude of 45° is either expressed in that manner, or as the 15th degree of Taurus, which is thus written ♉ 15°, or as $1^s\,15^\circ$, the 1^s, or one sign, being taken merely as a mode of expressing the quantity of longitude 30°. So, also, the longitude of 118° 30′ is equally well expressed as ♋ 28° 30′, or as $3^s\,28^\circ\,30'$: and 217° 19′ 10″ of longitude is written either ♏ 7° 19′ 10″, or $7^s\,7^\circ\,19'\,10''$: the character ♋ or ♏

expressing the sign, *in* which the remaining degrees, &c. are to be measured; the number 3^s or 7^s expressing the number of signs, or spaces of 30°, which are to be added to the remaining degrees, &c., to make up the whole longitude *. These signs, or portions of the ecliptic, continue to be measured, at intervals of 30° each, from the actual position of the vernal equinox; and the consequence is, that the equinox being removed back by precession, the signs or constellations of the zodiac no longer correspond with them. The constellation of the Ram is now near the sign ♉ of the ecliptic; that of the Lion, near ♍; that of the Waterman, near ♓.

We see therefore the importance of attending to precession, in reducing observations made at different times, for the purpose of comparison with each other. If we failed to do so, a star really at rest would be spoken of as having moved through the whole space, by which in reality the equinox has receded from it. But there is another not less important application of the doctrine. Our present observations must also frequently be corrected by considerations derived from precession. For instance, all observations which involve the right ascension as one of their elements, are made by observing the time by a clock adjusted to sidereal time. It is therefore necessary that we should be able to ascertain the correctness of this clock: and this may be done by trying whether it accurately marks the time at which a star, whose place is well known, comes to the meridian. Thus, if the right ascension of a particular star is known to be 90°, or a quarter of a circle, its time of coming to the meridian would be 6 hours, or a quarter of a day, after the first point of Aries had done so; and the clock would be correct, if at that instant it marked 6^h of sidereal time as that interval. If however the star's right ascension had been ascertained some time before, it would have been altered in the interval by the effect of precession; let us suppose it to have been increased by such an arc, as would correspond to 8 seconds of time †. The consequence would be that if, as we have supposed, at the instant when the star is on the meridian, the clock marked 6^h of sidereal time, it would be too slow by 8 seconds, and all observations made by it would require to be corrected accordingly.

Again, we have already seen (p. 17) that the North polar distance of any body can be ascertained by observations of its altitude when upon the meridian, if the elevation of the pole, or its distance from the zenith be known. Conversely, if the declination, or North polar distance of the body be known, and its altitude when on the meridian of any place observed, the elevation, or zenith distance, of the Pole there may be ascertained. It is accordingly the easiest way of ascertaining the elevation of the Pole at any particular place, to observe the meridian altitude of a star whose declination is known; and thence to compute the elevation of the Pole. But in this case also, if some time has elapsed since the declination of the star observed was ascertained, that declination will have been altered by precession; and the elevation of the Pole will therefore be erroneously deduced, unless we previously correct the registered declination, so as to have the accurate value of that element before we use it in our computation.

These observations naturally lead us to consider the subject of the *corrections* necessary to deduce from the observed place of a body its true one. These corrections are no less than five in number: precession, refraction, parallax, nutation, and aberration. We have already sufficiently considered the subject of precession.

Section II.—*Of Refraction—Its nature, variation, and amount—Oval appearance of Sun and Moon near the horizon—Dim appearance of objects near the horizon—Twilight.*

The different branches of science are intimately connected with each other,

---

* The word sign is even used sometimes merely to designate 30°, and applied in this manner to right ascension, with which, in its primary sense, it had no connexion. Thus the right ascension 289° 12′ 3″ is sometimes written 9^s 19° 12′ 3″. The symbols however of the particular constellations are never thus applied.

† The effect of precession though different for different stars is, for considerable periods of time, very nearly uniform in respect of the same star. If we know, for instance, that the annual effect of precession on the right ascension of a given star is 36″.5 in space, or $2^s.433333$ in time, we know that the effect in 100 days is $\frac{100}{365}$ of either of these quantities, or 10″ in space, or $\frac{2}{3}$ of a second in time; and if three years and 100 days had elapsed since the epoch for which the right ascension was ascertained, the present right ascension would exceed its registered value by $3 \times 36''.5 + 10''$, or 1′ 59″.5 in space; or $3 \times 2^s.4333 + \frac{2}{3}{}^s$, or $7^s.9666$ in time.

The rate of precession however, even for the same star, has some variation; and, for purposes of great accuracy, there is a further correction requisite on this account; but this need not be any further noticed here.

and few of them can be explained without occasionally having recourse to facts ascertained by, and properties resulting from, others. In explaining the manner in which a correction for refraction becomes necessary, and in which it is to be made, we must have recourse to ascertained facts and principles, derived from the sciences of Optics and of Hydrostatics, or rather that part of the latter which is known under the name of Pneumatics.

From the latter we learn that the atmosphere continually lessens in density, as it is more elevated above the surface of the earth; and that, with some allowance for local and accidental causes, it does so in nearly the same manner everywhere. The earth therefore being considered as spherical (for its variation from an accurately spherical form may here be neglected), the atmosphere will consist of concentric strata or layers of air, of densities continually diminishing; and the surface of the extreme stratum will also, in conformity with the general principles of the equilibrium of fluids, be nearly spherical also.

From the discoveries of Optics we collect that a ray of light always moves in a straight line through any transparent medium of equal density throughout. If it falls perpendicularly on the surface of such a medium, its course is not affected by it, but it passes on its original direction: but if it falls on it obliquely, its direction is changed, or the ray is *refracted*, at the point of incidence, and it passes on through the medium in a line making a certain angle with the former course of the ray, but which is itself a straight line if the medium be of uniform density. In this case, the sines of the angles of incidence and refraction (the angles which the ray, before and after meeting the surface, makes with a perpendicular to that surface at the point where the ray meets it) bear always, whatever be the amount of the angle of incidence, the same proportion to each other, as long as the medium is the same. These proportions are different in different media, and the ratio is generally higher, or the ray is more bent out of its course, as the density of the medium is greater; in passing from a rarer to a denser medium, the ray is generally bent *towards* the perpendicular. If the ray passes through several media bounded by parallel plane surfaces, the effect produced by refraction is the same in amount as if it had fallen originally on the last of these media at the same angle as that at which it fell on the first. If the medium is uniform, its parts on each side of the ray where it enters correspond, and there is no cause which can turn the ray to one *side* of its original course rather than the other, though the medium acts so as to bend the ray from its original course, either towards or from the perpendicular: the deflection, therefore, is either directly towards, or directly from, the perpendicular; and, consequently, the course of the ray after deflection is in the plane which passes through its course before deflection and the perpendicular, or in the same plane as before; and the deflection itself therefore is entirely in that plane.

In investigating the law of the *correction* to be applied for refraction, it is plain that the amount of the deflection in different circumstances is what we want to discover; for that is the difference between the observed and the true place of the body.

To ascertain this, we will first make the simple supposition that the surface of the earth is plane, and that of the atmosphere plane also, and parallel to it; and we will further suppose in the first instance that the density of the atmosphere is uniform, and equal to that which we observe at the earth's surface. The surfaces of the earth and atmosphere being thus supposed to be parallel, all perpendiculars to the surface of the atmosphere must be parallel to those to the surface of the earth. A ray, refracted at the exterior surface of the atmosphere, will pass thence in a straight line through the atmosphere, as its density is supposed to be uniform; and when it arrives at the surface of the earth, the angle which it makes with a perpendicular to that surface must be equal to that which it made with the perpendicular to the surface of the atmosphere at the point of refraction, since those perpendiculars are parallel to each other. The perpendicular to the earth's surface passes through the zenith, and consequently the angle between the refracted ray, or the apparent direction of the star, and that perpendicular, is the apparent zenith distance of the star; it is also, by the definition of the term, the angle of refraction; and consequently, the angle of refraction is equal to the apparent zenith distance of the star. The refracted ray lies in the same plane with the two parallel lines which it joins; and that plane, passing through a perpendicular at the surface of the earth, is necessarily a vertical plane there: the effect of refraction therefore takes place entirely in a vertical plane.

Now the whole amount of refraction produced by the passage of a ray of light from a vacuum into air of the density usual at the earth's surface is very small. According to the table given in the treatise on Optics, p. 6, the ratio of the sine of incidence to the sine of refraction is less than 1.0003 to 1. It follows, by a very easy computation, that the deflection increases very nearly in the same proportion as the tangent of the angle of refraction, or of the apparent zenith distance*. It is to be observed also that the refraction takes place on entering a denser from a rarer medium, and consequently the refracted ray is nearer the perpendicular, or more elevated, than the incident ray. The effect of refraction therefore is to cause the observed altitude of a body to be greater than its true altitude, and its observed zenith distance to be less than its true zenith distance.

The same conclusions will be true, although the ray may pass through strata of air of different densities, if the surfaces of all these strata are parallel to each other; for we have already mentioned that in that case the amount of refraction is the same as if the ray fell originally on the last medium. Let us now see in what manner the spherical form of the earth and atmosphere affects the result.

For this purpose we will again begin with the simplest supposition, that of an uniform density. We have already mentioned that the surface of the earth and of the atmosphere will be concentric spheres, and consequently a line drawn from the centre of the earth to meet either of them will be perpendicular to their surfaces where it meets them. The line therefore, drawn to the surface of the earth at any particular place, will pass through the zenith of that place. In *fig.* 8, let O represent the centre of the earth, and the concentric circles the earth and the outside of the atmosphere, S R T the course of a ray refracted at R,

Fig. 8.

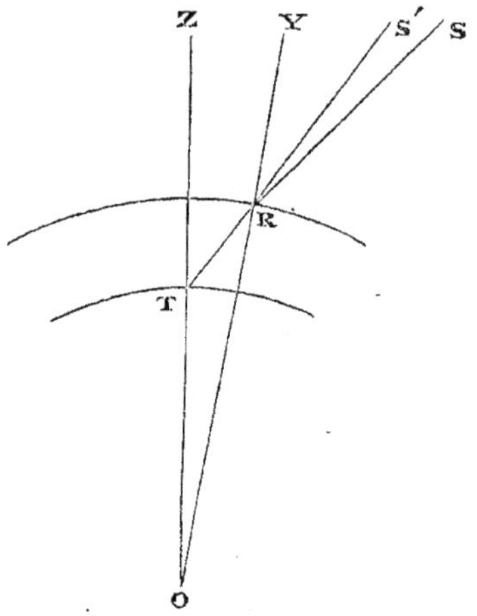

and passing to the eye of an observer at T. Join O T, O R, and produce them; the line O T will pass through the zenith of the observer at T, and any plane passing through it must be a vertical plane. Now the line T R necessarily is in the same plane with O R, O T Z, which it joins; it is therefore in a plane vertical at T; and so is the line R S, for the course of the ray before and after refraction is in the same plane. As in the case of a plane surface therefore, the deflection takes place in a vertical plane; and, as before, it raises the apparent place of the body above the real place, the deflection in this case also being towards the perpendicular.

Thus far the results correspond on the two suppositions. The amount however of the deflection is different both in quantity and in the law of its variation. It continues indeed to vary as the tangent of the angle of refraction, the angle T R O, or S′ R Y; but this angle is no longer equal to the apparent zenith distance S′ T Z, for S′ T Z, the exterior angle of the triangle R T O is equal to the sum of the interior and opposite angles T R O and T O R. The amount of the deviation too is diminished, for a ray falling parallel to S R upon a refracting surface at T, would

---

* [Let I represent the angle of incidence, R the angle of refraction, r the amount of deflection, m the index of refraction for the particular medium.

Then, by the general principles of Optics, $\sin. I = m \sin. R$.

But as the ray is bent *towards* the perpendicular, or R is less than I, $I = R + r$, and consequently the former equation becomes $\sin. (R + r) = m \sin. R$, or $\sin. R. \cos. r + \cos. R. \sin. r = m \sin. R$.

But as r is necessarily very small, $\cos. r = 1$ and $\sin. r = r$ very nearly. The equation therefore is equivalent to

$$\sin. R + r. \cos. R = m \sin. R,$$

$$\text{or } r. \cos. R = (m - 1) \sin. R,$$

$$\text{or } r = (m - 1) \frac{\sin. R}{\cos. R} = (m - 1) \tan. R.]$$

in the same manner make a greater angle with O T Z than S R does with O R Y; and the angle of incidence being thus increased, the angle of refraction and the deflection would be increased also. The actual deflection therefore is less than it would be if the ray were incident immediately at T.

We may now proceed to the case of several concentric strata of air, each of uniform density through its whole thickness, and the nearer to the earth always more dense than that more remote from it. These may be represented by the concentric circles in *fig.* 9, and the

Fig. 9.

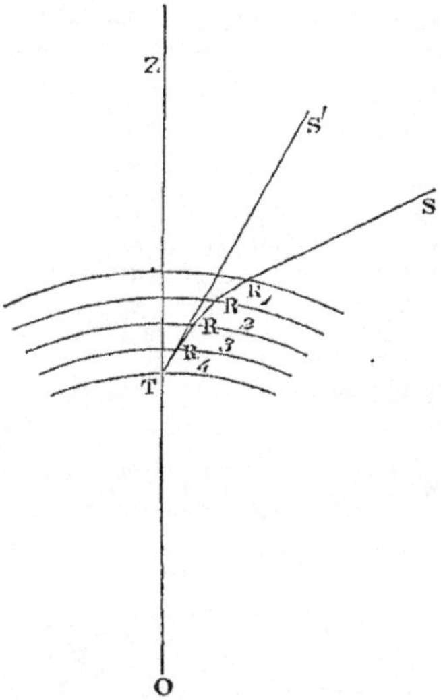

line O T Z, as before, will pass through the zenith of an observer at T. Let $S R_1 R_2 R_3 R_4 T$ represent the course of a ray, which will of course be refracted at the points R_1, R_2, R_3, R_4. In the same manner as before, the line $T R_4$ will be in a plane vertical at T, and the line $R_3 R_4$, the course of the ray before refraction at R_4, in the same plane with $R_4 T$, its course after refraction; and similarly, $R_2 R_3$ in the same plane with $R_3 R_4$, $R_1 R_2$ in the same plane, with $R_2 R_3$, and $S R_1$ in the same plane with $R_1 R_2$, and, consequently, with $R_4 T$, or in a plane vertical at T. The difference between their directions therefore will be in the same vertical plane; or in this case also the deflection takes place in a vertical plane; and as each successive refraction is in passing from a rarer to a denser medium, the ray will at each time be drawn nearer to the perpendicular; and consequently, as before, the body will be raised by refraction.

In this case however, as before, the amount of deflection will be less than if the ray were incident immediately on a refracting surface at T, and the law of its variation will differ from that of the tangent of the apparent zenith distance. If we suppose lines drawn from O through R_4, R_3, R_2, R_1, each of these lines will be perpendicular to the spheres at the points R_4, R_3, R_2, R_1, and pass through the zeniths of those points respectively. Taking therefore the case of the extreme stratum $R_1 R_2$, the results there will be exactly the same as those previously deduced for a single stratum of uniform density, for it is one; and then we have already seen that the deflection at R_2 will be less than if the ray were immediately incident there. In the same manner, taking the different strata in succession, the deflection of a ray incident at R_2 in the direction $R_1 R_2$ will be less than that of a parallel ray immediately incident at R_3; that of a ray incident at R_3 less than that of a parallel ray incident immediately at R_4; that of the ray $R_4 T$ less than that of a parallel ray incident immediately at T; and consequently the whole deflection, being the sum of these deflections each of which is diminished by the spherical form of the atmosphere, will itself be diminished by it. The amount of each particular deflection will be determined by the tangent of the angle of refraction there*. We have already seen, that in each particular stratum this angle does not accurately correspond with the apparent zenith distance there; and of course we have no reason to suppose that the aggregate of these deflections will correspond with the tangent of the apparent zenith distance at the ultimate place of observation.

* [In deducing the variation of the deflection as the tangent of the angle of refraction, we considered *m* to represent the index of refraction from a vacuum into air of the density usual at the earth's surface. The proof however was independent of this particular value, and only depended on the whole amount of refraction being very small. Of course the difference of the densities of the successive strata of air, since that next to the earth's surface is supposed the densest, must fall short of the difference between the density of that stratum and a vacuum; and consequently the amount and index of refraction in each case be less than in that already demonstrated. The same result therefore, that the deflection will *vary* as the tangent of the angle of refraction, will apply in these cases also, and with yet more accuracy from the yet smaller value of the quantities neglected.]

These arguments are evidently independent of the number of strata, and of their thickness. If therefore we conceive the atmosphere to be divided into an indefinite number of strata, and the thickness of each of course to be indefinitely small, the same conclusions will hold good. The different strata, being still successively denser as they approach the surface of the earth, the ray will be bent towards the perpendicular at each; but the difference of density being very small indeed, and the points at which the successive bendings down take place brought very close to each other, the path of the ray will hardly differ from a curve of a continuous curvature; that is to say, a line continually bending, and not changing its course, as in the figure we have drawn, by abrupt angular deviations at distinct points. The thinner and more numerous the strata, the more completely will the line assume this character. Now the air (speaking without reference to accidental circumstances, which may produce trifling exceptions, which even where they occur do not materially affect the truth of our conclusions) continually diminishes in density as its distance from the earth's surface increases; and consequently, the smaller the thickness which we suppose each stratum to have, the more nearly do we approach to a true representation of the atmosphere. Diminishing the thickness of each stratum indefinitely, we have a true representation of the atmosphere; and in this case we have seen that the line becomes a curve. Every deflection is towards the perpendicular; they are all therefore towards the earth, and the curve described by the ray is concave towards the earth.

In this case all the results which we have deduced will apply. The ray, after each successive bending, will be in the same vertical plane as before: there will therefore be no turning to one side or the other, but the ray, when it enters the eye of the spectator, will come in the direction of a point higher than the situation of the object from which it really proceeds. The whole refraction will also be less than it would be if it took place at once, when the air has acquired its final density.

This being the case, we have no longer any just ground to conclude that the refraction at different zenith distances will vary accurately as the tangent of the apparent zenith distance, though the difference thus occasioned is too small to make the law of the variation very far different in the existing from what it would be in the supposed case. The whole refraction is less than it would be on the supposition on which that result was deduced; but whether it is diminished at all altitudes in the same proportion, in which case the law of its variation would remain unaltered, or whether it is affected differently under different circumstances, we do not as yet see. And it is a subject of more intricate calculation than it is desirable here to introduce. The result however is, that the diminution of refraction occasioned by the spherical form of the earth and atmosphere varies as the sum of a certain series involving the cube, the fifth power, the seventh power, &c. of the tangent of the apparent zenith distance. The terms beyond the cube may, from the smallness of their coefficients, be safely neglected in all cases where the zenith distance does not approach 90°, and even the term involving the cube of the tangent, is very small in comparison with that involing the tangent itself merely. As a rough approximation therefore, the variation in the proportion of the tangent may, in this case also, be used*.

* [For scientific purposes we are in want of a more accurate estimate of refraction than the simple one given in the text. The formula given by La Place is

$$R = (a + \frac{2}{3} a^3 - a\,b) \tan. Z - ab. \tan.^3 Z,$$

where R is the refraction, Z the apparent zenith, $a = 57''.2$ and $b = .00125251$. This formula is generally used for computation, but does not apply accurately at zenith distances exceeding 74°, or at altitudes less than 16°. Another formula, of a simpler form, though not quite equally correct within the limits to which that already given properly applies, is found to give very accurate results to a greater extent, namely to the extent of about 80° from the zenith, or within 10° of the horizon. In this formula, the refraction is supposed to vary, not as in the rough approximation in the text, according to the variation of the tangent of the apparent zenith distance, but to that of the tangent of that distance diminished by three times the amount of the refraction, as computed by that imperfect approximation. If therefore A represent the amount of the refraction at a given zenith distance P, R the true amount of the refraction at any other zenith distance z, and a and r the amounts of the refraction at those zenith distances respectively, as computed on the original supposition of its varying exactly as the tangent of the zenith distance, we shall have this proportion—

$$R : A :: \tan. (z - 3r) : \tan. (P - 3a)$$

$$\text{or } R = \frac{A}{\tan. (P - 3a)} \tan. (z - 3r).$$

Now the tangent of 45° = 1; if, therefore, P be so assumed that $P - 3a$ be equal to 45°, $R = A \tan. (z - 3r)$, A being then the amount of refraction which corresponds to that value of P. A is found by observation, or it may be deduced from

The refraction increases as the zenith distance of the body increases; for the tangent of any angle increases as the angle increases. The tangent increases slowly while the angle is small, continually more rapidly as it becomes larger, and with excessive rapidity when the angle approaches 90°, at which time also it becomes very great in absolute amount, and when the angle is of 90° its tangent is of infinite magnitude. All the small inaccuracies therefore, which may be neglected while the angle is small, cannot be so when it approaches 90°, as they then affect a quantity of great amount, and very rapid variation. None of the formulæ deduced therefore can safely be relied on, except when applied to bodies whose zenith distance falls considerably short of 90°, or which are considerably above the horizon. None of them are considered to be practicably applicable if the zenith distance exceeds 80°, or the body observed is within 10° of the horizon.

This however is of little importance; for none of the most material observations are made on bodies below that altitude. Below it, the amount of refraction, from causes hitherto unascertained, seems irregular; the refraction indeed continues to increase, and that rapidly, as the zenith distance increases; but it is different at different times, and in different places, and its actual amount, and the manner in which it is affected by the state of the atmosphere, have not hitherto been accurately ascertained.

Having thus explained the manner in which refraction varies, we have still to see how it affects our observations of heavenly bodies, and to explain some remarkable phenomena which it produces.

In the first place, it is of importance to remark that all heavenly bodies are alike affected by it. The effect is produced merely by the operation of the atmosphere; and consequently every object, whether sun, moon, or star,

theory, to be of the value of 57″,* when the mercury in the barometer stands at the height of 29.6 inches, and Fahrenheit's thermometer stands at 50. The value of A (which is very nearly the same as that of a) being so small, P itself is very nearly equal to 45°.

We have mentioned the height of the barometer and thermometer at the time when the value of A is ascertained. It is evidently necessary that these should be taken into the account. The density of air increases in proportion as the weight pressing on it increases, or the density of the air at the earth's surface increases in proportion to the weight of a column reaching through the whole atmosphere. But to this weight, the weight of the column of mercury in the tube of the barometer is equal. The density of the air therefore increases in proportion as the height of the mercury in the barometer increases, if the density of the mercury in the barometer remains the same: its refractive power also, which increases, as its density increases, is found to do so in the same proportion or very nearly so. The whole amount of refraction therefore, which, when the barometer stands at 29.6, is A tan. $(z - 3r)$, will be greater or less than this quantity as the barometer stands above or below that height: or if h represent the height of the barometer, our formula will become

$$R = \frac{h}{29.6} \text{A tan. } (z - 3r).$$

The effect of an increase of temperature, on the other hand, is to increase the elasticity of the air, and also to diminish the density of the column of mercury in the barometer, and thus to make an equal height of the barometer correspond to a less weight of air. On this latter account, the formula already stated as involving the height of the barometer, requires a little alteration. If t be the height of Fahrenheit's thermometer at the time of the observation, the correcting fraction $\frac{h}{29.6}$ will have to be increased when t is less than 50, and diminished when it is greater; the correct value is found to be $\frac{h}{29.6}$ $(1.005 - .0001\,t)$.

A separate correction is required for the increased elasticity of the air, for the air in consequence, without having its whole weight diminished, is spread over a larger space; and its density, and consequently its refracting power, is diminished. The precise proportion in which this effect is produced however is not well ascertained. According to Bradley, the formula given above is made to include the effect of temperature in modifying the amount of refraction by multiplying it by $\frac{400}{350 + t}$, where t is the height of Fahrenheit's thermometer at the time of the observation: according to Dr. Brinkley, the fraction by which it is to be multiplied would be $\frac{500}{450 + t}$.

Using the latter fraction, and Dr. Brinkley's value of A, the formula becomes

$$\frac{500}{450 + t} \times \frac{h}{29.6} (1.005 - .000st) \times 57''.72, \text{tan. } (z - 3r).$$

This formula evidently corresponds with the general principle, that the refraction will diminish as the temperature increases; for as the temperature does so, the denominator of the fraction increases, and the value of the fraction is diminished, the value of the quantity .0001 t in the numerator being too small to affect the truth of this result. Still however some uncertainty remains as to the exact amount of the refraction at different temperatures. It is not found that any other circumstances with respect to the state of the atmosphere affect the amount of refraction.

There is a very slight additional correction to the formula, required for purposes of extreme accuracy, arising from the variation of gravity at different points of the earth's surface. The density of the air depends on the pressure upon it of the higher strata of air, and this is evidently greater as the gravity is greater. We shall hereafter see that the force of gravity is continually greater as we recede from the equator and approach the poles of the earth: the refraction therefore will be so also. But the variation arising from this cause is exceedingly small: even where it is greatest it will not cause the refraction to differ by 1-350th from the value which it would otherwise have.]

* 56″.9 according to Bradley; 57″.5 according to the French Tables; 57″.72 according to Brinkley.

will have its apparent place differ from its true one in the same manner, since the same causes work upon it, under the same circumstances, and for the same space. We have already seen that the only effect of refraction is to raise the place of the body; and consequently, that the correction of refraction is to be subtracted from the observed altitude, or added to the observed zenith distance.

The correction however is greater as the zenith distance is greater, and less as it is less. Now at every place, except where the pole is in the zenith, all heavenly bodies, except the circumpolar stars, rise from the horizon towards the east, return to it towards the west, and are continually at different elevations during their progress from the one point to the other, being highest upon the meridian. The circumpolar stars themselves are similarly affected, being at their highest and lowest points when they pass the meridian; and succes sively at different elevations at every instant between them. The correction for refraction therefore will be different at every different instant; and if the real path described by each star in its daily course is a circle, its apparent course will be unequally raised above this circle in different parts, and will not be accurately circular. And so we find it: the apparent path which, in the early part of this treatise, we have spoken of as circular, is not accurately so; but when the proper correction for refraction is made in each particular instance, it becomes so: or rather, to speak more correctly, it becomes more nearly so; for there are other corrections still to be made, before it becomes accurately so. All the corrections however are small, so that the apparent path does not very considerably differ from the true one; it will therefore to all common observation appear circular; and the results deduced in our early chapters will be arrived at, though not with strict accuracy, in the manner there pointed out.

There are however some important and curious conclusions, which may be at once deduced from the doctrine of refraction.

A ray of light which, passing out of a vacuum, falls on the surface of any medium, at an angle of 90° from the perpendicular to the surface at the point, will not enter the medium. The consequence would of course be, that if the external surface of the atmosphere were plane, a ray of light proceeding from a star 90° from the zenith would never enter it; and thus every heavenly body, as a star, or the sun, would never be seen when just upon the horizon. The shape of the atmosphere however is spherical, and we shall find that this prevents the consequence just pointed out.

In *fig.* 10, let O represent the centre

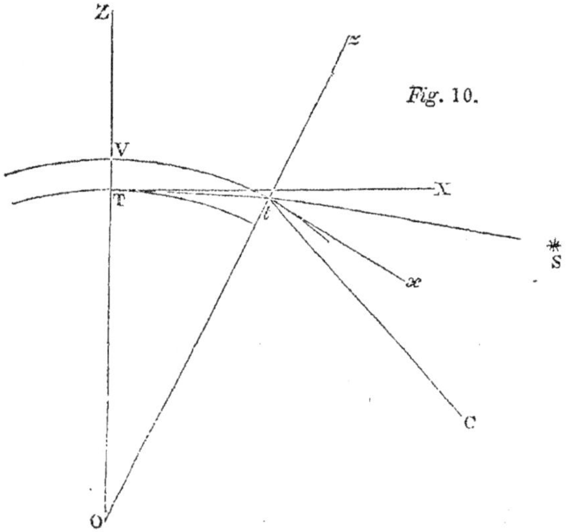

Fig. 10.

of the earth; the circular arc nearest it, part of its surface, that farthest from it, part of the outer surface of the atmosphere. Let T be the position of the observer, Z his zenith: O T Z, therefore, is a straight line, and if T X be drawn a tangent to the circle at T, it will be perpendicular to O T Z, and will be in the horizon of the observer. Independently of refraction, no heavenly body situated below this horizon could be seen by the observer at T, for part of the opaque earth would be between them. Let a ray, however, proceed from a heavenly body S, and meet the exterior surface of the atmosphere at the point t; through t draw O t z, a line passing through the centre of the earth, and t x perpendicular to that line.

If S be any heavenly body depressed below the horizon at T, but so little that a ray from it falls on the atmosphere at t above the line t x, that ray will fall upon it at an angle S t z, less than a right angle; and will consequently be refracted into the atmosphere, and follow the usual law of refraction, being continually bent downwards*. Different situations may be assigned to the point t, and the star S, each of which will transmit the ray to the observer at T. As the figure is drawn, the ray arrives at his position in the direction of the tangent T X, or the star would appear to be upon the horizon; if both S and t were taken higher, the curve would pass above that tangent, and the ray enter the eye in consequence from a higher point: in that case therefore a star, still really below the horizon, would appear above it. In the same manner, a ray proceeding from a body in the line T X, and falling on the atmosphere above that line, might be refracted to T; and thus the spherical form of the atmosphere would render a body, really exactly upon the horizon, visible: the consequence to which reference has been made above.

The amount of the horizontal refraction, or the real depression below the horizon of a body which appears to be just upon it, is not very accurately ascertained; it varies much in different climates, and from local and accidental circumstances in the same. Vapours and exhalations continually affect it; and the account given in the latter part of the treatise on Optics, p. 56, &c., of the phenomena of the mirage, abundantly shews how uncertain and irregular may be the refraction of a ray proceeding through a large portion of the lower strata of the atmosphere. Generally speaking, however, the average horizontal refraction in this country is about 33′.

It is obvious that the effect of this horizontal refraction is to lengthen the period during which every heavenly body appears to be above the horizon. It is seen as if upon the horizon when it is really 33′ below it; it then appears to rise above it, and continues apparently above it during its whole course, until it again becomes 33′ below it, when it apparently returns to the horizon, and then sinks below it. The time therefore of its appearing above the horizon is greater than that of its really being so, by the time which it takes to rise through an altitude of 33′ on the Eastern, and to sink through an altitude of 33′ on the western side of the heavens. Conversely, as the time during which a heavenly body is really above the horizon is readily computed when its place is ascertained, and that during which it appears above the horizon is easily observed, the difference between these two may be ascertained; and from it, the arc through which the body is apparently raised, or the horizontal refraction, may be calculated.

The sun, as well as every other heavenly body, being raised by refraction, the period during which he appears above the horizon is increased, or the day is lengthened by that means. In the same manner, as we have already observed, p. 28, the effect of refraction in places where at one season of the year the sun never appears to rise, and at another never to set, is to increase the length of time during which he is continually above, to diminish that during which he is continually below the horizon.

The doctrine of refraction furnishes an explanation of another remarkable fact. Every one has observed that the sun and moon, when very near the horizon, appear of an oval shape, their width horizontally seeming greater than their height vertically. The variation of re-

* [Since t x, T X are perpendicular to the lines O z, O Z respectively, on the same side of those lines, they are inclined to each other at the same angle as those lines, or at the angle T O t (Treatise on Geo. I. 18); or, as the line t x evidently tends downwards from T X, the line t x is depressed below T X by that angle; it therefore will fall below any star which is not depressed below T X by an angle equally great: or, if S be a star depressed below the horizon T X by an angle less than T O t, and S t a ray proceeding from it, the ray will fall on the atmosphere at t at an angle S t z, less than x t z, or less than a right angle.]

fraction at altitudes little above the horizon cannot be correctly computed from the formulæ given above, but it will still have some correspondence with them. Now they depend on the tangent of the zenith distance, or of an angle only a little less. The tangents of angles differing very little from right angles increase very rapidly for small alterations in the amount of the angle. It is plain then that the amount of refraction must vary very rapidly also. The whole amount of the refraction near the horizon is considerable: its changes therefore are considerable also. The apparent diameter of the sun is of about 32′; the variation of refraction very near the horizon, occasioned by a variation of altitude of 32′, is generally not less, or is even greater than 4′. Now distinct rays proceed from the different parts of the sun's body: a ray therefore proceeding from his upper limb, or rather from the highest point of it, will be raised by refraction 4′ less than one proceeding from the lowest point, or the apparent vertical diameter will be 4′ less than it ought to be, or than 32′, and will therefore be of 28′ only. The apparent horizontal diameter will hardly be at all affected: for as refraction takes place only in a vertical direction, all the points of the horizontal diameter will be raised very nearly parallel to each other*, and being at the same altitude, they will be equally raised. The apparent horizontal diameter therefore will continue of 32′; and the sun will appear to have its diameters in the proportion of 28 and 32 to each other, and will therefore assume a perceptibly oval form. The same observations will apply to the moon in similar situations. It is hardly necessary to observe, that the numbers here taken are only used for the purposes of illustration, and that the degree in which the apparent shape becomes oval, will continually vary in different conditions of the atmosphere. At greater elevations no such perceptible effect is produced, for there the refraction itself is small, and its variation also slow. For instance, at the altitude of about 45°, the refraction is only about 57″, and the variation of 32′ in altitude would not make the difference of more than about a fiftieth part of this quantity: an amount far too small to produce any perceptible effect on the general apparent form of so large an object*.

* [In strictness, the horizontal diameter will be a very little diminished: for the extreme points of the sun being raised in vertical circles, which converge to the zenith, they will be brought nearer to each other by reason of that convergence. The convergence however is exceedingly small, the circles being only about half a degree apart, and the situation of the points raised by refraction being nearly 90° from the zenith.]

Section III.—*Of Parallax—Its nature and variation—Mode of ascertaining its amount—Deduction from it of the distances, diameters, and magnitudes of Heavenly Bodies—Identity of sensible and rational horizon—Increase of Moon's apparent diameter as she approaches the zenith.*

Generally speaking, if different persons view the same object from different places, they will see it in different direc-

* In connexion with this part of the subject, though not strictly forming a part of it, or affecting the correctness of observation, we may explain (more fully than in p. 24) the cause of the comparatively dim appearance of objects seen in, or very near the horizon; and also that of the curious and important phenomenon of twilight.

When a ray of light passes through any medium whatever, however transparent it may be, a certain quantity of light is absorbed by the medium itself, and the quantity of course increases, as the ray passes through a greater space of the medium. The denser and less transparent also the medium, the greater is the amount of light absorbed. Now a ray proceeding from a body near or upon the horizon, passes through a much larger space of the atmosphere, than one from a body near the zenith. For instance, in the figure 10, a ray from the zenith passes only through V T, the ray from the star S passes through the longer space *t* T. Besides this, it is obvious, from mere inspection of the figure, that a larger proportion of the line *t* T is in the lower regions of the atmosphere, than of the line V T. The line V T passes directly through all the strata; the line *t* T passes through them much more nearly horizontally near T than near *t*, and consequently is longer in proportion among the strata nearest the earth's surface. But these are denser and less transparent than the higher strata. The quantity of light absorbed must therefore increase, both from the greater space of air which the ray traverses, and from the greater proportion of the denser, in comparison with the rarer, air through which it passes. The quantity of light absorbed being thus increased, that which remains is diminished, and the body appears more faintly. This effect is most strongly produced near the horizon: but the line drawn to the zenith is the shortest of any which passes through the atmosphere, and those more distant from it are continually longer as the distance increases. The absorption of light therefore is least in the zenith, and continually greater as the zenith distance increases, or the altitude diminishes; and a heavenly body will, although not very observably, except within a moderate distance of the horizon, be brighter the higher it is above the horizon.

The rays of light are capable of reflection, as well as of refraction or absorption. Thus, if, in figure 10, C *t* represents the course of a ray proceeding from a body depressed lower below the horizon than the line *t x*, the ray may meet a particle of air at *t*, and be in part reflected by it; and if the angle C *t* O be equal to the angle made by the direction of *t* T, at the point *t*, with the line *t* O, the course of the ray C *t* after reflection will be the same as that of the ray S *t*, after refraction: or it will come to the eye of a spectator at T. It will not, being reflected only from a single point, and a small part only of the light being reflected at all, produce an image of the body C: but though tions. Thus, if in figure 11, A and B represent the stations of two observers, each observing the same body X, and A B E be a line drawn through A and B

Fig. 11.

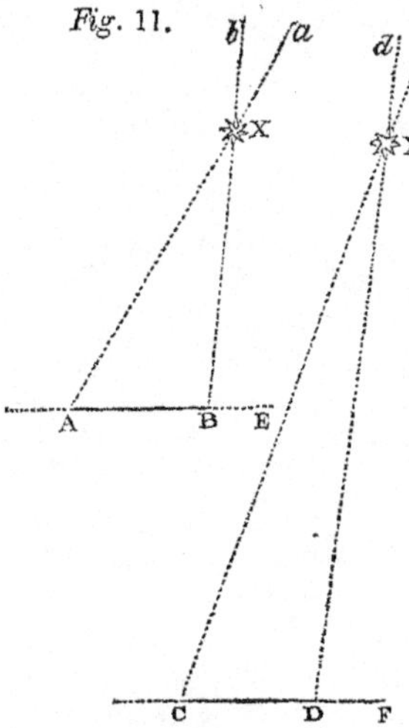

no distinct image is formed, light will be transmitted to the eye.

This however is not the only, nor the principal way in which light proceeds to the eye from a body depressed too far below the horizon for it to appear upon or above it by the effect of refraction. Light falls upon different particles of air presented to it in different situations, and under different circumstances, and is reflected in consequence, and dispersed through the atmosphere in a variety of directions. It is indeed this dispersion which occasions that general diffusion of light through the air to which we owe almost all our opportunities of sight, and without which nothing would be visible except the luminary itself, and those objects so situated with respect to it, as exactly to reflect to the eye rays proceeding from it. Without this dispersion we should see the sun at noon-day as a brilliant object surrounded by intense darkness; and its wide range cannot be better exemplified than by its making the whole vault of the sky visible, nor its variations in degree better than by the strong irradiation which pervades those parts of the sky near the sun, as compared with the mild blue of those more distant from that body. Rays thus dispersed may be successively reflected from one particle of air to another, and finally to the eye; and as there is nothing to determine the situation of the different particles with respect to the rays which fall on them, their reflections may take place in any possible direction; and there is in consequence no impossibility that a ray, once entering the atmosphere at any point whatever, may be thus transmitted to the eye of an observer situated anywhere soever on the earth. Wherever the sun may be, his rays fall on some part of the atmosphere and enter it: however far therefore he may be below the horizon, it is not impossible that some light may be transmitted from him to the eye, and it is even probable that some always is so.

The quantity of light thus received however will necessarily continually decrease as the sun sinks lower below the horizon. A greater number of reflections will be necessary to transmit the light to the eye at all; and thus not only the light of a single ray will be feebler, but, a greater complication of circumstances being necessary to transmit a ray accurately to the eye, a smaller number will arrive there. Altogether therefore as the sun sinks lower below the horizon, the light received from him is diminished; and we do observe correspondingly, that the light of day gradually fades away after sunset, or gradually increases before sunrise, and that there is a considerable interval between these periods and that at which the darkness is complete, or as nearly so as we ever observe it. This interval is called *twilight*: it is generally considered to last till the sun is 18° below the horizon, as it has been thought that the greatest observed darkness does not exist when he is less depressed. An element of this kind however, which depends on no principle of science, does not admit of any very accurate determination; it in fact differs in some degree in different climates; and the depressions assigned by different writers have accordingly been various.

The duration of twilight is different at different places of the earth's surface, and at different seasons of the year at the same place. It is least at those parts of the earth where the pole of the heavens is in the horizon; for there all the circles of daily rotation (their planes being always perpendicular to the line joining the poles, and that line being then in the horizon) are perpendicular to the horizon, and there is an end of twilight therefore, as soon as the sun has performed so much of his daily rotation as is equal in space to 18° of a great circle. Thus, when he is in the equator, 18° of his course are themselves 18° of a great circle, and the twilight ends when he has described 18° of his course after sinking below the horizon, or it is of $1^h\ 12^m$ duration: when he is at either of the tropics, 18° of his circle of daily rotation are equal to little more than 16½° of a great circle, and it takes upwards of 19½° of his daily course to be equal in space to 18° of a great circle; or the duration of twilight is somewhat more than $1^h\ 18^m$. Again, where the heavenly pole is in the zenith, as there is but one night, (p. 27) there are but two twilights in the year. At the North Pole there is night as long as the sun's declination is South; but whenever it does not exceed 18° South, the sun is never more than 18° below the horizon; or whenever this is the case there is twilight. From the autumnal equinox then, when the sun sinks below the horizon, till November 12th, when he attains the South declination of 18°, there is continual twilight; then night unrelieved by twilight until January 29th, when he returns to the declination of 18° South; and then again continual twilight until the vernal equinox, when he rises above the horizon, and makes the day of half-a-year. In intermediate points of the earth, the considerations regulating twilight are more complicated. When the pole is elevated above the horizon, but not in the zenith, the circles of daily rotation are oblique to the horizon, and cut it at different angles, and proceed below it in different directions, as they cut it at different points. The duration of twilight depends on the time at which the sun is 18° of a vertical circle below the horizon; and the actual space which it is necessary for him to describe that he may be thus far depressed, will evidently be greater the less rapidly he sinks below the horizon. Besides, the number of degrees of his diurnal circle, comprised in a given space of the heavens, will increase as his diurnal circle diminishes in magnitude; and on this account also, the length of twilight will be different at different times at the same place. It will also be different at the same time at different places; for the circles of diurnal rotation will cut the horizon at different angles, as the pole is more or less elevated above the horizon.

At London the North Pole is elevated 51½° above the horizon: the point, therefore, which is on the meridian 18° below the North point of the horizon, is 69½° distant from the North Pole, and therefore has 20½° of North declination, and is of course the lowest point of the parallel circle of that declination, or of the circle of diurnal rotation which the sun describes when his declination is 20½° North. At this time therefore, or on the 22d day of May, he never sinks more than 18° below the horizon, or there in the plane passing through A, B, and X, so as to furnish a standard to which to refer the apparent place of the body X, it is obvious that the angles X A E, X B E, made with the line by the body's apparent direction, are different in the two instances. In the same manner, if C and D represent the places of the two observers, supposed now more distant from the object, and C D F be a line drawn in the same manner as A B E before, the angles X C F, X D F, made with this line by the body's apparent direction, are again different from each other. The difference between the two angles, is, in each case, the angle A X B, or C X D, subtended by the distance between the places of the two observers at the position of the body X: for the exterior angle of every triangle being equal to the sum of the interior and opposite angles, X B E = X A E + A X B; or X B E − X A E (the difference of the apparent directions) = A X B. In the same manner, in the other case, X D F — X C F (the difference of the apparent positions) = CXD. It is evident, even from inspection of the figure, where the circumstances of the two cases, except as to the distance from the body X, are nearly similar, that this angle is larger in the case of the positions A, B, which are nearer to X, than in that of C, D, which are more distant from it. The manner however in which the amount varies deserves more minute consideration.

In the first place, it is evident that the figure represents the whole alteration produced by the change of position. It represents the alteration produced in the plane A X B; and as A, B, X, are all points in this plane, the apparent direction of X from both A and B will be a line in this plane, and its apparent position, *a* or *b*, a point in it. In the same manner, *c*, *d*, the apparent positions of X from C and D, will each be points in the plane passing through C, X, and D. The body therefore is seen in the same plane, both from A and B; or, in the other case, in the same plane both from C and D: the alteration therefore of its apparent position in that plane is the only alteration which it experiences.

[The manner in which the amount of this alteration varies may be at once deduced from the first principles of trigonometry. Confining ourselves for

---

is twilight for the whole period from sunset to sunrise; and this continues to be the case, as long as his North declination continues greater than, or equal to 20½°, or from May 22d to July 21st. From that time, he sinks farther than 18° below the horizon; at first for a very short period, so that as he then sets a little before eight o'clock, and the twilight does not end till he attains very nearly to his greatest depression, or till very near midnight, the duration of twilight is nearly four hours in the evening after sunset, and the same in the morning before sunrise. From that time the length of twilight continually diminishes till about three weeks after the autumnal equinox, when it attains its least value, which is about 1h 50m; it then increases again till the winter solstice, when it is 2h 7m; then again diminishes till about three weeks before the vernal equinox, when the sun being again at the same declination as at the former period of three weeks after the autumnal equinox, the twilight again has its least length of 1h 50m, and thence it again continually increases till the 22d of May, from which time it lasts during the whole night. The variations will be similar at other places, but the periods at which the twilight continues during the whole night, and at which it is shortest, and the actual length of its duration, will be different at every different latitude on the earth. When the latitude is less than 48½°, or the pole is less elevated above the horizon than 48½°, there will be no period during the year at which the twilight will last during the whole night; for in these cases the distance from the pole of a point in the meridian, 18° below the horizon, will be less than 66½°, and its declination therefore greater than 23½°, the greatest declination of the sun; and the sun consequently, when on the meridian below the pole, will be more than 18° below the horizon.

The duration assigned to twilight may seem longer to many readers than they would expect to find it; and they may be surprised at learning that it is not shortest in the depth of winter. This arises from the popular use of the word twilight. Astronomically speaking, it is the period during which it is thought that light comes *perceptibly* to the eye from the sun below the horizon: in common language, it is the time during which a considerable quantity of light, but decidedly inferior to that of day, so comes, a quantity sufficient to make a marked difference between the appearances presented, and those either of broad day, or of night; and thus we hardly begin to call it twilight till some little time after sunset, and cease to think about it as such, long before all rays cease to be perceptibly transmitted. This period therefore is very much shorter than that of astronomical twilight, and especially so in the winter, when much of the light that would otherwise come to the eye, and produce what is commonly called twilight, is lost and intercepted by a misty atmosphere, and the heavy clouds frequent at that season. The astronomical twilight of mid-winter does not exceed the shortest astronomical twilight by much more than a quarter of an hour: and these causes may easily make the difference between the astronomical and the popular twilight vary by more than this quantity, and if so, the twilight of winter will be called the shorter. On an evening of very heavy cloud indeed there may be very little perceptible twilight at any season of the year; so many of the rays reflected from the upper regions of the atmosphere, and which would otherwise produce it, being intercepted, and either absorbed, or reflected away from the earth, by the clouds in the lower regions of the air.

It is obvious that the same principles which produce twilight by the reflection of the sun's rays would produce a similar phenomenon, or a lunar twilight, by the reflection of those of the moon. The light of that body however is so inferior to that of the sun, that it never produces a twilight of practical importance and utility, and the consequence is, that it is never spoken or thought of. On a clear night nevertheless, the sky begins to brighten, especially in that part nearest to the moon, for some time before she actually appears in the horizon; and this is, strictly, a lunar twilight.

the future to the triangle A B X, we have these equations.

$$\sin. X = \sin. A. \frac{AB}{BX} \text{ and}$$

$$\sin. X = \sin. B. \frac{AB}{AX}.$$

If, therefore, the distances A B and B X remain constant, the sine of the angle expressing the change of position increases or diminishes in the same proportion as the sine of the angle made at the position of one observer, by the star and the position of the other observer; if this angle remain constant, the sine of the angle expressing the change of position increases with the increase of the distance between the two stations, and diminishes with the increase of distance of the body itself. Thus, in the figure as drawn, the distances A B, C D, and the angles A B X, C D X, are equal; the sine of the angle A X B, therefore, must exceed that of the angle C X D in the proportion of $\frac{1}{AX}$ to $\frac{1}{CX}$, or of C X to A X; or, as the figure is drawn, nearly in the proportion of 2 to 1; and if the angles are neither of them large, the ratio of the angles themselves will not be very different. If the distance A X or B X be indefinitely increased, while the line A B remains the same, it is evident that the sine of the angle X, or the angle itself, or the alteration of position, will become indefinitely small.]

The earth is a body of large magnitude; the distances therefore at which different observers may be placed from each other on its surface may be very great, and the differences between the apparent positions even of very distant bodies may in consequence be very considerable. It is necessary therefore to have some means of ascertaining the amount of these differences, before we can compare observations made at one place with observations made at another. This amount evidently varies for each particular pair of places, for it depends both upon their distance, and upon the direction of the line joining them with reference to the body observed. The computation therefore would be different for each case, and could not be made at all without knowledge of the distance of the two places, and the direction of one as measured from the other. These are both of them elements of very difficult ascertainment; so that if it was necessary to make these computations in every instance, it would be almost impossible to compare successfully the multiplicity of observations made by different observers at different places.

This difficulty however may be avoided. The centre of the earth is a point equidistant, or nearly so, from every point on its surface; it is also directly, or very nearly directly, below every point on its surface. An observer then, wheresoever situated, will know the direction of the centre of the earth: and its distance, even if not ascertained exactly, will yet be the same, or very nearly the same for all. It is also through the centre of the earth, not through the position of any particular observer, that we have considered the axis of the heavens, and all the great circles of the sphere to pass. On these accounts it has been agreed to regard that as the true place or direction of a heavenly body, which would be its place or direction, as seen by an observer placed at the centre of the earth, if the earth itself did not, by its opaqueness, intercept his vision. The apparent position at any point on the surface of the earth is different from this, and the difference is called *parallax*, from a Greek word, signifying *a change of position of one point with respect to another*. To ascertain therefore the true place of a body, that is to say, its place as if viewed from the centre of the earth, we must *correct* its observed place, by making an allowance for the effect of parallax. The manner in which this allowance is to be made, we shall now proceed to explain.

For this purpose, in *fig.* 12, let the

Fig. 12.

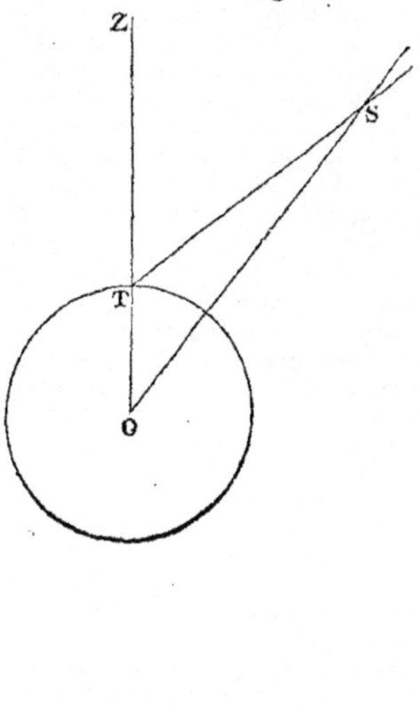

circle represent the earth, supposed to be accurately spherical, T the place of an observer, O the centre of the earth. O T therefore will be perpendicular to the horizon at T, and, if produced, will pass through the zenith; or it is a vertical line. Let S be the position of any heavenly body, and join T S, O S. The line T O being a vertical line, the plane T S O must necessarily be a vertical plane. Now the angle T S O is the difference between the apparent place of S, as seen from O and from T, or, by the definition of parallax already given, it is the parallax of S. Parallax therefore takes place in a vertical plane, and the angle Z O S, the interior angle of the triangle T O S, must necessarily be less than Z T S, the exterior angle of the same triangle, or the apparent is greater than the true zenith distance of the object. The body therefore is *lowered* by the effect of parallax. Parallax therefore, like refraction, acts entirely in a vertical direction; but, while refraction raises, parallax lowers the apparent place of the object; and they in some degree tend to counteract each other.

[The next object is to ascertain how parallax varies in different situations of the same body.

$$\sin. TSO = \sin. STO \, \frac{TO}{SO}: \text{or}$$

since $\sin. STO = \sin. ZTS$

$$\sin. TSO = \sin. ZTS \, \frac{TO}{SO}.$$

Now, in the case of the same body, if it continue always at the same distance from the centre of the earth, S O is always of the same value; and T O is always, in all cases, the same line. For the same body therefore, $\sin. TSO \propto \sin. ZTS$, or the sine of the parallax varies as the sine of the apparent zenith distance. When a body appears in the zenith, the apparent zenith distance is nothing; its sine therefore, and consequently the sine of the parallax is nothing, or the parallax itself is nothing; a conclusion which is obviously true, for observers at T and O would each see a body at Z, in the same direction O T Z. The sine of an angle continually increases as the angle itself does so, until it reaches the value of 90°; the greater therefore the apparent zenith distance, or the less the apparent altitude, of a body, the greater is the parallax, until the apparent zenith distance becomes 90°, or the body appears in the horizon, and then the parallax is the greatest possible. This greatest parallax, as it occurs when the body appears in the horizon, is called *the horizontal parallax* of the body. The sine of the horizontal parallax therefore, (since in this case, $\sin. ZTS = \sin. 90° = 1$ is equal to $\frac{TO}{SO}$, or to the radius of the earth, divided by the distance of the body from the earth's centre; or, if we call the horizontal parallax P, $\sin. P = \frac{TO}{SO}$.

We now proceed to point out the mode of ascertaining the amount of parallax. If P be the horizontal parallax, or the parallax when the apparent zenith distance is 90°, and p the parallax at any other apparent zenith distance, which we will call z; then as the sine of the parallax for the same body varies as the sine of the apparent zenith distance—

$$\sin. p : \sin. P :: \sin. z : \sin. 90°, \text{ or } 1.$$
$$\text{or, } \sin. p = \sin. P. \sin. z.$$

If p and P are both very small, which is found to be the case with all the heavenly bodies, $\sin. p = p$, and $\sin. P = P$, very nearly. On this supposition, therefore, $p = P. \sin. z.$ and $P = \frac{p}{\sin. z}$.]

If therefore we can find the parallax at any apparent altitude, we can find the horizontal parallax from it; if we can find the horizontal parallax, we can ascertain that for every other zenith distance. The parallax depending merely on the zenith distance, it is different in the same body at different periods of its diurnal course, and has a course of variation comprised within the compass of a day. It is therefore called also the diurnal parallax, to distinguish it from another element of the same nature, which we shall hereafter explain under the title of annual parallax. For the present, by parallax we mean diurnal parallax only.

Many methods have been proposed for determining the amount of parallax, some of considerable labour, and requiring observations made at distant places. Except in the case of the moon, the amount of parallax is in all cases exceedingly small, and therefore requires the utmost accuracy in its ascertainment; so that most of the methods suggested are insufficient, and it cannot be satisfactorily ascertained, except on the occurrence of phenomena, which very rarely happen, and cannot at present be explained. The parallax of the moon however may be correctly estimated by observations made at a single place; and it will be worth while to explain the principle of this method without going minutely into its details, especially as it is applicable to the case of refraction as well as to that of parallax.

In *fig.* 13, let H Q N represent the horizon, H E Z P N the meridian of any place, P the pole, Z the zenith of that place (both of them necessarily points

Fig. 13.

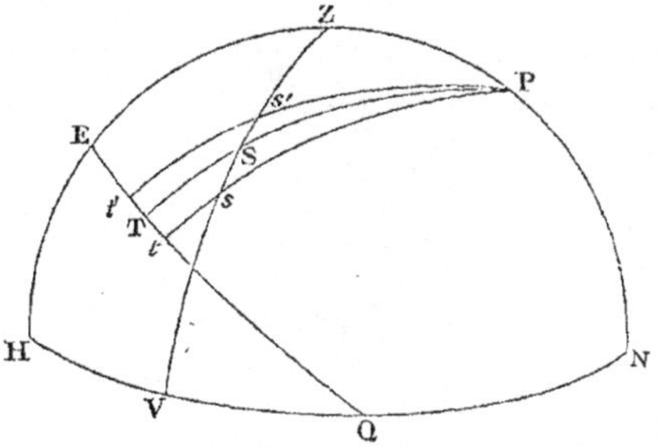

in the meridian), and S the true position of any heavenly body. Let Z S V be a circle drawn from the zenith through the body to the horizon: it is therefore a vertical circle; and both refraction and parallax will take place in it. Omitting for the present all consideration of refraction, parallax will depress the apparent place of the body; or some point *s*, below S, will represent the apparent place of the body. Let E Q be the equator, and draw from P, through S and *s*, to E Q, great circles P S T, P *s* *t*. The right ascension of the point T is evidently the same as that of the point S; that of *t* the same as that of *s*; and if E be the point of the equator upon the meridian, the difference of right ascension between E and S is E T, the difference of right ascension between E and *s* is E *t*; quantities differing by T *t*. If therefore S and *s* were the true places of two different heavenly bodies, they would be upon the meridian at different times, the body at S being upon the meridian before or after E by a time corresponding to the difference of right ascension E T, the body at *s* by a time corresponding to the difference of right ascension E *t*. When the body S really is upon the meridian, its apparent place is also on the meridian, for the meridian being a vertical circle, the parallax of a body upon it takes place entirely in it. The star S therefore, as soon as it really is upon the meridian, is seen there, though below its true place; or the time between the observation of the body apparently at *s*, and the observation of it on the meridian, is the time corresponding to the difference of right ascension E T, for it is the time during which the star is really transferred from its true place, S, to the meridian. This time therefore may be ascertained by observing, by the clock, the difference of time which actually intervenes between the two observations of the star. The place *s* also being observed, its right ascension may be computed, and thence the time in which a body really placed at *s* would be on the meridian ascertained; and the difference between this and the observed value of the time at which S comes upon the meridian, will be the difference in time corresponding to the space T *t* in right ascension; and hence T *t* may be found, and from it, by an easy process, the value of the parallax may be deduced. It is obvious that, by a similar process, the value of refraction may be deduced: the refraction, however, raising the object, if *s'* represent the apparent place of the body, as affected by refraction (now omitting parallax), *s'* will be above S, and drawing P *s'* *t'* as P *s* *t* was drawn before, E *t'*, the difference of right ascension between E and *s'* will be less than E T, the difference of right ascension between E and S, just as in the case of parallax E *t* was greater than E T.

We have now sufficiently explained the nature of parallax; and the importance of considering it for the purpose of reducing our observations to accuracy, is obvious. It is, however, necessary to observe, that if the figure of the earth is not accurately spherical, the distance T O will not be exactly the same for all places; nor will the line T O, when produced, necessarily pass exactly through the zenith. All our deductions concerning parallax therefore would on this supposition, which we shall hereafter find to be the more accurate one, require some qualification. This qualification is easily made; but we need not here enter into any details respecting it. So little does the earth differ in figure from a perfect sphere, that we may fairly treat it as one for the purpose of explaining some very important results which may be deduced from the observations and theory of parallax; and we proceed, therefore, in the next place, to detail and explain them.

From the expression already deduced for the horizontal parallax, of $\sin. P = \frac{T\,O}{S\,O}$, we learn that $S\,O = \frac{T\,O}{\sin. P}$, or that the distance of any body from the centre of the earth = the earth's radius, divided by the sine of the horizontal parallax; or the radius of the earth is contained so many times in the distance of the body, as the sine of the angle of horizontal parallax is contained in the radius of the circle on which it is measured. It is found by observation, that the horizontal parallax of every body which we have occasion to observe is very small, never much exceeding 1°, and, except in the case of the moon, falling very far short of that quantity. The sine of the horizontal parallax therefore will differ imperceptibly from the parallax itself. Now the length of the radius of a circle is equal to the length of 57.°29578 of that circle very nearly; and we consequently have this equation, that S O (the body's distance from the earth) $= \frac{1}{P} \times$ earth's radius $= \frac{57^{\circ}.29578}{P}$ × earth's radius.

We will exemplify, by one or two instances, the use of the equation above deduced. The horizontal parallaxes of the sun and moon vary: we have already seen that the distance of the former body is variable; we shall hereafter see that the distance of the latter is yet more so. We will therefore take their mean values. The mean value of the horizontal parallax of the moon is 57′ 4.″16844, according to La Place, or .951158 of a degree. The moon's mean distance, therefore, is $\frac{57.29578}{.951158}$. earth's radius, or 60.2379 × earth's radius; or very nearly 60 times the earth's radius. The mean value of the horizontal parallax of the sun is only 8.″75, or, according to another estimate, 8.″81; or, in the former case, .00243 of a degree, in the latter .002447 of a degree, and the distance consequently will be $\frac{57.29578}{.00243}$. earth's radius, or 23578 times the earth's radius, if we adopt the former value; $\frac{57.29578}{.002447}$ earth's radius, or 23414 times the earth's radius, if we use the latter. The difference of only six hundredth parts of a second in the value of the horizontal parallax, makes therefore a difference of 164 times the length of the earth's radius in the distance of the sun from the earth; and we may easily infer how very minute is the accuracy necessary in the determination of this element.

It may happen that the parallax of a heavenly body is too small to be observed; for in theory it must always exist. Let us suppose, for instance, that we are enabled to ascertain by observation, that the diurnal parallax of a fixed star does not exceed .36 of a second; or rather, that it appears to have no parallax whatever, but that we cannot rely upon the accuracy of our observations beyond this point. It would follow from this, as .36 of a second is .0001 of a degree, that the distance of the object could not be so small as $\frac{57.29578}{.0001}$ earth's radius, or as 572957.8 the earth's radius; a quantity considerably exceeding 2000 millions of miles. We shall hereafter see reasons for knowing that the distance of the nearest fixed stars is incomparably greater than even this vast quantity, and their diurnal parallax absolutely imperceptible.

Another important element which we may determine by means of the horizontal parallax is the actual radius of any heavenly body, as compared with that of the earth. The horizontal parallax $= \frac{\text{earth's radius}}{\text{distance of the body observed.}}$

The apparent radius of the body* $= \frac{\text{the body's radius}}{\text{distance of the body observed}}$, and as the denominators of each of these fractions are the same, the earth's radius : the body's radius :: horizontal parallax : apparent radius; or the body's radius, = earth's radius $\times \frac{\text{apparent radius of the body}}{\text{the horizontal parallax of the body}}$.

Thus the moon's mean horizontal parallax is 574.″16844, or .951158 of a degree: her mean apparent radius (or the half of her mean apparent diameter) is 1533.″ 8652, or .259407 of a degree, and consequently, her radius $= \frac{.259407}{.951158}$ earth's radius, or very nearly $\frac{3}{11}$ of the earth's radius, and their diameters are of course in the same proportion. In the same manner as we have stated the mean horizontal parallax of the sun at 8.″ 75, or .00243 of a degree, and as his mean apparent radius is 16′ 1.″ 6, or .26712 of a degree, his radius $= \frac{.26712}{.00243}$ · earth's radius, or 110 times the earth's radius nearly. The magnitudes of other bodies, whose parallax and apparent diameters can be observed, may be ascertained in the same way.

The fixed stars which have no observable parallax, have also no apparent diameter; that is to say, in the best observations, made with the most powerful telescopes, the stars still retain the appearance which they have to the naked eye, of mere luminous points. With respect to them therefore we can form no conclusions of this kind.

We have not yet however deduced all the results which follow immediately from the determination of this important element of parallax. Different spherical bodies (and the earth, the sun, and the moon are so nearly spheres, that they may for this purpose be treated as being such) have their actual bulk, or volume, in the proportion of the cubes of their radii: we may therefore determine the proportion of the volumes of these different bodies, whose apparent diameters and parallaxes we have observed, and estimate them also in terms of the volume of the earth. Thus the volume of the moon is to that of the earth, nearly as the cube of $\frac{3}{11}$, to 1, for that is nearly the proportion of their radii: or the volume of the moon $= \left(\frac{3}{11}\right)^3$ the volume of the earth; or $\frac{27}{1331}$ times the volume of the earth, or nearly $\frac{1}{49}$th part of it. Similarly the sun's radius being about 110 times that of the earth, his volume is about 110^3, or 133100 times the volume of the earth.

If there pass through the centre of the earth a plane parallel to the horizon of the observer, this is called the *rational horizon* of the observer, the actual horizon being called his *sensible horizon.* The perpendicular distance between these planes is the earth's radius, and this, as they are parallel, is their distance everywhere. They will therefore intersect the imaginary sphere of the heavens in lines at that perpendicular distance from each other. The angle there subtended by that distance is evidently equal to the horizontal parallax of the fixed stars; for the lines subtending the angle in question and the horizonal parallax are equal, and are both perpendicular to the same line, namely, the distance from the earth to a fixed star, which is supposed to be a point in this imaginary sphere. The angles they subtend at the extremity of this line are consequently equal. But the horizontal parallax of a fixed star is imperceptibly small: the angular distance therefore between these lines where they intersect the sphere of the heavens is imperceptible from the earth; or the rational and sensible horizon appear to intersect the heavens in the same line. It follows that the altitudes of heavenly bodies, which are observed with respect to the sensible horizon, are accurate also with respect to the rational horizon, after they have been corrected for parallax, where it exists*.

* See note, p. 18.

* [In connexion with these considerations of parallax, we may mention another circumstance also resulting from the different positions of a heavenly body, with respect to the horizon. The apparent diameter or radius of every body observed in the heavens being very small, it may be considered, at different distances, to vary inversely as the distance. If, therefore, referring to fig. 12, we take R to represent the apparent radius at the centre of the earth, r, that at the point T, R : r : : S T : S O : : sin. S O T : sin. S T O, or of S T Z,

SECTION 4.—*Of Nutation and Aberration—and general remarks on corrections.*

BESIDES the corrections already described, there are two others, called *nutation* and *aberration*. Their existence is discovered by observation; but it would be impracticable to explain and account for their existence, without assuming the possession of knowledge which we have not yet arrived at. The manner in which they affect our observations, however, may be shortly stated here.

We have already stated that the pole of the equator is found to describe a circle round the pole of the ecliptic in 25868 years, thus producing the precession of the equinoxes. On more minute investigation however, it appears that the place of the pole of the equator thus ascertained differs, though slightly, from the true one, that the real path of that pole, though never far from the circle supposed, does not coincide with it, nor indeed with any circle at all, but that the true place of the pole never differs more than about $9\frac{1}{2}''$ from its place as before ascertained. The true place is found to be successively on all sides of that deduced on the supposition of its being affected by precession only, and they are again in their original situation with reference to each other at the expiration of a fixed period, a little more than eighteen years and a half; and their relative positions then again go through the same order of change. We may therefore consider the place of the pole of the equator as determined on the supposition of its being affected by precession only, as its *mean place*; and, as in many other instances, we shall find it more convenient to ascertain the true place by reference to this, than independently.

A secondary to the equator drawn through the solstitial points is called the *solstitial colure*: it is perpendicular to the ecliptic also, and passes through the poles of both these circles. A secondary to the equator drawn through the equinoctial points is called the *equinoctial colure*: it intersects the solstitial colure at the pole of the equator, and makes an angle of 90° with it. The equinoctial and solstitial points and the pole of the equator receding by the precession of the equinoxes, these circles themselves move necessarily. There is a point, the nature of which will be explained in the next chapter, called the ascending node of the moon's orbit; this is always a point in the ecliptic, but it has a retrograde motion in that circle of about 19° 21′ in every year, and thus returns to its original situation in a little more than eighteen years and a half. The true position of the pole of the equator depends on the position of this point. It is thus determined. Let P in *fig.* 14 represent the mean position of the pole of the equator; A B C drawn through it, part of the solstitial

---

But S T Z is the apparent zenith distance, or z, using our former notation, and S O T is the true zenith distance, or the observed zenith distance diminished by parallax, or $z-p$. We have, therefore, this result R : r : : sin. $(z-p)$: sin. z; or, R sin. $z = r$ sin. $(z-p) = r$ (sin. z cos. p − cos. z sin. p) $= r$ (sin. $z - p$ cos. z), for p being very small, cos. $p = 1$, and sin. $p = p$ very nearly. But $p =$ P sin. z. We have therefore,

$$\text{R. sin. } z = r\,(\text{sin. } z - \text{P sin. } z.\ \text{cos. } z),$$

$$\text{or } r = \frac{\text{R}}{1 - \text{P. cos. } z}$$

very nearly: a quantity which will evidently increase, as cos. z increases, or the zenith distance diminishes. The apparent radius and diameter therefore increase as the zenith distance diminishes, and are greatest in the zenith. This general result indeed may be at once deduced, although not the precise law of the variation, from very simple considerations. The triangle S O T, wherever S be situated with respect to the zenith, has always, if S be considered as the same body, the two sides S O, O T, of the same value. The angle S O T included between them continually varies, and S T, the base of the triangle, will be greater as that angle is greater, and less as that angle is less. (*Euc. I.* 24, or *Treat. on Geom. I.* II.) The apparent radius varies inversely as the distance S T; it therefore increases as that distance diminishes, or as the true zenith distance S O T diminishes.

In the case of the moon, where P is about 1°, the difference is very perceptible; but as it depends on the magnitude of the horizontal parallax it will evidently diminish as that quantity diminishes. Even in the case of the moon it does not amount to much more than a sixtieth part of the apparent diameter, when the moon is in the zenith. The sun's horizontal parallax is only about 1-400th of that of the moon, and the difference between his apparent diameter in different situations will be diminished accordingly, and will obviously be too small to be noticed. The same remark will apply to the other bodies which are the subject of our observation; and the more because their apparent diameters are themselves very small, and any variation of so small a quantity is of course less observable than a similar variation in a larger one. The result which we have obtained is therefore of no practical importance except with reference to the moon.

With reference to the moon also, it is further necessary to remark, that her apparent diameter, as observed even in the horizon, differs from her *true apparent diameter*, if we use that term to express her apparent diameter as it would be seen from the centre of the earth. If we suppose the moon's distance from the centre of the earth to be sixty times the earth's radius, the distance of the observer from the point where she is in the horizon will be $\sqrt{3559}$ times that quantity; for that distance is the side of a right-angled triangle, whose hypotenuse is sixty times that quantity, and whose side is that quantity. For very nice purposes this difference slight as it is, sometimes becomes material.]

colure, which if produced in the direction A B C would pass through the pole of the ecliptic. If a portion of this circle, A B, be taken, extending 9″.648 on each side of P, and an ellipse A d B be described, of which A B is the transverse axis, and the semi-minor, or semi-conjugate axis P d is 7″.182, the true

Fig. 14.

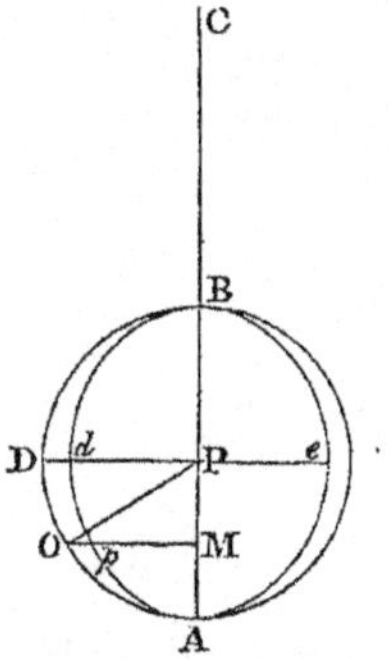

position of the pole will always be a point in the circumference of that ellipse. Its exact position is thus determined. When the ascending node of the moon's orbit coincides with the vernal equinox, the true position of the pole is at A, the extremity of the transverse axis of the ellipse most distant from the pole of the ecliptic. To determine its position at any other time, a circle A D B must be described with the centre P, and radius P A or P B. The ascending node of the moon's orbit has a retrograde motion: take the angle A P O, measured in a retrograde direction, or from East to West, equal to the arc through which the node has retrograded from the vernal equinox, and from O, the point where the radius P O cuts the circle, draw O M perpendicular to the transverse axis of the ellipse: the point p, where the line O M cuts the ellipse, will be the true position of the pole. It is obvious that when the node has retrograded through 90°, or coincides with the winter solstice, the pole will be at d, the extremity of the conjugate axis of the ellipse, for the arc AD being 90°, the perpendicular from D coincides with Pd. When the node has retrograded through 180°, or coincides with the autumnal equinox, the pole will be at B, the extremity of the transverse axis, which is nearest to the pole of the ecliptic. When the pole has retrograded through 270°, or coincides with the summer solstice, its place will be at e, the other extremity of the conjugate axis.

This variation of the true from the mean place of the pole is called *nutation*, as it may be supposed to arise from a *nodding*, or oscillation of the axis of the heavens, of which the pole is the extremity, about its mean position; or rather, as we shall hereafter see, it does actually arise from a similar motion of the axis of the earth, a line of which we have as yet taken no notice. This explanation however must be deferred to a later part of this treatise. The manner in which nutation affects the observation of the stars will be sufficiently understood from what has been already said about precession: of course the motion of the pole towards or from any particular star will affect its right ascension and declination, and as the same motion affects the position of the equinox also, the longitude also will be altered by it: the ecliptic and its pole continuing unmoved, the latitude will remain unchanged. Any detail of the precise manner in which these elements are affected by nutation is not within the scope of this treatise.

The nutation which we have described depends on the position of the node of the moon's orbit. There is also a nutation, exactly of the same kind, which depends on the position of the sun. Its amount however is so small, that it could never have been ascertained by observation only: it is indicated by theory, but a theory depending on considerations which we have not as yet approached. The ellipse by which the place of the pole, as affected by this nutation, may be determined, has its semi-transverse axis only of 0.″435, and its semi-conjugate axis only of 0.″399; and the angle by which the actual place of the pole is determined, is to be taken equal to twice the sun's longitude. When the sun's longitude is 180°, that angle is 360°; or the point P has gone quite round the circle; and this nutation therefore goes through all its values in the course of half a revolution of the sun.

There is one important observation to be made before we quit the subject of nutation. We have hitherto considered the obliquity of the ecliptic, or the inclination of the equator to that circle, to be always the same. This inclination is measured by the distance between their poles; and precession, although it changes the place of the pole of the equator, leaves its distance from the pole of the ecliptic unaltered; but nutation changes it. The mean place of the pole, P, remaining at the distance of what we must now call *the mean obliquity* from the pole of the ecliptic, the true place of the pole of the equator, when the ascending node of the moon's orbit coincides with the vernal equinox, is more distant from the pole of the ecliptic, by the space P A, or 9″.648, and the obliquity is therefore increased by that amount; and in the same manner, when the ascending node of the moon's orbit coincides with the autumnal equinox, the pole of the equator is at B, and the obliquity is diminished by the same amount, P B, or 9″.648. In other positions there will be corresponding alterations; and the nutation which depends on the position of the sun will of course produce effects similar in their nature, but less in amount.

The remaining correction, *aberration*, will be fully discussed and explained hereafter. Like refraction, it makes the apparent place of a star different from its true place; and it prevents us, except in the case of stars situated in the ecliptic, from ever seeing a heavenly body exactly in its true position or direction. All its effects are produced in the course of a year, and they are different from every different heavenly body, though they follow one law applicable to all. That law may be very shortly stated. In *fig.* 15, let S

Fig. 15.

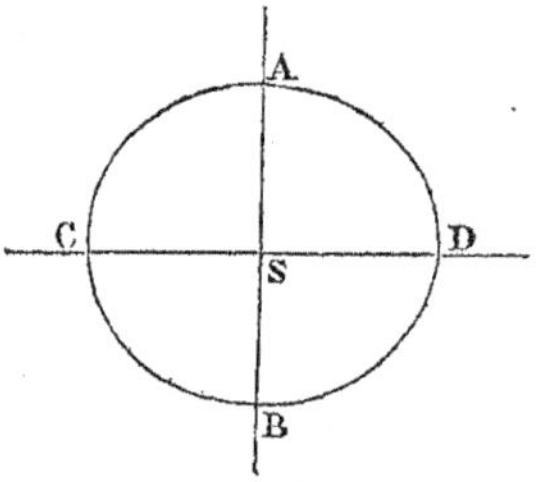

represent the true position of any heavenly body, and let A S B be a portion of a circle of latitude (a secondary to the ecliptic) passing through it. The apparent place of the star will always be a point in an ellipse of which the true place of the star, S, is the centre, and the conjugate axis is in the line A B; the transverse axis of course is in the line C D perpendicular to it. If A C B D, represent this ellipse, S C, the semi-transverse axis, is in all cases 20″.246 of a great circle. The semi-conjugate axis S A is to the semi-transverse axis S C, in the proportion of the sine of the star's latitude to the sine of 90°, or the radius. The ellipse therefore in which each star is to be found, is easily constructed. The apparent place at any particular time is very nearly found by drawing an arc of a great circle through the star to the point of the ecliptic which is 90° behind the place of the sun at the time: the intersection of this line with the ellipse described as before pointed out, is the apparent place of the star at the time of observation. We shall hereafter enter into further details with respect to aberration, and deduce from it some very important consequences: for the present this description of it will be sufficient.

We find then that we have no less than five corrections, as they are called, to apply to the place of a heavenly body, as immediately observed, before we can tell its true place: corrections for precession and nutation, which are rather the alteration of the language in which an observation is originally expressed, to make it correspond accurately with the language of a different period, than corrections of any real inaccuracy; corrections for aberration and refraction, to ascertain from the apparent situation of the star the real direction in which it is with respect to the observer; and a correction for diurnal parallax, which need not be applied in the case of the fixed stars, to deduce from the real direction with respect to the observer the true position as estimated with respect to the centre of the earth.

It will probably occur to the reader that there is a want of strict accuracy of reasoning in the processes which we have explained for ascertaining the amount of these corrections, and even the nature of some of them; because in estimating the one, we have not taken the others into consideration. For instance, taking the method explained in page 59, of determining parallax and refraction, we have shewn how the parallax would affect the value of E *t*, as compared with E T, on the supposition that *s* was the apparent place of the body, depressed by parallax below its true place S: and in the same manner we have shewn the effect of refraction, by supposing *s′* to be the apparent place of the body, raised by refraction above S. If both refraction and parallax operate, it is obvious that the apparent place of the body will be somewhere between s and s', and the same mode of proceeding will not give us the effect of parallax or refraction, but the combined effect of both. We know however from independent considerations, at least very nearly, the manner in which each of these elements varies with the height of the body above the horizon; and we can form a tolerably near, though by no means a sufficient estimate, of the amount of each at a particular height. We can tell that the amount of parallax of a fixed star must be very small indeed, before we can venture to say that it is absolutely too small to be at all estimated: for we know from other considerations, that the refraction varies very nearly as the tangent of the apparent zenith distance, while the parallax varies accurately as the sine of the same angle, or rather of the same angle increased by the amount of the refraction, but we find that the combined effect of the parallax and refraction in the case of the fixed stars, itself varies very nearly as the tangent of the apparent zenith distance; and we therefore conclude, that the parallax must necessarily be very small in comparison with the refraction. We also find that the difference between the angles really observed, and those estimated on the supposition that this refraction varies as the tangent of the apparent zenith distance, does not itself follow the law previously ascertained for the variation of the parallax, or one resembling it; and we consequently infer that this difference does not principally proceed from parallax, and that our supposition of refraction varying according to the variation of the tangent of the apparent zenith distance is itself inaccurate, although it furnishes an approximation to the truth. Knowing then that the parallax, if any, is very small in comparison with the refraction, we know that we may, with little error, consider the whole effect as produced by refraction only: and examining the law of refraction on this supposition, we find, (as we have already stated, in the note to p. 50,) that it is more accurately represented as varying, adopting the notation there used, as the tangent of $z-3r$. Still, if the fixed stars have any observable parallax, however small, it would be included in this quantity, and the real refraction would be greater than that thus computed, by the amount of that parallax; for the observed place would be above the true one, by the amount, not of the whole refraction, but of the refraction diminished by the parallax. To see whether this is so, we may resort to observations of the moon, whose parallax is considerable. The amount of refraction being very nearly, if not accurately, ascertained, the apparent zenith distance may be corrected from the error occasioned by refraction, or a very near estimate made of the angle Z T S, (in *fig.* 12.) although our observations are themselves affected by refraction, and therefore do not immediately give that angle. From these observations the amount of parallax may be computed. If it does not vary accurately as the sine of the corrected apparent zenith distance, the angle Z T S, it is evident that the refraction has not yet been rightly computed, but that some material parallax of the fixed stars is involved in the quantity which we have assumed to vary as the tangent of $z-3r$: and we should obtain only an approximate value for the parallax. With this approximate value of parallax we might however correct other observations of the same body, made under different circumstances. In these latter observations therefore we should know, approximately at least, how far the apparent place of the body would be higher than that actually observed, if it were not lowered by parallax, or what would be the apparent place of the body, if only affected by refraction; and from this computed place, and the known true place, as ascertained by the mode already referred to, the amount of refraction might again be more accurately ascertained than before, and a new series of results might be thus obtained and compared, so as to furnish a new and yet more accurate estimate, if necessary, of the mode in which refraction varies; and again, having thus a more correct estimate of refraction, we might repeat our observations on parallax and get a more correct estimate of that element also. And a similar process might be repeated alternately, as long as there was any material irregularity observed, or any discordance between theory and observation. In the case of refraction and parallax, these repeated corrections of one class of observations by another are little necessary, for the effect of parallax on the fixed stars being really quite imperceptible, our first observations on refraction do not require any correction on this account. We have however illustrated the manner in which the law of such corrections may be ascertained by the supposition of the necessity of pursuing the alternate correction further than is really necessary in this case, because the simplicity of the nature of these causes of error and the circumstance that we know independently the general principles which must regulate their variation, makes them furnish a simpler instance than otherwise could have been given of the manner in which such an investigation is to be pursued, and the steps by which errors, in the first instance inevitable, are gradually to be removed.

Where the existence of the correction itself, as in the case of precession, depends entirely on observation, the ascertainment of its exact amount and law is necessarily more difficult; and the number of alternate reductions and adjustments necessary to ascertain its value and the effect of any other source of inaccuracy with which it may be complicated, will be greater; and after all, until some cause in nature is discovered which will account for the being of such an irregularity, its existence may be considered as not completely established. The mode of ascertaining its value however will be of the same nature; and its existence may be assumed as probable, if we find all the phenomena of the heavens corresponding to that supposition. Thus, still taking the case of precession, we have already seen that upon the supposition of the retrograde motion of the equator on the ecliptic by $50''.1$ annually, the apparent places of every star in the heavens would be affected, but differently according to the different position of each star; and we find that the apparent places of each star do vary in the manner corresponding to that supposition, or very nearly so. It is therefore reasonable to believe that supposition true, even now, while as yet we have no further grounds for thinking it so. It is possible that each star may really have a motion which exactly corresponds to the effect which would be produced on its apparent place upon the supposition of this retrograde motion of the equinoxes, and thus that the equinox itself may be stationary; but it is not at all likely that an almost incalculable number of bodies should each of them have motions, all different in amount, and many even opposite in direction, and only corresponding to each other when referred to a point arbitrarily chosen with respect to them, and with which they seem naturally to have no more connection than with any other point in the surface of the heavens. Besides, even if these appearances were to be considered as corresponding to real motions in the stars themselves, and not to an alteration of the point and lines with reference to which they are observed, the importance of the correction would not be diminished; for it would then be ascertained that the stars had actually these motions with respect to that point and those lines, and their relative situation therefore would still be altered in the same manner, and our observations would still have to be corrected on account of that alteration, although the name of precession of the equinoxes would no longer be well chosen to express the cause of the necessary correction.

But in truth, whenever we can discover that all heavenly bodies are affected according to a certain *law*, that is to say, in a manner corresponding to their condition with reference to some certain object, we may fairly believe that the effect produced *depends* on their condition with reference to it. Thus if it were merely by observation that we found that parallax varied in all cases as the sine of the apparent zenith distance, and that all bodies, however different the absolute amount of their parallax, had its variations conformed to that rule, we might fairly conclude that the amount of parallax *depended* on the amount of the zenith distance; and in fact we know from theory that it is so. The same may be said of all the other corrections which have been mentioned: refraction, precession, nutation, aberration, all affect all bodies, however differently in appearance, according to their distances from certain points, lines, or objects; and we therefore conclude that the amount of these several corrections *depends* on these distances; and that, when we discover their causes, we shall find them such as would make the amount necessarily, and in its very nature, correspond to these distances. We shall hereafter see that it is really so with respect to precession, nutation, and aberration. We already know it with respect to refraction as well as parallax.

SECTION V.—*On the proper motions of the fixed Stars.*

WE have yet a further qualification to give to our original statements, with respect to the fixed and invariable position of the stars. We have already seen that the uniform position of which we spoke must not be considered to be that immediately observed, but only that deduced from it, after correction on all the different accounts we have mentioned. It is however found, after this is done, that many of the stars appear not to continue exactly in the same places, but to have a small motion still unexplained and unaccounted for. In this we do not speak of a small class of heavenly bodies, whose motions are very great and well understood, and of which we shall hereafter treat fully under the name of planets, but of many of that large class of bodies, which, from the very slight change which their position undergoes, and the appearance of unmovedness which they consequently present, are known by the name of fixed stars.

It is not necessary here to detail all the precautions necessary in observing these variations; they are very small in amount, the greatest known not being more than 4″ in the course of a year, and most of them not exceeding, or even equalling, 1″ in that time. It is by the comparison of observations made at distant periods that such minute motions can be best detected. If they accumulate for years, the aggregate will be so considerable that any inaccuracy in its ascertainment will bear but a small proportion to the total amount; and thus, if the whole amount be distributed equally over the whole space of time, the rate of motion will be ascertained, if not with absolute correctness, yet with a very near approach to it. It is thus found that many of the stars appear to have small motions which cannot be referred to any of the causes of variation which we have hitherto mentioned. If any law could be discovered, according to which these motions varied in each particular case, they would only furnish ground for introducing another general correction, to be applied to all stars. But this is not found to be the case: many, indeed far the largest portion of the stars, have not been found to have any such motion, and those which have it, have it of amounts and in directions not only very different in different cases, but the differences of which cannot be expressed by any assignable law. Thus of two stars situated very near each other, one will appear to have no motion whatever, the other a motion Eastward; while a third, also situated very nearly in the same part of the heavens, may perhaps have a motion directly Westward, or in some other completely different direction.

We cannot refer these apparent motions to any source affecting the accuracy of our observations, for any such would affect all observations, and would affect them in a manner, different perhaps in each particular instance, but corresponding and consistent in all. We therefore are reduced to the conclusion that they are produced by a real motion in the stars in question; or that some of those stars which are called, in common with those which really have no ascertainable motion, fixed stars, have really small continual motions of their own, or, as they are called, *proper motions*.

These proper motions of upwards of 400 stars, have been ascertained to exceed one-fifth of a second in a year. Many of the more brilliant stars are among those which have these proper motions. We may instance Arcturus, Sirius, Castor, Pollux, Procyon, Regulus, and the first star, or α, of the constellation of the Eagle. The proper motion of Arcturus is such as to diminish his North polar distance, or increase his declination by nearly 2″ every year, and to diminish his right ascension by a little more than 1″ in the same period. It is evident that these motions are not to be neglected in the use we make of observations connected with his place *.

For this reason the proper motions of several of the more remarkable stars which have them are inserted in astronomical tables. The proper motions, being in all directions, would, independently of their difference in amount, act differently in different cases, upon the right ascension, declination, &c. of the stars, increasing them in some instances, diminishing them in others, and requiring in each case to be separately calculated and estimated. The results of these computations however are themselves registered, or rather they are generally combined with

* We have already seen, (ch. ii. § 1.) instances of the manner in which variations in declination and right ascension, arising from precession, affect observations. Of course similar variations will do so in the same manner when they arise from any other cause.

the effects of precession, under the title of the *annual variation* of the particular star in question; so that for all purposes of practical utility the observer may at once refer to these computed results, and thus, by a single addition or subtraction, make the necessary correction for the effects both of precession and of the proper motion of the star observed.

We may now, for the present, quit the consideration of the stars. There are other cirumstances connected with them which will hereafter call for examination; but we have now deduced, from their general appearances, all that is necessary to fit us for the consideration of the more complicated phenomena of the moon and the planets; namely, the apparent general daily rotation of the whole sphere of the heavens, and the various corrections and allowances which must be made in every case for the purpose of deducing the true place from that actually observed. In speaking of the moon and planets, we shall always consider these corrections as made, and take no notice of them unless when they incidentally become material for the purpose of determining other elements, as, for instance, the parallax in discovering the distance of any of these objects. For the future therefore, when we speak of the place of a heavenly body, we speak of its place as it would be observed and estimated by an observer at the centre of the earth at the time of making the observation, and after correcting the observation from the errors occasioned by refraction, aberration, and nutation.

Having premised thus much, we proceed to the consideration of the lunar motions.

Chapter III.

Section I.—*Motion of the Moon in an Ellipse round the Earth—Motion of the Nodes and Apsides of her Orbit—Periodic time—Synodic period.*

It will be unnecessary to go into any detail of the manner in which we ascertain the nature of the moon's motions; they are found by observation of her places at different times, and of her relative distances at those times, just in the same manner as those of the sun. We thus find that the moon, like the sun, moves alternately from North to South, and from South to North, and continually from West to East amongst the fixed stars; and we find that her track, as well as his, may be represented by a great circle of the heavens, or by the intersection of a plane passing through the centre of the earth with the imaginary sphere of the heavens. Observing also her apparent diameter from time to time, we find that this continually increases during one, and continually diminishes through another portion of her course; and comparing these different observations with each other, in the manner already explained in the case of the sun, we find that she also appears to move in an ellipse about the earth, the earth being in one of the foci. By observation of her parallax we find that her mean distance from the earth is about sixty times the earth's radius; the greatest distance of the moon is about sixty-four times, and the least about fifty-six times the radius of the earth, quantities differing from each other in the proportion of 8 to 7: a much higher variation than that which subsists in the solar ellipse. The greatest apparent diameter of the moon is about 33′ 31″, the least 29′ 22″, quantities which obviously are very nearly in that ratio. We also find that her radius vector, like the sun's, describes equal areas in equal times. Another result also, deduced from observations similar to those already mentioned, is, that the apsides of the moon's orbit, the two extremities of the transverse axis, or her perigee and apogee, are not stationary, but move forwards at the mean rate of 40° 40′ 32″.2 in every year, or complete a circuit of the heavens in $3232^d\ 13^h\ 56^m\ 16^s.8$.

The orbit of the moon is not in the same plane with that of the sun, it is inclined to it at an angle of little more than 5°; but this inclination varies: it never falls short of 5°, it sometimes amounts to as much as 5° 18′. The moon therefore is seen sometimes to the North, sometimes to the South of the ecliptic; and as all great circles of a sphere bisect each other, an equal space of her course in the heavens will be to the North and to the South of that line. The circles intersect each other in only two points; and for the whole space from one to the other, the moon will either be always to the North, or always to the South of the ecliptic. The points where the moon's orbit intersects the plane of the sun's orbit are called *the nodes* of her orbit; that at which, having been South of the ecliptic, she passes to the Northward of it, is called *the ascending node*, the point which we have mentioned in treating of nutation; that at which, having been North of the ecliptic, she passes to the South of it, is *the descending node**. The nodes also are found to have a motion on the ecliptic, which carries back the intersection of the nodes with the ecliptic by about 19° 21′ 23″. 6, in every year; or makes these points complete the circuit of the heavens in $6793^d\ 10^h\ 6^m\ 29.^s952$; or 6793.42118 days.

We have said that the moon moves round the earth, from West to East. The sun moves in the same direction, but is much longer in completing an entire revolution than the moon. It is evident therefore that if the sun and moon are at some one instant in the same direction with respect to the earth, the moon will move more rapidly Eastward than the sun, and will consequently separate from him; and will continually outstrip him, until she returns to the same place at which we suppose her to have been when they were originally observed together. When however she arrives there, the sun will be there no longer, but at some distance to the Eastward of that position, and the moon will have to go on for some time longer before she overtakes him, and is again seen in the same direction with him. Her time of returning to this point in her orbit, from which we have supposed her to set out, or of making a complete revolution round the earth, is called her *periodic time;* her time of being again in the same direction with the sun, is called her *synodic period,* or *synodic revolution,* from a Greek word which signifies a *coming together.*

[If either of these quantities is known, the value of the other can be computed from it. Thus let s represent the synodic period of the moon, p her periodic time, P the period of a revolution of the sun; and let A represent the angle through which the sun has moved before the moon overtakes him; and let us also suppose for the present that the angular motions both of the sun and moon are uniform. If this be the case, as the sun moves through 360° in P, he will move through $360°\frac{s}{P}$, in the time s, or in the synodic period of the moon; or the angle A will be $360°\frac{s}{P}$. But the moon, before she overtakes the sun, will have moved through $360° + A$, or $360° + 360°\frac{s}{P}$; and as she takes the time p to move through 360°, she will take the time $\frac{360° + A}{360°}p$, to move through $360° + A$;

or $$s = \frac{360 + 360\frac{s}{P}}{360} \cdot p$$

or $$Ps = Pp + sp,$$

and hence $p = \dfrac{Ps}{P+s}$ or $s = \dfrac{Pp}{P-p}$: equations from which, if either quantity s or p be known, the other may be computed: for P, the length of the year, is known.

The angular motions however both of the sun and moon are variable, and this method therefore seems to give but an inaccurate mode of estimating the elements in question. We may however estimate in this manner their mean value. If we suppose the duration of an actual revolution of the moon, her periodic time, to be always the same, the length of a synodic revolution will only be affected by the different rates at which the sun and moon move in different parts of their orbits. It will necessarily increase when the angular velocity of the sun increases, for in that case, the angle which the moon has to traverse after returning to her original place before she rejoins the sun, increases also; and it will be diminished when the angular velocity of the moon herself in this part of her orbit is increased, for then the time of her describing that angle will be diminished.

Now the synodic period may be determined by observation; and the best mode of doing so, is by the observation of eclipses of the moon.* The nature of these phenomena will be hereafter explained: they are very easily observed; and the middle of the eclipse is very near the time at which the earth is directly between the sun and the moon, and that exact time may be easily computed from the observations made of the eclipse. From one of these times therefore to

* At the ascending node the moon's declination which has been more Southward than that of the sun, becomes more Northward, or she passes the meridian nearer to the North Pole: or, as in those parts of the world where astronomy has been most cultivated, the moon never passes the meridian to the North of the zenith, and the North Pole is above the horizon, she passes the meridian higher than the sun, having hitherto passed it below him. Hence this is called the ascending node; and, for the converse reason, the other the descending node.

* Eclipses of the sun may also be used, but more calculation is required to make the conclusions drawn from them equally accurate.

another, or from the commencement or close of one eclipse to those of another happening under similar circumstances (for then, in each case, the moon will be in the same position with respect to the earth), is necessarily either one synodic period of the moon, or some exact number of synodic periods: and if the time at which each takes place be observed, the interval between them, divided by the number of synodic revolutions, will give the length of the synodic period. If the eclipses are taken at the same, or very nearly the same period of the year, the moon and sun will each of them be nearly in the same position with respect to the earth at each observation, or each will have revolved a certain number of times round the earth; and consequently, as the principal inequalities of the motion of each are gone through in the space of one revolution, each will have gone through all the varieties of its motion a certain number of times, and may therefore be considered as having passed through the same space as if it had always moved with its mean motion. The synodic period therefore, thus deduced, will be the *mean* synodic period; except indeed that the motion of the moon's apogee being, as we have seen, considerable, must not be entirely left out of the account, as the rate of her motion in her orbit depends on her situation with relation to that point. To make the correctness of the result deduced complete therefore, we should add the further condition that the apogee of the moon should be about the same place at each observation. It would not however be easy to find observations so fully corresponding to each other. If however we take observations very distant from each other in time, the apogee will have revolved a certain number of times, and in each of these revolutions the moon will have had all her varieties of motion; there will besides be one incomplete revolution of the apogee, and one only, occasioning some deviation from the mean value. Still the error thus produced will be divided in the manner we have already illustrated, p. 40, among all the synodic periods included between the two observations, and will therefore produce very little effect on each; and on the same principle, if the time intervening be great enough, even the inequalities produced by observing at different seasons of the year may be neglected.

Now eclipses are phenomena so remarkable, that they have been very long observed; and the time of the occurrence of some has been recorded with sufficient accuracy, even before the Christian era. By comparing these with recent observations made at the same season of the year, the duration of the mean synodic period may be ascertained with very great accuracy; and it is thus found to be $29^d. 12^h. 44^m. 2^s.8$, or 29.530588 days*.

The sidereal year is of 365.256384 days: the mean periodic revolution of the moon therefore is

$$\frac{365.256384 \times 29.530588}{365.256384 + 29.530588} \text{ days, or}$$

27.321661 days, or $27^d\ 7^h\ 43^m\ 11^s.51$. This is evidently the *sidereal revolution* of the moon: her time of describing 360°, or of returning to the same position with respect to the stars. But while she performs her revolution, the equinox will have retrograded, and her return to the same position with respect to the equinox, or her *tropical period*, will be shorter. The tropical period may either be deduced from the sidereal period, or deduced in the same manner from a similar equation: for if q represent the tropical period of the moon, and Q the tropical year, s being the synodic period as before, $q = \frac{Q\,s}{Q + s}$, an equation exactly similar to the last. The considerations necessary to deduce this equation are rather more complicated than those involved in the proof of the former one; but it may nevertheless be worth while to subjoin them †. The length of

* This is the result deduced for the comparison of an eclipse observed by the Chaldeans 720 B. C. and one observed in Paris in 1771. The whole number of synodic revolutions in the interval was 30817; and any error produced by the inequalities of the moon's motion would be divided equally among all these revolutions.

† [Adopting the notation used in the text, let a = amount of precession in the time of the moon's tropical revolution, or in q; and b = precession in a tropical year, or in Q.

Then as the sun in a tropical year describes $360° - b$,

the angle described by the sun in the synodic period $s = (360 - b)\frac{s}{Q}$.

But the moon in the synodic period describes this angle and 360° besides, or $360 + (360 - b)\frac{s}{Q}$.

Now in q, the moon describes $360 - a$; and consequently the motions being supposed uniform,

$q : s :: 360 - a : 360 + \frac{(360 - b)s}{Q} :: 360\,Q - a\,Q : 360\,Q + 360\,s - b\,s$

and consequently $360\,Q\,q + 360\,s\,q - b\,s\,q = 360\,Q\,s - a\,Q\,s$.

But, if the precession be uniform, $a : b :: q : Q$ or $a\,Q = b\,q$ or $a\,Q\,s = b\,s\,q$.

Taking away therefore these equal quantities, $360\,Q\,q + 360\,s\,q = 360\,s\,Q$,

Or, $Q\,q + s\,q = s\,Q$, or $q = \frac{Q\,s}{Q + s}$: the result stated above.]

the tropical revolution therefore (the length of the tropical year being 365.242264) is $\frac{365.242264 \times 29.530588}{365.242264 + 29.530588}$ days, or 27.321582 days, or $27^d\ 7^h\ 43^m\ 4^s.685$.—]

Thus far our observations require little explanation; and we have thought it best to premise them, as the ascertainment of the shape and position of the moon's orbit depending on the same principles, and being deduced in the same manner as that of the sun, we may at once refer to it as so ascertained, without entering into any detail of the steps necessary to its investigation; and the difference between the synodic period and the periodic time follows so immediately from the motion of revolution, that it was necessarily introduced as a consequence of it. These results however do not at all explain the most remarkable appearances of the moon, or those to which the attention of a common observer is first directed: and we now proceed therefore to state the nature of those appearances, and to explain the causes from which they proceed.

Section II.—*On the Phases of the Moon—General account—Law of their Variation—Earthshine—Proportion of Moonlight at different seasons and places—Harvest Moon.*

The period of time during which the appearances now in question succeed each other, is the synodic revolution of the moon, or, as it is commonly called, *a lunar month.* There is a certain period during which the moon is not at all visible; and in the course of which, as we know from observation and computation of her course, she has the same right ascension with the sun, or comes on the meridian at the same time with him. It is some time after this before she becomes visible, and when she does so, she is seen in the West soon after sunset, with the appearance of a very thin crescent, the bright and visible part having the side nearest to the sun convex towards him, and apparently semicircular: the inner part, or the part farther from the sun, being elliptical, and convex in the same direction with the outer. From this time, as the moon's motion Eastward in the heavens is greater than the sun's, in the ratio of 13 to 1 nearly, their distance from each other continually increases, as the moon comes on the meridian continually at a longer interval after the sun; and while she does so, the breadth of the crescent continually increases, the outward, or Western line continuing to be circular, but the inner ellipse continually becoming less strongly curved, until, when the moon is distant about 90° from the sun, and comes upon the meridian about six hours after him, this inner curve is changed into a straight line, and the appearance is that known by the name of *half moon.* After this time the line becomes again elliptical, but has its convexity towards the side most distant from the sun, or bulges out in that direction, and the two lines which appear to bound the moon become concave to each other: in this condition the moon is called *gibbous,* and the side of the moon most distant from the sun continually becomes more and more strongly curved, and the apparent breadth of the moon consequently greater, until, when the right ascension of the moon is 180° different from that of the sun, and she comes on the meridian twelve hours after him, or at midnight, this side of the moon, as well as the other, is a semicircle; and the moon appears completely round in the heavens, or, as we say, it is *full moon.* From this period, although the moon still comes upon the meridian longer and longer after the sun, she approaches him in distance, for no points upon the sphere can be distant from each other more than 180°; and when the difference of right ascension, in the direction in which it is measured, exceeds this quantity, the distance measured backward must fall short of it: thus, if the moon comes on the meridian 15 hours after the sun, or the difference of their right ascension is 225°, the distance between them, measured in the opposite direction, is only 9 hours, or 135°, or she comes on the meridian only 9 hours before the succeeding noon. After the full moon therefore, the distance begins to diminish, and as it diminishes it is found that the appearances visible during its increase succeed each other in the contrary order; that the moon becomes gibbous, and her apparent breadth continually diminishes, the side now next the sun continuing apparently circular, the other becoming elliptical, and continually less and less strongly curved, until, about the time when the distance between them is 270° in one direction, or 90° in the other, the appearance of half moon is again presented; and from that time forward the moon appears as a crescent continually diminishing in breadth, until at length, when she has got as near the sun as she was when she first became visible at the beginning of the month, she disappears, and is not again seen till the corresponding period of the next month, when the same order of appearances recommences. The appearances during the period of her diminution are exactly the same with those during the period of her increase, and correspond to exactly the same distances in each case. Thus, if the shape of the moon be observed when her angular distance from the sun is 70° East, it is found that she has exactly the same shape when 290° East, or 70° West of him: the only difference is that the circular part, or *limb*, of the moon, which was turned in the former case towards the Western part of the heavens where he set, is turned in the latter towards the Eastern part where he is about to rise.

It is obvious, therefore, that these appearances, or *phases*, as they are called, of the moon, depend upon her angular distance from the sun, for they continually vary with the variation of that quantity; the visible magnitude of the moon increasing when that quantity increases, and diminishing when it diminishes; and, when that quantity is equal at different periods, these visible magnitudes being equal also. Nor is it difficult to perceive that the appearances presented correspond to those which would obtain, if the moon were an opaque body, giving forth no light of its own, but capable of reflecting the light received from the sun.

It is evident that the moon is not luminous of herself, for if she were she would always be visible when above the horizon, whatever were her position with respect to the sun. Considering her therefore as opaque, but capable of reflecting light, it is plain that one portion of the moon would always be light, namely, the whole portion which is turned towards the sun, and the rest would be dark. We should consequently see only that portion which the sun illuminated, and only so much of that portion as was on the side of the moon turned towards us. When therefore the moon was between us and the sun, she would be invisible, because the whole of her enlightened side would be turned from us; gradually, as she receded from this position, some part of her enlightened side would be within the view of an observer at the earth, and this part would continually increase as she got farther from the sun, until at length, when she was upon the opposite side of the earth, or the earth was between her and the sun, the same portion of the moon would be turned towards the sun and earth, or the whole of the enlightened portion would be visible to us. From this time she would again approach the sun, and some part of the enlightened portion would continually disappear; and as the quantity visible would depend merely on the relative positions of the sun, moon, and earth, the decrease would follow the same law as the increase, although in a reversed order, for the relative positions would succeed each other in this manner. These conclusions may be illustrated by inspection of *fig.* 16, where if T represent the earth, S the sun, (of which however

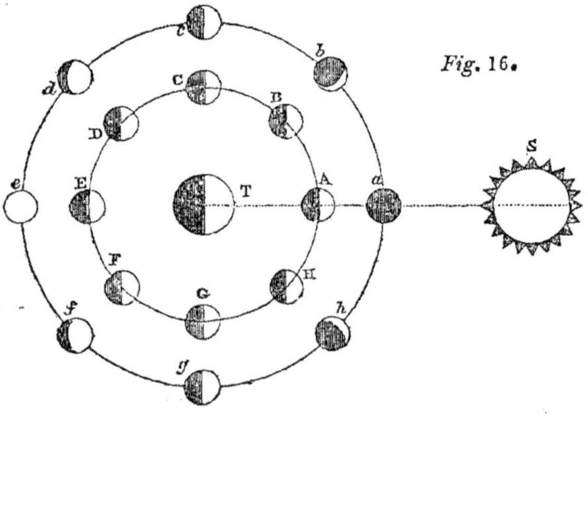

Fig. 16.

the distance must be taken to be very great, although it is not represented so for the convenience of the figure,) and A, B, C, D, E, &c., different positions of the moon (which we will suppose spherical) in her orbit, the enlightened portions of those circles will represent the enlightened part of the moon in each case; and, as the part of the moon turned towards the earth will be that, or very nearly that within the circle passing through A, B, C, &c., the part visible in each case will be only so much of the enlightened part as is within that circle. The figures in the outer circle, *a*, *b*, *c*, &c., will represent the appearances, or phases, of the moon in the corresponding situations, the light parts only being visible. Thus, at A, the whole of the enlightened part of the moon is turned from the earth, and none of it, in consequence, is visible: at B and H, a small part only, and that equal in both instances, is visible, namely, the light parts within the circle; and the appearances presented are represented by *b* and *h*, two similar figures, but with the convexity of the light part turned in each case towards S, and consequently in different directions with respect to T. To explain fully the correctness of the representation, the figures *b*, *c*, &c., should be considered as if they were perpendicular to the plane of the paper. The reason is obvious. Taking the figure B, the visible part is not a sector of a circle as it is unavoidably represented in the figure, which is drawn on a plane surface; but a portion of the surface of a sphere of which every part, except the external line actually drawn, is either above or below the plane of the moon's orbit, or the plane of the paper which represents it. The circle drawn through A, B, &c., will very nearly represent the line bounding the part of the moon visible from the earth; and this will be a circle described on the sphere very nearly perpendicular to the plane of the moon's orbit, and to the line joining the earth and moon, or the moon's radius vector; it will therefore be seen as a circle perpendicular to that same plane, or as the outer line of the light part of *b*. The dark line also, the boundary of the enlightened part of the moon visible from the earth, will also be a circle of the sphere, but this will be seen obliquely from the earth, and will therefore assume an oval appearance; it therefore will be represented by the inner boundary of the light part of *b*. These lines will evidently meet in a point, as they do in the figure, because the corresponding circles in B do actually meet so, and the visible light part of the moon itself consequently terminates in one.

The reader will have no difficulty in ascertaining in the same manner the correctness of the other delineations. We conclude therefore that the moon shines by light reflected from the sun; and we shall find a still farther proof of it when we treat of eclipses, for we shall see that even when she is opposite to the sun, at the time of the full moon, if the earth is directly between her and the sun, so as to intercept his light entirely, she then also becomes invisible.

The subject however is of too much importance for us to leave it with only this general sort of illustration. The law which determines the proportion of the surface of a spherical heavenly body, depending for its light on the reflection of light from the sun, which is visible at different times, according to its different situations with respect to that body, may be investigated without any difficulty by a reader very slightly acquainted with mathematics; and it is of the more importance to do so, because we shall not only find that it accounts for the various phases of the moon, but that it serves to explain the appearances of other heavenly bodies, on which we have not yet touched. We proceed therefore to investigate it. We confine ourselves to the case of a spherical heavenly body, because all those to which we shall have to apply our results, are very nearly of that form.

If the moon, or any other body which receives the sun's rays, be spherical, the boundary of the part on which they fall, or of the enlightened part, will necessarily be circular. This follows immediately from a very simple consideration. Let A (*fig.* 17.) represent the centre of any opaque sphere whatsoever, and B a

Fig. 17.

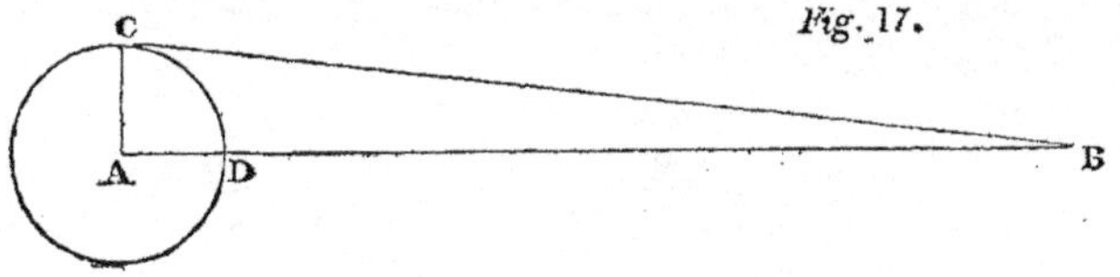

luminous object shining upon it; and let the line A B be drawn joining A and B, and passing through the sphere at D; and let B C be a tangent to the sphere. It is evident that C will be a point in the boundary of the illuminated part of the sphere, for no point farther from B than C is, can receive a ray of light from B, as some part of the opaque sphere will be between them; and every point between C and D must receive such rays, for there is nothing to interpose and prevent them from arriving there.

Now, all tangents drawn to a sphere from the same point are equal to each other; at every point therefore in the boundary of the illuminated part the value of C B is the same; A C, the radius of the spherical body, is equal at all points, and A B is always the same line: the value therefore of the angle D A C is the same in the case of every point in this boundary, for all the triangles, formed as A B C is, have all their sides equal, and consequently their angles equal also. Every point then in this boundary is at the same angular distance from the point B, or from D, and therefore in the circumference of a circle whose pole is D, and consequently whose plane is perpendicular to the line A B. This circle differs almost imperceptibly, in the case of the sun and any body on which he shines, from a great circle of the sphere; for, as the angle A C B is a right angle, being that made by a tangent with the radius which it meets, the angles C A B and C B A together are equal to a right angle; and as the sun's distance is exceedingly great in proportion to the radius of any heavenly body on which he shines, the angle C B A is very small indeed, or the angle C A B very nearly equal to a right angle *.

In drawing this conclusion, however, we have treated the light as proceeding from a single luminous point, B; and the result will require to be a little modified, as light does actually proceed from every part of a very large body, namely the sun. If the sun were of the same size as the moon, the extreme, or tangential rays, would be parallel; if smaller, they would continually diverge from each other; if larger, they would converge to a point. These results are obvious in themselves; or they will immediately appear by the inspection of *fig.* 18, where, if S represents

Fig. 18.

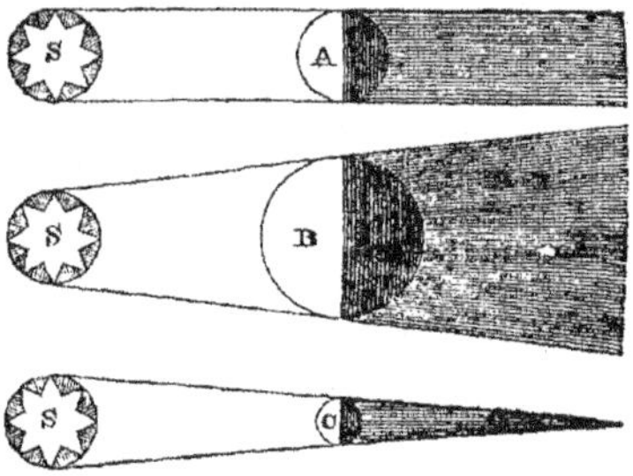

the sun, and A, B C, three bodies, the first equal to the sun in size, the second larger, the third smaller, the extremities of the shaded figures beyond them will evidently represent the course of the extreme rays, and the figures themselves the shadows cast by the bodies, which, in the two former cases, would be prolonged to an infinite distance; in the latter, they would terminate as in the figure. The third case represents that of the moon, which is smaller than the sun: the boundary of the illuminated part therefore will be determined by rays, not diverging from the illuminating body S, but converging to a point on the other side of the moon. But still, as they all meet in a point, the boundary of the illuminated part will be a circle, in the same manner as before; and as the difference of the diameters of the sun and moon is small in comparison with their distance, the rays will converge very slowly, and the extremity of the shadow, the point to which the extreme rays converge, will be very distant from the moon; and the boundary therefore, in this case also, will differ very little from a great circle. The whole part illuminated therefore rather exceeds half the sphere,

* [In the case of the moon, the mean value of the distance A B is very nearly equal to the distance of the earth from the sun (for the greatest distance exceeds that quantity by the moon's distance from the earth, and the least distance falls short of it in the same degree), or to about 23,500 times the earth's radius, (p. 60); the moon's radius on the other hand is only about $\frac{3}{11}$ths of the earth's radius (p. 61). C A therefore is to A B, or, which is very nearly the same, C B, in the proportion only of $\frac{3}{11}$ to 23500, or of 3 to 2585 00 and the angle A B C, which is very nearly equal to $57^{\circ}.29578. \frac{AC}{CB}$, is consequently only

$$\frac{3}{285800} \, 57^{\circ}.29578,$$

or to little more than 2″. The angle C A B therefore may be considered to differ imperceptibly from a right angle, or the arc D C from 90°, in which case the boundary of the enlightened part, in which C is situated, would be a great circle of the sphere.]

but so little, that we may say generally that half of any spherical heavenly body is illuminated by the sun, and that the boundary of the illuminated part is a great circle of the sphere, whose plane is perpendicular to the line joining the centre of the body with the centre of the sun.

Again, if we suppose B, in *fig.* 17, instead of being the position of an illuminated body, to be the position of an observer looking at the spherical body A, he will evidently see all between D and C, and nothing beyond C. C therefore, in this case, will represent a point in the boundary of the part of the sphere visible to the observer; and by the same reasoning as before, this boundary must be circular, and its plane perpendicular to A B: and further, it also may, if the distance A B is sufficient, be considered as a great circle of the sphere. In the case of the moon and earth, this distance is too considerable to be entirely neglected*. Still, even in this case, the supposition that the boundary is a great circle is so nearly correct, that we may use it to explain the manner in which the magnitude of the visible part of the object varies; and in the others to which we shall hereafter apply it, it differs quite imperceptibly from the truth.

Now in *fig.* 19, let E represent the situation of the earth, S that of the sun, M the centre of any spherical body reflecting the sun's light to the earth; the body itself being represented by the circle A B a b, which, however, is excessively magnified to admit of drawing the necessary lines within it. Draw the lines M S, E S, and E M, and produce E M to N. The angle M E S, subtended at E by the places of M and the sun, or their apparent angular distance from each other, is called *the elongation* of M, and the angle N M S is called the

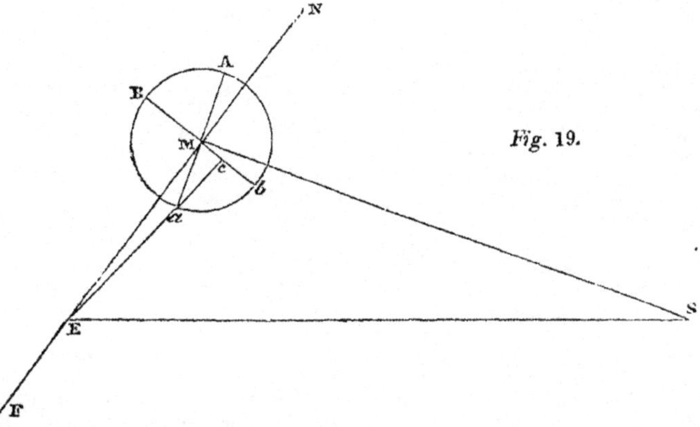

Fig. 19.

exterior angle of elongation. Draw A M a perpendicular to M S; then, by what we have already shown, A and a will each be points in the extreme boundary of the enlightened part of the body, or A b a will represent the enlightened part. In the same manner, if B M b be drawn perpendicular to E M, B and b will be points in the extreme boundary of the part of the body which is turned towards E, or B a b will represent that part: and a b consequently will represent the part actually visible at E by reflected light. Draw E a c through a to meet the line B b, and the points a and c would appear to the observer at E, in the same direction. If he could see the whole part B a b, which is turned towards him, he would see it occupying the same angle as the line B b; he sees a b occupying the same angle as the line b c: or the extreme

* [In this case A B = 60 times earth's radius nearly, and A C = moon's radius = $\frac{3}{11}$ earth's radius nearly. Consequently sin. A B C = $\frac{AC}{AB}$ = $\frac{3}{11 \times 60} = \frac{1}{220}$, a quantity so small that the angle and the sine may be considered to correspond, and consequently the angle A B C = $\frac{57^\circ.29578}{220}$ = .260435, or very little more than a quarter, of a degree, and the angle C A B falls short of a right angle only by that quantity. The same result may be obtained yet more simply thus; the angle C A B falls short of a right angle by the angle A B C; but A B C is the angle subtended at the earth by the radius of the moon, or it is half the apparent diameter of the moon, which we have seen, p. 68, to vary from 29′ 22″ to 33′ 31″: the mean value of the half will therefore be something more than 15′¾.]

apparent breadth of the visible part of the moon is to the apparent diameter of the moon in the proportion of bc to the real diameter of the moon: or the apparent breadth varies as bc. But ac is very nearly perpendicular to B b: for it is only in cases where the greatest possible angle subtended by any radius of the object (as M b) at E is very small, that we can take a *great* circle of the sphere as representing the boundary of the part turned towards E; and we have already stated that we may do so, even in the case of the moon, without any material inaccuracy*. The angle M c E, therefore, is very nearly a right angle: or bc is very nearly the versed sine of the arc ab, or the angle a M b: for by the definition of a versed sine, it would be that quantity, if the line ac were accurately perpendicular to B b. The visible breadth therefore of the illumined part of the body M varies very nearly as the versed sine of the angle aMb. But aM S, b M N, are right angles, and therefore equal: and taking the common angle bM S from both (or adding it to both in cases where the angle N M S is greater than a right angle) aMb=S M N the exterior angle of elongation. The visible breadth therefore of the illumined part of the body varies very nearly as the versed sine of the exterior angle of elongation.

[This is the extreme visible breadth. The whole visible part varies in the same manner; for the boundary of the illuminated part of the sphere is itself a circle, and the lines therefore which join E with every point of this boundary will form an oblique cone of which E is the vertex, and this boundary the base. The plane B b is perpendicular to the axis of this cone, and of course cuts it all round: the section therefore which it makes must (*Geom. App.* Prop. 24) be an ellipse; and the half of this section will be one of the boundaries of the visible part of the moon. The other boundary is that of the part turned towards the earth, which is a circle, and it is on this same plane: the visible part therefore will appear as a figure contained between a circle and an ellipse. The minor axis of this ellipse = Mc, the major axis = Mb. The area of the ellipse therefore, = 3.14159. M c. M b: and the area of the circle = 3.14159. M b^2: the area of the part between them therefore, = 3.14159. M b (Mb $\pm$ Mc) = 3.14159. Mb. bc, and therefore varies as bc, and of course, the difference or sum of the semicircle and the semiellipse varies in the same proportion.]

It is evident also that the whole of the exterior semicircle is, in every case, visible; for the boundary of the part turned towards the earth, and the boundary of the enlightened part of the moon, are each of them great circles, and therefore bisect each other; or the visible boundary is a complete semicircle, at whatever angle they meet. Whenever therefore the moon is visible, her extreme points, *or cusps*, are at the extremities of a diameter; and we may therefore, during all her phases, however small be the part of her surface which is actually visible to us, make those observations of her apparent diameter to which we have already referred for ascertaining her distance and the form of her orbit.

The deduction of the above propositions is not necessary for the purpose of satisfying us that the moon shines by light reflected from the sun, for her disappearance when between us and the sun, and whenever his light is intercepted from her, and the manner in which the light part of the moon is always turned towards the sun, are sufficient for that purpose. But these propositions are necessary to estimate at all accurately the part of the moon visible at different periods of the month. Thus, when the moon is between the earth and the sun, or as it is called *in conjunction with the sun*, the exterior angle of elongation is equal to nothing; for the line joining the earth and moon when produced is in the same direction with that joining the moon and the sun, and coincides with it: in this case therefore the versed sine of that angle is equal to nothing. When the earth is between the sun and moon, or the moon *in opposition to the sun*, the line joining the earth and moon when produced is exactly in the opposite direction from that joining the moon and sun, or it coincides with it, but is measured in the other direction, or it makes an angle of 180° with it; and the versed sine of this angle, or of the exterior angle of elongation, is the whole diameter. The whole face of the moon therefore is then visible. When the exterior angle of elongation is 90°, or when, as it is termed, the moon is *in quadratures*, the versed sine of

* At the extremity of the diameter B b, the difference between the angle made by the diameter itself with the line drawn from E to it, and a right angle, is only about a quarter of a degree, as shown in the last note: for every point nearer to M, the difference is of course less.

that angle is equal to the radius, or half the face of the moon is visible*. The elongation of the moon, together with the angle M S E, always are equal to this exterior angle; and consequently, in this case, the elongation = 90° − M S E, or very nearly = 90°; for, as M E is about sixty times the earth's radius, (p. 60) and E S nearly 24,000 times the same quantity, the angle M S E must necessarily be very small†. It is obvious that the observations which we have already detailed correspond with these results; and the uniform manner in which they do so, shows that the figure of the moon may, without material inaccuracy, be considered as spherical.

It is perhaps necessary to observe, that when we speak of the moon as reflecting light from the sun, we do not mean that she reflects it so as to present an image of the sun on one point of her surface, like a mirror; but that the light gets broken and diffused from part to part of her surface, and finally sent forward to us in such a manner as to render the whole surface of the enlightened part visible; just as light is diffused over bodies on the earth, which, if perfectly smooth, would only form an image of the sun, but do actually show, by reflected and broken light, the whole of their own surface, its form and colour.

It is evident, from the law of variation which we have deduced, that almost immediately after the moon and sun cease to be in the same line, there is some portion of the illuminated part of the moon turned towards the earth. It is however very small at first: the versed sine increases very slowly while the angle continues small, and it is therefore some time before its magnitude becomes considerable. During this period also the moon is apparently near the sun, and consequently in a very light part of the heavens; and it is therefore a good while before she really becomes visible, for she cannot be seen till the light which she reflects is sufficient to be distinguished from that which the sun spreads generally over the region of the atmosphere through which the rays which proceed from her must pass. The length of time therefore during which she is not actually seen, furnishes no exception to the correctness of our results.

There is a remarkable appearance presented by the moon when the visible part, according to the principles we have established, would be small, which this is the proper season for explaining. At these periods the whole of the moon's disk is frequently seen, part bright, and having its magnitude the same with that which we have explained as the whole visible magnitude of the moon; the rest visible by a pale and delicate light, and appearing, from the ordinary effect of brightness in augmenting the apparent magnitude of objects, somewhat smaller in its dimensions than the brighter part. The appearance, from this circumstance, and also as being oftenest observed in the evening, soon after the moon's first appearance, or after *the new moon*, when more persons have the opportunity of seeing it than in the early morning preceding the disappearance of the moon at the latter end of the month, has received, in common speech, the odd name of 'the old moon in the new moon's arms.' The French, with more accuracy of expression, have named it, from the pale colour of the greater part of the moon, *lumière cendrée*, or *ashy light*. The cause of this light is obvious: the earth, as well as the moon, reflects light, and consequently, the enlightened part of the earth, or so much of it as is turned towards the moon, will reflect light to that body. Some of that light will again be reflected back to the earth, and thus even that part of the moon which receives no light directly from the sun, may, by indirectly receiving it from the earth, become, as we see it, faintly visible. The appearance, being thus occasioned, has received the name of *earthshine*. The light thus indirectly supplied must necessarily be far inferior in quantity and brightness to that which the directly enlightened part of the moon receives immediately from the sun; and thus the great inequality of brightness in the two visible portions is accounted for. The only apparent difficulty arises

* If therefore we can observe the time when exactly half the moon is visible, or the inner line of light quite straight, we shall know that the exterior angle of elongation is 90°. We can observe the actual elongation at that time, and hence we may know the value of M S E, the other angle of the triangle. Hence the proportion of the sides may be ascertained, or of the sun's and moon's distances from the earth. We cannot make the necessary observation of the exact time of half moon with accuracy enough to make the result of any value, as we have now better means of ascertaining it; but this method is worth notice, as being the first ever employed for the purpose, having been adopted by Aristarchus, about 280 B.C.

† It follows hence that, as N M S = M E S + M S E always, in the case of the moon, N M S = M E S, very nearly. The visible part therefore will vary very nearly as the versed sine of M E S, or of the elongation itself, in this case.

from the circumstance that the appearance in question is only seen when the directly illuminated part is small. In reality however this seeming difficulty confirms the explanation given, for there are two obvious reasons for it. As the directly illuminated part increases, its light becomes greater, and the light diffused over that part of the atmosphere through which the moon shines, greater also: a stronger light therefore is required to be distinguishable. But this is not all; the light actually supplied to the moon from the earth diminishes. The earth being a spherical body, and reflecting light, appearances or phases will be presented by the earth to the moon similar to those which we, on the earth, observe in the moon; and all our results will be true for this case, as well as for that already examined. The order indeed will be different. Thus, when the moon is invisible to us, being between the earth and sun, the earth will turn the same part to the sun and moon, and will be visible to the moon with a full face: when we see the full moon, the earth is between the moon and sun, and therefore invisible to the moon. Without entering into any further detail of these appearances, as visible at the moon, it is evident that the general principles on which they were deduced apply equally to this case, and consequently that the part of the earth visible at the moon varies nearly as the versed sine of the earth's exterior angle of elongation there. The result will be less accurate than in the case of the moon, on account of the earth's greater magnitude *: but still it will be very nearly so. In *fig.* 19, if M E be prolonged to F, E M S is the earth's elongation, as estimated at the moon, and F E S is the earth's exterior angle of elongation. But N M S = M E S + M S E, and consequently N M S and F E S together, are equal to M E S + F E S + M S E, or to two right angles, + M S E, or very nearly to two right angles, as M S E is necessarily very small. The greater therefore N M S, or the moon's exterior angle of elongation, the less is F E S, or the earth's exterior angle of elongation; and as their versed sines increase and diminish when the angles do so, and the visible parts are in the proportion of the versed sines, the part of the earth visible from the moon diminishes as the visible part of the moon increases, and of course the quantity of light which the earth reflects to the moon diminishes also. The power therefore of distinguishing the moon by this light reflected from the earth, is diminished as the part visible by light directly reflected from the sun is increased; both because less light is thus transmitted to the moon, and because more is required before it can be distinguished.

* The real diameter of the earth being about $\frac{11}{3}$ times that of the moon, and their distance very considerable with reference to it, the apparent diameter of the earth at the moon will be about $\frac{11}{3}$ greater than that of the moon at the earth, or it will be about 115′, nearly 2°. The whole apparent magnitude will be greater than that of the moon to us in the proportion of the square of $\frac{11}{3}$ to 1, or of 121 to 9, or more than 13 to 1.

We have thus explained the manner in which the proportion of the moon which is visible at different periods of the month varies. This proportion however is not the only thing which we can observe with respect to her. Many singular marks and spots are apparent upon her, from which various and important conclusions may be drawn: but before we proceed to state these observations, and draw from them the inferences to which they lead, it will be worth while to pause, and deduce from the results already obtained some remarkable consequences, which materially tend to the convenience of mankind.

The most obvious practical service of the moon, as far as mankind are concerned, is the supply of light which she affords during the otherwise dark hours while the sun is below the horizon. But she herself is sometimes above, sometimes below the horizon: and as her declination is continually varying, her periods of continuance above the horizon continually vary also. Her light also is different at different periods of her course: sometimes none; sometimes little, and therefore of little practical utility; generally however enough to be of material service to the sailor, the traveller, and even to the husbandman; and this most when her light is the brightest, or at the full moon. It therefore matters little to man whether she is above or below the horizon at night, as long as her own light is little or nothing; but it is of much importance that she should be above the horizon at night when her light is considerable. Now this, by the conditions of her motion, she necessarily is. The quantity of light which she reflects is greatest when her distance from the sun is greatest, or when the difference of their right ascensions is 180°, or the time between their appearances on the meridian is twelve hours. When the moon's light is greatest therefore, she is on the meridian at midnight; and as her light is always greater as her distance from the sun, and consequently as the interval of time between their appearance on the meridian, is greater, the periods of her greater light will always bring her on the meridian nearer midnight, than those of her less brilliancy; and therefore, taking the whole course of one of her revolutions, she will be above the horizon during a larger proportion of the night, as her light, and consequently her power of being useful, are greater. The nights however are of different length at different periods of the year; and consequently, if the moon, when at her greatest brightness, were always at the same declination, and thus had the same proportion of her diurnal course above the horizon in all these cases, she would supply light through a much smaller proportion of the long and dark nights of winter, than of the short and comparatively light nights of summer. But this is not the case. For the purpose of illustration, we will suppose the moon to move in the plane of the ecliptic, and, as usual, that the North Pole is above the horizon. The full moon takes place when the difference of longitude of the sun and moon is 180°. Taking then the extreme cases of the two solstices, it is evident that when the sun is at the summer solstice, or in the tropic of Cancer, the full moon, being 180° distant, would be in the tropic of Capricorn; the sun therefore being at his greatest North declination, the moon would be at her greatest South declination: and throughout all that half of her orbit which is most distant from the sun, and where consequently her light is greatest and more than half her face visible, her declination would be South. In this case then, when the sun having his greatest North declination the day is longest, the moon would have her greatest South declination, and be least time above the horizon, when at the full: and during the whole time that more than half her face is visible, her declination would be South, and less than half of her diurnal course would be above the horizon. Exactly the contrary results would evidently follow when the sun is in the tropic of Capricorn, when he has his greatest South declination, or the days are shortest. In this case the moon would be at the full when in the tropic of Cancer, or at her greatest North declination; and throughout all that half of her orbit most remote from the sun, and when more than half her face is visible, her declination would be North; and consequently during this whole period she would be more than half her time above the horizon, and longest of all when her light is the greatest. In intermediate positions of the sun, the results would be intermediate; but it is not necessary to enter into any detail of them. The reader will have no difficulty in pursuing, if he is inclined, a similar course of argument with respect to them.

Again, the greater the elevation of the Pole above the horizon, the greater is the inequality of day and night, and the longer does a body with North declination continue above, or a body with South declination continue below the horizon. The more elevated the Pole therefore, the longer is the moon above the horizon when her declination is North; and as, during the winter, her declination is North, while her light is greatest, the proportion of time during which she continues above the horizon while her light is greatest, increases at that season as the latitude increases, or as the nights themselves are longer. Taking the extreme case, where the Pole is in the zenith the moon will never set while her declination is North; and at the winter solstice it will be so during the whole time that more than half her face is visible. There will therefore be a fortnight of the brightest moonlight at that season when the night, from the complete absence of twilight, would otherwise be darkest. There will always be a fortnight at a time, there, during which the moon will be above the horizon; but as the sun approaches the equinox, the light given by the moon during part of this fortnight will diminish. At this time however it may better be spared, for by this time, though the sun does not appear on the horizon, there will be a considerable twilight.

It is abundantly plain, that the same consequences will follow from South declination where the South Pole is above the horizon, as we have seen to follow from North declination where the North Pole is so: and thus that this beneficent provision, by which the greatest quantity of moonlight is afforded to those regions and at those seasons in which it is most wanted, is general over the whole earth.

The moon however moves not in the plane of the ecliptic, but in one inclined to it at an angle of a little more than 5°; but this difference will not materially affect the results we have obtained. It will indeed at different times, according to the place of the nodes of the moon's orbit, very materially affect the actual length of time during which she is above the horizon; but she never can be distant much more than 5° from the ecliptic, and so much only during a very small portion of her course. The ecliptic itself is, at its greatest distance, more than 23° from the equator. The distance of the moon from the ecliptic therefore, being so much smaller, cannot, except for a very small space comparatively, vary the nature of our results, though it will their amount; or make that declination North which would otherwise be South, or the contrary. It cannot even materially alter the point at which the declination would be greatest, though it may make that greatest declination differ, in different cases, by upwards of 10°. Thus, if we suppose the nodes of the moon's orbit to coincide with the equinoxes, the moon, when 90° from the node, will be about 5° North or South of the ecliptic at the tropic of Cancer, as the part of her orbit between the vernal and autumnal equinox is that which lies above or below the plane of the sun's orbit, or, in other words, as the node which is at the vernal equinox is the ascending or descending node. Her declination at this point therefore in the one case will be about 28° 30′ N., in the other only 18° 30′ N.; but in each case it will be the greatest North declination which during that revolution she attains. When however the node is not at the equinox and solstice, this will not necessarily be the case, but it can never differ much from it. We may therefore conclude our results to be generally true; and we thus see that the shape and motions of the moon, and the manner in which she reflects light to the earth, are so ordained as to make her most serviceable whenever and wherever her services are most wanted; an incidental consequence indeed of the general laws governing her motion, but one of the many remarkable instances in which, throughout the appearances of Nature, we see not only that the general scope and tendency of her operations are beneficial, but that collateral benefits continually flow from the manner in which those operations are conducted.

Another application of the same principles of reasoning, and, although in an inferior degree, a similar instance of a beneficial result arising incidentally from the operation of the general laws established in nature, will be found in the explanation of the phenomenon known by the name of the *harvest moon.* Whether we consider the moon as moving in the plane of the ecliptic, or in that of its real orbit, the variation of its declination will be most rapid where its orbit cuts the equator; and as no part of the real orbit is much more than 5° distant from the ecliptic, its intersection with the equator cannot be far distant from the intersection of the ecliptic with the equator. The moon therefore moves most rapidly Northward when near the equator, and also near the first point of Aries; most rapidly Southward when near the equator and the first point of Libra.

When the sun is near one equinoctial point, the moon, when full, is near the other, and necessarily near the equator also. The difference of their right ascension, estimated in time, is 12 hours, and this is also the length of the day at that period. The moon therefore will rise in the East, just as the sun sets in the West. As the moon completes her revolution round the earth in a little less than 28 days, her motion, if uniform, would be at the rate of about 13° a day; or, omitting any consideration of the manner in which the obliquity of her orbit would affect the results, she would come to the meridian of the place about 52 minutes later on each successive day. If therefore her declination continued the same, as she would always be an equal time above the horizon, she would rise and set about 52 minutes later every day: but when she is near the first point of Aries, her declination, and consequently the period during which she is above the horizon, rapidly increases; and consequently she is longer above the horizon before reaching the meridian, and the time of her rising is not retarded nearly to this average amount of 52 minutes. The time by which her setting is retarded is increased by as great a quantity as that by which the retardation of her rising is diminished. The degree of effect thus produced differs in different regions of the earth: here it is such that for two or three days the moon's rising is nearly as much accelerated by upon her surface, and any of these be her motion Northward as it is retarded by the continually later period at which she comes to the meridian: the consequence therefore is that she appears during this time to rise at very nearly the same hour, that is to say, almost at the instant of sunset. Bright moonlight therefore (when the moon is full near Aries, or the sun is near Libra) for two or three days immediately succeeds the disappearance of the sun; and as, by the general course of the seasons, this period is about the beginning of autumn, or the close of harvest-time, when the opportunity thus afforded of carrying on the works of husbandry after sunset is often very valuable, we speak in common language of the *harvest moon*, when we speak of her as rising at that season successively for two or three nights nearly at the same time.

About the vernal equinox, exactly a reverse operation must take place. The moon when full is then near the equator, but moving Southward, and her change of declination from day to day is the greatest. Her appearance on the meridian still continues to be later from day to day; but as her periods of being above the horizon diminish, her rising and setting will each be brought nearer to the time of her being on the meridian, and her rising is therefore additionally retarded, but the time of her setting is accelerated. Instead therefore of setting every day considerably later, her settings at this period are nearly at the same time for two or three days near the full moon; and for the same reason as before, her setting must, at the full moon, be just when the sun rises. For two or three days therefore at this period, the moon sets just about sunrise. The result is of no practical importance here; but it furnishes another instance of the application of the same principle, and is therefore inserted to familiarize the reader with it. Besides, although unimportant in this hemisphere, the inhabitants of the Southern half of the world, who have their autumn at the time of our spring, are thus furnished in their turn with the advantage of a harvest moon. Of course in every revolution of the moon there are periods when her rising and setting are thus affected, for they must be so whenever her orbit crosses the equator: but they excite little observation in other instances, not being then connected with the close or the beginning of day.

Section III.—*On the moon's rotation on an axis—Libration in latitude—Libration in longitude—Diurnal libration.*

We now proceed to the consideration of other observations which we can make on the appearances of the moon, besides those which we have already explained as to the proportions of her face, or *disk*, which we see at different times in the month. Every one is aware that the moon does not present (as the sun does, at least to the naked eye) an uniform face of uninterrupted light, but that she appears chequered and diversified with darkish spots and lines: indeed these have been so constantly the subject of observation, that in most countries fanciful resemblances have been imagined for them; and we still hear of the man in the moon, his bush, and his dog*. The probable cause of these appearances will be matter of consideration hereafter; but independently of any such speculations, they at once furnish us with the means of ascertaining a curious and important fact with respect to the motions of the moon.

As we have the means of making these observations upon the surface of the moon, we can tell by them what part of her surface is turned towards us. If these appearances are from time to time different, it would be natural to conclude that different parts of the moon are at different periods presented to us: if they are always the same, the same part of the moon is always turned towards us, unless indeed all her parts are marked so exactly alike that the one would be indistinguishable from the other. This however is not the case; for we see at the period of full moon half, or very near half, the surface of the moon, and the marks and spots with which it is diversified: and these, when examined through a telescope, are so different in their character, that any alteration in their position with respect to us would be immediately detected. We are thus able by observation to ascertain what part of the moon's face is presented to us: and we are so, whether the full moon or any smaller portion of her disk be presented to us, because, if we once learn to distinguish the marks

---

* *Moon.* All that I have to say, is, to tell you, that this lantern is the moon; I, the man in the moon; this thorn-bush, my thorn-bush; and this dog, my dog.
Midsummer Night's Dream, Act V. s. 1.

upon the part, however small, which is actually visible to us, we see what part of her disk that is, and consequently in what manner she is placed with respect to us.

The result of these observations is, that very nearly the same part of the moon is always turned towards us: there are some slight differences in the appearances presented at different periods of the month, but they are so small that, for the present, they may be left out of our consideration. Hence we may easily ascertain that the moon must herself have a motion of rotation, and that the period of her rotation must be the same as that of her revolution round the earth. Referring again to *fig.* 16, we may take the figures in the inner circle to represent different positions of the moon; and the shaded part of the figure, A, will represent the half turned to the earth in that position, the white part the part turned from it. As in the figure the boundaries of the white and shaded part in the figures of the inner circle are parallel in each case, the shaded part may in every case represent the side of the moon which was turned to the earth at A, if we suppose that the moon has no motion of rotation. The part of the moon actually turned towards the earth will in each case be very nearly represented by the part within the circle joining A, B, C, &c. It is obvious therefore from inspection of the figure, that, if the moon has no motion of rotation, a different part of her surface will be presented to the earth in every different position. Thus at A she presents one side to the sun, the opposite side to the earth, at E she would present the same side which she before presented to the sun, to the earth also; and in the intermediate positions, B, C, D, she would continually turn towards the earth less and less of the part turned towards it at A, as the positions, B, C, D, themselves successively became more distant from A. We find however that this is not the case, but that at every position very nearly the same part of her surface is turned towards the earth. This can only be done in one way. If the same part of the moon is to be turned towards the earth at B, which was so at A, or the shaded part in the figure is to coincide with the part within the connecting circle, it can only do so by a turning of the moon in the same direction, so as to bring it into the required position; and in the same manner the moon must have turned yet farther to produce the same effect at a greater distance, as C or D, and must have turned half round to make the same side face the earth at E, one extremity of the diameter A E, which had done so at A, the other. The same motion must evidently continue beyond E; and that the same part of the moon may again be presented to the earth on her return to A, the revolution must be completed at her return to A, and not sooner. There is then a revolution of the body of the moon, and it is completed during the space of a periodic month*; for it is obvious that the position of the sun has nothing to do with the part of the moon which is really turned towards the earth, though it determines that which is visible, and consequently that the time of rotation is the same with that of the moon's return to the point A, not as a point situated between the earth and the sun, but as a point in a given direction from the earth; or, in other words, that it is the same with the length of the periodic, not of the synodic revolution.

In the figure, as drawn, the moon is merely represented by a circle drawn on the plane of the paper; which obviously represents that of the moon's orbit, and the rotation deduced would be rotation in that plane, or round an axis perpendicular to that plane. If however the axis round which the moon turns is inclined at any angle to that plane, the appearances would be different. Let us call the extremity of the axis elevated above the plane of the paper, M, that depressed below it *m*; and let us suppose that the axis is everywhere parallel to itself, and that at the point A, the extremity M is inclined a little towards the earth, and of course the extremity *m* a little away from it. It is plain that, on this supposition, an observer at the earth would have turned towards him a part of the moon a little beyond M, and that, towards the other side, he would not see quite so far as *m*. When the moon arrived at E, these appearances would be reversed; the position of the axis continuing parallel to itself, but its situation with respect to the earth being reversed, the extremity *m* would now be inclined as much towards the earth, as

* We shall hereafter find that this remarkable law obtains, as far as we can discover by observation, in the case of all satellites, or bodies of the same order as the moon; that they all have a motion of rotation, and all perform it in the same time in which they move once round the principal planet which they attend.

in the former position M was; and as, in the former case, an observer would have had exposed to him parts beyond M, but not those extending to *m*, he would now have turned to him parts beyond *m*, but would not be able to observe those extending to M. There would therefore be a sensible difference in the parts which he could observe under the two circumstances. It is plain also that there would be similar and corresponding changes of appearance in the intermediate situations: it is not necessary to enter into detailed consideration of them. The facts actually observed however are found to correspond with the results deduced upon this supposition of a rotation on an axis moving parallel to itself, but not quite perpendicular to the plane of the moon's orbit. The points M and *m* are called, from their correspondence, which will hereafter appear, with the points called the poles of the earth, the *poles* of the moon; and the great circle perpendicular to the axis of the moon, is called for a similar reason the *equator* of the moon. The phenomenon which we have been explaining of the appearance of different parts of the moon's surface, differently situated with respect to these poles, is called the moon's *libration* (rocking or balancing) *in latitude*, because terrestrial latitude is reckoned by the distance from the terrestrial equator and its poles, and the corresponding measure on the moon's surface is therefore called lunar latitude. From the amount of this libration, the degree of inclination of the moon's axis to her orbit may be ascertained: the angle is $84° 51' 11''$; or it is $5° 8' 49''$ out of the perpendicular.

This however is not all. The boundary of the part of the moon presented to the earth is a circle of the moon, of which the plane is perpendicular to the radius vector. The angle therefore which the positions of these planes will make with each other at different points of the orbit, will be equal to the angle made by the radii vectores at these points. The moon however moves in an ellipse, and consequently her angular velocity is not uniform: the angle traced by the radius vector therefore will not increase in the exact proportion of the time. If then the rotation of the moon be uniform, as well as complete in the periodic time of the moon, she will describe 360° on her axis, in the same time that she takes to describe 360° round the earth: but while she takes exactly half this time to describe 180°, a quarter of it to describe 90°, and a tenth of it to describe 36° on her axis, in whatever part of her orbit this time be taken, she will not take accurately these same proportions of time to describe the corresponding angles of 180°, 90°, and 36° around the earth in every part of her orbit. At some periods therefore, when the moon's motion in her orbit is less than her mean motion, she will have turned farther on her axis than is necessary to keep the same face directly turned towards the earth; at other times, when the moon's motion exceeds the mean motion, she will not have turned sufficiently far on her axis for that purpose; and the consequence would be, that in the former case a small part of the moon on the side towards which her motion in her orbit takes place would be seen which had not been so in her original position: in the latter, a small part of that side of the moon away from which she moves: or as the motion of the moon round the earth is from West to East, speaking generally, in the former case some of the Eastern side of the moon, in the latter some of the Western, will be seen, beyond what was originally turned towards the spectator. The poles of the moon would be unaffected by the motion of rotation, and of course by its equality or inequality. These appearances again are actually found to take place, and are known by the name of the moon's *libration in longitude*, from an analogy between the lines and points on her surface, and those on the earth, like that to which we have already referred in speaking of the libration in latitude.

There is still another phenomenon of the same kind. The part of the moon presented to an observer at any place is bounded by a circle perpendicular to the line joining his place and the centre of the moon. To observers at different places, therefore, appearances in some degree different will be presented; for the moon is not so distant from the earth but that the lines joining her centre with different spots on the earth's surface may make a sensible angle; in the extreme case not less than twice the horizontal parallax of the moon, or nearly 2° on an average: and this angle will be that of the inclination of the planes bounding the part of the moon visible at each situation. A similar consideration will explain a further variation of the appearances of the moon, which every day presents to us,

When the moon rises in the East, an observer on the earth's surface will see a little more of her Western and then upper side, than he would do if placed in the centre of the earth; and when she is setting in the West, he will see a little more of her Eastern and then upper side. Thus, if in *fig.* 16 C represent the moon when rising, G when setting, to an observer at the point T, on the surface of the earth, and C B A H G the direction of its daily course, from East to West, the side turned towards A will be the Western side when the moon is rising, the Eastern side when it is setting; and it is obvious that the observer at T, being nearer to the point A than an observer at the centre of the earth would be, he will see rather more of those parts which are turned towards A. In intermediate points there will be corresponding varieties of appearance. This is called the *diurnal, or parallactic, libration.*

A certain degree of variation in the appearances presented by the moon is thus shown necessarily to exist, on the supposition of her uniform rotation round a fixed axis; and the appearances which actually exist are found very nearly to correspond with that supposition. But in this, as in almost every astronomical result, we are obliged to qualify our first conclusions in order to arrive at complete correctness. We find that the appearances presented so nearly correspond with those necessary on the supposition of uniform rotation round a fixed axis, that we are induced to conclude, in the first instance, that this supposition is strictly true; but more minute observation teaches us that it is not so. We have indeed no reason to doubt the uniformity of the motion of rotation; but the position of the axis itself is found not to be invariably the same. It is always inclined at the same angle, or very nearly so, to the lunar orbit; but its direction is different, for it is always perpendicular to the line joining the nodes of the orbit; and as they make a complete revolution in 6793.42118 days, the axis of rotation of the moon will in that time go through all its positions with respect to the heavens.

SECTION IV.—*Mountains of the Moon —Volcanoes—Atmosphere of the Moon.*

THE moon continually turning the same face, or very nearly so, towards us, we are able to observe it, by the aid of powerful telescopes, with great accuracy. The consequence has been, that maps of that part of its surface which is exposed to us have been constructed, and that we have obtained considerable knowledge of its constitution as well as of its motions.

On looking at the moon through a powerful telescope, we observe very different degrees of brightness in different parts of the surface, and especially some bright spots, which have beyond them, in the direction opposite to that of the sun, a comparatively dark shadow. The length and position of this shadow vary as the position of the spot with relation to the sun varies; the shadow being always opposite to the sun, but longer or shorter as the sun is more or less elevated above a plane touching the moon at that point, *or the horizon of that point of the moon.* It is obvious, from these appearances, that the dark part is really a shadow created by the interposition of the bright spot between it and the sun, or that the bright spot is an elevation above the general level of that part of the moon where it stands, or a mountain in the moon.

We are led to the same conclusion of the existence of mountains in the moon by another remarkable appearance. The inner or oval edge of the moon is the boundary of light and darkness there. If the moon were perfectly smooth, this boundary, as we have seen, would be a perfect circle, and its appearance would be that of an accurate ellipse; but if there are inequalities on the moon's surface, some of the higher points beyond the line which would be the boundary of the visible part if smooth, would project into the light, and would be visible sometimes even while the depressed parts between them and the boundary would be in shadow. Thus upon the earth's surface the tops of mountains continue to receive the light of the sun after he has set to the valleys or plains below them. Now these are exactly the appearances presented by the inner edge of the moon when seen through a telescope: whatever portion of the moon be visible, the edge is everywhere rough, with lines and points projecting beyond its general line, and with some insulated points completely beyond that line, and not connected with it by any lines of light.

It is evident that the former must be elevated ridges rising above the general surface of the moon; the latter, single points yet more elevated, and more distant from the part generally enlightened. We conclude therefore that the surface of the moon is everywhere rugged, without any great plains or large spaces, like the sea here, covered with water or any fluid, which, from the laws regulating the equilibrium of fluids, would necessarily have a smooth surface.

It may seem that corresponding appearances ought to be presented by the other limb of the moon, and that, just as the sun's rays enlighten prominent objects beyond that part of her surface which he generally shines upon, we ought also to be able to see prominent points beyond the part of the moon directly turned towards us. And it is so; but though such points are actually presented to our view, they are scarcely distinguishable to our senses, even when assisted by powerful instruments. It will be worth while to explain the cause of this difference, especially as it will show also how we are enabled to estimate the height of lunar mountains. For this purpose, in *fig.* 20, let

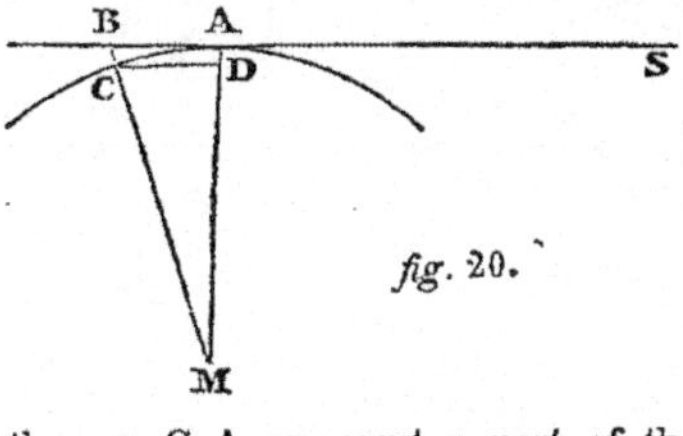

fig. 20.

the arc C A represent a part of the moon's surface, and S A B any straight line touching it; and let B be any point in that line, and join B with M, the centre of the moon. The part of the line B C, intercepted between B and the surface of the moon at C, is evidently the height of the point B above that surface; and whenever the angle A M C is small, B C is much less than A B*. Now the rays of light move in straight lines; S A B therefore may represent the course of a ray of light. A, where it touches the sphere, will be a point in the general boundary of light and darkness; but if B represent the top of a lunar mountain, whose height is C B, B will just receive the ray S A B, and will therefore be visible. The distance, A B, from the general boundary of light and shade, may be observed. It will not indeed be seen except in particular cases, perpendicularly to the line joining the observer and the moon; but from the observed distance, the real distance of A B may be computed, the inclination at which it is seen being known; and from the real distance, as computed, the angle A M B, and the height B C, may be ascertained. A B is necessarily greater than B C; but this, though it is one, is not the only reason why the distance of an elevated point from the boundary of the enlightened part of the moon is more easily observed than its actual prominence. We may now suppose S to represent the direction of an observer upon the earth; in which case it is plain that A will represent a point in the boundary of that part of the moon which, if she were perfectly smooth, would be visible from S. The observer at S, however, will be able to see the point B, if that be, as before, a point elevated above the general surface of the moon; but he will see it in the line S A B, or in the same direction with the point A, the extreme point of that part of the moon, which, independently of these prominences, would be presented to him: and he will not distinguish it as projecting beyond the general surface of the moon, though it is only by virtue of this projection that it is brought within his view at all. A similar mountain situated between the points A and C would indeed project beyond the line A B, and be partly distinguishable; and in the extreme case of its being situated exactly at A, it would project thus by its whole height; but even then, as we have already seen, the prominence is much less than A B.

It will naturally be supposed that the nicety of observation required to detect such small quantities as the altitude of lunar mountains at the distance of the moon must be very great; and their values may in consequence be considered as not very accurately ascertained. They were much overrated when first observed, and the observations then registered would correspond to surprising heights, as 15 miles, or thereabouts. Better observations however have reduced these extraordinary elevations; and Dr. Herschel considered few of

* [Drawing C D perpendicular to A M,
A B : C D : : A M : D M : : C M : D M : : B C : A D
or A B : B C . : C D : A D : : sin. M : ver. sin. M.
Whenever, therefore, sin. M is greater than ver. sin. M, which it is until M = 90°, A B is greater than B C, and very much so when M is small.]

these mountains to exceed half a mile in height. There are some however far higher; and one in particular, named after the philosopher Leibnitz, has been computed by M. Schroeter to be 25,000 feet, or nearly 5 miles high.

Besides these points near the illuminated edge of the moon, other bright points are occasionally seen in the dark parts of it, at so great a distance from the edge, that they cannot be accounted for in the same manner: for a mountain lofty enough to be enlightened in such a position could not escape our observation when in the generally enlightened part of the moon. These points therefore, when seen, must be luminous in themselves. They are seen occasionally, not always; they are therefore luminaries only occasionally: and the most probable account that has been given of them is, that they are volcanoes, and that when they are visible independently of reflected light, they are in a state of active eruption. Their appearance, when seen in the generally enlightened parts of the moon, corresponds with and confirms this supposition.

If there be an atmosphere surrounding the moon, at all analogous to that of our earth, there must be twilight there: and if there is twilight, there must be a partial and faint light beyond the boundary of that part of the moon which is fully and directly illuminated by the sun. Such a light has been actually observed, but very faint and of small extent; corresponding therefore to the supposition of there being such an atmosphere, but of little power in reflecting light. That its power is very small, is evinced also by another consideration. The same media generally are powerful both in refracting and reflecting light. Now we have complete proof that the atmosphere of the moon has very little power in refracting light. In the course of the moon's revolution through the heavens, she must necessarily sometimes be in the same direction with some of the stars, and, being much nearer to us than they are, pass between us and them, and conceal them from us. This is called the *occultation of the fixed stars by the moon*. Her motions being known, the time at which that part of her, which first coincides in position with a star, does so coincide, or the beginning of the occultation, may be computed; and so in the same manner may its end, and consequently its duration. This however is the duration independently of any refraction by the moon's atmosphere: for it is deduced from considering when the moon is exactly between the earth and the star, at which period it would intercept rays of light proceeding in a straight line from the star to the earth. If the moon have any refracting atmosphere, the duration would be shortened, for the rays would be bent towards * the moon in passing through it, and, as they would pass first on one side of the moon (that which first came in conjunction, or apparent contact with the star), and afterwards on the other (which last leaves it), they would be bent towards each other, and towards the earth, and consequently in one case would continue to reach the earth, after some part of the moon is between the earth and star, and in the other would begin to arrive at the earth before the whole of the moon had passed from between them. The observed duration of the occultation would be less than the computed duration. Such a difference is in reality scarcely observable: its amount has never been very completely ascertained, but it certainly does not exceed, in any case, 8 seconds of time; a diminution which is not greater than that which would correspond to a horizontal refraction not exceeding 2″. The horizontal refraction at the earth is not less than 33′; and as, on the supposition of the medium being similar, the refracting power increases with the density, it may be estimated from this, that the density of the lunar atmosphere must be nearly 1000 times less than that of ours. The identity of its nature however, and therefore the conclusion drawn from it, is to a certain degree conjectural.

One other circumstance respecting the lunar atmosphere may here be mentioned. It is clear that it is not loaded with heavy clouds, as the atmosphere of the earth so frequently is: for these would either themselves be visible to us,

* The constitution of the lunar atmosphere being supposed the same above different parts of the moon's surface, the ray will leave the atmosphere under circumstances corresponding to those under which it entered it. Its course therefore on quitting it will make the same angle with a perpendicular there, that its course before entering it did with a perpendicular at the point of its entrance. The angle between these two courses therefore, or the deflection occasioned by the ray's passage through the atmosphere, will be the same as the angle between the perpendiculars, and measured in the same direction, that is to say, towards the side of the moon opposite to that at which the ray enters, and consequently, the ray entering at some distance from the moon, towards the moon itself.

or would at least be discovered by the shadows they would cast in varying places upon the moon's surface. We can observe the shadows of her mountains, and should equally be able to observe those of her clouds. This observation comes in aid of that already deduced from other appearances, of the absence of any large spaces covered with water, or any evaporable fluid in the moon: for if there were any such, evaporation would take place, and clouds would be formed.

We have already said that many of our conclusions with respect to objects on the surface of the moon are to a certain extent conjectural; and it may possibly seem that the conjectures are impugned by the dissimilarity of appearances presented by the moon and earth. Objects on the earth are seen by reflected light, as well as those at the moon; yet with how much less brilliancy, except occasionally where a brightly reflecting surface is met with, and with how many shades of colour, instead of one uniform and almost white brightness, only diversified by its different degrees of intensity. And when we see objects melted together in distance, and almost of uniform colour, it is a pale bluish hue of faint lustre. Any such objection may easily be removed by these considerations, that the mixture of a great variety of colours will produce a white, or nearly a white light, and consequently that the blended light reflected from a large distant tract, ought in general to be nearly of that colour; that if the distant objects which we see on the earth are not so, it is because the rays which proceed from them pass through a long space of the lower and denser parts of our atmosphere, and the objects become, in consequence, tinged with its blue colour; that their faintness proceeds partly from this cause, and partly from their being visible only in the strong light of day; and that when the moon is seen so also, there is not that striking difference in her appearance and theirs which we suppose, when we contrast them with our notion of her, as derived from our common observations in the darkness of night. But it may be worth while to subjoin, as a still further answer to any objections of this nature, the description which a very accurate observer has given of the appearances presented by a part of the earth itself, where the brilliancy of the climate, and other accidental circumstances, diminished some of the causes of difference.

'On the 26th of May we sailed from Valparaiso, and proceeded along the coast to Lima. During the greater part of this voyage the land was in sight, and we had many opportunities of seeing not only the Andes, but other interesting features of the country. The sky was sometimes covered by a low dark unbroken cloud, overshadowing the sea, and resting on the top of the high cliffs which guard the coast, so that the Andes, and indeed the whole country, except the immediate shore, were then screened from our view. But, at some places, this lofty range of cliffs was intersected by deep gullies, called *quebradas*, connected with extensive valleys, stretching far into the interior. At these openings we were admitted into a view of regions, which, being beyond the limits of the cloud, and therefore exposed to the full blaze of the sun, formed a brilliant contrast to the darkness and gloom in which we were involved. As we sailed past, and looked through these mysterious breaks, it seemed as if the eye penetrated into another world; and had the darkness around us been more complete, the light beyond would have been equally resplendent with that of the full moon, to which every one was disposed to compare this most curious and surprising appearance.

'As the sun's rays were not, in this case, reflected from a bright snowy surface, but from a dark-coloured sand, we are furnished, by analogy, with an answer to the difficulties sometimes started, with respect to the probable dark nature of the soil composing the moon's surface*.'

Section V.—*On Eclipses.*

I. General nature of Eclipses.

II. Lunar Eclipses—Lunar ecliptic limits—Penumbra.

III. Solar Eclipses—Solar ecliptic limits—Total and Annular Eclipses.

IV. Number of Eclipses—Recurrence in about 18 *Years—Use in Chronology.*

I. We explained generally in the course of the last section the nature of an occultation. It is evident that, in the same manner as the moon was there represented to interpose between the earth and a fixed star, she may interpose also between the earth and sun, or some

* Captain Basil Hall's Extracts from a Journal, vol. i., p. 186.

part of the sun, and thus deprive us for a time of the whole or a part of its light. Again, when nearly opposite to the sun, she may be in such a position that the earth itself may be between her and the sun, and intercept the sun's light from the whole or a part of her surface. In this case, as the moon shines only by reflecting light proceeding from the sun, (for the dark side of the earth is then towards her, and consequently no light is supplied to her by earthshine) the whole or some part of her surface must become invisible. These phenomena are known by the name of *eclipses*, from a Greek word signifying *a failure* of light: the disappearance of part or the whole of the sun by the interposition of the moon being called *a solar eclipse;* the disappearance of the moon, or any part of her, by coming within the earth's shadow, being called *a lunar eclipse.*

If the moon revolved round the earth in the same plane as the sun, there would necessarily be a lunar and a solar eclipse every month. Referring again to fig. 16, if T represent the earth, S the sun, and A, B, C, D, &c. the moon's orbit, A is directly between the earth and sun, and the moon at A would eclipse the sun; T is directly between the sun and E, and the moon at E would therefore have the earth interposed between her and the sun, or she would be eclipsed. On this supposition therefore there would be a solar eclipse at every new moon; a lunar eclipse at every full moon. This however is not the case, and we proceed to shew more in detail why it is not so, and what circumstances must concur to produce an eclipse. In so doing we shall take the case of a lunar eclipse first, as being the most simple.

II. A lunar eclipse being occasioned by the interposition of the earth between the sun and moon, is an actual deprivation of light to the moon. She therefore becomes for the time really invisible, and disappears, not merely to a spectator at some particular place who is prevented from seeing her, but absolutely and universally. We have no occasion therefore to refer to any particular point on the surface of the earth, or to embarrass ourselves with any considerations of difference between observations made at different places: we have only to ascertain when the light transmitted to the moon from the sun actually fails.

For this purpose we must first ascertain the form and magnitude of the earth's shadow. Now we already know (p. 61) that the sun is larger than the earth. The earth's shadow therefore, as we have seen in p. 74, is a cone, and terminates. We can also tell, knowing, at least within a moderate distance of the truth, the relative magnitudes and the distance of the sun and earth, what the length of this shadow must be. The sun's radius is about 110 times greater than that of the earth: his distance nearly 24,000 times the same quantity. The length of the shadow then must be to its radius at the earth very nearly in the ratio of 24,000 to 110 − 1, or 109, or it must be $\frac{24000}{109}$ times, or about 220 times that radius: more accurately, its mean value is 216.531 times that radius; its value when the sun is in apogee being 220.238, when in perigee 212.896 times that quantity. The moon's mean distance from the earth is not more than 60 times that radius; she therefore passes far within the limits of the shadow, and may be eclipsed by it.

The next object is to determine what are the dimensions of the shadow at the distance where, if at all, the moon passes through it. The section, being one perpendicular to the axis of the cone, will necessarily be a circle, and we therefore only have to determine the proportion which the radius of this circle bears to that of the earth. Now the radius of this circle will bear to that of the earth, the same proportion as the parts of the axis of the cone intercepted between them and the vertex of the cone; or

rad. of shadow at moon's distance : rad. of earth :: 220 − 60 : 220 nearly
:: 160 : 220
:: 8 : 11.

The radius of the shadow being thus found to be about $\frac{8}{11}$ of the earth's radius, and that of the moon being only about $\frac{3}{11}$ (p. 61) of the same quantity, the shadow, at the distance where she may meet it, is considerably more extensive than the moon, and she may therefore fall entirely within it, and be totally eclipsed.

The apparent radius of the moon is about 16′: the angle subtended by the radius of the earth's shadow at the same distance must bear the same proportion to this quantity that the radii themselves bear; or the proportion of about 8 to 3; and consequently it would be about 42′ 40″. More accurately however, for the numbers we have taken are only a very rough approximation to the truth, its mean value is 41′ 8″.5; or the mean apparent diameter of the shadow is 82′ 17″. If therefore the moon pass through it along a diameter, and her apparent diameter be of 32′, when she is first completely within the shadow, she has a space of 50′ 17″ to move over before any part of her arrives at the opposite side of the shadow, and during the whole time in which she does so move, the eclipse will be total. The greatest possible length of a total eclipse of the moon may in this way be ascertained to be about two hours.

It will however be worth while to estimate the magnitude of the apparent diameter of the earth's shadow more accurately. For this purpose, in *fig.* 21, let the circle whose centre is S represent the sun, that whose centre is E represent the earth, and let RTP, rtP, be two lines drawn touching the sun and earth, and meeting in P. The line SE joining the centres of the sun and earth

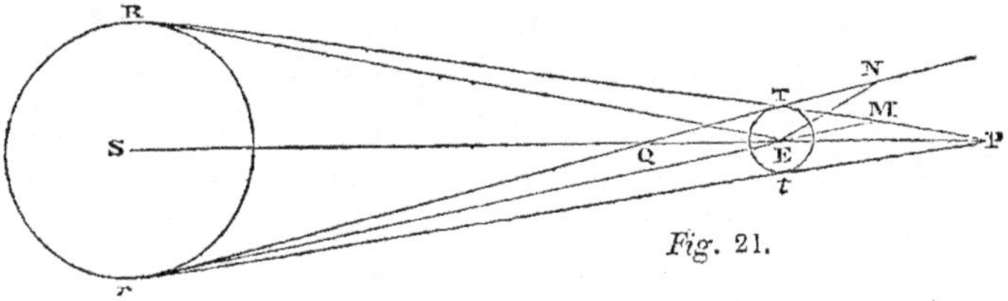

Fig. 21.

will pass through P also: for the line joining P with the centre of the circle S, will bisect the angle between the tangents PR, Pr; and in the same manner the line joining P, with the centre of the circle E, will bisect the angle between the tangents PT, Pt, and as these tangents are the same, the lines in question must coincide. Now let M be a point in the extremity of the earth's shadow at the moon's distance. The angle MEP therefore is the apparent radius of that section of the shadow. Now the angle

$$\begin{aligned} \text{MEP} &= \text{EMT} - \text{EPM} \\ &= \text{EMT} - (\text{RES} - \text{ERT}) \\ &= \text{EMT} + \text{ERT} - \text{RES}. \end{aligned}$$

Now RTM is a tangent to the earth at T; to an observer placed at T therefore the sun, and also the point M, would be in the horizon. The angle ERT therefore is the angle which the distance from the situation of an observer to the centre of the earth subtends at the sun, when he is in the horizon, or it is the horizontal parallax of the sun: the angle EMT is in the same manner the horizontal parallax of the point M, or of the moon; for the horizontal parallax of the moon, and of a point at the moon's distance, are the same. The angle RES is obviously the apparent radius of the sun. If therefore we call the horizontal parallax of the sun P, that of the moon p, and the apparent radius of the sun R, we have this equation; apparent radius of shadow at the moon's distance $=p+\text{P}-\text{R}$.

If we call the apparent radius of the moon r, then when the moon is just in contact with the shadow, but without it, the angular distance of her centre from the centre of the shadow, will be this quantity increased by her apparent radius, or $p+\text{P}+r-\text{R}$: at this distance therefore she will just be in contact with the shadow, or there will be that description of eclipse which is called an *appulse* of the moon; and if her distance be greater than this, she will not be eclipsed. In the same manner, when she is just in contact with the shadow, but within it, the angular distance of her centre from the centre of the shadow is the apparent radius of the shadow diminished by her apparent radius; or it is $p+\text{P}-\text{R}-r$. This therefore is her distance when she is just totally eclipsed.

The farther the sun is from the earth, the more slowly do the lines joining them, or the boundaries of the shadow approach each other, and the larger therefore is the shadow at the moon's distance: and besides this, the nearer the moon is to the earth, the larger, all other things continuing the same, is the part of the shadow through which she passes. On both accounts, the duration of an eclipse is greatest, when the moon is at the least, and the sun at the greatest distance*.

* [The same results may thus be more strictly deduced. The values of all the quantities, $p+\text{P}-\text{R}$, $p+\text{P}+r-\text{R}$, $p+\text{P}-\text{R}-r$, vary with the varying distances of the sun and moon from the earth; for both the horizontal parallax and the apparent radius vary inversely as the distance, and consequently their sum and difference do so also. Now p, the moon's horizontal parallax, is greater than r, her apparent radius: $p-r$ therefore is a positive quantity. P, the sun's horizontal paral-

Taking the extreme values, the greatest apparent radius of the shadow is of 45′ 52″.15, and the corresponding apparent radius of the moon is 16′ 5″.45; and the greatest distance from the centre of the earth's shadow, at which the moon can possibly come in contact with it, is the sum of these quantities, or 62′ 37″.65. In the same manner, the least apparent radius of the shadow is 36′ 42″.15, and the corresponding apparent radius of the moon is of 14′ 41″, and the least distance at which the moon can just be in contact with the shadow, and no more, is the sum of these quantities, or 51′ 23″.15. If, therefore, the moon, during her revolution, never comes so near as 62′ 37″.65 to the centre of the earth's shadow, there can be no eclipse; if she comes to that distance, or within it, there may: if she comes within the distance of 51′ 23″.15, there must*.

Now the centre of the earth's shadow is a point in the line which passes through the centres of the earth and sun; and as this line is in the plane of the ecliptic, the centre of the earth's shadow will be in the direction of a point in the ecliptic, directly opposite to the centre of the sun. This direction being thus ascertained, we have only to inquire whether this point be within the required angular distance of some point in the moon's orbit or not, to know whether there may or may not be an eclipse. It is evident that this will depend on the situation of the node of the moon's orbit. Thus, if the node happens to coincide with the centre of the earth's shadow, the moon, in passing through that part of her orbit, will pass through the very centre of the shadow, and be totally eclipsed; and such an eclipse is called a *central* eclipse of the moon. If, on the contrary, the node be 90° from this point, the moon, when in opposition to the sun, will be 90° from the node, or at her greatest distance from the ecliptic. Her centre in this case will be upwards of 5° from the centre of the shadow, and there will be no eclipse.

It is a known property of the sphere, that the shortest distance which can be measured on its surface from any point without a given great circle to that circle, is the arc of a secondary to that circle, passing through the point. But arcs of great circles measured from one point on the sphere to another, measure the angular distances between those points as seen at the centre. Let then, in *fig.* 22,

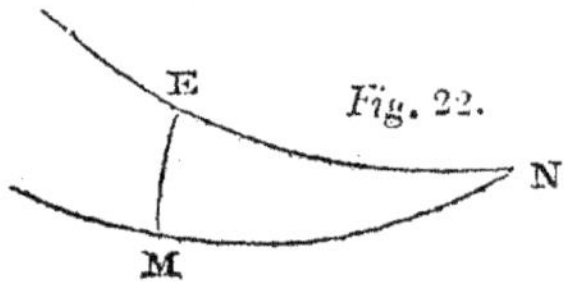

Fig. 22.

N M represent a portion of the moon's orbit, N E a portion of the ecliptic, N, of course, being the node; and let E be the position of the centre of the earth's shadow. Let E M be the arc of a secondary to N M, drawn through E; then M is the point in the circle N M which is nearest to E, and if the moon is not in contact with the earth's shadow when at M, she never will be so. If E M is greater than 62′ 37″.65, she never will be: if therefore we can ascertain what must be the value of N E to correspond with this value of E M, we shall ascertain how distant the node may be from the centre of the earth's shadow to admit of there being an eclipse of the moon. Now in the spherical triangle E M N, the angle E M N is a right angle; the angle E N M is known, for it is the inclination of the moon's orbit to the ecliptic, or 5° 17′ in its general value, and the side E M is 62′ 37″.65 by the supposition. The remaining sides and angles therefore may be computed; and the result is, that E N, on that supposition, is equal to 11° 25′ 40″ nearly*. The circles continually diverge from each other for 90°; if therefore E N be greater than this, E M will exceed 62′ 37″.65, and there will be no eclipse. If therefore the node be more than 11° 25′

lax, is less than R, his apparent radius: P−R therefore is negative. Each of the values deduced therefore is the sum of a positive quantity, p, $p+r$, or $p-r$, and a negative one P − R: and must increase with the increase of the positive one, which takes place as the moon's distance diminishes, and also with the decrease of the negative one, which takes place as the sun's distance increases. When the moon's distance is least therefore, and the sun's greatest, she will be at the greatest distance from the centre of the earth's shadow, both at the beginning and end of an eclipse, and also when it begins and ceases to be total: and the same quantities will have their least values when the moon's distance is greatest, and the sun's distance least.]

* We have here treated the distances and magnitudes of the sun and moon, their horizontal parallaxes and apparent magnitudes, as known, and have thence deduced the duration and circumstances of an eclipse. Conversely, if these be observed, the value of any one of the other elements involved may be computed from the same equations.

* [Rad. × sin. M E = sin. E N, × sin. E N M, or log. sin. E N = 10 + log. sin. E M − log. sin. E N M = 10 + log. sin. 62′ 37″.65 − log. sin. 5° 17′.]

40″ from the point of the ecliptic opposite to the sun, there will be no eclipse; and this quantity is called in consequence the *lunar ecliptic limit.* In the same manner, if we take E M = 51′ 23″.15, we shall have E N = 9° 20′ 29″ nearly; and as, if E N be less than this, E M must be less than 51′ 13″.1, because the circles converge towards the node, there *must* be an eclipse if the node be not more than 9° 20′ 29″ from the point of the ecliptic opposite to the sun. It is obvious that an eclipse may or must take place within these distances *on each side of the node.* In intermediate distances the non-occurrence or occurrence of an eclipse will depend on the question, whether or not E M is greater than $p+P+r-R$, the value of this quantity being determined from the horizontal parallax and apparent radius of the sun and moon at the time of the opposition.

Having thus ascertained the cases in which there may be a lunar eclipse, we might proceed to investigate the quantity of the moon which would be eclipsed under different circumstances, and the time during which such an eclipse would continue. Such an investigation, if conducted to its utmost extent, would be more elaborate than the purposes of this treatise admit: but for common purposes the difficulty may be in a great measure eluded by adopting an artifice of computation or representation. To simplify the conception of the subject as far as possible, let us suppose, in the first instance, that the paths of the moon and the shadow are straight lines intersecting each other, instead of great circles. Thus, in *fig.* 23, let M N represent the orbit of the moon, supposed to be a straight line, E N the path of the shadow, supposed also to be a straight line, and N their node or intersection; and let M and E be the respective places

Fig. 23.

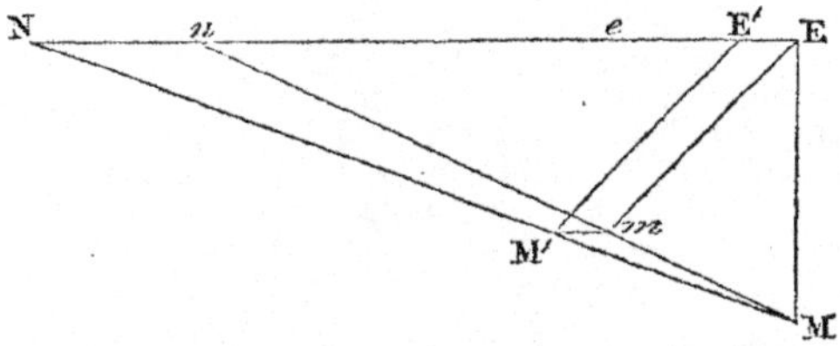

of the moon and the centre of the shadow at the time of opposition. Let it also be supposed that the motions of the moon and shadow are uniform; and that while the moon moves through the space M N, the shadow moves through E *e*. Take N *n* = E *e*, and join M *n*; the line M *n* is so drawn that if M′ represent the moon's place at any time, and E′ the place of the centre of the shadow at the same instant, if the line M′ *m* be drawn parallel to N E to cut the line M *n*, the point *m* is at the same distance and in the same direction from E that M′ is from E′; or in the same relative situation to E that the centre of the moon is to the centre of the shadow.

This will at once appear from the following considerations. The motions of the moon and shadow being uniform, whatever proportion of the space M N is described by the moon in a given time, the same proportion of E *e* is described in the same time: or

M M′ : M N :: E E′ : E *e*,
:: E E′ : N *n*,
for N *n* = E *e*;

But M N : M M′ :: N *n* : M′ *m*, for N *n*, M′ *m* are parallel; and compounding these proportions, E E′ and M′ *m* are equal; they are also parallel; and consequently E *m* is equal and parallel to E′ M′. This conclusion is obviously independent of any particular value of M M′; it is therefore generally true, that the situation of the point *m* relatively to E, is the same as that of the moon with relation to the centre of the shadow; and the phenomena of eclipses therefore, which depend merely on the relative distance and direction of the moon and shadow, and not on their absolute situation, would be the same if the shadow were at rest, and the moon moved in the supposed orbit M *n*, as they really are. The fictitious orbit M *n* is called the moon's *relative orbit*, as being her orbit with relation to the situation of the shadow considered as a fixed point E.

The angle M *n* E, at which the relative orbit cuts the ecliptic, may be very easily computed. M E is perpendicular to the ecliptic N E, for M and E have the same longitude; consequently

$$\tan. MnE = \frac{EM}{En} = \frac{EM}{EN - Nn} = \frac{EM}{EN - Ee}.$$

But since ME is perpendicular to the ecliptic, ME is the moon's motion in latitude, and EN her motion in longitude while she passes from M to N, and E e is the motion in longitude of the centre of the shadow, which is necessarily equal to the motion in longitude of the sun during the same time. If therefore the moon's motion in latitude be called n, her motion in longitude m, and the sun's motion in longitude s, $\tan. MnE = \frac{n}{m-s}$.

These conclusions however require to be modified, when applied to the case of the moon and shadow moving as they do in circular, not in rectilineal, orbits. The truth of our former conclusions depends on the uniform rate of the moon's motion in latitude as well as in longitude; and this is not the case in the real motions. The relative orbit therefore has to be ascertained with reference to the direction and circumstances of the motion at a particular period.

For a short time however, such as that of the continuance of an eclipse, the motions may be considered in this case also as uniform; and the angle at which the relative orbit is inclined to the ecliptic may still be ascertained in the same manner, if we only look to the rate of motion, instead of the actual space described in a considerable time; or if we consider n, m, and s, to represent the *horary* motions of the moon in latitude and longitude, and of the sun in longitude. Even this modification however does not allow for the variations which take place in the motions of the sun and moon, and in the manner in which that of the latter body is decomposed into motion in longitude and latitude, during a very short space such as an hour, or as the duration of an eclipse: and these must be taken into consideration to give complete accuracy to the computation. We need not however here enter into any detail of these minute corrections, but shall proceed to explain very shortly the manner in which the portion of the relative orbit enables us to represent the circumstances and compute the duration of an eclipse.

In *fig.* 24 let E represent the position of the centre of the earth's shadow at the time of the opposition of the moon, and Rr a portion of the relative orbit. A perpendicular EM_2, let fall on this line, is the shortest line which can be drawn from E to Rr; if this line there-

Fig. 24.

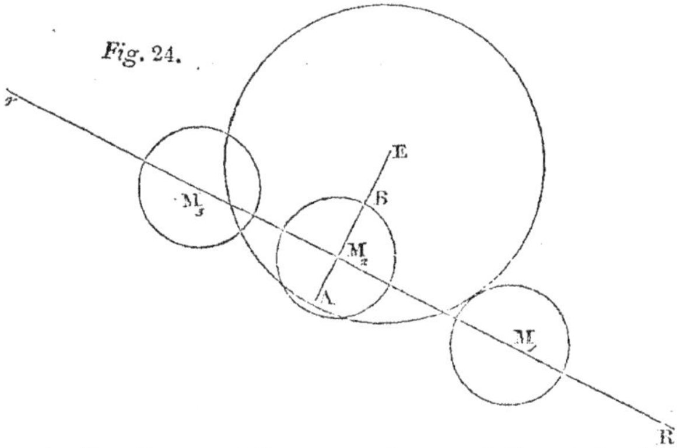

fore is greater than the sum of the apparent radius of the shadow at the moon's distance and the apparent radius of the moon, each estimated for the time of the opposition, there can be no eclipse at all: if there be an eclipse, the depth of the immersion must be greatest at the point M_2; or rather at the time corresponding to it, since M_2 is a point in the relative or imaginary, and not in the real, orbit of the moon. If we describe round E a circle whose radius bears the same ratio to the line EM_2, that the apparent radius of the shadow does to the nearest distance of E from the relative orbit, this circle will represent the earth's shadow, and if at the point M_1 we describe a circle touching the circle described round E, and whose radius bears the same ratio to the radius of that described round E, that the apparent radius of the moon does to that of the shadow, the point M_1 will be that at which the eclipse will begin: and a corresponding point on the other side of M_2 that at which it will end. The duration of the eclipse may hence be at once ascertained, for the motion in latitude is the same in the relative and in the real orbit; and the latitudes corresponding to the beginning and end of the eclipse being ascertained from the relative orbit, the time at which the moon is at each latitude, and consequently the interval between them, may be found from the tables of her motions: or if it be preferred, the time of describing a given arc of the relative orbit may be independently computed. If another circle, like that described about M_1, be described about M_2, it will represent the situation of the moon at the time of her greatest immersion; and if A be the extremity of the radius of the shadow along which E M_2 is measured, and B the point in which the inner limb of the moon cuts it, A B is the greatest depth of the immersion, which is generally estimated in *digits*, a term used to express twelfths of the moon's diameter. The part of the circle representing the moon, which is exterior to the circle described round E, will necessarily represent the part visible at the particular time; and any one who has seen an eclipse of the moon will at once recognize the figure, as correctly exhibiting at M_2 the manner in which the exterior limb sometimes falls short of, and at M_3 that in which it sometimes exceeds a semicircle, instead of being always, as in the phases of the moon, an exact semicircle, however broad or narrow the part of the disk exposed may be.

We have hitherto spoken of the shadow as conical; and it is true that the portion of space within which the earth will entirely conceal the sun is so. But it is clear, that there will be another portion within which part of the sun will be concealed. If a common tangent be drawn to the sun and earth, crossing the line which joins their centres, as the line in figure 21, which passes from a point very near r to another very near the point T, and thence to N, it is clear that the point r will be invisible at every spot between E P and the line T N, and that consequently through this whole space, some part of the sun will be concealed, or there will be a partial shadow even beyond the line T M P, within which the darkness is complete. This also will be the case alike on every side of the line S E P, and if the tangents drawn in the manner last described, cut the line S E in Q (for they all will evidently cut it in the same point), the *penumbra* or partial shadow will be a portion of a cone, whose vertex is Q.

[If N be the extreme point of the penumbra at the moon's distance, we may find in the same manner as before, the apparent radius of the penumbra, and deduce from it the extreme distance at which the moon can pass through the penumbra of the earth. Now, the angle

$$\begin{aligned} NEP &= ENQ + EQN \\ &= ENQ + SEr + QrE, \\ &= p + R + P, \end{aligned}$$

adopting the notation already used; for, as Q N is a tangent to both the sun and earth, T N Q is the horizontal parallax of N, a point at the moon's distance, or of the moon, Q r E or T r E, the horizontal parallax of the sun, and S E r, the sun's apparent radius. It is obvious that the same conclusions with respect to the effects of the varying distances of the sun and moon, will take place in this case and the former.]

These computations are of no very great importance, as we do not consider a lunar eclipse to take place, unless some part of the moon is entirely darkened. The lunar ecliptic limits therefore are not affected by them. But it is of importance to observe that the penumbra does exist, and that in proportion as the moon is nearer the absolute shadow, the proportion of the sun obscured to her is increased. The penumbra therefore increases in depth, or the brilliancy of the moon diminishes, as we approach the boundary of the dark part of the moon; and this appearance may actually be observed during a lunar eclipse, the brilliancy of the moon decreasing near the part which has disappeared.

There is however generally some light discernible on the whole face of the moon, even during a total eclipse. We have already seen that in the case of the occultation of a star by the moon, the rays of that star must, if the moon have any refracting atmosphere, be bent towards the earth. Exactly in the same manner, the rays of the sun in passing through the earth's atmosphere are actually bent towards the moon; and some of them reach it, even when the earth is directly interposed between her and the sun, so as to prevent her from receiving any light independently of this inflexion of the rays. The consequence is that in most eclipses the obscured part of the moon is more or less faintly visible. The degree in which this takes place depends much on the *general* state of the earth's atmosphere at the time. The red rays, having the greatest momentum (Optics, p. 65), are those which principally find their way to the moon. Thus, in the eclipse of September 2d, 1830, the moon, when she was not concealed by clouds, as was the case in London, appeared of a deep blood-red colour, even during the period of the greatest obscuration.

Having thus seen the general nature of the lunar, we proceed to the examination of solar eclipses. There are some considerations common to both, which will more fitly find a place after those peculiar to solar eclipses are gone through.

III. There is one very material distinction between these and eclipses of the moon. In an eclipse of the moon, as we have seen, her disk, or part of it, is absolutely deprived of light, and it consequently disappears alike to all observers. The sun however is a body luminous in itself, and does not cease to be so on account of the interposition of another body, as the moon: he is only partially or entirely concealed as the moon happens to interpose between the observer and part, or the whole, of his surface; just as he is concealed, to use a more familiar instance, behind a cloud which happens to pass over him, but concealed only at those places where the cloud does so. The sun's parallax is very small, and all observers therefore see him in very nearly the same point of the heavens: the moon's parallax is considerable, and different observers therefore see her in points considerably distant from each other. The appearances therefore will be different in different places, and the duration, extent, &c. of a solar eclipse will differ accordingly, and must be computed for the particular places at which the eclipse is to be observed. The difficulties of these computations are very considerable; but, omitting them, we may proceed to give the same sort of account as before of the limits within which a solar eclipse can take place, and to state some of the principal circumstances affecting it.

For this purpose, in *fig.* 25, let the circle, whose centre is S, represent the sun; that whose centre is E, the earth; and let R T be a common tangent to them, which does not cut the line, S E, which joins their centre. Let the circle, whose centre is M, and which touches the

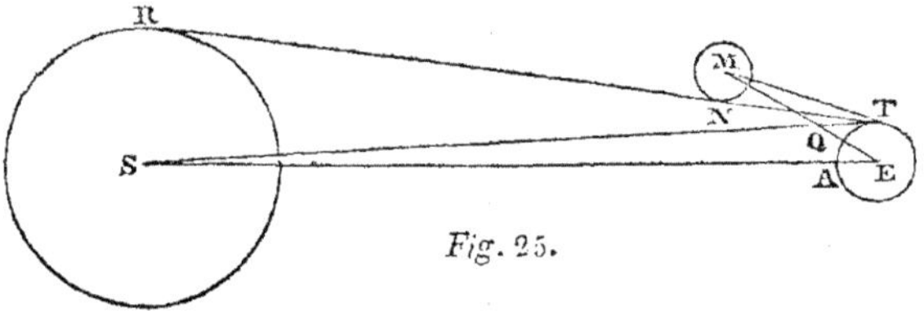

Fig. 25.

line R T on the outside with respect to S E, represent the moon. It is evident that this position of the moon will be that most distant from E S, at which a solar eclipse can take place; for at T the moon appears just to come in contact with the sun, or at T there is an *appulse* of the moon to the sun; but at every other point of the earth's surface the sun is seen entirely clear from the moon: if the moon be farther from S E than its position at M, then she is not in contact with the sun even at T, and clearly nowhere else: if she is nearer, then she obscures more of the sun from T, and part of it from portions of the earth between T and A. The angle M E S therefore represents the greatest angular distance at which the moon can be from the line joining the earth and sun, when a solar eclipse takes place; and if we can ascertain its value, we can find from it the *solar ecliptic limits*. Now the angle M E S = M Q S — Q S E

$= MTQ + QMT - QSE$

$= MTN + RTS + EMT - TSE$

$= r + R + p - P$ very nearly, adopting the notation formerly used: for M T N is evidently the apparent radius of the moon, and R T S that of the sun, as seen from T; E M T also is the parallax of the point M; and as R N T is a tangent to the earth at T, it is in the plane of the horizon there, and the altitude of the point M above the horizon is only the angle M T N, or the apparent radius of the moon; the parallax of M therefore will differ imperceptibly from the horizontal parallax of the moon: and the angle Q S E, or T S E, is evidently the parallax of the point S, a point only depressed below the horizon by the angle R T S, the apparent radius of the sun, and therefore differing imperceptibly from the horizontal parallax of the sun.

It is however necessary to observe, in using these results, that the apparent radius of the moon is not its true apparent radius, or that which it would have if seen from E, but that observed from T, a quantity differing from it as we have seen before (p. 62. *n*), in the proportion of the distances E M, T M.

It is evident that the value of the angle in question will be greatest when the distances both of the moon and sun are least; for then both the moon and sun occupy the greatest angle, and therefore interfere with each other at the greatest distance *. If we take the greatest values of all the quantities involved, we shall find the value of M E S to be 1° 34′ 23″.2 very nearly; and when the moon's centre therefore is at this distance from the line joining the earth and sun, there may be a solar eclipse: if we take their least values, we shall find M E S to be 1° 23′ 15″ nearly; and if the moon's centre be within this distance from the same line, there must be a solar eclipse. Between these limits there may or may not, as the case may happen.

Hence, in the same manner as before, we may find the solar ecliptic limits. The only difference will be that the point E (in *fig*. 22), instead of representing the place of the centre of the earth's shadow, as referred to the ecliptic, will now represent the sun's place in the ecliptic; and that the arc E M, if we want to ascertain the solar ecliptic limits, beyond which there can be no eclipse, will be 1° 34′ 23″.2; or if we want to ascertain the limit within which there must be an eclipse, it will be 1° 23′ 15″. The solar ecliptic limit, as computed on the first supposition, is 17° 21′ 27″; as computed from the second it would be 15° 14′ 47″: and in this case, as in that of lunar eclipses before, it is plain that these distances must be measured on both sides of the node, to find the whole space within which an eclipse can or must happen. Whenever therefore the sun is more than 17° 21′ 27″ from the node, there can be no solar eclipse; when he is between that distance and 15° 14′ 47″ from it, there may; when he is at or within 15° 14′ 47″ from it, there must.

Such are the circumstances on which the possibility of a solar eclipse depends. It would be attended with much more difficulty than would be consistent with the object of this treatise to go into any details with respect to the computations necessary to ascertain the length and magnitude of solar eclipses †; but

* [As before, (p. 89. *n*.) R—P is a positive quantity, and the whole expression therefore is made up of positive quantities, each varying inversely as the distance.]

† It is however of importance to point out generally the manner in which the possibility of an eclipse, at any particular place on the earth's surface, is ascertained, and also its duration.

For this purpose, in fig. 26, let the circles, whose centres are E and S, represent the earth and sun as before, and let T be the point on the earth's surface, with respect to which the computations

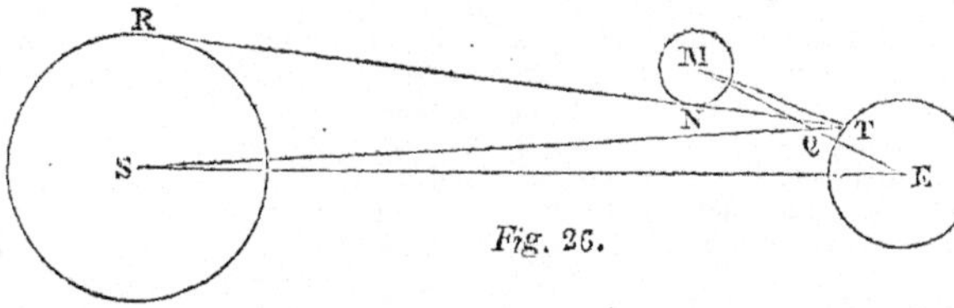

Fig. 26.

are to be made. Draw T R a tangent to the sun, and it is evident that the eclipse will commence at T, as soon as the moon, which is represented, as before, by the circle whose centre is M, comes in contact with the line T R. In this case, as before, the moon's distance from the line joining the earth and sun, or the angle $M E S = M T N + R T S + E M T - T S E = r + R + p' - P'$, if p' and P' represent the parallaxes of the moon and sun at the point T respectively, instead of the horizontal parallaxes. This angle being known, the situation of the moon at the time of the commencement of the eclipse may be ascertained. In the same manner, if we suppose a tangent drawn from T to the opposite side of the sun, the eclipse will be at an end at T when the moon has passed completely beyond, or, as the figure is drawn, below it; and her situation then will be computable in the same manner. Her position at each time being known, the arc which she describes in the interval is known also, and consequently the rate of her motion being known, the time which she takes to describe it, or the duration of the eclipse, is known.

We have no occasion to enter into any account of the manner in which these computations are made; they evidently depend ultimately on the apparent magnitudes of the sun and moon, quantities which may be observed, and the difference of their parallaxes.

a few more general considerations will be sufficient to illustrate the manner in which they vary, according to the different situations of different observers; and to show how materially their appearances may also vary, according to the different distances of the sun and moon.

For this purpose, let us refer again to *fig.* 21; but let us now suppose the circle, whose centre is E, to represent the moon instead of the earth. It is obvious that the moon's complete shadow will be represented, in this case, as the earth's was before, by the conical figure T P *t*, and that the line T N will represent the boundary of the penumbra of the moon, and there will of course be a similar boundary of penumbra at the same distance below the line S P. By the same considerations therefore, which we before used, at any point between the lines T P and P E the sun will be entirely, at any point between T P and T N he will be partially, hid by the moon. At any point therefore between T P and P E, there will be a total eclipse: at any point between T P and T N there will be a partial eclipse of the sun; for some part of his surface will be actually concealed from the observer, and that constitutes an eclipse. If we suppose the line S P prolonged beyond P, the central parts of the sun will be concealed from an observer situated in that prolongation; but it is evident that he will then see part of the sun above T, and below *t;* and as the figure may represent the section of the sun and moon made by any plane passing through the centres of both of them, the same result will follow in the case of every such plane, and there will therefore be a ring of light surrounding the moon. In this case, the ring will be bounded by two concentric circles, the centre of the sun and moon being in the same line as the observer; but there may be a ring when the observer is not accurately in the prolongation of the line joining these centres, if he be so little out of it, that he still is able to see part of the sun on every side of the moon, though more on one side than on the other. An eclipse taking place under these circumstances, is called *an annular eclipse*, from the Latin word *annulus*, a ring.

The actual magnitude of the sun and moon being invariable, the length of the moon's shadow depends on their distance, and is of course greatest when that distance is greatest, for then the boundaries of the shadow converge least rapidly; and least when it is least, for then they converge quickest. But this distance at the time of an eclipse is the excess of the sun's distance from the earth over that of the moon; for the moon is then between the sun and the earth: it is therefore greatest when the sun's distance is greatest, and the moon's least, or when the sun is in apogee, and the moon in perigee; and it is least when the sun is at his least distance, and the moon at her greatest, or the sun in perigee and the moon in apogee. The extreme values of the length of the shadow are 59.73 times the earth's radius, and 57.76 times the earth's radius. In the former case, when the moon is in perigee, her distance from the centre of the earth is only 55.9 of the earth's radius; the earth therefore is nearer to her than the extremity of her complete shadow, and consequently there may be some part of the earth within the shadow, and at that part there will be a total eclipse. In the latter case, when the moon is in apogee, her distance from the earth is 63.87 times the earth's radius. The earth therefore is at a distance greater than the extremity of the shadow, and therefore there can be no total eclipse, although there may be an annular one. It is evident that corresponding calculations must be made for intermediate distances of the sun and moon, to know exactly when a total eclipse is possible, and when it is not so. It is however clear that whenever the apparent diameter of the sun exceeds that of the moon, there can be no total eclipse; when it is less there can be no annular eclipse.

Again, taking the extreme case when the sun is in apogee and the moon is in perigee, the length of the moon's shadow is only 59.73 times the earth's radius; and as the distance of their centres is 55.9 of the same quantity, the distance of the earth's centre from the extremity of the shadow will be only 3.83 times the earth's radius, or the distance of the part of the earth most distant from that extremity will be 4.83 times the earth's radius. The radius of the shadow at that point therefore (which will be to the radius of the moon in the proportion of their distances from the point of the shadow) will be $\frac{4.83}{59.73}$ moon's radius, or less than $\frac{1}{12}$ of the moon's radius, and as the moon's radius is only about $\frac{3}{11}$ of the earth's, this quantity will be less than $\frac{1}{44}$ of the earth's radius. Its more accurate value is $\frac{1}{45.34342}$ or .022054 of the radius. Therefore the portion of the earth's surface at which the eclipse is total at the same moment cannot exceed a circle whose radius is .022054 of that of the earth, or about eighty-eight miles; if the part of the earth where the shadow falls be supposed to be that nearest to the moon, that is, when the axis of the shadow prolonged passes through the centre of the earth. But if the centre of the earth be not directly in the line joining the centres of the sun and moon, the point where this shadow meets the earth will be more remote from the moon than we have supposed, and the radius of the shadow there less. In this case, however, the section of the shadow by the surface of the earth will be oblique to the axis of the shadow; and the quantity of the earth's surface, from which the sun is completely hidden, will consequently also be increased. It is evident, however, that in this case it is only a small part of the earth's surface at which the eclipse can be total.

The whole region of the earth, however, where the eclipse may be total is greater than we have hitherto stated it to be. The moon is in motion, and the centre of her shadow consequently moves also; if it falls therefore upon the earth's surface, it will move along a certain line on the earth's surface, and the complete shadow will move along a zone or belt of that surface, and the eclipse will be total, though at different times, to the inhabitants of different parts within this belt. The whole of this belt however, for the reason above stated, can evidently be but a small portion of the entire surface of the earth.

We see then that the time at which a total eclipse takes place will be different at the different places where it is observable. Its duration also will be different, as the centre of the shadow, or only a more remote part of it, passes over the spot. It can never continue total, at any particular place, for more than $7^m 38^s$; nor be annular for more than $12^m 24^s$ *. Similar results may be deduced with respect to every other sort of eclipse, annular or partial. In the latter case, the place of observation is only within the penumbra of the moon, and a greater or smaller part of the sun will be hidden, as the place of observation is less or more remote from the centre of the shadow; the quantity of obscuration will therefore be different for different places.

Since so many circumstances are necessary to produce a total or an annular eclipse, their occurrence at any particular place is necessarily very rare. The last total eclipse which has taken place in London was in April, 1715: and there had not been a former one visible there since 1140. Dr. Halley, in the 29th volume of Philosophical Transactions, has given a curious and minute description of the eclipse of 1715; and there is also a valuable account, by Maclaurin, in the same work, of the phenomena of an annular eclipse, which he observed at Edinburgh on February 18, 1737. The darkness in each case was considerable; in the former so great, that the planets Jupiter, Mercury, and Venus, and the fixed stars Capella and Aldebaran, appeared: in the latter, only Venus was distinctly observed, though some observers thought they perceived other stars, and one that he recognized stars in the Great Bear. Maclaurin was watching the progress of the eclipse, and did not make any of these observations himself.

IV. Having thus explained generally the nature both of solar and lunar eclipses, we proceed to some considerations which affect both alike.

As the solar ecliptic limits exceed the lunar, there must be more eclipses of the sun than of the moon. But every eclipse of the moon is visible whenever the moon is above the hori zon at the time when it takes place; that

* The greatest possible duration of an annular eclipse necessarily exceeds that of a total eclipse. In the former case, the sun's apparent diameter is the greatest (32′ 35″), and the moon's the least possible (29′ 22″). The difference, or 3′ 13″, therefore, is the arc through which the moon may move while the eclipse continues annular: a quantity, however, which ought to be a little diminished, to allow the luminous ring to be distinctly formed all round the moon. The greatest duration of a total eclipse is when the sun's apparent diameter is the least (31′ 30″), and the moon's the greatest possible (33′ 31″). The difference, therefore, or 2′ 1″, is the arc through which the moon may move while the eclipse continues total. The duration of the greatest annular therefore exceeds that of the greatest total eclipse: and the more, because the moon in the former case being at the greatest, and in the latter at her least distance, her angular velocity will be less in the former than in the latter case.

is, to half the surface of the whole earth; and as she is above the horizon of each particular place as long, during the year, as she is below it, or very nearly so, she will everywhere, upon an average, be in any particular part of her orbit as often while above as while below the horizon, and thus half her eclipses, on an average, will be visible to an observer wheresoever situated. In the case of the sun, however, we have seen that it may very well happen, that there may be an eclipse, yet that it may not be visible at many places where he is above the horizon. Thus, taking the figure 25, at the point T there is that sort of eclipse called an appulse; and the sun is then on the horizon. He is above the horizon everywhere from T to A, and again beyond A to the point corresponding to T, or the other side: yet, to none of these places is he eclipsed. The same conclusions apply in different degrees to different cases of eclipses; and the consequence is, not only that half the eclipses of the sun, on an average, take place while he is below the horizon of any particular place; but that, of those which take place while he is above it, many are not visible there. Hence, though the whole number of solar eclipses exceeds that of lunar, the number visible at any particular place falls short of it.

The number of eclipses which take place in any particular year, and still more the number visible at any particular place cannot be accurately ascertained, except by calculation, for the particular year and place. But we can tell, by very simple considerations, what is the greatest, and what the least number of eclipses, which can happen in *any* year, and of what nature such eclipses must be.

It is plain that there must be at least one solar eclipse when the sun is near the node. The whole distance within which there must be one is 15° 14′ 47″ on each side of the node, or 30° 29′ 34″ in all, and the sun is more than a lunar month moving through this space; and consequently there must be one new moon, and may be two, while he is within 15° 14′ 47″ of the node, and thus there must be one solar eclipse, and there may be two. There need not be any lunar eclipse there, for the lunar ecliptic limits within which there can be an eclipse, being only 11° 25′ 40″ on each side of the node, or 22° 51′ 20″ in all, the sun is less than a month in moving through that space, and there need not, therefore, be any full moon while he is within 11° 25′ 40″ of his node: and if there is not, there can be no lunar eclipse, for the centre of the shadow is always exactly opposite the sun, and therefore at the same distance from one node that he is from the other. In the same manner there never can be two lunar eclipses near the same node, for if the moon is eclipsed at one full moon, the centre of the shadow cannot then be more than 11° 25′ 40″ either before or behind the node: at the succeeding or preceding full moon, the sun, and consequently the centre of the shadow, will have been 29° 6′ from this position, (for that is the space through which, at his mean rate of motion, he moves in the synodic period of the moon,) and will therefore be at least 17° 40′ 20″ from the node, or beyond the lunar ecliptic limit, and no second lunar eclipse therefore will take place. In the same manner, there cannot be three solar eclipses near the same node: for the first cannot take place till he is within 17° 21′ 27″ of the node, and as, before the next new moon, he must have moved through about 29° 6′, the second cannot take place till he is at least 11° 44′ 33″ beyond it; and the third new moon will not be till he is 29° 6′ beyond this place, or far beyond the solar ecliptic limits. There will therefore be no third solar eclipse. And of course, if on the occurrence of the first solar eclipse, he be nearer the node than we have supposed him, or beyond it, these conclusions are only strengthened. Again, when there are two solar eclipses near the node, there must be a lunar eclipse between them: for as the solar eclipse cannot take place unless the sun be within 17° 21′ 27″ of one node, at the full moon preceding or following it the sun will be 14° 33′ from that point, or within 2° 48′ 27″ of the node, and of course the centre of the shadow at the same distance from the opposite node, far within the lunar ecliptic limits: or there will necessarily be a lunar eclipse. In the same manner, if there be a lunar eclipse at the node, there must be at least one solar eclipse there also: for at the lunar eclipse, the moon must be within 11° 25′ 40″ of the node, and the sun within the same distance of the opposite node, and therefore, at the preceding or succeeding new moon, the sun will be 14° 33′ from that point, and consequently not more than that distance from the node, even if the lunar eclipse were in the node exactly. There never then can be more than three eclipses, or fewer than one, when either the sun or the centre of the shadow is near a given node: and if three, two must be solar: if one, it must be solar; if two, one must be solar, and the other lunar. We have spoken here only of the mean motion of the sun, but his extreme motions do not differ sufficiently from it to affect the truth of these conclusions. If however these results be true at one node, they must also apply to the other. Now the nodes, as we have seen, have a motion backwards upon the ecliptic of about 19° 21′ in every year: and the sun therefore having been in a given position with respect to one node, will be in the same position with respect to the other, not when he has gone through 180°, but through this quantity diminished by nearly half the annual retrogression of the nodes, or, taking its mean value, when he has gone through about 171°*; and in the space of six synodic revolutions of the moon, taking the mean values of the sun's motion and the synodic period, he goes through about 171° 18′. These values, in each case, may vary materially, but it is evident that they may easily happen so to correspond, that the sun during six synodic periods of the moon may have moved very nearly through that space of his orbit which was necessary to bring him into the same position with respect to one node, that at the beginning of them he occupied with respect to the other. In this case therefore, the eclipses will take place at each period, very nearly in the same order, and to the same amount; and there may, therefore, be three eclipses at each period, or there may be only one. Finally, the node in the course of the year moves through upwards of 19°: if therefore we suppose the first solar eclipse of such a series of three eclipses as we have supposed to happen in each case, to take place just at the beginning of the year, the node at the beginning of the year will not be more than 17°.21′ 27″ beyond the sun's place then; and as the sun's place at the beginning of the next year may be considered as the same, but the node will have retrograded 19° 21′, the node at the end of the year may be about 2° before the sun's place. Now twelve synodic periods are about 354⅓ days, and at the end of these the sun will be less than 11° short of his place at the beginning of the year, and consequently less than 9° from the node, or within the solar ecliptic limits: and it having been new moon at the beginning of the year, it will be new moon again at the end of the twelve synodic periods, and there will therefore be an eclipse of the sun. At the end of half a synodic period more, there will be a full moon, and an eclipse of the moon, but that will be more than 14½ days later, or more than 369 days from the first, and, consequently, not within the same year: and here, though the excess of 369 days above a year is small, the inequalities of the sun's motion and the synodic period will not be sufficient to bring this lunar eclipse within the year, for the sun's motion has all its values within the year, so that he will arrive at this point at the same time as if his motion were uniform.

The greatest number of eclipses then which can take place in a year, is seven: five solar, and two lunar. The least number is two, both of which will be solar. The greatest number of lunar eclipses which can take place in a year is three: for if a lunar eclipse takes place at the very beginning of the year, another may do so when the shadow comes near the next node: besides this, as the node cannot be more than 11° 25′ 40″ beyond the centre of the shadow, at the time of the first eclipse, or the beginning of the year, at the beginning of the following year it will, in the same manner as before, be 7° 55′ 20″ before it: at the end therefore of twelve synodic revolutions, the centre of the shadow, which will be about 11° from its former place, will be only about 4° from the node, and there will be a lunar eclipse at the end of the year, making three during its continuance.

The different phenomena of eclipses depending on the different situations of the nodes with respect to the sun, at the time of new and full moon, they must resemble each other whenever their situations are the same. Now the sun

* [The sun goes through 360° in a year, the node retrogrades through 19° 21′, or 19°.35 in the same time. Their mean motions therefore are in this proportion, or if the sun in a given time moves through x degrees, the node will move through $\frac{19.35}{360}x$. But when the sun returns to the same node, the sum of these two quantities (as the node has retrograded to meet him) makes up the whole circumference, or $x + \frac{19.35}{360}x = 360$, or $x = \frac{129600}{379.35} = 341°.63701$, or nearly 342°. This, therefore, is the angle described by the sun before he returns to the same node: the half of it, 171°, will be the angle described by him before he reaches the opposite node.]

returns to the same position with respect to the node sooner than the close of his own revolution, because the node has a retrograde motion on the ecliptic. The time in which he does so is $346^d\ 14^h\ 52^m\ 16^s.032$, or 346.61963 days: but this does not nearly correspond with the end of any exact number of synodic periods; for as each of these is of $29^d\ 12^h\ 44^m\ 2^s.8$, or of 29.530588 days, eleven of these periods make only 324.836468 days, and twelve of them make 354.367056 days, each of these quantities widely differing from that required to make the eclipses recur as at first. We may however find periods which will correspond. Thus 19 times 346.61963 days are 6585.77297 days, and at the end of this period the sun will be in the same place with respect to the node as at its beginning; and 223 times 29.530588 days are 6585.321124 days, a quantity differing by less than half a day or 12 hours from the other: at this latter period therefore the sun is very nearly in the same place as before with respect to the node, and accurately so with respect to the moon. From this period therefore eclipses will recur very nearly in the same order as before, yet not so accurately as to dispense with the trouble of computation: for the irregularities of the motion of both sun and moon make the positions, which are deduced from the mean values of the motion of the node and the synodic periods, inaccurate for any particular case; and besides this, the situations at the beginning and end of the period in question are not accurately and identically the same. This period of 223 months was early discovered by the astronomers of Chaldæa, and used by them for the purpose of foretelling eclipses.

Still, for general purposes, we may say that the same series of eclipses recurs in a little more than 18 years, for 6585 days exceed that time by 10 or 11 days, as there happen to be five or four leap years among the number. Even then however the same eclipses will not be visible at the same places as before. The complete period is of 6585.321124 days, or of $6585^d\ 7^h\ 42\frac{1}{2}^m$ nearly. The sun therefore will be in a different position with respect to the observer at the end from what he was at the beginning of the period; and the moon being either in conjunction with or in opposition to him will be so also. Even in the case of lunar eclipses therefore the moon will not be above the horizon of all the same places as before, and therefore they will not be visible at all the same places: and in the case of solar eclipses, besides this cause of apparent difference, the sun and moon, although similarly situated with respect to the centre of the earth, or to the earth in general, will be differently situated with respect to the particular point where the observer is placed; and an eclipse therefore which recurs in its regular order, and may be considered as identical in its circumstances with respect to the earth in general, will be of quite a different appearance to the observer at a given place in the two instances. Thus a total eclipse having been observed in London in April 1715, there must have been eclipses corresponding to it in 1733, 1751, 1769, 1787, 1805, 1823; but none of these have been visible as total eclipses in London*.

* It is worth while here to observe also that there must be a similar recurrence of the phenomena of occultations of the fixed stars by the moon. Her nodes continually retrograding, her course in each successive month will be through a part of the heavens a little different from the last: in different instances therefore her apparent place will coincide with that of different stars, between which and the earth she will pass, and there will be, according to the account already given of that phenomenon (page 86) *an occultation* of those stars. The time and place at which these phenomena are visible depend on considerations rather simpler than those relating to a solar eclipse; because, although the principles of the computation are very nearly the same, yet, the stars being motionless, we have no allowance to make for the motion of the body concealed, as in the case of the sun; and as they are too distant to have any observable parallax, we are also enabled to leave that element out of the calculation. These circumstances therefore depend merely on the moon's apparent diameter, parallax, place, and rate of motion. It is not necessary here to enter into any further detail of them.

It is plain that in this case, as well as in that of eclipses, these phenomena must succeed each other nearly in the same order, after a long interval. The stars themselves however, having no motion, the period in this case is more simply determined. The nodes complete their revolutions in 6793.42118 days, and at the end of this time the moon's orbit must coincide with its place at the beginning of it. There must therefore be very nearly the same succession of occultations. They will not however exactly correspond, because her situation with respect to the apogee or perigee of her orbit will not be the same in both cases; and consequently her distance from the earth being different, the elements involved in the computation of the occultation, her apparent diameter, parallax, and rate of motion, will differ also. Besides this, even if exactly the same phenomena were reproduced, they would appear differently, as in the case of solar eclipses, at the same place on the earth's surface; for the moon would be differently situated with respect to any particular spot on the earth's surface, although similarly with respect to its centre. These phenomena therefore require to be accurately calculated in each case, although a register of those which have hitherto taken place may enable us very nearly to tell what stars will be hidden, and when, from some part of the earth.

The maps of the stars, which have been published

Where the appearances themselves are so various, they might well continue long misunderstood. It requires very refined knowledge to foretel them; and, if unexpected, the disappearance for a time of one of the great luminaries of the heavens might not unnaturally occasion alarm among uneducated men. And accordingly we find in early historians instances of great terror thus produced. Eclipses were considered as omens of evil, and on one memorable occasion a battle between the Medes and Lydians is said to have been broken off, and a peace concluded, in consequence of an eclipse, which Thales, a philosopher of the time, had been enabled to compute*. We see, however, that in reality these remarkable phenomena are only the natural and necessary consequences of that uniform system of motion which we have described; that, however varied and complicated their appearances may be, they flow just as necessarily and directly from these simple laws as the common changes of the month and year: and that an eclipse is a phenomenon no more strange in its nature, or portentous in its occurrence, than the new or the full moon, or any other ordinary appearance which no one ever thought of making an object of alarm. In the present improved state of science, indeed, the very eclipses which, when they occurred, spread an ignorant terror among mankind, have been made subservient to the history of the times when they took place: for, by a more laborious application of the same principles which enable us now to compute the occurrence, the duration, and the amount of eclipses about to happen here, we can ascertain the same points with respect to those which did actually happen elsewhere long ago. For instance, the very day on which the great battle of Gaugamela or Arbela* was fought has been determined from its happening 11 days after an eclipse, the period of which has been computed; and the date of the battle already mentioned between the Medes and Lydians, of which even the year was not known, has been similarly investigated, for there was only one eclipse about that time which could be total in the part of Asia where it was fought. It is one of the most remarkable triumphs of science that it has been able to extract from phenomena so complicated and perplexing, the rules and principles on which they proceed, and from them to ascertain the dates of events almost lost in the confusion of early history, and removed by thousands of years from the time at which they have been investigated.

We have thus explained the main circumstances of the moon's motions, and may now proceed to the consideration of those presented by the remaining celestial bodies. In quitting the subject of the moon however for the present, it is of importance to remark that her motions are much more complicated than we have hitherto represented them to be. There are many apparent irregularities, which require corrections to be introduced into our computation, and no account of her phenomena can be considered accurate and complete without introducing and explaining these. Many of them are very small in amount, others are considerable; but at present it would only perplex the reader to enter into any detail of them. Hereafter when we explain the causes from which they proceed, we shall be better able to point out the manner in which some of them arise. It will be foreign to the purpose of this treatise, even then to attempt to explain them all, or even to give a complete list of them. Many of them cannot be explained except by means of the most laborious mathematical investigation, by which alone some of them have ever been discovered; and as tables are published for the purpose of enabling the sailor or the astronomer readily to compute the moon's place at any assigned period, in which all these corrections are included, it is to those elaborate works that every one will have recourse for purposes of actual investigation, in which, and in which only, these minute corrections have much importance.

by the Society for the Diffusion of Useful Knowledge, afford great facilities for discovering whether an occultation of any given star is likely to take place. For this purpose, it is only necessary to lay down two *places* of the moon in any given revolution from the Nautical Almanac upon one of the maps; and since the moon's path during a short time differs imperceptibly from a great circle, by joining these places *by a straight line* (for in these maps the projection of every great circle is a straight line), the path of the moon may be represented. If therefore this line nearly intersect the star, it may be inferred that the star must be occulted somewhere. It will be recollected that parallax depresses the moon; if therefore the line is rather below the star, it may be concluded that an occultation is not possible; if, on the contrary, the line is just above the star, in all probability it will take place. The fact, and the particulars of its occurrence, must then be ascertained by accurate calculations. It is unnecessary to take the effect of refraction into consideration, for that affects the moon and star equally.

* Herod., i. 74.

* Clinton, Fast. Hell. ann. 331. Hist. of Greece, 146.

Chap. IV.

Section I.—*Of the Planets generally—Distinction of superior and inferior Planets.*

We now enter upon the consideration of the motion of another set of bodies, which present some of the most remarkable appearances, and lead to the most remarkable conclusions of any which the heavens offer to our consideration. These are *the planets*. They appear in the heavens as stars, only distinguished from the rest, in the first instance, by the greater brightness and apparent magnitude of most of them, and by shining with a steady uniform lustre, somewhat like that of points upon the moon's surface, instead of the twinkling light of the stars in general. These are circumstances of distinction; but it is when we proceed to more minute inquiry,—when we examine their appearances through a telescope, and register their situations in the heavens at different times, that we discover ourselves to be regarding a class of objects widely different from any of which we have hitherto treated.

The peculiar appearances presented by each will be stated separately hereafter; but we may say at once, that some of them exhibit different phases; two, indeed, Venus and Mercury, presenting nearly all the phases of the moon; one other, Mars, sometimes very observably gibbous; the remaining planets, from circumstances hereafter to be explained, hardly perceptibly so. Those which present these phases, always turn their enlightened part towards the sun; and we conclude, therefore, with respect to them, in the same manner as in the case of the moon, that they shine only by the reflected light of the sun. We also find that their enlightened parts are bounded, like those of the moon, by elliptical and circular lines; and we thence infer, that, like her, they are nearly spherical. We shall hereafter see cause to form the same conclusions with respect even to those which have no perceptible phases.

Again, we are unable, by the most powerful instruments, to discover, at least with any certainty, that the brightest of the fixed stars subtend any assignable angle at the earth; we cannot, therefore, affix any value to their apparent diameter; but this is otherwise with respect to almost all the planets, their apparent diameters are easily observable, the greatest being not less than 61″, and the least which has been ascertained, not less than 4″*. A more important observation is, that these apparent diameters are found to vary in the same planet. The distance of the same planet from the earth, consequently, is variable also; and if we observe its direction from the earth at the time when these proportional distances are ascertained, we shall know, as in the former cases, what is the course it appears to describe with respect to the earth.

When we proceed, however, to make these latter observations, we shall not find the motions of the planets to be of that simple nature which those of the sun and moon, in their general outline, have appeared to be. They trace out no regular curve, as a circle or an ellipse; and, what is of still more importance, their radius vector (if we suppose a line drawn from them to the earth to be so called), does not describe equal areas in equal times. There are, however, some circumstances common to them all, which may be mentioned at once before we proceed to the more particular consideration of each. They all appear, although in different periods, to have a motion from North to South, and from South to North; they all appear, although in different periods, to have a motion, generally speaking, from West to East, and, consequently, to return again to the same right ascension which, at some former period, they had. But though these appearances correspond, in this general statement, with the appearances presented by the sun and moon, there is this remarkable difference between them besides those already mentioned: that all the planets, during the period which they take to return to their original right ascension, appear to have for a time a retrograde motion,—that is to say, a motion in the opposite direction from that of the sun, or from East to West. This is less in duration, and in its average rapidity, than their direct motion from West to East, and it is by the superiority of this latter, that they are finally brought round to their original right ascension; but the retrograde motion exists in all.

Another circumstance of similarity which was supposed to exist among these planets, was this, that they never appeared beyond a certain distance from

* However, the diameters of the small planets recently discovered are so small as hardly to admit of measurement. Vide *History of Astronomy*, part IV. p. 110.

the ecliptic. The moon's orbit we have already seen is inclined to the ecliptic at an angle of not much more than 5°, and her distance from it can never, therefore, exceed that quantity. The planets were supposed to be never observed beyond about 7° from it also; and a space of 8° on each side of the ecliptic was, in in consequence, supposed to include within it all the *varying* appearances of the year, and this space received the name of the *zodiac*. Within this space the twelve constellations, according to which we have already stated the different divisions of the ecliptic to have been named, are situated; and they have, in consequence, received the name of *signs of the zodiac*. At the time, however, when the zodiac was thus marked out, all the planets had not been discovered, and now that others have been so, the observation itself is no longer true. The planets then known were five, and each received the name of one of the divinities of ancient fable: they were Mercury, Venus, Mars, Jupiter, and Saturn. Modern discovery has added five to the list: Vesta, Juno, Ceres, Pallas, and Uranus. These are all named in conformity with the same system; but Uranus is also often spoken of as Herschel, having been so named after Dr. Herschel, who discovered it, and also as the Georgian, or Georgium Sidus, having been discovered in England during the reign of George III. Uranus, like the planets earlier known, is never seen beyond the limits of the zodiac; the other four new planets are.

For convenience in registering their observations, astronomers have agreed to represent these bodies by certain symbols, as in the following table:—

Mercury . .	☿	Ceres . . .	⚳
Venus . . .	♀	Pallas . . .	⚴
Mars . . .	♂	Jupiter . .	♃
Vesta . . .	⚶	Saturn . .	♄
Juno. . . .	⚵	Uranus . .	♅

Besides these, symbols have also been chosen for the sun, the moon, and the earth, namely:—

The Sun . . ☉ The Moon. . ☽

The Earth . ⊕, or ♁.

There are two remarkable distinctions between the appearances of Mercury and Venus, and those of the other planets. The other planets are seen at all distances or elongations from the sun, Mercury never more than 29° from him, Venus never more than 48°; and, besides this, Mercury and Venus are seen to pass occasionally between the earth and the sun, while the other planets never do so. Hence these two planets are called *inferior planets*, the rest *superior planets*; and the detail and explanation of their phenomena will be somewhat different in the two cases.

SECTION II.—*Of the Phases and apparent Motions of the two inferior Planets.*

THE appearances presented by the two inferior planets are exactly of the same nature. We may, therefore, detail them together; only mentioning afterwards the numerical values which in each case are to be assigned to the different elements of which we treat.

Neither of these planets is ever seen beyond a certain elongation from the sun. Within this distance it appears successively at all different angles from the sun, except, indeed, that when the planet comes very near the latter body, it becomes invisible in the bright light which, even while below the horizon, the sun spreads around its apparent neighbourhood. In this manner it is lost to view for some time, but it reappears again on the opposite side from that on which it disappeared; it gradually is seen further and further from the sun, until it arrives at a considerable distance from him; it then continues for a short time apparently at the same distance from him, and then again approaches him continually, until it is lost in the brightness of his beams, and afterwards reappears on the opposite side of him. Being never seen beyond a certain distance from him, it is evident that the planet's motions have, either by necessity or accident, a certain relation to his, which brings it alternately before and after him.

The place of these planets is never very distant from the ecliptic, and they, consequently, when on one side of the sun, set later; when on the other, they rise earlier than he does; and, as their elongation is confined within the limits we have mentioned, they can never, unless where the pole is very much elevated above the horizon, be visible all night; but, on the contrary, only for a certain period either after sunset, or before sunrise, according as they are to the East or to the West of the sun. Mercury is not an object of sufficient magnitude to be much noticed; but Venus is the brightest luminary of the heavens, after the sun and moon; and is spoken of in the one case as the evening, in the other as the morning star.

The motions of these bodies with respect to the stars have already, in common with those of the superior planets, been generally described. The period during which the motion of these planets is retrograde, takes place, of course, while they are passing from the Eastern to the Western side of the sun. The period of this retrograde motion, however, is very considerably less than that during which the planet passes from its greatest elongation East of the sun, to the greatest elongation West of him, and the extent of the retrograde motion also falls far within that of the direct. For some time after the Eastern elongation has attained its greatest amount, the planet's motion among the stars continues direct, or in the same direction with that of the sun, but its angular velocity becomes less than that of the sun, or the sun approaches it, and the elongation begins to diminish; after this, the planet for a short time becomes apparently stationary among the fixed stars, and, of course, the sun's motion continuing, he continues to approach it; the planet's motion then becomes retrograde, and continues so till the planet has passed the sun, and arrived at the same distance beyond it on the Western side which it had on the Eastern when it began its retrograde motion, during which whole time the sun, by virtue of his own motion, continues to gain upon the planet, first overtaking, and then passing it; the planet then becomes again stationary, the sun, of course, which is now beyond it, continuing to recede from it; and then the planet, recommencing a slow direct motion, with an angular velocity inferior to that of the sun, the sun continues to increase its distance from it, until the angular velocity of the planet gradually, but continually increasing, it becomes equal to that of the sun, and the planet is then at its greatest elongation West of the sun. From this time the motion of the planet continues direct, and its angular velocity rises above that of the sun; the planet, therefore, gains upon the sun, and continues to do so with an increasing velocity until it overtakes it, and then, after passing it, with a diminishing velocity, until the planet again reaches its greatest eastern elongation, when the angular velocities are again equal, and the same train of appearances recommences.

When a planet is in the same direction with respect to the earth as the sun, it is said to be *in conjunction* with the sun. We have already seen that, in the case of an inferior planet, this happens twice; once, while the planet is between the sun and the earth, and while its apparent motion is retrograde, which is called the *inferior conjunction;* and once, while the sun is between the earth and the planet, and the apparent motion is direct, which is called the *superior conjunction.*

During the retrograde motion, the planets are nearer to the earth than during their direct motion, a fact which we ascertain by observing the apparent diameter of these objects; for it is of sufficient magnitude to be ascertained by the telescope, and is found very materially to differ at different times, and to be greater during the period of the retrograde, than during the period of the direct motion. It is also found, during the retrograde motion, to be greater as the planet approaches the line joining the sun and earth; and, during the direct motion, to be less in proportion as it approaches the same line.

Another, and yet more important observation which we are enabled to make with the telescope, is this: that these planets go through all the phases of the moon, or very nearly so, in a constant order. When the planets are about their greatest elongation from the sun, they appear of the same shapes as the half moon; when the elongation diminishes, from its greatest amount Eastward, the part visible becomes a crescent, whose breadth continually diminishes as the planet approaches the line joining the sun and earth; and again, after we lose sight of it in the rays of the sun, when it reappears on the Western side of that body, it does so as a very thin crescent, gradually increasing in breadth as it approaches its greatest elongation Westward, about which period it again appears as a semicircle. Beyond this, as the elongation again diminishes, the planet becomes gibbous, and continually more and more so, until, before we lose sight of it in the sun's rays again, it shines very nearly with a full circular face, as it does when it reappears on the Eastern side of the sun, and then becomes gibbous, continually diminishing the luminous surface which it exhibits, till it returns to its greatest elongation, and again becomes a semicircle.

The general sum of these observations

is this. From the time that the planet first appears on the Eastern side of the sun, till it again disappears in its rays on the same side, it has first a direct motion, then becomes stationary, and then has a retrograde motion; and during the same period it is first seen merely as a circular orb, and continually shows a less proportion of its disk, showing half of it when about its greatest elongation Eastward, and only a very thin crescent before it disappears. During this whole time, also, its apparent diameter continually increases. When it reappears on the Western side of the sun, its motion is at first retrograde; the planet then becomes stationary, and then moves with a direct motion. During this time its apparent diameter continually diminishes, and the part of its disk visible continually increases, from a very thin crescent at its reappearance, to a semicircle about the time of its greatest elongation, and to very nearly a complete circle before it disappears again. The phenomena on the Eastern and Western sides are exactly the same, but in a reversed order of succession.

We are next to inquire what results we may deduce from these phenomena. It is obvious that, exactly in the same manner as in the case of the moon, we may conclude that these planets shine by light reflected from the sun, and are nearly spherical. The illumined portion of the disk must, therefore, vary, as the versed sine of the exterior angle of elongation, which must be equal to the radius at the greatest elongation, when half the disk is visible, and must be greater than the radius during all that part of the planet's course during which it appears gibbous, and less whenever it appears as a crescent. At the greatest elongation, therefore, the exterior angle of elongation must be a right angle; greater whenever the planet is gibbous, and less whenever it is a crescent.

Now in *fig.* 27, let S represent the sun's place, E the earth's, and SEZ the greatest angle of elongation of the planet. If SY be a perpendicular from S on EZ, Y will obviously represent the planet's place when at its greatest elongation; for we know by observation that half the planet is then visible, and therefore that the exterior angle of elongation is a right angle; and the planet is at that time somewhere in the line EZ, and SY is the only perpendicular which can be drawn from S to that line. SYE being a right angle is greater than any other angle in the triangle SYE, and consequently SE, the side opposite to it, is greater than EY, or the planet is then nearer to the earth than the sun is. Again, let EW be any line drawn between ES, EZ; it is therefore a line on which the planet will be seen twice; for it is twice at every elongation except the greatest, and if the points *v* V be taken, so that the angle S *v* W is acute, and SVW obtuse, they may represent positions of the planet; for at one of its positions on any line, the planet appears as a crescent, and its exterior angle of elongation is consequently acute; at the other it appears gibbous, and its exterior angle of elongation must be obtuse. When it appears as a crescent, the angle S *v* W being less than a right angle, the angle S *v* E is greater than one; it is therefore the greatest angle of the triangle S *v* E, and SE the side opposite to it is greater than E *v*. When the planet, therefore, appears as a crescent, it is nearer to the earth than the sun is. Finally, taking the case of the planet's appearing gibbous, the angle S V W will be greater, and therefore S V E less than a right angle. Its value may be ascertained by the degree of gibbousness, and the

Fig. 27.

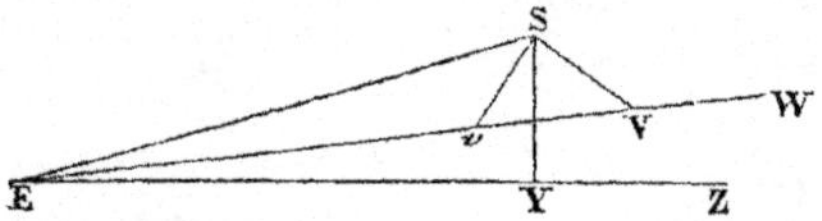

angle of elongation S E V may be observed; the value of the third angle E S V being the difference between the sum of these angles and 180° may thus be ascertained. If it is greater than E V S, E V is greater than E S, or the planet is more distant from the earth than the sun is. It is found by observations thus conducted, that it becomes so, not long after the planet has been seen at Y, and that it is continually more and more so, as the line

E W approaches to coincidence with E S*.

In addition, then, to the facts we have already stated, we find that during the course of phases which the planet goes through, it is during the whole period of its appearing as a crescent, nearer to the earth than the sun is; and that during part of the time of its appearing gibbous, it is farther from the earth than the sun is.

Combining this with the other phenomena of their appearance alternately on each side of the sun, it naturally follows that these planets must describe some sort of orbit round the sun: it remains, therefore, to examine whether this supposition will account for those retrograde and stationary appearances which we have detailed.

For this purpose we will first consider what would be the appearances presented if the planet were to describe any orbit round the sun, and the sun itself had no motion. We shall represent the orbit by a circle, not as considering that to be its precise shape, but as the most simple and familiar curve. We have already seen that the inferior planets occasionally pass between the earth and the sun: the radius of the circle which represents the orbit of one of them must therefore be less than the line representing the distance from the earth to the sun.

In *fig*. 28, then, let S represent the sun, E the earth, and Y_1 C Y_2 *c*, a circle described round S with a radius less than S E, the orbit of any inferior planet. Let EY_1, E Y_2, be tangents drawn from E to this circle, E v_1 V_1, E v_2 V_2, any two lines drawn through the orbit, one on each side of E S. Let also E S cut the orbit in the two points C *c*, and let us suppose the body to

Fig. 28.

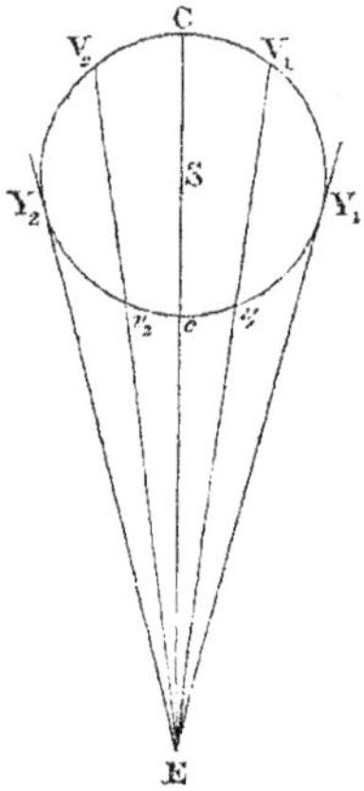

move in this direction; Y_1 C Y_2 *c*, and that this motion is from west to east round the sun.

We need not now trouble ourselves with considering, in any detail, the phases which would be presented by a body moving in the orbit Y_1 C Y_2 *c*; it is obvious that the exterior angle of elongation will continually increase from *c*, where it is nothing, to C where it is 180°, being 90° at Y_1, where the body is seen in the direction of a tangent to the orbit, and is consequently at its greatest elongation; and that it will again decrease in the corresponding manner from C to *c*, being again of the value of 90° at Y_2, the point where a tangent drawn from E meets the circle on that side of E S. The motion of the planet being from west to east, it will, throughout the semicircle, appear on the western side of the sun, throughout the semicircle C Y_2 *c* on the eastern*; and the portion of the disk visible will continually increase while it is in the former, and continually decrease while it is in the latter part of its course. These results correspond with those already mentioned as discovered by observation.

* The same results may be deduced by observation of the parallax of the planet at the different times in question; or more readily, perhaps, than in either way, by observation of its apparent diameter at different times. At any one of these, the angles of the triangle made by the sun, earth, and planet, being ascertained as above, the proportion of its sides may be known, and consequently the ratio of the planet's distance at that time from the earth to the sun's distance from the earth. The apparent diameter of the planet at any other time varies inversely as its distance; the apparent diameters, therefore, at different times being observed, the ratio of the corresponding distances will be known, and the proportion of one of these distances to the sun's distance having been previously ascertained, that of the others may be deduced from it. For instance, if in the case of the planet Venus, EY were found to be $\frac{5}{7}$ ES (which is very nearly its value), and the apparent diameter at any other point in its orbit were found to be half that at Y, the planet's distance from the earth would then be double its distance at Y, or it would be $\frac{10}{7}$ E S.

* The motion of the planet round the sun in the direction *c* v_1 Y_1 V_1 C being from West to East, it is obvious that Y_1 is West of V_1, and V_1 of C, with reference to the sun, and consequently that they are so also with respect to the earth at E, as the angles Y_1 E V_1, V_1 E *c*, are measured in the same direction with Y_1 S V_1, V_1 S C, although they are of inferior magnitude. As viewed from the earth, therefore, the semicircle *c* Y_1 C is on the West side of the sun.

Again, of all lines drawn from E to the circle, the least is E c, the greatest E C, and the lines are continually greater than each other as they recede from c and approach C; throughout the semicircle c Y_1 C, or the western semicircle, therefore the distances E c, E v_1, E Y_1, E V_1, continually increase, and the apparent diameters continually diminish; and throughout the semicircle C Y_2 c, the eastern semicircle, the distances continually diminish, and the apparent diameters continually increase. This again corresponds with the results of observation.

Lastly, it is evident that the motion of the planet being from west to east, while it moves through the arc Y_1 C Y_2, its apparent motion, which will be measured by the angle Y_1 E Y_2, will be in that direction also; or it will be *direct*. At the points Y_1, and Y_2, the motion in the orbit for a short period is in the direction of the tangent; the position of the body, therefore, will appear for a time unchanged, or the body will be *stationary*. Throughout the arc Y_2 c Y_1, the apparent motion of the body will be measured by the angle Y_2 E Y_1, measured in the opposite direction from Y_1 E Y_2, or from east to west; or the motion will then be *retrograde*.

These results have a considerable correspondence with those collected from observation. They give us direct and retrograde motions, and account for a stationary appearance. The motion, also, on the eastern side of the sun is at first direct from C to Y_2, then the body appears stationary, and then the motion becomes retrograde from Y_2 till it disappears near c; and on its re-appearance on the western side of c, the body still has a retrograde motion, till it becomes stationary at Y_1, and then has a direct motion between Y_1 and C. So far the results appear to correspond with those of observation; but they differ in two very material circumstances. On the supposition that we have made, the apparent direct and retrograde motions are measured by the same angle Y_1 E Y_2, measured in opposite directions; or they are equal, and the body appears stationary at Y_1, Y_2, the points of greatest elongation, and commences its retrograde motion as soon as it quits Y_2, and continues it till it arrives at Y_1. In reality, as we have already seen, the amount of retrograde is less than that of direct motion; and the body continues a direct motion after it passes Y_2, and recommences it before it arrives at Y_1, having its stationary points somewhere between those points, as at v_2 and v_1, and its retrograde motion confined to the arc between those points. Our supposition, therefore, though it gives us results corresponding to a considerable extent with those actually observed, does not at present represent them correctly. Indeed it is impossible that it should do so, for we have supposed the point S stationary, or that the motions of the planet are measured from S, the sun, considered as a fixed point; whereas S is itself in motion. Our results, therefore, drawn from *fig.* 22, cannot fully explain actual appearances until we take into account the motion of S.

The motion of the sun itself, as seen from the earth, is a direct motion; the motion given to the planet itself, therefore the supposition of its partaking of the sun's motion would be a direct motion also. This will at once appear from the inspection of *fig.* 29. Let A B C D represent any different bodies, to all of which an equal motion in the same direction is given; and let the amount and direction of this motion for a short time be represented by the equal and parallel lines A a, B b, C c, D d. Suppose, also, that E, a point in the line d D produced, is the situation of the observer; and that the apparent motion

Fig. 29.

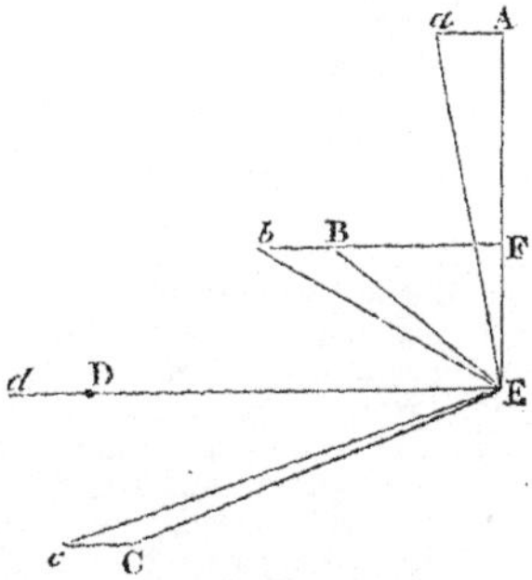

of the point A, which will be measured by the angle A E a, is a *direct* motion. It is obvious by mere inspection of the figure, that the angle B E b, which represents the apparent motion of the point B, will, if the point B be on the same side of the line E D that A is, be measured in the same direction that A E a is, or it will be a direct motion also; and that C E c, the apparent motion of C, a point on the opposite side of E D, is measured in the opposite direction, or it is a *retrograde* motion. The points D *d* are seen in the same line E D, and D therefore has no apparent motion at all. Now to apply these observations to the case of an inferior planet, we may suppose A to represent the sun, and A *a* its actual, or the angle A E *a* its apparent, motion. In this case, A *a* must be drawn perpendicular or very nearly so to A E, for the sun moves in an orbit very nearly circular, the earth being in the centre, and the direction of his motion is therefore very nearly perpendicular to his radius vector. For the sake of simplicity, we will suppose it accurately so. Now E D, the direction of the line in which there is no apparent motion, and beyond which the apparent motion becomes retrograde, is necessarily parallel to A *a*, the direction of the actual motion; in the case supposed, therefore, it is perpendicular to E A, for *a* A is so. But whenever the point B is on the same side of the line E D that A is, or whenever the angle A E B is less than A E D, that is, in the present case, less than a right angle, the apparent motion of B is direct, though it varies in amount according to its distance and direction. Now if we suppose B to represent a planet, A E B is the elongation, and in the case of an inferior planet, the elongation is never so great as 90°. If, therefore, an inferior planet partake of the sun's motion, the motion which it thus derives will appear to be direct in every position; and this apparent motion is to be combined with the apparent motion resulting from the planet's motion in its own orbit, before we can ascertain what the real apparent motion would be.

Now the apparent motion resulting from the planet's motion round the sun is sometimes direct, sometimes nothing, sometimes retrograde. In the first case, the real apparent motion will be the sum of two direct motions, and will of course be direct; in the second, it will be the direct motion occasioned by the sun's motion only; and in the last, it will be the difference between the direct motion occasioned by the sun's motion round the earth, and the retrograde motion occasioned by the planet's motion round the sun, and will therefore be direct or retrograde as the former or the latter of these two quantities is the greater. The direct motion might be so great as to exceed, in all cases, the utmost value of the retrograde, and the whole apparent motion would then always be direct. If, however, this be not so, it is yet evident that, for some time, the direct would exceed the retrograde motion; for the retrograde motion is evidently very little for some distance near the points Y_2, Y_1, in *fig.* 28, the points of greatest elongation, the curve at first departing very little from its tangent; and on the other hand, it is evident, also, that the retrograde motion is greatest at *c*, the point of inferior conjunction: for then the planet is nearest the earth, and its motion also being perpendicular to the line joining them produces the greatest effect. It may, therefore, well happen that for some time after the periods of greatest elongation, the direct may overcome the stationary motion, and the whole apparent motion, therefore, be direct; that they may then become equal, in which case, the planet would appear stationary; and that then the retrograde may exceed the direct motion, or the whole apparent motion become retrograde, and continue so until, after passing through inferior conjunction, the direct and retrograde motions again become equal, or the planet apparently stationary, before arriving at its greatest elongation; and then, the direct motion exceeding the retrograde, the whole apparent motion will again become direct till the planet arrives at its greatest elongation. At this period the retrograde motion first disappears, and then is converted into a direct one, and of course the whole apparent motion is direct until the same course of appearances recommences; and it is the course which we have already seen to obtain in nature.

We see, therefore, that the apparent motions of the inferior planets are not inconsistent with the supposition of their moving in an orbit round the sun. Their phases point strongly to the conclusion that they do so: and nothing could have prevented us from at once adopting that conclusion except the uncertainty, till the question were examined, whether it could be reconciled with their apparent motions.

It is, however, worth while to examine somewhat more minutely the law by which these apparent motions are regulated, for the purpose of deducing one or two results accurately corresponding with observation; and also for the sake of connecting the phenomena we are at present discussing, with some others to which we shall presently advert.

For this purpose we must again refer to *fig*. 29. The motion of the planet occasioned by its partaking of the sun's motion, being equal in all cases to that of the sun itself, is always the same, and may be represented by the line B b, of which the length will be known and constant, and the apparent motion by the angle B E b. Let b B be produced to meet A E in F: B F will, as we have seen before, be perpendicular to A E. Now, B b being very small, B E b must be very small also, and, consequently, the sine of B E b and B E b very nearly equal: and also B E and b E very nearly equal. But sin. B E b = sin. $b\,\mathrm{BE}\,\frac{\mathrm{B}b}{\mathrm{E}b}$ = sin F B E $\frac{\mathrm{B}b}{\mathrm{E}b}$ = cos. F E B $\frac{\mathrm{B}b}{\mathrm{E}b}$ (as F E B = 90°) = cos. E F B $\frac{\mathrm{B}b}{\mathrm{EB}}$ very nearly.

Therefore B E b varies as $\frac{\text{cos. FEB}}{\mathrm{EB}}$, or directly as the cosine of the elongation, and inversely as the distance.

Now, when the elongation corresponds on the opposite sides of E C, the distances are equal: thus in *fig*. 28, E Y_1 E Y_2, the two tangents are equal, and so are E v_1, E v_2, EV_1, EV_2, if the angles S E V_1, S E V_2 are so. The elongations, however, although equal, are measured in opposite directions, and therefore if one of them be considered as positive, the other will be negative. But the cosine of an arc is the same, whether the arc be positive or negative: the apparent motion B E b, therefore, which varies directly as the cosine of the elongation, and inversely as the distance, will be the same at equal elongations on each side of E A: for both its elements are equal in those cases. The direct motion, therefore, given to the planet by its accompanying the sun is the same on each side of the sun.

Again, the motion, whether direct or retrograde, produced by the planet's motion in its own orbit round the sun, is also the same on each side of the sun. In *fig*. 30 let S represent the sun, E the earth, and V W part of the orbit of an inferior planet, considered as circular. Let S E be joined, and let V W be two points in the orbit, having equal elongations on each side of S E. As before, therefore, V E, W E, the distances of the planet from the earth at the two positions are equal, and so also are the angles S V E, S W E, for the triangles E V S, E W S are equal in all respects. Now, let V v, W w represent

Fig. 30.

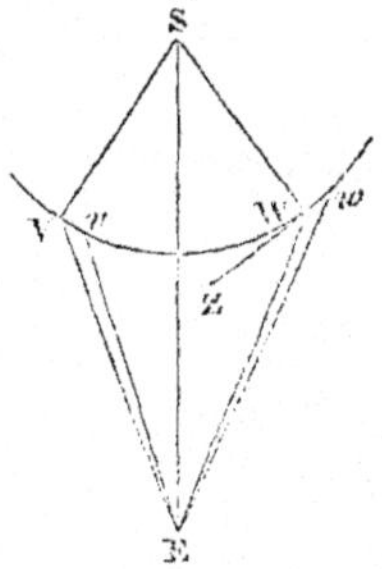

the portions of its orbit described by the planet in a very short time; these arcs will, if the motion is uniform, be equal. Besides this, for a very short space, the arc will coincide with the tangent, and may be considered as a straight line; if at W a tangent W Z be drawn in the opposite direction to the arc W w, W w may be considered as a continuation of Z W. Treating then the triangles V E v, W E w, as rectilinear,

$$\text{sin. VE}v = \text{sin. EV}v\,\frac{\mathrm{V}v}{\mathrm{E}v} = \text{sin EV}v\,\frac{\mathrm{V}v}{\mathrm{EV}} \text{ very nearly.}$$

$$\text{sin. WE}w = \text{sin. EW}w\,\frac{\mathrm{W}w}{\mathrm{E}w} = \text{sin. EWZ}\,\frac{\mathrm{W}w}{\mathrm{EW}} \text{ very nearly,}$$

(sin. E W Z = sin. E W w) and sin. V E v = sin W E w: for V v = W w, E V = E W: and E V v = E V S − S V v = E W S − S W Z = E W Z. The angles V E v, W E w, or the apparent motions, depending upon the motion of the planet round the sun, are therefore themselves equal in corresponding positions on different sides of the sun.

It is evident, therefore, that the whole apparent motions of the planet must correspond on each side of the sun: for these are always the sum, or the difference of the apparent motions derived from the planet's motion round the sun, and from the sun's motion round the earth: and as each of these is the same at equal elongations each way, their sum or difference, or the whole apparent motion, must be equal. We should therefore find the retrograde motion, before and after inferior conjunction, equal: the stationary points at the same elongation, the rate of motion at equal distances always equal—and so we very nearly do.

In these deductions, however, we have supposed the orbits circular, and the motion uniform: if either of these suppositions be inaccurate, our results will not accurately apply. With respect to the sun, we know them to be inaccurate, for his orbit is an ellipse, and his motion unequal, depending on the variation in his distance. Still, as the ellipse in which he moves does not differ much from a circle, and as the inequality of his motion is small, our conclusions, deduced on the erroneous suppositions we have adopted, will differ but little from the truth; and will sufficiently explain the manner in which his motion will affect the apparent motion of a planet revolving round him, though they will no longer be accurately correct as a representation of the amount to which it does so. In the same way, if we were to suppose the orbit of the planet some curve, differing not very considerably from a circle, the general effect of this motion in such an orbit would be sufficiently represented by the conclusions we have already deduced; though many particulars would cease to be accurately correct. For instance, the point of greatest elongation might not accurately correspond to that when exactly half the disk of the planet was visible; for the line drawn from the sun to that point might not be exactly perpendicular to the tangent then: the apparent motions at equal opposite elongations might not be exactly equal; for neither the distances, the amount of the actual motions, nor their direction, might accurately correspond. Still they would not differ much, and the particular conclusions we have come to would not be far from the truth: and the general principles by which direct and retrograde motion, and the existence and situation of the stationary points are accounted for, would evidently remain unaffected.

SECTION III. — *On the Phases and apparent Motions of the Superior Planets.*

THE phases of the superior planets present but little that is observable. Mars, indeed, is occasionally very perceptibly gibbous; but even he never assumes the appearance of a semicircle, or of a crescent: and the other superior planets are hardly seen to present less than their full face to an observer at the earth. This constitutes a marked difference between their appearances and those of the inferior planets.

Another marked difference consists in this: that, whereas the inferior planets are never seen beyond a certain distance from the sun, and occasionally pass between the sun and the earth, these planets are seen at all distances from the sun, and never pass between the sun and earth. They are, indeed, seen on the line joining the sun and the earth; but it is on the side opposite to the sun, or at the distance of 180° from him. At this time they are said to be in *opposition*. Their motions may also be ascertained, so as to enable us to compute that they are, at given times, in the line joining the earth and sun, on the same side as the sun, or *in conjunction*: but their distances from the earth are then greater than the sun's distance, or they are beyond him, just as the inferior planets are when in *superior conjunction*. The points of *opposition*, in the case of a superior planet, and of *inferior conjunction* in an inferior one, are in the same line with these, but on the opposite side of the sun: the points, therefore, of superior conjunction in the one case, and of conjunction in the other; and those of inferior conjunction in the one case, and of opposition in the other, seem to have a certain degree of correspondence.

They correspond also in another remarkable respect. The superior planets, as well as the inferior, appear sometimes to have a direct, sometimes a retrograde motion; and they also, like the others, are stationary for an interval between the two. In the case of the superior planets, the *conjunction* takes place during the period of direct motion, as we have already seen that the *superior conjunction* of an inferior planet does; and the *opposition* takes place during the period of retrograde motion, as we have already seen that the *inferior conjunction* of an inferior planet does. Besides this, we find the apparent diameters of these planets greatest in opposition, and continually diminishing as they approach conjunction, when they are least: another circumstance of the same kind of correspondence.

Taking the whole course of a superior planet from one opposition to another, we find its apparent motion at first retrograde, then at a certain elongation the planet becomes stationary, then its motion is direct, and continues so while it passes through conjunction, and till it arrives at very nearly the same elongation on the other side of the sun as it before had at its stationary point: then it again becomes stationary, and then its motion again becomes retrograde, till it is again in opposition. These appearances, as far as they go, exactly correspond with those of an inferior planet, substituting only the words *conjunction* for *superior conjunction*, and *opposition* for *inferior conjunction*. They are not, however, all the appearances of an inferior planet, for the remarkable circumstances of the points of greatest elongation, and of the recurrence of all the phases exhibited by the moon, are wanting.

The want of these makes it less easy, than in the case of an inferior planet, to divine the law of their motions; and, as these planets are seen at all angular distances from the sun, there is not the same obvious reason as before to conjecture that their motions depend upon his. Still there is a general similarity in the two sets of appearances, which makes it natural that we should inquire whether they cannot be explained in the same manner.

For this purpose, let us again take the case of a planet supposed to move in a circular orbit round the sun; and as before, let us omit for the present, for the sake of simplicity, all consideration of the sun's motion. The earth, being between the sun and the planet when the latter is in opposition, must then be at a point *within* the orbit of the planet. Let, then, in *fig*. 31, the circle represent the orbit of the planet; S, its centre, the sun, and E, the earth;

Fig. 31.

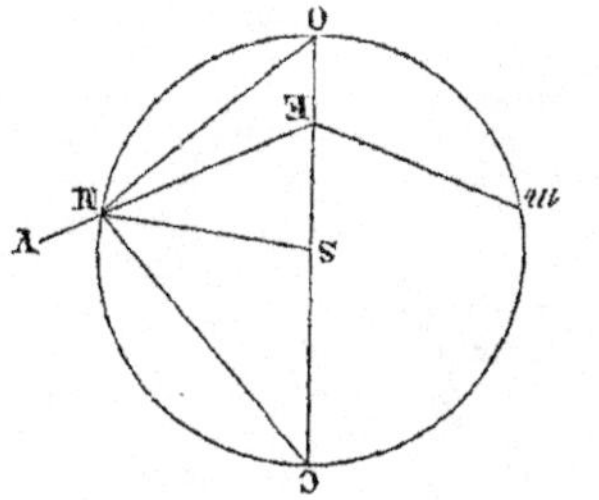

and draw the diameter C S E O through E; C, therefore, will evidently be the point of conjunction, and D the point of opposition.

Now E O is the shortest, and E C the longest line which can be drawn from E to the circumference of the circle; and all other lines so drawn are continually greater as they recede from E O and approach E C, and then again continually less as they recede from E C on the other side, and approach E O. The distance, therefore, of the planet from the earth is least in opposition, and continually increases thence till the planet is in conjunction, when it is greatest; and the apparent diameter is of course greatest in opposition, and continually less thence till the planet is in conjunction, when it is least. Again, if M m be points on opposite sides of E C, at equal distances from C, E M, E m, are equal; and of course the apparent diameter at M and m are so also. The same course of phenomena, therefore, succeed each other in a reverse order, at corresponding distances on the other side of C, and the apparent diameter continually decreases from opposition to conjunction. The phenomena, therefore, as in the case of inferior planets, correspond on this supposition with observation.

Supposing the planet itself to be spherical, the portion of the disk visible would vary (p. 76) as the versed sine of the exterior angle of elongation. At the points O and C, this angle is evidently 180°, or its versed sine is the whole diameter, and the full enlightened face of the planet would be turned towards the earth. At any intermediate point M, this will not be so: the sun and earth not being in the same line from the planet would have different parts of its surface turned towards them; and *phases* would be occasioned. The amount of these variations depends on the extent by which the angle S M Y differs from 180°, or it depends on the magnitude of the angle E M S (the angle subtended at the planet by the distance between the sun and earth, or the earth's elongation at the planet), for E M S and S M Y together make up 180°.

Now if we draw the lines C M, M O, the angle C M O, whenever the point M is taken between C and O, is the angle in a semicircle, and therefore a right angle. And it will obviously, in all cases, be composed of three parts, C M S, S M E, E M O; and, consequently, E M S, being one of these parts, will be less than a right angle. S M Y, therefore, will be in all cases greater than 90°, and the planet, though it may become gibbous, can never have half its face hidden, or become a crescent.

Again, $\sin. SME = \sin. SEM \frac{ES}{SM}$,

and as E S, S M are constant quantities, the sine of S M E, and consequently S M E itself, is greatest when sin. S E M is greatest, or when S E M is a right angle, or the planet *in quadratures*, as it is termed. The angle S M Y, therefore, is then least; and, consequently the visible part then differs most from a circle. And this, in observations of the planet Mars, on which alone they can be effectually made for this purpose, is found to be the case.

The sine of the angle S M E at this time $= \frac{SE}{SM}$; or it varies inversely as S M, the radius of the planet's orbit. The greater, therefore, this radius, the less is the extreme value of S M E, and the less, consequently, the utmost difference of the observed disk from the complete circle. We see, therefore, that the difference of phases must necessarily be less in the more distant planets; and, consequently, that their appearance with faces almost uniformly full, is the result which would follow from their revolving round the sun at distances far exceeding that of the sun from the earth.

It may be worth while to compute one or two numerical values to see how far this operates; versin. S M Y $= 1 +$ cos. S M E.

Let us suppose S M E $= 45^o$; in this case cos. S M E $= \frac{1}{\sqrt{2}} = .70\ 71$, &c., and the versin. S M Y $= 1.70\ 71$ &c. at its least value, or less than $\frac{3}{20}$ of the diameter are hidden when the planet is most gibbous. This, therefore, is the extreme value when $\frac{SE}{SM.} = \frac{1}{\sqrt{2}}$, or S E : S M :: 1 : 1.4142. In the case of Mars, the nearest of the superior planets, S M exceeds S E in a greater proportion than this, and the gibbousness will, therefore, even in this case, be less than that which we have deduced. In fact, in that instance, the greatest value of versin. S M Y is about 1.75, or seven-eighths of the diameter. Again, let us take the case of a planet revolving round the sun, at a distance five times as great as the sun's distance from the earth. The distance of Jupiter from the sun is rather more than this, and consequently his phases would be rather less considerable than those which will be deduced on this supposition.

In this case $\frac{SE}{SM} = \frac{1}{5}$, or .2, which therefore is the greatest possible value of sin. S M E, but cos. S M E $= \sqrt{1 - \text{sin.}^2 SME} = \sqrt{1 - .04} = \sqrt{.96}$ $= .9798$ nearly, and versin. S M Y $= 1 +$ cos. S M E $= 1.9798$ nearly, a quantity differing from the diameter by very little more than a hundredth part. The whole variation, therefore, of the phases of Jupiter would only be about a hundredth part, and he would, to all common observation, appear round at all times. The same remark would apply yet more strongly to planets still more distant from the sun, which we shall find to be the case of Saturn and Uranus. We see, therefore, that if the distances of the superior planets are sufficient, their want of phases corresponds with the supposition of their revolving round the sun, just as well as the variety of the phases presented by the inferior planets does.

It remains to see whether the same supposition will account for their direct and retrograde apparent motions. It is plain that the whole motion produced by the revolution of the planet round the sun, supposing the sun to be at rest, would be direct: for if, in *fig.* 31, the course of the planet's motion be O M C *m*, it is evident that the angle O E M will continually increase as the planet moves from O towards M and C, and that the motion round the sun, which is measured by the angle O S M, being direct, the apparent motion round the earth, which is measured by the angle O E M, will be direct also, that angle being measured in the same direction. The rate of this apparent direct motion will be different at different places; being greatest at O where the distance is least, and continually diminishing thence to C where the distance is greatest, and then increasing by corresponding degrees from C to O, being always equal at corresponding points on opposite sides, the distances and the angle made by the direction of the motion with the line joining the earth and planet being equal in those cases. This motion, then, being always direct, seems to furnish no explanation of the phenomena in question.

In the case, however, of the inferior planets, it was only by combining the motions produced by the revolution of the sun itself round the earth, with those produced by the revolution of the planet round the sun, that we arrived at a correct explanation of the phenomenon of the apparent motion. We proceed, therefore, to combine these motions now; and in doing so, we observe, that it appeared in p. 107 that the apparent motion produced by the motion of the sun round the earth would be direct, as long as a planet was on the same side of the earth as the sun was, or as long as its elongation did not exceed 90°, but would become retrograde as soon as ever it exceeded this quantity. In the case of an inferior planet it never does so; in the case of a superior planet it does. As soon, therefore, as this takes place, or whenever the planet is between its quadrature and its opposition, the apparent motion resulting from the sun's motion in its orbit becomes retrograde, and is opposed to that arising from the planet's motion round the sun. When this retrograde motion is equal to the direct motion, the planet will be stationary, when it exceeds it, the whole motion will be retrograde.

Now, as the amount of the apparent retrograde motion (p. 111) is equal at corresponding positions on each side of the point of opposition*; the same appearances of retrograde motion must, therefore, succeed each other in a reverse order, in the passing from opposition to quadrature, which had taken place in passing from the former quadrature to opposition. We have already seen that the direct motion produced by the planet's revolution round the sun is also equal, in corresponding situations, on both sides of the point of opposition. The whole apparent motion, therefore, which will always be the sum or difference of the apparent motions derived from these two causes, will be equal in corresponding positions, whether it be direct or retrograde; and the stationary points in the same manner, if there be any, will be between quadratures and opposition, and will be at equal distances from the point of opposition. Whether there will be any stationary points, or any retrograde motion, will depend, as in the former case, on the relative amounts of the absolute motions of the planet round the sun, and of the sun round the earth, and also on the relation of their respective distances: but it is obvious that, in this case also, these may be such as to account correctly for the apparent motions of the planets. The phenomena of their phases and apparent magnitudes we have already seen to correspond with the supposition of their motion round the sun. We see therefore, that all the appearances of the planets, whether superior or inferior, *may* be thus occasioned; and we proceed to inquire more minutely whether they are so.

Before entering, however, on that inquiry, it is necessary to observe, that the same degree of inaccuracy, and no more, will be introduced into the results deduced from superior planets, which we have already observed to exist in those obtained with respect to inferior ones, if their supposed orbits be not accurately circular, but differ little from circles. The general explanation will still be applicable, but the motions on each side of the point of opposition will not be precisely similar.

There is another remark also, which it may be worth while to subjoin. The whole apparent motion is combined of those produced by the revolution of the planet round the sun, and that of the sun round the earth. Of these, one is always direct, and one is alternately direct and retrograde; but it is remarkable that they change their characters as the planet is an inferior or a superior one; the motion of the planet round the sun producing an apparent motion alternately direct and retrograde in the case of an inferior planet, but always direct in that of a superior planet; and the motion of the sun round the earth producing an apparent motion, always direct in the case of an inferior, but alternately direct and retrograde in the case of a superior planet.

The whole apparent motion is retrograde, only when the retrograde part exceeds the direct part of the motion. In the case of an inferior planet the retrograde motion being then produced by the motion of the planet round the sun, this must be such that its effect to a spectator at the earth is greater than that of the sun's motion round the earth. On the other hand, in the case of a superior planet, the retrograde motion

* The amount of the retrograde motion at any particular point varies, as we have already seen, directly as the cosine of the elongation, and inversely as the distance (a fact which would by itself have sufficiently furnished us with the proof that it becomes retrograde when the elongation is greater than 90°, for then the cosine of the elongation becomes negative.) It therefore continually increases from quadratures to opposition, for the cosine of elongation is continually greater as the angle approaches 180°, and the distance continually diminishes as the body approaches O, the point of opposition. At equal distances on each side of the point of opposition, the retrograde motions are equal, as the cosines of the elongation are equal (for $\cos.(180° - A) = \cos.(180° + A)$) and the actual distances are equal also.

being produced by the motion of the sun round the earth, the motion of the planet round the sun must be such that its effect to a spectator at the earth may be less than that of the sun's motion.

We shall hereafter see that these various differences in the effects produced by the application of one principle under different circumstances, correspond accurately with a very simple and most important law.

Section IV.—*On the Motions of the Planets generally.*

The most important elements to be investigated with respect to the supposed orbits of the planets round the sun, are their figure, their magnitude, and the time which the planet takes in describing them, or, as it is termed, the periodic time of the planet. As our observations are only made at the earth, it is evident that there must be considerable difficulty in deducing from them all these elements, referred as they are to another point, situated at a vast distance, and itself in continual motion. It would be foreign to the purposes of this treatise to enter into the details of the necessary investigations; but it will be worth while to point out the possibility of conducting them, before we state the results to which they lead.

We can determine the apparent position of a planet at any instant by observations of its right ascension and declination; and knowing the sun's motion, we know the sun's position at the same time, although he may be below the horizon, and consequently not the subject of immediate observation; and consequently the angle between the apparent positions of the sun and planet at the time, or the angle made by the line joining the earth with each of them, may be ascertained. The distances, also, of the sun and of the planet from the earth at that time, may be ascertained by observation of their parallaxes; though a minute error in these observations is of so much importance that any estimate, so obtained, is liable to considerable inaccuracy. It may, however, be used as furnishing at least an approximate value; and the proportion which these distances bear to each other will be thus, to a certain degree, ascertained. If, therefore, in *fig.* 31, E represents the earth, S the sun, and M the planet, as before, the angle S E M may be deduced from observation of the apparent places of S and M, and the proportion of the sides E S, E M, from the proportion of their respective parallaxes. Now in every triangle, when three elements, one at least being a side of the triangle, are known, the rest may be computed. We have, therefore, the means of computing the value of the side S M, in terms of the side S E, and also the value of the angle E S M. We know the position of the line E S, or the apparent direction of the sun as seen from the earth; we know, therefore, the position of the line S E, or the apparent direction of the earth as seen from the sun, and knowing, also, the value of the angle E S M, and the plane in which it is measured (for this is the plane passing through E S and M), we know the exact direction of the point M as seen from S. We know, therefore, the position and distance of the point M with respect to S; or we ascertain one point in the orbit of the planet.

Some additional correctness may be given to these observations by a very simple consideration, which will also enable us to dispense in some cases with the observation of the parallax of the planet at all. The parallax varies inversely as the distances of the body; and so, for the same body, does the apparent diameter. The parallax, therefore, varies as the apparent diameter; and if several observations of the planet in different positions be made, and the apparent diameter and parallax ascertained in each, the relation of the parallax to the apparent diameter (which will be the same as the ratio of the real diameters of the earth and planet) will be ascertained by a comparison of these observations. One observation, if its accuracy could be relied upon, would be sufficient for this purpose; but the mean result of several is not likely to differ much from the truth, and its correctness will be the more to be relied upon in proportion as the individual observations more nearly approach to each other.

When this relation is once ascertained, we need not trouble ourselves further with the difficult and complicated observations necessary to ascertain the parallax. This element may be at once deduced from the apparent magnitude, which may be more easily observed, both because it is ascertained by a single observation instead of requiring the comparison of observations made at different times or places, and because in the planets whose parallax is smallest and most difficult of observation, the apparent magnitude greatly exceeds the parallax and its variations of course are greater also.

Considerations arising in a more advanced state of our knowledge, furnish much more accurate and ready methods of ascertaining the elements in question; but for the present it is sufficient to have explained the possibility of ascertaining them; and that sufficiently appears from the method above pointed out.

The same method may obviously be applied to the ascertainment of any number of points in the orbit of the planet; and consequently to the determination of the figure and magnitude of that orbit itself. Among the different results of such an investigation, we should at length discover one which would correspond in every particular to some former one; that is to say, we should find the planet again in the same position with respect to the sun in which we had ascertained it to be at the period of some previous observation. The interval between these two observations would, therefore, be the periodic time of the planet; and we thus see that we have the means of ascertaining all the elements in question, although subject to considerable inaccuracy from the unavoidable imperfections of observation.

The periodic time, however, is an element of the utmost importance, and there are other methods by which it may be found with so much more accuracy, and so much more ease, that it is worth while to indicate some of them, particularly as some of their circumstances correspond to other important results.

We have already stated that the planets are found at various distances north and south of the ecliptic. Their motions, therefore, are not in the plane of the ecliptic, for if they were, the earth being itself in that plane, they would always be seen somewhere in its intersection with the sphere of the heavens, or in the celestial ecliptic. Each planet, however, is found alternately above and below, or north and south, of the ecliptic. Its orbit, therefore, intersects that plane, and must have *nodes*, according to the explanation already given of that term. Again it is found by the observations which we have made that the sun is in the plane of the planet's orbit, and consequently a point in the intersection of that plane with the ecliptic, or in the line joining the nodes; and the planet passing round the sun, the nodes must be on opposite sides of that body, and must, therefore, with reference to that body, be 180° from each other. If, therefore, the motion of the planet round the sun were accurately circular and uniform, exactly half the orbit would be between each node, and the time which intervenes between the planet's appearance in each would be equal. If it is not so, the supposition either of circular, or of uniform, motion must be incorrect, or both may be so. If, however, the nodes themselves are fixed points, the whole time of the planet's passing from one node till it returns to the same again, will be the whole periodic time of the planet, with whatever inequality that time is divided at the other node; and if, therefore, this whole time can be ascertained, the periodic time is found also, if the node is a fixed point.

Now both the time at which the planet is in the node, and the position of the node itself, are very easily ascertained. The latitude and longitude of any body in the heavens may be deduced from observation; the latitude of the node, which is a point in the plane of the ecliptic, is nothing. To find its position, therefore, let us suppose observations of the planet made on the meridian on two successive days, on one of which it is north, and on the other south, of the ecliptic. Its positions at this time being ascertained, the arc of its orbit is ascertained also, and consequently the point at which it enters the ecliptic, or the position of the node; and as the motion of the planet during a single day may be considered as uniform, (for observation conclusively proves that its variation during that time must be very inconsiderable,) the exact time of its being in the node will be found also; for the interval between that time and the time of the first observation will bear the same proportion to the whole interval between the two observations that the distance of the node from the planet's place at the time of the first observation bears to the whole arc described by the planet in its orbit in the interval between the two observations.

The period, therefore, of the planet's being in the node, and the position of the node, may be ascertained. The same observations may be repeated when it is near the opposite node; and the exact time of its being there ascertained also; and again when it returns to the node at which we first observed it. It is found that the position of this node has undergone little or no alteration; the period which has elapsed, therefore, is the periodic time. It is also found, in most instances, that this time is not equally divided by the instant at which the planet was in the opposite node; the supposition, therefore, of uniform motion in a circle round the sun is inaccurate.

The method above explained is a very easy one of obtaining the length of a particular revolution. It may, however, happen, and it will be found to be the case, that this period is not always the same, though its variations are not considerable. If, however, we have the means of ascertaining the length of time occupied by many revolutions, the average length of one, or the *mean periodic time*, may be ascertained; and this may be done by means of early observations which have been registered, and are handed down to us.

There is another method, also, of ascertaining the periodic time which it is desirable to mention, though it is applicable only to the superior planets, whose periodic times, as we shall presently see, are longer than those of the inferior planets, and with respect to which, therefore, it is more important to have the means of comparing observations made at great distances of time, for the purpose of getting an average value. When the planet is in opposition, its direction, or its apparent place, as seen from the sun and from the earth, is the same. Referring to *fig.* 31, the planet when at O is seen both from S, the sun, and E the earth, along the same line S E O, and consequently in the same point in the heavens. If, therefore, two oppositions can be found at which the planet is seen at the earth in the same point of the heavens, it will be seen then at the sun also; and it will, therefore, be in the same relative position with respect to the sun, or it will have completed some exact number of revolutions. It is not likely, indeed, that the points where the planet is seen in opposition will exactly correspond; but this is not necessary. The mode of proceeding will be best explained by an example, in which, however, we will not take any real case, which might be embarrassed by considerable labour of computation, but merely illustrate the principle by easy numbers chosen for that purpose. Let us suppose, then, that by some of the former methods we had ascertained the periodic time of a planet to be somewhere about 3600 days; its mean motion, therefore, in its orbit would be 1° in 10 days. Let us, also, suppose that we had observations of different oppositions made at an interval of 720150 days from each other, and that the second opposition was at such a point of the heavens that the planet, at the time of this second opposition, would have 5° of its orbit to describe before it was exactly in the same position with respect to the sun that it occupied at the time of the first opposition. Our former observations enable us to say that it would be about fifty days describing these 5°; and consequently the whole period from the first observation to the planet's being in its original position with respect to the sun would be 720150+50 or 720200 days. Now this period must comprise an exact number of revolutions; it is not, however, divisible by 3600, and, consequently, that number cannot accurately express the *mean periodic time.* If, however, we divide 720200 by 3600, we find the quotient to exceed 200 by a small fraction only ($\frac{1}{18}$); we conclude, therefore, that 200 is the actual number of revolutions; and, consequently, that $\frac{720200}{200}$ or 3601 days is the *mean periodic time.* There are extant many early observations of the oppositions of the planets Jupiter, Saturn, and Mars; some of them considerably earlier than the Christian era; there are, therefore, ample opportunities, with respect to them, of making the computations explained above; and although the accuracy of these very early observations may not be very great, yet the error so produced, being divided among many revolutions, would be inconsiderable. For example, in the instance given, an error of 32″ in the observation would correspond to two days in time, divided among 200 revolutions, or the error in the mean periodic time would only be the $\frac{1}{100}$th of a day.

Without entering farther, at present, into the mode of ascertaining the important elements with which we have been concerned, we proceed to state the facts which have been discovered with respect to them. They were ascertained by the great astronomer Kepler, and are known by the name of Kepler's laws. They are three in number.

1. *The orbit in which each of the planets moves round the sun is an ellipse, of which the sun is in the focus.* The eccentricity of this ellipse is different in the different planets, but not great in any case. It does not, therefore, differ very considerably from a circle; and the explanation which we have given of the phenomena may apply to the case as it exists in nature.

2. *The areas described by the radius vector of the planet* (the line joining the planet to the sun), *are proportional to the times of describing them.* These two laws agree with those which we have already seen to obtain with respect to the motions of the sun and moon.

3. *The squares of the mean periodic times of different planets are to each other in the same proportion as the cubes of the axis major of their respective orbits.* For instance, if the axis major of Mercury's orbit be supposed to be 1, and that of Mars's orbit 4, the square of Mercury's periodic time will be to the square of Mars's periodic time as the cube of Mercury's distance to the cube of Mars's, or as 1 to 64; and the periodic times themselves in the proportion of the square roots of these quantities, or of 1 to 8. These numbers are not accurate, but they are not far from the relations subsisting between the motions of Mars and Mercury.

There are two very remarkable circumstances which require notice before we proceed further.

At the time when Kepler discovered these laws, the planets Uranus, Vesta, Ceres, Juno, and Pallas, were unknown. It is found, by observation, that they accurately follow the laws deduced by Kepler from comparison of the motions of the planets then known; their orbits are ellipses, the areas traced out by the radius vector of each are proportional to the times; and the periodic time of each is that corresponding according to the third law, with the magnitude of the major axis of its orbit. This furnishes very strong reason for believing that the laws are not the expression of a mere fortuitous coincidence, but the result of some fixed principle regulating these motions.

The periodic time of the sun round the earth is exactly that which, according to the third law, would be the periodic time of a planet moving round the sun in an ellipse whose major axis was equal to that of the ellipse which we suppose the sun to describe about the earth. This is, at least, an extraordinary coincidence, especially when we find no such relation existing with respect to the moon's motion round the earth, as compared with any of those which we have described. Hereafter we shall find this circumstance, among others, to lead to the most important conclusions.

We shall now revert, shortly, to the apparent motions of the planets as seen from the earth; for the purpose of showing that the third of these laws accounts for the phenomena of direct and retrograde motion, and makes it absolutely necessary that those phenomena should take place.

The square of the periodic time varies as the cube of the axis major of the orbit. If we suppose the orbits all similar, (and the difference in their figures is not sufficient to render the supposition materially untrue,) the circumference of the orbit will vary as the major axis. The mean linear motion in the orbit (or the actual motion, if it be supposed uniform) varies as the circumference divided by the time of describing it, which is the periodic time; or it varies directly as the axis major (for the circumference varies so), and inversely as the square root of the cube of the axis major (for the periodic time varies so); or, finally, it varies inversely as the square root of the axis major. The greater the axis major, the less is the linear motion or velocity.

We have already stated that the periodic time of the sun round the earth follows the same law as that of the planets round the sun. The inferior planets being always nearer the sun than he is to the earth; the axis major of the orbit of an inferior planet is less than that of his, and the linear motion of the planet, consequently, more rapid. A superior planet, on the contrary, has the axis major of its orbit greater than that of the sun's, and its own linear motion, consequently, less rapid than his.

Now at the period of inferior conjunction in an inferior planet, and of opposition in a superior one, the orbit of the planet, if it be supposed circular, is perpendicular to the line joining the planet to the earth; for this line, if produced, passes through the sun, the centre of the planet's orbit. The motion, therefore, of the planet in its orbit is, at this period, perpendicular to the line joining it to the earth, and produces the full effect of that amount of linear motion at that distance. Again, the motion of the planet, produced by the sun's motion in his orbit, is always in a direction parallel to the sun's motion; and if we suppose the sun's orbit also circular, the line of his motion, and, consequently, all lines parallel to it, will be perpendicular to the line joining him with the earth. But this line, or this line produced, passes through the points of inferior conjunction or opposition; the planet's motion thus produced, therefore, is perpendicular to the line joining the earth with these points, or with the planet when in them. At these points, therefore, this motion of the planet is perpendicular to the line joining it to the earth, and produces, like its motion in the orbit, the full effect of that amount of linear motion at that distance. The distance being the same for each, the one angular motion will exceed or fall short of the other, as the linear motion does so; and, consequently, the motion produced by the planet's motion in its orbit, will exceed that produced by the sun's motion round the earth in the case of an inferior planet; and will fall short of it in the case of a superior one. The planet's motion in its orbit produces retrograde motion in the case of an inferior planet; the sun's motion in its orbit produces retrograde motion in the case of a superior planet. The motion, therefore, which produces the appearance of retrograde motion, is always greater and more effective in opposition and inferior conjunction, than that which produces direct motion, and, consequently, there must necessarily, from the law regulating the periodic times, be retrograde motion at these seasons. These conclusions correspond with the remarks in page 113. For the sake of simplicity, they have been deduced on the supposition that the orbits are circular; but their deviations from the circular form, and from uniform motion, are not sufficient to render the conclusions untrue.

Another observation which we may now make, is this. The only line which can be drawn through the focus of an ellipse, so as to bisect the area of the curve, is the axis major. The line of the nodes passes through the sun, the focus of the ellipse; unless, therefore, it coincides with the axis major, it must divide the area of the orbit into unequal parts; and as the time taken to describe any area is proportional to the area described, the time of passing from one node to the other will be different from the time of returning to the former node again, unless in the single case of the coincidence of the line of the nodes with the major axis. Accordingly, we do, by observation, find that these times are unequal. In the case of Mars and Jupiter they are very considerably so; in the case of Saturn the difference is very inconsiderable. In the case, therefore, of Mars and Jupiter, the orbit must differ materially from a circle; in the case of Saturn it either must be very nearly circular, or the line of the nodes must nearly coincide with that of the major axis; and from accurate investigation of his motions, it appears that the latter is the case.

The interval of time between two successive conjunctions or oppositions of a planet is called, in the same manner as in the case of the moon, the synodic period of the planet. We shall find that relations exist between the synodic period and periodic time, in the planets, similar to those which we have already investigated in the moon. The formulæ deduced, however, are not exactly the same in the superior and inferior planets; and they must be separately investigated.

Taking the instance, then, of an inferior planet, let S represent the synodic period, P the periodic time of the sun round the earth, and p that of the planet round the sun. P is greater than p, by Kepler's third law; or the planet sooner returns to its original position with respect to the sun, than the sun with respect to the earth. Now if the sun were at rest, the planet, on completing its revolution round the sun, would be in the same position with respect to the sun and earth as before; but the sun having moved, it is not so. It will, therefore, have to pass through a further arc of its orbit before it arrives at this original position. Let this arc be A, then the planet will have described an arc $360°+A$, and the sun an arc A, when the planet is again in conjunction. But the time from one conjunction to another is the synodic period; hence the time of the synodic period = time in which the planet describes $360°+A$ round the sun, or $S=(360+A)\frac{p}{360}$, on the supposition that the motions are uniform. But $S=A.\frac{P}{360}$, (for it is the time in which the sun, whose periodic time is P, describes an arc of A°), and, consequently, $A = \frac{360\ S}{P}$. Substituting this value of A in the former equation,

$$S = \left(360 + \frac{360\ S}{P}\right)\frac{p}{360} = \left(1 + \frac{S}{P}\right)p;$$

or $PS = Pp + Sp$, and, consequently,

$$S = \frac{P\,p}{P - p}, \text{ and } p = \frac{P\,S}{P + S}.$$

In the case of a superior planet, if S′ represent its synodic period, and p' its periodic time, P, as before, being the periodic time of the sun, p' is greater than P by Kepler's third law; or the sun returns sooner to the same place with respect to the earth, than the planet with respect to the sun. If the planet were at rest, the relative positions of the sun, earth, and planet, would be the same as before when the sun had completed a revolution; but the planet having moved, it is not so. The sun will, therefore, have to pass through a further arc of its orbit before it arrives at its original relative position. Let this arc be A′; then in the synodic period, S′, the sun will have described 360 + A′, or $S' = (360 + A')\frac{P}{360}$, and the planet will have described A′, or $S' = A'.\frac{p'}{360}$, and, consequently, $A' = \frac{360\ S'}{p'}$. Substituting this value in the former equation,

$$S' = \left(360 + \frac{360\ S'}{p'}\right)\frac{P}{360} = \left(1 + \frac{S'}{p'}\right)P;$$

or $S'p' = Pp' + S'P$, and, consequently,

$$S' = \frac{P\,p'}{p' - P}, \text{ and } p' = \frac{P\,S'}{S' - P}.$$

In both cases, therefore, if either the synodic period or the periodic time of a planet be known, the other may be computed; and as the synodic period may be observed, this furnishes another, and a very easy method of ascertaining the periodic time, on the supposition of circular motion. The expressions, also, will be correct as to the *mean* synodic and periodic times, even in the case of elliptic motion.* It is at present, however, of more importance to observe, that the lengths of the synodic periods bear a certain relation to the proportion of their distances from the sun.

---

* The same conclusions may be deduced in the following manner, which has the advantage of showing more distinctly that the periods of conjunction or opposition investigated are those immediately succeeding each other:—

In *fig.* 32, let E represent the earth, and S S′ a position of the sun's orbit, supposed circular, and let the smaller circles described about S and S′ represent the orbit of an inferior planet. Let us suppose the planet, when the sun is at S, to be either in superior conjunction at C, or in inferior conjunction at *c*. If, then, the planet had no motion in its orbit, when the sun was at S′, the planet would be at A or *a*, (A *a* being a line parallel to

Fig. 32.

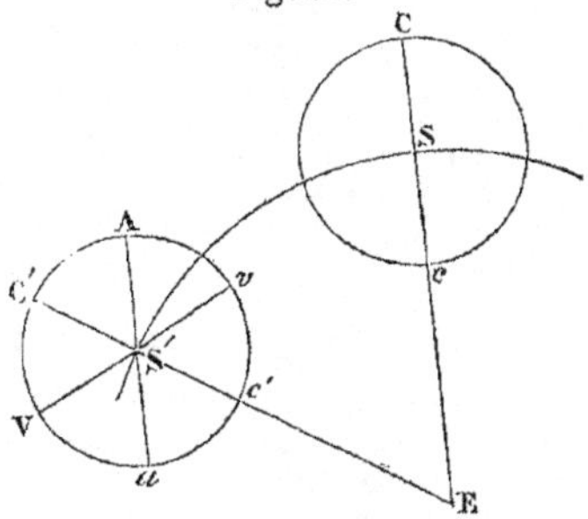

C *c*, or A, *a*, being the points in the orbit described round S′, corresponding to C, *c*, in the orbit described round S,) and, consequently, not in conjunction, for the points of conjunction are evidently C′, *c*′. The angle C′ S′ A, or its vertical angle *a* S′*c*′, is equal to S′ E S, for A S′ *a* is parallel to S E; if, therefore, the planet's angular motion in its orbit were equal to that of the sun, the planet would still be in conjunction when the sun was at any point whatsoever, S′, if it had been so when the sun was at S. But the angular motion of the planet is to that of the sun in the inverse ratio of their periodic times; for each describes a whole circumference in its periodic time. The angular motion of an inferior planet round the sun, therefore, being greater than that of the sun round the earth, the planet, instead of being at C′ or *c*′ when the sun is at S′, will have described some greater angle, A S′ V, or *a* S′ *v*, and will, therefore, be distant from the point of conjunction, C′ or *c*′, by the angle C′ S′ V, or *c*′ S′ *v*.

Let T = time of the sun's describing the angle S E S′, and let P, *p*, and S, retain their signification.

$$\therefore T : P :: S E S' : 360°$$
$$:: A S' C' \text{ or } a S' c' : 360,$$
$$\text{and } A S' C' \text{ or } a S' c' = 360° \frac{T}{P}.$$
$$\text{Again, } T : p :: A S' V \text{ or } a S' v : 360,$$
$$\text{and } A S' V \text{ or } a S' v = 360° \frac{T}{p}.$$

Hence the angle C′ S′ V or *c*′ S *v*′ = A S′V − A S′C′ or $a S' v - a S' c' = 360° \frac{T}{p} - 360° \frac{T}{P}$, a quantity which continually increases, as T increases; or the distance of the planet from the point of superior or inferior conjunction continually increases. When, however, this angle = 360°, (and not till then,) the situation of the body is the same as if the angle were nothing; or the planet is in superior or inferior conjunction; and consequently, in this case, T = S, the synodic period. The equation, therefore, becomes $360° = 360° \frac{S}{p} - 360° \frac{S}{P}$, or $Pp = PS - pS$, as before.

The same operations evidently apply to the case of a superior planet, except that its angular motion being less, the planet's place will *fall short* of the point of conjunction or opposition by the difference of the angles; and the angle by which it does so will be $360 \frac{T}{P} - 360 \frac{T}{p'}$:, and, therefore, $Pp' = p'S' - PS'$, as before.

In the case of an inferior planet $S = \frac{P p}{P - p} = p + \frac{p^2}{P - p}$, of which each term is smaller, as p is smaller. In this case, therefore, the shorter the periodic time, or the smaller the orbit, the shorter is the synodic period. The orbit of every inferior planet, however, being less than the sun's, the smaller the orbit of the planet, the more does it differ from that of the sun; and consequently among inferior planets, the greater the difference between the orbit of the planet and that of the sun, the less is the synodic period.

In the case of a superior planet $S' = \frac{P p'}{p' - P} = P + \frac{P^2}{p' - P}$, of which the first term is constant, and the second term diminishes as its denominator increases, and consequently as p' increases. In this case, therefore, the longer the periodic time, or the larger the orbit, the less is the synodic period. The orbit, however, of every superior planet is greater than the sun's; and the difference of course increases as the orbit of the planet increases: and consequently among superior planets as well as among inferior planets, the greater the difference between the orbit of the planet and that of the sun, the less is the synodic period.

We may also observe, that the retrograde motion, in the case of an inferior planet, is produced by the excess of the effect of its motion in its own orbit over that of the sun's motion in his. The greater, therefore, the motion of the planet in its orbit, the greater we should expect to find the proportion of the retrograde to the direct motion during a synodic period; during which all the phenomena of direct and retrograde motion take place. The planets, however, whose orbits are least, have the most rapid motion in their orbits: we should, therefore, expect to find that they had the greatest retrograde motion in proportion to their direct motion. In the case of a superior planet, on the contrary, the retrograde motion is produced by the excess of the effect of the sun's motion in his orbit over that of the planet's in its own. The less, therefore, the planet's motion, or the greater the planet's orbit, the greater should we expect to find the proportion of the retrograde to the direct motion during the synodic period. These results may be more fully and accurately deduced; but it is sufficient here to indicate them as we have done.

Another fact of the same nature may at once be deduced from the consideration of the respective distances of the different planets from the sun. The apparent diameter of a body varies inversely as its distance. Now, the greatest distance of an inferior planet from the earth, is when it is in superior conjunction; its least distance, when in inferior conjunction: the greatest distance of a superior planet is when it is in conjunction, its least when in opposition: the distance in the former cases being the sum in the latter, the difference, of the distances of the planet from the sun, and of the sun from the earth. Let D and d represent the greatest and least apparent diameters of a planet; R, the sun's distance from the earth, r, the planet's distance from the sun. Then in the case of an inferior planet when $R > r$

$$D : d :: \frac{1}{R - r} : \frac{1}{R + r} \text{ or } D = d.\frac{R + r}{R - r} = d.\left(1 + \frac{2r}{R - r}\right)$$

which necessarily increases as r increases.

In the case of a superior planet, where $r > R$

$$D : :: \frac{1}{r - R} : \frac{1}{r + R} \text{ or } D = d.\frac{r + R}{r - R} = d\left(1 + \frac{2R}{r - R}\right),$$

a proportion which necessarily increases as r diminishes *.

The greater r is, in the case of an inferior planet, or the less it is in the case of a superior planet, the more nearly does it approach to R. The more nearly, therefore, the planet's distance from the sun approaches to that of the sun from the earth, the greater is

* It is worthy of observation, that these equations also give us at once a direct method of ascertaining the proportion which the radius of a planet's orbit round the sun bears to that of the sun's orbit round the earth. Taking the case of an inferior planet $D = d\frac{R + r}{R - r}$, or $RD - rD = Rd + rd$, or $RD - Rd = rd + rD$, or $r = R.\frac{D - d}{D + d}$.

In the case of a superior planet $D = d.\frac{r + R}{r - R}$, or $rD - RD = rd + Rd$, or $rD - rd = RD + Rd$, or $r = R.\frac{D + d}{D - d}$.

the disproportion between the greatest and least apparent diameter.

Having deduced these theoretical truths from the laws discovered by Kepler, we proceed to give an account of the actual facts observed with reference to each planet, from a comparison of which with each other, it will be found that they accurately correspond with the results deduced. As this account will consist principally of a mere catalogue of numerical results, it will be most convenient to give it in the form of a table, (see following page, 122) subjoining any necessary explanations and remarks, and the statement of any circumstances peculiar to any particular planets.

It will at once be observed, on inspection of this table,* that it contains no notice of the phases of the planets, or of their greatest elongation. The table, however, is incomplete for Mercury and Venus, without the insertion of the latter: to the superior planets, which are seen at all distances from the sun, it has of course no application. In Venus and Mercury, and especially Mercury, whose motions, on account of the considerable excentricity of his orbit, are, as we shall presently see, very variable; they differ considerably in amount, according to the different positions of the planet with respect to the sun, and of the sun with respect to the earth. In the case of Mercury, the mean value of the greatest elongation is about 22½°, but it sometimes rises as high as 29°, and is sometimes nearly as low as 16°. In Venus, the variations are much less considerable, the least value being about 45°, and the greatest not exceeding 48°.

The table contains no account of the phases, because we have already given a very full detail of these appearances, and the results of observation accurately correspond with the phases we described. In Mercury, indeed, they are not easy of ascertainment, for he is generally so near the sun as to render observation difficult, on account of the great light diffused over that part of the heavens where he is situated; and his apparent magnitude is so small, that on that account also much delicacy of observation is required. In Venus, the observations are easy, and the discovery of her phases was one of the first fruits of the invention of the telescope in the hands of the illustrious Galileo.

One circumstance, however, is worth mention with respect to the phases of the planets. The actual brightness of a planet to the eye depends both on the quantity of illuminated surface visible, and on the distance. The brightness of a point, that is to say, the quantity of light which arrives from it at the eye, varies inversely as the square of the distance from the eye; this we know from the theory of optics. The brightness of the planet, therefore, varies directly as the quantity of illuminated surface, or as the versed sine of the exterior angle of elongation, and inversely as the square of the distance. In the case of a superior planet, the distance is least when the planet is in opposition, and the planet then also shines with a full face, and of course the brightness is then greatest; but in the case of an inferior planet, when the distance is least, the planet is in inferior conjunction, and the illuminated part is entirely turned away from the sun, or the planet is invisible. We must, therefore, have recourse to calculation, to discover the period of its greatest brightness, and we find it, in the case of Venus, to be when the elongation is about 40°, and the planet between its greatest elongation and its inferior conjunction. In the case of Mercury, the period of greatest brightness is between its greatest elongation and its superior conjunction, but as his brightness is never very great, the difference is not very observable; but we do practically find, in the case of Venus, that when she is near the situation above mentioned, and only then, her brightness is such, that for a considerable period she is visible before sunset and after sunrise; and that when the darkness of night has come on, she occasions a perceptible shadow. The first of these two phenomena seldom occurs without exciting observation and curiosity, simple as is its explanation, and common as it is; for it necessarily occurs twice in every synodic period, though it is only in the evening that it is much observed. It occurs in the case of Venus only; for even under less favourable circumstances, when she is not visible except at night, she exceeds all the other heavenly bodies in brightness.

A much more important circumstance remains to be mentioned with respect to

* The earth is inserted in the table for a reason which will hereafter appear. Many of the columns have no application to it: the applicability of those in which it is inserted will be discussed in the next chapter.

	Mean distance, that of the earth being 1.	Periodic time in days.	Eccentricity in 1801.	Synodic period in days.	Distance of stationary points from line joining earth and sun.	Time of retrograde motion in days.	Arc of retrograde motion.	Actual diameter, that of the earth being 1.	Greatest and least apparent diameters.	Apparent diameter of sun, as seen from planet.	Mean velocity in orbit, in miles per second.	Time of revolution on axis.	Inclination of orbits to the ecliptic in 1801.	Longitude of ascending node as seen from the sun in 1801.	Mean Longitude of perihelion seen from the sun in 1801.
Mercury	.3870981	87.96925804	.20551	115	18°	23	13½	.4	11″— 5″	82′½	30	24ʰ.5ᵐ½	7°. 0′. 1″	45°.57′.31″	74°.21′.46″
Venus	.7233323	224.70082399	.00685	584	29°	42	16°	.9	57″—10″	44′¼	23	23ʰ.21ᵐ	3°.23′.22″	74°.52′.39″	128°.37′
Mars	1.5236935	686.0761186	.09313	780	137°	73	16°	.8	29″— 6″	21′	15	24ʰ.40ᵐ	1°.51′. 3″	48°.14′.38″	332°.24′.24″
Vesta	2.373000	1335.205	.09322	503		83	13°			13′½	13		7°. 8′.46″	103°. 0′. 6″	249°.43′. 0″
Juno	2.667163	1590.998	.25494	474		99	12°			12′	12		13°. 3′.28″	171°. 6′.37″	53°.18′.41″
Ceres	2.767406	1681.539	.07835	466		99	12°			11′½	11		10°.37′.34″	80°.55′. 2″	146°.39′.39″
Pallas	2.767592	1681.709	.24538	466		99	12°			11′½	11		34°.37′. 8″	172°.32′.35″	121°.14′. 1″
Jupiter	5.2027911.	4332.5963076 or nearly 12 years.	.04818	399	115°	121	10°	11	40″—26″	6′	8	9ʰ.55ᵐ	1°.18′.51″	98°.25′.34″	11°. 8′.35″
Saturn	9.5387705	10758.9698400 or about 29½ years.	.05617	378	109°	139	6°	10	18″—15″	3′⅓	6	10ʰ.16ᵐ	2°.29′.35″	111°.55′.46″	89°. 8′.58″
Uranus	19.1833050	30688.7126872 or about 84 years.	.01667	369½	103½°	151	4°	4.3	4″	1′⅔	4		46′.26″	72°.51′.14″	167°.21′.42″
The Earth, see p. 121.	1	365.256384	.01685	—	—	—	—	1	—	32′	18½	23ʰ.56ᵐ	—	—	90°.30′. 5″

the two inferior planets. Being inferior planets, they pass between the sun and the earth; if, therefore, at the time of inferior conjunction, they are nearer the ecliptic than the distance of the sun's apparent radius, they will be in the same direction as some part of the sun itself, and will become visible upon his surface, and will appear to pass over it, exactly in the same manner as the moon does in the case of a solar eclipse. This appearance is called *a transit* (i. e. *passing across*) of Mercury or Venus. The general principles of its investigation are similar to those of a solar eclipse, and it is not necessary here to enter into its details. It is, however, of importance to point out the very important use to which the observation of them may be applied. It appears, that the duration of a solar eclipse, or transit, depends on the apparent diameters of the sun and of the body passing over it, and on the difference of their respective parallaxes at the place of observation. The former elements may be considered as nearly the same at the same time on every point of the earth's surface: the latter is different in different places, for the parallaxes themselves are so. The duration of the transit will therefore be different at different places, and if the parallaxes are accurately known, it may be accurately computed. Conversely, if the duration of the transit be observed, the difference of the parallaxes may be computed, and this with more accuracy, if the duration of the transit be observed in different places (chosen so that the phenomena of parallax should differ considerably at them), and the difference of the parallaxes ascertained at each. From the comparison of these observations, the actual parallaxes can be ascertained with much more accuracy than in any other way; and, as the sun's parallax is so small that the greatest nicety of observation is necessary to ascertain it within any tolerable limits of accuracy; and as its accurate ascertainment is of the highest importance in astronomy, great care has been taken by various governments to have these occurrences accurately observed in the most desirable situations. The last transit of Venus took place in 1769; and Cook made the first of his celebrated voyages for the purpose of conveying to Otaheite the astronomers sent out by this country to observe the transit there. The same transit was observed also at Wardhus in Norway, at Cajanebourg in Swedish Lapland, at Kola in Russian Lapland, at Petersburg, at Paris, at California, and at Hudson's Bay, &c.; and it is from a comparison of all these observations, that the value of the sun's horizontal parallax now received as correct has been deduced.

The last transit of Venus took place in 1769: there will not be another till 1874. The rare occurrence of these phenomena is easily explained: they can occur only when Venus is near one of the nodes of her orbit, and also in inferior conjunction. The periodic time of Venus, or time from leaving her node till she returns to the same node again, which is very nearly the same, is about 224 days: the synodic period, from inferior conjunction to inferior conjunction, is about 584. If, therefore, the planet were in its node at the time of one inferior conjunction, at the time of the next inferior conjunction it would be distant from that node by the space which it describes in 136 days; for $584 = 2 \times 224 + 136$. There would, therefore, be no transit, for the planet would be distant from either node: as the opposite node would be only about 112 days from the former. Five synodic periods, however, amount to about 2920 days, or very nearly eight years; and thirteen periodic times of Venus amount to 2912 days; so that at the fifth synodic period, or at the distance of eight years, Venus would again be very nearly in inferior conjunction, and a transit might take place. Accordingly, there was one in 1761, eight years before that in 1769. The correspondence of the periods, however, is not exact, and the difference accumulates. The consequence is that, although eight years after a transit another may occur, there will not be a third at the end of a second period of eight years: the difference of time being doubled, makes the distance from the node too great to admit of a transit. The next period at which a transit takes place at the same node is after 235 years; and then again in eight years. These results cannot be deduced without employing more accurate values of the synodic period and periodic time than those above used, for the purpose of explanation; and it is not worth while to occupy any time in obtaining them: there may, however, be transits at the opposite node. These will succeed each other in the same order, at intervals, alternately of 8 and of 235 years, at periods nearly midway between the transits at the opposite node; and there will accordingly be one in 1874, and another in 1882. The transits of 1761 and 1769 were at the descending node, and there will be a transit at that node again in 2004: the transits of 1874 and 1882 will of course be at the opposite, or ascending node. The transits of Mercury may be ascertained in the same manner, and they are of more frequent occurrence; they are, however, of less importance in ascertaining the sun's parallax, the one important element discovered by such observations. Mercury being much nearer the sun than Venus, his parallax is much smaller than hers; it is, therefore, less accurately ascertainable, and the difference of his parallax and the sun's parallax, which is the quantity involved in the observation of the transit, is less accurately ascertainable also.

After noticing these facts, of which the Table gives no intimation, we proceed to explain some columns of the Table itself, which our preceding remarks have not applied to. The first of these is the column headed excentricity. The excentricity of an ellipse is the proportion which the distance of the focus from the centre bears to the semi-axis major: and in the column which states it, the semi-axis major is supposed to be 1. The greater this excentricity is, the more, of course, does the ellipse differ from a circle: and the more, also, which is a very important result, does the rate of the planet's motion differ in different parts of its orbit. This will at once appear, when we remember the great law of all the planetary motions, that they describe equal areas in equal times round the sun in the focus of their orbits. The areas being equal, the angular velocity varies inversely as the square of the distance; and consequently the greater the variation in the distances, the greater is that of the angular velocity also. Now, the greatest and least distances of the planet from the sun are the two parts into which the major axis is divided; and of these, the former is the semi-axis major increased by the sun's distance from the centre of the orbit, and the latter the semi-axis major diminished by the same quantity. The greater the proportion which this quantity bears to the semi-axis major, or, in other words, as the sun is in the focus, *the greater the excentricity of the orbit*, the greater is the difference between the extreme rates of angular motion, and the more does the actual motion at different periods differ from the mean motion. The mean motion is that with which, were the angular velocity uniform, the orbit would be described in the time actually taken to describe it: the difference of the angle actually described at any time from that which, on this supposition, would be described in the same time, is termed the *equation of the centre*. The greatest value of this is of course greatest in those planets, of which the excentricity of the orbit is greatest.

The periodic time and the mean distance are not affected by the excentricity of the orbit; but most of the other elements stated in the Table are so. For instance, the synodic period is the period in which the sun passes, either, as in the case of an inferior planet, through an arc which falls short of the whole arc described by the planet by 360°, or, in the case of a superior planet, through the arc described by the planet increased by 360°. The time in which the sun does so will of course depend (omitting the consideration of his own inequalities of motion) on the magnitude of the arc described by the planet; and this will of course be affected by the rate of the planet's motion in that *part* of its orbit. The greater, therefore, the variation of this rate of motion, the greater may be the difference in the length of different synodic periods.

Again, we have already seen that the greatest elongation of Mercury varies very considerably; that of Venus but little. This is an immediate consequence of the excess of Mercury's excentricity over that of Venus: for the elongation is the angle subtended at the earth by the distance from the planet to the sun, and it varies, therefore, in corresponding situations, with the variation of the distance of the planet, both from the sun and from the earth. Both these are variable by reason of the excentricity of the orbit: and the greater, therefore, that excentricity, the greater will be their variations. We need not further pursue this subject: it is evident that wherever we have a mean value only, the greater excentricity, the greater will be the variation of the true from the mean value.

The next columns which require any explanation are those which state the apparent diameter of the sun, as seen from the planet, and the velocity of the planet in its orbit. These, of course, are not the subjects of observation; but they follow immediately as ascertained results from the other elements of the orbit: the mean velocity being known from the magnitude of the orbit, and the time of describing it; the apparent diameter of the sun bearing to its apparent diameter at the earth the inverse proportion of its distance from the two bodies. In making the former computation, the earth's distance from the sun, which is called 1 in the Table, is estimated at 93,726,900 miles. The light and heat of the sun depend, of course, not on the apparent diameter, but on the apparent magnitude of the sun; they vary, therefore, as the squares of the apparent diameter, or inversely as the squares of the distances. In Mercury, therefore, they are nearly seven times as great as at the earth; in Uranus nearly 370 times less.

The next column, which gives an account of the time of revolution on an axis, introduces us to a class of phenomena, of which we have as yet seen no instance, except in the case of the moon; we shall, however, find the class both extensive and important. By careful observation of the surface of several of the planets, spots and other marks may be perceived; and by continued examination of their situation, they are found to be transferred from place to place in the same manner as if the planet had a motion of rotation round an axis passing through its centre. This axis in different cases is inclined at different angles to the plane of the ecliptic, but in each planet it appears always parallel to itself. Similar observations lead to the same conclusion with respect to the sun itself; and his period of revolution is ascertained to be about $25^d\ 10^h$. Some account may here be given of the nature of the peculiar appearance observed on each of these bodies, from which these conclusions, and some others of considerable interest, are drawn.

The sun, when viewed through a telescope fitted up for the purpose, with a glass of dark colour to prevent the excessive brightness from disabling the eye from observation, is generally found to have several dark spots on his surface. These are different at different periods: sometimes numerous, sometimes few; sometimes small, sometimes so large as to be visible without a telescope. Some have been ascertained to be of four or five times the diameter of the earth. At other times, though rarely, the sun has presented none of these spots for several years together. Varying, however, as these appearances are, their variations are slow. The same spot may be observed, day after day, in the same situation with respect to other spots in its neighbourhood, so as to leave no doubt of its identity: but all these spots are seen successively more to the westward, till they finally disappear on the Western side of the sun, and then, after a considerable interval, re-appear on the Eastern side, and recommence the same order of appearances. By degrees these spots altogether vanish, but others which succeed them follow the same law in their motions; and by careful observation these motions are found to correspond with those which would be observed, if the sun were to revolve from West to East* in $25^d\ 10^h$, round an axis inclined at an angle of $82\frac{1}{2}^\circ$ to the plane of the ecliptic. If we call a great circle of the sun perpendicular to the axis, the *equator* of the sun, it is generally found that his spots are confined, for the most part, to a zone of his surface, not extending more than 31° on each side of this equator: some, however, have been seen about 40° from it. The light of the sun is found to be somewhat more intense towards the centre than towards the border; and it is supposed that this is occasioned by a dense atmosphere surrounding the sun, and weakening the effect of its rays. The rays proceeding from the centre of

* The motion of the sun from West to East round its axis, occasioning the transference of the spots from East to West, as viewed from the earth's surface. The surface of the planet which is exposed to the earth, is between the centre of the planet and the earth; any motion, therefore, of, or on, this surface will be seen in opposite directions at these different points, which are situated on opposite sides of the line of motion; just as, when a person standing with his face towards you, shifts anything from his right to his left hand, the direction of the motion, as referred to your own position, is from left to right.

In fact, this is an instance of retrograde motion, just as in the case of the inferior planets. Any point of the sun's surface may be considered as revolving round his centre, just as an inferior planet revolves round him; and its motion is *retrograde* when nearest the earth, or in *inferior conjunction*. We treat its motion, indeed, as retrograde throughout, which is not the case with an inferior planet, because, observing it only as to its situation with respect to the sun, and not making its position in actual space, we consider only the effect of its motion round the sun, uncombined with the sun's motion round the earth; and that we have already shown to be retrograde from one period of greatest elongation to the other, or as long as the point is visible.

the visible face of the sun to a spectator would of course pass through the least thickness of this atmosphere, and would, in consequence, be the least affected. It is also conjectured, that the zodiacal light, a faint light sometimes seen a little after sunset and before sunrise, especially about the time of the vernal equinox, is reflected by the atmosphere.

The observations on the planets which are found to revolve round an axis are of the same nature. The motion of each of them is found to be from West to East. In the case of Venus, luminous points are occasionally observable near the boundary of the unenlightened part; and from these, as well as from certain dark spots, her motion of rotation is ascertained. The equator of Venus—using the word in the same sense as that in which we have already applied it to the sun, namely, a great circle perpendicular to the axis of rotation—is inclined to the plane of the ecliptic at a considerable angle. Appearances detected by the same course of observations have led to the further conclusion, that there are very high mountains on the surface of Venus, and that she is surrounded with an extensive atmosphere, whose refractive power differs little from that of the atmosphere of the earth. These latter observations, however, have little certainty: the motion of rotation is completely established. Similar observations have been made, though with less certainty, upon Mercury. In the case of Mars, there is nothing particular to be detailed: his axis of rotation is inclined about 60° to the plane of the ecliptic.

The observations which are made on Jupiter are much more curious and important. Besides spots, which, as in the case of the other bodies, show him to revolve round an axis very nearly perpendicular to the ecliptic, there are several darkish belts observable on his surface, which cross it in lines perpendicular to his axis of rotation, and of course parallel to each other.

That planet is found also, by minute observation, not to be spherical, but to have its axis shorter than the equatorial diameter, in the proportion of thirteen to fourteen very nearly. We shall hereafter find this remark of importance. There are some irregularities in the motion of the spots from which Jupiter's motion of rotation has been ascertained, which give colour to a belief that they are not exactly upon his surface, but probably clouds in his atmosphere, and liable to much agitation from winds: but any such conclusions are, of course, involved in much uncertainty. These are the principal facts known with relation to Jupiter, except one of the highest importance—that he is accompanied by four small bodies which appear to revolve round him, and which are called his satellites. These bodies were discovered by Galileo as soon as the invention of the telescope gave him the means of making observations of so much delicacy. They are not visible to the naked eye; but a telescope of very moderate power is sufficient to exhibit them. Their motions, however, are of too great importance to be incidentally discussed here: and as the remaining planets are also attended by satellites, the description of their motions, and of the laws which govern them, will form the subject of a separate section.

Saturn, as well as Jupiter, has satellites, and also a very remarkable adjunct in a large ring which surrounds it, and a more particular account of which, as well as of his satellites, will also be found in the next chapter. For the present it is sufficient to say, that the centre of the ring coincides with that of the planet, and that its plane is inclined at an angle of a little more than 31° to that of the ecliptic. The ring casts an observable shadow on the planet. Besides these appearances, there are belts observable on the surface of the planet nearly parallel to the ring; and the planet itself is found to move from west to east round an axis which is perpendicular to the plane of the ring. As in the case of Jupiter, the period of rotation is very short, and as in the case of Jupiter also, we find the axis and the equatorial diameter very perceptibly different; the former being about one-eleventh shorter than the latter.

It will be observed that there is no period of rotation assigned to Mercury, Vesta, Juno, Ceres, Pallas, and Uranus. There are no observations which lead to the conclusion that these bodies do revolve. It is, however, probable that, as in all other respects they correspond with the remaining planets, they do so in this also. As, however, the motion of rotation is detected only by observation of particular marks and spots on the surface of the body, these observations must be more difficult as the body is more distant, and when its apparent diameter is very small, it is not to be expected that any such variations of its surface can be detected. Vesta, Juno, Ceres, and Pallas are so small that their apparent diameters have never been ascertained at all; and that of Uranus does not exceed 4″. It is not, therefore, surprising that observation should be unable to detect any motion of rotation; nor does the absence of such observations furnish any reason for disbelieving its existence. Even in the case of Mercury, though his extreme apparent diameter is not very much less than that of Saturn, the greater difficulty of making observations on his appearances, in consequence of the strong light diffused over the part of the heavens in which he is situated by the near neighbourhood of the sun, and the small part of his disk which is generally visible to the observer on account of his phases, will account for the impossibility of making the same observations with respect to him that have been made with respect to Saturn; and even in the case of Saturn the observations are of extreme difficulty, and Dr. Herschel was the first person who successfully made them, though the motion of rotation, and even its amount, had been conjectured before. Like Jupiter and Saturn, Uranus has satellites, an account of which will be given in the next chapter.

The remaining columns require very little explanation. The plane of the ecliptic being a known plane, the position of the plane of any other orbit will be ascertained, if we know the inclination of that orbit to the ecliptic, and the line in which it intersects it, or the line of the nodes. The position of the sun also is known, and it is a point in the plane of the ecliptic, and in the line of the nodes: the position of this line will therefore be ascertained, if the direction of the node with respect to the sun be known; and, as the node is in the plane of the ecliptic also, this direction will be at once ascertained by the *longitude* of the node as seen from the sun; that is, the distance from the first point of Aries of the point where a line drawn from the sun through the node meets the celestial ecliptic. Finally, we take the longitude of *the ascending* node, because the inclination may be measured either northwards, at the ascending node, or southwards at the descending node; and it is therefore necessary to specify which node it is, of which the longitude is ascertained. The other node is, of course, 180° distant from this. These two columns, therefore, ascertain the position of the plane of the planet's orbit. If the planet's orbit were circular, the sun being in the centre, a circle described round the sun in this ascertained plane at the planet's mean distance, would be its orbit. This is not the case. From the mean distance, and the excentricity however, we can exactly determine the figure of the planet's orbit; and, consequently, if we know the situation of its axis, we know the exact position of the whole orbit. Now, the points where the planet is at the least, and at the greatest distance from the sun, or, as they are respectively called, the *perihelion* and *aphelion* of the planet, are at the extremities of the axis major of the orbit; and a line, therefore, drawn through them, which will pass through the sun also gives the position of the axis. To ascertain this line, we want only the longitude, as seen from the sun, of the perihelion; for a plane drawn perpendicular to the ecliptic there will intersect the planet's orbit in the perihelion; and consequently this longitude being known, and also the inclination of the orbit, the exact situation of the perihelion may be discovered, and hence the position of the axis, and consequently of the whole orbit, ascertained*. The three columns, therefore, with those of the mean distance and the excentricity, furnish the means of completely ascertaining the position of the orbit.

It should, however, be observed that all these elements are subject to a very slow variation. The nodes of all the orbits have a retrograde motion, except that of Mercury which is direct; this, however, even in the case of Uranus, where it is greatest, does not amount to quite 1° in one hundred years. The longitudes of the perihelia and the inclinations of the orbits increase in some planets, and diminish in others: the utmost variation in the longitude of the perihelion, which is that of Saturn, is only of 32′ 17″ in one hundred years, in increase; and the greatest variation of inclination is that of Jupiter's orbit, the inclination of which diminishes 23″ in one hundred years. The excentricity of the orbits, also, is liable to a slight variation. These variations are not yet ascertained in the case of Vesta, Juno, Ceres, and

* The longitude of the perihelion and also of the node being known, their difference is known; the inclination also is known. The position, therefore, of the perihelion is ascertained by the solution of a right-angled spherical triangle, in which one side and one angle, besides the right angle, are known.

Pallas; with respect to which also it will be observed that many columns of the table are deficient. These planets are all very small, and only observable by telescopes of considerable power. Even through these they have no assignable apparent diameter, and of course we have no knowledge of their absolute magnitude, except that it is less than what would be sufficient to give them an ascertainable apparent diameter. Owing to their exceeding minuteness, they escaped observation till the beginning of the present century, when Ceres was discovered by Piazzi at Palermo on the very first day of January, 1801. Olbers discovered Pallas in 1802, and Vesta in 1807, and Harding Juno in 1803. The recent date of their discovery, as well as the difficulty of observing them, necessarily renders our knowledge of their motions imperfect. The same observation applies, although in a less degree, to Uranus; so far, at least, as to render our knowledge of its motions less accurate than those of the planets longer known. It is difficult to measure that body, both on account of its small apparent diameter, and the extreme faintness of the light which, at its great distance from the sun, it receives from that luminary, and reflects. The consequence was, that although Flamsteed observed it before the year 1700, and Mayer and Leckonnier afterwards, none of them detected its character, but considered it as a small fixed star; a mistake to which the extreme slowness of its motions would naturally conduce. It was not till 1781 that Herschel found it to be a planet.

Before quitting the subject of this table, it will be worth while just to refer to some of our previous conclusions, and to point out the correspondence of the table with them. The periodic time of a planet varies as the square root of the third power of its major axis, which is very nearly double its mean distance: the mean velocity varies inversely as the square root of the same quantity: the periodic times, therefore, increase, and the mean velocities in the orbit diminish, as the distance increases. On inspection of the table, it will be seen that they do so, and by calculation, it will be found that they do so in the proportions assigned. The velocities, indeed, are only rough approximations, and the conclusions from them will not be very accurate. A single instance of each, without the trouble of minute working, will illustrate these assertions. The mean distance of Jupiter is a little more than a fourth part of that of Uranus. Supposing it exactly a fourth part, his periodic time would be less than that of Uranus in the proportion of 1 to $\sqrt{4^3}$, or of 1 to $\sqrt{64}$, or of 1 to 8; and his velocity greater in the proportion of $\sqrt{4}$ to 1, or of 2 to 1. As his distance is somewhat greater than in this proportion, the periodic time of Uranus will be somewhat less than 8 times that of Jupiter, and so the table gives it; and its velocity rather more than half. The table represents it as half, this column not giving its values with any minute correctness.

We stated also that the synodic period was greater, in the case of an inferior planet, as the orbit was greater, and in the case of a superior planet, as the orbit was less; and on looking to the table we find the former of these conclusions verified by comparison of the values given in the cases of Mercury and Venus, the only inferior planets, and the latter in the case of all the others. Again, taking the case of the inferior planets, we saw that we might expect to find the retrograde motion bear the greatest proportion to the direct motion in those planets which had the smallest orbits. To compare these, we must ascertain the whole amount of direct motion during this synodic period. Taking the instance of Mercury, the synodic period being 115 days, and the sun's mean motion being very nearly 1° in a day, the position of the sun, and consequently the position of the planet, which is the same with respect to the sun at the end, that it was at the beginning, of the synodic period, will be, at the end of the synodic period, about 113° distant from what it was at its beginning. This quantity, therefore, must be the difference of the direct and retrograde motions, and the retrograde motion having been 13½°, the whole direct motion will have been 126½°, and the proportion of the retrograde to the direct motion will be 13½ to 126½, or nearly 1 to 9. In the case of Venus, the sun's mean motion during the synodic period of 584 days will have been about 576°; and as the planet is never more than about 45° distant from the sun, she must have accompanied him during this whole course; he cannot have left her, and returned to her again, performing a whole revolution without her. The difference, therefore, of her direct and retrograde motions is this quantity of 576°, and the retrograde motion being 16°, the direct motion will be 592°, and the proportion of the one to the other that of 16 to 592, or nearly 1 to 37. In the superior planets, also, on the other hand, the retrograde motion bears a greater proportion to the direct motion, the greater the orbit of the planet. In these, as the planet is in opposition, or 180° from the sun, during the synodic conjunction, and as the sun revolves more rapidly than the planet, the synodic period must always exceed a year, and the sun must make one revolution more round the earth than the planet round the sun in the period. This, indeed, appears directly from the manner in which the duration of the synodic period was obtained, by finding the arc described by the planet, while the sun described the same arc, increased by 360°. If, therefore, we take the case of Mars, whose synodic period is 780 days, the sun will have moved through about 769°, and the whole motion of the planet during that time will be 769°—360°, or 409°. This is the difference between the direct and retrograde motion, and the retrograde motion being 16°, the whole direct motion will be about 425°, or about $26\frac{1}{2}$ times the retrograde motion. Again, the sun's motion in 399 days, the synodic period of Jupiter, is about 394°, and the whole motion of the planet is 394°—360°, or 34°. The retrograde motion is 10°: the whole direct motion, therefore, is 34° + 10°, or 44°; or $4\frac{1}{2}$ times the retrograde motion. In Saturn, the sun's motion is 373°, the synodic period being 378 days: the planet's whole motion, therefore, is 13°; and the retrograde motion being 6°, the whole direct motion is 19°, or very little more than 3 times the retrograde motion. The same method of computation may be applied to the remaining instances with a similar result. Similar conclusions might have been deduced, although with less accuracy, from the column giving the time of retrograde motion: thus in the case of Mercury it is $\frac{23}{115}$, or $\frac{1}{5}$ of the synodic period, in that of Venus, $\frac{42}{584}$ or only about $\frac{1}{14}$ of it. In the superior planets the synodic period continually diminishes, and the time of retrograde motion continually increases as the orbits increase; of course, therefore, the proportion of the latter to the former increases also.

The only remaining observation respects the apparent diameters. The greatest diameter of Mercury is only about double of the least; the greatest diameter of Venus is nearly 6 times the least. Again, the greatest diameter of Mars exceeds the least in the proportion of 5 to 1 nearly: in the case of Jupiter the proportion is only of 3 to 2, in Saturn of 6 to 5; in Uranus the difference, in a great measure from the smallness of the quantities, cannot be appreciated. The orbit of Venus is more nearly equal in magnitude to that of the sun than that of Mercury is; the orbit of Mars is more so than that of Jupiter, of Jupiter than of Saturn, and of Saturn than of Uranus. In each case, therefore, the results correspond with those already deduced from theory—namely, that the difference between the greatest and least diameters is greater as the planet's distance from the sun more nearly approaches to that of the sun from the earth.

Section V.—*Of the Satellites of Jupiter, Saturn, and Uranus, and of the Ring of Saturn.*

When Galileo first applied his telescope to the observation of the heavens, the most brilliant of the planets, Venus and Jupiter, were naturally the objects of his particular attention. In Venus he almost immediately discovered her phases. In observing Jupiter, he found in his neighbourhood some very small luminous bodies, the existence of which seemed to have some relation to him. Jupiter having a proper motion of his own, if these bodies were fixed stars, his position with respect to them would alter from night to night; and the laws of his motion being known, the degree and manner of this alteration might be computed. On observation, however, the relative position of the planets and these bodies was found indeed to vary, but not in the manner in which it would have varied had the planet alone moved. On the contrary, the small stars were found to accompany him, sometimes on one side of him, sometimes on the other, but always very near him throughout the whole course of his motions. On minute observation of the motions of these bodies, they are found to revolve round the planet at different distances from him. They are called his *satellites*, a word derived from the Latin *satellites*, which signify the guards or attendants of a prince or officer, and which is applied generally to smaller bodies which thus accompany the course of any planet or moving heavenly body.

These satellites are four in number, and the orbit of each is nearly circular. The nearest to Jupiter is called the first, the next the second, the most remote the fourth satellite. The inclination of the orbit of Jupiter to the ecliptic is very small, and so also are the inclinations of the orbits of these satellites. The satellites, therefore, are always very near to the plane of the ecliptic, and consequently as the earth itself is in the same plane, the satellites are always seen very nearly in the line of the celestial ecliptic, which is the intersection of that plane with the plane of the heavens; or, as they are seen in a very small part of that line, they will appear very nearly in a straight line, which passes through the centre of the planet.

The orbits of these satellites are found to be elliptic; those of the first and second are very nearly circular; those of the third and fourth have observable excentricity, that of the fourth being the greatest, and the excentricity of that of the third varies in a remarkable manner. The squares of the periodic times are found to increase in the proportion of the cube of the major axis of the orbit, the same law as that observed by the planets in their motions round the sun; and like them also the areas described by the radius vector of each satellite, that is to say, by the line joining the centre of the satellite with the centre of Jupiter, the body round which it revolves, are, for each satellite, proportional to the times. The mean distances of the satellites are respectively, 5.81296, 9.24868, 14.75240, and 25.94686, the radius of Jupiter being considered as 1; and the periodic times of their revolution round Jupiter are respectively 1.76914, 3.55118, 7.15455, and 16.68877 days, quantities which will be found to be in the proportions above stated. The inclination and the position of the nodes of these respective orbits are found to vary, but we need not here make any statement of these particulars.

These satellites circulating round the planet in planes very little different from that of the planet's motion round the sun, the planet must frequently be between some of the satellites and the sun, and some of the satellites in the same manner between the sun and the planet. The periods at which these phenomena take place may be computed, for the positions of Jupiter and the sun being known, the position of the line joining them is known also. It is found that when the satellites pass within a certain distance of this line, at a greater distance from the sun than Jupiter is, or behind Jupiter and through his shadow, they lose all their light and become invisible. We conclude, therefore, that they shine only by reflecting light from the sun, as they disappear when the light of the sun is intercepted from them. The same conclusion may be deduced from observations made when the satellites pass between Jupiter and the sun, for then a shadow may be observed on the surface of Jupiter corresponding to the situation, and moving with the motion of the satellite, which therefore is evidently an opaque body, giving no light in itself, and capable of intercepting that of the sun. The satellites, therefore, when they pass behind Jupiter are *eclipsed*, just as the moon is when it passes behind the earth, and they eclipse the sun to Jupiter when they pass between the planet and the sun, just as a solar eclipse on the earth is occasioned by a corresponding situation of the moon. Another conclusion which follows from the same observations is, that Jupiter himself is an opaque body shining only by reflected light; for if he were luminous in himself he would enlighten the satellites while he was between them and the sun, and those parts of his surface between which and the sun a satellite was passing would continue to shine, although the solar light was intercepted from them. We have already seen, p. 112, that the absence of phases in Jupiter furnishes no argument against this conclusion, but corresponds with it.

The motion of these satellites, like those of all the bodies connected with the solar system which we have hitherto examined, is from West to East. The direction of their motion, its time, and their distances from the planet being known, their motions may be accurately computed, and consequently the time which they would take in passing through Jupiter's shadow, or during which they would cast a shadow on his surface, ascertained. The latter of these may be observed in the case of all the satellites, but probably not with sufficient accuracy to draw any material conclusion from it; the former cannot in the case of the first or second satellite, the entrance of which into the shadow, or else its departure from it, is always concealed from the earth by the planet. The third and fourth satellites, which are more distant from the planet, may both appear and disappear when the planet is not between them and the earth; and when they do so, the duration of the eclipse is found accurately to correspond with the calculation. All these circumstances prove the truth of our conclusions, that the satellites shine only by reflected light, and also confirm the accuracy of our knowledge of their motions.

The observation of these eclipses furnishes the best means of ascertaining the periodic time of the satellites. As they take place when the satellites and sun, seen from Jupiter, are in opposition, the interval between two successive eclipses is a synodic period, or some number of synodic periods; and the synodic period being thus ascertained by easy observation, the periodic time may be deduced from it, in the same way as in the case of the moon or of a planet.

Small as these bodies are, for their apparent diameter can scarcely be measured, it is hardly to be expected that any observations can be made, whereby to ascertain if they have any motion of rotation. In the fourth satellite, however, Maraldi observed a certain spot, from which he concluded that it revolved round an axis, and that the time of this revolution was the same as the periodic time round the planet. Dr. Herschel also observed, that the brightness of the different satellites varied at short intervals, and as they, like Jupiter himself, can have no observable phases at the earth, this variation of brightness could only proceed from some variation on their respective surfaces. By careful comparison of their respective brightnesses with each other, and by thus ascertaining when the brightness of each was at its greatest and least intensity, he finally arrived at the same conclusion with respect to all, that Maraldi had previously deduced for the fourth satellite—namely, that they all revolve round an axis, and that the time of rotation of each is equal to its periodic time.

Assuming the correctness of our knowledge of the motions of these bodies, we proceed to apply it to the ascertainment of two very important facts—the one purely of an astronomical nature, the other belonging more properly to the science of optics, but ascertained only by these astronomical researches, and of the highest importance also, as we shall hereafter see, to astronomical science.

We have stated in the table, in p. 122, the mean distance of Jupiter from the sun, and also his real diameter. Of course, if his distance from the earth can be ascertained, these values may be computed; the former directly from the triangle formed by the three bodies, of which two sides, the distances of Jupiter and the sun from the earth, would be known, and the included angle, the elongation of Jupiter might be observed; the latter from the comparison of his known distance from the earth with his apparent diameter which might be observed. But there is a difficulty in ascertaining his distance from the earth, for his parallax, the element generally used in such computations, is too small to be correctly ascertained. Taking the values given in the table, in p. 122, to be correct, Jupiter when nearest to the earth is more than four times the distance of the sun from it; his horizontal parallax, therefore, is more than four times less than that of the sun, and as the parallax of the sun is little more than 8″, that of Jupiter, even under these, the most favourable circumstances, will hardly exceed 2″, a quantity much too small to ground any calculations upon and still less on its variations. How, then, are these elements to be ascertained? Assuming, indeed, the correctness of the law that the periodic times vary as the square root of the cubes of the distances, Jupiter's distance from the sun may be ascertained from observation of the periodic times, and his distance from the earth, and consequently his real magnitude, deduced from this element; but it is obviously desirable to be able also to ascertain this independently. And the observation of the eclipses of his satellites enables us to do this. If we take an eclipse, of which the beginning and end can be observed, the time of the middle of the eclipse can be ascertained; and at that time the planet would be, with nearly complete accuracy, between the sun and the satellite. The line joining the sun and Jupiter is therefore at that time the same with that joining Jupiter and the satellite. Now the position of the line joining the planet and satellite may be ascertained from our knowledge of the motions of the latter, the position, therefore, of the line joining the sun and Jupiter may be known. That of the line joining the sun and earth at the same time is known also; and hence the angle between these two lines. Again, the angle of elongation of the sun and Jupiter, as seen from the earth at the same instant, may be observed; and consequently in the triangle formed by joining the sun, earth, and Jupiter, two angles, and of course the third are known. The ratio, therefore, of the sides is known; and the distance of Jupiter from the sun, in terms of the sun's distance from the earth, may thus be ascertained. In the same manner the distance of Jupiter from the earth at the same instant may be ascertained, and by comparing this with the apparent diameter, the real diameter of Jupiter may be ascertained also. We have already stated it to be about 11 times that of the earth; his actual magnitude of course exceeds that of the earth in the ratio of the cubes of their diameters, or of 11^3 or 1331, to 1.

When the motions of Jupiter and his satellites were sufficiently ascertained to enable astronomers to predict the periods of their eclipses, a singular irregularity was found to pervade the observations of them. They were found to occur very nearly at the times predicted; but not accurately. When Jupiter's distance from the earth was least, the eclipses occurred rather more than 8 minutes before the computed time; when greatest, about the same period after it; and at all intermediate distances the variation of the observed from the computed time was found accurately to correspond with the variation of the actual from the mean distance; the observed time preceding the computed time, whenever the distance of Jupiter from the earth was less than the mean distance, and following it, whenever it was greater. It is evident, therefore, that this variation *depends* on the distance of Jupiter from the earth, as it corresponds accurately with this element at all times, and under all circumstances. This discovery was made by Roemer, a Danish astronomer, in 1674; and he also suggested the true explanation. It clearly appears that the result exactly corresponds with that which would take place if the appearances in question were seen at different times by observers at different distances from the objects observed, and later as the distances increased; or, in other words, (as these appearances are only seen by the eyes receiving, or ceasing to receive light from the object observed,) if that light travelled gradually, instead of being propagated instantaneously, from the object to the observer. If this were so, and the rate of its motion uniform, the variation in time would be exactly in the proportion of the variation of distance; and as it is found to be so, we are authorized to conclude that the reason of its occurrence is that a certain time is necessary to the transmission of light. The utmost difference, or the sum of the extreme acceleration, and the extreme retardation, of the eclipses is $16^m\ 26^s$. The greatest distance of Jupiter from the earth is the sum of his distance from the sun, and the sun's distance from the earth; the least distance of Jupiter from the earth is the difference of these same quantities; and the extreme difference between his distances from the earth at different times is the difference between these values, or twice the sun's distance from the earth. The time, therefore, which light takes to traverse this distance is the before-mentioned period of $16^m\ 26^s$., or 986^s; and as the sun's distance is about 93,726,900 miles, or its double 187,453,800 miles, the motion of light in 1^s will be $\frac{187,453,800}{986}$ or about 190,000 miles.

The detail given of the motions of Jupiter's satellites and the mode of observing them will sufficiently explain the mode of proceeding with respect to the remaining planets. As they are more distant from the earth, the observation of their peculiar appearances is more difficult, and our knowledge respecting them more imperfect. The principles of investigation, however, are the same. Saturn is then found to have seven satellites, moving from West to East around the planet, in orbits nearly circular*, and generally inclined at angles of about 31° to the plane of the ecliptic. The seventh, or most distant satellite, however, has its orbit coinciding more nearly with the plane of the ecliptic. The mean distances of these satellites from Saturn are respectively 3.080, 3.952, 4.893, 6.268, 8.754, 20.295, 59.154 times the planet's radius; and their periodic times .94271, 1.37024, 1.88780, 2.73948, 4.51749, 15.94530, 79.32960 days, values which again correspond with the law that the periodic times vary as the square root of the cube of the axis major. In each of these satellites, also, as well as in those of Jupiter, the areas described by the radius vector are proportional to the time.

Like the satellites of Jupiter, these

* The orbit of the sixth satellite, however, is perceptibly elliptical.

have no ascertainable apparent diameter. The sixth, treating the satellite nearest to the planet as the first, and the most distant as the seventh or last, is the largest, and was first discovered. It was first observed by Huygens, and is thence called the Huygenian satellite. It is seen without much difficulty in a common telescope. The others are more difficult of observation; the first and second were never discovered till Dr. Herschel observed them; the others were discovered at different periods between the eras of Huygens and Herschel.

Difficult, however, as they are of observation, one of them, the seventh, furnishes a very remarkable result. Its light, whenever it is to the east of Saturn, becomes so faint that the satellite cannot be seen without great difficulty, much exceeding that presented at other times. As in the case of the satellites of Jupiter, this cannot proceed from any phases presented by the satellite to the earth; it must, therefore, depend on some variation of its surface. We are thus enabled to conclude that this satellite has a motion of rotation round an axis, and that this motion, like the corresponding motions of Jupiter's satellites, is performed in its periodic time. We have no means of making similar observations on the other satellites; but as far as our observations go, it appears to be a fact common to all satellites, that they do revolve on an axis, and that the period of their revolution is the same with their periodic time. This offers a remarkable analogy, and also a remarkable contrast, to the facts stated in the table in page 122, with respect to the planets; an analogy in the existence of motions of rotation, a contrast in the circumstance that the period of rotation appears in the case of satellites to be uniformly the same with the periodic time, while in the case of the planets it evidently is unconnected with it.

The most remarkable appearance presented by Saturn remains to be described. Galileo observed that Saturn did not appear spherical, but seemed to have two small bodies adhering to, or closely accompanying him; not moving round him like satellites, but continuing, at least for a considerable period of time, to hold the same position with respect to him. Some time afterwards, when the construction of telescopes was considerably improved, Huygens discovered that these were not separate bodies, but the two sides of a ring, which encompasses the planet. By minute observation, it has been ascertained that the plane of this ring is inclined at an angle of 31° 32′ to that of the ecliptic, and that it exactly corresponds with the plane of the equator of the planet itself. It is found that the thickness of the ring is very small, probably not more than 1000 miles, while its outer diameter is about 200,000 miles, (its apparent diameter at the mean distance of Saturn being 47½,″) and its inner diameter about 161,000; the width of the whole ring being, therefore, about 20,000 miles. On more minute observation, however, it is found that the ring is divided about 6700 miles from its outer edge, and that there is then an interval of about 3800 miles in width; so that it really consists of two concentric rings, an outer and an inner one, lying in the same plane. It has even been supposed from some faint appearances of division on its surface, that it consists of a greater number of these concentric rings. These minute observations are principally due to Dr. Herschel, who has also ascertained, by the observation of some brilliant points on its surface, that it has a motion of rotation from West to East round an axis perpendicular to its plane, and passing through the centre of Saturn, and consequently coinciding with the axis of rotation of Saturn himself. The period of this rotation is .437 of a day, very nearly the same as that of the rotation of Saturn himself. A more remarkable observation is this, that the period of rotation of the ring, is that which would be the period of rotation of a satellite whose orbit was the mean circumference of the ring, deducing that period from those of the actually existing satellites, according to the law that the periodic times vary as the square root of the cubes of the distances.

The plane of the ring being inclined to the ecliptic, and the earth being in the plane of the ecliptic, and Saturn never far distant from it, the plane of the ring can never be perpendicular to the line joining Saturn with the Earth. The ring, therefore, will always be seen obliquely. Now a circle, when viewed obliquely, at a considerable distance, assumes to the eye the appearance of an ellipse, an ellipse being the projection of a circle; and the more obliquely the circle is presented to the eye, or the smaller the angle made by the plane of the circle with the line joining the eye and its centre, the more excentric is the ellipse. When this line coincides with the plane of the circle, the circle assumes the appearance of a straight line. Now as the centre of the earth is a point in the ecliptic, it may evidently happen that the common section of the plane of the ring and the ecliptic, which is a line in the plane of the ecliptic, may pass through the earth; and in this case, as the same common section is a line in the plane of the ring also, the circle, which forms the ring, would be seen as a straight line. When this is the case, the edge of the ring only is presented to the spectator, and as that edge is comparatively very thin, it is not of sufficient magnitude to be visible except by telescopes of very great power. Herschel, however, succeeded in seeing it even under these circumstances, and found it, as it would naturally appear, in the form of a straight line. At different inclinations, the ellipse would of course appear of different width; but the inclination never much exceeding 30°, the minor axis of the ellipse will never be more than about half the major. These appearances may be advantageously represented by figures, of which *fig.* 33 represents Saturn with

Fig. 33.

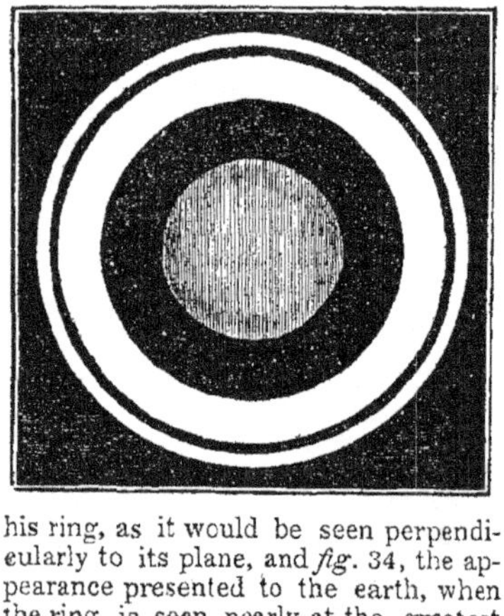

his ring, as it would be seen perpendicularly to its plane, and *fig.* 34, the appearance presented to the earth, when the ring is seen nearly at the greatest angle. Under these circumstances, there is a well-defined space between

Fig. 34.

the ring and the planet, through which stars have occasionally been seen.

Again, the sun is in the plane of the ecliptic; and therefore, on the same principle as before, the common section of the ring and ecliptic may pass through it. In this case, therefore, the rays of the sun will fall only on the edge of the ring. When they do so, the ring is found to disappear to common observation; but a line of shadow, corresponding with the position of its plane, is found crossing the disk of the planet. The conclusion is obvious, that both Saturn and the ring shine only by reflected light; for the ring becomes invisible when the rays of the sun do not fall on it, which it would not do if it shone by its own light; and Saturn is darkened where the rays of the sun are intercepted from it, which also would not be the case if it were itself luminous. The very edge of the ring, however, which in this case would receive the rays of the sun, would still be illuminated; but as in the case when the edge only is turned to us, it is invisible, except with telescopes of very great power. With these Dr. Herschel succeeded in observing it.

Besides these disappearances, however, which last for a comparatively short time, there are others of more importance. In *fig.* 35, let S represent the place of the sun, the circle described about it the orbit of Saturn, ♄ any position of Saturn, and A ♄ B (a line drawn through ♄) the intersection of the plane of the ring with the plane of the ecliptic. Let XY be drawn through S, parallel to A ♄ B, and meeting the orbit in X and Y. The plane

Fig. 35.

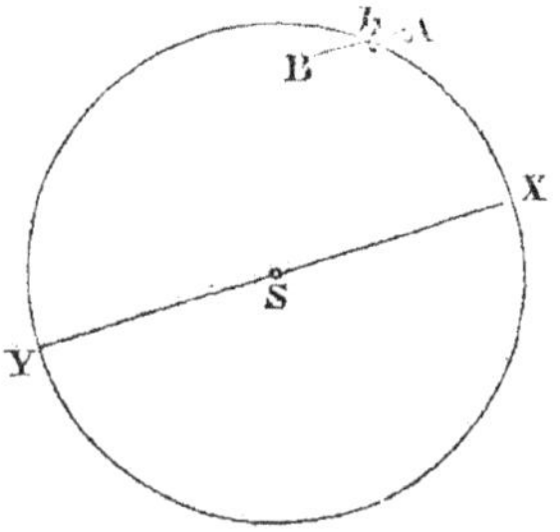

of the ring is always perpendicular to the axis of the planet; and consequently, as we have already stated that the axis of the planet is parallel to itself in all positions, the plane of the ring must be so also, and consequently the several intersections of that plane with the plane of the ecliptic will always be parallel. When, therefore, the planet is at X and Y, this intersection will coincide with the lines X S, Y S, joining the centre of the ring and the sun. This, therefore, must happen twice in each revolution of the planet round the sun; and on these occasions the disappearances of the ring, of which we have last spoken, will occur, the edge of the ring being then turned towards the sun. Besides this, it is plain that one side of the ring will be turned towards the sun throughout that half of the orbit which lies from X through ♄ to Y; and the other side, throughout the remaining half; and consequently that the opposite sides of the ring will be illumined during each of these periods; and that one side will always be light, and the other dark, except on the occasions when the edge of the ring only is presented to the sun, and both the sides are deprived of his light.

Now the earth is at a considerable distance from the sun, though that distance is small when compared with its distance from Saturn. It may easily, therefore, happen that the plane of the ring passes between the earth and the sun, and consequently that the opposite sides of the ring are presented to those bodies. The dark side, therefore, will be presented to the earth, and the ring will, on that account, be invisible. The very edge, indeed, of the ring would not be absolutely turned away either from the sun or the earth; it would therefore be illumined, and this illuminated edge might be visible; and here again Dr. Herschel, with the assistance of his very powerful instruments, has succeeded in seeing this small portion of the ring, when it was invisible to all other observers. Generally speaking, however, the ring may be considered as invisible during this period; and Galileo was led for a time to doubt the correctness of his discovery, from the disappearance of the ring from this cause after he had for some months observed it. Its nature and the laws of its motion were not then sufficiently ascertained to enable him to account for this disappearance, and the correctness of his original observations appeared doubtful, until the re-appearance of the ring, when the sun and earth again were on the same side of its plane, confirmed them.

The observation of the periods when the edge of the ring is presented to the earth, enables us to determine the direction of the line in which the plane of the ring intersects the ecliptic; for this plane then passes through the earth, and the line joining Saturn and the earth is the line of intersection required. This line always moves parallel to itself. When, therefore, it is ascertained by observation that the edge of the ring is presented to the sun, in which case we have already seen that this line of intersection must pass through the sun, the position of this line is known. The position of the line joining the earth and sun may be observed at the same period; and consequently the angle between these two lines known. The elongation, also, of Saturn from the sun, as seen at the earth, may be observed. As in the case of Jupiter, therefore, we know the angles of the rectilineal triangle formed by lines joining the sun, the earth, and Saturn; and we can, therefore, ascertain the distances of Saturn from the Sun, and from the earth; and from the latter, and the apparent magnitude at the time, we can ascertain the real magnitude. As in the case of Jupiter, the results so obtained correspond with those deduced from the law of the periodic times. His diameter is thus ascertained to be ten times that of the earth, and his whole bulk of course to exceed that of the earth in the proportion of the cube of 10, or of 1000 to 1.

Uranus is so distant from the earth that there is little opportunity of minute observation concerning him. He is, however, ascertained to have six satellites moving in orbits nearly circular, and with periodic times corresponding to their distances according to the usual law. Their mean distances are respectively 13.120, 17.022, 19.845, 22.752, 45.507, and 91.008 times the radius of the planet; and their periodic times, 5.8926, 8.7068, 10.9611, 13.4559, 38.0750, and 107.6944 days. There is one remarkable difference in their motions from those of the other bodies which we have considered. Their orbits are very nearly perpendicular to the ecliptic; and their motion, therefore, can hardly be considered from West to East, agreeably to the uniform rule found to obtain in all the other motions of the system. These satellites, as well as the planet which they accompany, were all discovered by Dr. Herschel; and he further suspected, from the result of his observations, that the planet was surrounded by two rings perpendicular to each other. This fact, if established, is a remarkable and very curious one; but it cannot at present be considered as at all certain. There are no means of ascertaining his distance and actual magnitude, like those mentioned in the case of Jupiter and Saturn; the determination of these elements must, therefore, rest on the observation of his periodic time, and the law by which the distances and periodic times are connected.

We have now gone through a statement of the motions of all the bodies which are commonly considered to form the solar system. There are, indeed, other bodies belonging to it which occasionally appear in the heavens, and excite much observation and curiosity when they do so, which are known by the name of *comets*. Their motions, however, are more complicated and difficult of ascertainment, and cannot well be explained without the aid of some theory to which we have not, up to the present period of this treatise, attained. These motions, also, are not suited to throw much light on certain very important considerations, to which we are now qualified to proceed; it will, therefore, be most convenient to postpone any consideration of them for the present, and proceed to examine the results which we have already obtained; and to deduce from them some very important modifications, or rather a complete reconstruction of a great part of our whole system of astronomy.

Chapter V.

Section I.—*First idea of the motion of the Earth.*

On comparing together all the facts which we have ascertained with respect to the planets and their satellites, we may consider certain general laws to be completely established. The proportion between the periodic times and the major axes of the orbits is found to subsist, both with respect to every planet in its revolution round the sun, and to every satellite of each particular planet in its revolution round that planet. Subsisting thus, at distances very far different from each other, it is at least a plausible conjecture that it does so generally, and, consequently, that if a new planet or satellite were discovered, its motions would follow this same law. The same remark will apply to Kepler's other laws, namely, that these revolutions take place in elliptic orbits round the sun, or the planet, in the focus; and that the radius vector of the revolving body describes equal areas in equal times.

Again, we have reason to conclude, though not with the same certainty, that all the bodies of the solar system have motions of rotation round an axis; the planets in times which do not appear to follow any particular rule; the satellites in times always equal to the periodic time of the particular satellite round its planet. We are not, indeed, able to ascertain these facts with respect to all the bodies with which we are acquainted, of either description; but we find the case to be so with respect to all which we can observe; and it is, therefore, a natural conjecture, that it is so in the others also, as they present no appearances inconsistent with the supposition. We should, therefore, be inclined to conjecture, that a new planet, if such should hereafter be discovered revolving round the sun, would have a motion of rotation round an axis; and that, if it were accompanied by any satellite, this satellite would itself revolve round an axis, and that its times of rotation on its axis, and of revolution round the planet, would be equal. All the other motions of the solar system are from West to East; we should, therefore, expect that these would be so also.

Now we have already stated, that the sun's periodic time is exactly equal to what would, in the supposition of the same law obtaining at that distance, be the periodic time of a planet revolving round the sun in an orbit of which the major axis was equal to that of the sun's orbit round the earth. If, therefore, for an instant we suppose that the appearances actually observed could equally well be explained on the supposition of the earth revolving round the sun at that distance at which we have hitherto considered the sun to revolve round the earth, the periodic time observed would correspond with that determined by the general law regulating the duration of such revolutions.

We are always able to observe the position of the moon with respect to the earth, and the supposition we are proposing would not be fit for adoption, if it were found inconsistent with her motions. But if it be not so, and if it be also true, the moon would evidently be a body revolving round the earth, and accompanying it in its course round the sun; or it would be a satellite of the earth. We should, therefore, be led to expect, that if the earth and moon were such a planet and satellite, they would each have a motion of rotation from West to East; and that the motion of rotation of the moon would be completed in the time of its revolution round the earth. In fact, we know that it is so; and this instance of correspondence, though by itself it furnishes no proof of the truth of the supposition, leads us at least to consider it as one worth investigation. The motions of the moon, also, both of revolution and rotation, are from West to East, like those of the other planets and satellites.

It would be easy at once to point out some arguments which, assuming that the supposition is not inconsistent with the appearances observed, would induce us to consider it the true one; but as no probabilities of this kind could warrant us in adopting the supposition, if it failed to correspond with these appearances, the preferable order will be first to examine whether it is capable of explaining them. If it proves so, it will then be time to consider whether it furnishes a more probable explanation of them than the supposition which we have hitherto adopted, and which observation originally leads us to, namely, that the earth is at rest, and that all the heavenly bodies revolve about it. We proceed, therefore, first to investigate the effects which might be produced by the rotation of the earth round an axis, and its revolution round the sun, supposing these to exist. We shall find that they correspond with the appearances actually presented. It is, however, a plausible objection, that these motions cannot exist, for that, if they did, we should be sensible of them. Our next object, therefore, will be to consider and remove this objection, and we shall, then, examine the probabilities which there are in favour of the supposition of the earth's motion; and, finally, point out one particular phenomenon which can only be explained on that supposition, and which, therefore, seems almost exclusively to demonstrate its truth. This is the phenomenon of the aberration of the fixed stars: it has been mentioned before, but its explanation deferred.

SECTION II.—*Identity of the appearances on the supposition that the Sun and Heavens are at rest, and the Earth describes an orbit round the Sun, and revolves on an axis with those actually observed.*

THE general principle which will lead us through all the considerations requisite to the discussion of this subject, is this: that wherever it is known that of two bodies one revolves round, or has any motion with respect to, the other, their relative motions may be equally well explained, whichever of the two we suppose to be at rest, and the other in motion.

The simplest case is that of two bodies only, considered without reference to any others, and on the supposition that the motion of the moving body takes place in a plane; and a very simple diagram, and very simple considerations, will suffice to show that their relative motions, whatever they be, may be equally well explained, whichever we consider to be at rest. Of course, when two bodies only are concerned, their *relative* motions will be the same on different suppositions, if their distances from, and directions with respect to each other, are the same on each; we are not concerned with their *actual* positions in space. Let then A and B in *fig.* 36, re-

Fig. 36.

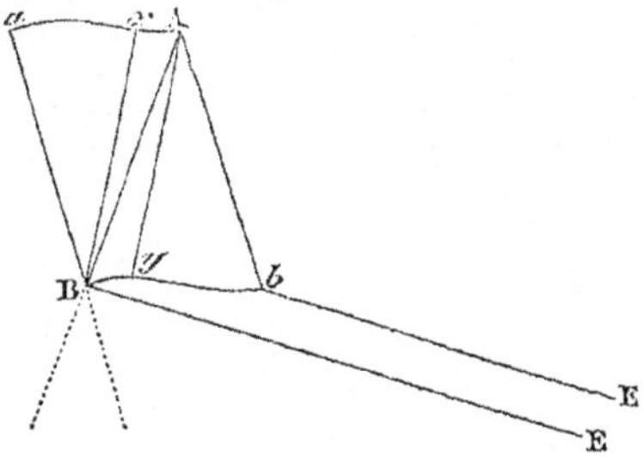

present any two bodies whatever, and let us first suppose B at rest, and A to move through any line A *a*, either curved or straight, and drawn in any direction whatsoever. Draw B E a line in any given direction whatever, as, for instance, Eastward, in the plane passing through A *a* B; then the relative positions of the two bodies at the beginning and end of the motion will be ascertained by the distances A B, *a* B, and the angles A B E, *a* B E, respectively. Let B *b* be a curve equal and similar to A *a*, and situated in the same manner with respect to the line A B, that the curve A *a* is with respect to the line B A; the relative position of the two bodies will be the same if A be supposed at rest, and B to move from B to *b*, that it was on the former supposition. To show this, let A *b* be joined, and *b* E be drawn parallel to B E; b E, therefore, is in the same direction as B E. Now A B a, B A b, are equal and similar figures; the corresponding sides and angles, therefore, are equal, or Ab=Ba, and the angle B A b=the angle A B a. Hence A b and B a are parallel, and b E B E are parallel also; the angles A b E, a B E, contained between these pairs of parallel lines, are, therefore, equal also. But b A measures the distance of A from b, and A b E its direction with respect to it estimated in the same manner as the direction of a with respect to B; these, therefore, are the same on the supposition of B's motion, that they were on that of A's; or the relative position is the same on each supposition. Again, the figures A B a, B A b, are equal and similar, and, of course, their areas are equal also; and if any similar portions of these areas A B x, B A y, be cut off, these, also, in the same manner will be equal to each other. It is evident that these conclusions are entirely independent of the particular figure assigned to the curve described.

The curve apparently described by the sun is an ellipse, of which the earth is in the focus. Their relative positions, then, will be the same, if the earth describes a similar ellipse round the sun. All the lines drawn from the different points in this ellipse to the sun will be equal to those drawn to the earth from the corresponding points in the ellipse supposed to be described by the sun. The angles between these corresponding lines will be equal also; and, consequently, the sun will occupy exactly the same position with respect to the ellipse supposed to be described by the earth, that the earth does with respect to that apparently described by the sun; or it would be in the focus.

Supposing the earth thus to move, her radius vector would describe equal areas in equal times, or areas always proportional to the time; for the areas described by the radius vector of the earth, (as the small area B A y, or the large one B A b, supposing B to represent the earth, and A the sun,) are always equal to the area described by the radius vector of the sun (as the small area A B x, or the large one A B a). These latter areas, however, we have seen to follow the law of Kepler; those described by the radius vector of the earth, therefore, will follow it also.

The motion of the earth, on the supposition thus made, will be from West to East. In the figure as drawn, A a representing the motion of the sun, is a motion from West to East; the angle A B a, therefore, is an angle measured in that direction. If the lines A B, a B, be continued by the two dotted lines, it is obvious that the radius vector B A, by continuing its motion in the same direction, will successively coincide with each of the two dotted lines, and describe the angle between them. But this angle is equal to the angle B A b, described by the radius vector A B, on the supposition of the earth's motion; for the dotted lines are parallel to the lines B A, A b, which contain it. The angle B A b, therefore, corresponds in magnitude and direction with that between the dotted lines, and it is, therefore, described like that angle, by a motion from West to East. The angular direction of the motion, therefore, is the same, although the linear direction is opposite.

We see, then, that the earth may be supposed to have a motion round the sun, which would make their relative situations always the same that they are observed to be; that this motion would be from West to East, like those of the other planets, and that the earth on this supposition would follow all Kepler's laws of planetary motion, having her orbit an ellipse round the sun in the focus, the areas described by the radius vector proportional to the time, and the periodic time bearing to that of all other planets the ratio of the square root of the cube of their respective major axes.

We observe, however, the situation of the moon, the planets, and the fixed stars, as well as that of the sun. We proceed, therefore, to examine whether these, also, are consistent with the same supposition.

In the case of the moon there is no difficulty. We have ascertained nothing with respect to its motions, except that it is always in a certain position with respect to the earth. If, therefore, the earth moves, the moon must partake of its motion, besides having a motion of her own round the earth. This motion round the earth is all which observers on the earth perceive; but the actual motion performed by the moon in absolute space will be, on this supposition, a curve of a very complicated nature. There is, however, nothing improbable in this result. It takes place in the case of all other satellites, and very simple instances may be adduced to show the manner in which these different motions of the moon round the earth, and of the earth round the sun, may be combined. If, for instance, we suppose the wheel of a carriage to have luminous points fixed on it, and the carriage to be driven on a dark night, a spectator standing still might observe the path of the luminous points. If they were on the edge of the wheel, the path would be a cycloid, if nearer its centre, a trochoid. These paths described in absolute space have a resemblance to a circle. We know, however, that the point is always at the same distance from the centre of the wheel, and, therefore, that they are actually described by the combination of two motions, that of the point round the centre, and of the centre itself along the road. We might, from mere calculation, ascertain from the observed path of the luminous point that it might be described by a combination of such motions; and we do, in fact, make a similar calculation in ascertaining the orbit of the satellite of a distant planet, when we observe the orbit in absolute space; having, however, this advantage, that we see, also, the planet itself, the centre round which one of its motions is performed, and the sun round which the satellite itself, with its planet, performs the other, and are thus led at once to refer its motions to their right centres. In the case of the moon, when we are ourselves at the centre of one of its motions, a different illustration will serve us best. Every one who has walked out in summer has observed insects buzzing about his head, and flying round and round it, accompanying him wherever he goes. The only motion which he at first observes is this circular motion about him; for it is the only change of their *relative* position. It is, however, evident, that their actual motion is not circular, for if it were, they would return to the same point of space, while he would have quitted it; they necessarily, therefore, follow him in all his motions, and partake of them, besides having a motion round him. Just the same, on the supposition of the earth's motion round the sun, would be the motion of the moon; the only difference is, that we are not conscious of the motion of the earth, we are of our own; in one case, we are slow to believe the supposition; in the other, we are aware of the fact. But the motion is similar in each case, and there is no real difficulty in the one, which does not equally apply to the other.

The manner in which the supposition of the earth's motion will apply to the observations of the motions of the planets, requires some rather more complicated examination.

Fig. 37.

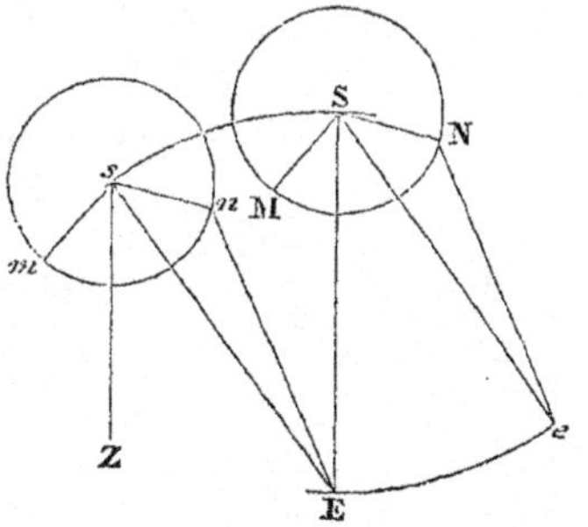

In *fig.* 37, let S represent the position of the Sun at any given time, E that of earth, and M that of any planet, which we will suppose, for the present, to move in the plane of the ecliptic. A curve, therefore, drawn in the plane passing through S, M, and E, will represent the orbit of the planet, and another curve drawn in the same plane, the orbit of the sun, if that luminary be supposed to move round the earth. Let S *s* represent a portion of the Sun's orbit, and let M N be the portion of the planet's orbit described round the sun in the same time. From the point *s*, draw *s* Z parallel to S E, and make the angles Z *s m*, Z *s n*, equal to E S M, E S N, respectively. And draw *m n* a curve similar and equal to M N. Then it is evident that the points *m*, *n*, are situated with respect to *s*, just as M, N, are with respect to S, for they are at the same distances from it, and their directions are equally inclined to lines *s* Z, S E, drawn in the same direction from the respective points *s*, S. But the planet was at M when the sun was at S, and would describe an arc M N, while the sun moved from S to *s*, the point *n*, therefore, represents the planet's situation after the sun has moved from S to *s*. Let us now suppose the earth to move round the sun, in the same manner as we before supposed the sun to move round the earth, and of course, a curve described in the plane of S M N E will represent the orbit of the earth: and if we make the angle E S *e* = S E *s*, and the curve E *e* similar to S *s*, and situated with respect to E S, as S *s* is with respect to S E, the curve E *e* will represent an orbit of the earth, which will make the relative situations of the earth and sun the same as they would be if the sun were supposed to move. On supposition, therefore, of the sun's motion, the respective places of the sun, earth, and planet, at the end of the given time, would be *s*, E, *n*, the planet's distance from the earth, *n* E, and its elongation *n* E *s*: on the supposition of the earth's motion round the sun, the places would be S, *e*, N, the distance N *e*, and the elongation N *e* S.

Now, *s* E = S *e*, for S E *s*, E S *e*, are equal and similar figures.

And *s n* = S N, for *m s n*, M S N, are equal and similar figures.

The included angles, also, E *s n*, *e* S N, are equal, for E *s n* = Z *s n* − Z *s* E = E S N − E S *e* = *e* S N,

the two triangles, therefore, are equal in every respect, and E *n* = *e* N, and *n* E *s* = N *e* S. The distances, therefore, of the planet from the earth are the same on the two suppositions; and so are its elongations from the sun. But the apparent position of the sun itself is the same on each supposition; and consequently the apparent position of a point at the same elongation from him is the same, if its distance is equal in each case. The apparent position of the planet, therefore, is so; and all the appearances of the planetary motion will be exactly the same on the one supposition that they are on the other.

As yet we have only considered the case of a planet moving in the plane of the ecliptic. It is, however, obvious that the same conclusions will apply also to the case of a body which moves out of that plane. If from every point of the orbit perpendiculars be let fall on the plane of the ecliptic, they will trace out a certain curve on that plane; and this curve, wherever it is situated, will always be similar and equal to itself; and, consequently, as in the former case, each point of this curve so constructed will be at the same distance, and in the same direction with respect to the earth, on either of the suppositions in question. The distance of these points from the earth being equal on each supposition, and the actual perpendicular distance of the planet from the plane of the ecliptic being the same in each case, the angle subtended at the earth by this perpendicular distance, or the planet's apparent distance from the ecliptic, that is, his latitude, is the same on each supposition, and his actual distance is equal also; for it is the hypothenuse of equal right-angled triangles: and the direction of the lines joining the earth and planet being the same, the vectors passing through these lines perpendicular to the ecliptic will be parallel, or run in the same direction; and thus the longitude also will be the same. In this case also, then, the latitude and longitude, and the actual distance being all the same, all the appearances of planetary motion will be exactly the same on the supposition of the earth's motion, that they are on that of the sun's.

Finally, the identity of these appearances has been ascertained only by showing that the actual distances are the same, and the angles made with certain lines drawn in fixed directions, and therefore parallel to each other, the same on each supposition. The position, however, of the sun and planets is observed with respect to the fixed stars also; and if, therefore, these parallel lines would point to observably different points in the heavens, when drawn from different situations of the earth on the supposition of its motion, the situations of the sun and planets, being the same with respect to these points, would be observably different with regard to the heavens in general. If, however, the sphere of the heavens be considered as immeasurably distant, the interval between the points where these parallels met them, being always equal to the distance of the situations of the earth from which they are drawn, would become, notwithstanding the great magnitude of this distance, too small, in comparison, to be at all observable; just as we have already seen, that, in fact, the distance of any two points on the earth's surface is too small to produce any distinction between the relative positions of the fixed stars as seen at them. There is, therefore, no impossibility which, on this account, attaches to the supposition of the motion of the earth; for any difficulty which would arise may be removed by increasing the distance at which we conceive the fixed stars to be placed from us. We shall hereafter consider the probability of this supposition. For the present, it is sufficient to have shown that, if admissible, it would obviate the only remaining difficulty; and, consequently, that the supposition of the earth's revolving round the sun is consistent with all our observations of the apparent motions of the sun, moon, and planets, and the lunar and planetary theory deduced therefrom. We conclude, therefore, that the earth *may* be a planet, subject to the same laws as the others. But if the earth be a planet, we should conjecture that it would have a motion of rotation upon an axis, and that this motion would take place from West to East. We proceed, therefore, to inquire what would be the apparent effect produced by such a motion.

Let us, then, suppose the earth to revolve round an axis passing through its centre. The centre of the earth being supposed to coincide with that of the sphere of the heavens, the axis will be a diameter of the sphere, and any plane passing through it will intersect that sphere in a great circle. The poles of the heavens are at the extremities of a diameter of the sphere. Let us suppose the axis of rotation of the earth to coincide with this diameter; and every great circle made by the intersection of any plane passing through this axis with the heavens will pass through the poles of the heavens; or it will be a *meridian*. Let us now suppose this plane to move from West to East, and in every position its intersection with the heavens will still be a meridian, but a meridian continually more to the East than before. If, therefore, it coincided at one hour with the meridian of certain stars, it will coincide an hour after with the meridian of certain stars more to the East; and which, therefore, at the former observation, were to the East of the then intersection. Again, if the motion of this plane be uniform, its approach to these stars will be uniform also; and as the angle between two great circles of a sphere is the same as the angle made by their planes at their intersection; if we suppose the plane to move from West to East at a certain rate, as 15° an hour, its intersection with the heavens will coincide from time to time with meridians distant by corresponding arcs; or distant by 15° for each hour of the motion. If, therefore, we suppose the earth to revolve uniformly round its axis in twenty-four sidereal hours, the intersection of any such plane will, in the course of these twenty-fours, have coincided with every meridian; and the motion having been uniform, the time elapsed in passing from one meridian to another will be in the proportion of the angular distances between these meridians, or of the differences of right ascension, and will be an hour for every 15° of right ascension. The intersection of this meridian with the earth's surface will always at each place be the same: all observations, therefore, made there, and referred to this intersection, will be consistent with each other. If we suppose the earth to be a solid of revolution, that is to say, that any section perpendicular to the axis is always a circle: a plane passing through the axis and any point on the earth's surface will be perpendicular to the plane touching the earth at that point, or to the horizon at that place; and it will therefore coincide with the plane of the meridian of the place. The meridian of the place, therefore, will travel Eastward at the same rate of 15° in an hour; and if, while it really did so, we were to suppose it fixed, and instead of observing its motions with respect to the stars, observe the positions of the stars with respect to it, we should find that the stars which were on it at one instant were to the West of it at the next, it having really moved to the East of them; and they would therefore appear to have moved from East to West, exactly at the same rate as the meridian really moved from West to East, or 15° in a sidereal hour.

These are exactly the appearances described in the former part of this treatise as occurring during a sidereal day. We come, therefore, to this conclusion —that the diurnal appearances of the heavens *may* be explained on the supposition that the heavens are at rest, and that the earth moves from West to East in twenty-four sidereal hours, round an axis coinciding with the line drawn from the centre of the earth to the poles of the heavens. We have already seen that the peculiar motions of the sun, moon, and planets are consistent with the supposition, that the sun also is at rest, and that the earth moves round him.

If the earth has such a motion as that now ascribed to it, the circle, the plane of which passes through its centre, and is perpendicular to its axis, will be its *equator*, according to the use previously made of this term. We will, therefore, use this term for speaking of that circle, although without at present affirming the actual existence of such a motion of rotation, and therefore only considering the axis as that diameter which, when produced, passes through the poles of the heavens.

SECTION III.—*Objections to the supposition of the Earth's motion removed.*

THE circumference of a great circle of the earth, considering that body as spherical, is about 25,000 miles. If, therefore, the earth revolves on an axis in 24 sidereal hours, a point at the equator will be transferred through 25,000 miles in that time, or it will move at the rate of more than 1000 miles in an hour, or about 17⅓ miles in a minute. The rate of motion of points situated elsewhere will fall short of this, in the proportion of the respective magnitudes of the circles which they describe.

Again, supposing the orbit of the earth to be circular, and its radius 93,726,900 miles, (which is about the mean distance, and the excentricity of the orbit will make very little difference in the result,) the length of the orbit will be 588,902,993 miles, which being described in a year, will make the average velocity of the earth in its orbit nearly 19 miles in a second.

Of these motions, if they exist, we are altogether unconscious; and it is a very plausible objection to the supposition of the earth's possessing them that we are so. They are motions far more rapid, especially that of the earth in its orbit, than any which are the subjects of our common observations; and knowing by experience the violent effects of all rapid motions observed round the earth, it is difficult to conceive that such motions as these, in which we ourselves partake, can possibly elude our observation.

To discover whether there is really any improbability in this supposition, it is necessary to consider in what manner it is that we become aware of the existence of any motion which we perceive in ourselves; and the effects by which we perceive it seem to be referable to two classes; change in the relative positions of objects, and the actual feeling of motion.

We have already seen that the change in the relative positions of objects may be the same on the supposition of the motion of the earth, as on that of the motion of the sun and heavens. In other words, the change in the relative position of objects only proves the existence of motion somewhere: it does not at all show which is the moving body.

We are, however, often fully conscious that we are ourselves in motion; and this consciousness is produced in several distinct ways; by ourselves performing some functions which produce it, as walking or running; by being carried by some body or machine, of which we either see the action which produces the motion, as in riding on horseback, or being drawn in a carriage, or else know from other sources that the motion must be in ourselves, as in sailing in a vessel in a high wind; by the inequalities of the motion, even when we do not observe the manner in which it is performed, as in driving in a carriage over a rough road, or in sailing on a troubled sea; by the sensation of passing rapidly through the air, when we see the leaves and every light body at rest, and therefore know that the sensation we perceive cannot arise from wind blowing upon ourselves at rest. There seems to be no evidence of actual motion which may not be referred to one of these classes. The last cannot possibly apply to the motion of the earth, if we suppose the atmosphere to partake equally of that motion; for then the motions being equal, an observer and the air about him would continue in the same relative position, or be relatively at rest; just as a traveller moving in the same direction and at the same rate as the wind, or shut up in a close carriage, and thus taking his atmosphere along with him, is unconscious that the air is not perfectly calm. It is obvious, also, that we cannot disprove the earth's motion, from our inability to observe or explain the exact manner of its production; for although such actual observation or discovery would prove its existence, the want of it can never disprove it. If, however, a certain inequality or roughness is incident to all motion which we know, so as to make us conscious of its existence, the absence of this consciousness would be a strong argument against it; and the existence of such an argument seems to depend on that question of fact.

The answer to this question is, that we are, in fact, often unconscious of the existence of motions which we ourselves perform, and that, while we continue so, we do in fact attribute the observed change in the relative positions of objects to a motion in them, and not in ourselves. We can only instance motions which we are capable of positively detecting; but even in these, attention to our own motion continually ceases.

Thus, while on a rough road or sea, the motion of a carriage or vessel continually reminds us of its existence, on a smooth one we often forget it; and whenever we do, the objects which we are passing appear in motion in the opposite direction to our own. This observation is within the experience of almost every one; and it furnishes by itself complete proof of the possibility of being unconscious even of a motion which, when our attention is called to it, we can easily discover. Perhaps a still more remarkable illustration is afforded by an experiment which every one can easily try. If there is any motion of which it would appear difficult to be unconscious, it would be one performed by our own active exertions, as that of walking; and, consequently, no one as he walks along attributes the general changes in the positions of objects to anything but his own motion. If, however, a man walks close along the side of a wall or railing, nearly of his own height, so that the outline of it may be about on a level with his eye and near it, he will see it continually dancing up and down as he moves, and he will find the *sensation* produced to be that of a vertical motion in the wall or railing itself. Reflection, indeed, will at once teach him that this cannot be so; and a little consideration will inform him that, as at every step he rises on the ball of his foot, and sinks again as he sets his foot down, the eye is at different elevations at different periods, and consequently the relative elevation of the outline different. The experiment may best be tried with a wall or paling of the height mentioned, because from its nearness to the eye the variations of the relative elevation are greater, and therefore more striking than in any other case; and a darkish night is, perhaps, best suited for the trial, because then we perceive the outline distinctly, and yet are not diverted by the presence of other observable objects. The natural observation to be made is this. Our progress forward is laborious in itself; we are, therefore, fully conscious of its existence, and we habitually consider that, in leaving objects behind us, we pass them, and not they us. But the vertical motion is only incidental to this; it therefore escapes our notice entirely till our attention is called to it by some foreign circumstance; and while it does so, we find that a motion which we absolutely perform by the muscular action of our own bodies, eludes our observation, and that the appearance is of motion in the objects at which we look. It is easy to vary our experiments, always with similar results; and we conclude that we are insensible of motion, unless made aware of it by mechanical obstructions or personal exertion.

Now it is obvious that when the question is whether the earth and all that is upon it move, no evidence can be derived from personal exertion; and no mechanical obstructions can be shown to exist. There is, therefore, no reason why we should not suppose its motion, if there be any, to be perfectly smooth and uninterrupted; and if so, experience leads us to believe that we should be unconscious of this motion, because the more nearly we can approach experimentally to such a motion, the more difficult do we find it to perceive the existence of any, and the more apt are we to attribute the effects of our own motion to the existence of motion in other objects.

It was, indeed, once imagined that the consequence of supposing the earth to move would be to show that bodies near it would continually have motions with respect to it materially different from those which would exist on the supposition of its being at rest; for instance, that if a body were let fall from the top of a tower, it would fall in a line joining its place with the earth's centre at that instant, and continue to move in this line; that the bottom of the tower, however, a point on the earth, would, by virtue of the earth's motion, move away from that line; and consequently that, when the body had fallen through the given height, the base of the tower would be found to be far from the line along which the body fell; while, in fact, it was found to coincide with it. The fallacy of this reasoning, however, is obvious, although it was long and much relied on as an objection to the theory of the earth's motion. If the body were at rest relatively to the earth before it began to fall, in which case only we find it fall at the bottom of the tower, it would only be so by having itself the motion of the earth. Its actual motion, therefore, when it fell, would be compounded of this motion and its motion of falling, and would be a curve in consequence; but the motion of falling is the only one which produces change of situation relatively to the earth, and this only therefore could we observe. The combination of the two, however, prevents the earth from leaving the body behind, and it falls, therefore, at the foot of the tower; just as a ball let fall from the top of the mast of a ship in motion shares in the motion of the ship, and falls at the foot of the mast exactly as it would have done had the ship been at rest.

These are the objections which have been urged against the supposition of the earth's motion; and we conclude that there is no reason to disbelieve it, either on the ground that we are unconscious of that motion, or that the *observable* motions of bodies on the earth's surface and near it are the same that they would be on the supposition that the earth is at rest. This supposition, then, is an *admissible* one; for the appearances of the heavens may be explained by it, and there is no evidence whatever to contradict it.

The next stage of the inquiry, therefore, as this supposition and that of the motion of the sun and sphere are equally admissible, is which of the two has the stronger arguments in its favour.

Section IV.—*Probability of the Earth's Motion.*

The first argument in favour of the earth's motion is derived from the much greater simplicity of such a supposition, and the less amount of motion thus introduced into the system.

In explaining this argument, we will begin with the diurnal motions. We have seen that these may all be explained by supposing the earth to revolve round an axis in twenty-four sidereal hours. The motion in this case is of one body, and its extreme velocity about 25,000 miles in this period. If, on the other hand, the earth is at rest, all the bodies visible in the heavens must have a motion of revolution in twenty-four hours; and the circles of their revolution will exceed that of a point on the earth's surface in the proportion of their radii, or of the distances of the bodies from the earth's axis. Instead, therefore, of the motion of one body, we have to suppose that of an incalculable number, scattered at all distances from the moon at sixty times the earth's radius to the sun at nearly 24,000 times the same quantity, to the remoter planets at yet greater and varying distances, and to the fixed stars at distances altogether inappreciable on account of their magnitude. Many of these bodies, again, are of magnitude very far surpassing the earth; and all these bodies, so vast and so immeasurably distant, are to be supposed to revolve round a comparatively small body, with which they do not appear to have any other assignable connexion whatever. The force of the argument will, perhaps, appear more strongly by an instance. The sun is nearly 24,000 times more distant from the axis of the earth than a point on its equator is: the daily circle, therefore, that it would describe, when in the equinoctial, would be 24,000 times greater than that described by such a point. Its magnitude, also, is about 1,331,000 times greater than that of the earth; if, therefore, we were to suppose all the earth to revolve at the same rate as a point on its surface, yet the whole motion of the sun, if it moves, would be greater than that of the earth, if it revolves, in the proportion of 24,000 × 133,100, or 3,194,400,000, to 1. The motion of a point in the equator is greater than that of any other in the earth, and consequently the whole motion of the earth is much less than here represented: on the other hand, the sun's density is probably very far inferior to that of the earth, and the actual proportion of the motions on the two suppositions is not very far from the truth. Monstrous, however, as the notion of so enormous a motion seems, it is but a small part of the difficulty involved in the supposition of the diurnal revolution of all the heavenly bodies round the earth; for the sun is only one, and one of the very nearest of their number. Nature, however, as far as we can observe her, works always frugally, and employs no more exertion than is necessary; and the mind at once recognizes the superior probability of the comparatively trivial motions supposed in the earth, over the inconceivably great and rapid motions which must otherwise be supposed to exist in all the heavenly bodies.

The same argument applies, and really with equal force, though the disproportion between the motions introduced on the two suppositions is not quite so overwhelming in support of the theory, to the supposition that the earth moves round the sun. If only the earth and sun were in existence, the whole motion thus introduced would be less than that involved in the supposition of the sun's motion in the proportion of their respective masses, or of 1 to 133,100; or, after allowing for the difference of density, probably about 1 to 30,000; a difference quite sufficient to induce us to prefer the former hypothesis. But, besides this, if the sun moves round the earth, he is accompanied by ten planets, and seventeen satellites attending them, all partaking in his motion, and consequently moving round the earth also. Some of these bodies, although very inferior to the sun, are themselves of vast magnitude. The earth, on the contrary, is attended only by one satellite, the moon. Her motions, indeed, are to be added to those of the earth; but the additional motion thus introduced is very inferior to that introduced by the motion ascribed even to one of the larger planets. and vastly less, indeed, than that of all the bodies implicated in the supposition of the sun's motion; and thus, also, the superior simplicity of the supposition of the earth's motion is evinced.

In all our other observations of the heavenly bodies, the smaller seems uniformly to attend and depend on the larger: the moon on the earth, the satellites on their respective planets, the planets on the sun. If the earth is a planet, and moves round the sun, her motions are an instance of this observation: if the contrary supposition is adopted, the motion of the sun is an exception to it.

This introduces the mention of what may, perhaps, be considered almost as a general law of the human mind; the readiness to believe in uniformity. Whenever we observe a series of things agreeing with each other in all the circumstances presented to our notice, we feel inclined to conjecture that they will agree, also, in other circumstances; and whenever we find that a fact, of the nature of which we are ignorant, can be reduced to some class of facts with which we are well acquainted, we are impelled to refer it to that class, and consider it to be occasioned in the same manner. These conclusions are often drawn too hastily, but the number of such errors which have been committed, only shows the more strongly the propensity of the human mind thus to generalize and to abstract. This is not the statement of the mere rule, that like causes produce like effects; it is rather the principle on which all our notions of cause and effect depend; for the very belief that, when one event is found uniformly to follow another, the earlier *produces* the latter, is itself merely an instance of the principle in question.*

---

* There are some very curious instances of this propensity to generalize connected with the history of Astronomy. Kepler deduced his laws merely by it; he found them to subsist in the planets which he observed, and boldly announced them as general truths; but he was unable to demonstrate that they were necessarily true universally, if at all. This was reserved for Newton. The confirmation which his researches gave them, fixed them as undoubted laws of nature; till they received this, they were liable to be questioned and even exploded, like many other suppositions of their fanciful, though most ingenious author, which he propounded with equal confidence. In all instances, however, whether he was right or wrong, he acted on the same principle which we are now discussing, that of believing in the generality of rules which he found to obtain in a few instances.

Another very remarkable guess of the same nature is contained in a singular analogy which Professor Bode, of Berlin, found to subsist between the major axes of the different planetary orbits. *He* found that the following table, the mode of the construction of which is obvious, very nearly expressed their relative distances, taking that of the earth as 10:—

Mercury's distance,	.	.	= 4
Venus's	,,	=	$4+3\cdot 0 = 7$
Earth's	,,	=	$4+3\cdot 2 = 10$
Mars's	,,	=	$4+3\cdot 2^2 = 16$
Vesta, Juno, Ceres and Pallas		=	$4+3\cdot 2^3 = 28$
Jupiter's	,,	=	$4+3\cdot 2^4 = 52$
Saturn's	,,	=	$4+3\cdot 2^5 = 100$
Uranus's	,,	=	$4+3\cdot 2^6 = 196$

It will be found, on inspection, that these numbers very nearly correspond with those in the table in p. 122. It is a very remarkable circumstance in the history of this table, that it was formed before the discovery of the telescopic planets, and that the void thus occurring in the series had led some persons to conjecture the existence of a planet between Mars and Jupiter, just about the distance at which the telescopic planets were afterwards discovered. A similar conjecture, if the earth's planetary character were unknown, would fix a planet at the distance actually occupied by the earth; and this adds a circumstance of similarity to those stated in the text, although the force of an argument resting on so loose a foundation as this empirical law, cannot be very great.

The law itself has lately received a remarkable extension from Mr. Challis. This gentleman has shown (*Cambridge Philosoph. Transact.* vol. iii. p. 171), that it prevails not only between the distances of the planets from the sun, but between the distances of the satellites from their respective primaries. Thus in the case of the system of Jupiter, the respective distances of the satellites may be expressed without considerable error by the following law:—

	Empirical Values.	True Values.
7	= 7 .	. 6·91
7+4 . . .	= 11 .	. 11·0
$7+4\times 2\frac{1}{2}$ . .	= 17 .	. 17·54
$7+4\times (2\frac{1}{2})^2$ . .	= 32 .	. 30·86

Again, for the system of Uranus, we have:—

	Empirical distance.	True distance.
1283 . . .	= 1283 .	. 1312
1283+437 . .	= 1720 .	. 1720
$1283+437\times \frac{3}{2}$.	= 1934 .	. 1984
$1283+437\times (\frac{3}{2})^2$.	= 2259 .	. 2275
$1283+437\times (\frac{3}{2})^5$.	= 4601 .	. 4551
$1283+437\times (\frac{3}{2})^7$.	= 8749 .	. 9101

(The distances of the fourth, fifth, and seventh, have not been observed.)

The system of Saturn, however, offers a peculiarity; the first, second, third, fourth, and fifth satellites are ranged according to one series; the

If, however, it be an argument in favour of any explanation of a fact, that it reduces it into a class of known phenomena, the supposition of the earth's motion round the sun is most strongly recommended to our belief. The motion of the sun round the earth, if it exists, is a fact unlike any other with which we are acquainted. It is, indeed, an elliptic motion round the earth, as is, also, that of the moon; but there is not the proportion between their periodic times which we find to subsist in all other systems of bodies revolving round the same principal; and it is a motion of the larger round the smaller body, of which we find no other instance. In spite, therefore, of some circumstances of resemblance, it would have to be considered as a motion quite distinct from all others, and governed by different laws. The earth, on the other hand, if we suppose it to move round the sun, is at once included in a class of objects, the planets, whose motions are well known; and its motions are found to correspond in every particular with those of the other planets. The proportion of the distance to the periodic time is the same; the shape of the orbit the same; the direction of the motion, whether of rotation or in the orbit, is the same; the existence of a motion of rotation in the earth itself, and also in a satellite revolving round the earth, and the direction and duration of the satellite's rotation, all correspond to exactly similar phenomena observed in other bodies of the solar system, and furnish arguments in support of the supposition that the earth is really one of the same class. The conclusion is irresistible, that the supposition is true; for how great is the improbability that all these coincidences should happen by accident only, which they must do, if the earth is a body of a different nature from the planets.*

Chapter VI.

On Aberration.

The arguments already adduced are probably abundantly sufficient to induce the reader to adopt the supposition of the earth's motion. If, however, the earth moves, its situations at different periods of the year are very distant from each other; and we might therefore expect to find the apparent situations of the bodies we observe, different in consequence. Those of the planets are so, for the effects of this variation of situation are included in the account already given of their apparent motions. But we might expect to find also a perceptible difference in the situation of the fixed stars, which, although too distant to be affected by parallax, considered as the variation produced by the distance of places on the earth's surface from its centre, might well be so by this far greater change of position; and any such effect, if produced, would be of the same nature, and would therefore be accurately described as *parallax;* and going through all its changes in the course of a year, it may properly receive the name of *annual parallax*. In fact, however, none such can be detected, at least with any certainty, in any of the fixed stars. Astronomers have been much divided in opinion on this question; but the amount of this parallax, if any such is observable, is at least ascertained to be exceedingly small. This, however, furnishes no reason for disbelieving the existence of the earth's motion. It may be accounted for, as in p. 60, by supposing the distance of the stars to be indefinitely great, in comparison even with the diameter of the earth's orbit; and our conclusions in this case, with respect to the distance of

first, fifth, sixth, and seventh, according to another:—

	Empirical distance.	True distance.
336	= 336	335
336+82	= 418	430
336+82×2	= 500	528
$336+82\times(2)^2$	= 664	682
$336+82\times(2)^3$	= 992	958
$336+82\times(2^3)\times3$	= 2304	2208
$336+(82)(2^3)\times3^2$	= 6240	6436

With the exception here noticed, it may then be affirmed, that the planets and satellites arrange themselves about their primaries at mean distances, which observe approximately this progression, a, $a+b$, $a+rb$, $a+r^2b$, &c. The value of r is always one of the terms of the series 1, $1\frac{1}{2}$, 2, $2\frac{1}{2}$, 3, &c. We may add, that this ratio of b to a is generally expressed by very simple numbers: thus, for the planets it is $\frac{2}{3}$ nearly; for the system of Jupiter $\frac{7}{4}$; for that of Saturn, $\frac{1}{4}$; of Uranus, $\frac{1}{3}$, nearly.

* In these reasonings we have treated the motion of rotation of the earth as already established, and have used it as one of the circumstances of correspondence to prove the planetary nature of the earth. If, on the other hand, we consider it to be sufficiently established without resorting to this argument, that the earth is a planet, its motion of rotation would itself be confirmed by the analogy of the other planets which have corresponding motions. This argument, indeed, may have some weight, without introducing the question of the earth's planetary character, as all the bodies, whose magnitude and nearness give us the means of ascertaining their motions, have a similar motion.

the fixed stars, will have to be increased in proportion to the increase of the possible distance between the two places of observation. Thus we find that the fixed stars, supposed to have an horizontal parallax of 0·″36, must be at the distance of 572957·8 times the earth's radius, or more than 2,000,000,000 of miles. Now this distance, if the annual parallax be confined within the same limit, must be increased in the proportion of the radius of the earth's orbit to the earth's radius, or in the proportion 47968 : 1. But as some astronomers admit a small perceptible annual parallax of 1″ or 2″, we find by the principles just laid down (taking the parallax at 2″) for their distance 499,702,352 times the earth's radius, which quantity again has to be multiplied by 3962 to turn it into miles. Such a distance surpasses all our powers of imagination; but we have no reason whatever for disbelieving that the fixed stars are really so far off, though the belief is eminently calculated to excite wonder at the vast extent thus attributed to the visible creation.

The question, however, whether any sensible annual parallax exists, is one which would naturally excite much curiosity; and many observations have been made for the purpose of determining it. In the course of these a very remarkable phenomenon was discovered by Dr. Bradley, which he named the *aberration*. We have hitherto deferred the explanation of its nature; we shall now be able fully to elucidate it, and to draw from it a very strong proof that the earth really does move. For this purpose, however, it will be convenient, before we give any account of the phenomenon observed, to enter into some preliminary investigations.

Fig. 38.

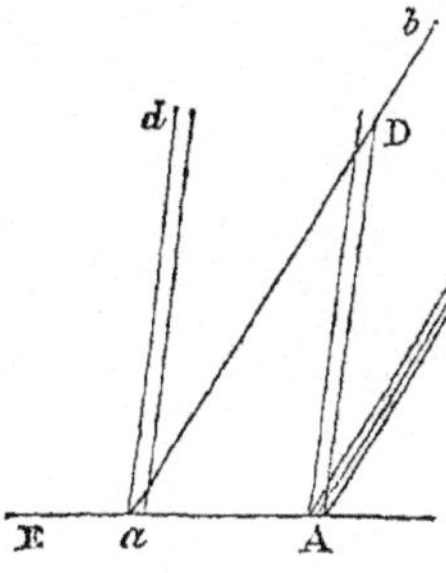

Let us suppose that the parallel oblique lines B A, *b a*, in fig. 38, represent the course of drops of rain falling towards the ground. If an observer be placed at A, and continue at rest there, he might point a tube, represented by the double lines A C, in the direction A B, and the drop would descend along the axis of the tube, which would coincide with and represent the direction in which the drop fell. But if, instead of continuing at rest, he were to move forward in the direction A E, carrying the tube parallel itself, the drop would no longer descend along the axis, but the back of the tube, being carried forward into and through the position formerly occupied by the axis, would come in contact with the drop, and either stop it, or change its course. If, however, we suppose A D to be another position of the same tube, it is obvious that it may be so taken, that the drop which falls along the line *b a* shall be at D a point in the axis of that tube, when the observer is at A; and we may further suppose the drop to have attained its terminal velocity, and its motion in consequence to be uniform, and that the observer also moves uniformly from A to *a*, while the drop falls from D to *a*. If, therefore, he continues to keep his tube parallel to itself, its situation will be *a d*, and the drop, arriving at *a*, will still be at a point in the axis of the tube. In the same manner, the motions of the drop and the observer being supposed uniform, and the tube always parallel to itself, the position of the drop at each instant will be a point in the axis of the tube in its corresponding position; and the drop will, therefore, when it arrives at *a*, have descended along the axis of the tube, which will then be in the position *a d*. It will, therefore, seem to come in the direction *d a* instead of *b a*; and it is obvious, from inspection of the figure, and consideration of the mode in which the motions take place, that the angle *d a* E, made by the apparent direction of the drop's motion with the direction in which the spectator moves, must be less than *b a* E, or B A E, the angle made with the same line by the real direction of the drop's motion.

Now, this is exactly what an astronomer does in observing the heavenly bodies. Of course, the apparent direction in which he sees them is the same, whether we introduce the supposition of pointing a tube towards them, or not: that supposition only facilitates explanation. But, in fact, he does so: he points a telescope to receive a ray of light from the heavenly body, just as we have supposed him to point a tube to receive the drop of rain: and it is necessary, in order that a correct image may be formed at the eye, that the ray also should proceed along the axis of the tube of the telescope. If, therefore, the spectator have any motion, the direction in which the telescope must be pointed, so that light will be received from any heavenly body observed, will not be accurately that of the body itself, but one nearer to the direction in which the observer moves.

We proceed to determine more accurately the amount of the deviation which would be thus produced. We have already seen that the direction of the tube must be so fixed, that A*a* may be to D*a* in the same proportion as the velocity of the observer bears to the velocity of the ray, and consequently, as the angles *d a* D, *a* D A, are equal.

$$\text{sin. } d\,a\,\text{D} = \text{sin. } a\,\text{D A} = \frac{\text{A}\,a}{a\,\text{D}}\,\text{sin.}\,a\,\text{A D}$$

$$= \frac{\text{A}\,a}{a\,\text{D}}\,\text{sin. } d\,a\,\text{E}$$

$$= \frac{\text{velocity of observer}}{\text{velocity of ray or drop}}\,\text{sin. of the ap-}$$

parent angle between the ray and the direction of the observer's motion (A), and this quantity will, of course, be greatest when the sine of this angle is the greatest, or the angle itself equal to 90°. In that case,

$$\text{sin. } d\,a\,\text{D} = \frac{\text{velocity of observer}}{\text{velocity of the ray or drop.}}$$

If, therefore, the velocity of the ray or drop be very great in proportion to that of the observer, this angle of deviation must be very small: if it be indefinitely great, in which case the motion may be considered as instantaneous, the angle of deviation must become imperceptible.

We have already seen that the velocity of light is not less than 190,000 miles in a second, a velocity so enormous, that, for most purposes, its motion may be considered as instantaneous. But every observer on the earth's surface must partake of the earth's motion in its orbit, if it have any; and we have seen (p. 142) that this motion, if it exists, is not less, in its average value, than 19 miles in a second, the 10,000th part or very nearly so, of the velocity of light. This is a very great disproportion; but still $\frac{1}{10,000}$ the value in this case of the fraction $\frac{\text{vel. of observer.}}{\text{vel. of ray,}}$ is the sine of a very appreciable angle. The sine of a very small arc may be considered as equal to the arc itself. The arc which is equal to the radius is 57°·2957795: the required sine or arc therefore

$$= \frac{57^\circ{\cdot}2957795}{10000} = \frac{206264{\cdot}8''}{10000} = 20''{\cdot}62648.$$

To make these computations more exact, more accurate value of the different velocities in question should be employed; and using them, the value of the angle thus deduced is found to be 20″·246*. This, therefore, is the greatest value which this deviation, or, as it is called, the *aberration*, can assume on this supposition; and this is so small, that the angle between the apparent direction of the body and the direction of the observer's motion, will in all cases be very nearly the same as the angle between the real direction of the body and the direction of the earth's motion: and thus the equation (A) will become in this case

sin. angle of aberration = 20″·246 × sin. angle between body's real place and the direction of the earth's motion†.

Besides this amount of aberration, there must be some produced by the earth's motion of rotation about its

---

* The time which light takes in traversing the diameter of the earth's orbit is found by observation of the eclipses of Jupiter's satellites to be $16^{m}\ 26^{s}$. If therefore the radius of the earth's orbit be r, the velocity of light will be $\frac{2\,r}{16^{m}.\,26^{s}.}$

The whole circumference of the earth's orbit, supposing it to be circular is $2\,\pi\,r$ and, this is described in a year, or in 365·25638 days, (for this is the length of a sidereal year, and consequently of a complete revolution of the earth,) the mean velocity of the earth is $\frac{2\,\pi\,r}{365d.\,25638}$

$$\frac{\text{The greatest aberration}}{\text{arc} = \text{radius}} = \frac{\text{Vy. of earth}}{\text{Vy. of light}}$$

$$= \frac{2\,\pi\,r}{365d.\,25638}\cdot\frac{16^{m}\,26^{s}}{2\,r}.\ \text{rdd.} = \frac{\pi.\,986^{s}}{365d.\,25638}$$

$$= \frac{3.14159.\ \ 986}{365{\cdot}25638.\ \ 24\times60\times60}$$

∴ The greatest aberration

$$= \frac{3{\cdot}14159\times986\times206264{\cdot}8}{31548151 5} = 20''\,246$$

† The same conclusion may be thus obtained. Let A represent the aberration, and S the real angle between the star's place and the direction of the earth's motion; then S—A will represent the apparent angle, and the equation will become

$$\text{sin. A} = 20''.246.\ \text{sin. (S—A)}$$
$$= 20''.246\ (\text{sin. S cos. A—cos. S sin A})\ \text{and}$$

as A is necessarily very small, (less than $\frac{1}{10000,}$) the last term, cos. S. sin. A may be neglected in comparison with sin. S. cos. A; and cos. A may be considered as equal to 1.

The equation, therefore, finally becomes sin. A = 20″·246, sin S, axis, if any such motion exist: the greatest velocity which can be thus produced is that of a point in the equator, which we have already seen to be about $17\frac{1}{2}$ miles in a minute; and the greatest aberration, therefore, which can be produced by this cause will fall short of the greatest aberration which can be produced by the earth's motion in its orbit in the proportion of their respective velocities, or of $\frac{17\frac{1}{2}}{60} : 19$. It cannot therefore exceed

$$20''{\cdot}246 \times \frac{17\frac{1}{2}}{60{\cdot}19} \text{ or } {\cdot}3108 \text{ of a second.}$$

Taking more accurate values of the motions, it is found that it cannot exceed ·3084 of a second; a quantity so small that we may, for the purposes of the present treatise, neglect it, and consider the whole aberration to be that which would be produced by the earth's motion in its orbit: and this of course will be the same to an observer, wherever on the earth he is situated; for being borne along with the earth, he partakes of all its motion.

We have already seen that the fixed stars have no observable annual parallax: it follows of course that their distance, or the radius of the imaginary sphere of the heavens in which they appear to be set, is indefinitely great in comparison with the diameter of the earth's orbit. Lines therefore drawn from the sun and earth parallel to each other may be conceived to meet this sphere in the same point of its surface. Now the earth moving in the plane of the ecliptic, the direction of its motion must always be towards some point in the celestial ecliptic, namely that point in which a tangent to its orbit at its then place meets the celestial ecliptic, for the celestial ecliptic is the intersection of that plane with the sphere of the heavens. This point, therefore, may be determined by drawing a tangent from the earth's place, or by drawing a line from the sun, which is considered as the centre of the sphere, parallel to that tangent, to meet the celestial ecliptic. Again, the orbit of the earth being very nearly circular, this tangent will always be very nearly perpendicular to the line joining the earth and sun. Now if the line joining the earth and sun be produced each way to meet the celestial ecliptic, the earth's apparent place from the sun, or her place as referred to, the celestial ecliptic will be at one extremity of this diameter of the sphere, and the sun's apparent place from the earth will be at the other, and there will be 180°, or a semicircle between them. Thus when the sun appears in Aries, the earth is in Libra. Of course, a line drawn through the sun very nearly perpendicular to this diameter will meet the circumference very nearly midway between or about 90° from its two extremities. But such a line will meet th point of the celestial ecliptic towards which the earth moves; this point therefore is about 90° from the earth's apparent place in the celestial ecliptic; and as the earth moves towards it, it is 90° before the earth's apparent place: and in the same manner it is 90° behind the sun's apparent place.

Finally the aberration takes place in the plane passing through the line joining the earth and star, and also the line of the direction of the earth's motion. The earth, however, may be considered as in the centre of the sphere, for it is distant from it only by its distance from the sun, which is indefinitely small in comparison with the radius of the sphere. This plane therefore passes through two radii of the sphere, and the angle between the two radii of the sphere is measured by the great circle of the sphere which joins their extremities. If, therefore, an arc of a great circle be drawn from the star to the point of the celestial ecliptic 90° before the earth's place, this arc will measure the angle contained between the direction of the earth's motion and the line joining the earth and the star; or it will measure the angle or the amount of which the quantity of the aberration depends.

Fig. 39.

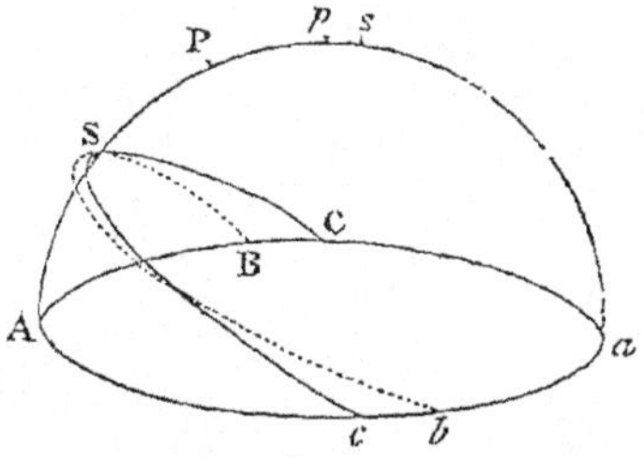

We may now explain and ascertain the manner in which all the phenomena of aberration would succeed each other during the whole course of the year, on the supposition of the earth's motion. For this purpose, let A B *a c*, in fig. 39, represent the celestial ecliptic, S the position of any star, and A S *a*, the half of a great circle drawn through the star perpendicular to the plane of the ecliptic: and let B, C, be other points in the celestial ecliptic, and B S *b*, C S *c*, arcs of great circles drawn through the star and these points respectively, which will of course be semicircles, as all great circles bisect each other. If then, A, B, C, *a*, *b*, *c*, represent different positions of the point towards which the earth moves (or the point 90° before the earth's place) S A, S B, S C, S *a*, S *b*, S *c*, will represent the arcs, to the sines of which the amount of aberration is proportional, and in the direction of which it takes place. Now, the earth being in every point of the ecliptic in the course of a year, every point of the celestial ecliptic must be 90° before its place in the course of the same period, and consequently the point S must be joined with every point of the celestial ecliptic to give all the arcs which determine the magnitude and direction of the aberration during the year. Of these, the least is A S, and the greatest is S *a*: the one being greater and the other less than 90°: and in passing from A towards *a*, the corresponding arcs must pass through all intermediate values, increasing as they approach *a*. Of course, among these, there must be one which is of 90°: let S C be this; and the aberration in the direction of the line S C will have its greatest value, and will of course be 20″·246. The arcs C S, S *c*, together make a semicircle, and S C being 90°, S *c* will be 90° also; of course therefore the aberration in the direction S *c* will also be 20″·246: and the extreme distance between the two apparent places as affected by the aberration in these opposite directions will be 40″·492. If again, B S *b* represent any other great circle passing through S, the aberration in the direction S B will be proportional to the sine of S B, and that in the direction of S *b* will be proportional to the sine of S *b*. But the arcs S B, S *b* together make up a semicircle, or S B is the supplement of S *b*: and as the sine of an arc and of its supplement are equal, the aberrations in the directions S B, S *b*, are equal.

In the same manner the aberrations in the directions S A, S *a*, are equal: and as S A is the least possible arc drawn from S to the celestial ecliptic, these are the least values of the aberration. The greatest values of the aberration are in the directions S C, S *c*; and as the arcs S C, S *c*, are of 90° each, the circle C S *c* cuts the circle A S *a* at right angles; or the directions of the greatest and least aberrations are perpendicular to each other, the least aberration taking place in a direction perpendicular to the ecliptic, or affecting only the latitude of the star. On investigation of the precise amount of the aberration in every direction, on the supposition that it is the effect of the earth's motion, we shall find that the apparent place of the star is always in the periphery of an ellipse, of which the centre is the true place of the star; the minor axis is in the direction of a great circle passing through the star perpendicular to the ecliptic, the major axis = 40″·492; and the proportion of the minor to the major axis, that of sin. ✳'s. latitude : radius *. Of course, therefore, the star is never seen in its true place, except in one case which we shall presently mention.

Now it is found by observation that the apparent places of a star do actually differ at different periods of the year, and that their variations accurately cor-

---

* Let the angle ASB $= \theta$, SA $= a$, SB $= b$.
$\therefore \cos.\ \theta = \tan.\ a \cot.\ b.$ or $\cos.^2 \theta = \tan.^2 a \cot.^2 b.$
and $\sin.^2 \theta = 1 - \cos.^2 \theta = 1 - \tan.^2 a \cot.^2 b =$

$$1 - \frac{\tan.^2 a \cos.^2 b}{\sin.^2 b}$$
$$= \frac{\sin.^2 b - \tan.^2 a\,(1 - \sin.^2 b)}{\sin.^2 b}$$
$$\therefore \sin.^2 b = \frac{\tan.^2 a}{(1 + \tan.^2 a) - \sin.^2 \theta} = \frac{\tan.^2 a}{\sec.^2 a - \sin.^2 \theta} = \frac{\sin.^2 a}{1 - \cos.^2 a \sin.^2 \theta}$$

Let ϱ = amount of elevation in the direction S B then $\varrho = 20''\cdot246 \sin.$ S B

$$\therefore \varrho^2 = (20''\cdot246)^2.\ \sin.^2 b = \frac{(20''\cdot246)^2 \sin.^2 a}{1 - \cos.^2 a \sin.^2 \theta}$$

the polar equation to an ellipse, whose semi-major axis = 20″·246, and semi-minor axis = 20″·246, sin. *a*.

It is obvious that this ellipse is the same as that produced by the projection upon the surface of the celestial sphere of a circle drawn parallel to the plane of the ecliptic, whose radius is 20″·246. The circle being so small, the projection may be considered as orthographic, and the surface of the sphere as a plane perpendicular to the line drawn from the observer to the star. The orthographic projection of a circle is an ellipse, whose minor axis : major axis : : cos. inclination of the plane : radius. In this case, the inclination of the supposed plane to the surface of the sphere = complement of, the star's latitude, for if the radius of the sphere be produced, the right angle between the radius so produced, and the surface of the sphere, is made up of that inclination, and the exterior angle formed by the outer of two parallel lines, and a line intersecting them, which is equal to the interior and opposite angle; that is, to the angle at the centre, or to the star's latitude. The cosine of the inclination, therefore, is equal to the sine of the star's latitude; and the minor axis = 20″·246 sin. ✳'s latitude, as before.

respond with those computed on the supposition of the earth's motion; that is to say, if all the apparent positions are registered, they are found to be points in the periphery of an ellipse, in which the magnitude and the proportion of the arcs are the same as those already deduced; and assuming the centre of that ellipse to be the true place of the star, then the apparent place is always found to be in the great circle joining that true place with the point of the ecliptic towards which the earth, on the supposition of its revolving round the sun, is moving at the time. The manner in which these observations are made will better appear when we have further considered the effects of the aberration, supposing it to exist: but we may at once observe, that the discovery of the existence of such a variation in the apparent place of the stars, at different times, was made before the theory, by which we have explained it, was invented; and consequently that the coincidence of theory with observation is not liable to the suspicion which often attaches to the correspondence imagined to exist between the results of observation, and those deduced from a theory previously adopted. Where the observations, indeed, are perfectly correct, and the data from which the theoretical calculations proceed perfectly ascertained, it is of little importance which came first in the order of time, for their coincidence, if the two really correspond, must be perfect. But this is not the case with the phenomena of aberration: the time by which the eclipses of Jupiter's satellites are accelerated or retarded is not known with sufficient accuracy to enable us to state with perfect certainty the exact magnitude of the apparent ellipse indicated by theory; nor is the accuracy of observation sufficient to enable us to determine with the minuteness necessary for such a purpose, the precise variations in the observed positions of the star; indeed, the exact amount of the greatest aberration is still a matter of controversy*: it is, therefore, of importance to know that observation preceded theory, and therefore has not been accommodated to it; and that the theory (which follows, indeed, as a necessary consequence from the supposition of the earth's motion) was not originally and independently deduced from that supposition, but adopted only as a mode of explaining facts to which it was experimentally found to correspond.

It is obvious that this coincidence between the results of theory, supposing the earth to revolve round the sun, and of observation, furnish the most convincing proof of the reality of the earth's motion*. It is, indeed, possible that the earth may be at rest, and that the varying positions of each star, during the year, may be attributable to a real motion in the star, performed in the space of a year, and following in every instance the same laws; but the improbability of such a supposition is manifest. On the supposition of the earth's motion all these variations are accurately accounted for, both in magnitude and direction; they necessarily all take place towards the same point, (that towards which the earth moves,) and in amounts depending, as they are found to depend, on the star's distance from that point. On the supposition of the earth's being at rest, and the star's having an actual motion of its own, there is no reason why any one star should have its position in any way determined with reference to the particular point in question, rather than to any other point in the heavens, and of course the improbability of its being so, even in the case of any particular star, is very great: the improbability of finding that all stars should have their positions so regulated (as we practically find them to be) of course altogether defies calculation.

* See a paper by Mr. Richardson in *Ast. Soc. Trans.* vol. iv., which assigns the value 20″·505 to the greatest aberration. Bradley considered the value to be 20″; M. Zach, deducing the value from Bradley's observation, 20″·232. M. Bessel, forming his computations from the same sources, 20″·68; Dr. Brinkley, 20″·37; M. Ludenau, 20″·61; M. Struve, 20″·35; Professor Woodhouse, 20″·246. M. La Place, 20″·25. All these results differ by quantities falling decidedly within the possible limits of error arising from the inaccuracies of observation, if they be deduced directly from it, or from the possible error in the computation of the velocity of light, if they be deduced from the theory of the earth's motion.

* If, on the other hand, we consider the earth's motion as sufficiently established by the other arguments which exist in support of it, we may use the phenomena of aberration for the purpose of ascertaining the velocity of light; for, whatever be that velocity, the effects of aberration will be of the same nature, and follow the same laws; and from their observed amount the velocity of light will be ascertained. The process will be very simple, and the reverse of that already adopted: as before, if A = greatest amount of aberration, $A = \frac{\text{Vy. of the earth}}{\text{Vy. of light.}}$, and, consequently, velocity of light = A, velocity of the earth, in which equation the value of A may be ascertained from observation, and the close agreement of the value thus deduced with that derived from the observation of the eclipses of Jupiter's satellites, furnishes the strongest confirmation of the accuracy of each.

Considering the fact of the existence of this aberration as established, we shall point out a few instances of the peculiar effects which it produces. In the first place, as it necessarily takes place in great circles drawn to intersect the ecliptic in every possible point, it takes place in every possible direction; and among those directions there must be some perpendicular, and some parallel to the equinoctial and the ecliptic respectively. When the aberration is in lines perpendicular to the equinoctial, it takes place, of course, wholly in declination; when in lines parallel to the equinoctial, it takes place wholly in right ascension; when in any other direction, it affects both the declination and the right ascension. It is not necessary, here, to investigate the laws by which the effect produced on the declination and right ascension is determined: it is sufficient to state that they are of easy and of certain computation, and that the observed results accurately correspond with them. Observations of right ascension and declination are those most easily made; and it was from such observations that Dr. Bradley detected the existence of the apparent irregularity in question, and investigated the laws which regulated it; and thus was led to the discovery of the cause from which it proceeded.

We proceed to point out some particular results affecting heavenly bodies in certain specific positions. In the first place, let us suppose the case of a star placed exactly in the pole of the ecliptic. In this case, the arc drawn from the star's place to every point in the ecliptic is exactly 90°, and its sine, therefore, always equal to the radius: and, consequently, the star will be seen in a circle always 20″·246 distant from its true place, and the amount of the aberration will always be equal, and always the greatest possible. The same result would follow from considering the apparent curve as the projection of a circle parallel to the plane of the ecliptic; for, in this case, such a circle would be seen perpendicularly, and would therefore appear circular. The next instance which we will take, will be that of a star situated in the ecliptic. In this case, the arc drawn from the star to the point of the ecliptic, towards which the earth is moving, will always be a portion of the ecliptic itself; and the whole aberration, therefore, will be in the plane of the ecliptic, or it will take place entirely in longitude. The magnitude of this arc, also, will have every value from 0° to 180°: of course, therefore, the aberration will, at two points, where the arc is 90°, have its greatest value of 20″·246. And at other two, where the arc is 0° or 180°; that is to say, when the earth is moving directly towards or away from the star, the aberration will be nothing. The same conclusions evidently follow from the results already deduced: for the star's latitude being nothing, the minor axis of the ellipse in which it is seen becomes nothing also, and the ellipse itself becomes a straight line, in which the star appears to oscillate backwards and forwards; of course passing through the centre, or having its apparent coincide with its true place in the course of its passage each way.

Another set of conclusions, of much importance in the history of the discovery of aberration, are deduced from considering its effect on stars situated in the solstitial colure; the meridian which passes through the pole of the ecliptic. In this case, therefore, the aberration in latitude and in declination are the same, and this aberration takes place entirely in latitude, or entirely in declination at the same time; that is to say, when the earth is about 90° behind either of the solstices, or about the time of the two equinoxes. At this time, also, the aberration in declination is greatest; for, although the whole aberration is least when it takes place in the direction S A, or S a (*fig.* 39), and consequently entirely in latitude, the aberration estimated in latitude is then greatest*. Now (in *fig.* 39), let P represent the pole of the heavens, p the pole of the ecliptic, S, s, two stars situated in the solstitial colure, on opposite sides of the pole, whose north polar distances (and of course their declinations) are equal, that is to say, $PS = Ps$: it is obvious that they will have their greatest

---

* The whole aberration $\propto \sqrt{1 - \cos.^2 a \sin.^2 \theta}$ (see note, page 150). But the aberration in latitude will evidently be : the whole aberration :: cos. θ : rad., or aberration in latitude $\propto \dfrac{\cos. \theta}{\sqrt{1 - \cos.^2 a \sin. \theta}}$. The square of the aberration ∴ $\propto \dfrac{\cos.^2 \theta}{1 - \cos.^2 a \sin.^2 \theta} \propto \dfrac{1 - \sin.^2 \theta}{1 - \sin.^2 \theta + a \sin.^2 a \sin.^2 \theta}$ $\propto 1 - \dfrac{\sin.^2 a \sin.^2 \theta}{1 - \cos.^2 a \sin.^2 \theta}$, which is evidently the greatest when the second or subtractive term vanishes, or when $\theta = 0°$ or $180°$: or the aberration is entirely in latitude.

aberration in latitude and declination at the same time,—namely, when the points A, c, are those towards which the earth is moving: but that the effect of aberration at S will be to increase the north polar distance, or diminish the declination; at s it will be to diminish the north polar distance, or to increase the declination. If these effects took place to an equal amount in each case, they might plausibly be attributed to an actual variation in the position of the pole itself, which, in moving towards the one, would move as much away from the other; and this was the first supposition of Dr. Bradley, when he discovered (by observation on γ Draconis, and 35 of Camelopardalus, two stars of equal north polar distances, situated very near the solstitial colure, and on opposite sides of the pole) the variation of declination. He found, however, that they were not equal, and consequently that this supposition could not be admitted: and we have already seen that the aberration in the one case would be $20''\cdot246$, sin. S A, in the other $20''\cdot246$, sin. s A: two quantities which are necessarily different. Bradley's original notion, however, was, that by these observations he should be able to detect an annual parallax in the fixed stars; and it is necessary to point out, that these same observations were inconsistent with the supposition, that the irregularities observed were the effect of parallax.

For this purpose it will be necessary to revert to the consideration of the annual parallax, and to ascertain in what manner it will affect the apparent place of any fixed star. The sun being considered as the centre of the system, the true place of the star will be that in which it is seen from the sun. Now, parallax (p. 56) always takes place in a plane passing through the object observed, and the different positions at which the observation takes place; that is to say, in the present case, through the star, the earth, and the sun. The line joining the earth and sun will, when produced, pass through the earth's place as referred to the celestial ecliptic, and the intersection of the plane just mentioned with the sphere of the heavens will, as it passes through the sun, the centre of that sphere, be a great circle of the heavens. Parallax, therefore, will appear to take place along this great circle; a great circle, namely, joining the star and the place of the earth as referred to the celestial ecliptic at the time of observation. It will, therefore, in the course of the year, take place, like aberration, in every possible direction, for it will take place in great circles joining the star's place with every point of the celestial ecliptic; and the circles S A, S B, S C, &c., in *fig.* 39, may represent lines in which, at different times, the effect of parallax takes place. This is not the only resemblance between the effects of parallax and aberration. If P represent the greatest possible parallax of any fixed star, the parallax in any particular situation, as when the earth is at B, will be P sin. S B: the greatest parallax will be in the same lines S C, S c, which are each arcs of 90°, in which the greatest aberration took place, and the least will be in the lines S A, S a, when the aberration is least. Besides this, the value of the parallax is P sin. S B; a quantity exactly of the same form as the value of the aberration $20''\cdot246$ sin. S B. By the same process of reasoning, therefore, as that before adopted, the effect of parallax will be to make the star apparently move in an ellipse, exactly similar to that occasioned by aberration, and differing from it only in magnitude, in the proportion of the constant coefficients P and $20''\cdot246$. These coefficients, indeed, have one remarkable difference: the coefficient $20''\cdot246$ being derived from the consideration of the velocity of light is the same for all stars; the coefficient P may be different in the case of each particular star, and if their distances are different, it will be so.

Still, with this general identity between the nature of the effect of aberration and of parallax, how do we distinguish that it is by aberration and not by parallax that the places of the stars are sensibly affected? The answer is very simple: the effects of parallax and aberration, though each in the course of the year tends in every possible direction, takes place at each particular instant towards different points: the former towards the earth's place, the latter towards a point in the ecliptic 90° distant from it. Thus when the earth is at c, the effect of parallax takes place in the line S c, that of aberration in the line S A. The arc c S C is bisected by A P a, and, therefore, they cut each other at right angles; for a short space about the point S, therefore, such as those within which the effects of parallax and aberration are confined, the arc c S C is parallel to

c A C. In this position, therefore, the whole effect of parallax will be in a direction parallel to the ecliptic, or it will affect the longitude of the star only, and not its latitude. At the same time the aberration, if any exists, will take place in the line S A, perpendicular to the ecliptic, and it will therefore affect the latitude only. If therefore we find, at this time, the latitude of the star to be affected, as we actually do, the inference is necessary, that it is affected by aberration, not by parallax. In the same manner, when the earth's place is at A, the parallax would take place in the line A S P, therefore entirely in latitude; but the aberration would take place in the direction S C, A C being 90°, and, consequently, in a direction parallel to the ecliptic, or entirely in longitude. Now at this time we do find that the star's longitude is affected, which it cannot be by parallax; this, therefore, must be the effect of aberration. The separation of the effects of the two causes is a little more complicated in intermediate situations; but they always take place in different directions, and may always, in consequence, be distinguished from each other; and we are thus clearly enabled to say, that the inequalities observed are referrible to aberration as their cause. Whether there be any sensible annual parallax at all is still a matter of dispute, and it is sufficient here to have pointed out the nature of its effects, supposing them to exist, and the impossibility of their being confounded with or included in those of aberration. It is not pretended that it can be detected, except in a very few stars; and in none of these is its amount supposed to exceed $2''$.

Dismissing the consideration of parallaxes, which evidently can be of little practical importance, we return for a short time to the consideration of aberration. It is evident that the apparent positions of the sun, moon, and planets, must be affected by it in the same manner as those of the fixed stars; for it arises only from the gradual propagation of light, which is the case as much in the one case as in the other.

Now the sun being always in the celestial ecliptic, and at the distance of 180° from the earth, he is always 90° before the point towards which the earth is moving; the aberration, therefore, in his case, has always its greatest value, of $20''{\cdot}246$. It always affects the longitude only, being in the plane of the ecliptic, and it always diminishes the apparent longitude, because it brings the sun nearer the point towards which the earth is moving. We always, therefore, by the effect of aberration, imagine the sun to be in a point $20''{\cdot}246$, behind the true direction of the ray by which we see him, or behind his real place at the time when the ray quitted him. But the ray quitted him $8^m\ 13^s$ before it arrived at the earth, for that is the time which light takes to pass from the sun to the earth, and, consequently, what we see is the apparent position of the sun $8^m\ 13^s$ before; and as we want to ascertain the sun's true place, we have not only to correct the apparent place of the error occasioned by aberration, but to add to it also the amount of the motion of the earth or sun during the space of $8^m\ 13^s$, that is to say, the arc $20''{\cdot}246$; for this is also the amount of the sun's mean motion in $8^m\ 13^s$.* The sun, therefore, will be seen $40''{\cdot}492$ behind his true place, in consequence of the progressive motion of light; $20''{\cdot}246$ by the effect of aberration, and $20''{\cdot}246$ from the space through which he moves while his light is coming to the earth.

The apparent places of the planets will be affected in an analogous manner. The following considerations will show how to estimate the quantity by which these places are altered. The true place of any planet at the time of observation will differ from the observed place by the arc that the planet describes in the time that a ray of light takes to pass from it to the earth. As the distance from the planet to the earth is always known, or may be calculated from the tables,† and since the time taken by light to traverse the diameter of the earth's orbit is ascertained, we may find by a simple proportion how long light will take to describe the distance in ques-

* The space described by the earth or sun, during the passage of light from the sun to the earth, is necessarily equal to the greatest aberration, whatever the amount of aberration may be. The space described by the earth or sun in any time is to the space described by light in the same time in the proportion of their respective velocities: if, therefore, to represent this time, S the space described by the earth or sun in t, and r the distance of the earth from the sun, which light, therefore, will describe in t, $S : r ::$ Vy. of ⊕ or ☉ . Vy. of light; or $S = r\,\dfrac{\text{Vy. of} \oplus \text{ or } \odot}{\text{Vy. of light}}$, the same expression as that deduced in another page, for the greatest aberration.

† The planetary tables generally give a table expressly for finding this distance, from the commutation; which is the name applied to the angle subtended at the centre of the sun, by the respective centres of the earth and planet.

tion; and again, knowing this time, we may deduce from the planet's geocentric horary or diurnal motion, the arc corresponding to it.

The astronomical formulæ for aberration are generally founded upon the supposition that the motion of the earth in the ecliptic is circular and uniform. This supposition is not strictly true; and to be mathematically exact, we should take into account the variation of the earth's velocity, due to its motion in an ellipse. The velocity in question not being constant, neither will the coefficient which we have assumed equal to $20''\cdot246$ have always that value, but it will vary, being greatest when the earth is in its perihelion, or nearest the sun; and least, when it is in its aphelion, or farthest from the sun. These variations are, however, confined within very small limits, owing to the very trifling excentricity of the ellipse in which the earth moves; in fact, they do not exceed $0''.003$, as the excentricity of the terrestrial orbit is only about 0,016,853, the semi-axis major being unity. We might, if necessary, allow for them, but this would be a refinement as unnecessary as taking into account the effects of the earth's rotation, or the diurnal aberration to which we have alluded above.

As the application of the correction for aberration is of constant recurrence in astronomy, it has been found desirable to construct tables, which shall give the effects of aberration both upon the right ascension and declination of any given star, without compelling the astronomer to make use of the algebraical formula in each case. To explain how these tables are constructed, we must revert to the principles already laid down; namely, first that the aberration = $20''\cdot246 \times$ sine of the angle between the body's real place, and the direction of the earth's motion:* and, secondly, that if an arc of a great circle be drawn from the star to the point of the celestial ecliptic, 90° before the earth's place, this arc will measure the angle contained between the direction of the earth's motion and the line joining the earth and star, being, in fact, the angle on which aberration depends.

Now, suppose (*fig.* 40) that E be the earth, E I O the ecliptic, and E O an arc of 90°, let S be the place of the star, then O S will be the great circle in which the aberration takes place, and let I S s' be any plane the effect of aberration in which it is required to measure:

Fig. 40.

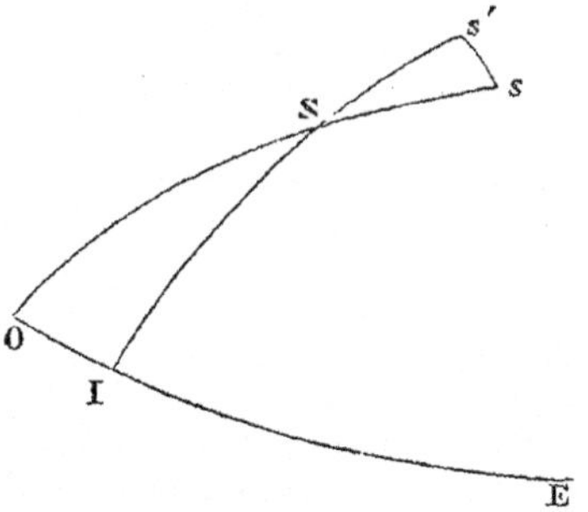

produce O S, and take on it S s equal to the whole aberration, and draw from s, $s\,s'$ perpendicular to I S s', then $s\,s'$ will measure the deflection from the plane I S s' caused by aberration. The triangle S s s' being extremely small, we may solve it as rectilinear, and then we get

$$s\,s' = \text{S}\,s.\ \sin.\ s'\,\text{S}\,s = \text{S}\,s\ \sin.\ \text{I S O}$$

but S s = $20''\cdot246$. sin. S O, by the principles already laid down

$$\therefore\ s\,s' = 20''\cdot246.\ \sin.\ \text{S O}\ \sin.\ \text{I S O}$$
$$= 20''\cdot246.\ \sin.\ \text{O I}\ \sin.\ \text{S I O}$$
$$= 20''\cdot246.\ \cos.\ \text{I E}\ \sin.\ \text{S I O}$$

should E I be greater than E O, the expression is the same, but must be taken with a negative sign.

Now, let S I (*fig.* 41), be a circle of declination meeting the equator ♈ s s' in s, and the ecliptic ♈ I E in I, and let S S′ be the aberration perpendicular to I S. Draw through S′ the cricle of declination p S′ s': call the angle E ♈ s',

Fig. 41.

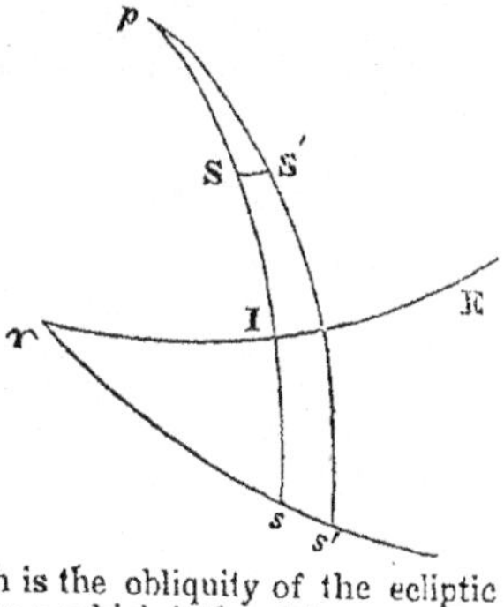

which is the obliquity of the ecliptic ω, and ♈ s, which is the right ascension of the star, α; and let ♈ I = ϕ: we have,

* It is to be observed, that a very small arc may be considered equal to its sine.

Υ I s being a right-angled spherical triangle, $\cot. \phi = \cos. \omega. \cot. a$, and hence ϕ may be calculated. Now, by what has preceded

$$S S' = 20''{\cdot}246. \cos. I E, \sin. S I E.$$

but, $\sin. a = \sin. \phi. \sin S I E.$

$$\therefore \sin. S I E = \frac{\sin. a.}{\sin. \phi.}$$

$$\text{and } S S' = 20''{\cdot}246, \cos. I E, \frac{\sin. a.}{\sin. \phi.}$$

Again, $I E = \Upsilon E - \Upsilon I.$
= Earth's heliocentric longitude $- \phi$.
$= \oplus - \phi.$

$\therefore \cos. I E = \cos. (\oplus - \phi)$, but $\oplus = 180^\circ + \odot$.

$\odot$ being the geocentric longitude of the sun.

$$\therefore \cos. I E = \cos. (180^\circ + \odot - \phi).$$
$$= - \cos. (\odot - \phi.)$$

$\therefore$ substituting

$$S S' = - 20''{\cdot}246. \frac{\sin. a. \cos. (\odot - \phi).}{\sin. \phi.}$$

but $s s'$, which measures the aberration in right ascension $= \frac{S S'}{\cos. \delta}$, δ being the declination of the star

$$\therefore s s' = 20''{\cdot}246. \frac{\sin. a. \cos. (\odot - \phi).}{\sin. \phi. \cos. \delta.}$$

This, then, is the expression for the aberration in right ascension.

Again, let the plane in which the aberration is to be measured be perpendicular to the circle of declination s D (*fig.* 42), and let the equator Υ Q meet these circles in M and D, then M is

Fig. 42.

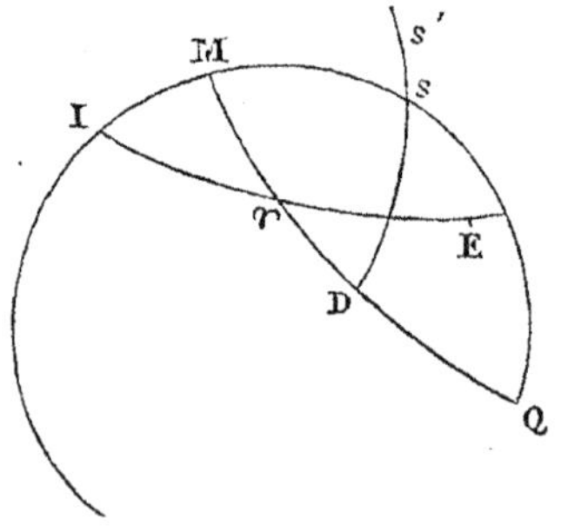

the pole of s D, and the angle s M D $= s$ D $= \delta$: therefore, in the triangle M I Υ, we have

$$M \Upsilon = 90^\circ - a$$
$$M \Upsilon I = \omega$$
$$I M \Upsilon = 180^\circ - \delta$$

hence I Υ, which we shall call θ, may be easily calculated.

Now $s s'$, which is the aberration in declination,

$$= - 20''{\cdot} 246. \cos. E I. \sin. s I E,$$

$$\text{but } \sin. I = \frac{\sin. M \Upsilon}{\sin. \theta}$$

$$\therefore s s' = - 20''{\cdot}246. \frac{\sin. M \Upsilon}{\sin. \theta} \cos. E I$$

$$= 20''{\cdot}246. \frac{\sin. M \Upsilon.}{\sin. \theta} \cos. (180^\circ + \odot + \theta)$$

$$= + 20''{\cdot} 246 \frac{\sin. M \Upsilon. \cos. (\odot + \theta)}{\sin. \theta}$$

$$\text{but } \sin. M \Upsilon = \cos. a \sin. \delta$$

$$\therefore SS' = 20''{\cdot}246. \frac{\cos. a. \sin. \delta. \cos. (\odot + \theta)}{\sin. \theta}$$

Now if we observe the form of these expressions for the aberration in right ascension and declination, we shall see that for any given star, the angles ϕ and θ, and the coefficients

$$\frac{20''{\cdot} 246. \sin. a}{\cos. \delta. \cos. \phi} \text{ and } \frac{20''{\cdot} 246. \cos. a. \sin. \delta}{\sin. \theta}$$

respectively, may be considered as constant: the variations of a and δ, on which they depend, being so slow as to be of no importance; on the other hand, $\odot$ varies from day to day. In order, then, to form our tables, we have only to calculate for each star ϕ, θ, and the efficients just mentioned, or rather their logarithms; and to tabulate the results. When the aberration of any star for a particular day is required, we take from the Nautical Almanac, or the Solar Tables, the angle $\odot$, we add to it ϕ for the right ascension, θ for the declination, which are given in the Tables of Aberration. To the logarithms of the sines of the angles thus obtained, we add the logarithms of the respective coefficients taken from the same tables. The sum of these logarithms, in each case, will be the logarithm of the aberration required.

It is easy to see that the aberration will be greatest when the sine of $\odot - \phi$, in one case, and of $\odot + \theta$, in the other, are greatest; that is, when they are respectively equal to unity: in this case the aberration is expressed simply by the coefficients

$$\frac{20''{\cdot} 246. \sin. a}{\cos. \delta. \cos. \phi.} \text{ and } \frac{20''{\cdot} 246. \sin. \delta. \cos. a}{\cos. \theta.}$$

consequently these coefficients are the maxima of aberration in right ascension and declination. We have mentioned this circumstance, because it is under this name that they are generally entered in the tables.

We have already explained the principles upon which the aberration of a planet may be calculated. The time employed by light to describe the radius of the earth's orbit is $493^s \cdot 2$. If R be the radius of the orbit of the planet, then the time taken by light to describe the distance R : $493^s \cdot 2$ :: R : radius of the earth's orbit or unity. ∴ the time in question = $493'' \cdot 2 \times R$. Now the time taken by the planet to describe this arc, : a day or 86400 seconds :: as the arc itself : planet's motion in a day : ∴ the arc in question

$$= \frac{\text{planet's diurnal motion} \times 493^s \cdot 2 \times R}{86400^s.}$$

$$= \frac{\mu \times 493 \cdot 2 \times R}{86400.}$$

μ being the diurnal motion, expressed in seconds of arcs,
or aberration $= 0'' \cdot 0057083\, \mu R$ *.

This expression is the same for right ascension, declination, longitude, or latitude, provided we take, in each case, the respective values of μ. As long as the planet's motion is *direct*, the aberration in right ascension and longitude is *negative*, and *positive* in the other case. The aberration in declination and latitude is *negative* when the motion is towards the *north pole*, and *positive* in the other case.

Chapter VII.

On the Precession of the Equinoxes, and the Nutation of the Earth's Axis.

We have already mentioned that Bradley's first notion, when he discovered the annual variation in the apparent position of the fixed stars, was to attribute it to a motion in the pole of the heavens itself. Further observation led him to abandon this theory, but he had occasion to resume it afterwards for the purpose of explaining another apparent irregularity which he discovered by a yet longer and more laborious course of observation. It will be convenient, however, before we proceed to explain the nature of this irregularity, to refer to one already explained, and to see whether the notions we have now formed of the existence of motion in the earth do not furnish a more easy and intelligible account of it than we have hitherto been able to give.

This is the phenomenon of the precession of the equinoxes. We have stated that the intersection of the equator and ecliptic continually varies, receding in such a manner that the pole of the equator describes a circle of 23°. 28′ radius (nearly) round the pole of the ecliptic. Now we have recently seen that exactly the same apparent effects will be produced by the rotation of the earth round an axis coinciding with the axis of the heavens, as by the rotation of the heavens themselves. If, therefore, this supposition is to be adopted, the axis of rotation of the earth must continually vary its position, to coincide with the varying positions which the axis of the heavens is found by observation to assume: the stationary points of the earth, the extremities of this axis are found never to vary; the motion of the earth, therefore, will be, on this supposition, a motion of rotation round a constant axis, the position and inclination of which offer a slow change. A familiar instance may serve to explain the sort of motion in question. A top when spinning is often seen to incline to one side, and to perform a sort of conical motion, the point of the peg continuing in the same position. The top turns round many times in the course of one of these conical revolutions; it has therefore a compound motion, a rapid motion of rotation round its axis, of which the point of the peg is the one extremity and the centre of the top is the other, and a slower conical motion of the axis itself upon its fixed point, with which the motion of rotation is combined. The parallel is not a very close one; the motion of the axis in the top is rapid, in the earth almost inconceivably slow (completing a revolution in about 25,000 years) and the fixed point in the top is one extremity of the axis, fixed chiefly by the resistance afforded to it by the ground on which it rests. The fixed point in the earth is the centre of the axis, which retains its position in the orbit independently of any such mechanical obstruction: and the axis thus projecting on each side of the fixed point, the parts of it on each side of that point describe similar cones vertically opposed to each other.* Still

* μ and R in this expression may be found from the planetary tables.

* The illustration may be carried a step farther The top, if spun on a hard and smooth surface, will often traverse it with considerable regularity and velocity, the motions already mentioned remaining unaltered, the whole top thus moving in space. and having combined this motion, a rapid motion of rotation round a fixed axis, and a slower conical motion of the axis itself. The motion of the top

the conception of the motion may be rendered easier by the illustration; and it will be seen that the elements of which it is made up are in no way inconsistent with each other. And if they are not, this explanation of precession furnishes one of the most striking instances which we have met with of the superior simplicity introduced into the system of the heavens by the supposition of the earth's motion. It is sufficiently difficult to imagine that the sun, the moon, and all the stars and planets, at distances almost infinitely various, should all revolve round the earth, even for a single day in tracks corresponding to the supposition of rotation round an axis with which they have no apparent connexion: but it is incomparably more so, to suppose that this axis is continually changing its position, and that all their motions, separate and independent as they are, are continually changed also, so as always to conform to that same supposition, which for a single day and a single position of the axis seemed so improbable. If, on the other hand, we suppose the earth to revolve, and the position of its axis of revolution to vary, all these apparently independent motions are explained, and become the necessary consequences of this one supposition; which involves nothing improbable; and which we shall hereafter see to be itself the necessary consequence of certain laws to which all the appearances of nature conform themselves.

We have already shewn how Dr. Bradley was induced to abandon the notion that the appearances really produced by aberration were the result of a motion of the pole. After he had done so, however, and had deduced from his observations the theory of aberration, he found that there still remained another irregularity to be accounted for, and that, on comparing together the observations of a considerable length of time, the north polar distances of different stars were still found to vary in a manner not accounted for by the effects of refraction, aberration, parallax, and precession. It is remarkable that the observation of the same stars, γ of the Dragon, and 35 of Camelopardalus, which led him to attribute aberration to its true cause, and to reject the supposition of a motion of the pole to explain it, led him now to adopt the supposition then rejected. On comparison of observations made at different times, he found that one of these stars became nearer the pole, while the other became more distant from it; and that in this case, unlike the former, the approach of the one exactly equalled the recess of the other. These results obviously corresponded with the supposition of a motion in the pole itself, which being situated exactly, or very nearly, between the two stars, would necessarily, if it changed its position, approach the one by the same space by which it receded from the other. If, however, the pole really shifted its place, it would affect the right ascension and declination of every star in the heavens, for these depend on the situation of the pole; and the effects of its motion upon these elements could be discovered by computation. If, on investigating these effects, and comparing them with the observed phenomena of a great variety of stars, taken in all different situations, the results were found to correspond, there could be no doubt that they were really occasioned by a motion of the pole; if they failed to do so, then some other cause must be sought for them. Now it was found that they did accurately correspond with the supposition that the pole, besides its motion occasioned by precession, had another, which we shall presently more fully explain, but which made its true place occasionally on one side, and occasionally on the other of its mean place, the mean place being that which it would occupy if effected by precession alone. The place of the pole we have seen to be determined, on the supposition of the earth's rotation, by the position of the earth's axis: the earth's axis therefore must have a motion, sometimes to one side, sometimes to the other of its mean position, or it must have a slight rolling or nodding motion combined with the continued motion of precession; and this motion is called *nutation*. The apparent irregularity in question therefore is known by the name of *nutation*, or more fully, *the nutation* of the earth's axis.

It is found that the motion of the pole, thus ascertained, makes its true place in all cases different from its mean place; and that the true place may always be represented by supposing it to describe an ellipse round the mean place, which itself proceeds in a circle round the pole of the ecliptic, in the manner already

over the surface may represent the motion of the earth in its orbit; the spinning, its rotation in 24 hours; the conical motion of the axis, the motion of the earth's axis which occasions precession.

explained in treating of precession. This ellipse is found to be described in rather more than $18\frac{1}{2}$ years, and in exactly the time in which the nodes of the lunar orbit perform a complete revolution on the ecliptic*. In fact, there is an intimate connexion between these phenomena, the nature of which will be explained when we come to treat of the Theory of Universal Gravitation. We will only observe here, in order to give some idea of the nature of this connexion, that the earth not being a perfect sphere, the attraction exercised by the moon upon it will vary with different positions of that planet with regard to the earth's equator. Now these positions depend upon the position of the lunar orbit itself, the nodes of which describe the whole circumference of the ecliptic in the time above stated. At the expiration of this period of $18\frac{1}{2}$ years, the orbit has very nearly returned to its primitive position: the attraction then, and inclination of the earth's axis consequent upon it, return to the same state too.

Having then acquired some idea of the cause of Nutation, and having ascertained that its period coincides with a revolution of the moon's nodes, let us now proceed to examine more accurately the laws of the phenomenon, and show how the nature of the orbit described by the true, round the mean pole of the earth, has been ascertained. On the one hand, the comparison of the effects produced on the right ascension and declination of a great many stars; on the other, the theory deduced from the principle of gravitation, show that the variation in the obliquity of the ecliptic produced by this cause, may be represented by the formula $9''.63.\cos.\ \Omega$, where Ω stands for the mean longitude of the moon's node: and the variation of the place of the equinox on the ecliptic by the formula $-18''.\sin.\ \Omega.\cot.\ 2\,\omega$, where ω represents the obliquity of the ecliptic. Now it is easy to show that the effects produced on the right ascension and declination of a star by a *small* simultaneous variation of the longitude, and the obliquity, may be expressed as follows. Let a and δ represent the right ascension and declination respectively; and $d\,a$ and $d\,\delta$ the variations required. Take ψ for the variation of the longitude and ϕ for that of the obliquity. Call the longitude λ, the latitude θ, then

* See page 69.

$$d\,\delta = \psi.\sin.\ \omega.\cos.\ a + \phi.\sin.\ a$$
$$d\,a = \psi\,\{\cos.\ \omega + \sin.\ \omega.\ \sin.\ a.\tan.\delta.\} - \phi.\cos.\ a.\tan.\ \delta.$$

But, $\psi = -18''.\sin.\ \Omega.\cot.\ 2\,\omega.$
$\phi = 9''.63.\cos.\ \Omega.$

substituting these values, and assuming $\omega = 23°.27'.52''$, we get

$$d\,\delta = 8''.4\ \sin.\ (a - \Omega) + 1''.23.\ \sin.\ (a + \Omega).$$
$$d\,\alpha = -16''.51\sin.\ \Omega - \{8''.4.\cos.\ (\alpha - \Omega) + 1''.23.\cos.(\alpha + \Omega)\}\ \tan.\ \delta\,*.$$

* In order to demonstrate the expressions for $d\,\alpha$ and $d\,\delta$ given in the text, let us call as before the obliquity of the ecliptic ω, the right ascension α, the declination δ, the longitude β, the latitude θ. By spherical trigonometry—

$$\sin.\ \delta = \sin.\ \omega.\cos.\ \theta.\sin.\ \beta + \cos.\ \omega.\sin.\ \theta$$

$$\tan.\ \alpha = \frac{-\tan.\ \theta.\sin.\ \omega + \sin.\ \beta.\cos.\ \omega}{\cos.\ \beta}$$

The effect of nutation is to produce a small augmentation in the longitude of each star, and a small variation in the inclination of the equator to the ecliptic. These changes being in themselves extremely small, we may, by a well-known mathematical principle, find their united effect, by calculating separately their respective effects, and adding these together for the whole result. Let us consider then first a small variation in the longitude, in which case β becomes β', and let us find the correspondent changes in α and δ.

Recurring to the expressions for α and δ, which now become α' and δ', we have $\sin.\ \delta' = \sin.\ \omega.\cos.\ \theta.\sin.\ \beta' + \cos.\ \omega.\sin.\ \theta \therefore \sin.\ \delta' - \sin\delta = \sin.\ \omega\ \cos.\ \theta\,\{\sin.\ \beta' - \sin\beta\}$

or by trigonometry—

$$\sin.\frac{1}{2}\left(\delta' - \delta\right).\cos.\frac{1}{2}\left(\delta' + \delta\right) = \sin.\ \omega.\ \cos.\ \theta.\left\{\sin.\frac{1}{2}\left(\beta' - \beta\right).\cos.\frac{1}{2}\left(\beta' + \beta\right)\right\}$$

The difference of the longitudes or $\beta' - \beta$ is in fact the quantity which we have called ψ; $\delta' - \delta$ is the nutation in declination which we have called $d\,\delta$. Hence $\sin.\frac{d\,\delta}{2}.\ \cos.\left\{\delta + \frac{d\,\delta}{2}\right\} = \sin.\ \omega.\cos.\ \theta.\sin.\ \frac{\psi}{2}.\cos.\left\{\beta + \frac{\psi}{2}\right\}$

But since the arcs ψ and $d\,\delta$ are very small, we may substitute for the ratio of the sines that of the arcs themselves, and confining ourselves to the first powers of these arcs, we get

$$d\,\delta = \psi.\frac{\sin.\ \omega.\cos.\ \theta.\cos.\ \beta}{\cos.\ \delta.}$$

To simplify this expression, let us suppose, in figure 43, P′ to be the pole of the ecliptic, P

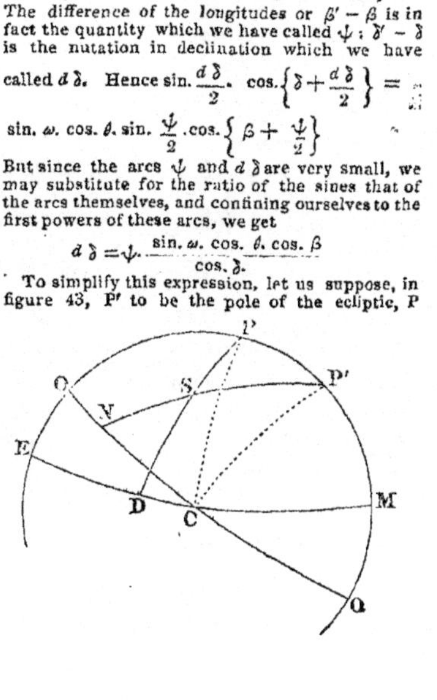

The formulæ for the effects of the solar nutation are exactly similar: but the values of ϕ and ψ are different. In this case,

---

that of the equator, S the place of any star whose changes of position we are considering; E C M the equator, O C Q the ecliptic, P S D a circle of declination passing through the star, P′ S N a circle of latitude; the great circle E O P P′ will be the solstitial colure, C the equinox (suppose it the vernal); let P C, P′ C be arcs of great circles respectively. Now, in the spherical triangle P′ P S the side P S is the complement of the declination, P′ S that of the latitude; the angle P′ P S = P′ P C + C P S = 90° + α; the angle P P′ S = P P′ C − C P′ S = 90° − β. Now,

$$\frac{\sin. PS}{\sin. P'S} = \frac{\sin. PP'S}{\sin. P'PS}$$

$$\therefore \frac{\cos. \delta}{\cos. \theta} = \frac{\sin. (90^\circ - \beta)}{\sin. (90^\circ + \alpha)}$$

$$= \frac{\cos. \beta}{\cos. \alpha}$$

$$\therefore \cos. \beta = \frac{\cos. \alpha \cos. \delta}{\cos. \theta}$$

But we have found

$$d\,\delta = \frac{\psi. \sin. \omega. \cos. \theta. \cos. \beta}{\cos. \delta}$$

substituting them for cos. β its value, we get

$$d\,\delta = \frac{\psi \sin. \omega. \cos. \theta}{\cos. \delta} \times \frac{\cos. \alpha \cos. \delta.}{\cos. \theta}$$

$$= \psi. \sin. \omega. \cos. \alpha \quad . \quad . \quad . \quad . \quad (a)$$

Let us now consider the effects produced on the right ascension, by the motion parallel to the ecliptic, of which we have been treating. By spherical trigonometry, we have

$$\tan. \alpha = \frac{-\tan. \theta. \sin. \omega + \sin. \beta. \cos. \omega}{\cos. \beta}$$

Suppose now, that β becoming β', and the latitude remaining constant, that α becomes α', then

$$\tan. \alpha' = \frac{-\tan. \theta \sin. \omega + \sin. \beta'. \cos. \omega}{\cos. \beta'}$$

consequently

$$\tan. \alpha' - \tan. \alpha = \frac{-\tan. \theta. \sin. \omega \left\{\cos. \beta - \cos \beta'\right\} + \cos. \omega. \sin (\beta' - \beta)}{\cos. \beta \cos. \beta'}$$

$$\text{or, } \frac{\sin. (\alpha' - \alpha)}{\cos. \alpha. \cos. \alpha'} = \frac{-2 \tan. \theta. \sin. \omega. \sin. \frac{1}{2}(\beta' + \beta) \sin. \frac{1}{2}(\beta' - \beta) + \sin. (\beta' - \beta) \cos. \omega.}{\cos. \beta. \cos. \beta'}$$

$\alpha' - \alpha$ is the nutation in right ascension; it is extremely small, as is also $\beta' - \beta$ which we have previously called ψ: we may then substitute the ratio of these small arcs for that of their sines. We ma also in every term which is multiplied by these small factors, suppose $\beta = \beta'$, and $\alpha = \alpha'$; hence

$$\alpha' - \alpha \text{ or } d\,\alpha = \frac{\psi. \cos.^2 \alpha \left\{-\tan. \theta. \sin. \omega. \sin. \beta. + \cos. \omega\right\}}{\cos.^2 \beta}$$

From this expression we must eliminate β and θ, in order to have terms involving only the right ascension and declination. For this purpose let us substitute in the denominator on the right hand side, for cos. β, its value, $\frac{\cos. \alpha. \cos. \delta}{\cos. \theta.}$, we obtain

$$d\,\alpha = \psi \left\{\frac{-\sin. \theta. \cos. \theta. \sin \omega. \sin. \beta + \cos.^2 \theta. \cos. \omega}{\cos.^2 \delta}\right\}$$

Recurring now to the fundamental equations with which we set out, it appears that $\sin. \omega. \cos. \theta. \sin. \beta = \sin. \delta - \cos. \omega. \sin. \theta$

substituting this value, we obtain

$$d \quad = \psi \left\{\frac{-\sin. \theta. \sin. \delta + \cos. \omega}{\cos.^2 \delta}\right\}$$

To eliminate θ, it is easy to see, by applying the formulæ of spherical trigonometry to the triangle S P P′ fig. 43, that

$$\sin. \theta = \cos. \omega \sin. \delta - \sin. \omega. \cos. \delta. \sin. \alpha$$

substituting this value of sin. θ

$$d\,\alpha = \psi \left\{\frac{-\sin. \delta (\cos. \omega. \sin. \delta - \sin. \omega \cos. \delta. \sin. \alpha) + \cos. \omega}{\cos.^2 \delta.}\right\}$$

$$= \psi \left\{\frac{-\cos. \omega. \sin.^2 \delta + \sin. \omega. \sin. \delta. \cos. \delta. \sin. \alpha. + \cos. \omega}{\cos.^2 \delta.}\right\}$$

$$= \psi \left\{\frac{\cos. \omega (1 - \sin.^2 \delta)}{\cos.^2 \delta} + \sin. \omega. \tan. \delta. \sin. \alpha\right\}$$

$$= \psi \left\{\cos. \omega. + \sin. \omega. \tan. \delta. \sin. \alpha\right\} \qquad (b)$$

The reader will observe that, in what precedes, we have only considered the effects produced by a small change in the longitudes, the obliquity remaining constant. Let us now consider the effect of a small change in the obliquity of the ecliptic, the longitude remaining constant. For this purpose, let us take the equations

$$\sin. \delta = \sin. \omega. \cos. \theta. \sin. \beta. + \cos. \omega. \sin. \theta$$

$$\tan. \alpha = \frac{-\tan. \theta. \sin. \omega + \sin. \beta. \cos. \omega}{\cos. \beta}$$

which we have already employed.

Assume $\tan. \chi = \frac{\sin. \beta}{\tan. \theta}$

then $\sin. \delta = \frac{\sin. \theta. \cos. (\chi - \omega)}{\cos. \chi}$

and $\tan. \alpha. = \frac{\tan. \beta. \sin. (\chi - \omega)}{\sin. \chi}$

Now let ω become $\omega + \phi$, β and θ remaining the same, χ will not vary, but δ and α will change, and become δ' and α'. Thus for the first of these quantities, we shall have

$$\sin. \delta' = \frac{\sin. \theta \cos. (\chi - \omega - \phi)}{\cos. \chi}$$

$$\therefore \sin. \delta' - \sin. \delta = \sin. \theta \frac{\left\{\cos. (\chi - \omega - \phi) - \cos. (\chi - \omega)\right\}}{\cos. \chi}$$

$$\varphi = 1''.34. \cos. 2\,\Lambda\,*$$
$$\psi = -1''.34. \sin. 2\,\Lambda. \cot. \omega.$$

In very accurate calculations it is necessary, having found the effects of the solar nutation in right ascension and declination respectively, by substituting these values of φ and ψ in our general formulæ, to add the expressions for the variations thus obtained, to the respective variations caused by the lunar nutation. It will, however, be seen that these terms must be very small: when they are taken into account, the nutation thus corrected is called the luni-solar nutation. As the object of this Treatise is rather to explain general methods than to go into the niceties of astronomical calculation, we shall for the future consider the effects of lunar nutation only.

Let us now determine by means of our formulæ, the motion of the equinox in right ascension caused by nutation, or as it is sometimes called the equation of the equinoctial points. In the case of the equinox $\alpha = 0$ and $\delta = 0$, and therefore $d\,\alpha = -16''.51. \sin. \Omega$

This then is the difference in right ascension, between the true and the mean place of the equinox. The equation in declination will be

$$d\,\delta = -7''.17 \sin. \Omega.$$

The same general formulæ will also serve us, to show how the apparent pole of the equator describes a small ellipse round the place of the mean pole.—From what has been already said about nutation, it appears that the ellipse in question is of very small dimensions: we may then, without sensible inaccuracy, suppose its projection on the concave sphere of the heavens to be the same as if it were projected on a plane surface: and as the oscillation takes place round the mean place of the star, we shall take this mean place for the origin of the co-ordinates. Draw through this point an arc of a great circle perpendicular to the meridian, and take the tangents at this point to the arc just mentioned, and to the meridian respectively, as axes of rectangular co-ordinates. These co-ordinates consequently will be the difference of declination, $d\,\delta$, and the difference of right ascension, measured on the great circle perpendicular to the meridian, which passes through the place of the star; or, in algebraical language,

$$x = d\,\delta$$
$$y = d\,\alpha. \cos. \delta.$$

$$\sin. \tfrac{1}{2}(\delta' - \delta). \cos. \tfrac{1}{2}(\delta' + \delta) = \frac{\sin. \theta. \sin. \frac{\varphi}{2} \sin. (\chi - \omega - \frac{\varphi}{2})}{\cos. \chi}$$

substituting for $\dfrac{\sin. \frac{(\delta' - \delta)}{2}}{\sin. \frac{\varphi}{2}}$, $\dfrac{\frac{\delta' - \delta}{2}}{\frac{\varphi}{2}}$, and putting in the other terms of the equation $\delta' = \delta$, $\alpha' = \alpha$, and $\varphi = 0$, we obtain

$$\delta' - \delta = \frac{\varphi. \sin. \theta. \sin. (\chi - \omega)}{\cos. \delta. \cos. \chi}$$

But $\quad \sin. (\chi - \omega) = \dfrac{\tan. \alpha\,.\,\sin. \chi}{\tan. \beta}$

$$\therefore \delta' - \delta = \frac{\varphi \tan. \alpha. \sin. \theta. \tan. \chi}{\cos. \delta. \tan. \beta}$$

Again $\quad \sin. \theta \tan. \chi = \sin. \beta. \cos. \theta.$

$$\therefore \delta' - \delta = \frac{\varphi \tan. \alpha}{\cos. \delta}. \cos. \theta. \cos. \beta.$$

But $\quad \cos. \theta. \cos. \beta = \cos. \alpha. \cos. \delta$

as may be seen by referring to the triangle S P P′, consequently

$$\delta' - \delta = \varphi \sin. \alpha \quad . \quad . \quad . \quad . \quad . \quad (c)$$

This then is the effect produced on the declination.

Now for the right ascension

$$\tan \alpha' = \frac{\tan. \beta. \sin. (\chi - \omega - \varphi)}{\sin. \chi}$$

$$\therefore \tan. \alpha' - \tan. \alpha = \tan. \beta \frac{\sin. (\chi - \omega - \varphi) - \sin. (\chi - \omega)}{\sin. \chi}$$

$$\sin. (\alpha' - \alpha) = \frac{-2. \tan. \beta \cos. \alpha \cos. \alpha' \sin. \frac{\varphi}{2} \cos. (\chi - \omega - \frac{\varphi}{2})}{\sin. \chi}$$

Using the same substitutions as before

$$\alpha' - \alpha \text{ or } d\,\alpha = \frac{-\varphi \tan. \beta. \cos.^2 \alpha \cos. (\chi - \omega)}{\sin. \chi}$$

But $\quad \cos. (\chi - \omega) = \dfrac{\sin. \delta \cos. \chi}{\sin. \theta.}$

$$\therefore d\,\alpha = \frac{\varphi \cos.^2 \sin. \delta. \tan. \beta}{\sin. \theta. \tan. \chi}$$

As before $\quad \sin. \theta. \tan. \chi = \sin. \beta. \cos. \theta$

$$\therefore d\alpha = \frac{-\varphi \cos.^2 \alpha. \sin. \delta}{\cos. \theta. \cos. \beta}$$

Again as before $\quad \cos. \theta. \cos. \beta. = \cos. \alpha \cos. \delta.$

$$\therefore d\,\alpha = -\varphi \cos. \alpha. \tan. \delta. \quad . \quad . \quad . \quad (d)$$

We have now calculated the values of $d\,\alpha$ and $d\,\delta$, supposing the longitude and obliquity of the ecliptic to vary successively: to get the effects of their simultaneous variations, we have only, as these are very small, to add together the expressions already obtained for $d\,\alpha$ and $d\,\delta$ respectively, that is, (a) to (c), and (b) to (d), and we get

$$d\,\delta = \psi \sin. \omega. \cos. \alpha. + \varphi. \sin. \alpha$$

$$d\,\alpha = \psi \left\{\cos. \omega. + \sin. \omega. \sin. \alpha. \tan. \delta\right\} - \varphi \cos. \alpha. \tan. \delta$$

* Λ is the longitude of the sun.

Now, if we refer to page 159, we shall find that these expressions become

$$x = 9''{\cdot}63.\ \sin.\ \alpha.\ \cos.\ \Omega$$
$$- 7''{\cdot}17.\ \cos.\ \alpha.\ \sin.\ \Omega$$
$$y = - 16''{\cdot}51.\ \sin.\ \Omega\ \cos.\ \delta$$
$$- \{ 9''{\cdot}63\ \cos.\ \alpha.\ \cos.\ \Omega + 7''{\cdot}17.\ \sin.\ \alpha.\ \sin.\ \Omega \} \sin.\ \delta.$$

If between these two equations we eliminate the angle ☊, we shall have the orbit described by the apparent place round the mean place generally. Our object is to find the orbit described by the apparent pole of the equator round the true: consequently in this case, $\delta = 90°$, $\cos.\ \delta = 0$, $\sin.\ \delta = +1$: and, further, as the choice of the meridian on which x is to be counted is arbitrary, let us take that which passes through the poles both of the equator and ecliptic; consequently $\alpha = 90°$ $\sin.\ \alpha = 1$, and $\cos.\ \alpha = 0$; our expressions now become

$$x = 9''{\cdot}63.\ \cos.\ \Omega$$
$$y = - 7''{\cdot}17\ \sin.\ \Omega$$
$$x^2 = (9''{\cdot}63)^2.\ \cos.^2.\ \Omega$$
$$y^2 = (7''{\cdot}17)^2.\ \sin.^2.\ \Omega$$
$$\cos.^2\ \Omega = \frac{x^2}{(9''{\cdot}63)^2}$$
$$\sin.^2\ \Omega = \frac{y^2}{(7''{\cdot}17)^2}$$
$$\therefore 1 = \frac{x^2}{(9''{\cdot}63)^2} + \frac{y^2}{(7''{\cdot}17)^2}$$

or $(9''{\cdot}63)^2\ (7''{\cdot}17)^2 = (9''{\cdot}63)^2.\ y^2 + (7''{\cdot}17)^2.\ x^2$.

which is the equation of an ellipse referred to its centre. Now, as we have taken the place of the mean pole for the centre of the co-ordinates, and as these co-ordinates give the place of the apparent pole referred to it, it follows that the apparent pole describes an ellipse round the mean, as stated in page 63. The major axis is in the direction of the axis of x, and is a tangent to the meridian passing through the centre of the ellipse; its value is $19''{\cdot}36$. The minor axis in the direction of the axis of y, and is perpendicular to the former; its value is $14''{\cdot}34$. This minor axis is a tangent to the parallel of the ecliptic, which passes through the mean pole of the equator. If we wish to express it in parts of this parallel, we must divide $14''{\cdot}34$ by the cosine of the distance from the pole of the ecliptic, or, what is the same, by $\sin.\ \omega$. With regard to the major axis, it is to be observed that it is situated on the solstitial colure, for it will be recollected that, in taking the axis of y, we have made $\alpha = 90°$.

To obtain the place of the apparent pole in its ellipse at any given moment, we must recur to our original values of x and y, putting in these $\cos.\ \alpha = 0$, $\sin.\ \alpha = 1$, $\cos.\ \delta = 0$, $\sin.\ \delta = 1$, we get

$$x = 9''{\cdot}63.\ \cos.\ \Omega$$
$$y = - 7''{\cdot}17.\ \sin.\ \Omega.$$

Referring now to the figure and construction in page 63, $PM = PO.\ \cos.\ APO$: but $APO = \Omega$, or the longitude of the moon's node; $PO = PB = 9''{\cdot}63$ $\therefore PM$, which is the abscissa of the point $p = 9''{\cdot}63.\ \cos.\ \Omega$; but this is the value of the abscissa of the place of the apparent pole: p then will, as has been stated, be the place of that pole.

The Tables of Nutation are similar in their form to those of Aberration. It is necessary, as in the latter case, to effect a transformation in the formula by the introduction of an auxiliary angle, which we will call η. The formulæ already obtained for the nutation in right ascension and declination may be thus expressed,—

$$d\,\alpha = - m\,.\cos.\,\alpha\,.\tan.\,\delta\,.\cos.\,\Omega - (n + p\,.\sin.\,\alpha\,.\tan.\,\delta)\sin.\,\Omega$$
$$d\,\delta = m\,.\sin.\,\alpha\,.\cos.\,\Omega - p\,.\cos.\,\alpha\,.\sin.\,\Omega,$$

when m, n, and p stand for the different numerical constants employed in page 159.

Let us assume $$\cot.\,\eta = \frac{n\,. + p\,.\sin.\,\alpha\,.\tan.\,\delta}{m\,.\cos.\,\alpha\,.\tan.\,\delta}$$

$$\therefore d\,\alpha = - m\,.\cos.\,\alpha\,.\tan.\,\delta \{\cos.\,\Omega + \cot.\,\eta\,.\sin.\,\Omega\}$$
$$= - m\,.\cos.\,\alpha\,.\tan.\,\delta \frac{\{\cos.\,\Omega\,.\sin.\,\eta + \sin.\,\Omega\,.\cos.\,\eta\}}{\sin.\,\eta}$$
$$= - m\,.\cos.\,\alpha\,.\tan.\,\delta\,.\frac{\sin.\{\eta + \Omega\}}{\sin.\,\eta}$$

Now we may remark here, as we did in the case of Aberration, that the angle φ and the coefficient $- m\,.\cos.\,\alpha\,.\tan.\,\delta$ are constant for the same star, at least for many years, as the changes of the right ascension and declination caused by precession are very small. We may also remark that the nutation in right ascension will be greatest when sin. $(n + \Omega) = 1$, in which case

$$d\alpha = -\frac{m \,.\, \cos.\, \alpha \,.\, \tan.\, \delta}{\sin.\, n}$$

The coefficient then of our general expression for $d\,\alpha$ is identical with the maximum value of $d\,\alpha$.

Let us now take the expression for $d\,\delta$. It is

$$m \,.\, \sin.\, \alpha \,.\, \cos.\, \Omega - p \,.\, \cos.\, \alpha \,.\, \sin.\, \Omega.$$

Assume $$\cot.\, n' = \frac{p}{m} \,.\, \cot.\, \alpha$$

$$d\,\delta = m \,.\, \sin.\, \alpha \,.\, \left\{\cos.\, \Omega - \frac{p}{m} \,.\, \frac{\cos.\, \alpha}{\sin.\, \alpha} \,.\, \sin.\, \Omega\right\}$$

$$\therefore d\,\delta = m \,.\, \sin.\, \alpha \,\{\cos.\, \Omega - \cot.\, n' \,.\, \sin.\, \Omega\}$$

$$-\frac{m \,.\, \sin.\, \alpha}{\sin.\, n'} \sin.\, \{\Omega - n'\}.$$

Here, as before, n' is constant for the same star; and the coefficient $\frac{m \,.\, \sin.\, \alpha}{\sin.\, n}$ represents the maximum of the nutation in declination.

In order, then, to form Tables of Nutation for the principal fixed stars, we calculate for each star the two angles n and n', the logarithms of their sines, as well as those of the two maxima, and arrange them in tables. The calculator who wishes to determine the nutation of any star in right ascension, for example, finds in the table n, to which he adds Ω (taken from the Nautical Almanac); in the tables of logarithmic sines, he finds the logarithm of the sine of $\{n + \Omega\}$; to this he adds the logarithm of the maximum which he finds in the tables of nutation, and subtracts the logarithm of the sine of n, which he finds in the same tables. He proceeds in a similar manner for the declination. By this means, the calculation of nutation is most essentially abridged; and, consequently, several Astronomers have published tables on the principles just laid down, of the nutation of a large number of fixed stars. The most remarkable of these are the Tables of Nutation and Aberration for 1404 stars, by the Baron de Zach. But it is to be borne in mind of these, and all similar tables, that they will only serve for a limited number of years, as in fact the tangents of n and n', as well as the maxima, are not constant, since the right ascensions and declinations which they involve are not constant. The length of time for which they may be considered as invariable, is different for different stars: near the Pole the variation is very considerable, and the tables will not be serviceable for more than a few years, while near the Equator they may be used without alteration for half a century. The reason of this difference will be explained when we come to treat of the formulæ for precession. Under these circumstances de Zach, and other Astronomers, have given special tables of nutation for the Polar star, and β Ursæ Minoris, stars much used in astronomy for the determination of latitudes and other purposes.

Some difference exists among astronomers as to the exact value of the principal constant of nutation, or the semi-axis major of the ellipse, of which we have frequently spoken. Bradley estimated it at 9″·0; Mayer at 9″·65; Maskelyne 9″·55; von Lindenau from observations of the polar star during three revolutions of the lunar nodes, at 8″·977; Dr. Brinkley, by the observation of 1618, different stars at 9″·25. Observations of the pendulum bring it down to 8″·6. According to La Place the chances are 21405 to one that this coefficient is not less than 9″·31, or greater than 9″·94.

To give an idea of the way in which the constant of nutation may be determined, we must recur to the expression for the nutation in declination, page 159; it is

$$\psi \,.\, \sin.\, \omega \,.\, \cos.\, \alpha + \phi \,.\, \sin.\, \alpha.$$

Now the theory of gravitation shows that $\psi = -2m \,.\, \sin.\, \Omega \,.\, \cot.\, 2\omega$, and $\phi = m \,.\, \cos.\, \Omega$, where m is an unknown constant. As for any particular star, the right ascension is found from the catalogues; and as the obliquity of the ecliptic is known, or may be determined by independent methods, every thing is known in the expression except the coefficient m: let us call then the whole expression m Q, where Q is a known quantity. Let us suppose now that having calculated Q for a given star, we observe the meridian altitude of that star; this meridian altitude, corrected for refraction, will give us the apparent declination. We will also correct the apparent declination for precession and aberration, which are supposed known. Then the true declination will equal the apparent declination thus corrected,—nutation in declination. Call the true declination Δ, the apparent declination corrected for precession and aberration δ, then $\Delta = \delta - m\,Q$.

Suppose now the same star be observed again on another day, the true declination remains the same. Let the apparent declination, corrected as before, be δ', and Q' the coefficient, then

$$\Delta = \delta' - m\,Q'$$

$$\therefore \delta - \delta' = m\,.\,(Q - Q')$$

$$\therefore m = \frac{\delta - \delta'}{Q - Q'}.$$

By repeating these observations we may multiply indefinitely equations of this form, and the mean will give us m with all desirable accuracy. It was by observations of this kind on the pole star that von Lindenau found for the constant of nutation $8''\cdot977$.

It may be as well to explain here how the arc ☊ is calculated. The lunar theory teaches us that the node retrogrades annually $19^\circ\,.\,20'\,.\,30''$, or in other terms $19^\circ\cdot34166$. We also know that on the 1st of January, 1830, the longitude of the node was $5^\circ.23'.1''\cdot18$. By subtracting from this arc the annual retrogradation taken 1, 2, 3 . . . . times, we shall have the longitude of the node at the commencement of the successive years 1831, 1832, 1833 . . . . and we may find the longitude for any day of these years by subtracting again so many times the daily retrogradation, or $3'\,.10''\,.64 = 0^\circ\cdot052955$. It is found, however, in calculation, that addition is always more easy than subtraction. As the longitude of the node decreases, the supplement of this longitude to 360° increases; it is then more convenient instead of subtracting the motion of the node in longitude, to add the supplement of the node's longitude; consequently it is always the supplement which is given in the tables and ephemerides. It is also more convenient to divide the circle into 1000 than into 360 parts; a simple proportion enables us to turn degrees, &c., into thousandths of a circle. We find in this way that the supplement of the longitude of the node for the beginning of 1830 is 519·38; the annual motion 53·7269, the diurnal 0·147098. Thus, if we call the number of years from 1830, t, and the number of days from the beginning of the year, f; then the supplement of the longitude for a given day, expressed in thousandths of the circumference, will be

$$519\cdot38 + 53\cdot7269\,.\,t + 0\cdot14798\,.f.$$

For years anterior to 1830, t must be taken with the negative sign.

Having sufficiently discussed the subject of Nutation, we shall now resume the consideration of Precession, and show how its effects may be calculated and allowed for. In what has been said hitherto, we have taken no notice of a slow change in the obliquity of the ecliptic which complicates the results. In speaking of precession we have considered merely the general motion of the stars in longitude, while their latitudes were supposed to remain unaltered. But the attraction of the planets upon the terrestrial spheroid causes a displacement of the ecliptic, which would not take place were the earth perfectly spherical. The angle at which the ecliptic is inclined to the equator experiences a slow diminution, the effects of which are certainly small, but which it is necessary to take into account and to allow for in very accurate calculations. In fact, this diminution is so small that its existence was long contested, but it is now established as well by theory as direct observation. The problem we shall first consider, however, is to calculate the effects produced on the right ascension and declination in a few years merely, and for this purpose we may suppose the obliquity of the ecliptic constant. We may also suppose the increase of longitude, or the absolute quantity of precession constant, though in fact this quantity varies very slowly from year to year. We may now proceed as in finding the nutation, by taking the formulæ which express the right ascension and declination in terms of the longitude and latitude, and ascertaining the changes produced in them by a small variation in the longitude only. Thus we found from

$$\sin.\ \delta = \sin.\ \omega.\ \cos.\ \theta.\ \sin.\ \beta + \cos.\ \omega.\ \sin.\ \theta$$

by supposing δ to become δ', when β became β', that

$$\delta' - \delta \text{ or } d\,\delta = \psi.\ \sin.\ \omega.\ \cos.\ \alpha$$

and again from

$$\tan.\ \alpha = -\frac{\tan.\ \theta.\ \sin.\ \omega + \sin.\ \beta.\ \cos.\ \omega}{\cos.\ \beta}$$

that

$$d\,\alpha = \psi.\ \{\cos.\ \omega + \sin.\ \omega.\ \sin.\ \alpha.\ \tan.\ \delta\}$$

In these expressions ψ, and the angle ω are supposed constant: let us call $\psi.\ \cos.\ \omega = m$, and $\psi.\ \sin.\ \omega = n$, then

$$d\,\alpha = m + n.\ \sin.\ \alpha.\ \tan.\ \delta\ {}^{*}$$
$$d\,\delta = n\ \cos.\ \alpha$$

According to the best modern determinations, if we call t the number of years elapsed since 1750,

$$m = 46''\cdot 02824 + 0''\cdot 0003086450\cdot t$$
$$n = 20\ \cdot 06442 - 0\ \cdot 0000970204\cdot t$$

The smallness of the coefficients of t shows sufficiently, that for at least ten years, (even for purposes where the greatest exactness is required) that m and n are sensibly constant. In researches of less nicety, a longer period may be assumed: in fact, in thirty years the difference will not exceed a hundredth of a second: so that, for the ordinary purposes of astronomy, we may fairly assume m and n as constant for half a century.

It is easy to see how m and n may be deduced from observation. Suppose the right ascension and declination of any star to be observed at epochs distant by an interval of ten years: the tenth part of the variations observed will give the respective changes for one year. Thus $d\,\alpha$ and $d\,\delta$ are known in our expressions for the annual variation, given above: and every thing is known in our two equations except m and n, which may be found by the ordinary methods. It is sufficiently evident, that by employing a considerable number of stars, these quantities may be found with great accuracy. But $m = \psi.\ \cos.\ \omega$ and $n = \psi.\ \sin.\ \omega$; hence the annual precession and the obliquity of the ecliptic may be found: but it is the former quantity alone that is usually the subject of these researches.

The constant ψ thus determined, gives us the effect of the whole or general precession: that is, the motion which results from the combined effects of the variation of the obliquity, and of the general motion of the stars in longitude. The former is caused by the attraction of the planets upon the spheroidal figure of the earth; the latter by the attraction of the sun and moon only; it has thence obtained the name of Lunisolar precession. It is easy to see how the variation of obliquity modifies the lunisolar precession.

Fig. 44.

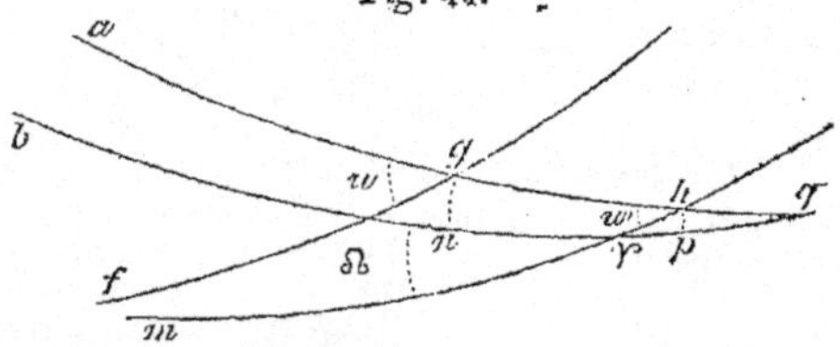

* For stars near the pole, whose declination is very considerable, tan. δ increases so rapidly with small changes in δ, that the annual variation in right ascension cannot be regarded as constant even for a small period of years. This variation must either be calculated from year to year with corresponding new values of α and δ, or it must be found by taking into account the second differences in the variations. If we differentiate the expressions already obtained for $d\,\alpha$ and $d\,\delta$, and multiply by sin. 1″, to get the result in seconds of arcs, we shall have for the second differences

$$d^2\alpha = \left\{ d\,\alpha.\ \tan.\ \delta + \frac{n.\ \sin.\ \alpha}{\cos.\ \delta} \right\} d.\ \delta.\ \sin.\ 1''$$
$$d^2\delta = -n \sin.\ \alpha,\ d\,\alpha.\ \sin.\ 1''$$

It is farther to be observed, that in the catalogues of stars, and the right ascension and its variations are generally given in *time*, not in space; the circumference being divided into 24 hours instead of 360 degrees. The values then of $d\,\alpha$ and $d^2\alpha$ must be expressed in time, which is done by dividing them by 15: since 15° are equivalent to one hour.

Let $a\,q$ be the ecliptic (fig. 44), $f\,g$ the equator at a given epoch. At the end of t years from this time, the vernal equinox g has moved to h : $m\,h$ will be the second position of the equator: $g\,h$ will be the lunisolar precession : the effects of nutation are supposed to have been already allowed for, and are, therefore, not taken into account here, as we are only comparing the *mean* positions of the equinox and equator. Now, the action of the planets displaces the ecliptic $q\,a$ by a small quantity, and at the end of t years has brought it into the position $b\,q$. The new equator $h\,m$ cuts the new ecliptic under the angle ☊, and the old ecliptic $a\,g$ under the angle ω', which differs by a small quantity from the old angle ω. Thus, ♈ is the new vernal equinox, and ☊ the new obliquity. The longitudes and right ascensions were originally counted from g towards a and towards f; they are now counted from ♈ towards b and towards m. On $q\,b$ take $q\,n$ equal to $q\,g$, ♈ n will be the *total* precession, which is evidently less than $g\,h$, the *lunisolar* precession.

In comparing the positions of stars at very distant periods of time, it is absolutely necessary to allow for the change in the obliquity as well as that in longitude. The method of doing this is as follows. The position of the star at some given epoch, as, for example, the year 1750, is known ; that is, its right ascension and declination are given in the catalogue. The obliquity of the ecliptic for the same year is also known. With these data we can calculate the longitude and latitude of the star for the epoch in question. This is done by means of the formulæ—

$$\sin.\ \theta = -\sin.\ \omega.\cos.\ \delta.\sin.\ \alpha + \cos.\ \alpha.\sin.\ \delta$$

$$\tan.\ \beta = \frac{\tan.\ \delta.\sin.\ \omega + \sin.\ \alpha.\cos.\ \omega}{\cos.\ \alpha.}$$

Fig. 45.

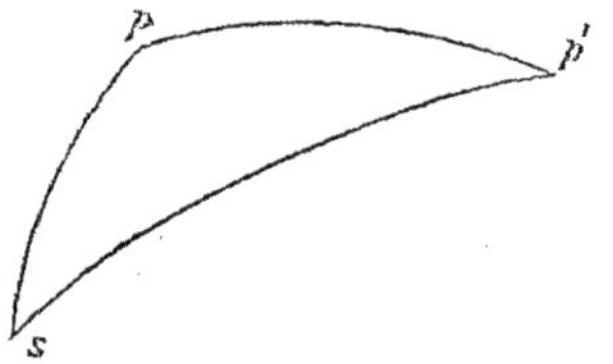

The reader may satisfy himself of the accuracy of these formulæ by considering the spherical triangle $p\,p'\,s$ where p represents the pole of the equator, p' the pole of the ecliptic, s the place of the star: $p\,s = 90^\circ - \delta$, $p'\,s = 90^\circ - \theta$, $p\,p' = \omega$, $p'\,p\,s = 90^\circ + \alpha$ (it will be recollected $p\,p'$ is necessarily in the solstitial colure), $p\,p'\,s = 90 - \beta$.

$$\text{Assume } \tan.\ \chi = \frac{\sin.\ \alpha}{\tan.\ \delta}$$

Then by substituting for sin. α. in the preceding equations, the value tan. χ. tan. δ, we get

$$\sin.\ \theta = \frac{\sin.\ \delta.\cos.\ (\chi + \omega)}{\cos.\ \chi}$$

$$\tan. = \frac{\tan.\ \alpha.\sin.\ (\chi + \omega)}{\sin.\ \chi}$$

Having thus found the longitude and latitude for the epoch, β and θ, we add to β the arc ψ, which is the precession for the given interval along the ecliptic of 1750 considered as invariable, and corresponds to $g\,h$ in our figure. According to the best authorities,

$$\psi = 50.''37572.t - 0''0001217945t^2$$

where t expresses the number of years counted from 1750. We must now calculate the new obliquity of the equator to this same ecliptic, or the angle ω' in our figure. Now

$$\omega' = \omega + 0\cdot 00000984233t^2$$

and ω which is the obliquity of the ecliptic for 1750 = 23° 28′ 18″. Then, with the longitude $\beta' = \beta + \psi$, the latitude θ, and the obliquity ω' we must calculate the right ascension and declination of the star with reference to the new position of the equator $m\,h$. We will call these new coordinates α' and δ'. They are obtained from formulæ exactly analogous to those which we have already had occasion to employ.

$$\sin.\ \delta' = \sin.\ \omega'\cos.\ \theta.\sin.\ \beta' + \cos.\ \omega'.\sin.\ \theta.$$

and

$$\tan.\ \alpha' = \frac{-\tan.\ \theta.\sin.\ \omega' + \sin.\ \beta'.\cos.\ \omega'}{\cos.\ \beta'}$$

These formulæ are deduced, exactly as those above, from consideration of the spherical triangle $p\,p'\,s$ between the poles of the equator and ecliptic and the place of the star. To adapt them to logarithmic computation, assume

$$\tan. \chi = \frac{\sin \beta'}{\tan. \theta'}$$

$$\sin. \gamma' = \cos. \theta \left\{ \sin. \omega'. \sin. \beta' + \cos. \omega'. \tan. \theta \right\}$$

$$= \cos. \theta \left\{ \sin. \omega'. \sin. \beta' + \frac{\cos. \omega'. \sin. \beta'}{\tan. \chi'} \right\}$$

$$= \cos. \theta. \sin. \beta' \left\{ \frac{\sin. \omega.' \sin. \chi' + \cos. \omega'. \cos. \chi'}{\sin. \chi'} \right\}$$

$$= \frac{\cos. \theta. \tan. \chi'. \tan. \theta \cos. (\chi' - \omega')}{\sin. \chi'}$$

$$= \frac{\sin. \theta. \cos. (\chi' - \omega')}{\cos. \chi}$$

Again substituting in the second formulæ for tan. θ its value,

$$\tan. \alpha' = -\frac{\dfrac{\sin. \omega'. \sin. \beta'}{\tan. \chi'} + \sin. \beta' \cos. \omega'}{\cos. \beta'.}$$

$$= \sin. \beta' \left\{ - \frac{\sin. \omega'. \cos. \chi' + \cos. \omega.' \sin. \chi'}{\cos. \beta'. \sin. \chi'} \right\}$$

$$= \frac{\tan. \beta'}{\sin. \chi'} \sin. (\chi' - \omega')$$

We know in this way, α' and β' the right ascension and declination referred to the new equator $m\ h$. But the former is counted from the equinoctial point h, while the equinox is really at ♈. The arc ♈ h represents the motion of the equinoctial point in right ascension. It is necessary to subtract this arc from the right ascension of the star just calculated to get its true value. Now the arc ♈ h is thus estimated. From h draw $h\ p$ perpendicular to $q\ b$; the triangle ♈ $h\ p$ may be considered as rectilinear; for the angle at q is very small, even after the lapse of a great many centuries. Solving, then, this triangle as rectilinear, we have

$$♈ h = \frac{♈ p}{\cos. \Omega}, \text{ but } ♈ p = \frac{h g - ♈ n}{\cos. \Omega}$$

$$= \frac{\psi - \psi'}{\cos. \Omega}$$

when ψ represents the lunisolar, and ψ' the general precession. According to Bessel,

$\psi = 50.''37572t - 0.''00012\ 17945 \cdot t^2$
$\psi' = 50.\ 21129t + 0.\ \ 00012\ 21483 \cdot t^2$
t being the number of years counted from 1750.

$\Omega = \omega - 0.''52114t - 0.''00000\ 272295t^2$

Call ♈ h, ϵ, then $\alpha' - \epsilon$ will be the right ascension of the star counted from the equinox ♈: the declination is γ': with these quantities and the obliquity Ω, we may, if necessary, calculate the longitude and latitude of the star by the formulæ already employed for such purposes. In all the foregoing expressions, it is necessary to attend to the sign of t, making it negative for years anterior to 1750, positive for years posterior to that time.

An example will illustrate the principles here laid down, and at the same time show how the theory of precession may be applied to test the authenticity of ancient observations. According to certain Chinese authors, it appears that Tcheou-Kong, regent of China about the year 1100 before Christ, observed the position of the winter solstice, and found that its right ascension exceeded, by two Chinese degrees, the right ascension of the constellation Nû, which constellation begins with the star that the moderns have named ϵ Aquarii. Now, two Chinese degrees are equivalent to 1° 58′ 17″ of our division of the circle. The right ascension of the winter solstice is 270°: if we subtract from this 1° 58′ 17″, we shall have 268° 1′ 43″ for the right ascension of ϵ Aquarii as observed by Tcheou-Kong. Let us compare our observation with the real right ascension, as determined by our formulæ.

According to modern catalogues, at the beginning of the year 1750, the longitude of ϵ Aquarii was

$\beta = 308° 14' 10''$

Its latitude $\theta = 8° 6' 20''$

To find the precession on the fixed ecliptic of 1750, we must put in the expression for ψ, $t = -2850$, times anterior to 1750 being considered negative: ψ thus becomes $-40^\circ\ 2'\ 43''$. In fact, it is evident that for times before the epoch selected, the precession must be subtracted from, instead of added to, the longitudes. Consequently $\lambda + \psi = 278.\ 11.\ 27$: with this new value of the longitude, with the latitude θ, and the obliquity ω', calculated by our formulæ, we shall find the value of the right ascension $268^\circ\ 8'\ 31''$. We must now calculate ψ', and $\frac{\psi - \psi'}{\cos.\ \Omega} = -0^\circ\ 42'\ 41''\cdot 5$. This being the motion of the equinoctial point is to be subtracted from the right ascension above obtained, and we shall ultimately get the right ascension of ϵ Aquarii for the year 1100 before Christ, referred to the real equinox $= 268^\circ\ 51'\ 16''$. This surpasses by $49'\ 33''$ the determination attributed to Tcheou-Kong. The difference is very trifling, considering, in the first place, the difficulty of making the observation, and in the second, the uncertainty in which we are as to the exact time in which Tcheou-Kong flourished. Were we to carry this date half a century farther back, the whole difference would be removed.

Other observations made by the same prince give for the obliquity of the ecliptic, as observed by him, $23^\circ\ 53'\ 47''$. Now, according to our formulæ for the law of the diminution of the obliquity, the value in the time of Tcheou-Kong, supposed to be 1100 B.C., was $23^\circ\ 49'\ 2''$: the difference is $4'\ 5''$: which is really surprisingly small, considering the nature of the instrument employed, a vertical gnomon of eight feet.

Thus also, it is stated, that Eudoxus, a Greek astronomer, who lived about four centuries before Christ, had observed that in the celestial sphere there was a star which corresponded to the pole of the equator. It could not have been the star now called the polar star, which at that time was far removed from the pole. From what we have already said about precession, the reader is aware that the pole of the equator moves in a circle parallel to the ecliptic, round the pole of the latter. We must then look along this parallel to see what stars are situated near it: we shall find that there is but one sufficiently brilliant to have been observed in those days, when the use of telescopes was unknown. This star is κ Draconis. According to our catalogues, the longitude of this star, on the 1st of January, 1800, was $133^\circ\ 26'$. But if it were really in the pole of the equator at the time of Eudoxus, the circle of latitude which passes through it must at that time have passed through that pole: consequently, the star must have been in the great circle passing through the poles of the equator and the ecliptic, which great circle is called the solstitial colure. The longitudes being counted from the equinox, which is 90° from the solstice, it follows that the longitude of the star, when in the pole, must have been 90°; the difference between this and $133^\circ\cdot 26$ is $43^\circ\ 26'$. To find out when this took place, we must divide this quantity by the amount of the yearly precession, or $50''\cdot 23$. This division gives us 3110 years for the epoch in question, or, as these are to be counted from the year 1800 after Christ, 1310 years before Christ. We know, however, from other sources, that Eudoxus lived about 400 B.C.: it appears then that he could not have described the state of the heavens as it appeared in his time, but that corresponding to a period anterior by nine centuries. It is true that our calculations have been of a very rough kind, as we have supposed the precession uniform, while, on the contrary, we know it to be subject to a small secular variation; but, on the other hand, any greater degree of nicety would be misplaced, as by the naked eye the position of the pole could not be determined very exactly.

In fact, it appears that the star κ Draconis never has been in the pole at all, as may be easily shown. To do this, we must ascertain its declination when in the solstitial colure; for it is only when in this colure that it could have been in the pole, and its declination at that moment should be 90°. Now we may suppose the latitude of the star in the year 1310 B.C. to be sensibly the same as at present, that is, $= 61^\circ\ 44'$. At this time, the circle of latitude, being on the colure, and consequently perpendicular both to the equator and ecliptic, was a circle of declination also. If, then, we add to the latitude the obliquity of the ecliptic, we shall get the declination required. Now the obliquity of the ecliptic at that epoch was $23^\circ\ 55'$, very nearly, adding this to $61^\circ.44'$ we get $85^\circ\ 39'$; the star, consequently, when nearest to the pole, was considerably more than four degrees distant from it.

Generally speaking, calculations founded on the motion of the stars caused by precession are of great use in chronological inquiries. They afford means, as also do solar and lunar eclipses, of verifying many important dates. The ancients, who did not enjoy the advantage of an uniform chronological system, such as we possess, frequently had recourse to astronomical phenomena. They were in the habit more particularly of referring to what are called the heliacal, cosmical, and acronycal risings of certain stars; and not merely are these used by historians, (as when Thucydides refers an event to the heliacal rising of Arcturus,) but the poets teem with allusions to them; so much so that many passages must be unintelligible without we know the times at which certain stars or constellations, more particularly Capella, Arcturus, the Pleiades, Orion, rose and set heliacally and acronycally at those times and for those countries. But it is necessary to explain the terms of which we have just been making use.

A star is said to rise or set cosmically when it rises or sets in the morning, at the instant of sunrise.

It is said to rise or set acronycally when it rises and sets the instant of sunset. It is said to rise heliacally when it appears in the morning on the horizon a little before the sun; and to set heliacally when it sets in the evening a very little after him. It may be estimated that, in general, stars are not visible till the sun is at least 12° or 15° below the horizon at the instant of their rising or setting; consequently, the cosmical rising precedes by 12 or 15 days the heliacal rising (since the sun moves through about a degree a day); and the heliacal setting precedes the acronical setting by about the same time. It is the heliacal rising which seems to have been more observed than any other. Of this we may give one memorable instance in the fact that the Egyptian rural year was determined by the heliacal rising of Sirius. Thus, the age of Hesiod, a disputed point, may be determined by the fact that he mentions that Arcturus rose heliacally sixty days after the winter solstice. It becomes, then, a matter of some interest to know how to calculate these phenomena for past ages.

The problem proposed comes to this, to find the point of the ecliptic, which is about 12° below the horizon when a given star is rising.

Fig. 46

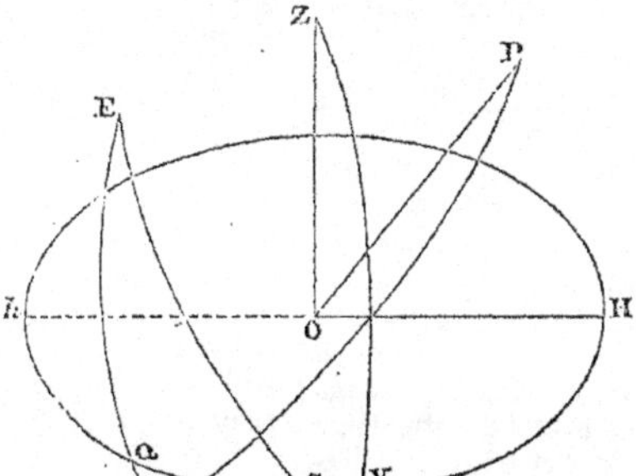

Let E be the vernal equinox, whence the right ascensions and longitudes are counted. Z the zenith, P the pole of the equator E A, and E C S′ the ecliptic. Let H V Q *h* be the horizon, the plane of which is supposed perpendicular to O Z, O being the centre of the celestial sphere. Let S be the star rising heliacally, while S′ is the position of the sun at the same moment; E A is the right ascension of the star; A S its declination; E S′ the longitude of the sun; we will suppose the sun to be 12° under the horizon; the arc V s' = 12°. Of course the latitude of the place is known, and consequently the angle made by the equator with the horizon, which is its complement; we shall call this angle ϕ: the right ascension E A, and declination A S of the star are supposed known, we shall call them α and δ*: the angle A E S′ = the obliquity of the ecliptic = ω.

* α and δ must be determined by taking their values from a modern catalogue, and deducing from these their values for the epoch desired. If the epoch be, as it always is, a very distant one, we must allow for the inequalities of precession, and the variation of the obliquity of the ecliptic. This will be done by following the methods already given for that purpose. See page 166.

The first thing to be done is to find the point of the equator Q, which is in the horizon at the same time with the star S: in the triangle A Q S, the angle at A is a right angle: A Q S = E Q h = ϕ : A S = δ

$$\therefore \sin. A Q = \frac{\tan. S A}{\tan. A Q S} = \frac{\tan. \delta}{\tan. \phi}$$

Q E = α — A Q, and is consequently known; it is the right ascension of the point of the equator which rises with the star.

We must now determine the point C of the ecliptic which rises simultaneously with the point Q of the equator. Now in the triangle E Q C we know the angle at E = ω: we know the angle at Q = 180° — ϕ, and the side E Q, which we shall call α'. Calling E C, β, we have by the formulæ of spherical trigonometry,

$$\cot. \beta = \frac{- \cot. \phi. \sin. \omega + \cos. \omega. \cos. \alpha'}{\sin. \alpha'}$$

The angle at C is determined by means of the formula

$$\cos. C = \cos. \alpha' \sin. \phi. \sin. \omega + \cos. \phi. \cos. \omega.$$

Now the angle E C Q, which we have called C = V C S'; and the angle C V S' is a right angle, V S' = 12°, consequently

$$\sin. C S' = \frac{\sin. 12^{\circ}}{\sin. C}$$

Calling C S', β', and adding β' to β we get the whole arc, E S', which is the longitude of the sun for the instant of the heliacal rising of the star for the epoch we are considering. Now this longitude being known, the solar tables will give us the time of year corresponding to it.

The problem is thus completely solved, supposing we know the year or nearly so in which the heliacal rising took place, and only wish to find the time of the year, indicated by that rising*. But it often happens that we know the time of year corresponding to the heliacal rising, and wish to deduce from this the date of events connected with it, as in the case of Hesiod, abovementioned, who has given us the time of year at which Arcturus rose heliacally, and from this circumstance it is wished to deduce the time at which the poet lived. The latitude is supposed known (in the case of Hesiod it is that of Bœotia), but if the date be altogether unknown, we must calculate the heliacal rising (of Arcturus) from century to century for the given latitude, making the proper allowances for precession. In the case just mentioned, we shall find the phenomenon comprised between the years 900 and 1000 B.C. We must then calculate for 950, which we shall find nearly to correspond with the conditions. A great nicety is not to be expected in these calculations, as the phenomenon itself of the heliacal rising cannot be susceptible of being observed with great precision. It must evidently depend in part upon a number of local and accidental causes, among which the transparence of the air and the extent of the horizon are sufficiently obvious.

* As when Thucydides tells us, that in the second year of the Peloponnesian war, (a well-known date,) the investment of Platæa was completed about the heliacal rising of Arcturus.

Chapter VIII.

Annual Parallax—Motion of the whole solar system in space—Double and multiple stars—Variable stars—Nebulæ.

It is our intention to treat in this chapter of several real or presumed motions of the fixed stars, which, from their extreme minuteness, have only been established in later years, and some of which still remain enveloped in doubt. We shall begin with

Section I.—*Annual Parallax.*

The nature and effects of annual parallax have already been explained in page 153; and it has been remarked that the subject is not one of practical importance. It is, however, one of very great interest in a merely scientific point of view. The existence of some parallax, were our instruments fine enough to appreciate it, cannot be doubted: yet, from the days of Galileo, the unremitting efforts of astronomers have led to no decisive and satisfactory result. We no longer feel this as a fatal objection to the Copernican system; yet it cannot be denied, that in earlier times it must have been felt as a very great difficulty. In searching for parallax, Bradley fell upon the discovery of aberration and nutation; and since his time, from the perfection of instruments, numerous small motions have been discovered, yet the question of parallax remains nearly where it was. However, we, perhaps, need not yet despair of final success, when we consider the improvements made in instruments, and in the methods of observation, during the last fifty years. Should our progress continue to be as rapid, ere the close of the century we may hope to ascertain, by actual observation, the existence of a phenomenon which, mathematically speaking, cannot be disputed. It is of importance for this purpose to be in possession of the formulæ which represent the effects of parallax either in longitude and latitude, or in right ascension and declination.

The difference between the effects of annual parallax and aberration on the place of a star, has already been pointed out: it will become still clearer on comparing the formulæ for both of these cases. It is evidently unnecessary in the case of parallax to take into account the small ellipticity of the earth's orbit; we shall therefore suppose it circular. Call the distance from the sun to the earth r, from the star to the earth r'; and the longitude of the earth, as seen from the sun, ϕ: the centre of the sun being the origin of the rectangular coordinates, let the axis of x and that of y be in the plane of the ecliptic, the axis of x passing through the equinox, that of z perpendicular to that plane. The coordinates of the earth, x, y, z, will evidently have for values

$$x = r. \cos. \phi \qquad y = r. \sin. \phi \qquad z = 0.$$

Let now the coordinates of any star be x', y', z': let its heliocentric longitude and latitude be θ' and λ', and call the projection of its radius vector on the plane of the ecliptic ϱ': then evidently

$$x' = \varrho' \cos. \theta', \qquad y' = \varrho' \sin. \theta', \qquad z' = \varrho'. \sin. \lambda'.$$

If now r' represent the distance from the sun to the star, then

$$' = r'. \cos. \lambda'$$

consequently, eliminating ϱ', we get

$$x' = r'. \cos. \lambda'. \cos. \theta'$$

$$y' = r'. \cos. \lambda'. \sin. \theta'$$

$$z' = r'. \sin. \lambda'$$

and also we may deduce the geocentric longitude and latitude of the star, θ and λ, by the formulæ

$$\tan \theta = \frac{y' - y}{x' - x}$$

$$\sin. \lambda = \frac{z'}{r'}$$

$$= \frac{z'}{\sqrt{(x - x')^2 + (y - y')^2 + (z - z')^2}}$$

Taking the first of these two equations, and substituting in it for x, y, x', y', their respective values, we obtain

$$\tan. \theta = \frac{r'. \cos. \lambda'. \sin. \theta' - r. \sin. \phi}{r'. \cos. \lambda'. \cos. \theta' - r. \cos. \phi}$$

$$= \tan. \theta'. \frac{1 - \frac{r}{r'}. \frac{\sin. \phi}{\cos. \lambda'. \sin. \theta'}}{1 - \frac{r}{r'} \frac{\cos. \phi}{\cos. \lambda'. \cos. \theta'}}$$

Now, if we call the greatest annual parallax ϖ, we shall have $r = r'. \sin. \varpi$ or $\frac{r}{r'} = \sin. \varpi$: for it is clear that, in this case, a line drawn from the star to the earth will be perpendicular to the radius of the terrestrial orbit.

$$\therefore \tan. \theta = \tan. \theta'. \frac{1 - \frac{\sin. \varpi. \sin. \phi}{\cos. \lambda'. \sin. \theta'}}{1 - \frac{\sin. \varpi \cos. \phi}{\cos. \lambda'. \cos. \theta'}}$$

Now, if the greatest annual parallax ϖ be imperceptible, then $\varpi = 0$, sin. $\varpi = 0$, and $\theta = \theta'$, or the geocentric longitude of the star is equal to the heliocentric, as must obviously be the case. If ϖ exists, we know that it must be excessively small: in α Lyræ, which has been much observed with a reference to this particular object, it certainly does not exceed two seconds, though it, perhaps, in some other stars, may be something more. Consequently sin. ϖ must be an excessively small fraction; and we may confine ourselves to terms involving its first power only. If then we divide out by the denominator of the expression for tan. θ, confining ourselves to such terms, we obtain

$$\tan.\ \theta = \tan.\ \theta'.\left\{1 + \frac{\sin.\ \varpi.\ \sin.\ (\theta' - \varphi)}{\cos.\ \lambda'.\ \sin.\ \theta'.\ \cos.\ \theta'}\right\}$$

$$\sin.\ (\theta - \theta') = -\frac{\sin.\ \varpi.\ \sin.\ (\varphi - \theta').\ \cos.\ \theta}{\cos.\ \lambda'.\ \cos.\ \theta'}$$

$\theta - \theta'$ and ϖ being, as has just been explained, extremely small angles, we may substitute their ratio for that of their sines; and also on the right-hand side of the equation, we may suppose $\theta = \theta'$. Hence

$$\theta - \theta' = -\frac{\varpi.\ \sin.\ (\varphi - \theta')}{\cos.\ \lambda'}$$

Proceeding similarly with regard to the expression for sin. λ, we get by the substitution for x, y, x' and y'

$$\sin.\ \lambda = \frac{r'.\ \sin.\ \lambda'}{\sqrt{r'^2 - 2\,r\,r'.\ \cos.\ \lambda'.\ \cos.\ (\varphi - \theta') + r^2}}$$

$$= \frac{\sin.\ \lambda'}{\sqrt{1 - 2\ \sin.\ \varpi.\ \cos.\ \lambda'.\ \cos.\ (\varphi - \theta') + \sin.^2\ \varpi}}$$

If now we neglect, as before, terms involving powers of sin. ϖ, beyond the first, and if we divide out on the right-hand side, we get

$$\sin.\ \lambda = \sin.\ \lambda'\ \{1 + \sin.\ \varpi.\ \cos.\ \lambda'.\ \cos.\ (\varphi - \theta')\}$$

$$2\ \sin.\left(\frac{\lambda - \lambda'}{2}\right).\ \cos.\left(\frac{\lambda + \lambda'}{2}\right) = \sin.\ \varpi.\ \sin.\ \lambda'.\ \cos.\ \lambda'.\ \cos.\ (\varphi - \theta')$$

$\lambda - \lambda'$ is a very small angle of the same order as ϖ; putting as before the ratio of the arcs for that of the sines, and supposing $\lambda = \lambda'$ in the terms which are multiplied by ϖ

$$\lambda - \lambda' = \varpi.\ \sin.\ \lambda'.\ \cos.\ (\varphi - \theta')$$

It is evident, from these expressions, that the annual parallax in longitude is nothing when sin. $(\varphi - \theta') = 0$, and $\varphi - \theta' = 0$ or $180°$: that is, when the star is in conjunction or opposition with the sun. In this case, the expression for the parallax in latitude acquires its greatest value, ϖ. sin. λ. This latter expression, on the other hand, becomes nothing when cos. $(\varphi - \theta') = 0$ or $\varphi - \theta' = + 90°$: that is, when the arth, seen from the sun, is in quadratures with the star, and at that time the parallax in longitude attains its greatest value $-\dfrac{\varpi}{\cos.\ \lambda'}$.

To compare the expressions we have obtained with those for the aberration in longitude and latitude, it is necessary to refer the former to geocentric coordinates, as we have done with the latter. Now, here it is to be remarked that the geocentric longitude and latitude differ from the heliocentric only by the parallax itself; consequently we may use them indifferently one for the other in the expression for that parallax, as we are confining ourselves simply to the first power of ϖ. But instead of the longitude of the earth, as seen from the sun φ, we must introduce the longitude of the sun, as seen from the earth $\odot$. Now $\odot = 180° + \varphi$, or $\varphi = \odot - 180°$; and substituting for φ, we have

$$\theta - \theta' = \frac{\varpi.\ \sin.\ (\odot - \theta')}{\cos.\ \lambda'}$$

$$\lambda - \lambda' = -\ \varpi.\ \sin.\ \lambda'.\ \cos.\ (\odot - \theta')$$

let us call $\theta - \theta'$, $d\,\theta$, and $\lambda - \lambda'$, $d\,\lambda$: the corresponding expressions of $d\,\theta$, and $d\,\lambda$, as caused by aberration, are

$$d\,\theta = -\frac{20''.\ 25.\ \cos.\ (\odot - \theta')}{\cos.\ \lambda'}$$

$$d\,\lambda = -\ 20''\!\cdot\!25.\ \sin.\ \lambda'.\ \sin.\ (\odot - \theta').$$

Now, comparing these with the formulæ for parallax, we see that they are just the inverse one of the other: the parallax, either in longitude or latitude, being greatest when the aberration is least, and the converse: and generally as the one increases, the other diminishes. To make them agree, we must change, in the formulæ for aberration, $\odot$ into $\odot + 90^\circ$. This is in accordance with what has been stated, page 153. By this consideration, we may get the effect of parallax in right ascension and declination at once: we have only to substitute in the corresponding expressions for the aberration $\odot + 90^\circ$ instead of $\odot$. We shall thus have

$$d\,\alpha = \frac{\pi.\ \sin.\ \alpha'}{\sin.\ \psi.\ \cos.\ \delta'}\ \sin.\ (\odot - \psi)$$

assuming

$$\tan.\ \psi = \frac{\tan.\ \alpha'}{\cos.\ \omega}$$

And

$$d\,\delta = \frac{\pi.\ \cos.\ \alpha'.\ \sin.\ \delta'}{\cos.\ \psi'}.\ \sin.\ (\odot - \psi')$$

assuming

$$\tan.\ \psi' = \frac{\sin.\ \alpha'.\ \cos.\ \omega - \cot.\ \delta'.\ \sin.\ \omega}{\cos.\ \alpha'}$$

It must be recollected, that in these formulæ θ', λ', α', δ', represent the heliocentric elements. These may be obtained by taking for θ' the geocentric longitude observed at the syzigies, that is, at the instant of conjunction or opposition with the sun: and for λ' the geocentric latitude observed when the star is in quadratures with the sun, that is, at 90° from the syzigies. In fact, we have seen that in the first of these situations $\theta = \theta'$; and in the second, that $\lambda = \lambda'$. With these values of θ' and λ', and the obliquity of the ecliptic ω, which is known, we must calculate α' and δ'.

The parallax in question may become much more apparent in one of these elements than another, on account of the constant factors which enter into the expressions, and augment or diminish their influence. Thus the parallax in latitude being generally $-\pi.\ \sin.\ \lambda.\ \cos.\ (\odot - \theta')$ it is greatest for the same star when $\odot - \theta' = 90^\circ$, in which case it becomes $-\pi.\ \sin.\ \lambda'$: and then for different stars, it is greatest when $\lambda' = 90^\circ$, in which case it becomes π. But this limit it never can exceed. On the other hand, the parallax in longitude having $\cos.\ \lambda'$ for a divisor, may exceed π very much: it increases with the latitude. But we must not conclude that it would become infinite when $\lambda' = 90^\circ$, that is, when the star is in the pole of the ecliptic. Our formulæ are only approximate, and they have been deduced on the supposition that $\theta - \theta'$ is very small. That is no longer true for a star so near the pole of the ecliptic, that the annual parallax π becomes a quantity comparable to $\cos.\ \lambda'$. In this case, θ can no longer be taken $= \theta'$ in our approximations. It results from these remarks, that it is desirable to observe stars at a distance from the ecliptic; the effects of the parallax on the longitude being then very considerable.

There may, however, be other circumstances to determine our choice: it is natural to presume that those stars whose proper motion is greatest are nearest the earth, and consequently have the most perceptible parallax; and these stars, it is important to remark, are by no means always among the brightest. This remark is the more necessary, that many have thought that there was a strong *à priori* probability that the brightest stars were nearest to the earth. Such observations as have been hitherto made, however, seem unfavourable to this opinion, as we shall presently show. On the other hand, it is desirable to avoid stars that do not rise sufficiently high above the horizon to be beyond the effects of the uncertain and variable refractions that take place at low altitudes.

The ordinary instruments and methods of observation certainly appear insufficient to settle the long agitated question of parallax. The greatest amount of this quantity does not exceed $2''$; yet we can hardly be certain of the place of a star to one or two seconds. That such is the case is sufficiently proved by the difference of opinion between the Astronomer Royal and Dr. Brinkley on the existence of a parallax of $2''$ in α Lyræ. When two of the most able observers of the day, furnished with instruments of the most splendid kind, thus disagree, it is pretty evident that recourse must be had, if possible, to some methods of still greater nicety for the detection of the phenomenon in question. For stars passing very near the zenith of the observer, the instrument called a zenith sector, which will be hereafter described, may be advantageously employed; but its use is limited to such stars which may not happen to be desirably situated for such observations, according to the remarks made above to guide us in our selection. Another method originally employed by Galileo has been improved upon, and used with much success lately, by a French naval officer, Count D'Assas-Montdardier. On a mountain at a considerable distance from his observatory, Galileo placed, at a few feet from the surface, a horizontal beam. He observed with his telescope certain stars, which, when on the meridian, appeared just to graze this beam. From the great distance between the observer and the beam the least change in the altitude of the star would make it appear either to pass above the beam without touching it, or to become eclipsed by it; and Galileo satisfied himself, by observing the sun in this way, that a change of less than one second could easily be appreciated.

For the horizontal beam of Galileo M. D'Assas de Montdardier substitutes a triangle in a vertical plane formed of three beams, or rather three bands of iron, the base of which is about fifteen times its length. This triangle he places in the meridian on a mountain distant more than 600 French metres from his observatory. It is evident that if the height of the star varies in the least, that a great variation will be produced in the time it takes to traverse the triangle, and in this way the most minute changes in the meridian altitude, and consequently in the declination, may be appreciated. The method pursued by M. D'Assas is to compare the variations (obtained by this method) of declination in the star he is examining with those of 12 or 15 small stars lying very near it *. These stars being very small and faint, may be supposed to have neither proper motion nor annual parallax. If these comparisons all indicate the same change of declination in the larger star, while the small ones keep the same relative positions with regard to one another, we may be certain that these latter are really invariable, and that the former alone is in motion. This motion may be the result of two combined causes, parallax and the proper motion of the star. Parallax having an annual period, the proper motion may be distinguished from it by comparing observations made at the same time of the year in different years *. This motion once determined, the effect produced by parallax on the declination may be easily ascertained; and thus $d\delta$ being known in our formulæ, ϖ or the whole annual parallax may be easily found. By applying this method to the star called Keid, or 29 Eridani, M. D'Assas found its annual proper motion in declination 4″13; its absolute parallax about 2″. This result is confirmed by the observations of the celebrated Piazzi, who found by totally different methods for this remarkable star a proper motion in declination of 3″·6. Rigel was found by M. D'Assas to have an absolute parallax of 1″·43; Sirius of 1″·24. According to these results the distance of Keid from the earth is about one hundred thousand times the radius of the earth's orbit round the sun, which radius itself = 95 millions of miles. The distance of Rigel is about 140,000 times that radius; of Sirius about 160,000. Keid is the one of the nearest to us of all the fixed stars; yet light, which goes from the earth to the sun in about eight minutes, takes a year and a half to pass from that star to the earth; and we actually might see the star for that time after it had ceased to exist. Well might La Place say that astronomy shows man his own greatness in the smallness of the base which has served him to measure the universe.

Section II.—*On the supposed motion of the Solar System in Space.*

It does not seem probable that while, as it would seem, all the heavenly bodies are in motion, the sun itself, which, from analogy, we may class with the fixed stars, should be at rest. Mechanical considerations also, which it is unnecessary to go into here, make it probable that as we know him to have a motion of rotation on his own axis, so

* By this comparison the errors arising from the uncertainty and variableness of refraction are eluded, as, whatever the amount of refraction may be at the moment, it will be the same for stars observed very nearly at the same altitude and same time.

* In all these calculations it is hardly necessary to observe, that due allowance must be made for precession, nutation, and aberration.

he has also a motion of translation in space. It is of course to be understood that the whole solar system accompanies him in this translation, the planets preserving their relative motions about his centre unaltered, just in the same way as the satellites of Jupiter accompany that planet in his periodical revolutions. The celebrated Herschel conceived that such a motion was actually perceptible by the change of place in certain fixed stars, and concluded that the sun and whole solar system was in motion towards a point in the constellation Hercules.

To comprehend how the existence of such a motion can be ascertained, we may begin by remarking that the motion in question, if it exists at all, being certainly very slow with regard to the immense distance of the fixed stars, we may consider the small arc described in a short time (as for instance half a century) as rectilinear. Not, however, that we suppose the revolution of the sun to take place round any given fixed star, but rather round the common centre of gravity of the system of stars, of which our sun is part. It is easy to form to ourselves an idea of the effects produced on the apparent places of the stars by a motion such as we have supposed. The stars in that part of the heavens towards which the sun is in motion will appear to open out and recede from each other, diverging on all sides from the point to which the motion is directed, which point we shall call the pole of translation. Take in the heavens the point diametrically opposed to this pole; the stars in that part will appear to close together, and converge to this opposite pole. Such will be the general effects produced; but to ascertain the existence of the motion, and, if that be established, its magnitude and direction, we must investigate formulæ for the effects on the right ascension and declination of any given star.

Suppose in fig. 47 that P is the pole of the equator, p the pole of translation, S the place of the given star.

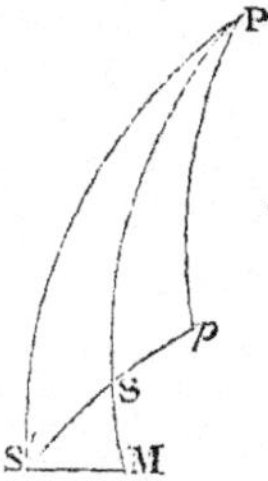

Fig. 47.

P S, the north polar distance of the star, we shall call Δ; P p the north polar distance of the pole of translation Δ'; the angle S P p will be the difference between the right ascension of the star α, and that of the pole of translation α'; it will then be $\alpha - \alpha'$. Now by spherical trigonometry we obtain the angle at S by the formula

$$\text{Cot. S} = \frac{\text{cot. } \Delta' . \text{ sin. } \Delta - \text{cos. } (\alpha - \alpha') . \text{ cos. } \Delta}{\text{sin. } (\alpha - \alpha')}.$$

Now if we prolong the arc of a great circle p S which passes through the pole of translation and the first position of the star S, it must also pass through the second position S'. This is evident, since the apparent motion of the star supposed immoveable must necessarily be in the plane determined by the lines joining the pole of translation with the earth, and with the star. Produce P S and let it meet the parallel of declination passing through S' in M. The small triangle S S' M may be solved as rectilinear from what has been said above, and we have

$$\text{tan. S} = \frac{\text{S}'\text{M}}{\text{S M}}.$$

Now S' M is the variation in right ascension measured on the parallel of the star, or $d\alpha$ sin. Δ; S M is the variation in polar distance, or $d\,\Delta$

$$\therefore \text{tan. S} = \frac{d\,\alpha . \text{sin. } \Delta}{d\,\Delta}$$

$$\therefore \text{cot. S} = \frac{d\,\Delta}{d\,\alpha . \text{sin. } \Delta}.$$

Equating this value of cot. S to the value before obtained, we get

$$\frac{d\,\Delta}{d\,\alpha . \text{ sin. } \Delta} = \frac{\text{cot. } \Delta' . \text{ sin. } \Delta - \text{cos. } (\alpha - \alpha') . \text{ cos. } \Delta}{\text{sin. } (\alpha - \alpha')}$$

$$\therefore d\,\Delta . \text{ tan. } \Delta' . \text{ sin. } (\alpha - \alpha') = d\,\alpha \{\text{sin.}^2 \Delta - \text{sin. } \Delta . \text{cos. } \Delta . \text{tan. } \Delta' . \text{cos. } (\alpha - \alpha')\}$$

By means of this formula, when α' and Δ' are known, we may calculate the ratio $\frac{d\Delta}{d\alpha}$ of the two motions in right ascension and declination, and then comparing these with the motions actually observed, verify the truth of our hypothesis as to the position of the pole of translation. By repeating the same operation with different stars, we shall see whether they agree in giving the same result for the place of the pole, or whether the results obtained from them are in contradiction with each other, which latter circumstance would prove that no such motion towards a pole really exists.

Sir William Herschel conceived the pole of translation to be situated at 245° 52′ of right ascension, and 49° 38′ of north declination, near the star 34 Herculis. Assuming these values of α' and δ', we might calculate $\frac{d\alpha}{d\delta}$; or taking $d\alpha$ from observation, calculate $d\delta$, and compare it with the observed value of that quantity. Such an examination does not seem to support Sir W. Herschel's ideas. Bessel, who has examined this question with his usual care and ability, states that many stars indicate a point in the heavens very remote from that assigned by Sir W. Herschel; and that, in fact, there is no one point in particular towards which he can discover any such tendency.

A general consideration will completely set the matter at rest. If there exist a polar motion towards any point whatsoever, stars near each other must have motions very nearly parallel, so that the ratio of the motion in right ascension to that in declination will be nearly the same. But, on examining the catalogues of stars, it is found that this is not by any means the case. M. Biot observes, that α Herculis and α Ophiuchi differ only 5° in right ascension and 2° in declination; yet the variation of right ascension in forty-two years has been, for the former, 2″·26, for the latter, 8″·54; of declination, — 10″·8 for the former, and + 2″ for the latter. Similarly, the three stars, α, β, and γ Aquilæ, differ only 2° in right ascension, and 5° in declination. Their respective variations in the former direction for forty-two years have been + 30″, + 5″, and + 2″; in the latter, — 21″·8, — 8″·5, and + 14″·9 respectively.

It may be concluded, from these remarks, that if a motion of translation of the solar system in space really exists, as is *à priori* not improbable, that it is masked by the more considerable proper motions of the fixed stars themselves. Under these circumstances we must leave the problem to be solved by future astronomers, who shall possess very accurate observations, with a sufficient interval between them for such determinations.

One remarkable circumstance has been noticed by Bessel;—that a great proportion of those stars which have a proper motion, are double stars. A singular case of this binary system, as well as the greatest in the amount of its proper motion, is 61 Cygni. The two stars which form this double star are 15″·4 asunder, and have a motion, round their common centre of gravity, of 0°·73 annually; while the annual proper motion of the double star in right ascension is 5″·46 of space; and in declination 3″·19. Other stars in the neighbourhood of 61 Cygni do not appear to have any proper motion at all. On the other hand, A Ophiuchi and 30 Scorpii, two stars distant from each other 13′ in space, and which have no revolving motion round each other, are evidently travelling together through space, and leaving the neighbouring stars behind.

SECTION III.—*Double and Multiple Stars.*

On applying good telescopes to the examination of the heavens, it appears that many of those luminous bodies which strike us simply as bright points, are of a more complicated nature than appears to the naked eye. Of the various appearances they present we shall, in the first instance, notice the separation of many stars into two or more others, thus offering instances of really compound stars. Double stars, as they are called, are sufficiently frequent in the heavens; instances of greater composition occur, but are much less common. The star called θ Orionis is quadruple*. Sometimes these double stars are composed of one bright star, accompanied by another of much smaller magnitude, as in the case of α Lyræ, α Herculis, Rigel, Polaris, and others; sometimes of two stars, quite or nearly equal

* In fact it is quintuple: but the fifth star is a variable star, being only visible at certain states of its periodical brightness.

in brightness, as in the case of γ Arietis, θ Serpentis, γ Virginis, and many others. It was for a long time supposed that the contiguity of these stars was merely an optical effect, arising from both being placed nearly on the same visual line drawn from the earth. Modern discoveries have, however, taught us that in many instances, there is a more intimate connexion between the bodies forming these double stars than was at first supposed.

The pursuit of annual parallax seems fated to lead to curious discoveries. Galileo in the first instance, and subsequently Herschel, had proposed a plan for determining the existence of parallax by the observation of double stars. Supposing, as these astronomers did, a double star to be composed of two independent stars at a great distance from each other, and consequently at very different distances from the earth, it is sufficiently evident, that the quantity of annual parallax will be very different for the two stars. Consequently, their apparent places being very unequally affected by this cause, they will appear at one time to recede from, at another to approach towards, each other according to laws dependent on their relative positions and distances. The instruments of Galileo were not sufficiently good to enable him to put in practice this method of observation, a method evidently requiring very delicate admeasurements: but the telescopes of Herschel being unrivalled in excellence, he was enabled to undertake the research with reasonable hopes of success. While pursuing this investigation, and measuring with care the angles of position and distances of these compound stars, he was led to a curious and unexpected discovery. He found that several of these are not stars that appear double from a fortuitous juxtaposition, but in reality are intimately connected, forming binary systems in which either one star revolves round the other, or both round their common centre of gravity*. Mathematically speaking, there can be no doubt that the latter is always the case; but when one star is very much larger than the other, the common centre of gravity will lie so near the centre of the former, that at the great distance at which we are, the difference between the two centres will be insensible. It is thus that the planets of the solar system revolve round the common centre of gravity of that system; but their mass, compared with that of the sun, is so small, that, for all practical purposes, we suppose these two centres to coincide. When, on the other hand, the two stars are nearly of equal magnitude, each will sensibly revolve round the imaginary point, the common centre of gravity. Among the most remarkable instances of such binary systems in the heavens are—

Castor.—One of the two stars in this beautiful system describes round the other in about 253 years an ellipse, of which the following are the dimensions:—Semiaxis major = 8″·086; excentricity = 0·7582. But as this ellipse does not lie in a plane perpendicular to the visual ray, but in a plane inclined to it about 70°, the *apparent* semiaxis major is only 5″·34; while the *apparent* semiaxis minor is 2″·72.

γ Virginis.—This fine star, composed of two of equal magnitude, has a periodic time of about 629 years. The semiaxis major of the ellipse = 12″·09; the excentricity = 0·8335. The plane of the orbit is inclined to the tangent plane to the celestial sphere at that point about 67°.

ξ Ursæ Majoris.—This is a very remarkable double star, (composed of two of nearly equal magnitude,) having a periodic time of only 60 years. The semiaxis major = 3″·278; excentricity = 0·3777; inclination of the orbit 56°·6.

ξ Bootis has a period of 117 years: semiaxis major = 12″·56; excentricity = 0·5937; inclination to the plane of the heavens = 80° 5′.

γ Leonis is a very remarkable star; it appears at first composed of two reddish stars of unequal magnitude, but with very good telescopes it is shown quadruple. The two principal stars certainly form a binary system with a relative angular motion of about 0°·30 annually, the distance being about 3″·3.

η Coronæ.—This star has already gone through more than a complete revolution since its discovery, as a double star, by Sir W. Herschel, in 1781. Its excentricity is only 0·2603, which hardly exceeds that of the orbits of Mercury, Pallas, and Juno, in our own system: the axis major is only 0″·8325; the periodic time rather more than 44 years.

We have selected these instances to give some idea of the nature of the

* In either case the theory of gravitation shows that the orbit is an ellipse.

motions in those stars in which they are most evident. Many other systems are known to exist, of which we cannot take notice at present, it not being our object to give in this place a catalogue of multiple stars: we shall only just notice the singular contrasts of colour offered frequently, though not invariably, by the two stars of a binary system. Sometimes it seems these colours in the minor star are complementary of those in the larger; when the former are seen by themselves they appear colourless: this, however, is not universally the case; indeed there is an extraordinary difference in the colour of the simple stars, as the naked eye shows in many instances, and as is still more visible in the telescope. On the causes of these differences it would be idle to speculate. As, however, it is always desirable to compare facts, we may remark that a new star appeared suddenly in the year 1572 in the constellation of Cassiopeia. On its first appearance it was of a brilliant white, and surpassed in splendour even Jupiter; its brightness gradually diminished, till, in sixteen months, it had completely disappeared. During this time its colour experienced considerable changes: from brilliant white it passed to a reddish yellow, like that of Mars or Aldebaran, and thence to a leaden white, like that of Saturn. It is unnecessary to point out the analogy between these changes and those produced by terrestrial bodies at different degrees of heat.

Section 4.—*Variable Stars.*

Under this designation we include a class of stars whose brightness is subject to certain periodical variations, as the star we have already noticed in the group which forms the compound star θ Orionis. One of the most remarkable of these is the star Algol, in the constellation Perseus, which, every sixty-nine hours, experiences a diminution of splendour, which reduces it, in three hours and a half, from the second to the fifth magnitude: it then takes about the same time to return to its original state. Another remarkable changing-star is Mira, in the neck of the Whale, which has a period of about 334 days; being, at its brightest, of the second or third magnitude; and when least bright, invisible to the naked eye. A similar star exists in the neck of the Swan, with a period of about 397 days. Another star in the breast of the same constellation has a period of about 15 years, during ten of which it is apparent, and during five invisible. The constellation Hydra possesses a star which passes from the fourth magnitude to being invisible: its period is 404 days. δ Cephei has a period of five days; η Antinoi of seven; but it is unnecessary here to enumerate all of this description.

Several explanations of these changes have been offered, on the relative probabilities of which it is hard to decide. Some astronomers have supposed that the variable stars are, like all the other luminous bodies, revolving on their own axes, after the manner of our sun; but that their surface is partially covered by large obscure spots, (such, in fact, as we see on a small scale in the sun,) which, by the rotation of the star, are at certain times turned towards us. Others have attributed these variations to large opaque bodies revolving round those stars, and periodically intercepting their light. Lastly, some have explained the facts by attributing to the stars themselves an extremely flattened form—such, to use a familiar illustration, as that of a mill-stone; in which case, the brightness of the star would evidently depend upon the position in which it happens to be seen. Time must decide which of these suppositions is nearest to the truth.

We have already alluded to the bright star which appeared suddenly in the constellation Cassiopeia, in the year 1572. On its first appearance it surpassed in splendour even Jupiter and Venus, and might be seen on the meridian in broad day. Its brightness gradually diminished, till in sixteen months after its first apparition it disappeared, without having changed its place in the heavens. The changes in its colour, as its brightness diminished, have been already noticed. Those changes, and the suddenness of its appearance, seem to point forcibly to a vast combustion as the origin of this extraordinary phenomenon. A similar phenomenon occurred in the year 1604, in the constellation Serpentarius. It is said that a similar circumstance led Hipparchus to form his catalogue of the fixed stars, about a century and a half before the birth of Christ.

Section V.—*Nebulæ.*

The term nebula is applied to certain

irregular spots of pale light and ill-defined figure, which occur frequently in the heavens. Some of these are nothing more than clusters of small stars, so near each other as not to be separated by the naked eye. The telescope, however, shows their real nature. Of this kind is the nebula in Cancer, called Præsepe Cancri, composed of about forty very small stars, easily separated in a telescope; and such are many others that need not be enumerated here. There are others, however, that are not wholly resolvable into separate stars, such as the beautiful nebula in the sword of Orion, the great nebula of Andromeda, and others. The Milky Way itself appears to the eye as a vast nebula traversing the celestial sphere: when, however, it was examined by the powerful telescopes of Herschel, it was resolved into an incredible number of small stars. To give some idea of their number, we may state, that in a zone of fifteen degrees in length by two in breadth, he has observed more than fifty thousand. It is supposed that our sun and the brightest of the fixed stars form a part of this great nebula. We must conceive it a vast stratum, whose depth is immense, (about a thousand times the distance from Sirius to the earth,) yet very inconsiderable with respect to its other dimensions. When we look in the plane of the nebula itself, in which we are situated, the stars appear so thickly clustered together as to form, apparently, a confused mass of light: if, on the other hand, we look in a direction perpendicular to this plane, it is evident that we shall see the stars thinly scattered over the surface of the heavens; the depth of the stratum in which we are placed, enormous as it is, being trifling compared with its length. In fact, if we suppose the average distance of each fixed star from that nearest it to be the same as the distance of Sirius from the earth, the depth of the stratum, one thousand times that distance, will not appear so very considerable.

It has already been said, that all nebulæ are not to be resolved into clusters of stars closely packed together. Some resemble rather planetary bodies, forming distinct masses of equable pale light. These have been called by Herschel planetary nebulæ. Others appear to be composed of one or more bright stars situated in a less luminous mass. It has been remarked, that in this case the immediate neighbourhood of the star appears much darker than the rest of the nebula. An analogous fact is, that in the neighbourhood of a nebula there is generally an absence of stars. Herschel, observing these facts and the varieties of appearance presented by the different nebulæ, was led to the conclusion, that they were all parts of a luminous substance disseminated generally over the heavens, which accumulates in certain points, either from mutual attraction or from that of a neighbouring star. He thought he could distinguish, by the greater or less degree of sphericity, and the brilliancy of the central nucleus as compared with the surrounding nebulosity, the progress in condensation and the relative ages of the different nebulæ. The first stage is that of an uniformly nebulous mass; the second, that of a similar mass slightly condensed round one or more faintly luminous nuclei: these nuclei gradually become brighter; then the nebulous atmospheres of each separating by the effects of a farther condensation, there result compound nebulæ formed of brilliant centres very near each other, and surrounded respectively by their separate atmospheres. Sometimes the luminous matter, by a more uniform condensation, forms the planetary nebulæ of which we have spoken *. Lastly, a higher degree of condensation transforms the nebulæ into groups of stars thickly set together. In confirmation of these ideas, it certainly appears that the fine nebula in the sword-hand of Orion, and that in the girdle of Andromeda, have undergone evident changes since the times of Huyghens and Simon Marius, who first observed them. These changes seem to indicate a contraction and condensation of the nebula, analogous to that supposed by Herschel. Such changes, however, are of course so very slow in their progress,

---

* A catalogue and description of the nebulæ, as they exist at present, has been undertaken, and in a considerable degree completed, by Sir J. F. W. Herschel. It has not as yet been made public, but some idea of the interesting results it contains may be formed from the following words of the author. 'I have already determined, with as much accuracy as the nature of such observations permits, the places, and obtained sufficient descriptions of the physical peculiarities, of between two and three thousand of these wonderful objects, a great part of them by many repeated observations; and made careful drawings of the most remarkable for their shape, size, or structure. Among these are objects so surprising, that I shall earnestly desire to see my observations verified by the powerful instruments (if sufficiently so) which are now become common in the hands of observers.'—Mem. Ast. Soc. vol. V. p. 47.

that, in all probability, it will take centuries of observation to establish the theory of Herschel by direct evidence.

Chapter IX.

Comets.

Of all the phænomena presented by the heavens, there is, perhaps, none which excites such general interest among all classes of society as the appearance of a comet. In general the motions of the heavenly bodies are characterized by a periodicity and regularity which render any unusual appearance the more striking. To which may be added, that a great comet is in itself an object well calculated to impress every beholder with astonishment and awe. The last great comet was that of 1811, which may be recollected by many of our readers; but brilliant as it was, it sinks into insignificance when compared with other apparitions of the same kind. Comets have been known whose tail has extended from the zenith to the horizon, while the disk of the body itself was equal in size to the full moon*. Such appearances are well calculated to impress the human mind with awe, as magnificent deviations from the usual unvarying regularity of the heavens. Ever prompt by a singular superstition to connect the aspects of the stars with the destiny of man, the belief of the multitude, up to these later times, has attributed to them malevolent influences over the fate of empires, or regarded them as announcing to the human race some impending scourge, as 'shaking from their horrid hair pestilence and war.' To these idle terrors have succeeded fears, perhaps not much more reasonable; the dread of a shock between one of these bodies and the earth. To say nothing of the extreme improbability of such an event, there is good reason for believing that, were it to take place, it would not be productive of such disastrous consequences as are usually anticipated.

The term Comet is derived from the appearance frequently presented by these bodies, that of being accompanied or rather followed by a large mass of luminous matter which we call the tail, but which the ancients likened to the appearance of hair, or a beard. The tail, however, is not a necessary appendage to a comet, as many well-authenticated instances exist where a comet has presented no such appearance. The essential character of a comet is derived from its motion, not from its constitution or external appearances, which seem liable to great variation. Most frequently a comet appears to be composed of a bright nucleus, partially or completely surrounded by a paler light, which spreads more particularly on one side, constituting the tail or hair. In other comets the nucleus is extremely faint, or even positively wanting; and in a third class there can neither be said to exist nucleus nor tail, as the comet presents the appearance of a very small nebulous body, which can only be distinguished from other nebulæ by its motion. As the circumstance of its motion distinguishes a comet from the fixed stars and nebulæ, so the law of the motion and the nature of the curve described distinguishes it from a planet. Another distinctive character might have been taken in former times from the fact that the paths of comets are not confined to the zodiac, nor to the direction from west to east, like the old planets, but take place in planes inclined at all angles to the ecliptic, and as well with a retrograde as with a direct motion. But this character has now less value since some of the newly discovered planets are known to have orbits inclined more than 30° to the ecliptic, and the satellites of Uranus are said to revolve perpendicularly to that plane. One character is common to all the planets and satellites of our system, the revolution in ellipses of very small eccentricity; on the other hand, the cometary ellipses are, without any exception, extremely excentric, and so much so, that in all but a very few cases they are sensibly identical with the parabola.

The tails of comets are evidently formed of highly rarefied matter, as is sufficiently indicated by their extreme transparence, which permits the smallest stars to be distinguished through them; and there is reason to believe that the nucleus, though in a greater state of condensation, is very far, at least in some instances, from being solid. Astronomers have occasionally distinguished fixed stars of no great bright-

* Still more extraordinary accounts are given in ancient authors, which have not been alluded to in the text, as they may be suspected of exaggeration.

ness through the *nuclei* of comets *. Indeed when we consider the enormous heat to which many comets are exposed when near their perihelion, it is difficult to conceive that any part of them can escape complete vaporization. Thus the comet of 1680, at its perihelion, was 166 times nearer the sun than the earth. If we suppose, as there is every reason for believing, that the intensity of the solar heat varies, like the intensity of light, inversely as the square of the distance, it appears that the comet must have been subjected to a heat 27,556 times as great as that received by the earth, or 2,000 times as great as that of red-hot iron. On the other hand, it must be confessed that, in their aphelion, comets experience a degree of cold of which we cannot form any calculation or conception.

The tail generally begins to appear as the comet draws near the sun; its length increases with its proximity, but does not acquire its greatest extent till after the perihelion passage. Its direction is always from the sun †, forming a curve rather concave towards that body, as in Fig. 48, where A represents the

Fig. 48.

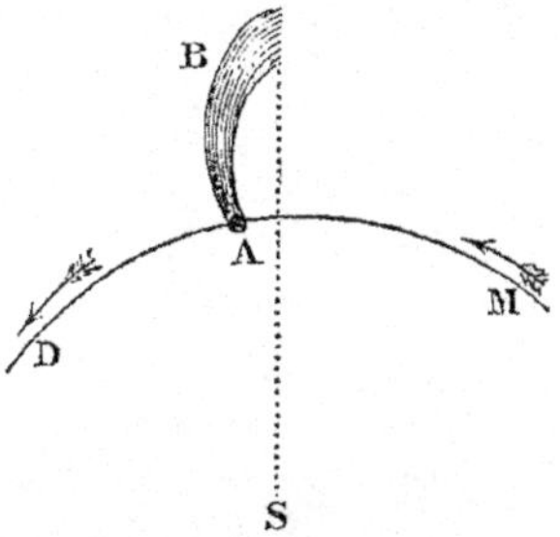

nucleus of the comet, B the tail, S the sun, MAD the comet's orbit in the direction from M to D.

The position and form of the tail indicate plainly its real cause, which is the vaporization produced in the body of the comet by excessive heat * : how great that heat must be, has been already explained. The tail acquires its greatest size after the perihelion, when the comet has been thoroughly heated, just as the earth does not attain its highest temperature till after the summer solstice. There are, however, considerable anomalies in the appearances presented by the tail, not merely with regard to its size, but also with regard to its shape and general appearance. Sometimes the tail is divided, or even the comet has two tails in different directions; sometimes, as in the comet of 1769, the tail has a double curvature of this form. Occasionally the tail is of immense length, and narrow; at other times broad and fan-shaped, as in that of 1744, which was 17° long by 130° broad. Sometimes the tail has a waving or undulating motion †, at others an instantaneous increase and decrease has been observed. The multiplicity of tails and the fan-shaped appearance which results from it, are probably caused by a more than ordinary velocity of the comet, in which case the curvature of the orbit in a given time becoming more sensible, the columns of vapour that arise deviate more and more from the direction of the original tail, and spread over a larger portion of the heavens. It is also to be recollected that the tail being situated in the plane of the comet's orbit, the appearance presented to the earth will depend very much on the position of that plane with regard to the earth. The curvature of the tail, for example, could never be visible to us if the comet were moving in the plane of the ecliptic.

The idea which prevailed for a long time with regard to the nature of comets was, that they were meteors of temporary duration, engendered in the atmosphere of the earth. Some circumstances certainly led to this view—the suddenness, in many cases, of their appearance

* A strong proof how trifling the mass of a comet usually is, may be found in the fact that the comet of 1770 passed between Jupiter and his satellites, without deranging at all by its attraction the motions of those minute bodies. The same comet approached us so nearly, that, according to the calculations of La Place, had its mass equalled that of the earth, the sidereal year would have been increased 2h 28m. Yet, probably, had the sidereal year varied a second, we should have perceived it. The comet's mass certainly could not have been a 5,000th part of that of the earth.

† The tail is always on a prolongation of the straight line which joins the comet to the sun: thus, if the comet be to the east of the sun, and set after him, the tail takes an easterly direction; but a westerly, if the comet be to the west of the sun, and set before him.

* The vaporization mentioned would not produce, however, a tail, without we suppose the comet to move in a resisting medium, which, as we shall presently see, is, from other causes, highly probable.

† It has been stated that, in the comet of 1811, such undulations were seen to pass from the comet to the extremity of the tail in 2 or 3 seconds; a distance of 4 millions of leagues. This velocity surpasses even that of light. A similar fact has been stated of the comet of 1807.

and decline, the transparence of their tails, and the apparently small density of their nuclei. But it was found by an accurate comparison of observations, (and we owe this discovery to Tycho Brahé,) that their parallax is so small as to put them far beyond the region of the moon. This point, once ascertained, decided that they could not be vapours generated in our atmosphere, and gave a strong probability to the opinion advocated of old by the Chaldæans, and supported by Seneca, that they were bodies permanent as the planets of our system, and reappearing at certain intervals depending on their peculiar orbits. The discoveries of Hevelius, Dörfel, and Newton, showed, in confirmation of this, that comets move in an orbit sensibly parabolic, the sun being in the focus. It was proved by Newton to be a necessary consequence of the theory of universal gravitation, that bodies attracted by the sun must describe round him one of the conic sections.

It is probable that the orbit described round the sun by comets is always an ellipse; at least we know it to be so in some instances: but the ellipse being of great excentricity, and a small part of the orbit only being visible to us, namely, that described near the perihelion, it is sensibly identical with a parabola. The first conjecture of an elliptic orbit is due to Halley, who, seizing the analogy established by the Newtonian theory between the motions of comets and planets, instituted a comparison between the elements of the orbits of the comets then known, and, perceiving a great coincidence between those of the years 1531, 1607, and 1682, with an equality of intervals, did not hesitate to throw out the conjecture that it was one and the same comet which had reappeared three successive times, and that it would reappear a fourth time in the year 1758 *. It may easily be supposed, that astronomers were eagerly on the look-out to verify this prediction; and accordingly, on the 25th of December in that year, it was again perceived. Thus the permanent nature and periodic returns of comets were for the first time established, and the truth of the Newtonian theory with regard to them placed beyond all dispute. More recent times have afforded us two additional instances of periodic comets, which have, however, this remarkable point of difference from that of Halley,—that the period in both is strikingly shorter. In the latter, the period is about seventy-six years, while in one of the two former it amounts to only six years and a fraction *; and in the other to between three and four †. In fact, these bodies may be said, in every sense, to belong to our system, as the orbit of the latter lies within even that of Jupiter, and it is in reality only distinguished from the planets by the great excentricity of its orbit.

It is unnecessary to explain in this place how Tycho Brahé was enabled to ascertain that the comet he observed had no sensible parallax, and consequently was far beyond, not merely the earth's atmosphere, but the region of the moon. On this head we must refer to the chapter on Parallax. The method pointed out in page 59, though it would be insufficient to determine with nicety a very small parallax, would be sufficient to show that the parallax of a body thus examined was very small; in fact, much smaller than that of the moon, which is sufficient for the purpose in question. The idea of Kepler was, that the orbit was a straight line, which would have precluded the possibility of a comet ever reappearing: Hevelius proved by observation, that the orbit was sensibly curvilinear; but Dörfel first distinctly announced the parabolic motion round the sun, about which there can be no doubt, as it has been verified by the calculation of the orbit of every comet that has appeared since.

* This comet had previously appeared in the year 1006: it is then said to have been four times as large as Venus: another of its apparitions was in 1456, and when it passed near the earth, its tail at that time occupied an arc of 60°, and was curved like a sabre. It excited great alarm, as coinciding with the capture of Constantinople by the Turks. Its subsequent apparitions have been much less brilliant. It is expected to reappear in 1835.

* This comet is called, from the name of its discoverer, von Biela's comet: it is supposed to be identical with the comet of 1772: it certainly appeared in 1806 and 1826; in which latter year it was first recognized as a periodic comet.

† This comet is called the comet of Encke, as that astronomer first recognized it in the year 1818, as a periodic comet, which had already appeared in the years 1786, 1795, 1801, 1805: its extreme smallness, perhaps its proximity to the sun, or other causes, prevented its being observed in 1808, 1812, and 1815. It has been repeatedly observed in its re-apparitions since 1818, and its orbit is now well determined. This remarkable body, which certainly forms a part of our solar system, has the major axis and inclination of Ceres; its sidereal revolution is 46 days less than that of Vesta; its perihelion falls within the orbit of Mercury; its aphelion between Jupiter and the new planets.

We cannot in this place go into an account of the Newtonian theory of gravitation, which belongs to the department of Physical Astronomy; it will be sufficient for our purpose to take it as a matter established by observation, that comets in general move round the sun in a parabola, of which that body occupies the focus. As, however, the parabola is not a closed curve, the fact of a comet's returning periodically, which is now established with regard to three, proves that the orbit of reappearing comets is, in fact, an ellipse; but the ellipses of comets being excessively excentric, the visible portion of them, which is always small, will not be distinguishable from a parabola. This is generally true: there are, however, some cases in which it becomes absolutely necessary to calculate the orbit as an ellipse; as, for example, in the case of the two comets of short periods of which we have spoken. These two remarkable bodies have opened us to a new view of the nature of comets; and that of Encke, in particular, (the comet of 3⅓ years,) bids fair to lead us to some remarkable discoveries. It has long been supposed that the space in which the planets move is filled with an extremely subtle fluid, the existence of which, hitherto, has been merely conjectural, or founded upon an optical theory, which of late years has gained many partisans, but perhaps can hardly be considered as demonstratively proved. The revolutions of the planets which have now been so carefully observed for many years, show pretty clearly, that if such a fluid exists, it is too rare to have any sensible effect on their motions. But as we have good reason for supposing the comet of Encke to be of incomparably less density than any planet or satellite of our system, it might happen that the fluid would cause a sensible resistance to the comet, while it was imperceptible for another body. Accordingly it has been found, that the magnitude of the axis major of the orbit decreases from revolution to revolution, and consequently, the periodic time diminishes. This result seems now pretty well established, and it is precisely the effect which would be produced by a resisting medium. A few more revolutions will probably put the matter beyond doubt, if, indeed, it be not considered as already established.

The discovery of three periodic comets, with the great probability of discovering more, has given additional interest to the problem of determining from observation the elements of the orbit, namely, the longitude of the perihelion, the longitude of the node, the inclination of the plane of the orbit to the ecliptic, its distance from the sun at the perihelion, and the time of its passage through that point. As soon as a new comet is discovered, astronomers hasten to determine, by means of three observations, these elements. Having found them, they compare the first four with the corresponding elements of former comets, as registered in the catalogues of those bodies: should they be fortunate enough to find a prior appearance of the same comet, the comparison of the perihelion passages will give its periodic time.

It is no long time since the calculation of a comet's orbit from observation was considered a most laborious operation; but the progress of astronomical science has so much simplified the problem, that it has ceased to be one of great labour or difficulty. As these calculations are becoming daily more general and more interesting, we shall give here the most simple and elegant mode of determining the orbit from three observations; a method which we owe to Olbers.

We begin by supposing the right ascension and declination of the comet to have been observed on three different days, the intervals between the observations being *small* and *nearly equal*. The problem is, from these three given positions, to determine the elements of the parabola described round the sun. The first step is to convert these right ascensions and declinations into longitudes and latitudes: let us call the longitude and latitude corresponding to the first observation λ' and θ' respectively: to the second λ'' and θ'': to the third λ''' and θ'''. Take from an Ephemeris the longitudes of the sun for the three moments of observation, and call them Λ', Λ'', Λ''' respectively. Let now, in fig. 49, S represent the sun; A, B, C, the three places of the comet; *a*, *b*, *c*, the places of the earth, at the time of the observations: the method rests on the following principle,—that the times between the observations being small and nearly equal, the middle radii vectores S B and S *b* cut the chords A C and *a c* in D and *d* in the ratio of the times: that is, calling the interval between the first and second observation t'; between the second and third t''; then that $ad : dc ::$ A D : D C : : $t' : t''$. This supposi-

Fig. 49.

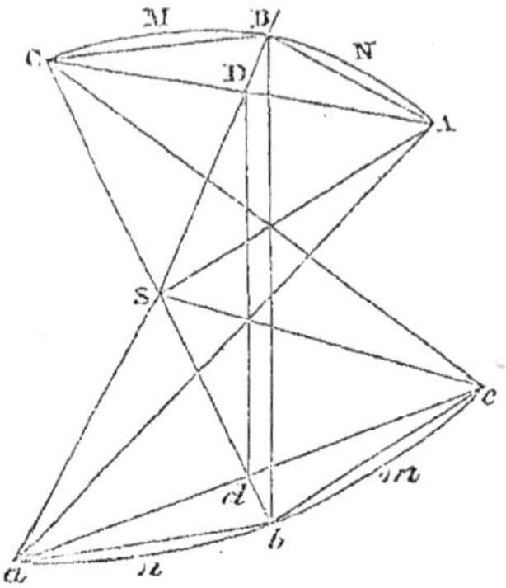

tion is not mathematically true; but when the arcs A C, *a c* are small, it is very nearly so. We know, by the theory of centripetal forces, that the times are as the parabolic and elliptic sectors, A N B S, B M C S, *a n b* S, *b m c* S: while the segments of the chords are as the triangular sectors A B S, C B S, *a b* S, *c b* S. The difference between these will be the small segments A N B A, *a n b a*, B M C B, *b m c b*. If the arcs and sectors are small magnitudes of the first order, these segments will be of the third order. Besides, it has been proved by Lambert, that for every parabolic and elliptic arc there is a radius vector which will cut the chord exactly in the ratio of the time, when the small segment A N B A, B M C B will be exactly in the ratio of A D to D C, and this will be the case, approximately, in the parabola, when the arcs are *small* and the times nearly equal, as we have supposed. The same will take place in the earth's path for intervals very nearly equal, since this path differs so little from a circle. Our method, then, which is only an approximate one, cannot lead to great errors. We may here observe, that all methods for determining, in the first instance, the elements of a comet's orbit, are approximate: the approximate values of the elements once found, more accurate values are obtained by a method of correction, for which we must refer to the work of Olbers*, the Astronomy of Delambre †, or the Mécanique Céleste ‡.

We now proceed to find the apparent place which the comet would have had at the time of the second observation, supposing that the comet had really been at D and the earth at *d*. The apparent places A D C, seen from *a d c*, lie in a great circle of the sphere: *b d* S D B lie all in one plane; consequently all the points of the line *d* D, seen from a given point in the line *b* S, lie in one and the same great circle.

To find the position of the line *d* D, we must determine the point of intersection of these two great circles. The first passes through the observed places of the comet at the first and third observation; the second, through the middle observation and the place of the sun at the same time. Take

$$\cot.\ \varpi = \frac{\tan.\ \theta'''}{\sin.\ (\lambda''' - \lambda').\ \tan.\ \theta'} - \cot.\ (\lambda''' - \lambda')$$

Then the angle $\alpha' - \varpi$ will give the point where the great circle drawn through both extreme places of the comet cuts the ecliptic: the angle η at which it cuts it is given by the equation

$$\tan.\ \eta = \frac{\tan.\ \theta'}{\sin.\ \varpi}$$

The longitude of the point where the other great circle cuts the ecliptic is evidently $= \Lambda''$, or the longitude of the sun at the middle observation: its inclination ι is given by the equation

$$\tan.\ \mathrm{I} = \frac{\tan.\ \theta''}{\sin.\ (\Lambda'' - \lambda'')}$$

Taking now

$$\cot.\ \sigma = \frac{\tan.\ \eta}{\tan.\ \mathrm{I}.\ \sin.\ \{\Lambda'' + \varpi - \lambda'\}} + \cot.\ \{\Lambda'' + \varpi - \lambda'\}$$

then the longitude c'' of the point of intersection $= \lambda'' - \varpi + \sigma$. and the latitude γ'' comes out.

$$\tan.\ \gamma'' = \tan.\ \eta.\ \sin.\ \sigma$$

According to our supposition, the chord of the comet's path A C and the chord of the earth's path *a c* are cut by the lines of sight A *a*, *d* D, *c* C in the ratio of the times; now the same ratio will hold for all orthographic projections of these chords and lines of sight.

* Abhandlung, sec. iv.
† Astron., vol. iii. chap. xxiii.
‡ Vol. i. page 227.

Suppose in Fig. 50, that C D A is the projection of the chord of the comet's path on the plane of the ecliptic, $a\,c\,d$ the chord of the earth's path as before: a A, d D, c C the projections of the corresponding lines in Fig. 49.

Fig. 50.

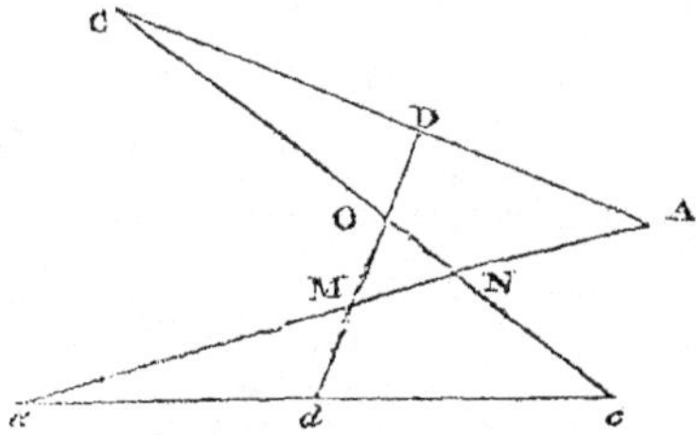

Now,

$$\mathrm{C\,O} : \mathrm{A\,M} :: \frac{\mathrm{C\,D}}{\sin.\,\mathrm{C\,O\,D}} : \frac{\mathrm{A\,D}}{\sin.\,\mathrm{D\,M\,A}}$$

$$c\,\mathrm{O} : a\,\mathrm{M} :: \frac{c\,d}{\sin.\,\mathrm{C\,O\,D}} : \frac{a\,d}{\sin.\,\mathrm{D\,M\,A}}$$

Since

$$c\,d : d\,a :: \mathrm{C\,D} : \mathrm{A\,D} :: t'' : t'$$

$$\text{and } \mathrm{C}\,c = \mathrm{C\,O} + c\,\mathrm{O}$$

$$\mathrm{A}\,a = \mathrm{A\,M} + a\,\mathrm{M}$$

$$\therefore \mathrm{A}a : \mathrm{C}\,c :: \frac{t'}{\sin.\mathrm{D\,M\,A}} : \frac{t''}{\sin.\,\mathrm{C\,O\,D}}$$

$$\text{But } \mathrm{D\,M\,A} = c'' - \lambda'$$

$$\mathrm{C\,O\,D} = \lambda''' - c''.$$

Also A a, C c are the curtate distances of the comet from the earth at the first and third observations: let us call them ρ' and ρ'''.

$$\therefore \rho' : \rho''' :: \frac{t'}{\sin.\,(c'' - \lambda')} : \frac{t''}{\sin.\,(\lambda''' - c'')}$$

$$\therefore \rho''' = \rho'.\,\frac{t''.\,\sin.\,(c'' - \lambda')}{t'.\,\sin.\,(\lambda''' - c'')} = \mathrm{M}\,\rho'.$$

Now since c'' has been already found, M is a known quantity, and therefore the ratio of ρ''' to ρ' is known.

There are, however, cases in which this mode of finding M cannot be employed, as when the apparent motion of the comet is nearly perpendicular to the ecliptic. Then the arcs $c'' - \lambda'$, and $\lambda''' - c''$ are evidently very small, and M cannot be determined from them with accuracy: and then the more general method, which we shall presently give, must be adopted. On the other hand, this latter method cannot be used when the comet's apparent motion is slow and nearly parallel to the ecliptic; in this case we must have recourse to the one already given. This method also has great conveniences when the intervals of time are very small. We may then, instead of calculating c'', assume it equal to λ'', and consequently

$$\mathrm{M} = \frac{t''.\,\sin.\,(\lambda'' - \lambda')}{t'\,\sin.\,(\lambda''' - \lambda')}.$$ This is the same thing as supposing the lines B b, D d, in Fig. 49, to be parallel: a supposition which cannot be far wrong when the arcs $a\,c$, A C are small, and consequently the lines $b\,d$, B D very small.

Now since all orthographic projections of the lines of sight cut the orthographical projections in the same ratio, let us project these lines on a plane perpendicular to the ecliptic, and so placed that the middle radius vector of the earth is perpendicular to it. Take

$$\tan.\,b' = \frac{\tan.\,\theta'}{\sin.\,(\mathrm{A}'' - \lambda')}$$

$$\tan.\,b'' = \frac{\tan.\,\gamma''}{\sin.\,(\mathrm{A}'' - c'')}$$

$$\tan.\,b''' = \frac{\tan.\,\theta'''}{\sin.\,(\mathrm{A} - \lambda''')}$$

then b', b'', b''' will be the angles which the lines of sight in the projection make with the projected chord of the earth's path. But

$$\frac{\tan.\ \gamma''}{\sin.\ (\Lambda''-c'')}=\frac{\tan.\ \theta''}{\sin.\ (\Lambda''-\lambda'')}$$

wherefore in this case it is unnecessary to calculate γ'' and c''.

Call the projected distance at the first observation d, and at the third N d, then, since in this case also the chords are cut in the ratio of the times

$$N=\frac{t''.\ \sin.\ (b''-b)}{t'.\ \sin.\ (b'''-b'')}$$

Now

$$\rho'=\frac{d.\ \cos.\ b'}{\sin.\ (\Lambda''-\lambda')}$$

$$\rho'''=M\rho'=\frac{N\,d.\ \cos.\ b'''}{\sin.\ (\Lambda''-\lambda''')};$$

consequently

$$M=\frac{\cos.\ b'''.\ \sin.\ (\Lambda''-\lambda')\ .\ \sin.\ (b''-b').t''}{\cos.\ b'\ .\ \sin.\ (\Lambda''-\lambda''')\ .\ \sin.\ (b'''-b'')\ t'}$$

$$=\frac{\sin.\ (\Lambda''-\lambda')\ (\tan.\ b''-\tan.\ b').t''}{\sin.\ (\Lambda''-\lambda''')\ (\tan.\ b'''-\tan.\ b'').t'}$$

$$=\frac{\{\tan.\ \theta''.\ \sin.\ (\Lambda''-\lambda')-\tan.\ \theta'.\ \sin.\ (\Lambda''-\lambda'')\ \}\ t''}{\{\tan.\ \theta'''\ \sin.\ (\Lambda''-\lambda'')-\tan.\ \theta''.\ \sin.\ (\Lambda''-\lambda''')\}\ t'}$$

Put now

$$m=\frac{\tan.\ \theta''}{\sin.\ (\Lambda''-\lambda'')}$$

$$\therefore\ M=\frac{\{m.\ \sin.\ (\Lambda''-\lambda')-\tan.\ \theta'\}\ t''}{\{\tan.\ \theta'''-m.\ \sin.\ (\Lambda''-\lambda''')\ \}\ t'.}$$

Hence the ratio of the curtate distances of the comet from the earth at the first and third observations is known; to find from these the distances themselves, we must proceed as follows: call S a. S c the distances of the earth from the sun at the first and third observations, R', and R''': call S A, S C the corresponding distances of the comet from the sun, r', r'''.

Now,

$$r'^2=R'^2-2R'\rho'.\ \cos.\ (\Lambda'-\lambda')+\rho'^2.\sec.\ \theta'$$

$$r'''^2=R'''^2-2R'''\ M.\rho'\ \cos.\ (\Lambda'''-\lambda''')+M^2\rho'^2.\sec.^2\ \theta'''$$

Now, calling the chord A C, k''.

$$k''=\sqrt{(x'''-x)^2+(y'''-y')^2+(z'''-z')^2}$$

x, y, z, being rectangular coordinates, whose centre is at the sun, the axis of z perpendicular, to the plane of the ecliptic; and that of x passing through the vernal equinox. Developing

$$k''=\sqrt{r'^2+r'''^2-2x'x'''-2y'y'''-2z\ z'''}$$

It is easy to see that

$$x'=\rho'.\ \cos.\ \lambda'-R'.\ \cos.\ \Lambda'$$
$$y'=\rho'.\ \sin.\ \lambda'-R'.\ \sin.\ \Lambda'$$
$$z'=\rho'.\ \tan.\ \theta'$$
$$x'''=M'.\ \cos.\ \lambda'''-R'''.\ \cos.\ \Lambda'''$$
$$y'''=M\rho'\ \sin.\ \lambda'''-R'''.\ \cos.\ \Lambda'''$$
$$z'''=M\rho'.\ \tan.\ \theta'''$$

Substituting these values we get

$$k''^2=r'^2+r'''^2-2\,R'R'''.\ \cos.\ (\Lambda'''-\Lambda')+2\rho'R'''.\ \cos.\ (\Lambda'''-\lambda')$$
$$+2M\rho'R'.\ \cos.\ (\Lambda'-\lambda''')-2M\rho'^2.\ \cos.\ (\lambda'''-\lambda')-2M\rho'^2\ \tan.\ \theta',\ \tan.\ \theta''$$

or shortly,

$$k''^2 = F + G\varrho' + H\varrho'^2$$

the co-efficients F, G, and H, being known quantities, and ρ' unknown, as well as k''.

The most convenient way in practice is to find ρ' by successive approximations, which may be easily done in this way. If k'' were known, ρ' might be thus determined: take $F' = K^2 - F$ and

$$\text{assume tan. } \psi = \frac{2\,H}{G} \cdot \sqrt{\frac{F'}{H}}$$

$$\text{then } \rho' = \text{tan. } \frac{\psi}{2} \sqrt{\frac{F'}{H}}\text{*}$$

As long as F′ is positive, there is only one positive value of ρ'; when G is negative, the angle ψ is greater than 90°.

But to find in this way an approximate value of ρ', we must have an approximate value of k''. We may get this from the following considerations: F is the square of the chord of the earth's path; now, as long as both chords are small, we have very nearly $k''^2 = \frac{4\,F}{r' + r'''}$. It is true that r' and r''' are unknown; but $r' + r'''$ cannot be less than 1, as long as the apparent angular distance from the comet to the sun is greater than 30°; and on the other hand, $r' + r'''$ is almost always smaller than 3, because nearly all visible comets are within the orbit of Mars. Hence 2 is always an approximate value for $r' + r'''$: consequently, one may always assume at first

$$F' = K^2 - F = \frac{4\,F}{r' + r''} - F$$

$$= \frac{4\,F}{2} - F = F.$$

If, however, in the equation for the chord the angles $(A - \lambda)$ are greater than 90°, and consequently, the coefficient of ρ' positive, then $r' + r'''$ is necessarily greater than 2 R. In this case, instead of assuming F′ = F, we should take

$$F' = \frac{4\,F}{2.4} - F = \frac{2}{3} F.$$

With one of these two suppositions for F′, we find by the formulæ given above, an approximate value of ρ', and this being known, we may substitute it in the values of r'^2 and r'''^2, and thus find r' and r'''. The best way, however, is to take some simple fraction that comes near to the value of ρ' which we have just found, and to calculate r'^2 and r'''^2 roughly, thus:

$$\text{supposing } r^2 = 1 + b$$

$$\text{then } r = 1 + \frac{b}{2 + \frac{1}{2}b} \text{ approximately.}$$

Substituting these approximate values of r' and r''' in the equation

$$F' = \frac{4\,F}{r' + r'''} - F$$

we now get a more accurate value of F, and again from that, of ρ'. With this second value of ρ' we must now proceed to calculate rather more accurately r' and r'''. This having been done, if T be the interval between the first and third observations, put $\tau = 4\,m\,T$, where log. $m = 8.5366114$, then approximately

$$\text{†}F' = \frac{\tau^2}{r' + r'''} + \frac{\tau^4}{(r' + r''')^2} \cdot \frac{1}{12(r' + r''')^2} + 4\,\frac{\tau^6}{(r' + r''')^3}\left(\frac{1}{12\,(r' + r''')^2}\right)^2 - F$$

Calling the three first terms on the right-hand side of the equation A, B, C, respectively

$$A = \frac{\tau^2}{r' + r'''}$$

$$B = \frac{A^2}{12\,(r' + r''')^2}$$

$$\text{‡}\,C = \frac{4\,B^2}{A}.$$

With this more accurate value of F′ we calculate again ϱ', which now begins to be near its exact value. If we repeat the last process once more, (which is not always necessary,) we shall find ϱ' with all desirable accuracy. As soon as ϱ' is determined, the elements of the orbit are found with ease. We must calculate exactly

$$r',\ r''',\ \varrho', \text{ and } \varrho''' = M\,\varrho'.$$

---

* F′ is rarely negative: it can only be so when $r' + r'''$ is greater than 4. In this case take

$$\sin.\ n = \frac{2\,H}{G} \cdot \sqrt{\frac{F'}{H}},$$

then,

$$\rho' = \tan. \frac{n}{2} \cdot \sqrt{\frac{F'}{H}} \quad \text{or,} \quad = \cot. \frac{n}{2} \sqrt{\frac{F'}{H}}.$$

Both these values of ρ may be positive, but the last is always the true one.

† This may easily be deduced from Lambert's Theorem—

$$T = \frac{\left(\frac{r' + r''' + k'}{2}\right)^{\frac{3}{2}} - \left(\frac{r' + r''' - k'}{2}\right)^{3}}{3m\sqrt{2}}.$$

‡ In most cases C will be too small to make it necessary to take it into account.

The equation for the r's is of this form:—

$$r = \sqrt{f + g\,\varrho + h\,\varrho^2}$$

$$= \sqrt{f\left\{1 + \left(\frac{g}{h} + \varrho\right)\frac{h}{f}.\,\rho\right\}}$$

When $\frac{g}{h} + \rho$ is positive put $\left(\frac{g}{h} + \rho\right)\frac{h}{f}.\,\rho = \tan^2\varphi.$

———————— negative ———————— $= \sin.^2\varphi.$

In the first case $r = \frac{\sqrt{f}}{\cos.\varphi}$

—— second —— $= \sqrt{f}.\cos.\varphi.$

Knowing now r', r''', ϱ', and ϱ''', we get the elements of the orbit in this manner. Call the heliocentric latitudes at the first and third observations γ' and γ''', we have

$$\sin.\ \gamma' = \frac{\tan.\theta'.\,\rho'}{r'}$$

$$\sin.\ \gamma''' = \frac{\tan.\theta'''\,\rho'''}{r'''}.$$

Call the heliocentric elongations from the earth at the same time ε', ε''', then

$$\sin.\ \varepsilon' = \frac{\varrho'.\sin.(\Lambda' - \lambda')}{r'.\cos.\lambda.}$$

$$\sin.\ \varepsilon''' = \frac{\rho'''.\sin.(\Lambda''' - \lambda''')}{r'''\cos.\lambda'''}.$$

Hence the two heliocentric longitudes β' and β''' are known. If we call now ξ the angular distance of the comet from its ascending node, measured along the ecliptic, then $\beta' - \xi$ is the longitude of the node. ξ is found from the formula

$$\tan.\left(\xi + \frac{\beta''' - \beta'}{2}\right) = \frac{\sin.(\theta''' + \theta')}{\sin.(\theta''' - \theta')}.\tan.\left(\frac{\beta''' - \beta'}{2}\right).$$

Hence ξ and the longitude of the node $\beta' - \xi = \Omega$ is known.

The inclination of the orbit to the plane of the ecliptic is found by the formula

$$\tan.\ \iota = \frac{\tan.\ \theta'}{\sin.\ \xi},$$

ι being the inclination in question.

To get the heliocentric distances of the comet from its node measured in the plane of its orbit υ' and υ'''

$$\tan.\ \upsilon' = \frac{\tan.\ \xi}{\cos.\iota.}$$

$$\tan.\ \upsilon''' = \frac{\tan.(\xi + \beta''' - \beta')}{\cos.\ \iota.}$$

Now $\upsilon''' - \upsilon'$ is the difference of the true anomalies at the first and third observations. Let $\upsilon' - \upsilon''' = \chi$, and take

$$\tan.\ x = \sqrt{\frac{r'''}{r'}}$$

$$\tan.\left(\frac{\varphi}{2} + \frac{\chi}{4}\right) = \frac{\tan.(45^\circ - x)}{\tan.\frac{x}{4}}$$

where φ is the true anomaly at the first observation. But knowing φ, we know the longitude of the perihelion*.

The perihelion distance d is found by the formula

$$d = r'\cos.\frac{\varphi^2}{2}.$$

All the elements of the orbit are now found, except the time of the passage through the perihelion; this is easily found by the theory of parabolic motion, which gives us, calling t the time from the first observation to the perihelion passage,

$$t = \frac{d^{\frac{3}{2}}.\,T}{\pi\sqrt{2}}\left\{\tan.\frac{\upsilon'}{2} + \tfrac{1}{3}.\tan.^3\upsilon\right\}$$

π representing a semicircumference to radius = 1, and T being the duration of a sidereal revolution of the earth*.

This expression is very inconvenient for finding the anomaly when t is given,

* υ' and υ''' represent the heliocentric distances of the comet from the node measured in the plane of its path; adding υ to φ we get the distance of the perihelion from the node; project this distance on the ecliptic, and add it to the longitude of the node, and we shall have the longitude of the perihelion.

* The law of the equable description of areas which holds in all cases of bodies acted upon by centripetal forces, gives us the equation

$$r^2\,d\,\upsilon = c\,d\,t.$$

By the properties of the parabola

$$r = \frac{d}{\cos.^2\frac{\upsilon}{2}}$$

where d is the perihelion distance.

Substituting this value of r, and remembering that $c = \frac{\upsilon}{T}$, we shall get, upon integration, the expression for t, given in the text. It must be remembered that the time is counted from the perihelion, or $t = 0$ when $\upsilon = 0$.

as we have to solve a cubic equation. Astronomers have eluded this difficulty by calculating the time corresponding to every degree of anomaly, for a comet whose perihelion distance is equal to unity, or the mean distance from the sun to the earth: then the time for any given anomaly for another comet is found by taking from the table the time corresponding to the given anomaly, and multiplying it by $d^{\frac{3}{2}}$. Such a table as that we speak of is usually called table of the comet of 109 days, because a comet whose perihelion distance is unity, takes nearly 109 days to describe 90° of anomaly from the perihelion. If the comet be well known to recur, as, for example, that of Halley, it may be desirable to correct the parabolic elements by allowing for the ellipticity of the orbit. The best way of doing this is as follows. The semi-axis major is known, the periodic time being known; for, by Kepler's third law, the squares of the periodic times are as the cubes of the major axis: the periodic time of the earth is known, and its semi-axis major. Now, the semi-axis major being known, the excentricity may be found by subtracting from it the perihelion distance. Hence $a\,e$, and consequently e, is known, e being the ratio of the excentricity to the semi-axis major. Now, the anomaly in a very excentric ellipse may be deduced from the corresponding anomaly in a parabola, by adding to the parabolic anomaly υ a small angle φ, which is determined by the equation

$$\sin.\varphi = \frac{1}{10}.\tan.\frac{\upsilon}{2}\left\{4 - 3\cos.^2\frac{\upsilon}{2} - 6\cos.^4\frac{\upsilon}{2}\right\}.$$

We find then, by the tables of comets, the parabolic anomaly υ corresponding to the given time: we add to it, and $\upsilon + \varphi$ will be the elliptic anomaly. The elliptic radius vector may be calculated by the expression

$$r = \frac{d}{\cos.^2\left(\frac{\upsilon+\varphi}{2}\right)}\left\{1 - \frac{(1-e)}{2}.\tan.^2\left(\frac{\upsilon+\varphi}{2}\right)\right\}.$$

The following example, taken from Olbers, will contribute to make the subject more intelligible, while, at the same time, it will show the simplicity and shortness of the method. From three observations of the comet of 1769, the three corresponding longitudes and latitudes have been deduced, as follows:—

	days	hours	λ	θ
September 4	14	0	80° 56′ 11″	17° 51′ 39″
8	14	0	101° 0′ 54″	22° 5′ 2″
12	14	0	124° 19′ 22″	23° 43′ 55″

The tables of the sun give us for these three moments—

Λ	log. R.
162° 42′ 5″	0·003132
166° 35′ 31″	0·002665
170° 29′ 20″	0·002184

$$\text{Hence } t' = t'' = 4 \text{ days} \therefore \frac{t''}{t'} = 1 \text{ and } T = 8 \text{ days.}$$

The first thing is to determine M: for this purpose, we have to calculate the expression

$$m = \frac{\tan.\theta''}{\sin.(\Lambda'' - \lambda'')}$$

Hence
$$M = \frac{m.\sin.(\Lambda'' - \lambda') - \tan.\theta'}{\tan.\theta''' - m.\sin.(\Lambda'' - \lambda''')}$$

and then the two equations

$$r'^2 = R'^2 - 2R'\rho'.\cos.(\Lambda' - \lambda') + \rho'^2\sec.^2\theta'$$
$$r'''^2 = R'''^2 - 2R'''M.\rho'.\cos.(\Lambda''' - \lambda''') + M^2\rho'^2.\sec.^2\theta'''.$$

In our case, these equations become

$$r'^2 = 1·01453 - 0·28854\,\rho' + 1·10393\,\rho'^2$$
$$r'''^2 = 1·01011 - 1·21482\,\rho' + 0·90869\,\rho'^2$$

The equation for the chord is

$$k^2 = r'^2 + r'''^2 - 2\,R\,R'''.\cos.(\Lambda''' - \Lambda') + 2\,R'''.\cos.(\Lambda''' - \lambda')\,\varrho' + 2\,R'.\cos.(\Lambda' - \lambda''')\,M\,\varrho' - 2\,M.\cos.(\lambda''' - \lambda')\,\varrho'^2 - 2\tan.\theta'.\tan.\theta'''.\,M\,\varrho'^2$$

or $$k^2 = 0{\cdot}01868 - 0{\cdot}10954\,\varrho' + 0{\cdot}49702\,\varrho'^2.$$

In order to solve this equation by approximation, we must have recourse to the formulæ

$$\tan.\psi = \frac{2\,H}{G}.\sqrt{\frac{F'}{H}} \text{ and } \varrho = \tan.\frac{\psi}{2}.\sqrt{\frac{F'}{H}}$$

We have $F = 0{\cdot}01868$ $G = -0{\cdot}10954$ $H = 0{\cdot}49702.$

If we suppose now $F' = F$, we find $\varrho' = 0{\cdot}3332$, take then $\varrho' = \frac{1}{3}$ the formulæ

$$r^2 = 1 + b$$

$$r = 1 + \frac{6}{2 + \frac{1}{2}6}$$

give us $r' = 1{\cdot}02$, $r''' = 0{\cdot}83$ $\therefore r' + r''' = 1{\cdot}85$, and $F' = \frac{2{\cdot}15}{1{\cdot}85}.F.$

With this value of F' we calculate ϱ' over again, and find for it $0{\cdot}3465$. With this value of ϱ', we now calculate accurately the values of r' and r''', and we get $r' = 1{\cdot}02326$, $r' = 0{\cdot}83565$ $r' + r''' = 1{\cdot}85891$. To determine F' from this value of $r' + r'''$, we employ the series

$$F' = A + B - F, \text{ where } A = \frac{\tau^2}{r' + r'''}$$

$$B = \frac{A^2}{12.(r' + r''')}$$

$$\tau = 4\,m\,T$$

C being in this case insensible. Hence we obtain $F' = 0{\cdot}222112$, and from this $\varrho' = 0{\cdot}34817$: as this value of ϱ' is so near the former $0{\cdot}3465$, it is unnecessary to pursue the approximation any farther. ϱ' being thus found, the elements of the orbit are easily determined by the equations in page 57. We find these

1. The heliocentric latitudes
   $\gamma' = 6\ 17\ 34$ $\gamma''' = 9^\circ\ 12'\ 19''$
2. The heliocentric elongations from the earth
   $\iota' = 19\ 47\ 4$ $\iota''' = 15^\circ\ 25'\ 16''$
3. The heliocentric longitudes
   $\beta' = 2^\circ\ 29'\ 52''$ $\beta''' = 5^\circ\ 54'\ 36''$
4. Angular distance of the comet from the node on the ecliptic
   $\xi = 7^\circ\ 11'\ 45''$
5. Longitude of the descending node
   $\beta' - \xi = 355^\circ\ 18'\ 7''$ . . . . (I.)
6. Inclination of the orbit $= 41^\circ\ 21'\ 30''$ . . . (II.)
7. $\upsilon''' - \upsilon' = \chi = 4^\circ\ 27'\ 46''$
8. True anomaly of the comet at the third observation $= \varphi = 135^\circ\ 52'\ 24''$
9. Longitude of the perihelion $= 145^\circ\ 11'\ 11''$ . . (III.)
10. Perihelion distance $= 0{\cdot}11872$ . . . . . (IV.)
11. Time from the third observation to the perihelion passage
$= 24^d\ 20^h\ 22'$ $\therefore$ time of the perihelion passage is Oct. 7. $10^h\ 22^m$ (V.)

Chapter X.

Construction of the Tables of the Sun, Moon, and Planets.—Equations of Condition.—Method of least Squares.

To avoid tedious calculations, astronomers have formed tables by which the places of the sun, moon, and planets for any given instant may be found. Of the utility of such tables, it is unnecessary to say anything: the reader will have already perceived that such calculations must be of frequent recurrence in astronomy, and therefore it is of the utmost importance to simplify them as much as possible. As the tables of the sun are of more frequent utility than the others, we shall commence by giving some account of their construction.

Section I.—*Tables of the Sun.*

It is hardly necessary to premise, that what we call the tables of the sun are in reality tables of the earth; since it is the latter body which is in motion round the former. As, however, in practice, the object is to determine the apparent place of the sun, as seen by a given observer; and as it is immaterial whether his motions are real or apparent, we shall continue to speak of him as if he were the moving body. The reader will bear in mind that all the inequalities which we shall presently mention are inequalities in the earth's own orbit. We begin then by supposing the earth to be in the focus of the ellipse described round it by the sun, which ellipse of course lies in the plane of the ecliptic. The problem is to determine what point of its ellipse the centre of the sun occupies at any given instant.

Now evidently, if we know the longitude of the sun, his position in the celestial sphere is determined, since the inclination of the ecliptic to the equator is known; and we may, by the solution of a right-angled spherical triangle, obtain his right ascension and declination, which determine his apparent position on the celestial sphere.

Fig. 51.

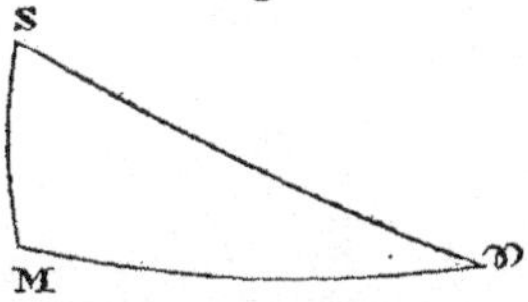

Let ♈ be the vernal equinox; ♈ S, the ecliptic; ♈ M, the equator; S the place of the sun; S M an arc of a great circle at right angles to ♈ M: then ♈ S will be the sun's longitude. If we know this, and the angle M ♈ S, which is the obliquity of the ecliptic, we can calculate ♈ M, the right ascension, and M S, the declination of the sun. We shall explain subsequently how the obliquity of the ecliptic is determined: for the present we suppose it known: all then depends upon finding the sun's longitude for the instant required. The sun's apparent position in the heavens, the longitude being known, may then be determined as we have just shown: but if we wish to have his absolute position in space, we must also know his distance from the earth, or the radius vector of his orbit at that moment. The longitude fixes the direction of the radius vector; the length of the latter, the absolute position of the sun on it.

Let us begin by showing how the longitude may be found. The first approximation is to neglect the ellipticity of the orbit, and to suppose the motion of the sun in his orbit circular and uniform. This motion will be proportional to the time; and the place of the fictitious sun thus moving will give what is called the *mean longitude:* this it is necessary subsequently to correct to get the *true longitude.* To find the mean longitude, it is necessary to know the mean velocity of which we have spoken, and the longitude of the sun, at some one time, which is called the *epoch.* Knowing these, we may find the *mean longitude* at any other time; since, as the velocity is supposed uniform, this longitude will be proportional to the time. The determination of the epoch is a matter entirely of observation. It is usual to take the beginning of some particular year for epoch*, and to determine the place of the sun for that time by a great number of observations made shortly before and afterwards, and reduced to that instant by the mean motion, which may always be supposed sufficiently well known to serve for short intervals.

The mean motion is ascertained by comparing two epochs, determined carefully by observation at very distant periods; for example, at an interval of 50 years. As the inequalities of the sun's motion return to nearly the same value within the course of a year, they will disappear in such a comparison as that just mentioned, and we shall obtain the mean motion of the sun in this interval with great accuracy: if we divide this by 50, we get the mean motion for one year. Now this being known, let us suppose that the longitude for the be-

---

* The solar tables of Delambre take for epoch the midnight (mean time at Paris) which separates one year from the preceding; namely, that between the 31st of December and the 1st of January. The reason of this is, that the astronomical day of the French is counted, like the civil day, from midnight to midnight, while the astronomical day of English astronomers is counted from midday to mid-day. It is important to bear in mind this difference.

ginning of the year 1800 has been carefully observed; by adding to it once, twice, three times the motion for one year, and so on, we get the longitudes for the years 1801, 1802, 1803, &c., successively. These longitudes, being entered into a table, form what is called a table of epochs. But knowing the mean motion for one year, simple division will give us the mean motion for a day, an hour, and a minute. To save trouble, the mean motion is given in the solar tables ready calculated for every day in the year, and every hour and minute of the day. To find, then, the mean longitude of the sun for any given moment, as, for example, 23d Sept. 1805, at 9 minutes after 4, mean Paris time, begin by taking from the *Table of Epochs*, the mean longitude for the beginning of the year; this is, by Delambre's tables, 280° 11′ 2″·8; by the same tables the mean motion from the beginning of the year to the 23d Sept. is 261° 11′ 47″·4; add this to the epoch, and subtracting a whole circumference, we get 181° 22′ 50″·2. To this we must add, from the tables, the mean motion for 4 hours, which is 9′ 51″·4, and that for 9 minutes, which is 22″·2; so that finally we obtain for the mean longitude of the sun, at the instant abovementioned, 181° 33′ 3″·8.

To understand how the elliptic motion of the sun is calculated, we must refer to the following figure:

Fig. 52.

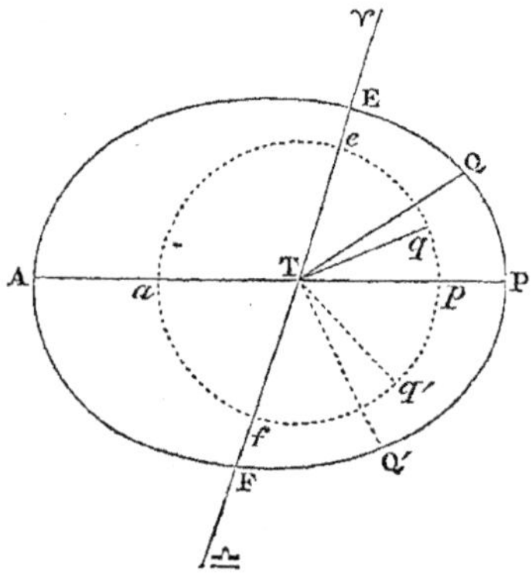

Let P E A F be the apparent ellipse described by the sun round the earth T, which occupies the focus. Let F E be the line of the equinoxes, on the prolongation of which, in the celestial sphere, are the points ♈ and ♎; let *afpe* be a circle described round the centre T, and conceive a fictitious sun S′, which describes this circle with an uniform motion, and its velocity such that it is on the radius T P drawn to the perigee P, when the sun S is at P, and it returns to P with S at the end of each revolution. At the point P, S has its greatest velocity, and immediately gets beyond S′ which moves with a uniform velocity. At a given time from their being both on the line T P, S will be, as it were, at Q, while S′ is only at *q*. But on the other hand, as S approaches its apogee A, its velocity diminishes, while that of S′ always remaining the same, it finally comes up to S′, and reaches it when both are on the line T A. The velocity of S is then the least, and it is passed by S′; but the velocity of S gradually increasing, it finally reaches S′ on the line T P. It appears from this, that from P to A, S is before S′, while from A to P, S′ is before S. Suppose at a given moment *q* to be the place of S′, and Q that of S; the angle *q* T P is called the mean anomaly, Q T P the true anomaly; the arc E A P Q will be the true longitude; *eapq* the mean longitude; *eafp* will be the longitude of the perigee. It is easy to see that in all cases the mean anomaly = the mean longitude of the sun − the longitude of the perigee.

We must here remark, that the solar perigee is not a fixed point in the heavens, but moveable according to certain laws, the investigation of which belongs to the department of Physical Astronomy. It may be stated, however, as a result of observation, that it has a motion to the eastward of nearly 12″ annually*. Of this, however, it is unnecessary to say more, as the solar tables, by the side of the column of the epochs of the mean longitude, give the epochs of the longitude of the perigee from year to year. A simple subtraction accordingly will give the epoch of the mean anomaly. In the same way, by the side of the column giving the motion in mean longitude for months, days, hours, and so on, is given the motion of the perigee for the same, so that another subtraction gives the mo-

* This motion, referred to the ecliptic, will appear to be about 62″ annually, in consequence of the precession of the equinoxes.

tion of the mean anomaly for any required fraction of a year *.

We now proceed to determine the elliptic place of the sun. For this purpose, as the real or elliptic sun is at Q, while the mean sun is at q, we must find the angle qTQ, and add it to the mean longitude $e\,a\,p\,q$. This angle is called the *equation of the centre*. Theory gives us the equation of the centre in terms of the mean anomaly; and in the solar tables there is a table giving this equation for every ten minutes of mean anomaly.† The equation of the centre is subject to a very slow secular variation, which tends gradually to diminish it; that is, in fact, to make the solar orbit more nearly circular: this change is very small; the tables give a column in which, by the side of each value of the equation, is found the corresponding variation for 100 years, counted from 1810. A proper fractional part of this variation must be taken, according to the year for which we calculate: thus, if we are finding the equation of the centre corresponding to 3° 20′ of mean anomaly in the year 1840, we find in the first column the equation of the centre = 6′ 6″·4; then under the head

---

* To follow the example already given by Delambre's Tables ……………… { Longitude of the perigee for Jan. 1st, 1805 = 279° 34′ 13″; Motion of the perigee up to Sept. 23 = 45″; Longitude of the perigee Sept. 23, 1805 = 279 34 58 }

Now we found, for the same time, mean longitude of the sun = 181° 33′ 3″·8

Difference = mean anomaly = 261 58 5·8

It is important to remark, that in the Solar Tables of Delambre, which are those most used, and which we are now following, the anomaly is counted from the perigee; in the case of comets, which are always invisible at their aphelion, this is absolutely necessary; but in many tables of the sun and planets the anomaly is counted from the apogee.

† Kepler's first law gives us the equation

$$r^2\,dv = c\,dt$$

$$\text{when } c = \sqrt{\mu}\,\sqrt{a\,.\,(1-e^2)}$$

μ being the sum of the masses of the sun and planet.

The equation to the ellipse gives

$$r = \frac{a\,.\,(1-e^2)}{1+e\,.\,\cos.\,v}.$$

Substituting for r^2 in the first equation, we get

$$dt = \frac{a^{\frac{3}{2}}.(1-e^2)^{\frac{3}{2}}}{\sqrt{\mu}}\,.\,\frac{dv}{(1+e\,.\,\cos.\,v)^2}.$$

To integrate this expression, take another variable u, such that

$$\cos.\,v = \frac{\cos u - e}{1 - e\,.\,\cos.\,u}.$$

$$\text{Hence } t + C = \frac{a^{\frac{3}{2}}}{\sqrt{\mu}}\,.\int du\,(1 - e\,.\,\cos.\,u).$$

Taking the time to run from the perihelion, $t = 0$ when $u = 0$; and putting $\frac{a^{\frac{3}{2}}}{\sqrt{\mu}} = \frac{1}{n}$ we get

$$nt = u - e\,.\,\sin.\,u \quad \ldots\ldots\ldots\ldots \quad (a)$$

nt is the mean anomaly; the auxiliary angle u is technically termed the eccentric anomaly, v the true anomaly. It will be seen, by referring to the value of cos. u, that the following relation holds

$$\tan\frac{v}{2} = \sqrt{\frac{1+e}{1-e}}\,.\,\tan.\,\frac{u}{2} \quad \ldots\ldots \quad (b)$$

From equation (b) we get the series

$$v = u + e\,.\,\sin.\,u + \frac{e^2}{4}\,.\,\sin.\,2u \quad \ldots\ldots\ldots\ldots \quad (c)$$

and from equation (a)

$$u = nt + e\,.\,\sin.\,nt + \frac{e^2}{2}\,.\,\sin.\,2nt \quad \ldots\ldots\ldots \quad (d)$$

Substituting in (c) the value of u from (d) we get

$$v = nt + 2e\,.\,\sin.\,nt + \frac{5}{4}\,.\,e^2.\,\sin.\,2\,nt \quad \ldots\ldots \quad (1)$$

Now, the equation the centre = true anomaly − the mean anomaly,

$$= v - nt$$

$$= 2e\,.\,\sin.\,nt + \frac{5}{4}\,.\,e^2\,.\,\sin.\,2nt + \ldots\ldots \quad (e)$$

secular variation $1''\cdot04$; the hundredth part of this, ·0104, will be the variation for one year; we must multiply this by 30, the number of years from 1810, which gives ·312; it will be observed, by looking to the top of the column, that the secular variation is negative; this, then, must be subtracted from the equation of the centre above found, and we finally get its value for the time required $6'\ 6''\cdot08$.

The equation of the centre being added to the mean longitude of the sun, we have now got his elliptic longitude, but we have not yet got his apparent place, as there are still some trifling inequalities to be considered. In the first place, we have already shown that, in consequence of the nutation of the earth's axis, the real place of the equinox oscillates about the mean place; and as we want the sun's longitude counted from the real equinox, we must add to his mean longitude the motion of the equinox on the ecliptic. We have seen, page 159, that this is represented by the expression $-\ 18\ .\ \sin.\ \Omega\ .\ \cot.\ 2\ \omega$, where ☊ is the longitude of the moon's node, and ω the obliquity of the ecliptic. This formula has been reduced into a table, where for every ten * parts of the supplement of the longitude of the moon's node the motion of the equinox on the ecliptic is given. This only gives us the lunar nutation: if we wish to be very accurate, we must take into account the solar nutation; this depends directly upon the sun's mean longitude †, and another small table gives the effect on the equinox from five to five days throughout the year.

Besides all this, the attractions of the larger planets, Venus and Jupiter particularly, disturb the orbit of the earth, and prevent that body from describing exactly an ellipse. The calculation of these perturbations belongs to the highest department of mathematical science; it is enough for our purpose in this place to mention, that the effects of these perturbations, as well on the epochs as on the motions for days, hours, &c., are given in the tables in columns headed respectively A, B, C, and so on. These numbers A, B, C, ... represent the angular distances of the different planets from the earth, as seen from the sun, the circumference being supposed divided into a thousand parts; they are preceded by a column headed N, which gives the supplement of the moon's node, which is necessary for the calculation of nutation. These columns give us the respective quantities (technically called *arguments*) with which we enter another table, which gives the corresponding effects on the longitude of the sun. All these must be added together to get the total quantity of the perturbations, and then applied to the sun's longitude.

Lastly, it would be necessary to take into account aberration. The solar orbit being very nearly a circle, the aberration of the sun is sensibly constant, and always equal to $20''\cdot25$: it is found, in consequence, more convenient to include it in the epochs of the table: it has been, in forming the epochs, already taken into account. Strictly speaking, the sun's orbit being an ellipse, the aberration varies a little in different parts of the orbit, and a small table shows the quantity to be added for every three degrees of true anomaly: but as the maximum of the variation is only $0''\cdot34$, we need take no further notice of it in this place.

Resuming all that has been said, we find the

True longitude ☉ = mean longitude ☉ + equation of the centre + nutation + perturbations.

The sun's latitude may always be supposed = 0, though, in fact, the perturbations produce a very small latitude, which never amounts to $1''$.

We have now found the sun's apparent position in the heavens, or the direction in which he is viewed: the elliptic theory gives us his distance or radius vector, or rather its logarithm, in terms of the mean anomaly. This will be found tabulated for every degree of mean anomaly, and other tables give the effect of the perturbations upon this logarithm. It must be observed, that here also is a secular variation, forming a column by the side of the logarithms of the radius vector, to be applied, as we have explained, when speaking of the secular variation of the equation of the centre.

The sun's mean horizontal parallax is about $8''\cdot5$: the parallax in altitude may be deduced from it by the usual formulæ. The horizontal parallax varies inversely as the distance from the earth:

* The whole circumference is supposed divided into 1000 parts.
† The formula is $-\ 1''\cdot34\ .\ \sin.\ 2\,A\ .\ \cot.\ \omega$. See page 161.

but its variations are so small as to be quite insensible.

It is of importance to know the sun's apparent semi-diameter: this also varies inversely as the distance from the earth, that is, as the radius vector: it is given in the tables for every degree of mean anomaly (the radius vector itself being a function of the mean distance and mean anomaly): the hour angle subtended by the semi-diameter is equal to the semi-diameter divided by 15 × cosine of the declination. This is placed in an adjoining column.

The motion of the sun in longitude in any given time varies inversely as the square of the distance of the sun from the earth *. This variation is not considerable, as the solar ellipse differs not much from a circle; still, however, it is necessary to take into account the effect produced by the variation of distance. The tables give the motion in an hour, or the horary motion, corresponding to every different degree of mean anomaly; the distance from the earth, or radius vector, being a function of the mean anomaly, and it being more convenient to express the horary motion in terms of the latter than the former quantity.

The longitude of the sun being once known, we may, by the method already explained, calculate his true right ascension and declination. Astronomers suppose a fictitious sun, S″, which describes the equator with an uniform velocity, exactly in the same time that the true sun, S, describes his orbit; both being supposed to start together from the real vernal equinox, and to return to it again at the same moment. This sun, S″, is not to be confounded with the other fictitious sun, S′, before-mentioned, which describes the ecliptic with uniform velocity, but there is this relation between them, that the right ascension of the *mean sun*, as it is called, S″, is always equal to the longitude of S′, which is, as we have before stated in other terms, the mean longitude of the sun. By referring to what has been said, page 193, about S′, it will be seen that this relation must subsist: as both S′ and S″ start from the same point, and describe the one the equator, the other the ecliptic, with an uniform velocity in the same time. The daily returns of S″ to the meridian determine the interval called the mean solar day, and mean time depends absolutely on the right ascension of S″, the mean sun, or rather is identical with it. True time, on the contrary, is measured by the true right ascension of the sun, and the difference between true and mean time, called technically the *equation of time*, is equal to the difference of these two right ascensions.

The equation of time, however, may be represented in another form. It evidently depends upon two causes; the unequal motion of the sun in the ecliptic, and the inclination of that orbit to the equator. The first effect is represented by the equation of the centre; the second by the difference between the longitude measured on the ecliptic and the same when reduced to the equator. Consequently, (neglecting the perturbations,) the equation of time = equation of the centre + the reduction of the true longitude to the equator. As to the reduction to the equator, it may be expressed in terms of this kind—

$$\tan.^2 \frac{\omega}{2} . \sin. 2 \odot - \tan.^4 \frac{\omega}{2} . \sin. 4 \odot$$

where $\odot$ is the true longitude of the sun †.

As the obliquity of the ecliptic is a most important element of the solar theory, it is desirable to explain how it may be determined with the requisite

---

* $r^2 \, dv = c \, dt$

$$\therefore \frac{dv}{dt} = \text{velocity of the earth in its orbit} = \frac{c}{r^2}$$

† If, in Fig. 51, we call ♈ M which is the projection of ♈ S on the equator, α; and ♈ S or the longitude λ, the angle S ♈ M or the obliquity being ω, we have the equation;

$$\tan. \alpha = \cos. \omega . \tan \lambda$$
$$= n . \tan. \lambda \quad \text{calling } \cot. \omega, n,$$
$$\tan. (\lambda - \alpha) = \frac{\tan. \lambda - \tan. \alpha}{1 + \tan \lambda . \tan. \alpha}$$
$$= \frac{(1 - n). \tan. \lambda}{1 + n . \tan.^2 \lambda}$$
$$= \frac{(1 - n). \sin. \lambda . \cos. \lambda}{\cos.^2 \lambda + n . \sin.^2 \lambda}$$

but by trigonometry we have the following series:

$$= \frac{(1 - n). \sin. 2\lambda}{1 + n + \cos. 2\lambda - n . \cos. 2\lambda}$$
$$= \frac{\frac{1-n}{1+n} . \sin. 2\lambda}{1 + \frac{1-n}{1+n} . \cos. 2\lambda}$$
$$= \frac{m . \sin. 2\lambda}{1 + m . \cos. 2\lambda}$$
$$\text{putting } m = \frac{1 - n}{1 + n}$$

degree of precision. If at the instant of the solstice the sun were on the meridian, we should only have to observe the meridian altitude of his centre, correct it for refraction and parallax, then adding or subtracting this altitude from the height of the equator above the horizon, (that is, the complement of the latitude of the place,) according as we are observing the summer or winter solstice, we should have half the obliquity in question. If we observe both solstices, we have only to subtract the meridian altitude at the winter solstice from that at the summer, to get immediately the obliquity. As, however, it will very rarely happen that the sun is on the meridian of the observer at the instant of the solstice, we must have recourse to other methods.

Measure the meridian altitude of the sun daily for several days before and after the solstice; as the latitude of the place of observation is supposed known each of these observations will give us the declination of the sun at the moment of observation: the problem is to conclude from all these declinations, the declination at the instant of the solstice.

Referring to the right-angled spherical triangle ♈ S M, Fig. 51; let S be the place of the summer solstice, ♈ that of the vernal equinox, ♈ S the sun's longitude, S M the sun's declination when near the solstice, S ♈ M the obliquity of the ecliptic. Call S M, Δ; ♈ S, Λ; and S ♈ M, ω. Then $\sin. \Delta = \sin.$ ♈ $\sin. \omega$: and each separate observation gives us an equation of this form. We shall show that from each of these equations the value of the declination at the solstice may be found. We shall thus get several values of this declination, the mean of all of which will be necessarily very exact.

Let Δ be an observed declination which we wish to reduce to the solstice: that is, from which we wish to conclude the declination at the instant of the solstice. Let δ be the difference between them: then the declination at the instant of the solstice, or

$$\omega = \Delta + \delta$$

$$\therefore \omega - \delta = \Delta$$

$$\text{but } \sin. \Delta = \sin. \Lambda \,.\, \sin. \omega$$

$$\therefore \sin. (\omega - \delta) = \sin. \Lambda \,.\, \sin. \omega$$

$$\text{or taking } \lambda = 90^\circ - \Lambda\,^*$$

$$\sin. (\omega - \delta) = \cos. \lambda \,.\, \sin. \omega$$

where, as we are now near the solstice, λ must be a small arc †. Developing this equation

$$\cos. \delta - \cot. \omega \,.\, \sin. \delta = \cos. \lambda = 1 - 2 \,.\, \sin.^2 \tfrac{1}{2} \lambda$$

$$\cot. \omega \sin. \delta = 2 \sin.^2 \frac{\lambda}{2} - 2 \sin.^2 \frac{\delta}{2}$$

$$\therefore \sin. \delta = 2 \tan. \omega \,.\, \sin.^2 \tfrac{1}{2} \lambda - 2 \tan. \omega \,.\, \sin.^2 \frac{\delta}{2} \quad \ldots\ldots (a)$$

Now, as near the solstice δ is very small, the last term of this equation may be neglected, and

$$\sin. \delta = 2 \,.\, \tan. \omega \,.\, \sin.^2 \tfrac{1}{2} \lambda.$$

It is true that this value of δ involves ω, which is the very quantity we are trying to find; but it is to be remarked, that ω is always known very nearly indeed, and we are only trying to get it with the greatest accuracy: as δ itself is a very small quantity, no sensible error can arise from a small uncertainty in the value of ω. However, this value of $\sin. \delta$ is only approximate: to be more exact, we must substitute it for $\sin. \delta$ in the last term of equation (a): or taking

$$\sin. \tfrac{1}{2} \delta = \tfrac{1}{2} \sin. \delta = \tan. \omega \,.\, \sin.^2 \tfrac{1}{2} \lambda$$

we get

$$\lambda - \alpha = m. \sin 2 \lambda - \frac{m^2}{2} . \sin. 4 \lambda + \frac{1}{3} m^3 \,.\, \sin. 6 \lambda \quad \ldots\ldots$$

$$\text{But } m = \frac{1 - \cos. \omega}{1 + \cos. \omega} = \tan.^2 \frac{\omega}{2}$$

$$\therefore \lambda - \alpha = \tan.^2 \frac{\omega}{2} \sin. 2 \lambda - \frac{\tan.^4 \,.\, \frac{\omega}{2}}{2} \,.\, \sin. 4 \lambda \quad \ldots\ldots$$

or expressing the reduction in seconds

$$\lambda - \alpha = \frac{\tan.^2 \frac{\omega}{2}}{\sin. 1''} \sin. 2 \lambda - \frac{\tan.^4 \,.\, \frac{\omega}{2}}{2 \sin. 1''} . \sin. 4 \lambda \quad \ldots\ldots$$

* At the winter solstice we must take $\lambda = 270^\circ - \Lambda$.

† λ should never be greater than 12°.

$$\sin.\delta = 2.\tan.\omega.\sin.^2\tfrac{1}{2}\lambda - 2.\tan.^3\omega.\sin.^4\tfrac{1}{2}\lambda.$$

To simplify this expression, we may put δ instead of sin. δ, and instead of sin. $\frac{1}{2}\lambda$, we may put $\frac{1}{2}\lambda - \frac{1}{48}.\lambda^3$

$$\therefore \delta = \tfrac{1}{2}\tan.\omega.\lambda^2 - \tfrac{1}{24}.\tan.\omega(1 + 3.\tan.^2\omega)\lambda^4 + \&c. \ldots\ldots$$

To express the arcs δ and λ in seconds, we must multiply the terms of the series respectively by sin. 1″. Since ω varies very slowly, we may consider the coefficients of λ^2, λ^4, and so on, as constant: and calling them A, B . . . we have $\delta = A\lambda + B\lambda^2 + \ldots\ldots$ This is the expression for the reduction to the solstice;* ω thus determined is only the apparent obliquity: to get the mean obliquity Ω we must apply the correction for nutation, as has been already explained when treating of that inequality.

The next element of importance to be ascertained is the position of the equinox, and the quantity of its annual motion on the ecliptic. This is easily done when the obliquity of the ecliptic is known. For some days before and after the equinox, his meridian zenith distance must be observed, and hence the declination found daily, just as in the observations for the solstice. Now each observation gives us an equation of the form

$$\sin.\alpha = \frac{\tan.\delta}{\tan.\omega}$$

α being the right ascension of the sun. It is to be observed, that the declination δ must be necessarily very small, and α will be determined with exactness, even if ω be not quite precisely known. Knowing α, or the right ascension of the sun at mid-day, add to it the time of culmination of some given star, then the right ascension of the star fixes the place of the equinox. Several observations of this kind will give the place of the equinox with great exactness, as referred to the fixed star†: when they are repeated after an interval of a few years, it will be seen how much the equinox has changed its place in consequence of precession. Since a knowledge of the latitude of the place is involved in the determination of δ, it is necessary to observe both equinoxes, that errors arising from this and other sources may compensate each other ‡.

The next element to be found is the position of the major axis of the solar ellipse, or of the perigee and apogee of the orbit. These may be found roughly, by observing the diurnal motion, as at the perigee the sun's velocity is greatest, and least at the apogee. But to find these points more exactly, we must remark, that they are exactly 180° from each other, and that the sun takes exactly half a year to pass from one to the other. The union of these two properties characterises exclusively the perigee and apogee of the orbit: the two points in the sun's orbit which satisfy this condition are the points required. For if we draw through the centre of the earth any straight line which is not the axis major of the orbit, this line will intersect the sphere of the heavens in two points 180° from each other; but the time taken by the sun to pass from one to the other, will never be exactly half a tropical year.

We must now determine the excentricity of the orbit. This may be done roughly by observing the greatest and least apparent diameters of the sun: since, the apparent diameter varying in-

---

* Differentiating the equation $\sin.\Delta = \sin.\omega.\sin.A$, with regard to ω and Δ, we have $d\Delta.\cos.\Delta = d\omega.\cos.\omega.\sin.A$: dividing the second equation by the first

$$d\Delta.\cot.\Delta = d\omega.\cot.\omega$$

$$\text{If } d\omega = -1''$$

$$\text{then}.\ d\Delta = -1''.\cot.\omega.\tan.\Delta.$$

We thus get the variation produced in the declination by a diminution 1″ in the obliquity.

† In the course of a few days the change of apparent place of the star from aberration, &c., would be quite insensible; yet, if necessary, it might be allowed for.

‡ To estimate the effect produced on the equinox by a small error on the declination, we must differentiate with regard to α and δ the expression

$$\sin.\alpha = \frac{\tan.\delta}{\tan.\omega}, \text{ and we get}$$

$$\frac{d\delta}{d\alpha} = \tan.\omega.\cos.\alpha.\cos.^2\delta.$$

At one equinox α very nearly = 0, at the other it nearly = 180°; consequently in these two cases cos. α will have different signs, while tan. ω and cos. δ will retain the same sign.

versely as the distance, these observations will give us the ratio of the greatest and least distances from the earth, and hence we have the excentricity of the orbit. But the nature of such observations, and the smallness of the variations in the semi-diameter, prevent this method from having any accuracy. The way in which we must proceed is this. The theory of elliptic motion gives the excentricity in terms of the greatest equation of the centre, that is of the greatest value which the equation of the centre attains *. The problem is then

---

* Call the greatest equation of the centre E, and put $\frac{E}{R} = \epsilon$, where R is the radius expressed in seconds; and let the excentricity be e, then

$$e = \frac{1}{2}\,\epsilon - \frac{11}{768}\,\epsilon^3$$

e being expressed in seconds.

To demonstrate this series, we must set out from the principle, that at the moment of the greatest equation of the centre the true angular motion of the sun is equal to his mean angular motion. Now the latter is proportional to the time described; call it v, and the time, from the origin t, let T be the time of a whole revolution, then evidently

$$\frac{v}{2\pi} = \frac{t}{T} \text{ or } v = 2\pi\,\frac{t}{T}.$$

With regard to the true motion, we know that the arcs described are proportional to the times of describing them. Now if the time be very small, the area traced out by the radius vector is very nearly $\frac{r^2\phi}{2}$; r being the radius vector, and ϕ its angular motion in the short time t. The smaller ϕ is, the more exact does this expression become, so that by diminishing it more and more, the error might be rendered less than any given quantity. Call S the whole surface of the ellipse, T the time employed to describe it; then

$$\frac{r^2\phi}{2\,S} = \frac{t}{T} \text{ or } \phi = \frac{2\,S\,t}{r^2\,T} = \frac{2\pi\,a^2\sqrt{1-e^2}}{r^2\,T}\,.\,t$$

a being the semi-axis major, e the excentricity of the ellipse.

Now ϕ is equal to v at the moment of the greatest equation of the centre, or

$$\frac{2\pi\,t}{T} = \frac{2\pi\,a^2\sqrt{1-e^2}}{r^2\,T}\,.\,t$$

$$\therefore\ r^2 = a^2\sqrt{1-e^2}$$

$$r = a\,(1-e^2)^{\frac{1}{4}}.$$

This equation not involving t, will be equally true, whatever value we attribute to that quantity; and this expression, which we have deduced from a certain supposition with regard to t, will not be merely approximate, but rigorously exact.

Now the polar equation to the ellipse is

$$r = \frac{a\,.\,(1-e^2)}{1+e\,.\,\cos.\,\theta}.$$

Substituting for r the value we have just found, we get

$$1 + e\,.\cos.\,\theta = (1-e^2)^{\frac{3}{4}} \quad\ldots\ldots\ldots\ldots \quad (a)$$

The equation (c) in the note to page 194, gives us,

$$\tan.\,\frac{\theta}{2} = \sqrt{\frac{1+e}{1-e}}\,.\,\tan.\,\frac{u}{2}.$$

This equation may easily be transformed into the following,

$$1 - e\,.\,\cos.\,u = \frac{1-e^2}{1+e\,.\,\cos.\,\theta}.$$

In the particular case we are now treating

$$1 - e\,.\,\cos.\,u = (1-e^2)^{\frac{1}{4}}$$

$$\therefore\ \cos.\,u = \frac{1-(1-e^2)^{\frac{1}{4}}}{e}$$

Put $u' = 90^\circ - u$

$$\sin.\,u' = \frac{1-(1-e^2)^{\frac{1}{4}}}{e},$$

Recurring to equation (a) we get from it,

$$\cos.\,\theta = -\frac{\{1-(1-e^2)\}^{\frac{3}{4}}}{e}$$

and putting $\theta' = \theta - 90^\circ$

$$\sin.\,\theta' = \frac{1-(1-e^2)^{\frac{3}{4}}}{e}.$$

Since e is very small, θ' and u' are small quantities of the same order, we may then, not going beyond terms involving e^2, get reduced to finding this greatest equation of the centre. Now if we refer to *fig.* 52, when the true sun S leaves its perigee P, (simultaneously with the fictitious sun S′, moving uniformly in the ecliptic,) it has at that time its greatest velocity, in consequence of which it soon gets before S′, and continues to gain on it, till its velocity, having reached a maximum, begins again to decrease. It is easy, however, to see, that it will continue to gain on S′, until its velocity has decreased so as to be equal to the velocity of S′, after which S′ will begin in its turn to gain on S. It is then at the moment that the velocity of S equals that of S′, that the angular distance between them will be the greatest. This angular distance is the angle Q T q, or the equation of the centre. The greatest equation of the centre corresponds then to the moment when the velocities of S and S′ are equal: and evidently there will be two such greatest equations, Q T q, and Q′ T q', in every revolution of the sun, one on each side of and symmetrically situated with regard to the major axis; and a little consideration will show, that in both cases the true sun is farther from the perigee than the fictitious sun. If we knew the angles Q T Q′, q T q', their difference would give us Q T q + Q′ T q = 2 Q T q, since the orbit is perfectly symmetrical on each side of its axis. But Q T Q′ is easily known, for supposing ♈ T ♎ to be the line of the equinoxes, it is the difference of the true longitude E T Q′, E T Q observed at the moments of the greatest equation of the centre, which moments are known by the velocity of the true sun being then equal to the uniform velocity. Similarly q T q' is the difference of the mean longitudes. By subtracting then the two true longitudes from the two mean longitudes corresponding to the instants before mentioned, we get twice the greatest equation of the centre.

SECTION II.—*Tables of the Moon.*

The motions of the moon are altogether much more complicated than the apparent motions of the sun. The moon revolves in an ellipse, of which the earth occupies the focus; while the latter body revolves round the sun, it carries along with it the moon, whose elliptic motion round the earth, remains unaltered by this circumstance: but the attraction of the mass of the sun produces great disturbance in the lunar motions. Thus the plane of the orbit oscillates about its mean inclination to the ecliptic, which is about 5°: at the same time the axis major of the ellipse turns in its own plane, at the rate of about 40° annually, while the line of the nodes travels completely round the ecliptic in eighteen years and a half.

The place of the moon is found from the tables in much the same way as that

---

$$\sin.\ \theta' = \frac{3}{4}\ .\ e + \frac{3}{32}\ e^3,\quad \sin.\ u' = \frac{1}{4}\ e + \frac{3}{32}\ e^3.$$

$$\therefore\ \theta' = \frac{3}{4}\ e + \frac{21}{128}\ .\ e^3\ \ldots\ u' = \frac{1}{4}\ e + \frac{37}{384}\ .\ e^3\ \ldots\ldots$$

but $nt = u - e\ .\ \cos.\ u = 90^\circ - u' - e\ .\ \cos.\ u'$

$$\theta - nt = \theta' + u' + e \cos.\ u'$$

$\theta - nt$, in the case which we have been considering, is the greatest equation of the centre: call it E, then

$$E = \theta' + u' + e - 2e\ .\ \sin.^2 \frac{1}{2}\ u\ \ldots\ldots\ldots\ (b)$$

The term $\sin.^2 \frac{1}{2}\ u$ is a very small quantity of the order e^2, and it is multiplied by e: we may then put for it $\frac{1}{4}\ \sin.^2 u' = \frac{1}{16}\ e^2$.

Substituting, then, in equation (b) this value, and those of θ' and u' found just above, we get

$$E = 2e + \frac{88}{384}\ e^3$$

and hence

$$e = \frac{1}{2}\ E - \frac{44}{384}\ e^3.$$

In the second term, put for e, $\frac{E}{2}$, then

$$e = \frac{1}{2}\ E - \frac{11}{768}\ E^3.$$

of the sun. We begin by supposing the satellite to have an uniform and circular motion; that is, we take from the tables the mean longitude for the epoch, and the mean motion for the number of days, hours, minutes, and seconds required. The mean place now ascertained, we must proceed to correct it first for the excentricity, and then for various inequalities, of the most remarkable of which we shall proceed to give some account.

We shall begin by considering those which are called the secular inequalities, to distinguish them from those called periodic, because they have a period sufficiently short, to have been described a great many times since astronomical observations were first made. We have seen the existence of such secular variations in the position of the perigee and the excentricity of the earth's orbit. But the mean motion of the earth is subject to no such acceleration: it is a result of the laws of gravitation, that the major axes, and consequently the periodic times of all the primary planets, are invariable. But this is not the case with regard to the moon. It is found that, when we compare ancient eclipses of the moon, observed upwards of seven hundred years before Christ by the Chaldeans, with recent observations, that they give the moon's mean motion much less than that which results merely from the comparison of modern observations with each other*. The same fact results from the comparison of the Arabian observations with those of modern astronomers. La Place has shown that this acceleration is caused by the secular diminution of the excentricity of the earth's orbit; as that diminution will, in the course of many ages, attain a limit, after which it will begin again to increase, consequently the acceleration of the moon's motion will find its limit at the same time, and be changed into a retardation. But an immense period of time, probably millions of years, will elapse ere this takes place.

The motion of the lunar perigee is subject to a secular *retardation*, depending on the same cause as that which produces the *acceleration* of the mean motion. Consequently the mean anomaly, which is equal to the mean longitude of the moon, minus the mean longitude of the perigee, is subject to a secular equation, which is equal to the difference between the two former.

The motion of the nodes is subject, like that of the perigee, to a secular retardation. Generally, if we call the acceleration of the mean motion 1, that of the perigee will be represented by $-3 \cdot 00052$, and that of the node by $-0 \cdot 73452$.

Of the periodic inequalities the most remarkable is the evection. Its effect is to diminish the equation of the centre in the syzigies, that is when the moon is in conjunction or opposition with the sun; and to increase it in quadratures, that is, when there are 90° of angular distance between these bodies. But the evection does not depend solely upon this angular distance; it also depends upon the distance of the moon from the perigee of her orbit. This equation may be represented generally by the following expression:—

$$1^\circ\, 16'\, 29'' \cdot 6 \,.\, \sin.\{2\,(☾ - ⊙) - A\} + 31'' \cdot 2 \sin.\, 2\,\{2\,(☾ - ⊙) - A\}\dagger$$

where ☾ is the mean longitude of the moon, ⊙ that of the sun, A the mean anomaly of the moon. In the syzigies, when the moon at the same time is in its perigee or apogee, 2 (☾ − ⊙) will be = 0 or 180°, as also will A; in this case the evection will be nothing. It will attain its greatest value when 2 (☾ − ⊙) = 90° or 270°; that is, when the moon is in quadratures, and when at the same time A, or the mean anomaly, = 90°. In this case, it = 1° 16′ 29″·6.

* As Ptolemy, who records these eclipses, gives the day and hour at which the principal phases (the beginning and end) were observed, we may determine the instant when the moon was exactly 180° in longitude from the sun. Now we can easily find the sun's longitude at that time, and hence we can get that of the moon. This is the true longitude; if we correct her place for the periodic inequalities, we get her mean longitude. Take now the mean longitude for some given moment in the modern tables; the difference between these mean longitudes gives us the moon's mean motion on the ecliptic in the interval between them, adding to it the proper number of whole circumferences. The mean motion may also be found by comparing two mean longitudes of modern times, and in this case the mean motion will be found to be much more rapid.

† It may be seen from this formula, that generally in the conjunctions the evection will have a contrary sign to the equation of the centre; for then 2 (☾ − ⊙) being equal to nothing, the argument is reduced to − A, which gives a negative sign if the anomaly is less than 180°, and a positive sign if it is greater; but in the first case, the equation of the centre is to be added to the mean longitude, in the second to be subtracted. It is always then of a contrary sign to the evection. In the oppositions, since 2 (☾ − ⊙) = 360°, the argument is reduced as before to − A, and the same consequence follows. But in the first quadrature, for example, ☾ − ⊙ = 90°, 2 (☾ − ⊙) = 180°, and the argument becomes sin. (180° − A) = sin. A; and so on for the other quadratures.

The next great periodic inequality is the variation, which disappears in syzigies and quadratures, and attains its greatest value when the moon is in octants, that is at 45°, 135°, 225°, or 315° of angular distance from the sun. This inequality depends then on this angular distance; and its period has been discovered to be half a synodical revolution. It may be represented generally by the formula

$$+ 39' 30'' \sin. 2 (☾ - ☉)$$

There is also an inequality known by the name of the annual equation, which has for its period a solar anomalistic year. It is of exactly the same form with the equation of the centre of the sun, but it has an opposite sign. Its analytical expression is $11' 13''\cdot 7$ sine of the mean anomaly of the sun. This inequality has a great analogy with the secular equation, since it is produced by the inequality of the earth's orbit, while the secular equation is caused by the secular variation in that excentricity. This equation might be entirely suppressed if we modified in a suitable way the equation of time, by making the necessary change in the solar equation of the centre, and then calculating the place of the moon with the equation of time thus modified*. It was in this way that the equation was first presented by Tycho Brahe, its discoverer. It is also remarkable that the apparent effect of the annual equation in eclipses is to increase the solar equation of the centre by its own value; and that the Indians, in consequence, who determined the inequalities of the sun and moon merely with reference to eclipses, made the solar equation of the centre too large by the whole value of the annual equation.

The last inequality we shall notice is the equation of long period. This has its name from its period being of 184 years. Its analytical expression is

$$47''\cdot 51 \sin. \{2 ☊ + \varpi - 3 \varphi\}$$

where ☊ is the longitude of the moon's node, ϖ that of her perigee, φ that of the sun's perigee.

Resuming then what has been said, we find that the moon's true longitude = mean longitude + equation of the centre + evection + variation + annual equation + equation of long period + perturbations + nutation.

The moon's latitude is found much in the same way as her longitude. We begin by supposing the inclination of the orbit to the ecliptic invariable. In this case, calling the difference between the true longitude of the moon and that of her node ψ, the latitude θ, and the obliquity ι, then

$$\tan. \theta = \tan. \iota . \sin. \psi$$

or, since θ and ι are very small,

$$\theta = \iota \sin. \psi$$
$$= 5° 8' 59''\cdot 8 \sin. \psi.$$

This is the first term of the expression for the latitude; we must add to it the various inequalities.

One of the most remarkable of these is an inequality which is proportioned to the sine of the true longitude. This equation is produced by the attraction of the terrestrial spheroid, and consists in a nutation of the lunar orbit, which corresponds exactly to the nutation of the earth's axis, one of these being in fact the reaction of the other, and it has given mathematicians a method of determining the compression of the earth's elliptic figure, independent of all local irregularities. This inequality diminishes the inclination of the lunar orbit to the ecliptic when the ascending node coincides with the vernal equinox; it augments it when this node coincides with the autumnal equinox.

The greatest inequality in latitude is one which has for its argument twice the distance of the sun from the node of the lunar orbit, and is proportional to the cosine of that angle. The motion of the lunar node has an inequality depending upon the sine of the same angle. These two inequalities together may be represented by supposing the true pole of the lunar orbit to oscillate round the mean pole in a small ellipse, which it describes in half a revolution of the sun with regard to the lunar node, that is in $178^{d}\cdot 30996$.

The attraction of the sun on the moon being different at different distances causes variations in the lunar orbit of which we have already seen an instance in the annual equation. In fact the nearer the sun is to the moon, its attraction on that body increasing in intensity tends to diminish the action of the earth,

---

* The expression for the solar equation to the centre is $1° 55' 27''\cdot 0$ sin. mean anomaly of the sun; the expression for the annual equation is $-11' 13''\cdot 7$ sin. mean anomaly of the sun. Combining these two, we have 1° 44′ 1[illegible]″·6 sin. mean anomaly of the sun for our fictitious equation of the centre, with which we must calculate the fictitious equation of time. But there is no practical advantage in this complicated mode of proceeding. The annual equation is now tabulated and applied in the same way as all the other corrections.

and to dilate as it were the orbit of the moon; and again, as the sun recedes, that orbit is contracted. It is in this way that the excentricity of the sun's orbit produces the evection, and at the same time it must produce corresponding variations in the radius vector of the lunar orbit. We might take a mean value of the radius vector, and determine by theory the perturbations by which it is affected; but as the horizontal parallax is the immediate result of observation, and the radius vector may always be deduced from it by a simple proportion, it is found more convenient to take a mean value of the parallax, and to calculate the variations produced in it by the change of distance.

This mean value of the parallax, or constant of the parallax, may be determined (supposing the earth's mean radius known) by a knowledge of the mean motion of the moon, and of the intensity of gravity at the earth's surface, which is found by observations of the pendulum. Now as gravity varies inversely as the square of the distance from the centre of the earth, we may calculate at what distance the moon must be from that centre, to have such a mean motion in her orbit as we find she really has. This distance gives us the mean parallax, which is thus found to be 57′ 0″·9. We add to this mean parallax the perturbations, in order to get the horizontal parallax Π at any required time; and this being known, we have the parallax for any given zenith distance ζ equal to $\Pi \sin. \zeta$. It has already been observed (see p. 60) that the earth not being exactly spherical, not only the line drawn from the centre of the earth to the observer will not generally pass through the zenith, but the horizontal parallax is different in different terrestrial latitudes. It is only in the case of the moon, which is so near us, that the variation in the horizontal parallax arising from this cause can be perceived; in the case of that body, however, it is absolutely necessary to take it into account.

Call the horizontal parallax at the equator Π, the radius of the earth at the equator R, D the distance of the moon from the centre of the earth: let the corresponding quantities at the latitude ϕ, be Π′, R′; the distance D will evidently be the same for both. Now

$$D = \frac{R}{\sin. \Pi} = \frac{R'}{\sin. \Pi} \therefore \frac{R}{R'} = \frac{\sin. \Pi}{\sin. \Pi'}$$

or, as Π, and Π′ are small, putting the ratio of the arcs for that of their sines

$$\frac{R}{R'} = \frac{\Pi}{\Pi'} \therefore \Pi' = \Pi . \frac{R'}{R}.$$

Now it may be shown, supposing the earth to be an ellipsoid, that

$$R' = R\left\{1 - p. \sin^2 \phi\right\},$$

where p designates what is called the compression, or the ratio of the difference of the two axes to the axis major. Consequently

$$\Pi' = \Pi. \left\{1 - p. \sin.^2. \phi\right\}$$

Π′, it will be observed, is always less than Π; that is, the equatorial horizontal parallax is the greatest. Knowing this, which we have before stated to be equal to 57′ 0″·9, we can, by the formulæ just given, find the horizontal parallax in any other latitude.

The apparent semi-diameter of the moon may be expressed in terms of the horizontal parallax. For, in the first place,

$$\sin. \Pi' = \frac{R}{D}$$

and calling the apparent semi-diameter Δ, we have

$$\sin. \Delta = \frac{R'}{D}$$

where R′ is the moon's radius. Hence

$$\frac{\sin. \Delta}{\sin. \Pi'} = \frac{R'}{R}$$

or

$$\frac{\Delta}{\Pi'} = \frac{R'}{R}$$

$$\therefore \Delta = \frac{\Pi'. R'}{R} = 0{\cdot}2725. \Pi'$$

This, it will be recollected, is the expression for the horizontal semi-diameter; for, from causes which have been explained in page 61, the apparent semi-diameter increases, the nearer the body approaches to the zenith. It appears from the formulæ there given, that calling the horizontal semi-diameter Δ, that at zenith distance ζ, Δ′, and the horizontal parallax Π′; that

$$\Delta' = \frac{\Delta}{1 - \Pi'. \cos. \zeta}.$$

Hence

$$\Delta = \Delta\left\{1 + \Pi'. \cos. \zeta\right\} \text{ approximately;}$$

$$\Delta' - \Delta = \Delta\left\{1 + \Pi'. \cos. \zeta - 1\right\}$$

$$= \Delta . \Pi'. \cos. \zeta$$

But $\Pi' = \dfrac{\Delta}{0{\cdot}2725}$

$$\therefore \Delta' - \Delta = \frac{\Delta^2 . \cos. \zeta}{0 \cdot 2725}$$

This, then, is the expression for the augmentation of the apparent semi-diameter as the moon rises towards the zenith.

Section III.—*Tables of the Planets.*

We begin in the planetary tables by supposing the motion of the planets round the sun to be circular and uniform; and thus we get, as in the case of the sun and the moon, the epochs, and the mean motions for days, hours, &c. The longitude, too, of the perihelion, with its secular variation, being known, we get the mean anomaly, by subtracting it from the mean longitude. The next step is to calculate the equation of the centre, which is given in terms of the mean anomaly; and its secular variation. We have, then, the true heliocentric longitude of the planet measured in the plane of its own orbit. To get the longitude on the ecliptic, we must add the reduction to the ecliptic (see page 196); and as the planetary orbits are all very slightly inclined to that plane, the first term of the reduction, viz.:—

$$\tan.^2 \frac{\iota}{2} . \sin. 2 (\lambda - \Omega)$$

will always be sufficient; in which expression, ι is the inclination of the orbit, λ the true longitude of the planet measured in its orbit, Ω the longitude of the node measured in the same plane. The latitude is given by the equation

$$\sin. \theta = \sin. \iota . \sin. (\lambda - \Omega).$$

We have now the true heliocentric longitude and latitude, that is, the longitude and latitude as seen from the centre of the sun. It is generally necessary to find the geocentric longitude and latitude; that is, the longitude and latitude as seen from the centre of the earth. To do this, we must know the radius vector of the planet corresponding to the moment for which we have calculated λ and θ. The radius vector is given in the tables in terms of the mean anomaly *, with its perturbations and secular variation, and is therefore known. To find now the geocentric coordinates, we must proceed as follows.

Fig. 53.

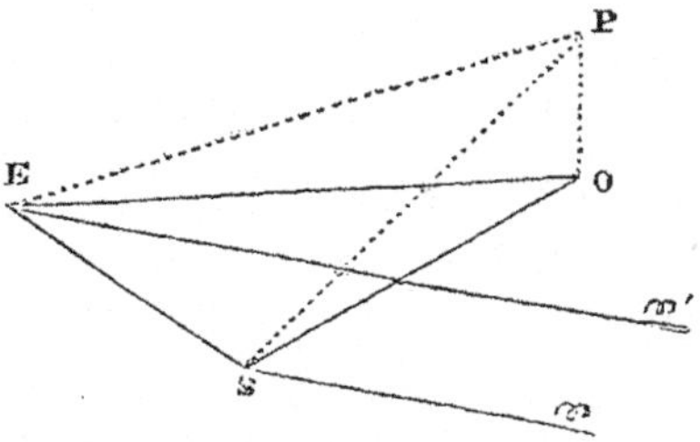

In *fig.* 53, let P be the place of the planet, S that of the sun, E that of the earth, O the projection of the place of the planet on the plane of the ecliptic: then P O is perpendicular to O E and O S, which are called the *curtate distances* of the planet from the earth and sun respectively. Let S ♈ be the line of the equinoxes: a line drawn from E to the place of the equinox in the celestial sphere will be parallel to S ♈, since the distance from the earth to the sun vanishes when compared with the distance of the fixed stars. ♈ S O will be the heliocentric longitude λ; γ' E O the geocentric longitude λ'; O S P the heliocentric latitude θ; O E P the geocentric latitude, θ'. The angle O S E is technically called the *commutation;* the angle S O E the *annual parallax;* O E S the *elongation.*

* By the equations of elliptic motion, page 193,

$$\frac{r}{a} = 1 - e . \cos. u$$

u being the excentric anomaly.

Now $$u = n t + e \sin. n t + \frac{e^2}{2} . \sin. 2 n t \; . \; . \; . \; .$$

Hence $$\frac{r}{a} = 1 + \frac{e^2}{2} - e . \cos. n t - \frac{e^2}{2} \cos. 2 n t \; . \; . \; . \; .$$

Now O S E = ♈ S E − ♈ S O = difference of the longitudes of the earth and planet. But the longitude of the earth = 180° + longitude of the sun = 180° + ☉

∴ commutation = 180° + ☉ − λ

in which equation ☉ and λ are known, and hence the commutation is known.

Now, in the triangle O E S, we know the sides S E = R, S O = S P. cos. $\theta = r$. cos. θ, and the comprised angle S, or the commutation. By Napier's analogies,

$$S O + S E : S O - S E :: \tan. \tfrac{1}{2}(E + O) : \tan. \tfrac{1}{2}(E - O)$$

$$\text{But } S + O + E = 180^\circ \qquad \therefore \tfrac{1}{2}(O + E) = 90^\circ - \frac{S}{2}$$

$$\therefore \frac{\tan. \frac{1}{2}(E - O)}{\cot. \frac{S}{2}} = \frac{r. \cos. \theta - R}{r. \cos. \theta + R}$$

Assume
$$\tan. \chi = \frac{r \cos. \theta}{R}$$

then
$$\frac{\tan. \frac{1}{2}(E + O)}{\cot. \frac{S}{2}} = \frac{\tan. \chi - 1}{\tan. \chi + 1}$$

$$= \tan. (\chi - 45^\circ)$$

$$\text{Assume } \xi = \tfrac{1}{2}(E - O) \text{. whence } E = 90^\circ + \xi - \frac{S}{2}$$

$$\therefore \tan. \xi = \tan. (\chi - 45^\circ). \cot. \frac{S}{2} .$$

but S and χ being known by what precedes, we thus find tan. ξ, and knowing ξ, we have E; for since $\frac{1}{2}$ (E − O) and $\frac{1}{2}$ (E + O) are now known, E and O are easily found.

But
$$\lambda' = ♈' E O = S E O - S E ♈' = E - S E ♈'$$
$$= E - 180^\circ + \text{longitude of the earth}$$
$$= E - 180^\circ + \oplus$$
$$= E - 360^\circ + ☉$$

$$\therefore \lambda' = E + ☉$$

Now
$$P O = E O . \tan. \theta' = S O . \tan. \theta$$

$$\frac{\tan. \theta'}{\tan. \theta} = \frac{S O}{E O}$$

$$= \frac{\sin. E}{\sin. S}$$

$$\therefore \tan. \theta' = \tan. \theta. \frac{\sin. E}{\sin. S}$$

We shall now proceed to say something on the methods by which the elements of the orbit of a planet are determined. These elements are five in number. 1. The inclination of the plane of the orbit to the ecliptic. 2. The longitude of the node of the orbit. 3. The longitude of the perihelion. 4. The excentricity of the orbit. 5. The axis major.

To determine the longitude of the node, it is necessary to proceed as follows. The right ascension and declination of the planet being known, we deduce from them its (geocentric) longitude and latitude. From such observations we determine the instant when the latitude is equal to nothing, and the geocentric longitude corresponding to that moment.

Fig. 54.

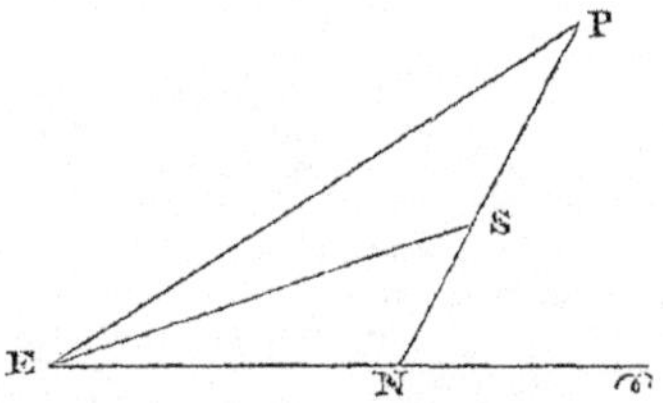

At the moment that the planet is in the plane of the ecliptic, that is, in its node, let its place be P, that of the sun S, of the earth E; then P S N will be the line of the nodes. Calling ♈ the equinoctial point from which the longitudes are counted, ♈ E S will be the geocentric longitude of the sun, which we shall call $\odot$; ♈ T P the geocentric longitude of the planet, which we shall call λ; ♈ N S the heliocentric longitude of the node, which we shall call ν. Call also S P, r; S E, R; then

$$R. \sin. (\lambda - \odot) = r. \sin. (\nu - \lambda)$$

Let us wait now till, after one revolution, the planet passes again through the same node: now the motion of the perihelion being very slow, may be neglected in the time of one revolution (at least in a first approximation); r then, corresponding to the same node, may be supposed to correspond with the same point of the orbit and to be constant. The distance, however, from the sun to the earth will be different; this is known by the tables; call it R′: we have then

$$R'. \sin. (\lambda' - \odot') = r. \sin. (\nu - \lambda')$$

Hence

$$\frac{R'. \sin. (\lambda' - \odot')}{R. \sin. (\lambda - \odot)} = \frac{\sin. (\nu - \lambda')}{\sin. (\nu - \lambda)}$$

$$\therefore \tan. \nu = \frac{R'. \sin. \lambda. \sin. (\lambda' - \odot') - R. \sin. \lambda'. \sin. (\lambda - \odot)}{R'. \cos. \lambda. \sin. (\lambda' - \odot') - R. \cos. \lambda'. \sin. (\lambda - \odot)}$$

To determine ν or the longitude of the node, we have made use of two passages through the same node; but if the orbit be supposed circular, which in a first approximation is allowable, we might employ two passages through the two opposite nodes: in this case as before r may be supposed constant. We have then

$$R. \sin. (\lambda - \odot) = r. \sin. (\nu - \lambda)$$

by the first observation. At the second observation

$$R'. \sin. (\lambda' - \odot) = r. \sin. (\nu' - \lambda')$$

But since the nodes are opposed to each other $\nu' = \nu + 180^\circ$

$$\therefore R'. \sin. (\lambda' - \odot) = r. \sin. (\nu - \lambda')$$

$$\therefore \tan. \nu = \frac{R'. \sin. \lambda. \sin. (\lambda' - \odot') + R. \sin. \lambda'. \sin. (\lambda - \odot)}{R'. \cos. \lambda. \sin. (\lambda' - \odot') + R. \cos. \lambda'. \sin. (\lambda - \odot)}$$

Either this equation, or the former, gives us a value of ν: repeating the operation several times, we get several values of this angle, and we shall find it, in fact, pretty nearly constant. The nodes of the respective planetary orbits have very small motions, which may, in determining approximately the elliptic elements, be altogether overlooked.

In the case of the two inferior planets Venus and Mercury, which from the shortness of their revolutions pass frequently through their nodes, it is desirable to select those passages which take place, when at the same time the planet is in inferior conjunction. At these times, the planet is seen to pass over the sun's disk, and it is easy to determine the latitude at the instant of conjunction; but this latitude and the inclination of the orbit being known, the solution of a right-angled spherical triangle will give us the difference of longitude between the node and the sun's centre: and hence we deduce the longitude of the node itself.

To recur to the superior planets: it will in reality hardly ever happen that we observe the planet exactly at the moment that it has no latitude; but in the present state of astronomy, the motions of the planets can always be considered (like the rest of their elements) as very nearly known; our object, in fact, is to determine the elements with greater exactness. The error in a few days certainly cannot be sensible; with then the motion as approximately known, we reduce to the ecliptic the observations made very near the node, each of which will give us a value of the longitude of the node, and the mean of all these longitudes will give pretty exactly the true value.

The longitude of the node being determined, the next thing is to find the inclination of the orbit. For this purpose it is necessary to wait for the moment that the sun passes through the node of the planetary orbit. Suppose at this moment the sun to be at S, which is on the line E N passing through the node: let P be the place of the planet, M its projection on the

Fig. 55.

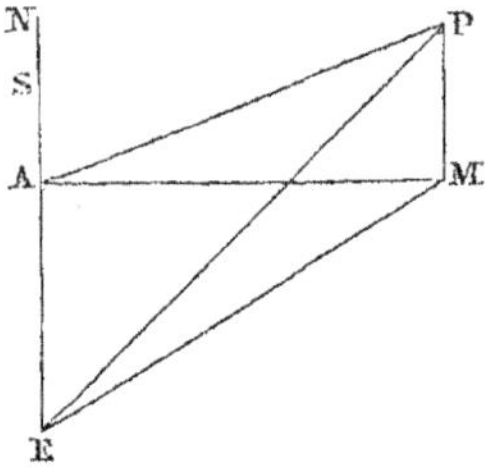

plane of the ecliptic: draw M A perpendicular to E N, and join E M, E P, A P. The angle P A M will represent the inclination of the orbit; call it ι: A E M will be the difference of longitudes between the planet and the node, which difference is known; call it φ:

$$PM = AM . \tan. \iota = EM . \tan. \theta$$

θ being the angle P E M, or the geocentric latitude of the planet.

$$\therefore \tan. \iota = \frac{EM}{AM} \cdot \tan. \theta$$

But $$AM = EM . \sin. \varphi$$

$$\therefore \tan. \iota = \frac{\tan. \theta}{\sin. \varphi}$$

It is, of course, impossible to seize exactly the instant of the sun's passage through the node; but as the motion of this body in longitude is pretty exactly known, it is very easy, if we observe him several times near the node, to find exactly by interpolation the instant at which he must have been in that point. The planet having been observed in the same way, we determine by interpolation the longitude and latitude it must have had at the same instant. It is desirable, as much as possible, to observe the planet in quadratures: at this time φ is nearly 90°, and a small error on its value will very little affect ι.

The longitude of the node, and the inclination of the orbit, determine the position of the plane in which the planet moves: we have now to determine the elements of the ellipse, lying in that plane, which it describes. This ellipse will be determined if we know the axis major, (or, what is the same thing, the periodic time*,) the longitude of the perihelion†, and the excentricity. The first of these, the periodic time, is found by observing the interval of time that elapses between two consecutive passages of the planet through the same node: or still better, by taking two passages through the same node at very considerable intervals of time, embracing several complete revolutions. Of course, in this case, it is necessary to make the requisite allowances for the change of position in the ecliptic, and also for the small secular variation in the place of the planet's node. The periodic inequalities having compensated themselves mutually several times within this interval, what remains of them becomes insensible when spread over so large an interval. This method is exactly similar to that by which the length of the tropical year is determined, from the passage of the sun through the equinoxes.

We have now to find the longitude of the perihelion and the excentricity of the ellipse. The method most frequently pursued is, to observe the planet in its oppositions and conjunctions. Let us consider the oppositions, which are now generally used for the superior planets.

* By the third law of Kepler, the squares of the periodic times are as the cubes of the greater axes.

† Or aphelion. Some tables count the anomaly from one of these points, and some from the other.

Fig. 56.

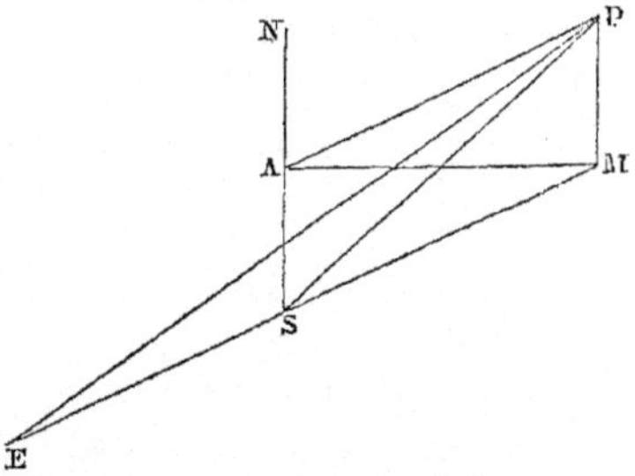

Let P be the planet in opposition, M its projection on the ecliptic, E the earth, S the sun, S N the line of the nodes, M A perpendicular to S N. At this moment the heliocentric longitude of the planet is equal to 180°—its geocentric longitude; and the latter being known as well as the longitude of the node, we have the angle A S M = φ

Call S P, the radius vector, r; and the angle N S P or the angular distance on the orbit of the planet from its node, ψ, and the geocentric latitude of the planet θ; its heliocentric latitude P S M, γ:
then

$$PM = SM . \tan. \gamma = AM . \tan. \iota$$

$$\therefore \frac{AM}{SM} = \frac{\tan. \gamma}{\tan. \iota}$$

But

$$AM = SM . \sin. \varphi$$

$$\sin. \varphi = \frac{\tan. \gamma}{\tan. \iota}$$

$$\text{or, } \tan. \gamma = \tan. \iota . \sin. \varphi$$

Hence γ is known.

Again,

$$PM = SM . \tan. \gamma = EM . \tan. \theta$$

$$\therefore \frac{SM}{EM} = \frac{\tan. \theta}{\tan. \gamma}$$

$$\therefore SM = EM . \frac{\tan. \theta}{\tan. \gamma}$$

$$= \frac{R . \tan. \theta}{\tan. \gamma}$$

Calling E M, or the distance from the sun to the earth which is known, R;

$$\text{But } SP . \cos. \gamma = SM$$

$$\text{or } r . \cos. \gamma = SM$$

$$= \frac{R . \tan. \theta}{\tan. \gamma}$$

$$r = R . \tan. \theta . \sin. \gamma$$

Hence r is known.

Again,

$$AS = SM . \cos. \varphi = SP . \cos. \psi$$

$$\therefore \frac{SM}{SP} = \frac{\cos. \psi}{\cos. \varphi}$$

$$\therefore \cos. \gamma = \frac{\cos. \psi}{\cos. \varphi}$$

$$\therefore \cos. \psi = \cos. \gamma . \cos. \varphi$$

And hence ψ, or the angle A S P, is known.

By means then of these calculations, each observed opposition of the planet gives us the radius vector at that moment, and the distance of the planet from its node, measured on the orbit. As the successive oppositions take place in different parts of the heavens, and consequently, in different parts of the planet's orbit, these oppositions will give us three observed values of the radius vector, and the three angles made by the radius vector with the line of the nodes.

These three angles would be the heliocentric longitudes reduced to the orbit, if the longitudes were counted from the line of the nodes. Now, instead of that, they are counted from a line drawn through the centre of the sun to the vernal equinox, supposed at an infinite distance on the celestial sphere. Suppose a plane perpendicular to the orbit to pass through the sun and the equinox: the intersection of this plane with the orbit will be the line whence the heliocentric longitudes on the orbit are counted. The angle between this intersection and the node is the hypothenuse of a right angled spherical triangle, of which one of the sides is the longitude of the node ν, and the acute angle adjacent to ν, is ι, or the inclination of the orbit to the ecliptic. Hence, calling the hypothenuse β, we have

$$\tan. \beta = \frac{\tan. \nu}{\cos. \iota}.$$

This constant angle β is to be added to each of the distances of the planet from its node, and then we shall have the three heliocentric longitudes reduced to the plane of the orbit. We shall call them υ, υ', υ''.

Let us call the longitude of the perihelion ϖ; then $(\upsilon-\varpi)$, $(\upsilon'-\varpi)$, $(\upsilon'''-\varpi)$ will be the three true anomalies at the three observations. Applying the inverse method of series or La Grange's Theorem to the equation (1) in the note to page 194, we have, limiting ourselves to the first power of the excentricity,

$$nt = \upsilon - \varpi - 2e \,.\, \sin.\,(\upsilon - \varpi)$$

Similarly $nt' = \upsilon' - \varpi - 2e \,.\, \sin.\,(\upsilon' - \varpi)$

and $nt'' = \upsilon'' - \varpi - 2e \,.\, \sin.\,(\upsilon'' - \varpi)$

Hence $n \,.\, (t' - t) = \upsilon' - \upsilon - 2e\{\sin.\,(\upsilon' - \varpi) - \sin.\,(\upsilon - \varpi)\}$

$$n \,.\, (t'' - t) = \upsilon'' - \upsilon - 2e\{\sin.\,(\upsilon'' - \varpi) - \sin.\,(\upsilon - \varpi)\}$$

Here, be it observed that $n = \frac{2\varpi}{T}$ where T is the periodic time and is known: that $(t'-t)$, and $(t''-t)$ are the intervals of time between the first and second, and first and third observations. Let us call then

$$n\,(t'-t) = p \qquad n\,(t''-t) = p' \qquad \upsilon' - \upsilon = \sigma \qquad \upsilon'' - \upsilon = \sigma'$$

$$p = \sigma - 2e\{\sin.\,(\sigma + \upsilon - \varpi) - \sin.\,(\upsilon - \varpi)\}$$

$$p' = \sigma' - 2e\{\sin.\,(\sigma' + \upsilon - \varpi) - \sin.\,(\upsilon - \varpi)\}$$

$$p - \sigma = -2e\{\sin.\,\sigma \,.\, \cos.\,(\upsilon - \varpi) + \cos.\,(\sigma - 1) \,.\, \sin.\,(\upsilon - \varpi)\}$$

$$p' - \sigma' = -2e\{\sin.\,\sigma' \,.\, \cos.\,(\upsilon - \varpi) + \cos.\,(\sigma' - 1) \,.\, \sin.\,(\upsilon - \varpi)\}$$

$$\therefore \frac{p - \sigma}{p' - \sigma'} = \frac{\sin.\,\sigma - 2 \,.\, \sin.^2 \tfrac{1}{2}\sigma \,.\, \tan.\,(\upsilon - \varpi)}{\sin.\,\sigma' - 2 \,.\, \sin.^2 \tfrac{1}{2}\sigma' \,.\, \tan.\,(\upsilon - \varpi)}$$

$$\therefore \tan.\,(\upsilon - \varpi) = \frac{(p' - \sigma') \,.\, \sin.\,\sigma - (p - \sigma)\,\sin.\,\sigma'}{2\,(p' - \sigma') \,.\, \sin.^2 \tfrac{1}{2}\sigma - 2\,(p - \sigma) \,.\, \sin.^2 \tfrac{1}{2}\sigma'}$$

this formula gives us $(\upsilon - \varpi)$, and since υ is known, ϖ, or the longitude of the perihelion.

For the excentricity

$$2e = -\frac{p' - \sigma'}{\sin.\,(\sigma' + \upsilon - \varpi) - \sin.\,(\upsilon - \varpi)}$$

$$\text{or } e = -\frac{p' - \sigma'}{4\sin.\,\tfrac{1}{2}.\,\sigma' .\, \cos.\,(\tfrac{1}{2}\sigma' + \upsilon - \varpi)}$$

Knowing e and $\upsilon - \varpi$, we have

$$nt = (\upsilon - \varpi) - 2e \,.\, \sin.\,(\upsilon - \varpi)$$

where all is known but t. But t being thus determined gives us the time of the passage through the perihelion.

By the methods above given, the different elements of a planet's orbit are determined. Astronomers, however, are not satisfied with one single determination of this kind, however accurate. They are in the habit of constantly making observations, in order, by means of those observations to correct the value of the elements already found. It is to this continual revision and amelioration that the planetary tables owe their present great exactitude.

The manner in which a number of observations are combined in order to correct simultaneously all the elements of the orbit, deserves a particular explanation.

It will be observed, that though, in the methods hitherto followed, we have endeavoured, as much as possible, to determine each element at a moment when it is least affected by any error on the magnitude of the others, yet that we have not been able completely to isolate any one of them. Thus, the inclination of the orbit is affected to a certain extent by an error in the longitude of the nodes; and again, a mistake in this longitude, or in the inclination, would affect the place of the perihelion and the excentricity of the ellipses. All these elements and their secular variations, exercise a reciprocal influence upon each other, and it is only by correcting them all simultaneously, that this influence can be destroyed. Another important condition is, that, in the correction of the elements, a great number of observations should be employed, as thus the inevitable errors of these observations compensate each other, and the results deduced will possess the greatest accuracy.

Let us suppose now, that the place of a planet for a certain moment is calculated, with the approximate value of the elements which we possess; let us suppose also, that the place of the planet at the same moment is actually observed,—there will, of course, be between the calculated and observed place a difference, which difference is called the *error of the Tables.* Now, as we are certain that there is no mistake in the form of the tables, that is, in the mathematical formulæ on which they are founded, the error must arise entirely from mistakes in the values of the different elements. This error then is a function of all the errors on the elements; but as these latter are very small, we may make use of a principle that we have already employed, and consider the total error as the sum of all the partial errors produced by mistakes on the value of the elements*. We calculate then, separately, the effect produced by a small indeterminate error on each of them; we add all those terms together, and equate their sum to the observed error of the Tables. We thus get an equation in which there are as many unknown quantities as there are indeterminate errors, that is, as there are elements to be corrected.†

A second observation gives us a second equation of the same kind, the unknown quantities remaining the same, and in the same way we may obtain any number of equations that we may want. Strictly speaking, we need not have more equations than elements; but, for the reasons above stated, it is desirable to employ a great many more. The question then arises, what is the most advantageous mode of combining them.

The most exact and generally useful method for this purpose, is that called the method of *least squares.* Call e the error of the tables given by one observation; e' that given by a second, e'' by a third, and so on. Let $x, y, z,$ be the unknown errors of the elements. Let $a\,x, b\,y, c\,z \ldots$ represent the effects produced separately upon the place of the planet by the respective errors, $x, y, z \ldots$ in the first observation: $a'x, b'y, c'z \ldots$ in the second; and so on. $a, b, c, a', b', c' \ldots$ will be known coefficients, then

$$e = a\,x + b\,y + c\,z \ldots$$
$$e' = a'\,x + b'\,y + c'\,z \ldots$$
$$e'' = a''\,x + b''\,y + c''\,z$$
$$\cdot \quad \cdot \quad \cdot \quad \cdot \quad \cdot$$

each observation will give an equation of condition, in which the coefficients $a, b, c \ldots$ will be different, but the error on each element remaining the same, $x, y, z \ldots$ are constant.

Square each of these equations, and then add them all together, we have, writing down merely the terms involving x, since the others are exactly of the same form.

$$e^2 + e'^2 + e''^2 + \ldots = (a^2 + a'^2 + a''^2 \ldots)\,x^2 + 2\,x\,y\,(a\,b + a'\,b' + .) + 2\,x\,z\,(a\,c + a'\,c' + \ldots$$

Now this equation is of the form $S = M\,x^2 + N\,x + P, +$ &c., to make it a minimum, its first differential coefficient must be equal to nothing, $\therefore M\,x + N = O$. Thus, considering x alone,

$$x\,(a^2 + a'^2 + \ldots) + y\,(a\,b + a'\,b' + \ldots) + z\,(a\,c + a'\,c' + \ldots) + \ldots = O$$
$$\text{or } a\,(a\,x + by + c\,z \ldots) + a'\,(a'\,x + b'\,y + c'\,z) + \ldots = O$$

That is, to form the equation that gives a minimum for any one of the unknown quantities, multiply each equation of condition by the coefficient of the unknown quantity in that equation, taken with its proper sign, and then add together all these products. For each unknown quantity we must proceed in the same way*, so that we get ultimately

* The total variation arising from a number of *small* simultaneous variations is equal to the sum of these variations taken separately. It is thus that in investigating the formulæ for precession and nutation, page 159, we have calculated, separately, the effects produced by a change in the obliquity, and a change in the longitude, and then added our results together to get the effect of a simultaneous variation in those quantities.

† This equation is called an *equation of condition.*

* The theory of the maxima and minima of several variables, shows us that the condition for the minimum is to be satisfied separately with regard to each.

just as many equations as unknown quantities, from which equations the latter must be found by the ordinary methods of elimination.

To explain more completely this important subject, we shall give an example of the application of the method of equations of condition, and of their combination by the principle of least squares, to the simultaneous correction of the longitude of the perihelion, and eccentricity of the orbit of a planet *.

Let us calculate from the tables the following quantities, Λ, the mean longitude of the perihelion, reduced to the orbit, ϕ the mean anomaly counted from the perihelion; then $2\,e \sin.\,\phi$ will be the equation of the centre (limiting ourselves to one term). Now, calling λ the longitude in the orbit,

$$\lambda = \Lambda + \phi + 2\,e \sin.\,\phi$$

Let x be the unknown correction to be applied to the epoch, z that to the eccentricity; then the real elements will be

$$\Lambda + x;\ 2\,(\epsilon + z) \sin.\,\phi.$$

These elements, substituted in the general expression for λ, will give us the real value of the heliocentric longitude, or λ'. Let us suppose that λ' is observed: we have calculated λ, consequently we know $\lambda' - \lambda$; let us call it e. But, on the other hand,

$$\lambda' = \Lambda + x + 2\,(\epsilon + z) \sin.\,\phi$$
$$\lambda = \Lambda + 2\,e.\sin.\,\phi$$
$$\therefore \lambda' - \lambda = e = x + 2\,z.\sin.\,\phi$$

It will be observed, that in this expression for e, everything is known except x and z.

A second observation gives us the equation,

$$e' = x + 2\,z \sin.\,\phi'$$

where, as before, ϕ' is known. In this way we may collect any number of equations to determine x, z. Let us suppose, for example, that the other elements of the orbit, excepting the longitude of the perihelion and the excentricity being known exactly, that four oppositions of Jupiter have been observed, and that from each opposition his heliocentric longitude has been deduced, as explained before in page 207. Let us suppose that, for each of these four instants of time, his longitude has been calculated from the tables, and compared with the observations, so that the four respective errors of the tables are known. In this way let us suppose that the four following equations have been formed:

$$x + 0{\cdot}98\,z = -\,11''{\cdot}8$$
$$x - 0{\cdot}99\,z = -\,12''{\cdot}4$$
$$x + 0{\cdot}57\,z = -\,14''{\cdot}7$$
$$x - 0{\cdot}85\,z = +\,15''{\cdot}4$$

We shall proceed to combine these equations by the method of least squares. The coefficient of x in each equation is unity; we have then only to add all these equations together for x, and we get

$$4\,x -.\,29\,z = -\,54''{\cdot}3$$
$$\therefore x -.\,07\,z = -\,13''{\cdot}7 \quad \ldots (1)$$

To get the equation for z, we must multiply each equation by the coefficient of z contained in it. Hence we obtain

$$+\,0{\cdot}98\,x + 0{\cdot}96\,z = -\,11''{\cdot}5$$
$$-\,0{\cdot}99\,x + 0{\cdot}98\,z = +\,11''{\cdot}2$$
$$0{\cdot}57\,x + 0{\cdot}32\,z = -\,\ \ 8''{\cdot}3$$
$$-\,0{\cdot}85\,x + 0{\cdot}72\,z = +\,13''{\cdot}0$$
$$-\,0{\cdot}29\,x + 2{\cdot}9\ \ z = +\,\ \ 4''{\cdot}4$$
$$\therefore x - 10\,z = -\,18''{\cdot}1 \quad \ldots (2)$$

We have now to eliminate by the ordinary methods x and z between the equations (1) and (2), and we get

$$x = -\,13''{\cdot}38$$
$$z = +\,\ \ 0''{\cdot}45$$

Hence we find that the longitude of the perigee is to be diminished by $13''{\cdot}38$, and the excentricity to be increased by $0''{\cdot}45$.

The method of equations of condition is of continual application in astronomy, nor is it confined merely to correcting the solar or planetary tables. To give an example of another kind, we may refer to what has been said, page 164, on the method of determining the constant of nutation by observations of the pole star, supposing the aberration known. But it is possible to determine simultaneously by these observations the constant of aberration, as well as that of nutation; and this process is preferable to the other. There can be no

* In this example, which is given merely as an illustration of the method, we suppose known, not merely the longitude of the node and the inclination of the orbit, which only fix the position of the plane of the orbit, but also the mean longitude and the mean motion. To proceed correctly, we ought to correct the two latter simultaneously with the longitude of the perigee and the excentricity. As, however, this would lead to long calculations, we have avoided it, wishing only to give an idea of the nature of the method.

doubt about the correctness of the form of the expressions for the nutation and aberration in declination, and the places of the stars are always sufficiently well known for calculating these small quantities, but there may be a trifling uncertainty on the amount of the constant coefficients. If n be the constant of aberration, and P the factor by which it is multiplied to express the effect on the declination, then each observed altitude of the pole star gives us an equation of the form $\Delta = \delta - m\,Q - nP$, and subsequently

$$\delta - \delta' = m\,.\,(Q - Q') + n\,.\,(P - P')$$

we can get any number of equations of this form, and then combine them by the method of least squares, whence we determine simultaneously m and n.

$$P = \frac{\cos.\,\alpha\,.\sin.\,\Delta}{\cos.\,\phi}\,.\,\sin.\,(\odot + \phi)$$

where ϕ is an auxiliary angle involving only the right ascension, declination, and obliquity of the ecliptic, and may therefore be considered constant. It is desirable that $P - P'$ which becomes ultimately a divisor, should be as large as possible, to diminish the effect of any unavoidable error on δ and δ'; since ϕ is constant, this will be the case when $\odot = \odot' + 180°$, or nearly so, that is when the observations are made at an interval of six months. This can only be done on stars not too far from the pole: on the other hand, the altitude must be considerable to avoid any uncertainty on the refraction: these considerations limit us either to the pole star itself, or circumpolar stars on their superior passage above the meridian.

When we determine the constants of aberration and nutation simultaneously, it is better to select the pole-star, for this reason, $Q - Q'$ must be as great as possible; now Q is of the form

$$n\,.\,\frac{\sin.\,\alpha}{\sin.\,\phi'}\,\sin.\,(\Omega - \phi)$$

where ϕ is a constant angle; we must then, to determine the nutation most favourably, observe when $\Omega' = \Omega + 180°$, or at intervals of half a revolution of the lunar nodes, while the aberration requires other observations at intervals of six months. Now the polar star may always be observed both above and below the pole; and in Europe its altitude is sufficiently great to avoid any errors on the refraction.

We shall give one more illustration of a different manner of employing a number of equations, which will serve at the same time to illustrate the manner in which the constant of refraction is determined. From optical and physical considerations exclusively, La Place has deduced an expression for the refraction in terms of the zenith distance, which has already been given in page 50. He has also presented this expression in a form more convenient for our present purposes *. Call r the refraction, n the unknown constant to be determined, ζ the zenith distance, h the height of the mercury in the barometer, (corrected for the effects of the temperature on the mercury,) q a known constant expressing the expansion of the air for every degree of Fahrenheit's thermometer †, t the number of degrees above 50 marked by the thermometer,

$$r = \frac{n\,.\,h\,.\,\tan.\,\zeta}{\overset{\text{inch.}}{29{\cdot}6}\,\{1 + q\,t\}} + \frac{\tfrac{1}{2}\,n^2\,.\,h^2}{\{\overset{\text{inch.}}{29{\cdot}6}\,(1 + q\,t)\}^2}\,.\,\frac{(1 + 2\,\cos.^2\,\zeta)\,\tan.\,\zeta}{\cos.^2\,\zeta}$$

$$- \frac{n\,.\,h\,.\,0{\cdot}00125254\,\tan.\,\zeta}{\overset{\text{inch.}}{29{\cdot}6}\,\cos.^2\,\zeta}$$

When the zenith distance of a star has been observed, all in this expression is known but n, which may be determined by taking the meridian altitudes of circumpolar stars above and below the pole, as follows:—The expression for the refraction we shall call shortly $A\,n + B\,n^2$; A and B being known coefficients. The true zenith distance equals the apparent zenith distance + the refraction $= \zeta + A\,n + B\,n^2$. Let us suppose that ζ is the zenith distance observed at the superior meridian passage of a circumpolar star; call ζ' the apparent zenith distance at the inferior passage, then the true zenith distance on this latter occasion

$$= \zeta' + A'n + B'n^2.$$

But half the sum of the true zenith distances will give the true co-latitude; call this ϕ; then

* See Mecanique Céleste, vol. iv. p. 271.

† For the value of q, see page 51.

$$\zeta + \zeta' + (A + A')\,n + (B + B')\,n^2 = 2\phi \quad \ldots \ldots \quad (a)$$

Now if ϕ were known n might be found by the solution of a quadratic equation, but in fact ϕ cannot be known *quite exactly* without a knowledge of the refraction; it is then desirable to determine at once n and ϕ. Now every circumpolar star that we observe will give us an equation similar to (a). Rigorously speaking, then, two equations would give us the values of n and ϕ. It is desirable, however, to observe a greater number, if possible; and each star gives an equation of the form

$$M + Pn + Qn^2 = 2\,\phi,$$

where M, P, and Q, are known. Now n is always very nearly known, and the coefficient of n^2 is extremely small; the approximate value then of n will produce no sensible error on n^2. Substituting this approximate value, then, of n in n^2, our system of equations will be one of linear equations between n and ϕ, which quantities may be determined by the methods above explained. If we wish to be very accurate, we may substitute in n^2 the value of n thus obtained, and recommence our calculations, but this would be an unnecessary refinement.

Chapter XI.

Terrestrial Longitudes and Latitudes.

Section I.—*Determination of Terrestrial Latitudes.*

The determination of terrestrial longitudes and latitudes by means of astronomical observations, forms the most important application of the science. It is unnecessary to dwell upon the utility of these determinations in navigation and geography; we shall proceed at once to consider the various methods by which the latitude may be found. Generally, if we observe at any time the altitude of a known star, or of the centre of the sun or a planet, the latitude may be found by the solution of a spherical triangle, in which, besides the zenith distance, the north polar distance is known, as well as the hour angle, which is opposed to the zenith distance; we have to determine from these the third side, which is the distance from the zenith to the pole, or the co-latitude. But it is desirable, in practice, always to select those moments for observation which tend to give the most accurate result. Now when a star is near the meridian, its altitude varies very slowly, and consequently the altitude on the meridian may be found very accurately; when this altitude is known, we have no occasion to go through the process of solving a spherical triangle, we shall obtain the latitude by simple addition or subtraction. The latitude is then always most simply found by observations made on or very near the meridian. There are several modes of doing this, which we shall discuss *seriatim*.

1. By simply observing the altitude, when *on* the meridian, of the sun, a planet, or a fixed star. The altitude thus obtained is the apparent altitude to which we must apply the necessary corrections. In the case of a fixed star, the only correction necessary is for the refraction. The refraction, we know, varies as the sum of the first and third powers of the tangent of the zenith distance, each multiplied by certain coefficients involving a known constant, and the height of the barometer and thermometer at the time of observation. But as it would be troublesome to calculate this formula in each particular case, it has been put into tables, where we find, first, the refraction for any given zenith distance when the barometer is at 29·6 inches, and Fahrenheit's thermometer at 50°; and afterwards the corrections to be applied for the differences between these normal heights, and the really observed heights of the barometer and thermometer. A great variety of such tables exist. Having corrected for the refraction, we find the co-latitude at once, or the height of the equator, by subtracting the star's declination from its altitude, if the declination is north, and adding it if it is south*. It is important to observe, that we must employ the apparent declination, that is, the declination as affected by precession, nutation, and aberration: the Nautical Almanac gives these apparent declinations for 100 principal fixed stars for every ten days throughout the year. Nothing, then, can be more simple than the determination of the latitude in this way.

* This supposes that the star passes the meridian to the south of the pole; if it passes to the north of the pole, add its north polar distance = 90° − declination to the altitude, and we get at once the height of the pole, that is, the latitude.

If the sun or a planet be observed, other corrections are necessary. As the centre of the disks of these bodies cannot be distinguished by the eye, we are obliged to observe the altitude either of the upper or lower limb, and then to subtract or add the apparent semidiameter *. This gives the apparent altitude of the sun or planet's centre; we must now apply the correction for refraction, and then that for parallax. The formula for the latter is so simple that it may be directly calculated; it will, however, be found tabulated among the solar tables. We now apply, as before, to the altitude, the declination of the sun or planet's centre; which declination is found in the Nautical Almanac, calculated for every day from the tables in the way which has been explained in the preceding chapter.

II. A second and more accurate method is the following. Instead of determining the meridian altitude by a single observation on the meridian, we take several altitudes very near the meridian and on both sides of it: from these we conclude the meridian altitude. To show how this is done; suppose in the annexed figure P to be

Fig. 57.

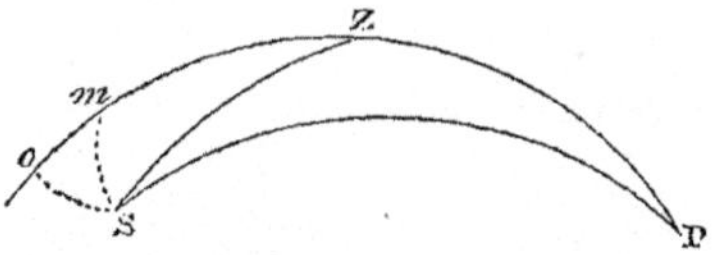

the place of the pole, Z the zenith, S a star or planet observed very near the meridian: suppose the altitude to have been corrected for refraction, and, if necessary, for parallax; then 90° − altitude = Z S = ζ. Knowing ζ we wish to deduce the zenith distance Z m when the star is on the meridian. It is to be observed that we know the time of the observation, and, consequently, the hour angle Z P S = h; and the declination of the star, = δ = 90° − P S. Z P is the co-latitude = 90° − ϕ. Take Z O = Z S = ζ, join S O, S M by arcs of great circles. Now by spherical trigonometry

$$\cos.\zeta = \sin.\phi.\sin.\delta + \cos.\phi.\cos.\delta.\cos h$$
$$= \sin.\phi.\sin.\delta + \cos.\phi.\cos.\delta - 2\cos.\phi.\cos.\delta.\sin.^2\frac{h}{2}$$
$$= \cos.(\phi - \delta) - 2\cos.\phi.\cos.\delta.\sin.^2\frac{h}{2} \quad \ldots\ldots$$

$$\text{But } Z\,S = Z\,o = Z\,m + o\,m$$
$$= Z\,m + \psi \quad \text{calling } o\,m,\ \psi$$

$$\text{and } Z\,m = P\,m - P\,Z$$
$$= P\,S - P\,Z \text{ (the declination being supposed invariable.)}$$
$$= \phi - \delta$$

$$\therefore \cos.Z\,S = \cos.\zeta = \cos.(\phi - \delta)\cos.\psi - \sin.(\phi - \delta)\sin.\psi$$
$$= \cos.(\phi - \delta)\left(1 - \frac{\psi^2}{2}\right) - \psi.\sin.(\phi - \delta) \quad \ldots\ldots(2)$$

$$\text{For } \cos.\psi = 1 - \frac{\psi^2}{2} + \ldots\ldots$$
$$\sin.\psi = \psi - \frac{\psi^3}{3}$$

and the arc ψ being very small, we take only the first term of the series in each case.

Equating now the expressions (1) and (2), we get

$$\tfrac{1}{2}\psi^2\cos.(\phi - \delta) + \psi.\sin.(\phi - \delta) = 2\cos.\phi.\cos.\delta.\sin.^2\frac{h}{2}$$

* As the sun's apparent semidiameter is not constant, its value is given for every day, in the Nautical Almanac.

To find ψ from this equation, we begin by neglecting the term involving ψ^2, and

then $$\psi = \frac{2.\cos.\phi.\cos.\delta.}{\sin.(\phi - \delta)}.\sin.^2\frac{h}{2}$$

or expressing ψ in seconds, $$\psi = \frac{2\sin.^2\frac{h}{2}}{\sin. 1''}.\frac{\cos.\phi.\cos.\delta}{\sin.(\phi - \delta)} \quad \ldots\ldots(3)$$

When we do not observe too far from the meridian, this term is always sufficient*. Each altitude observed near the meridian will give a value of ψ, which, being added to it, will give a value of the meridian altitude; the mean of all these meridian altitudes must be taken. Or it is better to observe an equal number of altitudes on each side of the meridian, and to take their mean; then to take the mean of all the arcs ψ corresponding to them; and apply this mean correction to the mean observed altitude, to deduce the true meridian altitude †. We may here remark that the factor $\frac{\cos.\phi.\cos.\delta}{\sin.(\phi - \delta)}$ remains constant for all the observations, while $\frac{2\sin.^2\frac{h}{2}}{\sin. 1''}$ is different for each. Astronomers have constructed tables of this last term for every second of time up to half an hour. In finding, then, the mean correction, we take the mean of all these terms as given by the table: we then multiply it by the factor $\frac{\cos.\phi.\cos.\delta}{\sin.(\phi - \delta)}$. Should we aim at great precision, and be observing the sun or a planet, it will be necessary to allow for the change of declination during the interval of the observations, as we have hitherto supposed the declination invariable. In the first place, we may observe, that the motion in declination being very small, it may be supposed uniform during the whole interval of the observations; and the variation will be proportional to the time from the meridian transit, that is, to the hour angle: also, the corrections will evidently be of different signs before and after the meridian transit; for the declination goes on either increasing or diminishing throughout, while the zenith distances diminish on one side of the meridian, and increase on the other. Take the sum of the hour angles (these angles being expressed in minutes of time) on one side of the meridian, then that on the other: divide the difference of these sums by the number of observations, and multiply the result by the motion in declination of the planet for a minute. The product is the correction to be applied to the meridian altitude determined by the methods above given, on the supposition of the declination remaining invariable.

The method above given for the determination of latitudes supposes that we know at each observation the hour angle. Now the hour angle depends on a knowledge of the time at which the star comes to the meridian, which time is found by a knowledge of its right ascension, and of the moment at which any given zenith distance is observed: the difference of these two times is the hour angle. Now the watch or clock that we use will never follow exactly either the motion of the mean sun (if we are employing mean solar time) or the motion of the star (if we follow sidereal time): but in either case we may determine pretty exactly the daily loss or gain of the clock, which loss or gain we suppose not to vary from day to day. Let us suppose the clock to lose daily a number of seconds, which we shall call r; (if it gains r will have a

* If, afterwards, we wish to take into account ψ^2, we must calculate ψ^2 with the value of ψ found by the first approximation: substitute this value of ψ^2 in equation (3), and then determine ψ; it will thus be given with great accuracy; but it is perhaps better only to employ observations within 10 or 15 minutes of the meridian, when the second term may be neglected. If taken into account the reader may easily satisfy himself that it will be of the form

$$-\frac{2\sin.^4\frac{h}{2}}{\sin. 1''}.\quad\cos.(\phi - \delta).\left\{\frac{\cos.\phi.\cos.\delta}{\sin.(\phi - \delta)}\right\}^2$$

† It is to be observed, that the latitude is always known approximately, or, if not known, may be determined approximately by a single altitude near the meridian, for then the altitude changes very slowly; it is with this approximate value of ϕ, that we calculate the correction

$$\frac{\cos.\phi\cos.\delta}{\sin.(\phi - \delta)}.$$

negative sign). Then, in 24 hours, the clock, instead of indicating 86400 seconds, will indicate $86400^s - r$, the hour angle h, consequently in our formulæ of reduction will not be the real hour angle. Call h' the real hour angle, then

$$\frac{h'}{h} = \frac{86400^s}{86400 - r}$$

$$\therefore h' = h.86400 + \frac{r.h.86400}{86400 - r}$$

Let $h.86400^s = \text{H}$: and $\frac{r}{86400 - r} = r'$, then

$$h' = \text{H} + \text{H}\, r'$$

$$\therefore \sin.\tfrac{1}{2} h' = \sin.(\tfrac{1}{2}\text{H} + \tfrac{1}{2}\text{H}\, r')$$

$$= \sin.\tfrac{1}{2}\text{H}.\cos.\tfrac{1}{2}\text{H}\, r' + \cos.\tfrac{1}{2}\text{H}.\sin.\tfrac{1}{2}\text{H}\, r'$$

In squaring this expression, let us confine ourselves to the first power of sin. $\frac{1}{2}$ H r' and take $\cos.\frac{1}{2}\text{H}\, r' = 1$

$$\therefore \sin.^2\frac{h'}{2} = \sin.^2\frac{\text{H}}{2} + \frac{\sin.\text{H}.\sin.\text{H}\, r'}{2}$$

We may substitute in the second term, H and H r' being very small, $2\sin.\frac{\text{H}}{2}$ instead of sin. H, and $2\, r' \sin.\frac{1}{2}\text{H}$ for sin. H r'. This gives us

$$\frac{\sin.\text{H}.\sin.\text{H}\, r'}{2} = 2\, r' \sin.^2\frac{\text{H}}{2}$$

$$\therefore \sin.^2\frac{h'}{2} = (1 + 2\, r')\sin.^2\frac{\text{H}}{2}$$

We have, then, in the constant factor common to all the reductions, an additional term to introduce; namely, the multiplier $(1 + 2\, r')$. To exemplify this method, let us suppose that at a certain spot, eight altitudes of α Polaris near the meridian (at the superior transit) have been taken, four on each side of it: let the apparent declination of the star for the day in question be δ . . . 88° 17′ 41″
Let the approximate value of the latitude be φ . . . . 51° 2′ 5″

Then φ − δ . . . . = 37° 15′ 36″

Hence $\log.\left\{\frac{\cos.\delta.\cos.\varphi}{\sin.(\varphi - \delta)}\right\} = 8.4900862$

The clock loses daily on sidereal time $69^s{\cdot}5$

$$\therefore \log.(1 + 2\, r') = 0.0006986$$

$$\therefore \log.\left\{\frac{\cos.\delta.\cos.\varphi\,(1 + 2\, r')}{\sin.(\varphi - \delta)}\right\} = 8.4907848$$

this is the logarithm of the constant factor by which we must multiply the mean of the reductions.

Now let us suppose that it has been found, from a knowledge of the right ascension of α Polaris, that it will pass the meridian at its superior transit at $0^h. 24^m. 44^s$. By comparing this time with the time of the respective observations, we get the several hour angles, and taking from the tables the corresponding reductions, they appear as follows:—

Sidereal Time. h. m. s.	Hour Angle.	Reduction to the Meridian.
23 18 57	5·47	65·7
22 23	2·21	10·8
22 59	1·45	6·0
23 43	1·1	2·0
29 19	4·35	41·2
41 59	17·15	583·9
45 54	21·10	879·0
46 36	21·52	938·1
Sum of reductions	=	2616·7

log. 2616.7	= 3.4177539
log. constant factor	= 8.4907848
	1.9085387
∴ there are eight observations ∴ subtract log. 8	= 0.9030899
∴ log. mean reduction	= 1.0054488
mean reduction	= 10″.13
Now the mean observed zenith distance, we suppose	= 37° 15′ 1″
Refraction (by the tables)	= 1′ 46″
	37° 15′ 47″
Reduction to the meridian	= − 10
Meridian zenith distance of Polaris	= 37° 15′ 37″
North polar distance of Polaris	= 1° 42′ 19″
Zenith distance of pole = co-latitude	= 38° 57′ 58″
∴ latitude	= 51° 2′ 2″

Lastly, we shall remark that the declinations of the stars may always be considered as sufficiently well known for the most exact determinations in navigation and geography. But in very nice geodetical measures, such as the measure of an arc of the meridian, it is desirable to determine the latitude independently of any value of the declination. This is done by observing the meridian altitudes of a circumpolar star, both above and below the pole: half the difference of those altitudes will give us the altitude of the pole, that is, the latitude. It is hardly necessary to add, that in this case each meridian altitude is to be found by the method, and with the reductions just explained. In this way we are rendered independent of any knowledge of the declination of the star employed, and, if necessary, that declination may be determined with considerable exactness *.

III. The last method which we shall notice, is that of Professor Littrow, for finding the latitude by observations of the Pole star off the meridian. This is a method of great practical utility, as we are enabled to find our latitude with great exactness, at any moment, without being obliged to wait for the transit of the star, or of the sun or a planet. Let a series of four or six consecutive altitudes be observed at very small intervals of time, and take the mean of these altitudes as corresponding to the mean of the times of observation. To find the meridian altitude of Polaris we proceed as follows: let P in the annexed figure be the place of the pole; Z that of the zenith:

Fig. 58.

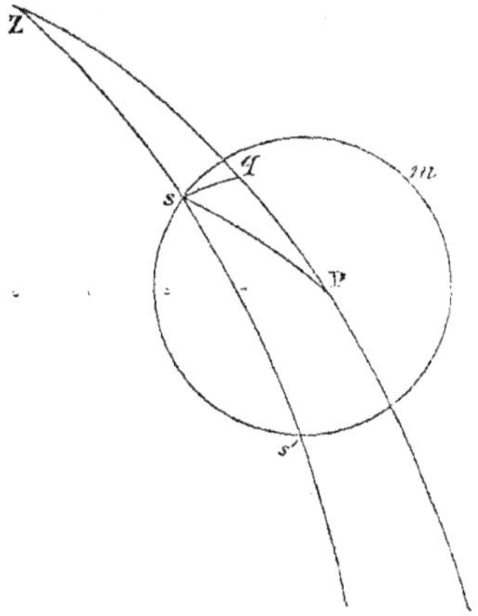

s' s′ m the circle described by the Polar star round P, and s the place of the star at the time of observation. Call P s, the north polar distance of the star, Δ: Z P = 90° − φ, φ being the latitude: Z S = 90° − H, H being the observed altitude. Our object is to find the difference between the arcs Z P, and Z s; as then Z s being known, we may conclude from it Z P or the co-latitude.

Now

$$ZP - ZS = H - \varphi = \psi \therefore \varphi = H - \psi$$

ψ it will be recollected, must always be small, as P s is not more than 1° 40′. Now let p be the hour angle Z P s

* The circumpolar stars usually employed for the determination of latitudes are Polaris and δ Ursæ Minoris, both of which, in European latitudes, are sufficiently high at their lower transit to avoid any uncertainty on the refraction; and, at the same time, from their small north-polar distance, move so slowly that observations may be made half an hour on each side of the meridian. In the Nautical Almanac, the apparent right ascensions and declinations of these two stars for the upper transit at Greenwich are given for every day throughout the year.

$$\cos. Z s = \cos. P s . \cos. P Z + \sin. P s . \sin. P Z . \cos. p$$

$$\sin. H = \cos. \Delta . \sin. (H - \psi) + \sin. \Delta . \cos. (H - \psi) . \cos. p .$$

$$\therefore 1 = \cos. \Delta (\cos. \psi - \sin. \psi . \cot. H) + \sin. \Delta . (\cos. \psi . \cot. H + \sin. \psi) \cos. p$$

$$\therefore 1 = \cos. \psi . (\cos. \Delta + \sin. \Delta . \cot. H . \cos. p) - \sin. \psi (\cos. \Delta . \cot. H - \sin. \Delta . \cos. p .)$$

Put $a = \cos. \Delta + \sin. \Delta . \cot. H . \cos. p$

$b = \cos. \Delta . \cot. H - \sin. \Delta . \cos. p$

$$\therefore \quad 1 = a . \cos. \psi - b . \sin. \psi \quad \ldots \quad (1)$$

Now, $a = 1 + \Delta . \cos. p . \cot. H - \frac{1}{2} \Delta^2 - \frac{1}{6} . \Delta^3 . \cos. p . \cot. H$

Δ being small

$$b = \cot. H - \Delta . \cos. p . - \frac{1}{2} \Delta^2 . \cot. H + \frac{1}{6} \cdot \Delta^3 . \cos. p$$

Suppose, on the other hand, ψ developed according to the powers of Δ, we may put $\psi = A \Delta + B \Delta^2 + C \Delta^3 \quad . \quad . \quad . \quad (2)$

There is no term in this expression independent of Δ, for if $\Delta = 0$, we should clearly have $\psi = 0$.

Now, $\cos. \psi = 1 - \frac{1}{2} A^2 \Delta^2 - A B \Delta^3$

$\sin. \psi = A \Delta + B \Delta^2 + (C - \frac{1}{6} A^3) \Delta^3$

Substitute now these values of sin. ψ and cos. ψ in equation (1), and at the same time the values above found for a and b, we then get, arranging our terms according to the powers of Δ,

$$\begin{aligned} 1 = 1 + \cos. p . \cot. h . \Delta . - \tfrac{1}{2} \Delta^2 - \tfrac{1}{6} \Delta^3 . \cos. p . \cot. H \\ - \tfrac{1}{2} A^2 \Delta^2 - \tfrac{1}{2} A^2 \Delta^3 . \cos. p . \cot. H - A B \Delta^3 \\ - A . \cot. H . \Delta + A . \cos. p . \Delta^2 + \tfrac{1}{2} A \Delta^3 . \cot. H + B \Delta^3 . \cos. p \\ - B . \cot. H . \Delta^2 - (C - {}^1) A^3 . \Delta^3 . \cot. H \end{aligned}$$

This equation is identical. The terms then which involve the same power of Δ must be separately $= 0$: whence we obtain

1. $\cos. p . \cot. H - A . \cot. H = 0 \therefore A = \cos. p$
2. $- \frac{1}{2} - \frac{1}{2} A^2 + A . \cos. p - B . \cot. H = 0$

or, $- \frac{1}{2} + \frac{1}{2} \cos.^2 p = B . \cot. H$

$\therefore \quad B = - \frac{1}{2} \sin.^2 p . \tan. H$

3. $- \frac{1}{6} . \cos. p . - \frac{1}{2} A^2 . \cos. p + \frac{1}{2} A - (C - \frac{1}{6} A^3) = 0$

$\therefore C = \frac{1}{3} . \cos. p . \sin.^2 p$

Let us now substitute these values in equation (2)

$$\therefore \psi = \Delta . \cos. p - \tfrac{1}{2} \sin.^2 p . \tan. H \Delta^2 + \tfrac{1}{3} . \cos. p . \sin.^2 p . \Delta^3 .$$

To express ψ and Δ in seconds of arcs we must change each into ψ . sin. 1″, and Δ sin. 1″: put now $m = \frac{1}{2}$ sin. 1″, $n = \frac{1}{3}$. sin.^{2} 1″, and we get finally

$$\psi = \Delta . \cos. p - m . (\Delta . \sin. p)^2 . \tan. H + n . (\Delta \cos. p .) (\Delta . \sin. p)^2 .$$

The arc ψ being applied with its proper sign to the observed altitude, gives us the latitude: since $\varphi = H - \psi$ *.

Let us suppose that on a certain day, we find

$\delta = 88° 23' 27''$ $\quad p = 18° 55' 59''$ $\quad H = 49° 2' 38''$

$\Delta = 1° 36' 33'' = 5792''·7$

* In this expression the first term Δ . cos. p is found tabulated in the Nautical Almanac for 1834 for every ten minutes of sidereal time, with a mean constant value of Δ''. The second term is given in a table of double entry, of which the arguments are the approximate latitude for every five degrees, and the sidereal time for every thirty minutes. We may always take H for the approximate latitude, for it can never differ from the real latitude two degrees. This second term thus found is corrected by a table of double entry, of which the arguments are the sidereal time of the observation for every two hours, and the month of the year. This term involves Δ^2, and as Δ is not rigorously constant throughout the year, the term will differ a little in different months. The third term of our series, being necessarily always very small, is neglected.

$$
\begin{array}{llll}
\log.\ \Delta = & 3\cdot762\ \ 8795 & \ldots\ldots & 3\cdot76288 \\
\log.\ \cos.\ p = & 9\cdot383\ \ 6587 + & \log.\ \sin.\ p = & 9\cdot98691 \\
1401''\cdot3\cdot & 3\cdot146\ \ 5382 + & & 3\cdot74979 \\
 & 7\cdot499\ \ 58 & & 2 \\
\log.\ \tfrac{1}{3}.\ \sin.^2 .\ 1'' = & 12\cdot89403 & & 7\cdot49958 \\
\log.\ 0''\cdot35 = & 1\cdot54915 & \log.\ \tan.\ \mathrm{H} = & 0\cdot06151 \\
 & & \log.\ \tfrac{1}{2}.\ \sin.\ 1'' = & 6\cdot38454 \\
 & & 1'.28''\cdot2 & 1\cdot94563
\end{array}
$$

$$
\begin{array}{lr}
\text{Hence } \psi = \Delta\ .\ \cos.\ p & = 23'\ 21''\cdot3 \\
\quad - m\ .\ (\Delta\ .\ \sin.\ p)^2\ .\ \tan.\ \mathrm{H} & -\ 1'\ 28''\cdot2 \\
\quad + n\ .\ (\Delta\ .\ \cos.\ p)\ (\Delta\ .\ \sin.\ p)^2\ . & +\ 0''\cdot4 \\
 & 21'\ 53''\cdot5
\end{array}
$$

$$
\begin{aligned}
\therefore\ \phi = \mathrm{H} - \psi &= 49^\circ\ 2'\ 38'' - 0^\circ\ 21'\ 54'' \\
&= 48^\circ\ 40'\ 44''
\end{aligned}
$$

Section II. *Longitudes—Eclipses of Jupiter's Satellites—Marine Chronometer—Lunar distances—Transits of Moon Culminating Stars—Occultations—Eclipses.*

I. We now come to the determination of terrestrial longitudes, a subject of much greater complexity. It will of course be understood, that we only speak here of the astronomical methods employed for this purpose. The earliest and most obvious of these methods was that proposed and employed by Ptolemy, in the observation of lunar eclipses. The physical instant at which the moon's disk enters or quits the earth's shadow, is of course the same for all parts of the earth from which it is visible; and were the observation susceptible of accuracy, it would be a good means of determining terrestrial longitudes. But this is not the case, and the causes are sufficiently obvious. The penumbra of the earth causes the immersion of the moon's disk into the pure shadow to be a phenomenon, the precise instant of which it is impossible to determine with precision, and as an error of a minute in this observation would cause an error of a quarter of a degree on the longitude, while in fact the error of the observation is necessarily much greater, it will easily be understood why this method was soon abandoned.

II. The discovery of Galileo of the system of the satellites of Jupiter, led to the adoption of another and a much more accurate method. These satellites are subject to eclipses similar to those of our moon, but the periods of some of them are much shorter, and from this cause the uncertainty is much less; consequently, these eclipses were for a long time much used in the determination of longitudes. The principle here is the same as in that of a lunar eclipse: the phenomenon takes place at the same physical instant for all observers, but the time which they will each count at the observations will depend upon the difference of the meridian, and the difference of the times will be exactly proportional to that difference, that is, to the longitude.

The eclipses of Jupiter's satellites (with, perhaps, the exception of the first) are not now much used for the purpose in question. The reason is the same as that which has caused lunar eclipses to be abandoned, the difficulty of determining exactly the instant of immersion or emersion. This will vary considerably according to the goodness of the telescope employed, and other accidental causes. It is much less for the first satellite, which performs its sideral revolution in 1 day $18^{h}\ 28^{m}$, than for any of the others: and as, on the other hand, it is eclipsed much more frequently than the rest, it is desirable to confine our observations, if possible, to this satellite. However, the Nautical Almanac gives the eclipses for the first three satellites; the fourth moves too slowly to be of any utility in such researches.

Nothing can be simpler than the method of finding the longitude by these eclipses. Let us suppose, for example, that the immersion of the first satellite into Jupiter's shadow has been observed on a certain day at $7^{h}\ 17^{m}\ 10^{s}$; while another observer at Greenwich has observed it at 6^h 53^m 5^s. The difference of the times is 24^m 5^s, which is also the difference of meridians; that is to say, the longitude of the observer is 24^m 5 = 6° 1′ 15″ west of Greenwich. If, however, there is no corresponding observation made under the meridian of Greenwich, we take from the Nautical Almanac the calculated time of that place at which the eclipse will happen, and we regard this time as an actual observation.

III. The observation of the eclipses of Jupiter's satellites being generally impracticable at sea, navigators have been obliged to have recourse to other methods. The great improvements in the art of watch-making, principally owing to Harrison, have caused the general adoption of the marine chronometer. It is sufficiently evident, that if while we are at a place whose longitude is unknown, we can, by any means, find to what instant of Greenwich time a given instant corresponds, we have at once the difference of longitudes between that place and Greenwich. Now the time reckoned by the observer is easily determined by the altitudes, or the transits of the heavenly bodies: the Greenwich time is found by transporting from Greenwich to the place of observation a chronometer which has previously been carefully adjusted to Greenwich time. If it were possible to have a chronometer which should neither gain nor lose in the least, a comparison of this chronometer with a watch indicating the time counted at the place of observation would be all that was necessary for our purpose. But this is not possible, nor is it necessary; all that is essential is that the chronometer should either gain or lose uniformly a certain quantity in a certain time; this uniform gain or loss is called the rate of the chronometer. If, for example, the chronometer loses 5 seconds daily, nothing can be more easy than to make the proper allowance for this at the end of any given time. The rate is always carefully determined before the vessel leaves some known meridian; but it is always desirable whenever the vessel remains sufficiently long in any port to verify the rate, in order to be perfectly sure that it has not undergone any variation, and if it has varied, to determine the new rate. There are various astronomical methods which we shall explain below, for finding the time at any given moment; this being done for several consecutive days, we obtain the rate of the chronometer. Thus, suppose that on the 26th of May we find by altitudes of Arcturus, that the sidereal time is 19^h 48^m 7^s, while the watch marks at the same instant 19^h 37^m 2^s, we know that the watch is too slow 11^m 5^s. Let us suppose that the next day, by altitudes of the same star, the sidereal time is 19^h 54^m 12^s, while the watch marks 19^h 43^m 8^s, it will then be too slow 11^m 4^s; it has thus gained a second in the interval of the observations. As the observations, however, are not 24 hours distant from each other, we find, by a simple proportion, what the gain is in 24 hours. Another observation on the 28th, compared with that of the 27th, will give us the gain in the second 24 hours; and by a series of observations of this kind, we ascertain whether or not the chronometer preserves an uniform rate, and the amount of that rate.

The most simple and exact method of finding the time at a given spot, is by observing the instant at which the centre of the sun, or a star of known right ascension, is on the middle wire of a transit instrument properly adjusted. At that moment the star or sun is on the meridian; the right ascension of the star at that moment, expressed in hours, minutes and seconds, is the sidereal time*. The instant of the sun's centre being on the meridian, is the instant of *apparent* noon; but as all clocks and watches are necessarily regulated on *mean* solar, and not apparent time, we must add the equation of time with its proper sign, to get the instant of *mean* noon.

No method is equal, in accuracy, to this method of transits; but as it can hardly be employed except in fixed observatories, it is necessary to adopt some other in the determination of longitudes. Let us suppose that by means of a sextant, or other instrument, the altitude of a known star has been taken, the time being carefully observed, while the star is at a considerable distance from the meridian: let us wait now till the star has passed the meridian, and attains exactly the same altitude on the other side; then also note the time carefully. Let us suppose, now, that the state of the atmosphere has not varied the least in the interval of the observations; then since the apparent altitudes are equal, the refractions also will be equal, and

* Sidereal time is counted from the transits of the first point of Aries.

the true altitudes equal. But since the declination of the star has not perceptibly varied in the interval, true equal altitudes will correspond to equal hour angles on opposite sides of the meridian; or, in other words, the instant of the star's being on the meridian will exactly bisect the interval of the observations, which interval is known. Add then half the interval in question to the instant of the first observation, and we get the instant at which the star was on the meridian; and its right ascension being known, we may get the sidereal time, as before.

The refractions are the same at the same altitude, if the barometer and thermometer have not varied; but as it is highly improbable that this will be the case, we must introduce a small correction of this nature. Let r, r', be the respective refractions calculated for the first and second altitudes; we may calculate without difficulty the time taken by the star to describe the vertical arc $r - r'$: the time thus calculated must be applied with its proper sign to the interval of the observations. To find the time in which the star describes a small vertical arc, let us consider the two triangles Z P S, Z P S′, where S and S′

Fig. 59.

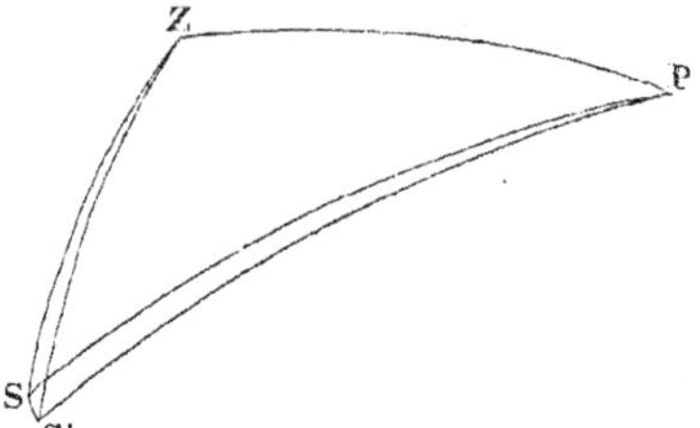

are the successive positions of the star, and the arc S S′ is very small; Z is the zenith, S the pole: call the hour angles p and p', the latitude ϕ, $\therefore$ Z P $= 90^\circ - \phi$; the declination, δ $\therefore$ P S $=$ P S′ $= 90^\circ - \delta$; call the altitudes h and h', then

$$ZS = 90^\circ - h \quad ZS' = 90^\circ - h',$$

then

$$\sin. h = \sin. \phi . \sin. \delta + \cos. \phi . \cos. \delta . \cos. p$$

$$\sin. h' = \sin. \phi . \sin. \delta + \cos. \phi . \cos. \delta . \cos. p',$$

$$\therefore \sin. h' - \sin. h = \cos. \phi . \cos. \delta . (\cos. p' - \cos. p),$$

$$\therefore \sin. \tfrac{1}{2}(h' - h) . \cos. \tfrac{1}{2}(h' + h) = \cos. \phi . \cos. \delta \sin. \tfrac{1}{2}(p - p') \sin. \tfrac{1}{2}(p + p');$$

and since h and h' differ very little,

$$(h' - h) . \cos. h = \cos. \phi . \cos. \delta (p - p') \sin. p;$$

$$\therefore p' - p = \frac{(h' - h) . \cos. h}{\cos. \phi . \cos. \delta . \sin. p}.$$

This is the expression for the variation in the hour angle produced by a small given variation in the altitude. The latter variation, in our particular case, is the difference of the refractions; substituting this in our formula, $p' - p$ is known, as h, p, ϕ, and δ, are all known quantities.

This method, which is called that of *corresponding altitudes*, is susceptible of great precision, when all the proper precautions are taken. It will easily be understood that it is advisable to determine the instant of the meridian transit, by the mean of several pair of observations. The star should always be observed as far from the meridian as it can with convenience. When it is the sun which we observe, it will be recollected that it is the moment of apparent noon which is found; to get the mean noon, we must apply the equation of time with its proper sign.

The sun is more frequently used in the observation of corresponding altitudes than stars. It is then absolutely necessary to know how to make the proper correction for the change in declination during the interval of the observations. For this purpose let us resume the equation

$$\sin. h = \sin. \phi . \sin. \delta + \cos. \phi . \cos. \delta . \cos. p.$$

Supposing, then, in this equation, δ and p alone to vary, we have

$$d\delta . \cos. \delta . \sin. \phi = \cos. \phi \cos. \delta \, dp . \sin. p . + \cos. \phi . \cos. p \, d\delta . \sin. \delta$$

$$\therefore d\delta (\cos. \delta . \sin. \phi - \cos. \phi \cos. p . \sin. \delta) \; . \; . \; . \; . = dp . \sin. p . \cos. \phi . \cos. \delta;$$

$$\therefore dp = d\delta \left(\frac{\tan.}{\sin. p} - \cot. p . \tan. \delta\right)$$

Where dp is the variation of the hour angle corresponding to a very small variation in the declination, $d\delta$ is the variation in declination during the interval of the observations. Let ψ be the diurnal variation in declination, (which is given in the Nautical Almanac,) then $d\delta : \psi ::$ half the time between the observations : 24 hours. Hence $d\delta$ is calculated. It is better, however, to take for ψ the mean of the two daily variations of the two consecutive astronomical days between which the noon in question lies. Several astronomical tables give ready calculated these means, or rather their logarithms for every day throughout the year. When this is done, the whole formula may be easily calculated, by dividing it into two terms, the factor of each being found by a separate table. The first term is $d\delta . \frac{\tan. \phi}{\sin. p}$; the second $- d\delta . \cot. p \tan. \delta$. or $A \psi \tan. \phi$ and $- B \psi \tan. \delta$. In Mr. Baily's astronomical tables, the logarithms of A and B are given for every two minutes of interval from two hours up to twenty-three. We must then add to log. A taken from the tables, log. ψ log. tan. ϕ, for the logarithm of the first term; and to log. B, log. ψ, and log. tan. δ, for the logarithm of the second term: the difference of these two terms is the correction. Let us suppose, for example, that on the 17th of October, 1827, corresponding altitudes of the sun have been taken in a place whose latitude we know to be 48° 41′. The interval of the observations is $7^h\ 40^m$: the sun's declination is given by the Nautical Almanac $-$ 9° 4′ 15″.

By the same ephemeris ψ determined by the mean of the preceding and consecutive daily variation $= - 22{\cdot}4'' = - 1324''$.

$$\log. d\delta = \log. 1324'' + \log. 3^h\ 50^m - \log. 24^h$$
$$= \log. 1324 + \log. 57^\circ.30' - \log. 360^\circ$$
$$= 3{\cdot}12205 -$$
$$0{\cdot}58399$$
$$3{\cdot}44370$$

$$\log. d\delta = 1{\cdot}14974 -$$
$$\log. \tan. \phi = 0{\cdot}05595 +$$
$$1{\cdot}20569 -$$
$$\log. \sin. p = 9{\cdot}92627 +$$
$$1{\cdot}27942 -$$

$$\therefore d\delta . \frac{\tan. \phi}{\sin\ p} = - 19^s{\cdot}0$$
$$- d\delta . \tan. \delta . \cot. p = - 1{\cdot}4$$
$$\therefore \text{whole correction} = - 20{\cdot}4$$

$$\log. d\delta = 1{\cdot}14974 -$$
$$\log. \tan. \delta = 9{\cdot}20317 -$$
$$\log. \cot. p = 9{\cdot}80314 +$$
$$0{\cdot}15605 +$$
$$- d\delta . \tan. \delta . \cot. p - 1^s{\cdot}4$$

The semi-interval $= 3^h\ 50^m\ 0$
Correction $= - \quad 20{\cdot}4$

Corrected semi-interval $= 3^h\ 49^m\ 39^s{\cdot}6$

According to this, the watch, at the first observation, should have indicated 12 hours $- 3^h\ 49^m\ 39^s{\cdot}6 = 8^h\ 10^m\ 20^s{\cdot}4$. Suppose it really indicated $8^h\ 6^m\ 3^s$, then we perceive that it must have been $43^m\ 36^s{\cdot}4$ too slow.

The time may also be found, though with less accuracy, by a single altitude of a star. But in this case it is necessary to select the moment when the star changes its altitude most rapidly in a short given time. We know, by what has preceded, that its variation of altitude is least when near the meridian: it is easy to show that it is greatest when near the prime vertical, that is, when its azimuth is 90°, counting from the meridian. Let ζ be the zenith distance, δ the declination, ϕ the latitude of the observer, p the hour angle, then

$$\cos. p = \frac{\cos. \zeta - \sin. \phi . \sin. \delta}{\cos. \phi . \cos. \delta}$$

$$dp . \sin. p = \frac{d\zeta . \sin. \zeta}{\cos. \phi . \cos. \delta}$$

Now call the angle P Z S, which is the azimuth, A,

$$\sin. A \,.\, \sin. \zeta = \sin. p \,.\, \cos. \delta,$$

$$\therefore dp = \frac{d\zeta}{\cos. \phi \,.\, \sin. A}$$

$$\frac{d\zeta}{dp} = \cos. \phi \,.\, \sin. A.$$

But $\frac{d\zeta}{dp}$ represents the variation of the zenith distance in a short time; this variation, then, will be greatest when sin. A is greatest; that is, when sin. A = 1, or A = P Z S = 90°.

As one altitude of a star would not give the time with sufficient accuracy, it is desirable to take a group of six or eight altitudes very near each other, and then to consider the mean of the altitudes as corresponding to the mean of all the times of observation. Knowing, then, in the triangle Z P S, the three sides Z P, Z S, and P S, we determine the angle at P, or the hour angle by the usual trigonometrical formulæ; call, for example, these three sides θ, ζ, and χ, respectively; then putting $2k = \theta + \zeta + \chi$,

$$\tan. \frac{p}{2} = \sqrt{\frac{\sin. (k - \theta). \sin. (k - \chi)}{\sin. k. \sin. (k - \zeta)}}$$

We may of course employ the altitude of the sun or a planet instead of that of a star: but in this case it is necessary to know, not only the latitude of the place of observation, but also an approximate value of the longitude, as this latter is necessary, in order to calculate the declination for the moment in question. This is more particularly necessary for the sun, whose daily motion in declination is so considerable. We are likewise obliged for the sun to correct the apparent altitude observed, not only for refraction, but parallax. This is unnecessary for Jupiter and Saturn, whose parallaxes are insensible. We may either always observe the same limb, whether upper or lower, and then correct for the apparent semi-diameter; or, which is more convenient, we may observe alternately the upper and lower limb: then at the end of an *even* number of observations the semi-diameter is eliminated when we come to take the mean.

Let us, as an example, suppose that the four following zenith distances of the sun's upper and lower limbs have been observed alternately, and that their sum = 252° 10′ 22″: and let the latitude of the place be 48° 40′ 50″: suppose the mean of the four instants of observation to be $7^h\ 30^m\ 37^s{\cdot}5$. This is the time by the watch, and therefore mean time: add to it the equation of time $3^m\ 7^s$ to get the true time of observation = 7° 33′ 44″·5.

		°	′	″
4 zenith distances	=	252	10	22
¼th	=	63	2	35·5
Refraction — parallax	=		1	46·4
ζ	=	63	4	21·9
Co-latitude = θ	=	41	9	10
North polar distance of the sun's centre = χ	=	74	36	18·2
$\therefore 2k = \zeta + \theta + \chi$	=	178	59	50·1
$\therefore k$	=	89	29	55·0
$k - \theta$	=	48	20	45·0
$k - \chi$	=	14	53	36·8
$k - \zeta$	=	26	25	33·1

$\therefore$ log. sin. $(k - \theta)$ = 9.8734194
log. sin. $(k - \chi)$ = 9·4099675
9·2833869
9·6483813
9·6350056

log. sin. k = 9.9999833
log. sin. $(k - \zeta)$ = 9·6483980
9·6483813

$$\text{log. tan. } \frac{p}{2} = 9{\cdot}8175028$$

$$\therefore \frac{p}{2} = 33\ 18\ 5 = 2^h\ 13^m\ 12^s{\cdot}3$$

$$\therefore p = 4^{h}\ 26^{m}\ 24^{s}\cdot 6$$

Complement to 12 hours $= 7\ 33\ 35\cdot 4$

This is the time that the watch should have marked at the mean of the observations, while in reality it marked $7^{h}\ 33^{m}\ 44^{s}\cdot 5$: it is then $9^{s}\cdot 1$ too fast.

The sun's declination and the equation of time are given for every day at noon at Greenwich: knowing approximately the longitude of the place of observation, we find to what time at Greenwich $7^{h}\ 30^{m}$ corresponds. Suppose the approximate longitude 22° 30′ W.; then $7^{h}\ 30^{m}$ corresponds to 9^{h} at Greenwich. To find the sun's declination and the equation of time at that moment, find how much each of these quantities vary in 24 hours: $\frac{1}{24}$th part will be their variation in one hour; and nine times that the variation in nine hours. Apply this variation with the proper sign to the declination at noon, and to the equation of time, we shall have these quantities for the instant of observation. Suppose our observation made on the 2nd of May: by subtracting the declination as given for Greenwich noon on the 2nd, from that for Greenwich noon on the 3rd, we get the diurnal variation, and $\frac{1}{24}$th part is the horary variation: suppose we find this $= 44''\cdot 5$; then nine times the horary variation $= 440''\cdot 5 = 7'\ 20''\cdot 5$. Suppose at Greenwich noon the declination was 15° 16′ 21″·3; then at nine hours it will be 15° 23′ 41″·8.

When we wish merely to determine the rate of a watch or chronometer, and not the absolute time, it is sufficient, having taken the altitude of a star not far from the prime vertical, to observe some days afterwards the moment at which it comes to the same altitude. In this time the changes of apparent place produced by aberration, precession, and nutation will be quite insensible; and consequently at the same *sidereal* time the star will return to the same altitude. Allowance however must be made for the changes of the atmosphere in the mean time both in temperature and density: these changes must be allowed for in the manner which has been previously explained. When the watch or chronometer marks, as is very usual, mean solar time, it is necessary in this kind of observation to allow for the daily acceleration of the fixed stars on the sun, which acceleration $= 3^{m}\ 55^{s}\cdot 91$. Suppose, for example, 9th June, Arcturus attains a certain altitude at $8^{h}\ 33^{m}\ 17^{s}$: and on the 14th, the star attains the same altitude at $8^{h}\ 7'\ 3''$. Then

		m s
difference in 5 days	=	26 14
in 1 day	=	5 14·8
Constant	=	3 55·9
Watch loses daily		1 18·9

If the watch went exactly by mean solar time, it ought to lose daily $3^{m}\ 55^{s}\cdot 9$ on sidereal time: it really loses $5^{m}\ 14^{s}\cdot 8$. The difference between these quantities then is its real diurnal loss.

IV. Having now sufficiently explained how to determine the longitude by means of the chronometer, we now proceed to examine the other methods most generally in use. In the first rank, in point of general utility, is the method of lunar distances. The original principle of this method is identical with that of lunar eclipses, of the occultations of Jupiter's satellites, and so on. The object is to find some celestial phenomenon which shall serve, as it were, for a signal, which may be observed under different meridians, and by which the two observers can compare the times they reckon at the same physical instant. Nor is it necessary that the phenomenon should be observed under both meridians: the hour at which the phenomenon must happen may be calculated for one of them, and observed at the other: the comparison of the observed time at one, with the calculated time at the other, will be sufficient. It is evident that there is only one particular instant at which the moon can be at a certain distance from any fixed star*: if we ascertain then at any moment this distance, and if we determine from the tables what time was counted at Greenwich at the moment she had the same difference, the longitude is found ex-

* The reason why the moon alone is employed lies in the extreme rapidity of her motion. In consequence of this, her distance changes so rapidly, that there is much less uncertainty about the precise moment corresponding to a given distance.

actly as in the case of the eclipses of Jupiter's satellites. The problem then divides itself into two parts; first, to ascertain the distance of the moon's centre from one of the principal planets or fixed stars at a certain moment; secondly, to find the Greenwich time to which, according to the tables, that distance corresponds.

The first part, namely, the observation of the distance, requires some corrections, which we proceed to explain. We begin by observing consecutively, and with as little interval as possible, four or six, or even sometimes eight, distances from the star to the nearest part of the moon's limb, noting at each observation the corresponding time: we take the mean of all these distances and consider it as answering to the mean of all the times; we thus get a single distance corresponding to a certain known instant of time. We add the apparent semi-diameter to get the apparent distance from the star to the moon's centre, and we have from this to find the *true* distance. The star is elevated by refraction, the moon is elevated by refraction, and depressed by parallax: to allow for these combined effects, it is necessary to measure the apparent altitudes both of the moon and star. Suppose, in the spherical

Fig. 60.

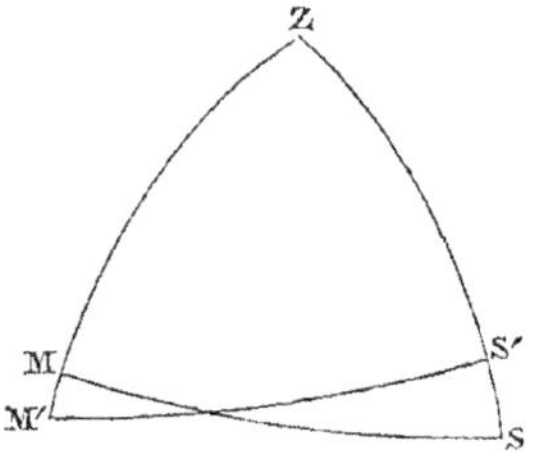

triangle Z M′ S′, Z to be the zenith, M′ the observed place of the moon's centre, S′ of the star: M′ S′ the observed distance. Suppose the true place of the moon corrected for parallax—refraction to be at M, and that of the star corrected for refraction to be at S; then M S will be the true distance which we wish to find. Now, since the apparent altitudes h and h' have been observed, Z M′ = 90° − h, and Z S′ = 90° − h' are known, and Z M, Z S may be found by calculation: M′ S′ = Δ' has been observed. Now

$$\cos. Z M S = \frac{\cos. \Delta' - \sin. h. \sin. h'}{\cos. h. \cos. h}$$

But calling M S, Δ

$$\cos. Z M S = \frac{\cos. \Delta - \sin. H. \sin. H'}{\cos. H. \cos. H'}$$

H and H′ are the true altitudes

$$\therefore \frac{\cos. \Delta' - \sin. h. \sin. h'}{\cos. h. \cos. h'} = \frac{\cos. \Delta - \sin. H. \sin. H'}{\cos. H. \cos. H'}$$

$$1 + \frac{\cos. \Delta' - \sin. h. \sin. h'}{\cos. h. \cos. h'} = \frac{\cos. \Delta - \sin. H. \sin. H'}{\cos. H. \cos. H'} + 1$$

$$\therefore \frac{\cos. \Delta' + \cos. (h + h')}{\cos. h. \cos. h'} = \frac{\cos. \Delta + \cos. (H + H')}{\cos. H. \cos. H'}$$

To simplify the expression, put $h + h' + \Delta' = 2\,m$

$$\therefore \frac{2. \cos. m. \cos. (m - \Delta')}{\cos. h. \cos. h'} = \frac{2. \cos.^2 \frac{H + H'}{2} - 2 \sin.^2 \frac{\Delta}{2}}{\cos. H. \cos. H'} *$$

$$\therefore \sin.^2 \tfrac{1}{2} \Delta = \cos.^2 \tfrac{1}{2} (H + H') - \frac{\cos. H. \cos. H'}{\cos. h. \cos. h'}. \cos. m. \cos. (m - \Delta')$$

$$* \; 1 + \cos. A = 2 \cos.^2. \frac{A}{2}$$

$$1 - \cos. B = 2 \sin.^2 \frac{B}{2}$$

$$\therefore \cos. A + \cos. B = 2 \cos^2. \frac{A}{2} - 2 \sin.^2 \frac{B}{2}$$

To put this equation in a form adapted for logarithmic computation, take

$$\sin. \phi = \frac{\sqrt{\frac{\cos. H. \cos. H'}{\cos. h. \cos. h'}. \cos. m. \cos. (m - \theta)}}{\cos. \frac{1}{2}. (H + H')}$$

$$\therefore \sin. \tfrac{1}{2} \Delta = \cos. \tfrac{1}{2} (H + H'). \cos. \phi.$$

The process then is, first, to correct the observed altitudes for parallax and refraction, in order to get the true altitudes. Knowing H, H′, h, h' and θ, we calculate $2\,m = h + h' + \theta$: then the auxiliary arc ϕ, and lastly Δ, or the true distance by the equation just given. It is unnecessary to observe, that in taking the distance from the moon to a planet, we must add the planet's semi-diameter to the observed distance.

The operation we have just explained is technically called clearing the distance. When this is done, we turn to the Nautical Almanac, which gives us for every three hours the distance of the moon's centre from the four larger planets and the principal fixed stars, and we look for the greater and less than that which has been observed: we find the change of distance in three hours; and hence by interpolation the moment of Greenwich time which corresponds exactly to the distance observed. If we suppose the variation of distance in the interval of three hours to be uniform, this interpolation offers no difficulty, as it is accomplished by a simple proportion. Suppose the distance observed to be 41° 2′ 3″, and that the Nautical Almanac gives the distance at 6 hours, 39° 44′ 26″; at 9 hours, 41° 29′ 6″; then the difference 1° 44′ 40″ is the change of distance in 3 hours: subtract from 41° 2′ 3″, the next smaller distance in the Nautical Almanac, 39° 44′ 26″; the difference is 1° 17′ 37″. We have then the proportion 1° 44′ 40″ : 3^h :: 1° 17′ 37″ : the time required. Hence this time $= 3^h \frac{1° 44' 40''}{1° 17' 37''}$. The Nautical Almanac gives us the proportional logarithm of the numerator 2343 : the proportional logarithm of the denominator is 6021 : the difference = 3678 corresponds to $2^h\ 24^m\ 18^s$, which, added to 6 hours, gives $8^h\ 24^m\ 18^s$ for the Greenwich time, corresponding to the moment of observation. Suppose the observer had counted at this instant $8^h\ 3^m\ 7^s$, the longitude is $21^m\ 11^s$ W.

It is, however, often productive of very considerable inaccuracy to suppose, as we have done, the moon's variation of distance in three hours uniform: and consequently, instead of interpolating as we have done, we must take into account the second differences. To comprehend the principle on which this is done, we must refer to the theory of finite differences, which teaches us that if we call the first difference between the two distances Δ', the difference between the first differences or the second difference Δ^2, and so on, then that the distance at any time t from the nearest value in the Ephemeris is equal to that nearest value $+ \frac{t}{3^h}. \Delta' + \frac{t\,(t-3^h)}{18^h}. \Delta^2 \ldots$ We shall suppose the second differences constant; in this case the third differences vanish, or $\Delta^3 = 0$; our series, then, is limited to the three first terms. Put $\theta = \frac{t}{3^h}. \Delta' + \frac{t.(t-3^h)}{18^h}. \Delta^2$, then θ is the difference between the nearest distance in the almanac, and that actually observed. Δ' is found by taking the two distances in the table immediately preceding and following the nearest, subtracting the first from the second, and the second from the third, or the reverse, and then putting Δ' equal to the sum of the two differences thus obtained divided by two. Δ^2 is the difference of these two differences: all, then, is known in our equation but t. We obtain from it

$$\frac{t}{3^h} = \frac{\theta}{\Delta' - \frac{1}{2}\Delta^2. + \frac{1}{6}. \Delta^2. t}.$$

If we were to neglect Δ^2, we get $t = \frac{3^h. \theta}{\Delta'}$; which is the same thing as supposing the variation of the distance uniform, and is in fact the process we have already followed in our example. To correct the result, substitute in the denominator this approximate value of t; and then calculate t over again by the formula we have just given.

The best practical way of proceeding is to employ the table of the correction for second differences in the Nautical Almanac in the following way. In the example we have already been considering, the approximate value of t is $8^h\ 24^m\ 18^s$.

The proportional logarithm of 6^h is 2355, of 9^h 2343; the difference is 12: entering the table of second differences with 12 at the top of the column, and $2^h\ 20^m$ at the side, we get for the correction 2^s; since the proportional logarithms are *decreasing* the correction is to be *added*, and we obtain for the correct moment of Greenwich time corresponding to given lunar distance $8^h\ 24^m\ 20^s$.*

When the proportional logarithms decrease, Δ^2 is positive, and the correction is positive: it is negative in the contrary case. It may be as well to remark, that the proportional, or, as they are sometimes called, logistic logarithms, are in this case the logarithms of the fraction $\frac{3^h}{t}$, and are given in tables for every minute and second of t, t being a fraction of 3 hours: of course when $t = 3^h$ $\log. \frac{3^h}{t} = 0$.

When the sun or moon or both are observed near the horizon, it is necessary to make a correction on the semi-diameter. Since refraction acts in a vertical plane, it will elevate the upper limb of the sun or moon less than the lower, so as to give the disk the form of an ellipse. The vertical semi-diameter in consequence will not have the same value as the horizontal semi-diameter, and the differently inclined semi-diameters will have different values according to their inclination. As the star is generally brought into contact with the moon's disk at the extremity of an inclined semi-diameter, it is necessary to know how to correct this source of error. The tables give us the horizontal, whereas we wish to know the inclined semi-diameter. Generally let a and a' be the true altitudes of the upper and lower limb: r and r' the corresponding refractions; then $a + r$, $a' + r'$ will be the apparent altitudes: their difference $a - a' - (r - r')$ will be the vertical semi-diameter $= 2\ (d - \varrho)$ calling d the horizontal semi-diameter, and putting $r + r' = 2\,\varrho$. The disk will have the form of an ellipse whose semi-axes are d and $d - \varrho$. The compression $\mu = 1 - \frac{d - \rho}{d} = \frac{\varrho}{d}$. Hence the radius d', which makes with the horizon the angle θ

$$= d\ (1\ \mu\ \sin.^2\ \theta)$$
$$= d\ -\ \varrho.\ \sin.^2\ \theta.$$

As the correction in question is very small, it is not necessary to know θ very exactly; and, indeed, except within 10 degrees of the horizon, this correction may be neglected altogether.

However, in estimating the moon's apparent semi-diameter, it is quite necessary to take into account the augmentation produced by her altitude above the horizon. This subject has already been examined in the note to page 61; and the expression for the augmentation is there given: we shall only observe here that it may be found tabulated in the lunar tables, and several astronomical works.

Lastly, we may observe, that when a sextant is employed, and particularly at sea, that it is difficult to observe the altitudes of the two bodies with much precision: besides this method requires three observers, one for the altitude of the moon, a second for that of the sun or star, and a third for the distance. It is then in many instances preferable not to observe but to calculate the altitudes of the two bodies. This, it is true, requires a knowledge of the longitude, which is what we are seeking, but practically it is always known approximately, and the error on the calculated altitude would probably be less than the inevitable error of the observation. Calling t the hour angle, δ the declination, φ the latitude, a the altitude,

$$\sin. a = \sin. \varphi . \cos. \delta - \cos. \varphi . \sin. \delta . \cos. p;$$

when a has been calculated by this formula, we must apply the proper correction for refraction according to the existing state of the atmosphere, and also the correction for parallax, to the moon, in order to get the apparent altitude.

We shall conclude this subject with the following example of clearing the distance by the formula above given. Let us suppose that the distance of Regulus from the moon's centre has been observed at $14^h\ 59^m\ 59^s{\cdot}7$, and that it was 58° 25′ 36″; and that the calculated apparent altitudes of the star and moon's centre at that moment were 70° 34′ 9″ and 48° 0′ 49″, while the true altitudes were 70° 33′ 49″, and 48° 40′ 38″, respectively.

---

* Since the variation of the proportional logarithms is very small, it may be supposed proportional to the time.

$\therefore H + H' = 119^\circ\ 14'\ 27''$	$\theta = 58^\circ\ 25'\ 37''$	
$\frac{1}{2} \quad = \ 59\ \ 37\ \ 13{\cdot}5$	$h = 48\ \ \ 0\ \ 49$	$\therefore$ log. cos. $= 9{\cdot}8253962$
log. cos. $\frac{1}{2}(H+H') = 9{\cdot}7039156$	$h' = 70\ \ 34\ \ \ 9$	log. cos. $= 9{\cdot}5220119$
log. cos. $\phi \quad = 9{\cdot}9801947$	$177\ \ \ 0\ \ 34$	$-\ 9{\cdot}3474081$
log. sin. $\frac{1}{2}\Delta \quad = 9{\cdot}6841103$	$m = 88\ \ 30\ \ 17$	log. cos. $8{\cdot}4165499$
$\frac{\Delta}{2} = 28^\circ\ 53'\ 36''$	$m - \theta\ \ 30\ \ \ 4\ \ 41$	log. cos. $9{\cdot}9371885$
$\Delta = 57\ \ 12\ \ 12{\cdot}4$	$H = 48\ \ 40\ \ 38$	log. cos. $9{\cdot}8197415$
	$H' = 70\ \ 33\ \ 49$	log. cos. $9{\cdot}5221313$
		$18{\cdot}3482031$
		$\frac{1}{2} = 9{\cdot}1741016$
		log. cos. $\left(\frac{H+H'}{2}\right) = 9{\cdot}7039156$
		log. sin. $\phi = 9{\cdot}4701860$
		$\therefore\ \phi = 17^\circ\ 10'\ 20''{\cdot}6$

We owe the formula for clearing the distance which has been given above to the Chevalier Borda, a distinguished French navigator and astronomer, and it usually bears his name: it is the most simple of all the *accurate* formulæ which have been proposed for this purpose. But it is found very useful in navigation, to employ *approximate* formulæ, which, though not mathematically exact, are sufficiently true for practical purposes, and at the same time possess the advantage of a form, that permits their being put into tables. It is unnecessary to insist upon the advantage of this, not merely to navigators destitute of mathematical knowledge, but often to scientific men, when the number of observations that they have to reduce is considerable. One of the most accurate and convenient of these formulæ is the following, due to Simonoff, a Russian astronomer.

Adopting the same notation as before,

$$\begin{aligned} \text{Cos. } H \,.\, \cos. h' \,.\, \cos. \Delta &= \cos. H \,.\, \cos. H' \,.\, \cos. D' \\ &+ \tfrac{1}{2} \,.\, \sin. (H + h) \,.\, \sin. (H' - h') \\ &+ \tfrac{1}{2} \,.\, \sin. (H' + h') \,.\, \sin. (H - h) \end{aligned}$$

In this expression the two last terms are always very small, and may both be taken without interpolation from a small table, such as that which follows.

Expressing, as we may always do, $H' - h'$ and $H - h$ in minutes of degrees, instead of their sines, we have only to tabulate the value of $\frac{1}{2} \,.\, x \,.\, \sin. 1''$ for all the values of x. Thus, for example.

x	$\frac{1}{2} \,.\, x \,.\, \sin. 1'$.	x
1°	0·000002538	179°
2°	0·000005076	178°
3°	0·000007612	177°
4°	0·000010146	176°
5°	0·000012676	175°
etc.	. . . . . .	etc.

This table should be calculated for every ten minutes up to $x = 90^\circ = 180^\circ$, and then no interpolation will be required. The second and third terms are found by this table, entering it with the values of $H - h$ and $H' - h'$ respectively; the first must be calculated trigonometrically; and then, as well as the other two terms, divided by cos. h . cos. h', to get the value of cos. Δ, the cosine of the real distance.

We might, and this would be rather more accurate, construct for the two latter terms of our expression, a table of double entry; the argument at the head of the table would do for either $H' - h'$, or $H - h$, and should be carried up to 62′ or 63′, the utmost amount of parallax combined with refraction. The argument down the table would do either for $H + H'$ or $h + h'$, and should be carried up to 90°. The table may be constructed for greater or less intervals of space, according to our desire of accuracy, and wish to avoid interpolation. We may further remark, that if such a table be wanting, the two latter terms may be calculated trigonometrically. Lastly, we shall observe that the formula is as exact as that of Borda.

Let us take the following example:—

$$
\begin{aligned}
\Delta' &= 36^\circ\ 50'\ 22''\\
h' &= 26\ \ 47\ \ \ 4\\
H' &= 26\ \ 45\ \ 16\\
h &= 24\ \ 33\ \ 44\\
H &= 25\ \ 20\ \ 38
\end{aligned}
$$

Hence we get

$$
\begin{aligned}
H + h &= +\ 49^\circ\ 54'\ 22''\\
H - h &= +\ \ 0\ \ 46\ \ 54\\
H' + h' &= +\ 53\ \ 32\ \ 20\\
H' - h' &= -\ \ 0\ \ \ 1\ \ 48
\end{aligned}
$$

Consequently

$$
\begin{aligned}
\text{log. cos. } \Delta' &= 9 \cdot 9032630\\
\text{log. cos. } H &= 9 \cdot 9560507\\
\text{log. cos. } h &= 9 \cdot 9508242\\
\hline
& \ 9 \cdot 8101379 \qquad \therefore \text{ First term} = +\ 0 \cdot 64589.
\end{aligned}
$$

$$
\begin{aligned}
\text{Log. sin. } (H + h) &= 9 \cdot 8836558\ +\\
\text{Log. sin. } (H' + h') &= 6 . 7189986\ -\\
\hline
& \ 6 \cdot 6026544\ -\\
\text{Log. } \tfrac{1}{2} &= 9 \cdot 6989700\ +\\
\hline
& \ 6 \cdot 3016244\ - \qquad \text{Second term} = -\ 0 \cdot 00020.
\end{aligned}
$$

$$
\begin{aligned}
\text{Log. sin. } (H' + h') &= 9 \cdot 9053967\ +\\
\text{Log. sin. } (H - h) &= 8 \cdot 1348855\ +\\
\text{Log. } \tfrac{1}{2} &= 9 \cdot 6989700\ +\\
\hline
& \ 7 \cdot 7392522\ + \qquad \text{Third term} = +\ 0 \cdot 00549.
\end{aligned}
$$

$$
\begin{aligned}
\therefore \text{ Sum of the three terms} &= 0 \cdot 65115\\
\text{Log. sum} &= 9 \cdot 8136778
\end{aligned}
$$

$$
\begin{aligned}
\text{Log. cos. } h &= 9 \cdot 9588077\\
\text{Log. cos. } h' &= 9 \cdot 9507096\\
\hline
& \ 9 \cdot 9095173
\end{aligned}
\qquad
\begin{aligned}
& 9 \cdot 9095173\\
\hline
\text{log. cos. } \Delta &= 9 \cdot 9041605\\
\therefore\ \Delta &= 36^\circ\ 40'\ 50''
\end{aligned}
$$

which is exactly what is given by the formula of Borda*.

By Transits of the Moon and Moon-culminating Stars.

Let us suppose that one observer under the meridian of Greenwich observes the sidereal time at which the moon's centre is on the meridian, while a second observer, at a place whose longitude is unknown, observes the sidereal time of the same phenomenon on his own meridian. Were the moon immoveable, each observer would count exactly the same instant of sidereal time when the moon was on the meridian,

---

* To demonstrate the formula of Simonoff for clearing the distance, we must recur to the figure in page 22. Calling the angle at Z, θ, we have, by the fundamental equation of spherical trigonometry,

$$\cos. \Delta = \sin. H \,.\, \sin. H' + \cos. H \,.\, \cos. H' \,.\, \cos. \theta$$

$$\cos. \Delta' = \sin. h \,.\, \sin. h' + \cos. h \,.\, \cos. h' \,.\, \cos. \theta$$

$$
\begin{aligned}
\therefore \cos. h \,.\, \cos. h' \,.\, \cos. \Delta &= \cos. H \,.\, \cos. H' \,.\, \cos. \Delta' + \sin. H \,.\, \sin. H' \,.\, \cos. h \,.\, \cos. h'\\
&\quad - \cos. H \,.\, \cos. H' \,.\, \sin. h \,.\, \sin. h'\\
&= \cos. H \,.\, \cos. H' \,.\, \cos. \Delta'\\
&\quad + \tfrac{1}{4} \left\{ \sin. (H + h) + \sin. (H - h) \right\} \left\{ \sin. (H' + h') + \sin. (H' - h') \right\}\\
&\quad - \tfrac{1}{4} \left\{ \sin. (H + h) - \sin. (H - h) \right\} \left\{ \sin. (H' + h') - \sin. (H' - h') \right\}\\
&= \cos. H \,.\, \cos. H' \cos. \Delta'\\
&\quad + \tfrac{1}{2} \,.\, \sin. (H + h) \sin. (H' - h')\\
&\quad + \tfrac{1}{2} \,.\, \sin. (H' + h') \sin. (H - h)
\end{aligned}
$$

for the sidereal time of transit is the difference in time between the transits of the moon and the first point of Aries, or, in other words, it is the right ascension of the moon; consequently, did the right ascension remain invariable, the two sidereal times of which we have spoken would be the same. But the moon's right ascension is constantly increasing; if we suppose the second observer to be west of Greenwich, in the interval between the moon being on the meridian of Greenwich, and on his meridian she will have increased her right ascension, and, consequently, the sidereal time of transit, which is equivalent to it: and if we suppose the motion in right ascension uniform, the difference of the sidereal times of transit will be proportional to the difference of meridians.

Nothing can be simpler in principle than this method, but it requires, in practice, several precautions. In the first place, should there be any error in the position of the transit instrument, or in the rate of the clock, the whole of these errors will fall upon the longitude. This is eluded by both observers observing the transit of a star of very nearly the same declination as the moon, and not differing much from it in right ascension, and then each noting the difference in sidereal time between the two transits. This difference, from the moon's motion in right ascension, will not be the same under the different meridians; the variation in it will give the longitude as before. The advantage we have gained is, that we are dependent upon the clock only for the very short interval of time between the two transits of the star and moon: and as the former has very nearly the same declination with the latter, it is seen in the field of view without altering the position of the telescope: we are thus rendered independent of any error in the position of the transit instrument, as this will affect both star and moon equally. The stars which may be thus observed are called, in the Nautical Almanac, moon-culminating stars, and their apparent places are given in that Ephemeris for every day of the year, except the four days preceding and following new moon.

We have hitherto reasoned with regard to the centre of the moon; now practically this cannot be observed—what is done, in reality, is to observe the transit of the bright limb. If the moon's semidiameter be supposed constant during the interval of the observations, it is unimportant whether we observe the centre or the limb, which in both cases is equidistant from it. If the two places which are compared are not too far distant from each other, the variation of the semi-diameter in the interval will be insensible. But if this be not the case, it is necessary to correct the difference of the intervals of transit as follows. The increase of semidiameter of the moon arising from its altitude above the horizon does not alter the time employed by its semi-diameter in passing the meridian. At the two instants of observation let r and r' be the horizontal semi-diameters; δ and δ' the moon's declinations of the moon's centre at the times of transit; then $\frac{r}{15.\cos.\delta}$ $\frac{r'}{15.\cos.\delta'}$ will be respectively the distances in right ascension of the moon's centre from the meridian, when the limb is on the meridian; the difference of these two quantities is the correction to be applied to the longitude previously obtained. It is to be added, when the western limb has been observed, and subtracted if the eastern. We have supposed the accentuated letters to refer to the eastern station. As one of the meridians is supposed known, there will be no difficulty in calculating for that r and $\cos.\delta$: but for the other, we must take an approximate value of the longitude which does not require to be known accurately for this purpose.

When the meridians are not above one or two hours distant, we might (except where very great accuracy is required) neglect the above correction, as well as that we are just going to explain; it is, however, already allowed for in the column in the page of moon-culminating stars in the Nautical Almanac, which gives us the variation of right ascension corresponding to one hour difference of longitude: having ascertained, then, the variation of right ascension in our particular case, we may conclude by a simple proportion our difference of longitude. But if great accuracy be required, then take the variation in right ascension corresponding to the middle of the times between the observations, which may be obtained by interpolation between the numbers in the Nautical Almanac.

We have hitherto supposed that corresponding observations are actually made on the two meridians, but this is not necessary, though more exact. If we have at a given station observed the interval of transits between a moon-culminating star and the moon's limb, we may compare it with the *calculated* interval that takes place at Greenwich which may be deduced from the Nautical Almanac.

To illustrate more fully this method, let us suppose, that the apparent times of the moon's culmination (as determined by previous calculation approximately, to the nearest minute) for the two meridians are M and M′; then compute for these times, r, δ, r', δ'; and let t and t' represent the difference (in sidereal time) between the transit of the moon's limb and star at the places of observation. We shall suppose throughout the accentuated letters to refer to the eastern observatory. The observed difference of the right ascension of the moon's centre for the time elapsed between the two observations $= \Delta = (t - t') \pm \frac{r}{15 \cos. \delta} \pm \frac{r'}{15 \cos. \delta'}$ in which expression the signs are to be used, according to the rule above given.

Δ measures the increase of right ascension: if the moon's motion were uniform, it would be proportional to the difference of longitudes, but this is not the case. In order to proceed with accuracy, it is necessary to find the true solar Greenwich times of the culminations, which we shall call H and H′; and then to calculate from the Nautical Almanac the increase of right ascension corresponding to the interval H − H′: let $\alpha - \alpha'$ be the difference of right ascensions; convert this into solar time, and do the same for Δ; it is evident that $\alpha - \alpha'$ thus expressed ought to be equal to Δ similarly expressed. But this will generally not be the case; as in order to calculate H, we have employed the difference of longitudes deduced on the supposition, that the moon's motion was uniform. The whole error of the result is $(\Delta) - (\alpha - \alpha')$; it being understood that these quantities are now expressed in true solar time.

Let x be the error on the moon's motion; find, from the Nautical Almanac, the moon's motion in 24 hours of true solar time and expressed in time, and call it m; let s be the length of the true solar day: then

$$x : \Delta - (\alpha - \alpha') :: s : m$$

$$\therefore x = \{\Delta - (\alpha - \alpha')\} \frac{s}{m}$$

then x is the correction, which, added to L, the presumed difference of longitudes, will give us the true difference.

The operations will be better understood by an example. On December 5th, 1824, Lieutenant Forster observed the differences in the culmination of the moon, and the two stars 62 and 95 Tauri at Port Bowen; the presumed longitude being $5^h\ 55^m\ 40^s$ W. from Greenwich. These differences, in sidereal time, were as follows:—

	Greenwich.	Port Bowen.
62 Tauri	$t' = + 9^m\ 45^s\ 58$	$t = + 24^m\ 53^s\ 93$
95 . . .	$t = - 9\ \ 25\ \ 98$	$t = + 5\ \ 24\ \ 90$
The mean gives	$t' = + 0\ \ 9\ \ 80$	$t = + 15\ \ 18\ \ 44$

$$\therefore t - t' = + 15^m\ 8^s\ 64$$

Add supposed difference of longitudes $= 5^h\ 55\ 40$

Sum $= 6\ \ 10\ 48\ 64$

or in true solar time (roughly) $= 6^h\ 10^m$

This then is the approximate value of the time elapsed between the two culminations.

By the Nautical Almanac, the moon's first limb culminated at Greenwich at $11^h\ 34^m$: consequently (adding $6^h\ 10^m$) it culminated at Port Bowen at $17^h\ 44^m$ of Greenwich time. Calculating by the Nautical Almanac the declination and semi-diameter of the moon for those instants

at $11^h\ 34^m$	at $17^h\ 44^m$
$r' = 0^\circ\ 15'\ 42''$	$r = 0\ 15'\ 44''\ 39$
$\delta' = 23^\circ\ 39'\ 20''$	$\delta = 23\ 53\ 30$

$$\therefore \Delta = (t - t') + \frac{1}{15}\left(\frac{r}{\cos. \delta} - \frac{r'}{\cos. \delta'}\right) = 15^m\ 8^s\ 938$$

Calculate now the moon's right ascension for the times $H' = 11^h\ 34^m$ and $H = 17^h\ 43^m\ 41^s\ 67$

$$\begin{array}{rrrrr} \alpha = & 69 & 53 & 49 & 21 \\ \alpha' = & 66 & 6 & 29 & 93 \\ \hline \alpha - \alpha' = & 3 & 47 & 19 & 28 \\ \Delta \text{ (in space)} = & 3 & 47 & 14 & 07 \\ \hline \therefore \Delta - (\alpha - \alpha') = & & & -5 & 21 \end{array}$$

From the Nautical Almanac $\frac{s}{m} = \frac{24^h\ 4^m\ 22^s}{14^\circ\ 49'\ 24''}$

$$\therefore x = -5''\ 21\ \frac{s}{m} = -8^s\ 44$$

and the correct longitude $= 5^h\ 55^m\ 40^s - 8^s\ 44$
$= 5^h\ 55^m\ 31^s\ 56$

Longitude deduced from the Observation of the Occultations of Fixed Stars by the Moon.

The method we are about to explain is unfortunately seldom practicable at sea; a circumstance the more to be lamented, because it is the most exact of all astronomical methods, when the longitude is found by a single observation. It is, however, eminently useful in geography, and maritime surveying, and its importance will lead us to explain it in some detail. The only disadvantage which attaches to it is the length of the calculations by which the longitude is deduced.

In this method, the observer having ascertained, by previous calculations, within two or three minutes, the time at which a certain fixed star will disappear behind the moon's disk, or, as it is technically called, be occulted by the moon, places himself at the telescope, and waits carefully for the instant of its disappearance or immersion; the minute and second being carefully noted; the observer also endeavours to observe, if the circumstances admit of it, the reappearance of the star, or its emersion. The longitude may be found from either of these observations singly, but it is desirable to observe both, if possible. Let us suppose now, that a second observer, under a known meridian, has made a similar observation of the immersion: if the moon were so distant as to have no sensible parallax, the physical instant of the immersion would be the same for both observers, and the occultation would then be an instantaneous signal simultaneously observed under two different meridians, and the difference of the time counted by the observer would be the difference of longitudes. We may reason in the same way with regard to the emersion. But, in consequence of her parallax, the moon appears at the same instant to observers differently placed, to occupy different portions of the heavens, consequently the physical instant of immersion or emersion is different for these observers. We cannot then use an occultation as an instantaneous signal, but we may make it available for our purpose by deducing from it the true* position of the moon at a known physical instant, and the corresponding times counted under the two different meridians. It is evident, that since the moon's apparent semi-diameter may always be calculated, at the instant of occultation, we know exactly the distance of her centre from a certain fixed star, affected, however, by parallax and refraction; by proper corrections we may get the true distance, and the method becomes similar in principle to the method of lunar distances already explained.

Before entering more minutely into the nature of these reductions, we shall show how the instant of the expected occultation may be determined within the limits of two or three minutes. We require for this purpose an approximate knowledge of the longitude, but this is, in fact, what we always have. At sea, it is given by the log and compass; on land, by magnetic bearings and the distance of the day's marches, (supposing the country perfectly unknown;) or it may be found by an eclipse of one of the satellites of Jupiter, by a lunar distance, or other methods of less precision. Let

* The true position of the moon is her position as seen from the centre of the earth; this is always the same at the same moment, but will be reckoned differently according to the difference of meridian. This true position (concluded from the apparent) supplies the place of an instantaneous signal.

us call the apparent right ascension and declination of the fixed star to be occulted A and D; the true right ascension and declination of the moon α and δ; her apparent ditto α', and δ'; the equatorial parallax Π; horizontal semi-diameter ϱ; apparent semi-diameter ϱ'; μ the sidereal time; r the distance of the observer from the earth's centre; ϕ the latitude of the observer; ϕ' his latitude corrected for the ellipticity of the earth. The earth being supposed a spheroid of revolution, each terrestrial meridian is an ellipse, and, consequently, a normal to the surface at a given point of the meridian does not generally pass through the centre; the line drawn from the observer to his zenith makes an angle with the prolongation of the radius of the earth at that point, which is called the angle of the vertical. Let A P B D be a

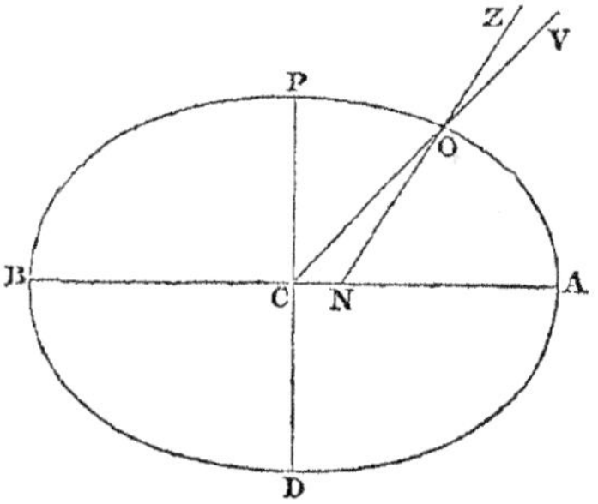

meridian, O, the place of the observer; O Z, a normal to the ellipse at O; C O V, a line from the centre C to the observer at O prolonged: then V O Z is the angle of the vertical. Prolong O Z till it meets the axis in N; the observed latitude is Z N A, while the reduced latitude will be V C A; their difference is the angle V O Z; from the observed latitude ϕ, then, we must subtract the angle of the vertical to get the geocentric or reduced latitude ϕ'. Now by the theory of the ellipse, calling the angle of the vertical θ, the compression p, where p = the ratio of the difference of the two axes to the axis major; and ϕ the observed latitude, then $\tan. \theta = p. \sin. 2\phi$.

It is also to be observed, that the horizontal parallax π is not the equatorial horizontal parallax, which we shall call Π, but that which corresponds to the latitude of the observer. For the earth being a spheroid, the horizontal parallax under different latitudes will be different. Let M be the distance from the centre of the earth to the centre of the moon: R the equatorial radius of the earth; then

$$M \sin. \Pi = R, \quad M \sin. \pi = r$$

$$\therefore \frac{\sin. \Pi}{\sin. \pi} = \frac{R}{r};$$

or since Π and π are small angles,

$$\frac{\Pi}{\pi} = \frac{R}{r}.$$

But by the theory of the ellipse,

$$r = R\,(1 - p. \sin.^2 \phi + \tfrac{5}{8}.p.^3 \sin.^2 2\phi \;\ldots)$$

$$\therefore \pi = \Pi\,(1 - p. \sin.^2 \phi. \;\ldots\ldots)$$

the first two terms of this series are generally sufficient. The equatorial horizontal parallax is given by the lunar tables; when it is known, we conclude from it the horizontal parallax corresponding to the latitude of the observer by means of the above series. If we take $p = \frac{1}{305}$, which is the value most usually assumed, then,

$$\log. p. = \bar{3}.5157002.$$

Let Σ be the angular distance of the star, and the centre of the moon, measured on a great circle; the angle made by this great circle with the circle of declination passing through the star, P; reckoning from 0 to 360° through P, so that P is between 0 and 180° if $\alpha' < A$, and between 180° and 360° if $\alpha' > A$, then

$$\sin \Sigma . \sin. P = -\cos. \delta' . \sin. (\alpha' - A),$$
$$\sin. \Sigma . \cos. P = \sin. \delta' . \cos. D - \cos. \delta' . \sin. D . \cos. (\alpha' - A).$$

But the formulæ which give the apparent place of the moon in terms of the true,* afford us

$$\Delta . \cos. \delta' . \sin. \alpha' = \cos. \delta . \sin. \alpha - r . \cos. \phi' . \sin. \pi . \sin. \mu,$$
$$\Delta . \cos. \delta' . \cos. \alpha' = \cos. \delta . \cos. \alpha - r . \cos. \phi' . \sin. \pi . \cos. \mu,$$

where Δ is the distance of the moon's centre from the place of observation. Substituting this above, we have

* These are known in astronomy as the formulæ of Olbers; the demonstration of them will be given in a subsequent part of this work.

$$\Delta . \sin. \Sigma . \sin. P = - \cos. \delta . \sin. (\alpha - A) + r . \cos. \phi' . \sin. \pi . \sin. (\mu - A),$$
$$\Delta . \sin. \Sigma . \cos. P = \sin. \delta . \cos. D - \cos. \delta . \sin. D . \cos. (\alpha - A),$$
$$- r . \sin. \pi \{\sin. \phi' . \cos. D - \cos. \phi' . \sin. D . \cos. (\mu - A)\}.$$

At the instant of immersion and emersion $\Sigma = \rho'$, and $\Delta \sin. \rho' = \sin. \rho$; also $\Delta . \sin. \Sigma = \sin. \rho$; whence Δ is eliminated, and we have

$$\sin. \rho . \sin. P = - \cos. \delta . \sin. (\alpha - A) + r . \cos. \phi' . \sin. \pi . \sin. (\mu - A),$$
$$\sin. \rho . \cos. P = \sin. \delta . \cos. D - \cos. \delta . \sin. D . \cos. (\alpha - A),$$
$$- r . \sin. \pi . \{\sin. \phi' . \cos. D - \cos. \phi' . \sin. D . \cos. (\mu - A)\}.$$

Put now $\sin. \rho = k . \sin. \pi$, ($k$ is a constant given in the lunar tables,) and divide out by $\sin. \pi$, then

$$k . \sin. P = - \frac{\cos. \delta . \sin. (\alpha - A)}{\sin. \pi} + r . \cos. \phi' . \sin. (\mu - A)$$

$$k . \cos. P = \frac{\sin \delta . \cos. D - \cos. \delta . \sin. D . \cos. (\alpha - A)}{\sin. \pi}$$
$$- r . \sin. \phi' . \cos. D - \cos. \delta . \sin. D . \cos. (\mu - A);$$

squaring each equation, and then adding them together,

$$k^2 = \left\{\frac{\cos. \delta . \sin. (\alpha - A}{\sin. \pi} - r . \cos. \phi' . \sin. (\mu - A)\right\}^2$$
$$+ \left\{\frac{\sin. \delta . \cos. D - \cos. \delta . \sin. D . \cos. (\alpha - A)}{\sin. \pi}\right.$$
$$\left. - r [\sin. \phi' . \cos. D - \cos. \phi' \sin. D . \cos. (\mu - A)]\right\}^2.$$

These equations are rigorous, but when our object is merely to approximate to the time of occultation, we may put

$$\frac{\cos. \delta . \sin. (\alpha - A)}{\sin. \pi} = \frac{\alpha - A}{\pi} \cos. \delta,$$

and
$$\frac{\sin. \delta . \sin. D - \cos. \delta . \sin. D. \cos. (\alpha - A)}{\sin. \pi} = \frac{\delta - D}{\pi}.$$

Suppose also $\quad r . \cos. \phi' . \sin. (\mu - A) \; . \; . \; . \; . = u,$

$$r . \sin. \phi' . \cos. D - r . \cos. \phi' . \sin. D . \cos. (\mu - A) = v.$$

Let us suppose now α, δ, π and μ calculated for a certain time T, which lies so near the time $(T + t)$ of the immersion or emersion that we are investigating, that the terms on the right hand side of the equation may be expanded in rapidly converging series according to the powers of t. Then for the time $T + t$

p becomes $p + p' t$
q ,, $q + q' t$
u ,, $u + u' t$
v ,, $v + v' t$

Here p' evidently $= \frac{\Delta \alpha}{\pi} . \text{cor.} \delta$

$$q' = \frac{\Delta \delta}{\pi}$$

where $\Delta \alpha$ and $\Delta \delta$ are the horary variations of the moon's right ascension and declination respectively.

In the expressions for u' and v', neglect t, then

$$u = r . \cos. \phi' . \sin. (\mu' - A)$$
$$v = r . \sin. \phi' . \cos. D - r . \cos. \phi' . \sin. D . \cos. (\mu' - A)$$
$$u' = r . \cos. \phi' . \lambda . \cos. (\mu' - A)$$
$$v' = r . \cos. \phi' . \lambda . \sin. (\mu' - A) . \sin. D$$

which we may put in this form:

$$u = a \qquad u' = b . \lambda$$
$$v = c - b . \sin. D . \qquad v' = a . \lambda . \sin. D ,$$

where λ is a known constant, $a = r.\cos.\varphi'.\sin.(\mu' - A)$

$$b = r.\cos.\varphi'.\cos.(\mu' - A)$$

$$c = r.\sin.\varphi'.\cos.D.$$

The values of p, q, p', and q' being known, our equation becomes

$$k^2 = \{p - u + (p' - u')\,t\}^2 + \{q - v + (q' - v')\,t\}^2$$

In this equation t is the unknown quantity, which being added to T will give us the time of immersion or emersion. To solve the equation conveniently, let us put

$$p - u = m.\sin.M \qquad p' - u' = n.\sin.N$$

$$q - v = m.\cos.M \qquad q' - v' = n.\cos.N$$

$$\therefore\ k^2 = m^2.\sin.^2(M - N) + \{m.\cos.(M - N) + n\,t\}^2$$

Let us put now $\dfrac{m}{k}.\sin.(M - N) = \cos.\psi$

$$\therefore\ t = -\frac{m}{n}.\cos.(M - N) \mp \frac{k}{n}.\sin.\psi$$

The upper sign is to be taken for the immersion, the lower for the emersion, provided we have taken $\psi < 180°$, which can always be done.

If $\dfrac{m}{k}.\sin.(M - N) > 1$,

then there is no occultation, the moon passes by the star without covering it. It is better, however, to try repeated approximations before we decide whether $\cos.\psi$ really is greater than unity, as this result may arise from the neglect of certain quantities in establishing our formula. It is perhaps more convenient, instead of calculating φ' from φ, to calculate at once

$$r.\cos.\varphi' = \frac{\cos.\varphi}{\sqrt{1 - e^2.\sin.\varphi}}$$

$$r.\sin.\varphi' = \frac{(1 - e^2).\sin.\varphi}{\sqrt{1 - e^2.\sin.^2\varphi}}\ ;$$

where e is the excentricity of the terrestrial meridian.

It is of importance, and more particularly with regard to the emersion, to be apprised beforehand of the part of the moon's disk where the star may be expected to disappear or reappear. If we do not know in observing an emersion to what part of the limb we ought to direct our attention, we are exposed to miss the observation altogether. The formula in page 233 may be put in this form

$$k.\sin.P = -m.\sin.M - n.\sin.N.t$$

$$k.\cos.P = -m.\cos.M + n.\cos.N.t$$

or substituting for t its value

$$k.\sin.P = -m.\sin.(M - N)\cos.N \pm k.\sin.N.\sin.\psi$$

$$k.\cos.P = -m.\sin.(M - N).\sin.N \mp k.\cos.N.\sin.\psi$$

and since

$$m.\sin.(M - N) = k.\cos.\psi$$

$$\sin.P = -\cos.(N \pm \psi)$$

$$\cos.P = -\sin.(N \pm \psi)$$

$$\therefore\ P = 270° - N \mp \psi$$

If we wish to reckon the place by means of the angle included between the two great circles drawn from the centre of the moon to the star and the north pole respectively, reckoned from the north to the left hand all round, then this angle Q very nearly $= 180° - P = N \pm \psi - 90°$. The calculation of Q, it may be observed, is hardly necessary except for an emersion: in this case the *lower* sign is to be taken.

Let us take the example given by Bessel, who first proposed this method, of the occultation of the star 82 Leonis on the 5th April, 1830. It is to be remarked, that in our expressions, one hour is taken for the unit of time: consequently t expresses a fraction of an hour. The occultation is reckoned for the meridian of Berlin; throughout the reckoning, only minutes and tenths of minutes are taken into account. It is found that

$$T = 7^h. \quad \mu' = 118°\,32' \quad \mu' - A = -50°\,42' \quad \pi = 54'\,18$$

$$\alpha = 168°\,37'\,9 \qquad \delta = 4°\,44'\,0 \qquad \Delta\alpha = 28{\cdot}5$$

$$\underline{A = 169\;\;14\;\;1} \qquad \underline{D = 4\;\;14\;\;1} \qquad \Delta\delta = -9{\cdot}1$$

$$\alpha - A = -36'\,2 \qquad \delta - D = +29\;\;9$$

$$p = -0{\cdot}6656 \qquad u = -0{\cdot}4717 \qquad \log. m.\sin. M = 9{\cdot}2876 -$$

$$q = +0{\cdot}5523 \qquad v = +0{\cdot}7591 \qquad \log. m.\cos. M = 9{\cdot}3156 -$$

$$p' = +0{\cdot}5242 \qquad u' = +0{\cdot}1014 \qquad \log. n.\sin. N = 9{\cdot}6261$$

$$q' = -0{\cdot}1683 \qquad v' = -0{\cdot}0092 \qquad \log. n.\cos. N = 9{\cdot}1892 -$$

$$M = 223°\,9' \quad \log. m = 9{\cdot}4525 \quad N = 110°\,5' \quad \log. n = 9{\cdot}6533$$

$$*\log.\frac{m}{k} = 0{\cdot}0171 \qquad \log. -\frac{m}{n} = 9{\cdot}7992 - \qquad \log.\frac{k}{n} = 9{\cdot}7821$$

$$\log.\sin.(M-N) = 9{\cdot}9638 \quad \log.\cos.(M-N) = 9{\cdot}5931 - \quad \log.\sin.\psi = 9{\cdot}4626$$

$$\psi = 16\cdot 52' \qquad \frac{-m}{n}\cdot\cos.(M-N) = 0{\cdot}247 \mp \frac{k}{n}\cdot\sin.\psi = \mp 0{\cdot}176$$

$$\therefore t = 0{\cdot}247 \mp 0{\cdot}176 = 0{\cdot}071 \text{ or } 0{\cdot}423$$

$$\therefore \text{ for the immersion } T + t = 7^h + 0{\cdot}071 = 7^h{\cdot}4^m \quad Q = 37°$$

$$\text{for the emersion } T + t = 7^h + 0{\cdot}428 = 7^h{\cdot}25^m \quad Q = 3°$$

The Berlin Ephemeris gives for every expected occultation the quantities p, q, p' and q' as also the angle $\mu' - A$: the four first quantities remain the same for any other meridian; the last is to be increased by the longitude counted from Berlin when to the east of that place, and to be diminished by that longitude when to the west; supposing that we make use of the Berlin Ephemeris. In the Nautical Almanac p, q, p' and q' are not given, but they may be very easily calculated, as in that work the moon's right ascension and declination are given for every hour. In both of these works nearly all the occultations worth observing which occur throughout the year are indicated; and a short calculation by the method above explained will show us whether the occultation takes place at all at the spot where we happen to be, and if it does, the times of immersion and emersion, and the angles from the north point.

Having now shown how the approximate time of the occultation, and the place of the disk where it is to take place, may be predicted, (which are necessary preparations for the observation,) we will now consider the converse problem; and supposing the occultation to have been observed, show how to deduce from it the longitude of the observer. The principle of the method is, as has been before explained, to determine from the observed instant of immersion or emersion the instant of the apparent conjunction of the moon and star. From the time of apparent conjunction we calculate the time of the true conjunction. If the occultation has been observed under a second meridian, we find in the same way the instant of true conjunction at the second place: now the physical instant of this is the same for both places; the difference of the times found for it is the difference of meridians, that is of longitudes. If the occultation has not been observed under a second meridian, we must calculate the time of true conjunction for Greenwich, and compare this time with that deduced from the observation; the difference will be the longitude from Greenwich counted in time. It may be as well to remind the reader that the moon is said to be in conjunction with a star, when she has either the same longitude or the same right ascension as the star. In the calculation of occultations it is the conjunction in right ascension which we shall consider. The true conjunction is the conjunction as seen from the centre of the earth; it differs from the apparent conjunction by the effects of the parallax in right ascension.

Let P be the pole of the equator, M the centre of the moon, S the star at the instant of occultation: S M will be the moon's apparent semidiameter; call it Δ': let S Q be a parallel of declination passing through S, A B the arc of the equator intercepted between the hour circles P S and P M prolonged; then M Q = the apparent difference of declinations $= \delta$; $\frac{SQ}{\cos. AS}$ = apparent dif-

* Log. $k = 9{\cdot}4353665$.

ference of right ascensions; call it α. Now the triangle S M Q being necessa-

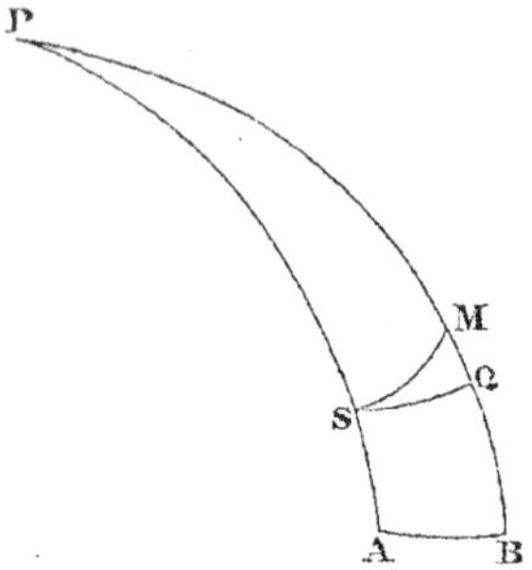

rily very small, we may solve it as a plane triangle, for S M, the greatest side, cannot exceed between 16 and 17 minutes. Consequently

$$SQ^2 = \Delta'^2 - \delta^2 = (\Delta' + \delta)(\Delta' - \delta)$$

$$\therefore \alpha^2 = \frac{(\Delta' + \delta)(\Delta' - \delta)}{\cos. d}$$

putting $d = AS$.

Let now ☽ and ☽$'$ be the true and apparent right ascensions of the moon, ✳ the right ascension of the star, ϖ the moon's parallax in right ascension.

Now ☽ = ☽$'$ $- \varpi$

and at the immersion α = ✳ − ☽$'$

∴ ✳ − ☽ = $\alpha + \varpi$

for the emersion ☽ > ✳, and

☽ − ✳ = $\alpha - \varpi$.

Generally we have for the difference of the true right ascensions, calling it c,

$$c = \alpha \pm \varpi,$$

taking the superior sign for the immersion.

We have now only to find the time taken by the moon to describe the arc c, to get the time from the true conjunction. Knowing the moon's horary motion, we find, by a simple proportion, the time she takes to describe the arc c; call this time t. Then

$$t = \frac{3600^s}{m} . c$$

$$= \frac{3600^s}{m} . (\alpha \pm \varpi)$$

when m is the arc described by the moon in one hour, or 3600^s.

Let T be the observed instant of immersion or emersion, then $T \pm t$ will be the moment of the true conjunction, taking, as before, the superior sign for the immersion.

It is necessary to know how to calculate the parallaxes in right ascension and declination.

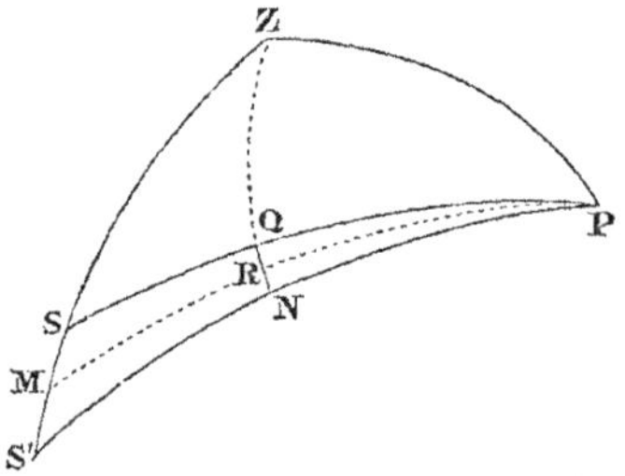

In the figure let us suppose S to be the place of the star as seen from the centre of the earth, S$'$ its apparent place, that is the true place depressed by parallax, Z the zenith, P the pole of the equator; then Z S S$'$ will be in a vertical plane; S S$'$ will be the parallax in altitude; the angle S P S$'$ will be the parallax in right ascension, or ϖ; the difference between P S and P S$'$ will be the parallax in declination, or π. When we know π, we must add it to the moon's true declination calculated for the moment T, to get her apparent declination: from this we must subtract the star's declination to get δ.

Let us call the *true* zenith distance ζ, the *true* declination D, the *true* right ascension AR; the corresponding *apparent* quantities ζ', D$'$, and AR$'$. From the spherical triangle S P S$'$, we have

$$\frac{\sin. SS'P}{\sin. PS} = \frac{\sin. SPS'}{\sin. SS'}$$

and from the triangle Z P S$'$

$$\frac{\sin. ZPS'}{\sin. ZS'} = \frac{\sin. SS'P}{\sin. PZ}$$

multiplying these two equations together, we have

$$\frac{\sin. ZPS'}{\sin. PS . \sin. ZS'} = \frac{\sin. SPS'}{\sin. SS' . \sin. PZ}$$

or calling the hour angle Z P S, h; the geographical latitude ϕ; and the parallax in altitude p;

$$\frac{\sin.(h+\varpi)}{\cos. D . \sin. \zeta'} = \frac{\sin. \varpi}{\sin. p . \cos. \phi.}$$

But if Π be the horizontal parallax, then

$$\sin. p = \sin. \Pi . \sin. \zeta',$$

$$\therefore \frac{\sin.(h+\varpi)}{\cos. D . \sin. \zeta'} = \frac{\sin. \varpi}{\sin. \Pi . \sin. \zeta' . \cos. \phi}$$

$$\frac{\sin.(h+\varpi)}{\cos. D} = \frac{\sin. \varpi}{\sin. \Pi . \cos. \phi} \quad . \; . \; . \; . \; . \; . \; . \; . \quad (1)$$

$$\sin. \Pi . \cos. \phi (\sin. h + \cos. h . \tan. \varpi) = \tan. \varpi . \cos. D,$$

$$\therefore \tan. \varpi = \frac{\sin. \Pi . \cos. \phi . \sin. h}{\cos. D - \sin. \Pi \cos. \phi . \cos. h}.$$

Assume $$\cos. \psi = \frac{\sin. \Pi . \cos. \phi . \cos. h}{\cos. D}$$

$$\therefore \tan. \varpi = \frac{\sin. \Pi . \cos. \phi . \sin. h}{\cos. D (1 - \cos. \psi)}$$

$$= \frac{\sin. \Pi . \cos. \phi . \sin. h}{2 \cos. D . \sin.^2 \frac{\psi}{2}}$$

a formula adapted to logarithmic computation, and from which the arc ϖ may be found.

When it happens that the moon's declination is not very great at the time of an occultation, it may be more convenient to calculate ϖ by a series.

Put $$q = \frac{\sin. \Pi . \cos. \phi}{\cos. D}$$

then $$\tan. \varpi = \frac{q . \sin. h}{1 - q \cos. h}$$

$$= q . \sin. h \{ 1 + q . \cos. h + q^2 . \cos.^2 h \; . \; . \; . \}$$

Put for $\tan. \varpi$, $\varpi \sin. 1''$ to express ϖ in seconds, and we obtain

$$\varpi = \frac{q . \sin. h}{\sin. 1''} + \frac{q^2 . \sin. 2h}{2 \sin. 1''} + \frac{q^3 . \sin. 3h}{3 . \sin. 1''} \; . \; . \; . \; . \; .$$

q is always very small, since it has for factor $\sin. \Pi$ which is always very small; consequently our series is, in general, sufficiently convergent to enable us to neglect the third term.

Again, $$\cos. P Z S = \frac{\sin. D - \cos. \zeta . \sin. \phi}{\sin. \zeta . \cos. \phi} = \frac{\sin. D' - \cos. \zeta' . \sin. \phi}{\sin. \zeta' . \cos. \phi}$$

$$\therefore \sin. \zeta' (\sin. D - \cos. \zeta . \sin. \phi) = \sin. \zeta (\sin. D' - \cos. \zeta' . \sin. \phi)$$

$$\sin. \zeta' . \sin. D - \sin. \phi (\sin. \zeta' . \cos. \zeta - \cos. \zeta' \sin. \zeta) = \sin. \zeta . \sin. D'$$

$$\therefore \sin. \zeta' . \sin. D - \sin. \phi . \sin. (\zeta' - \zeta) = \sin. \zeta . \sin. D'.$$

But $$\sin. (\zeta' - \zeta) = \sin. p = \sin. \Pi . \sin. \zeta'$$

$$\therefore \sin. \zeta' (\sin. D - \sin. \phi . \sin. \Pi) = \sin. \zeta . \sin. D'.$$

But we have $$\sin. P Z S = \frac{\cos. D . \sin. h}{\sin. \zeta} = \frac{\cos. D' . \sin. (h + \varpi)}{\sin. \zeta'}$$

$$\therefore \sin. \zeta . \cos. D' = \frac{\sin. \zeta' . \cos. D . \sin. h}{\sin. (h + \varpi)}.$$

Dividing the equation obtained just before by this last, we have

$$\tan. D' = \frac{\sin. D - \sin. \Pi . \sin. \phi}{\cos. D . \sin. h} . \sin. (h + \varpi).$$

This equation may be put under the form

$$\frac{\sin.\ \phi\ .\ \sin.\ \Pi}{\cos.\ D} = \tan.\ D = \frac{\sin.\ h\ .\ \tan.\ D'}{\sin.\ (h+\varpi)}.$$

Now, we have

$$\tan.\ D - \tan.\ D' = \frac{\sin.\ D}{\cos.\ D} - \frac{\sin.\ D'}{\cos.\ D'}$$

$$= \frac{\sin.\ D\ .\ \cos.\ D' - \cos.\ D\ .\ \sin.\ D'}{\cos.\ D\ .\ \cos.\ D'}$$

$$= \frac{\sin.\ (D - D')}{\cos.\ D\ .\ \cos.\ D'} = \frac{\sin.\ \pi}{\cos.\ D\ .\ \cos.\ D'}.$$

Adding now this equation to the preceding one, we have

$$\frac{\sin.\ \phi\ .\ \sin.\ \Pi}{\cos.\ D} - \tan.\ D' = \frac{\sin.\ \pi}{\cos.\ D\ .\cos.\ D'} - \frac{\sin.\ h\ .\ \tan.\ D'}{\sin.\ (h+\varpi)}$$

$$\frac{\sin.\ \phi\ .\ \sin.\ \Pi}{\cos.\ D.} = \tan.\ D' \left\{1 - \frac{\sin.\ h}{\sin.\ (h+\varpi)}\right\} + \frac{\sin.\ \pi}{\cos.\ D\ .\ \cos.\ D'}.$$

But $1 - \frac{\sin.\ h}{\sin.\ (h+\varpi)} = \frac{\sin.\ (h+\varpi) - \sin.\ h}{\sin.\ (h+\varpi)}$

$$= \frac{2\ \sin.\ \frac{\varpi}{2}\ .\ \cos.\ \left(h+\frac{\varpi}{2}\right)}{\sin.\ (h+\varpi)}.$$

[Multiplying above and below by $\cos.\ \frac{\varpi}{2}$, and recollecting that $2\ \sin.\ \frac{\varpi}{2}$. $\cos.\ \frac{\varpi}{2} = \sin.\ \varpi$].

$$= \frac{\sin.\ \varpi\ .\ \cos.\ \left(h+\frac{\varpi}{2}\right)}{\sin.\ (h+\varpi)\ .\ \cos.\ \frac{\varpi}{2}}$$

$$= \frac{\sin.\ \Pi\ .\ \cos.\ \phi}{\cos.\ D\ .\ \cos.\ \frac{\varpi}{2}}\ .\ \cos.\ \left(h+\frac{\varpi}{2}\right) \text{ by Equation (1) p. 237}$$

Consequently substituting above this quantity,

$$\frac{\sin.\ \phi\ .\ \sin.\ \pi}{\cos.\ D} = \tan.\ D'.\frac{\sin.\ \Pi\ .\ \cos.\ \phi}{\cos.\ D.\ \cos.\ \frac{\varpi}{2}}\ .\ \cos.\ \left(h+\frac{\varpi}{2}\right) + \frac{\sin.\ \pi}{\cos.\ D\ .\ \cos.\ D'}$$

$$\therefore\ \sin.\ \pi' = \sin.\ \phi.\ \sin.\ \pi\ .\ \cos.\ D' - \frac{\sin.\ D'\ .\ \sin.\ \Pi\ .\ \cos.\ \phi\ .\ \cos.\ \left(h+\frac{\varpi}{2}\right)}{\cos.\ \frac{\varpi}{2}}$$

Assume $\cot.\ \chi = \frac{\cot.\ \phi\ .\ \cos.\ \left(h+\frac{\varpi}{2}\right)}{\cos.\ \frac{\varpi}{2}}$

$$\sin.\ \pi = \sin.\ \phi\ .\ \sin.\ \Pi\ .\ (\cos.\ D'. - \sin.\ D' \cot.\ \chi)$$

$$= \sin.\ \phi.\ \sin.\ \Pi \left(\frac{\cos.\ D\ .\ \sin.\ \chi - \sin.\ D'.\ \cos.\ \chi}{\sin.\ \chi}\right)$$

$$= \frac{\sin.\ \phi\ \sin.\ \Pi}{\sin.\ \chi}\ .\ \sin.\ (\chi - D')$$

Take $n = \dfrac{\sin.\phi.\sin.\Pi}{\sin.\chi}$

$$\sin.\pi = n.\sin.(\chi - D + \pi)$$
$$= n.\sin.(\chi - D).\cos.\pi + \cos.(\chi - D).\sin.\pi$$
$$\tan.\pi = n.\sin.(\chi - D) + n.\cos.(\chi - D).\tan.\pi$$
$$\therefore \tan.\pi = \frac{n.\sin.(\chi - D)}{1 - n.\cos.(\chi - D)}$$
$$= n.\sin.(\chi - D)\{1 + n.\cos.(\chi - D) + n^2.\cos.^2(\chi - D)\ldots\}$$

or rather $\pi = \dfrac{n.\sin.(\chi - D)}{\sin.1''} + \dfrac{n^2.\sin.2.(\chi - D)}{2.\sin.1''} + \ldots\ldots$

it will be rarely necessary to take into account the third term.

We must begin then by calculating π, or the parallax in right ascension, then the angle χ, then n, and hence the parallax in declination π. This applied with its proper sign to the true declination as deduced from the tables, or the Nautical Almanac, gives us the apparent declination. The difference between this and the declination of the star is the quantity δ. In calculating the parallaxes of the sun and planets, the first terms of our respective series for the effects in right ascension and declination are sufficient; and it is not necessary to take into account the spheroidal figure of the earth. But in calculating these quantities for the moon, as in the instance of an occultation, it is absolutely necessary to allow for the ellipticity of the terrestrial meridians. Now this is done as has been explained in a previous part of the work by substituting for ϕ the astronomical latitude, ϕ' the geocentric latitude, which is deduced from ϕ, by the formula given above, page 232. We must also for Π, which is the equatorial horizontal parallax, substitute Π', the horizontal parallax corresponding to the latitude ϕ. The series for Π' in terms of Π has been already given.

We have already shown how the augmentation of the moon's apparent semidiameter may be calculated when we know the zenith distance. But in calculating an occultation we must find this augmentation without a knowledge of ζ or ζ'. In the triangle S P S′ (figure, p. 236) bisect the angle S P S′ by the arc S M; and draw from Z, Z N perpendicular to P M, and intersecting P S in Q and P M in R: it is evident that Q P N is an isosceles triangle, in which P Q = PN, and the angle at Q = the angle at N. Now the triangles P Z R, P Q R which have both a right angle at R, give us

$$\tan.PR = \tan.PZ.\cos.ZPR = \cot.\phi.\cos.\left(h + \frac{\varpi}{2}\right)$$

$$\tan.PQ = \frac{\tan.PR.}{\cos.QPR} = \frac{\tan.PR}{\cos.\frac{\varpi}{2}}$$

Assume $PQ = 90^\circ - \epsilon$

$$\therefore \cot.\epsilon = \frac{\cot.\phi.\cos.\left(h + \frac{\varpi}{2}\right)}{\cos.\frac{\varpi}{2}}$$

Now $SQ = PS - PQ = (90^\circ - D) - (90^\circ - \epsilon) = \epsilon - D$

$S'N = PS' - PN = \epsilon - D' = \epsilon - D + \pi.$

But since the angle at Q = the angle at N. sin. Z Q S = sin. Z N S; and P Z S is the same for the two triangles Q Z S, N Z S′.

$$\therefore \frac{\sin.ZS}{\sin.QS} = \frac{\sin.ZQS}{\sin.QZS} = \frac{\sin.ZS'}{\sin.S'N}$$

$$\frac{\sin.\zeta}{\sin.(\epsilon - D)} = \frac{\sin.\zeta'}{\sin.(\epsilon - D + \pi)}$$

$$\therefore \frac{\sin. \zeta'}{\sin. \zeta} = \frac{\sin. (\iota - D + \pi)}{\sin. (\iota - D)} = \frac{\sin. \Delta'}{\sin. \Delta}$$

$$= \frac{\Delta'}{\Delta} \text{ approximately}$$

$$\therefore \Delta' = \Delta . \frac{\sin. (\iota - D + \pi)}{\sin. (\iota - D)}$$

∴ the augmentation of the apparent semidiameter, or $\Delta' - \Delta$

$$= \Delta . \frac{\sin. (\iota - D + \pi) - \sin. (\iota - D)}{\sin. (\iota - D)}$$

$$= \Delta . \frac{2 \sin. \frac{\pi}{2} . \cos. \left(\iota - D + \frac{\pi}{2}\right)}{\sin. (\iota - D)}$$

$$= \Delta . 2 \sin. \frac{\pi}{2} . \left\{\cot. (\iota - D). \cos. \frac{\pi}{2} - \sin. \frac{\pi}{2}\right\}$$

$$= \Delta . \sin \pi . \cot. (\iota - D) - \sin.^2 \frac{\pi}{2}$$

The last term, never exceeding a quarter of a second, may be neglected, and substituting for sin. π, π sin. 1″, in order to get the whole expressed in seconds as Δ is already, then finally

$$\Delta' - \Delta = \Delta . \pi . \sin. 1'' \cot. (\iota - D)$$

The following example will illustrate the use of the formula and the method above given. Let us suppose that the immersion of Aldebaran has been observed at Paris on the 5th of October, 1830, at $10^h\ 21^m\ 9''$ mean time, and that we wish to find the time of the true conjunction. The position of the star is

$$✳ = 62^\circ\ 32'\ 36''\cdot 9 \qquad d = +\ 15^\circ\ 12'\ 39''\cdot 0$$

the positive sign being taken when the declination is north.

For the moon

☽ =	$61^\circ\ 43'\ 59''\cdot 26$	D =	$+\ 15^\circ\ 39'\ 27''\cdot 23$
h =	$-\ 72^\circ\ 21'\ 31''\cdot 3$	m =	$37'\ 13''\cdot 73$
Δ =	$16'\ 24''\cdot 2$	ϕ' =	$48^\circ\ 38'\ 27''\cdot 6$
Π =	$60'\ 13''\cdot 3$	Π' =	$60'\ 6''\cdot 62$

Hence we calculate, by the series above given, the parallaxes in right ascension and declination, and we find*

$$\varpi = -\ 39'\ 26''\cdot 36 \qquad \pi = 40'\ 31''\cdot 64$$

---

* The details of the calculation of the parallaxes are as follows:—

Parallax in Right Ascension.

log. sin. Π' = 8·2426531
log. cos. ϕ' = 9·8200536
8·0627067
log. cos. D = 9·9835776

log. q = 8·0791291 +	log. q^2 = 6·15826 +	log. q^3 = 4·2373 +
log. sin. h = 9·9790804 −	log. sin. $2h$ = 9·76163 −	log. sin. $3h$ = 9·7802 +
8·0582095 −	5·91989 −	4·0175 +
log. $\frac{1}{\sin. 1''}$ = 5·3144251 +	log. $\frac{1}{\sin. 2''}$ = 5·01340 +	log. $\frac{1}{\sin. 3''}$ = 4·8373 +
3·3726346 −	0·93329 −	8·8548 +
First term = − 39′ 18″·50	Second term = − 8″·58	Third term = + 0″·72.

∴ Finally ϖ = − 39′·26″·36.

Parallax in Declination.

ϕ' = 48° 38′ 27″·6 log. cot. ϕ' = 9·9446542

$h + \frac{\varpi}{2}$ = − 72 41 14·5 log. cos. $\left(h + \frac{\varpi}{2}\right)$ = 9·4736118

9·4182660

$\therefore$ the apparent right ascension and declination are respectively

$$☽' = 61^\circ\ 7'\ 32''\cdot 90 \qquad D' = 14^\circ\ 58'\ 55''\cdot 59$$

and the apparent semi-diameter $\Delta' = 16'\ 30''\cdot 9$ *.

Hence $d - D' = \delta = 13'\ 43''\cdot 41$

$$\Delta' + \delta = 30'\ 14''\cdot 38 \qquad \log.\ 30'\ 14''\cdot 38 = 3\cdot 2587163$$
$$\Delta' - \delta = 2'\ 47''\cdot 51 \qquad \log.\ 2'\ 47''\cdot 51 = 2\cdot 2240407$$
$$5\cdot 4827570$$
$$\therefore \log.\ \sqrt{(\Delta' + \delta)(\Delta' - \delta)} = 2\cdot 7413785$$
$$\log.\cos.\ d = 9\cdot 9845124$$
$$\therefore \log.\ \alpha = 2\cdot 7568661$$
$$\therefore \alpha = 571''\cdot 30 = 9'\ 31''\cdot 30$$
$$\varpi = 36\ \ 26\ \cdot 36$$
$$\therefore \alpha - \varpi = -\ 26\ \ 55\ \cdot 06$$
$$\log.(\alpha - \varpi) = 2\cdot 2081887 -$$
$$\log.\ 3600 = 3\cdot 5563025 +$$
$$5\cdot 7644912 -$$
$$\log.\ m = 3\cdot 3490307 +$$
$$\log.\ t = 2\cdot 4154605 -$$
$$\therefore t = -260^s\ 29 = -\ 4^m\ 20^s\cdot 29$$
$$T = 10^h\ 21\ \ 9\ \ 0$$
$$\text{time of true conjunction} = T + t = 10\ \ 16\ \ 48\ \ 7$$

Let us suppose now that the immersion has also been observed at Greenwich, and that in a similar way the instant of true conjunction has been found to be $10^h\ 26^m\ 7^s\cdot 7$: then Greenwich will be $9^m\ 21^s$ to the west of Paris. If the immersion or emersion has not been observed at Greenwich, then we must find by the Nautical Almanac the instant of Greenwich time at which the moon has the same right ascension as the star; but it is better, when it can be done, to employ the conjunction deduced from actual observation, as thus we are independent of the errors of the lunar tables in right ascension.

$$\frac{\varpi}{2} = -\ 19'\ 43''\cdot 18 \qquad \log.\cos.\ \frac{\varpi}{2} = 9\cdot 9999929$$
$$x = 75\ \ 19\ \ 10\cdot 0 \qquad \log.\cot.\ x = 9\cdot 4182631$$
$$D = 15\ \ 39\ \ 27\cdot 2$$
$$x - D = 59\ \ 39\ \ 42\cdot 8$$

Now
$$\log.\sin.\ n' = 8\cdot 2426531$$
$$\log.\sin.\ \phi' = 9\cdot 8753993$$
$$8\cdot 1180524$$
$$\log.\sin.\ x = 9\cdot 9855853$$
$$\log.\ n = 8\cdot 1324671 \qquad \log\ n^2 = 6\cdot 26493$$
$$\log.\sin.(x - D) = 9\cdot 9360409 \qquad \log.\sin.2(x - D) = 9\cdot 94045$$
$$8\cdot 0685080 \qquad 6\cdot 20538$$
$$\log.\frac{1}{\sin.\ 1''} = 5\cdot 3144253 \qquad \log.\frac{1}{\sin.\ 2''} = 5\cdot 01340$$
$$3\cdot 3829331 \qquad 1\cdot 21878$$
$$\text{First term} = +\ 40'\ 16''\cdot 9 \qquad \text{Second term} = +\ 16''\cdot 45$$
$$\therefore \text{finally}\ \pi = 40'\ 31''\cdot 64.$$

*
$$\log.\ \Delta = 2\cdot 9930834$$
$$\log.\ \pi = 3\cdot 3858993$$
$$\log.\sin.\ 1'' = 4\cdot 6855749$$
$$\log\cot.(x - D) = 9\cdot 767338.$$
$$\log.(\Delta' - \Delta) = 0\cdot 8318957$$
$$\therefore \Delta' - \Delta = 6''\cdot 7$$
$$\therefore \Delta' = 16'\ 30''\cdot 9$$

To explain how those tables may be corrected by an occultation, we must suppose the observation to be perfectly exact, but that the errors produce on the value of t, calculated as above shown, a small error $d\,t$. Reverting to the expression for t, and putting the constant factor $\frac{3600^s}{m} = a$.

$$d\,t = a\,(d\,\alpha \pm d\varpi)$$

But since $\alpha^2 = \Delta'^2 - \delta^2$

$$\alpha\, d\alpha = \Delta' d\,\Delta' - \delta d\delta = \Delta' d\,\Delta' - D' d D'$$

Substituting this value of $d\,\alpha$, in the expression for $d\,T$, we have

$$d\,t = \frac{a}{\alpha}\left\{\Delta' d\,\Delta' - D' d\,D' \pm \alpha\, d\,\varpi\right\}$$

Now our formula for the parallax in right ascension (page 237) may, considering that ϖ and Π are very small, be put in this form

$$\varpi = \Pi'.\frac{\cos.\,\phi.\,\sin.\,h}{2\cos.\,D.\,\sin.^2\frac{\psi}{2}}$$

we wish to find, supposing a small error on Π', the corresponding error on ϖ. The supposed small error on Π' will not affect sensibly the factor of that quantity in our expression,

$$\therefore d\varpi = d\,\Pi'.\frac{\cos.\,\phi.\,\sin.\,h}{2\cos.\,D.\,\sin.^2\frac{\psi}{2}}$$

$$= d\,\Pi'.\frac{\varpi}{\Pi'}$$

Again $D' = D - \pi$

$$\therefore d\,D' = d\,D - d\,\pi$$

and reasoning for $d\,\pi$ as for $d\,\varpi$

$$d\,D' = d\,D - \frac{\pi.\,d\,\Pi'}{\Pi'}.$$

$$\therefore\ d\,t = \frac{a}{\alpha}\left\{\Delta' d\,\Delta' - D' d\,D + \frac{\pi\,D' \pm \varpi\,\alpha}{\Pi}\,d\,\Pi'\right\}$$

$$= A\,d\,\Delta' - B\,d\,D + C\,d\,\Pi'$$

$d\,\Delta'$ is the error of the tables on the apparent semi-diameter, $d\,D$ on the true declination, and $d\,\Pi'$ on the horizontal parallax of the moon.

Now if the immersion is observed at a second place, at the time T', $d\,\Delta'$, $d\,D$, and $d\,\Pi'$ will be the same, the coefficients will be A', B', C', and the true time of conjunction

$$T' + t' + A'\,d\,\Delta' - B'\,d\,D + C'\,d\,\Pi',$$

while for the first place it was

$$T + t + A\,d\,\Delta' - B\,d\,D + C\,d\,\Pi'.$$

We have then for the difference of longitudes

$$T' - T + t' - t + (A' - A)\,d\,\Delta' - (B' - B)\,d\,D + (C' - C)\,d\,\Pi':$$

if upon calculating the co-efficients $A' - A$, $B' - B$, $C' - C$, we find them very small, the errors of the tables will produce no sensible effect on the longitude, and may be neglected.

But when the object is to determine the errors of the tables, then we must suppose the emersion at the first place to have been observed as well as the immersion; the time of true conjunction deduced from this should be the same as that deduced from the immersion, this will not generally be the case: suppose the error in the latter case to be $d\,t'$, then $d\,t + d\,t'$ will be the difference of the times above mentioned, which difference of course is a known quantity. Supposing

$$d\,t' = A_i\ d\,\Delta' - B_i\ d\,D - C_i\ d\,\Pi'$$

then $$d\,t + d\,t' = (A + A_i)\,d\,\Delta' - (B + B_i)\,d\,D + (C + C_i)\,d\,\Pi'$$

Each place, where both the immersion and emersion have been observed, will give us an equation of the same form, involving the same three unknown quantities $d\,\Delta'$, $d\,D$, and $d\,\Pi'$. Such observations then, made at three places, will serve us to correct the lunar tables, and subsequently the longitudes.

VII. We have already explained why the eclipses of the moon cannot be made available for the determination of longitudes. Yet as these phænomena always possess considerable interest, we shall say a few words on the method of calculating beforehand the instants of the principal phases. The magnitude of the lunar ecliptic limits has already been determined, page 90; and it has also been shown (page 89) that the apparent radius of the earth's shadow at the distance of the moon $= p + P - R$; where P is the horizontal parallax of the sun: p that of the moon; and R the apparent semi-diameter of the sun*.

This being premised, adopting the notation of page 92, if we call t the time from the conjunction to the phasis for which we are calculating, $s\,t$ will be the motion in longitude of the centre of the shadow in that time, $m\,t$ the motion in longitude of the centre of the moon, $n\,t$ the motion of the latter in latitude. Let c be the distance of the two centres at the time t; we may consider c as the hypothenuse of a right-angled triangle, of which one side is the difference of the motions in longitude of the centres of the moon and shadow, the other side the latitude of the moon's centre at conjunction + its motion since that time: or $(m - s)\,t$ and $\theta + n\,t$ respectively, where θ is the latitude at conjunction.

Hence $$c^2 = (m - s)^2\, t^2 + (\theta + n\,t)^2$$

or $$c^2 - \theta^2 = \left\{ (m - s)^2 + n^2 \right\} t^2 + 2\,\theta\,n\,t$$

or assuming $$\tan.\,\psi = \frac{n}{m - s}$$

$$n^2\,t^2 + 2\,n\,\theta \sin.^2 \psi\,.\,t = (c^2 - \theta^2)\,.\,\sin.^2 \psi$$

$$\therefore\, t = \frac{-\,\theta\,.\,\sin.^2 \psi\,.\, \pm \sin.\,\psi\,.\,\sqrt{c^2 - \theta^2\,.\,\cos.^2 \psi}}{n}$$

We have now only to put for c the different values of the distance of the centres which suits the beginning or end of the eclipse to get the corresponding times counted from the instant of conjunction; which last may be always found from the tables. Before going any further, we shall merely remark that the angle ψ is identical with the inclination of the relative orbit to the ecliptic.†

At the beginning or end of the eclipse, the distance of the centres is equal to the moon's apparent semi-diameter + p + P − R. Substituting this quantity for c, the two corresponding values of t will be the instants of the beginning and end of the eclipses. We might also, if necessary, find the radius of the penumbra; add to it the moon's apparent semi-diameter for c, and substituting this value, find the instants of immersion into and emersion from the penumbra. But this cannot be of any practical utility. At the instant of the middle of the eclipse the two values of t must evidently be equal; that is, the radical must vanish: consequently

$$c^2 - \theta.^2 \cos.^2 \psi = 0 \quad \text{or } c = \theta.\,\cos.\,\psi$$

$$\text{and } t = -\frac{\theta.\,\sin.^2 \psi}{n}$$

We have thus found the instant of the middle of the eclipse, and the distance of the centres at that moment, that is, the least distance of the centres. Add to this least distance of the centres θ cos. ψ, the moon's apparent semi-diameter, we shall have the distance from the exterior edge of the moon to the centre of the shadow: subtract from this the radius of the shadow, we have the whole part of the diameter not eclipsed: the remainder of course will be the quantity of the diameter eclipsed.

Let us suppose that for a certain day the following quantities have been calculated.

Instant of opposition			March 18d 0h 6m 12s
Moon's horary motion in latitude,	n	=	− 3′ 26″ ‡
. . . . longitude,	m	=	37 23 §
Sun's	s	=	2 29
Moon's apparent semi-diameter,	Δ	=	16 39
. . . horizontal parallax	p	=	61 0
Sun's apparent semidiameter	R	=	16 5
. . horizontal parallax	P	=	0 9

$$\text{Hence } \tan.\,\psi = \frac{n}{m - s} = -\frac{206''}{2094''}$$

$$\therefore\, \psi = -\,5^\circ\ 37'\ 7''$$

For the middle of the eclipse $$t = -\frac{\theta.\,\sin.^2 \psi}{n} = 0^h{\cdot}108047 = +\,6^m{\cdot}29$$

* It is found necessary to add to the quantity p + P − R, an empirical correction = 1′ 40″, in order to allow for the effects of the earth's atmosphere.

† See page 92.

‡ n is negative because the latitude is *decreasing*.

§ $m - s$ is always positive, because the moon always moves faster than the sun.

$\therefore$ the middle of the eclipse is at March $18^d\ 0^h\ 12^m\ 41^s$

Least distance of the centres $= \theta . \cos\psi = 38'\ 31''$

Add $\Delta = 16\ \ 39$

Distance of moon's exterior edge to the centre of the shadow $= 55\ \ 10$

Radius of shadow $= p + P - R + 1'\ 40'' = 46\ \ 44$

Breadth of the part of the disk not eclipsed $= 8\ \ 26$

Part eclipsed $= 24\ \ 52$

in digits $= 8^d\ 96$

For the instants of the beginning and end of the eclipse

$$c = \Delta - R + p + P + 1'\ 40'' = 63'\ 23$$

Hence the two values of t are

Beginning $t = +\ 0^h \cdot 108047 - 1^h \cdot 43546 = 1^h\ 19^m\ 39^s$

End $t = +\ 0 \cdot 108047 + 1 \cdot 43546 = 1\ \ 32\ \ 27$

Hence the instant of the beginning of the eclipse is

March 18^d	0^h	6^m	12^s	(the instant of opposition)
$-$	1	19	39	
= March 17	22	46	33	

That of the end of the eclipse is

March 18	0	6^m	12^s
+	1	32	37
= March 18	1	38	49

This is mean solar Paris time, and according to the custom of several French astronomers, counted from midnight. It is easy to see that the moon must have been above the horizon of Paris during the whole duration of the eclipse. The eclipse then was entirely visible to this town, and of course invisible in the opposite hemisphere.

We shall not say more on this subject, but proceed to one of greater interest and greater complexity, that of solar eclipses. But here the problem separates itself into two, according as we wish to determine the circumstances of the eclipse for the earth in general, or for a given spot on the earth's surface. The former problem becomes identical with that of lunar eclipses, if we suppose the moon to be in the place of the earth, and the converse; the general question of an eclipse of the earth by the moon may be treated by the same formulæ as those given for the calculation of a lunar eclipse, provided we modify properly the value of the radius of the shadow. Now the same figure and considerations (p. 89) by which the radius of the earth's shadow at the moon's distance were determined, show that the radius of the lunar shadow, as seen by an observer in the moon, is equal to the parallax of the sun with regard to the moon, plus the parallax of the earth, minus the apparent semi-diameter of the sun as seen from the moon. This parallax is the apparent semi-diameter of the moon as seen from the earth. Now as the parallax of the sun is extremely small, and may be neglected without producing more than half a second of error, we shall not take it into account, but simply state, that *the radius of the lunar shadow, as seen from the moon, is equal to the difference between the apparent semi-diameters of the moon and sun**.

---

* The rigorous expression is $(\Delta - R) . \dfrac{p}{p - P}$. This is easily shown: the lunar parallax of the earth is the moon's apparent semi-diameter, or Δ: the lunar parallax of the sun is $\dfrac{P . \Delta}{p - P}$; the lunar apparent semi-diameter of the sun is $\dfrac{R . p}{p - P}$: adding these three quantities together we get $(\Delta - R) . \dfrac{p}{p - P}$. It is easy to see that the lunar parallax of the sun will have the value we have assigned to it; for it will be equal to the terrestrial parallax of the sun, augmented in the ratio of the distances, and diminished in the ratio of the radii of the moon and earth. Call the radii of the earth and moon R and R′ respectively; their distance from the sun to the earth D, to the moon D′: then the distances of the earth and moon from the sun at conjunction will be D, and D − D′. The lunar parallax of the sun then will be equal to $P . \dfrac{R'}{R} . \dfrac{D}{D - D'}$. Now $\dfrac{R'}{R} = \dfrac{\Delta}{p}$ (page 61,) and $D = \dfrac{R}{\sin. P}$, $D' = \dfrac{R}{\sin. p}$ $\therefore \dfrac{D}{D - D'} =$

Thus too the radius of the lunar penumbra, as seen from the moon, will be equal to the sum of the parallaxes of the sun and earth, plus the apparent semi-diameter of the sun; all these quantities being calculated for the moon. Here, as before, we neglect the parallax of the sun, and take for the parallax of the earth the moon's apparent semi-diameter; we thus get *the radius of the lunar penumbra, as seen from the moon, equal to the sum of the apparent semi-diameters of the sun and moon.**

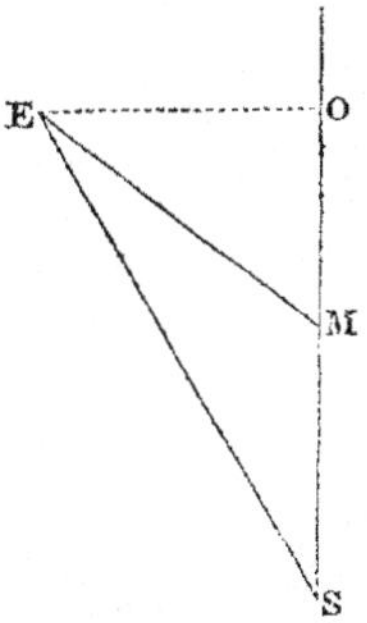

Let us suppose that, in the accompanying figure, S is the place of the centre of the sun, M of the moon, E of the earth: the angle O M E exterior to this triangle will represent the apparent distance of the centres of the moon's shadow and of the earth as seen from the moon; let us call it c: we shall have $c = S + E$ or $S = c - E$. From E draw E O perpendicular to the axis of the shadow; then we have $EO = EM . \sin. c = ES . \sin. S$. $\therefore EM . \sin. c = ES . \sin. S = ES . \sin. (c - E)$. But E S and E M, the distances of the sun and moon, are in the inverse ratio of the parallaxes of these two bodies, or $\frac{ES}{EM} = \frac{\sin. p}{\sin. P}$, or $= \frac{p}{P}$, the arcs p and P being very small. Hence $p . \sin. (c - E) = P . \sin. c$; or with the same approximation $p . (c - E) = P . c \quad \therefore E = c . \frac{p - P}{p}$.

Now E is the apparent distance of the centres of the sun and moon as seen from the earth. Now we may use, in calculating the general circumstances of a solar eclipse, the same formulæ as those which we have established for a lunar eclipse, provided we substitute for the angle c of page 243, the angle E which measures the apparent distance of the centres: but $E = c . \frac{p - P}{p}$; we have then merely to substitute in the formula already obtained $c \cdot \frac{p - P}{p}$ for c. We thus get (see page 243)

$$t = \frac{-\,\delta . \sin.^2 \psi \pm \sin. \psi \sqrt{c^2 \left(\frac{p - P}{p}\right)^2 - \delta^2 . \cos.^2 \psi}}{n}$$

We show, as before, that, at the middle of the eclipse, $t = \frac{-\,\delta . \sin.^2 \psi}{n}$ and $c = \left(\frac{p}{p - P}\right) \delta . \cos. \psi$; and we can determine as before the magnitude of the eclipse in digits.

The phases the most important to determine are the beginning and end of the eclipse. Now the eclipse begins or ends at the moment that the earth's disk enters the lunar penumbra; for as long as the earth is in that penumbra, a portion of the sun's disk must be obscured to some part of the earth's surface. To determine then the corresponding values of t, we must substitute for c, the apparent radius of the lunar penumbra as seen from the moon plus the parallax of the moon;† this is (by note to page 244) $p + (\Delta + R) \frac{p}{p - P}$. Substituting then this value of c, we get

---

$\frac{\sin p}{\sin. p - \sin. P} = \frac{p}{p - D}$, since the arcs p and P are very small. Hence, finally, the lunar parallax of the sun $= p . \frac{\Delta}{p} . \frac{p}{p - P} = \frac{p . \Delta}{p - P}$. It is easy, following the same reasoning, to see that the lunar semi-diameter of the sun $= R , \frac{p}{p - P}$.

* In this case, however, the parallax of the sun can hardly be neglected with safety. The exact value of the radius of the lunar penumbra, as seen from the moon, is $(\Delta + R) \frac{p}{p - P}$.

† It will be recollected that the parallax of the moon represents the apparent radius of the earth as seen from that satellite.

$$t = \frac{-\ell . \sin.^2 \psi \pm \sqrt{\{\Delta + R + p - P\}^2 - \ell^2 . \cos.^2 \psi}}{n}$$

Let us take for example the solar eclipse of 1764, which has been repeatedly calculated. Let us suppose

Instant of conjunction April 1^d 10^h 31^m 5^s. Apparent Paris solar time

Latitude of the moon at conjunction ℓ	=	+ 39′ 32″
Horary motion in latitude . . . . n	=	+ 2 24
. . . . longitude . . . m	=	29 39
Motion of the sun in longitude . . m'	=	2 27·7
Horizontal parallax of the sun . . P	=	8·8
. . . . of the moon* p	=	54 1·5
Apparent semi-diameter of the moon Δ	=	29 29
. . . . sun	=	31 32

Substituting these quantities in our expression for t, we obtain

Beginning of the eclipse $t = -0^h \cdot 14478 - 2^h \cdot 73898 = -2^h 53^m 1^s$
End . . . . $t = -0 \cdot 14478 + 2 \cdot 73898 = +2\ 35\ 39$
Middle . . . . $t = -0 \cdot 14478 \qquad = -0\ 8\ 1$

It will be recollected that the negative sign indicates times anterior to the conjunction: the positive sign, times subsequent to it.

The apparent distance of the centres of the earth and lunar shadow, as seen from the moon, at the instant of the greatest eclipse, is $\frac{p}{p-P} . \ell . \cos. \psi =$ 39′ 27″. Now this distance is less $p = 54' \ 1'' \cdot 5$, the moon's parallax, which represents the apparent semi-diameter of the earth as seen from the moon: consequently at this instant the axis of the shadow fell somewhere on the earth's surface. The eclipse then was *central* for some places at that moment, but not *total*; for the apparent semi-diameter of the sun was greater than that of the moon; it was *central and annular*, the sun's disk forming a ring of light round the dark surface of the moon. It is easy to see that, in other places, it must have been annular without being central: others again where it was partial without being annular or central: lastly, more than half the earth's disk was not covered at all by the penumbra, for the radius of the penumbra $(\Delta + R). \frac{p}{p-P} = 30'45''$, which is not only less than the lunar parallax 54′ 1″·5, but less than the shortest distance of the centres, 39′ 27″, so that even on the side where the shadow passes there is a portion of the disk = 8′ 42″ which is not eclipsed. The part eclipsed amounts to 45′ 19″·5, or in digits $5^d \cdot 3$.

In arriving at these conclusions we have supposed the horary motions of the moon uniform, and neglected the variations in her parallax and apparent semi-diameter; but it may be observed that the error arising from these circumstances is trifling, and that there is no utility in predicting the phases of an eclipse with extreme precision; it is quite enough to have a general idea of them to be prepared for the observation, which should be made with extreme accuracy as well as all the calculations founded upon them.

Having shown how to determine the phases of a solar eclipse for the earth in general, the next question that presents itself is to determine the points of the earth that will see these phases. This question, if completely gone into, is one of some length; for we should first determine what places see any eclipse; then those which see it annular, and those which see it central and annular; if the apparent diameter surpasses that of the sun, then there is no annular eclipse, and we must find those places which see it total, and those which see it total and central. When, in each of these cases, we have determined a certain number of places, we lay them down on a map of the earth's surface and join them by a curve. Thus also we may ascertain the points where the eclipse begins when the sun is rising, or ends when it is setting, or where it ends at the rising and begins at the setting. Again we might trace a line through those places which have the middle of the eclipse at rising, and those which have the middle at set-

* Calculated for the instant of conjunction.

ting. The limits of this treatise do not permit us to enter into these details: we shall content ourselves with showing how the two points of the earth's surface may be found which come first and last into contact with the penumbra.

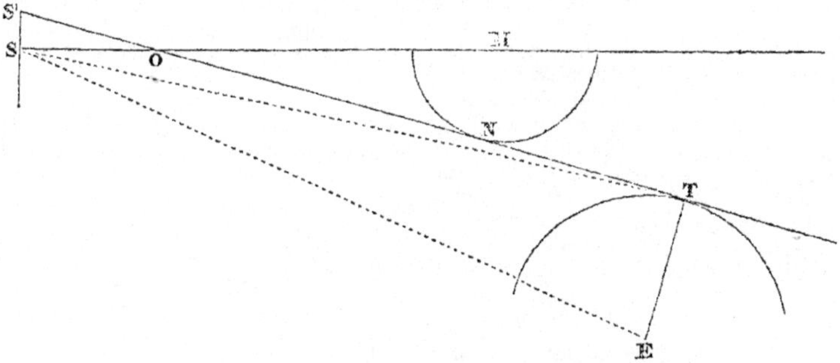

Let us suppose that, in the annexed figure, S, M, E are the centres of the sun, moon, and earth respectively: the circumferences described round the two latter represent the intersections of the moon and earth by the plane passing through the points S, M, E. S' O N T, a tangent to these two circumferences, is the extreme limit of the penumbra, which is always in this plane and must be a tangent to the terrestrial sphere at the instant of first and last contact; consequently the angle E T O is a right angle. Hence E T S = 90° − S T S' = 90° − R: also T S E = P ∴ E T S + T S E = 90° − R + P; and since S E T is the supplement of these two, S E T = 90° + R − P.

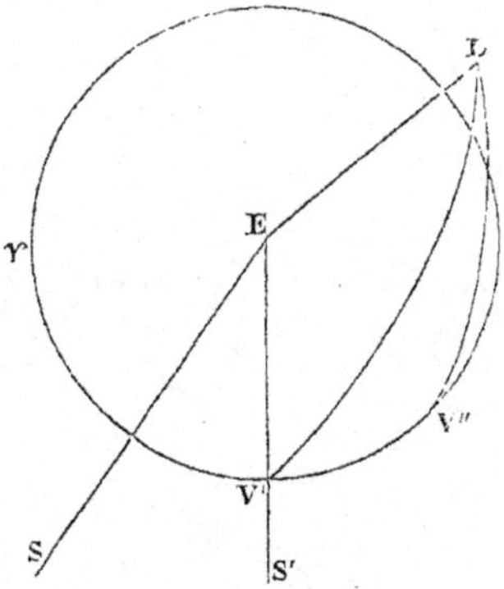

Let us now suppose that, in the adjoining figure, E is the centre of the earth, ♈ V V' V'' is the section of the earth's surface by the plane of the ecliptic: S, S' the successive positions of the sun in this plane at the instants of conjunction and of the last contact; lastly L (which we must conceive to be elevated above the plane of the paper) to be the point of the surface where this contact takes place. Join L E, the angle L E S' is known by what has just preceded. The first point is to find the angle ι which the plane L E S' makes with the ecliptic. This is easily done; for since the moon is in this plane at the instant of the last contact, which we will suppose to take place at the time t'', (counted from conjunction,) its latitude at that instant is $\lambda + n\,t''$; its difference of longitude from the sun $(m - m')\,t''$; and its apparent distance from the sun seen from the earth is $\Delta + R + p - P$*: consequently, as these arcs are very small,

$$\sin.\ \iota = \frac{\lambda + n\,t''}{\Delta + R + p - P}$$

$$\cos.\ \iota = \frac{(m - m')\,t''}{\Delta + R + p - P}$$

With these data we must calculate the latitude of the point M, and its difference of longitude from the sun. Draw from L on the ecliptic the circle of latitude L V'': the three points L, V', V'' on the earth's surface will form a spherical triangle right-angled at V'', in which we know the hypothenuse V' E L = 90° + R − P, and the angle L V' V'' or ι which we have just found.

$$\therefore \sin.\ LV'' = \cos.\ (R - P).\ \sin.\ \iota$$

$$\tan.\ V'V'' = \tan.\ (90° + R - P).\ \cos.\ \iota$$

We may simplify these expressions still further by putting cos. (R − P) = 1, and putting for the latter of the two,

---

* This is the same as the angle E, in page 245.

$$V'V'' = 90^{\circ} + \frac{(R - P)}{\cos. \iota}.$$

Add now to $V'V''$ the quantity $m' t''$, we shall evidently have the longitude of the point M, counted from V, the point of the ecliptic where the conjunction happened. Lastly adding ♈ V, which we shall call A, we have the longitude of M counted from the equinox. These two angles LV'', and ♈V'' are what are technically called the latitude and longitude of the zenith. Hence, by the appropriate formulæ, we may obtain the right ascension and declination of the zenith. The latter is identical with the geographical latitude of the place. The right ascension of the sun must be calculated for the instant of contact, and subtracted from the right ascension of the zenith. This gives us the hour angle of the sun, that is to say, the time counted at the place in question. As we have already calculated this time according to our first meridian (Paris, for example), we know the difference of times and consequently the longitude from Paris.

To apply these formulæ to the case of the eclipse of 1764, let us suppose that we have found the instant of the last contact with the penumbra, $t'' = 2^{h} \cdot 59420$.

At the instant of conjunction A = 12° 9′ 56″

Hence longitude of the zenith = 102 35 15

latitude . . . . = 33 27 35 N.

Similarly for the beginning of the eclipse

$$t' = -2^{h} \cdot 88376$$

Longitude of the zenith = 281° 45′ 48″

Latitude . . . . = 21 59 0

Hence we obtain for the point of last contact

Right ascension of the zenith = 109° 1′ 55″

Declination . . . . . . = 56 6 20

For the point of first contact

Right ascension of the zenith = 281° 53′ 57″

Declination . . . . . . = − 1 2 39

To get completely the position of the two points in question on the celestial sphere, we must determine the geographical longitude from the right ascension of the zenith. Operating in the manner explained above, we have

	First contact	Last contact
Longitude of the sun at conjunction	12° 9′ 56″	12° 9′ 56″
Reduction to the moment of the phasis	$m' t' = -7\ \ 7$	$m' t'' = +6\ \ 24$
Longitude of the sun at the moment of the phases	= 12° 2′ 49″	12° 16′ 20″
From these longitudes (supposing the obliquity of the ecliptic 23° 28′ 31″) the corresponding values of the right ascensions are	13° 5′ 53″	13° 20′ 28″
Subtracting these results from the respective right ascensions of the zenith as given above, we have the hour angles of the sun, at the moments of the phases	267° 48′ 4″	95° 41′ 27″
or in solar time	$17^{h}\ 51^{m}\ 12^{s}$	$6^{h}\ 22^{m}\ 46^{s}$

These angles are counted from the upper meridian and from east to west, from 0° up to 360°; counting in the same way, at Paris, the times were

	First contact	Last contact
at the instant of conjunction	$22^h\ 31^m\ 5^s$	$22^h\ 31^m\ 5^s$
Reduction to the instant of the phasis	$t' = -2\ 53\ 1$	$t'' = +2\ 35\ 39$
Paris time at the instant of the phasis	$= 19\ 38\ 54$	$1\ 6\ 44$
By the calculations above given the hour angles of the sun at the two places considered, are . . .	17 51 12	6 22 46
Hence the geographical longitudes of these two spots referred to the meridian of Paris, are . . .	1 46 52	18 43 58
The latitudes as given above* . .	$-1° 2' 39''$	$+56° 6' 20''$

We shall now proceed to the prediction of the phasis of a solar eclipse for a given place. For this purpose it is necessary to be able to calculate the apparent distance of the centres of the sun and moon at a given instant as seen from the point in question. We begin by calculating for the given instant the true longitudes ☉ and ☽, of the sun and moon, their horary motions in longitude m and m', their horizontal parallaxes p and Π, their semi-diameters R and Δ', and the *geocentric* latitude of the place ϕ'.

After this we deduce from Π and ϕ', the lunar horizontal parallax Π' by means of the formula in page 232. Now the parallaxes in longitude, from the form of their expression, are proportional to the horizontal parallaxes, and as the horizontal parallax of the sun does not exceed $8''\cdot 8$, we may take it into account by substituting for Π', $\Pi' - p$, or the differences of the parallaxes of the two bodies.

We now calculate the lunar parallaxes in longitude and latitude, and by adding them to the true longitude and latitude respectively, we get the apparent longitude and latitude for the instant required; we calculate also Δ', or the apparent semi-diameter of the moon, by the formula in page 240.

This being done, we proceed to calculate the apparent distance of the centres for the instant required. Now we may always, without sensible error, consider this distance as the hypothenuse of a right-angled triangle, of which the two sides are, respectively, the latitude of the moon's centre, and the difference of longitudes.

Calling the apparent longitudes of the centres ☉′ and ☽′, let us put ☉′ — ☽′ $= \lambda'$; and let us call the apparent latitude of the moon's centre θ'; then

$$c^2 = \lambda'^2 + \theta'^2.$$

This equation may be put more conveniently for logarithmic computation under this form. Assume $\tan. \psi = \frac{\lambda'}{\theta'}$, then

$$c = \frac{\theta'}{\cos. \psi}.$$

At the beginning or end of the eclipse, the distance of the centres, or $c = R + \Delta'$; if the eclipse be annular, at the moments of interior contact, $c = R - \Delta'$: both cases may be embraced under the general expression $c = R \pm \Delta'$*. At these moments then, c, as above calculated, ought to be equal to $R \pm \Delta'$. But as we do not know beforehand the instant of the phasis, which, indeed, is what we are seeking, we must calculate for every five minutes during an hour before and after conjunction, the values of c, and compare them with the value of $R \pm \Delta'$; by observing when c is greater and when less than this quantity, we shall have found, within five minutes, the time of the phasis, which may be found more nearly by a simple proportion, supposing the change of distance proportional to the time. As the whole variation of Δ' only amounts to $18''$, we may calculate this quantity for the instant of conjunction, and suppose it constant during the two hours. The lunar parallaxes, and the true longitudes and latitudes, should be calculated for every half hour, and hence the apparent motions determined; we may then, by interpolation, find λ' and θ' for every five minutes. As R and Δ' may be supposed constant during the whole interval of our calculations, the greatest eclipse will take place when c is least;

* The negative sign denotes a southern latitude; the positive a northern.

* Experience teaches us that, in these calculations, it is necessary to diminish the diameter of the sun by $3''\cdot 5$, and that of the moon by $2''$: the first of these corrections is attributed to an optical effect called *irradiation*; the latter to the *inflection* of the sun's rays at the surface of the moon.

but the equation $c^2 = \lambda'^2 - \theta'^2$ gives, for a minimum,

$$\lambda' d\lambda' + \theta' d\theta' = 0 \quad \therefore \lambda' = -\frac{d\theta'}{d\lambda'} \cdot \theta'.$$

We may take for θ' the apparent latitude at the middle of the eclipse*; $d\lambda'$, $d\theta'$, are the apparent motions which we have already found; thus λ' is known, and we may easily deduce the corresponding value of c; and c being known, the corresponding time may be found by interpolating between two known values of c for two successive epochs, five minutes distant.

The observer would not be able to observe the instant of immersion, unless he knew beforehand the point of the solar disk at which to look for the first appearance of the moon's dark limb. This is determined by means of what is called the parallactic angle, or the angle which the ecliptic makes with a vertical circle passing through the sun, at a given sidereal moment, s.

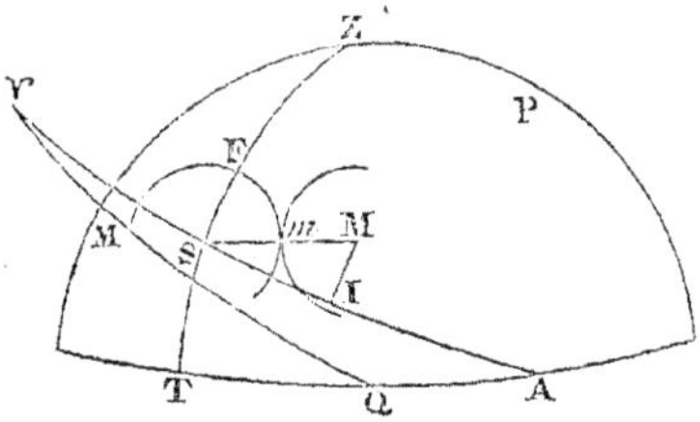

Suppose that, in the adjoining figure, P is the pole of the equator ♈ Q, P M the meridian, S the sun, Z T a vertical circle passing through its centre, M the centre of the moon, ♈ A the ecliptic, Z S A the parallactic angle, ξ. Now ♈ Q $=$ Q M $+$ M ♈ $= 90^\circ + s$; the angle at Q is ω, or the obliquity of the ecliptic: ♈ Q A $= 180^\circ -$ ♈ Q T $= 180^\circ - (90^\circ - \phi) = 90^\circ + \phi$; where ϕ is the latitude of the observer. Hence, by spherical trigonometry, we may calculate the angle at A, and the side ♈ A; but S A $=$ ♈ A $-$ ♈ S $=$ ♈ A $- \odot$. Consequently in the right-angled spherical triangle S T A, S A, and the angle at A are known, whence we get

$$\cot. \xi = \cos. SA . \tan. A,$$

at the instant of the contact at the point m, S M $= R + \Delta'$: M I $= \theta'$;

$$\therefore \cos. MSI = \frac{\theta'}{R + \Delta'}: \text{consequently}$$

we know $\xi -$ M S I $=$ F S m, or the arc of the solar disk comprised between the point of contact m, and the vertex of the sun, F.

We now come to the application of solar eclipses to the determination of terrestrial longitude. It will at once be perceived that this problem bears the greatest analogy to that of the determination of the longitude by lunar occultations of the fixed stars. In fact the problems are identical, with the exception of three points; these are

First, that, instead of determining the conjunction in right ascension, we determine the conjunction in longitude; that, in consequence, it is necessary to refer both bodies to co-ordinates of longitude and latitude, instead of those of right ascension and declination.

Secondly, the sun will have no latitude, which considerably simplifies the calculations; but the instant of conjunction will be determined by the *difference* of the motions in longitude of the sun and moon; and to find the sun's *apparent* longitude, his parallax in longitude must be calculated.

Thirdly, at the beginning or end of the eclipse, the apparent distance of the centres is equal to the sum of the apparent diameters of the two bodies. The latter of these is found from the tables or Ephemeris, and may be considered as constant for all altitudes above the horizon. Consequently, the same formulæ which we have already investigated for occultations will serve, provided we substitute for the difference of right ascensions the difference of longitudes; for the difference of declinations the latitude of the moon; for the moon's parallax in latitude, her parallax in declination; for her parallax in right ascension, the difference of the parallaxes of the sun and moon in longitude: provided also that we put the apparent dis-

* The middle of the eclipse is the middle of the interval between the two exterior contacts; it is always very near the instant of the greatest eclipse.

tance of the centres equal to the sum of the apparent semi-diameters. This having been done, we conclude the instant of apparent conjunction, and from that the instant of true conjunction, as before; and in the same way, from the instant of true conjunction, the longitude is determined.

Let λ be the apparent difference of the longitudes of the two centres; θ' the apparent latitude of the moon; δ and Δ' their apparent semi-diameters; then

$$\lambda^2 = (\Delta' + \delta)^2 - \theta'^2.$$

Now let ☽ and ☽′ be the true and apparent longitudes of the moon, ⊙ and ⊙′ those of the sun, ϖ and π the parallaxes of the moon and sun in longitude: then ☽ = ☽′ − ϖ. Now at the beginning of the eclipse,

$$\odot = \odot' - \pi,\ ☽ = ☽' - \varpi,\ \therefore \lambda = \odot' - ☽'$$
$$\therefore \odot - ☽' = \lambda + (\varpi - \pi);$$

at the end of the eclipse

$$☽ - \odot = \lambda - (\varpi - \pi);$$

or, generally, if the true difference of longitudes $= c$, then we have

$$c = \lambda \pm (\varpi - \pi),$$

taking the superior sign for the immersion.

The phases of the internal contacts in an annular eclipse may be employed in the same way, if we substitute the difference of the apparent semi-diameters for their sum.

It is necessary, as has been previously remarked, to diminish the sun's semi-diameter 3″·5, on account of irradiation, and that of the moon 2″, on account of inflexion.

In finding the time corresponding to the arc c, we regard the sun as immoveable, and attribute to the moon an horary motion equal to the difference of the motions in longitude. Let t be the time corresponding to the arc c; m and M the horary motions of the sun and moon;

$$\text{then } t = \frac{3600^s}{m - M}\{\lambda \pm (\varpi - \Pi)\}$$

We may dispense with the calculation of π, by calculating ϖ with the value $\Pi' - 8''$, instead of Π', to allow for the parallax of the sun; and we may then, in our formula, omit π. V. page 249.

All, then, that is required, is to know how to determine the parallaxes in longitude and latitude. The formulæ for expressing the effects of parallax in right ascension and declination may, by a slight modification, be made to serve for those in longitude and latitude. All that is necessary for this purpose is to substitute for the hour angle in the expression for the parallax in right ascension, the difference of longitudes between the moon and the zenith, and the latitude of the zenith for the geographical latitude in the expression for the parallax in declination. The longitude and latitude of the zenith are deduced from its right ascension and declination by the appropriate formulæ. To obtain these, we must consider the spherical triangle ZPP′ between the zenith, the pole of the equator P, and that of the ecliptic P′. In this triangle ZP is the colatitude of the place, Z P′ the complement of the latitude of the zenith; P P′ is the obliquity of the ecliptic.

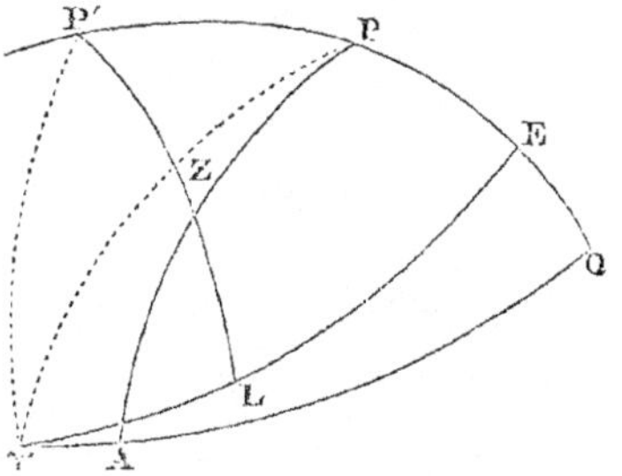

Let ♈ be the vernal equinox, ♈ E the ecliptic, ♈ Q the equator: it is evident that ♈ A is the right ascension of the zenith, ♈ L its longitude; the angle P′ P Z = P′ P ♈ + ♈ A. (P, it will be recollected, is the pole of the great circle ♈ Q, and therefore ♈ A measures the angle ♈ P A); but P′ P ♈ = 90°; for P ♈, being the equinoxial colure, is at right angles to P′ P Q, the solstitial colure; consequently P′ P Z = 90° + the right ascension of the zenith. Again, P P′ L = P P′ ♈ − ♈ L, (P′ being the pole of the great circle ♈ E); but P P′ ♈ = 90°, for we may consider ♈ as the pole of the great circle P′ P Q; consequently P P′ Z = 90° − the longitude of the zenith.

Let us then take

$$P P' = \omega$$
$$p' Z = 90° - \theta$$
$$P Z = 90° - \delta$$
$$p' P Z = 90° + \alpha$$
$$p P' Z = 90° - \lambda$$

Now applying the common formulæ of spherical trigonometry, and recollecting that cos. $(90° + \alpha) = -$ sin. α, we have

$$\cos.\,\theta = -\sin.\,\omega\,.\cos.\,\delta\,.\sin.\,\alpha + \cos.\,\omega\,.\sin.\,\delta$$

$$\tan.\,\lambda = \frac{\tan.\,\delta\,.\sin.\,\omega + \sin.\,\alpha\,.\cos.\,\omega}{\cos.\,\alpha}.$$

Assume $$\tan.\,\psi = \frac{\sin.\,\alpha}{\tan.\,\delta}$$

then $$\sin.\,\theta = \sin.\,\delta\,.\frac{\cos.\,(\psi+\omega)}{\sin.\,\psi}$$

$$\tan.\,\lambda = \tan.\,\alpha\,.\frac{\sin.\,(\psi+\omega)}{\sin.\,\psi}$$

It may be as well to observe, that what we have here called the longitude of the zenith, is often called the longitude of the nonagesimal; and what we have called the latitude of the zenith, is the complement of what is often called the altitude of the nonagesimal *. The quantity which we have termed the right ascension of the zenith, which is, in fact, identical with the sidereal time, is often called the right ascension of the mid-heaven.

To illustrate these formulæ, we shall take the following example:—

☽ =	152° 9′ 33″·00		
β = +	33 55·70		
Π =	54 6·75	p =	8″·42
Δ =	14 44·72	δ =	15′ 56·14
		ω =	23° 27′ 48″·1
		ϕ =	51 36 10·0

Sidereal time = $8^h\ 35^m\ 6^s{\cdot}17$ = 124 46 32·5 = s

Let the compression of the earth, or p, (see page 203,) be supposed $=\frac{1}{300}$.

Hence the calculations become

	log. p = 7·52288	Π = 54′ 6·75
log. $(1-p^2)$ = 9·9970999	log. Π = 3·51145	− 6·61
log. tan. ϕ = 0·1009945	log. sin.$^2\phi'$ = 9·78608	− p = − 8·42
log. tan. ϕ' = 0·0980944	log. 6″·61 = 0·82041	Π' = 53 51·72
ϕ' = 51° 24′ 59″		

Hence Π' and ϕ' being found, we proceed to calculate with these quantities the position of the nonagesimal,

log. cot. ϕ' = 9·9019056	log. sin. $(\psi+\omega)$ = 9·9151817	log. cos. $(\psi+\omega)$ = 9·7548371
log. sin. s = 9·8918781	log. tan. s = 0·0951316	—log. sin. ϕ' = 9·8930395
log. tan. ψ = 9·7937837	log. sin. ψ = − 9·7227655	log. cos. ψ = − 9·9289818
∴ ψ = 31° 52′ 52″·4	log. tan. λ = 0·2875478	—log. cos. θ = 9·7188948
ω = 23 27 48·1	λ = 117° 17′ 3″	∴ θ = 58° 26′ 4″
∴ $\psi+\omega$ = 55 20 40·5	☽ = 162 9 33	
	∴ ☽ − λ = 44 52 30	

* Calling the h the altitude of the nonagesimal, we have

$$\cos.\,h = \sin.\,\delta\,.\frac{\cos.\,(\psi+\omega)}{\cos.\,\psi}$$

$$= \sin.\,\phi\,.\frac{\cos.\,(\psi+\omega)}{\cos.\,\psi}$$

ϕ being the geographical latitude.

Let us refer then to the formula for the parallax in right ascension, (page 237) and substitute for ϕ, the altitude of the nonagesimal, or θ; for D, the altitude of the moon, which we shall call β: then take

$$q = \frac{\sin. \Pi . \cos. \theta}{\cos. \beta}$$

We must also substitute for the hour angle h, the difference in longitude between the moon and the nonagesimal, or ☽ − λ, calling ☽ the moon's longitude. Then if the parallax in longitude be ϖ, we have

$$\tan. \varpi = \frac{q . \sin. (☽ - \lambda)}{1 - q . \cos. (☽ - \lambda)}$$

$$\text{or} \qquad \varpi = \frac{q . \sin. (☽ - \lambda)}{\sin. 1''} + \frac{q^2 . \sin. 2 (☽ - \lambda)}{2 . \sin. 1''} + \&c.$$

$$\text{Again, assume } \cot. \chi = \frac{\cot. \theta . \cos. \left(☽ - \lambda + \frac{\varpi}{2}\right)}{\cos. \frac{\varpi}{2}}$$

$$\text{and } n = \frac{\sin. \Pi . \sin. \theta}{\sin. \chi} . \sin. (\chi - \beta)$$

then if π be the parallax in latitude, we have

$$\tan. \pi = \frac{n . \sin. (\chi - \beta)}{1 - n . \cos. (\chi - \beta)}$$

$$\text{or } \pi = \frac{n . \sin. (\chi - \beta)}{\sin. 1''} + \frac{n^2 . \sin. 2 . (\chi - \beta)}{2 \sin. 1''} + \&c.$$

Thus also calling the apparent semi-diameter, as given by the tables, Δ, and assuming

$$\cot. \varepsilon = \frac{\cot. \theta . \cos. \left(☽ - \lambda + \frac{\varpi}{2}\right)}{\cos. \frac{\varpi}{2}}$$

we have, for the augmentation of the apparent semi-diameter, the following expression,

$$\Delta . \pi . \sin. 1'' . \cot. (\varepsilon - \beta)$$

Chapter XIII.

On the use of Astronomical Observations in determining the Bearings of Terrestrial Objects, and the Variation of the Magnetic Needle.

It frequently happens that in the course of trigonometrical surveys, or those great geodetic operations which have been undertaken to determine the figure of the earth, it becomes necessary to determine the position of some given line, such as the side of a triangle, with regard to the meridian; or, in other words, it is necessary to determine the azimuth of this line. For this purpose it is absolutely necessary to have recourse to astronomical observations. This may be done in the following manner:—Let us suppose the observer placed at one end of the line in question to observe the difference between the azimuths, or the angle in a horizontal plane, comprised between a signal placed at the other end of the line, and some heavenly body, such as the Sun, if the observation be made in the daytime, or a fixed star if at night. Now the azimuth of the sun or star may be calculated, (supposing the latitude known,) provided we know the instant of the observation; and thus the azimuth of the signal is immediately found. In showing how to calculate the azimuth of the heavenly body, we shall begin by remarking, that since refraction and parallax both act in a *vertical* circle, they do not change the azimuth, and therefore may be neglected in these calculations.

Let us suppose that in the spherical triangle Z P S, between the pole of the equator, the zenith, and the sun or star, that Z P or the colatitude $= \phi$, P S or the north polar distance $= \delta$, the hour angle Z P S $= h$, the azimuth P Z S = A, and the angle Z S P, which is technically called the angle of variation, $= \xi$. In this triangle, ϕ, δ, and h are known, and it is required to find A. Now the formulæ of spherical trigonometry, known by the name of Napier's Analogies, give us the following equations:

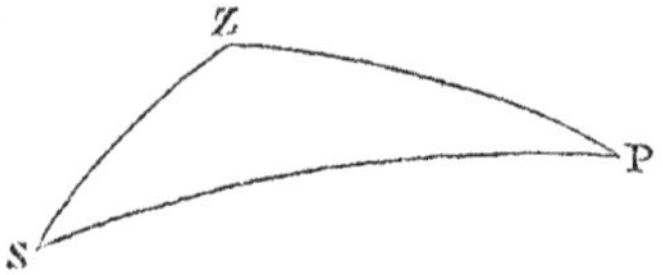

$$\tan.\ \tfrac{1}{2}\,(A + \xi) = \cot.\ \frac{h}{2} \cdot \frac{\cos.\ \frac{1}{2}\,.\,(\delta - \phi)}{\cos.\ \frac{1}{2}\ (\delta + \phi)}$$

$$\tan.\ \tfrac{1}{2}.(A - \xi) = \cot.\ \frac{h}{2} \cdot \frac{\sin.\ \frac{1}{2}\ (\delta - \phi)}{\sin.\ \frac{1}{2}\ (\delta + \phi)}$$

In the case of the sun, the hour angle h is the true time, expressed in degrees; in the case of a star, it will be the difference between the right ascension of the meridian and the right ascension of the star, or in other words, the difference between the sidereal time and the right ascension of the star: but the sidereal time is equal to the apparent solar time + the right ascension of the sun. From the sum, then, of these latter quantities, we must subtract the right ascension of the star to obtain h: this difference must have the negative sign when the star is to the east of the meridian, and the positive when to the west of it.

Also, if the sun's azimuth be employed, it is necessary to add or subtract his semidiameter, as the case may require*: or if we take the mean of a series of observations, this correction may be eluded by observing alternately the opposite limbs.

Let us proceed, as an example of the formula above given, to find the azimuth of α Aquilæ (to the east of the meridian) on the 9th June, 1830, at $9^h\ 18^m$ mean time on the meridian of Paris: ϕ is supposed $= 41^\circ\ 19'\ 10''$. The right ascension of the star is supposed to be $19^h\ 42^m\ 31^s{\cdot}33$; and its declination $8^\circ\ 25'\ 45''{\cdot}24$.

Now mean time	$= + 9^h$	18^m	0^s	
mean AR ☉	$= + 5$	29	14	
	$+14$	47	14	
AR ✶	$= -19$	43	31·3	
∴ − hour angle	$= - 5$	15	10·19	
or hour angle	$= + 5$	15	10·19	
Change of AR ☉ in $9^h\ 18^m$	$=\ -$	1	31·42	
∴ h =	5	13	38·77	∴ $\frac{h}{2} = 39^\circ\ 12'\ 21''$

$\delta = 81^\circ 34'\ 14''{\cdot}8$

$\phi = 41\ \ 19\ \ 10\ {\cdot}0$

log. cot. $\frac{h}{2}$ = 0·08846 = 0·08846

$\delta - \phi = 40\ \ 15\ \ \ 4\ {\cdot}8$ ∴ $\frac{1}{2}(\delta - \phi) = 20^\circ\ 7'\ 32''{\cdot}4$ log.cos. = 9·97264 log.sin. = 9·53666

$\delta + \phi = 122\ \ 53\ \ 24\ {\cdot}8$ $\ \frac{1}{2}(\delta + \phi) = 61\ \ 26\ \ 42\ {\cdot}4$ log.cos. = 9·97943 log.sin. = 9·94367

0·38167 9·68145

∴ log. tan. $\frac{1}{2}$ (A + ξ) = 0·38167 ∴ $\frac{1}{2}$ (A + ξ) = $67^\circ 26' 55''$

log. tan. $\frac{1}{2}$ (A − ξ) = 9·68145 ∴ $\frac{1}{2}$ (A − ξ) = 25 39 7

∴ A = 93 6 2

* Strictly speaking, we should employ for this purpose not the semidiameter of the tables, but the *azimuthal* semidiameter, which is equal to the former divided by the cosine of the sun's altitude. But this nicety is unnecessary, except in the most delicate operations.

This, then, is the azimuth of 'α Aquilæ at the time mentioned, counted from north to east.

The polar star is used very advantageously in this way at the moment of its greatest digression, that is, at the moment when it is farthest from the meridian: it is towards this time that its motion in azimuth is the least, and, indeed, perfectly insensible during the time occupied in observing a set of distances. It is necessary to calculate then the instant of greatest digression: let h be the hour angle at this instant, ϕ the complement of the latitude, δ the north polar distance of the star; at the instant of greatest digression we have a spherical triangle between the zenith Z, the pole P, and the star S, which is right-angled at S. Hence

$$\cos. h = \tan. \delta . \cot. \phi.$$

From a knowledge of the hour angle the corresponding mean solar, or sidereal time, may be deduced.

Lastly, the azimuth of Polaris is given by the formula

$$\sin. A = \frac{\sin. \delta}{\sin. \phi}.$$

Thus, let it be required to calculate the time of the greatest digression of Polaris, on a given day, for which the apparent place of the star, that is, its place corrected for precession, aberration, and nutation, is as follows:—

$$AR = 1^h\ 0^m\ 22^s \cdot 36 \qquad \delta = 1^\circ\ 35'\ 20''\cdot 27$$

and at the place for which this digression is to be calculated,

$\phi = 46^\circ\ 52'\ 40''$

$$\begin{aligned} \log. \cot. \phi &= 9\cdot 9715130 \\ \log. \tan. \delta &= 8\cdot 4431331 \\ \hline \log. \cos. h &= 8\cdot 4146462 \\ h &= 88^\circ\ 30'\ 40''\cdot 55 \\ &= 5^h\ 54^m\ 2^s \cdot 70 \end{aligned}$$

$$\begin{aligned} \log. \sin. \delta &= 8\cdot 4429658 \\ \log. \sin. \phi &= 9\cdot 8632617 \\ \hline \log. \sin. A &= 8\cdot 5797041 \\ A &= 2^\circ\ 10'\ 38''\cdot 51 \end{aligned}$$

is the azimuth at the digression to the N. W.

AR ✲	= 1 0 22 ·03
Mean AR ☉	= −17^h 2^m 57^s·48
	13 51 27 ·25
Acceleration of the fixed stars	− 2 16 ·21
	13 49 11 ·04 mean time of the digression to the W.

From what has preceded, it will easily be seen how the deviation of the magnetic needle from the true meridian may be ascertained by an observer on land. In fact, he has only, by means of a properly constructed compass, to take the magnetic bearing of some signal, and then to determine the true azimuth of that signal by some one of the methods above explained. The true azimuth, compared with the magnetic azimuth, will give the deviation of the compass. But the observer at sea must obviously have recourse to other means. It is to be borne in mind that for the purposes of navigation, it is unnecessary to determine the magnetic deviation with mathematical accuracy: it is more important to employ methods that are simple, and of easy practical application. It is usual at sunrise or sunset to observe the magnetic azimuth of the sun's centre*; which, compared with the calculated true azimuth, gives the deviation. This true azimuth is easily calculated; as we have a spherical triangle of which one side is the *apparent* zenith distance = 90° + refraction − parallax = ζ: the other two sides are the colatitude of the observer, ϕ, and the north polar distance of the sun's centre Δ. Hence

$$2\psi = \zeta + \Delta + \phi$$

$$\cos.^2 \frac{\alpha}{2} = \frac{\sin. \psi . \sin. (\psi - \Delta)}{\sin. \zeta . \sin. \phi}$$

* By observing alternately the opposite limbs.

We may, with sufficient accuracy for our purpose, regard the horizontal refraction as constant, and $= 33' \, 45''$, and the parallax $= 8''$; consequently ζ always $= 90^\circ \, 33' \, 37''$.

Navigators are in the habit of employing, instead of the azimuth of the rising or setting sun, its amplitude, which differs from the azimuth merely in being counted from the east and west points, instead of from north and south. Hence this method is often designated by the name of that of *ortive* and *occasive* amplitudes.

Suppose the magnetic azimuth of the centre of the rising sun has been found $= 83^\circ \, 21' \, 14''$ E. at a place whose co-latitude $= 59^\circ \, 56' \, 40''$; and that we have found, from the solar tables for the corresponding instant,

	$\Delta = 74^\circ \; 13' \; 38''$	
then	$\phi = 59 \;\; 56 \;\; 40$	$\log. \sin. \phi = 9{\cdot}9372872$
	$\zeta = 90 \;\; 33 \;\; 37$	$\log. \sin. \zeta = 9{\cdot}9999792$
	$2\psi = 224 \;\; 43 \;\; 55$	$\log. \sin. \phi . \sin. \zeta = 9{\cdot}9372664$
	$\psi = 112 \;\; 21 \;\; 57{\cdot}5$	$\log. \sin. \psi = 9{\cdot}9660347$
	$\psi - \Delta = 38 \;\; 8 \;\; 19{\cdot}5$	$\log. \sin. (\psi - \Delta) = 9{\cdot}7906848$
		$\log. \cos.^2 \frac{\alpha}{2} = 19{\cdot}8194531$
	$\therefore \frac{1}{2}\alpha = 35 \;\; 40 \;\; 37$	$\log. \cos. \frac{\alpha}{2} = 9{\cdot}9097265$
	$\alpha = 71 \;\; 12 \;\; 14$	

This is the true azimuth from N. to E.; subtract it from the magnetic azimuth $83^\circ \, 21' \, 14''$, and we obtain for the deviation of the magnetic needle from the true meridian 12° W.

HISTORY OF ASTRONOMY.

Chapter I.

Oriental Astronomy.—The Chinese.

Astronomy is in all probability the most ancient, as it is unquestionably the most perfect of the physical sciences. But the very antiquity of its origin, coæval it would seem with the earliest civilization of the East, throws an obscurity round it, which the present state of our knowledge does not enable us to dispel. Thus some have been induced to look to the Egyptians or Chaldæans, some to the Indians, and others to the Chinese, as the inventors of the science; but the truth is, that notwithstanding all the learning and acuteness that have been shewn in discussing these questions, the facts in our possession are too few and unconnected to guide us with any certainty to a conclusion. That in some of the nations mentioned, observations of the heavens, though perhaps rude and incomplete, have been made from the earliest times, appears unquestionable; but this is nearly all that can safely be affirmed. Whether any of these nations borrowed from the others, or all from a common source; or whether, on the other hand, there was no scientific communication between them, is a point upon which we may speculate with more or less probability, but which we can never hope to establish beyond controversy.

In a fine climate and a level country, in the plains of Chaldæa, or the valley of the Nile, the spectacle of the heavens, everywhere so striking, must have forcibly arrested the attention of a people just beginning to emerge from barbarism. But it is principally to the superstitious ideas of the inhabitants of the East that we must look for the motives which induced them to follow with so much care the varying phenomena of the celestial sphere. The Chaldæans have been celebrated in all ages for their attachment to judicial astrology; the Chinese, from time immemorial, have considered solar eclipses and conjunctions of the planets as prognostics of importance to the empire, and the observation of them has been made a matter of state policy. Traces of a similar belief may be found in Egypt and India; and there can be little doubt that to a superstition vain and degrading in itself, we owe the early observations made in China and Babylon, the zeal with which the Arabs embraced the science of Ptolemy, and the revival of astronomy in modern Europe.

In treating of the early Oriental astronomy, we are induced to give the priority to the Chinese, without meaning to affirm that the science is of more ancient date among that people, than among the Babylonians and Egyptians. But it can scarcely be disputed that they possess the oldest authentic observations on record, and consequently have well-founded claims to our earliest notice. It is also to be remarked that the long residence of the Jesuit missionaries in China, and the peculiar opportunities they enjoyed of examining the records, afford great comparative facility in the investigation of this part of the subject. For a long time the office of president of the tribunal of mathematics was filled by members of this order: their assistance having become indispensable to the Chinese for the correction of their astronomical tables and methods of calculation. The Jesuits did not fail to profit by this opportunity of studying the history, antiquities, and scientific monuments of the country; and we owe to their researches a great mass of curious and interesting facts. Father Gaubil has written a treatise professedly on the history of Chinese Astronomy, which with another treatise on the same subject in the 26th volume of the *Lettres Edifiantes* comprise nearly all that is known on the subject. But after all it must be recollected that the learning and diligence of the missionaries have been able to collect nothing beyond detached observations, and fragments of science, which could convey no information except to those who were already versed in astronomical calculations. The reader will not expect a connected history of the origin and progress of astronomy in the east; this we can trace but imperfectly for Greece, a country with which we are so well acquainted: with regard to China and India we must be contented with establishing a few facts, which, at some future time, increased knowledge may perhaps enable us to connect.

It appears that the Chinese, whose annals are in some particulars of more than doubtful veracity, carry up the foundation of the Empire to a prince named Fou-hi, as early as nine and twenty centuries before Christ; and they refer to the same period the institution of their cycle of sixty years, and the composition of certain mysterious figures, of which we shall say more presently. Though the cycle just mentioned is purely civil, it is sufficiently ingenious to merit farther notice in this place. In order to give each year a name that shall indicate at once its place in the cycle, the Chinese have taken two series, the first composed of ten, the second of twelve monosyllables. In forming then the names of the respective years, they begin by combining the first word of the first series with the first of the second to make the name of the first year in the cycle; for the name of the second year they combine the second of the first with the second of the second series, and so on: after ten of these combinations, the first word of the first series answers to the eleventh of the second, the second of the first to the twelfth of the second, the third of the first to the first of the second, and so they proceed till the first word of the first series corresponds again to the first of the second*. This happens at the end of sixty years. There can be no question that this cycle is extremely ancient; it is quoted in the Chou-king†; an historical work of which some parts are nearly as old as the time of the Emperor Yao; that is, about 2,300 B.C.; but there is some difficulty in fixing precisely its origin; the Chinese tribunal of mathematics places the first year of the first cycle at the eighty-first year of the Emperor Yao‡.

The first phenomenon recorded in the Chinese annals is a conjunction of five planets, in the reign of the Emperor Tchuen-hiu, which lasted from about 2514 to 2436 B.C. The conjunction is said to have taken place just beyond the constellation Che, which occupies about 17° of longitude, and the centre of which is in 6° Piscium; on the same year spring began before the first day of the first moon§. According to the calculations of Father De Mailla all these circumstances are verified in the year 2461.—In this year, the spring, which begins in China with the passage of the sun into 15° Aquarii, fell on the 4th of February, the new moon on the 6th, and on the 9th, the four planets Saturn, Jupiter, Mars, and Mercury, were, with the moon, comprised within an arc of about 12°, from 15° to 27° Piscium.* But, in the first place, we must remark that these calculations cannot inspire a great confidence; they were made with the tables of Lahire, now quite antiquated; and even had they been founded on the best modern tables, for example, those of Delambre, Burckhardt, and Lindenau, they would be open to much doubt; as small errors in the elements will, in the course of forty centuries, produce deviations of considerable magnitude. Besides, a conjunction of five planets is recorded, whereas it appears, from De Mailla's own calculations, that only four were in conjunction; the Chinese text requiring, according to Gaubil, that the moon be not included among the five in question†. According to the last mentioned astronomer, this conjunction was supposed, in order to serve as an epoch for the tables. For it appears that for a long time the Chinese always took for epoch a fictitious general conjunction of the planets; the date of which they fixed by calculating backwards with the respective mean motions they supposed to belong to each planet, till they found the conjunction they sought‡. It is remarkable enough that the Indians fixed the epoch of their tables in a similar way; nor is this the only point of coincidence in the astronomy of the two nations. Both divided their zodiac nearly in a similar way, the latter into twenty-seven, the former into twenty-eight constellations; to which number they were probably led by observing the moon's revolution to take place in something between twenty-seven and twenty-eight days; subsequently both seem to have become acquainted with a division of the zodiac into twelve

* The words of the first series are, Kia, Y, Ping, Ting, Vou, Ki, Keng Sin, Gin, Quey; of the second, Tse, Tcheou, Yn, Mao, Chin, Se, Ou, Ouey, Chin, Yeou, Su, Hay. The first year of the cycle would be called Kia-Tse, the second Y-Tcheou, and so on; for example, the fifteenth would be Vou-Yn. V. Souciet, Recueil d'Observations faites à la Chine, p. 174.

† Souciet, vol. iii. p. 14.

‡ Gaubil, in Souciet, vol. ii. p. 137.

§ The literal translation of the Chinese is, 'Hoc anno primæ lunæ primâ die præcesserat ver: quinque planetæ convenere in cœlo transmissâ constellatione Che.' De Mailla, Hist. de la Chine, vol. i. p. clv.

* On the 9th of February 2461 B.C., Gregorian style, at half-past seven P.M., mean time on the meridian of Paris, the longitudes of the planets were as follows:

☿ = ♓ 14.56.16
♂ = ♓ 26.45.11
☽ = ♓ 23.18.21
♃ = ♓ 17. 2.12
♄ = ♓ 24.39.47

De Mailla, vol. i. pp. clv. vi. vii. viii.

† Souciet, vol. ii. p. 149.

‡ Souciet, vol. ii. p. 16. Vide also Gaubil, Hist. de l'Astron. Chin. in the Lett. Edifiant. 1811. Toulouse, vol. xxvi. p. 212.

signs; but this division was purely mathematical, the constellations of the zodiac remaining unchanged.

The circumstance above-mentioned, that under the Emperor Tchuen-hiu, spring began with the passage of the sun into 15° Aquarii, appears to Bailly another proof of the connexion between the astronomy of India and China. The Indians who made use of a sidereal year, fixed the beginning of their year at the entry of the sun into the beginning of their zodiac, which, as we have just seen, was determined entirely by the fixed stars; now, in consequence of the phenomenon called the precession of the equinoxes, the beginning of this zodiac, which was marked by a certain fixed star, appeared to move, from year to year, at the rate of about 50″ annually, (the Indians supposed 54″); and if, at some remote time, the beginning of the year coincided with the winter solstice, some centuries later it would coincide with the vernal equinox. Now, the Indians suppose, that, in the year 3102 B. C., the beginning of their zodiac was in 6° Aquarii: in the time of Tchuen-hiu, it would be pretty near 15° Aquarii, and the Indian year at that time would begin with the passage of the sun through that spot in the heavens. This, as we have just seen, was the moment at which Tchuen-hiu fixed the beginning of the Chinese year. The explanation of Bailly is ingenious and plausible; but it is difficult to admit the extreme antiquity which the reality of this determination by the Chinese necessarily implies.

Though Gaubil rejects the conjunction of Tchuenhiu, he seems to think that the beginning of civilization in China, and the earliest observations of the stars, are at least as old as this emperor, though later than the time at which Fou-hi is placed. For it would seem that the stars Tay-y and Tien-y of the Chinese catalogues have been successively observed in the pole. It is not easy to identify either of these stars with any contained in the modern European catalogues; but, according to Gaubil, the first was in the pole about the year 2259, the second about 2669 B. C. But about 2850, the star α Draconis was polar star; this, however, is not marked as having been such in the Chinese sphere, and hence Gaubil deduces that their observations are subsequent to this time.

According to the opinion of the astronomer just mentioned, the authentic history of China begins with the reign of the Emperor Yao, about three-and-twenty centuries before the Christian era,—an antiquity still too great to be easily admitted. We have seen that the beginning of the cycle of sixty years, which the common tradition refers to Fo-hi, is fixed by the Tribunal of Mathematics more than eighty years after the Emperor Yao; and we may probably bring down to an age posterior to this prince, the date of the Ko-tou and the Koua, certain mysterious figures, supposed by the Chinese literati of all ages, to contain important astronomical truths. Unfortunately, these figures, if they have any meaning, have long ceased to be intelligible, and the commentary of Con-Fu-Tso on them is equally obscure*. The most eminent literati of the empire have in vain tortured their imagination to decipher these enigmatical records; and, even in Europe, they have exercised the ingenuity of a celebrated philosopher. Leibnitz imagined that he had found in the Koua, which are groups of straight lines, some continuous, others disjoined in the middle, (the straight and broken lines being combined in a variety of ways,) a system of binary arithmetic. This is a conjecture, perhaps rather specious, but difficult to establish by any solid reasons. After all, it is little better than a waste of time to employ it in speculating on the meaning of characters which have long been unintelligible to those who enjoyed the best opportunities for deciphering them. There are more interesting records of the time of Yao, to be found in the Chou-king, an extremely ancient work, one chapter of which, called the Yao-tien, is said by Gaubil to have been composed either in the time of this emperor, or very shortly afterwards†. From a passage of this chapter, it very clearly appears that the solar year was fixed at the length of 365¼ days, and a method of intercalation adopted to reconcile the motions of the sun and moon. Unfortunately, the intercalation used is not explained; but Gaubil thinks that it was the insertion of seven months in nineteen solar years,—a period which, according to him, has been known in China from the most remote antiquity ‡. In this he is confirmed by the Chinese annals, which attribute directly the intercalation in question to the Emperor Yao §. It is re-

* Souciet, vol. iii. p. 2, *et seq.*

† Souciet, vol. iii. p. 6.

‡ Lett. Edif. vol. xxvi. pp. 66 and 106.

§ In this period the months were alternately of twenty-nine and thirty days; each month bore the markable that this period of nineteen years seems to have been known at an early time to most of the eastern nations. We shall find it in the tables of India and Siam; and we have the testimony of Geminus to its being used among the Chaldæans.

According to Gaubil, the invention of the Chinese zodiac, divided into twenty-seven constellations, is to be referred to Yao*; and so it would seem is the estimation of the obliquity of the ecliptic at twenty-four Chinese degrees †. It is necessary to explain, that the Chinese degrees differ slightly from ours: instead of dividing the circumference of the circle into 360°, they have divided it into 365¼ parts; the object of which rather singular division seems to have been to facilitate the calculation of the sun's longitude, corresponding to each day of the year ‡. For, as they supposed the motion of the sun in its orbit to be uniform, and as they had fixed the length of the solar year at 365¼ days, it is evident that the sun would describe daily exactly one Chinese degree. It is evident, from the text of the Chou-king, that in the time of Yao, not only could the Chinese distinguish the equinoxes and solstices by the length of the days and nights, but that these phenomena were marked by a reference to certain stars. The text does not mention how these stars mark the passage of the sun through the colures; with regard to the winter solstice, and perhaps the two equinoxes, it appears that they designated them by their passage over the meridian about six in the evening: but as the stars are not visible at six at the summer solstice, Bailly supposes, that to mark this season, they took Antares, which then passed the meridian about eight in the evening §.

The reign of the Emperor Tchong-kang, the grandson of Yao, is memorable in Chinese history from the observation of a solar eclipse, which is interesting in itself, and still more so from the circumstances which are recorded as attending it. It is said that the emperor was so irritated against two great officers of state, who had neglected to predict the eclipse, that he put them to death on this account; though some writers have insinuated that he was influenced by political motives, for which the neglect of the prediction merely served as a pretext. It would be a most curious and important fact, could it be clearly made out, that in the time of Tchong-kang, the Chinese had methods for the prediction of eclipses: as to the method itself, there would be little doubt that it was by means of the period of nineteen years. Unfortunately there is some difficulty in verifying this eclipse, as there is an uncertainty about the date of Tchong-kang, amounting to twenty or thirty years. Father Gaubil, who has written a dissertation expressly on this eclipse, fixes it in 2155 B. C.*: the Chinese annals, translated by De Mailla, in 2159 †: most of the Chinese astronomers, particularly Cocheou-king, the greatest of them all, in 2128. According to Gaubil, the eclipse was observed in the constellation Fang, (corresponding to Scorpio,) in the ninth moon: though it is to be noticed, that the literal translation of the text, as given by De Mailla, places it simply in autumn. From a comparison of all the authorities, it is evident that we may fix it near the autumnal equinox, in the fifth year of the reign of the Emperor Tchong-kang. The question is, to find exactly the date of this emperor. Now, the text of the ancient book, the Chou-king, does not enable us to do this with precision; but the Tcheou-tchou, a work of less antiquity, though still very old ‡, refers the eclipse in question to the year 2128, and even fixes the very day on the first of the ninth moon §. This date, as Gaubil admits, is preferable to any other; but he rejects it in favour of the year 2155, because the tables of Halley give, on the former, only a very small eclipse of not more than a digit and a half. But this argument cannot be decisive, as, in cal-

name of the sign into which the sun entered at its end. When a month finished without the sun's entering into the sign of which it bore the name, then they intercalated a month.

* The Chinese astronomers in general fix the beginning of their zodiac in the constellation Hiu (from 19° to 29° Aquarii): now, in the time of Yao, the winter solstice was certainly in this constellation: hence Gaubil considers Yao as the founder of the Chinese astronomy, it being highly probable that originally the winter solstice coincided with the origin of the zodiac.

† Equal to 23° 38′ 11″ of our division.

‡ Carlini, Tavole del Sole, p. 1. Milano, 1810.

§ In this case, the star Ho, mentioned by Yao as marking the summer solstice, would correspond to the modern constellation Sing, and not Fang, as it is explained by the Chinese interpreters. V. Bailly, Hist. de l'Astron. Ancienne, Suppl, p. 349; and Gaubil in Souciet, vol. iii. p. 8.

* Souciet, vol. ii. p. 140.

† Vol. i. p. 137.

‡ Certainly not later than 279 B. C.

§ 13th Oct. 2128 B. C. It is to be noticed, that the text of the Chou-king is very differently translated by Gaubil and by De Mailla (vol. i. p. 137.) According to the latter, the eclipse was not merely near the equinox, but on the very day. This, if true, would quite exclude the eclipse of 2128, and force us to remount to that of 2159; but we prefer following the interpretation of Gaubil, who was not merely a good astronomer, but profoundly versed in the Chinese and Mandchou languages.

culating for such remote periods from the best modern tables, (and, *à fortiori*, from those of Halley,) it is impossible to be certain of the magnitude of the eclipse, the uncertainty may amount to some digits; and in fact the tables of Lemonnier give an eclipse of more than four digits *. These considerations seem pretty decisive in favour of the year 2128, if indeed such an observation ever were made; to which its extreme antiquity, and the long silence as to any similar observation which follows it, form very weighty objections †.

Indeed, admitting the eclipse of Tchong-kang, it seems quite inexplicable that, during the ten centuries following, not a single observation or fact connected with astronomy is to be found in the old Chinese histories ‡. This circumstance will probably appear to many conclusive against the early science of the Emperor Yao, and his immediate successors; and any scepticism on this point must be considered as very justifiable. But in the regency of Tchou-kong, about eleven centuries before Christ, we meet with observations, of the authenticity of which little doubt can be entertained, and which are among the most interesting transmitted to us by antiquity. It appears from a memoir of Gaubil, first printed in the *Connoissance des Temps* for 1809, that at the town of Loyang, now called Hon-an-fou, Tchou-kong found the length of the shadow of the gnomon § at the summer solstice equal to one foot and a half, the gnomon itself being eight feet in height. According to a tradition not quite so certain, the shadow at the winter solstice was thirteen feet. The exact date of these observations is not known; but we may suppose them, without any sensible error, to have been made about the year 1100 B. C.‖ Applying, then, to the data here given, the proper corrections for the sun's semi-diameter *, refraction, and parallax, we get for the respective zenith distances observed, 10° 53′ 7″ and 58° 40′ 46″. Half the sum of these quantities will give us the latitude of Loyang,—half their difference, the obliquity of the ecliptic observed. The former, then, we find 34° 46′ 55″, and the latter, 23° 53′ 47″. Now, the latitude, as determined by the observations of the missionaries, is 34° 46′ 15″ †: the obliquity of the ecliptic, calculated for the year 1100 B.C. from the formulæ founded on the theory of universal gravitation, is 23° 49′ 42″. The agreement of the latitude and obliquity deduced from the observations of Tchou-kong, with those just given, is really remarkable; and it is easy to show that this agreement could not be the result of artifice; for at the time Gaubil wrote, not only was the law of the diminution of the obliquity unknown, but the very fact of such a diminution was much doubted: he could not then have forged observations to represent it. But indeed such a suspicion is not likely to enter the minds of those who are familiar with his works. The good faith of the Chinese is much more doubtful; but the argument we have just used with regard to the Jesuit will apply with much more force to them.

We have also some observations of Tchou-kong on the position of the winter solstice, with regard to the fixed stars. This he placed at two Chinese degrees‡, within the constellation Nu, which begins with the star called ε Aquarii §. Hence it appears that he made the right ascension of that star about 268° 1′ 44″. If we calculate the place of the star for the year 1100 B. C., from the most accurate formulæ, we shall find, for its right ascension, 268° 47′ 14″. The error is not great, considering the uncertainty as to the exact year of the observation, and the difficulty of referring the solstice to the fixed stars. We do not know the method employed for this purpose by Tchou-kong. The ancient Chinese certainly used clepsydræ for the division of time; and the obvious inaccuracy of such an instrument would account for a large part of the error.

A curious fragment of the Tchou-pey —a work written more than three cen-

* Lett. Edif. v. xxvi. p. 279. Paris.

† Compare La Place (Syst. du Monde, liv. v. chap. 1.) Delamb. Ast. Anc. vol. i. p. 352.

‡ A solstice marked as having been observed in the year 1643 B. C., appears to be a calculation of much later astronomers, and interpolated from a period proposed by the astronomers of the Hans, a little before the Christian era. For a long series of similar supposititious solstices, see *Con. des Temps* pour 1809. Addit. They all seem to have been calculated from the periods of nineteen and seventy-six years.

§ The gnomon is essentially composed of some vertical object, as an obelisk, pillar, &c., the shadow cast by which is received on a carefully levelled horizontal plane: the trigonometrical tangent of the sun's apparent altitude is equal to the length of the shadow divided by the length of the gnomon.

‖ Fréret, in accordance with Gaubil, places the regency of Tchou-kong from 1104 to 1098 (Conn. des Temps pour 1811. Add. p. 432). The Chinese annals make it begin in 1115. V. De Mailla, vol. i. p. 279.

* Till the time of Cocheou-king, A.D. 1280, the Chinese only measured the length of the pure shadow.—V. Gaubil, *Connoissance des Temps*, 1809.

† V. Conn. des Temps, 1809, p. 394.

‡ Each degree = 59′ 8″ of our division.

§ Lett. Edifiant. vol. xxvi. p. 100.

turies before Christ—has been supposed to attribute to a philosopher contemporary with Tchou-kong a knowledge of the famous property of the right-angled triangle. In this fragment, the philosopher mentioned concludes, that if the two sides of a right-angled triangle are respectively equal to 3 and 4, the base will be equal to 5. Here we must remark, that the numbers 3 and 4 are not taken at hazard, but selected from some mysterious connexion supposed by Con-fu-tso to exist between these numbers and the universe *. Gaubil concludes, that in the time of Tchou-kong, the Chinese had methods for the resolution of right-angled triangles, though spherical trigonometry was unknown to them till the time of Cocheou-king, twelve centuries after Christ.

The history of Con-fu-tso, which extends from 720 to 481 B. C., records several eclipses, of which a good many have been verified by modern astronomers; others appear to have been marked in the wrong month; from the rough way in which all are given, they can only serve to fix the dates of Chinese chronology, and to show the assiduity with which these phenomena were noticed at so early an age. The most interesting observation of this period regards the position of the winter solstice, which was placed in the beginning of the constellation Nieou, the first star of which was β Capricorni. Now, as there can be no doubt that the astronomers who found this position of the solstice, were acquainted with the observations of Tchou-kong, who placed it at 2° from ϵ Aquarii, nearly 9° of longitude distant, it seems evident that they must have perceived the apparent retrogradation of the solstices and equinoxes. Such is the conclusion drawn, and apparently with reason, by Gaubil † and La Place‡; but in this case, it seems strange that the astronomers of the Hans, two centuries before Christ, should have been ignorant of the effects of precession.

Subsequently to this time the Chinese astronomy ceases to possess the same interest for us: as it has always remained in so rude and imperfect a state, that nothing but the antiquity of the determinations to be found in their books could make them worthy of attention. Complete astronomical treatises, as early as the year 108 B.C., are still in existence; but the reader who wishes to investigate their tables and methods, is referred to the treatises of Gaubil on this subject, in the second and third volumes of the collection of Souciet. We shall only notice here the method of establishing their epochs, followed by all the Chinese astronomers, anterior to Cocheou-king, above mentioned; and we will take, as an example, the treatise of Lieou-hiu, the most ancient of those now extant. The epoch of these tables is a general conjunction of the sun, moon, and planets, the moon being on the ecliptic, about 143127 years, before the year 104 B.C.* This latter seems to be the real epoch of his tables: the other, it is scarcely necessary to observe, is obtained by calculating back from the year 104 B.C., with the mean motions found for the different planets, till a general conjunction was obtained. This, as we have noticed, was the method usually followed by the Chinese; but sometimes, in order to avoid such large numbers as that just given, they would content themselves with a very rough approximation to a general conjunction, and neglect the errors arising on the mean motions, as too small to be noticed; which, if the epoch were at all distant, they would really be. The pretended general conjunction in the reign of Tchuen-hiu, above noticed, is, in all likelihood, an epoch obtained in this way. In the astronomy of Lieou-hiu we find the cycle of nineteen years very clearly explained; and, indeed, not only this, but the period of seventy-six years proposed in Greece by Callippus, was known in China before the Christian æra †.

It seems that, about the year 164 after Christ, the Chinese began to have communication with subjects of the Roman empire ‡; and it is worthy of notice, that very shortly afterwards some important reforms were made in their astronomy. They now ascertained the eccentricity of the solar orbit, the principal inequality of the moon, and a more exact value of the solar year; and we now find, for the first time, a distinct account of precession; a phenomenon, however, with which we believe they must have been acquainted long before. That some of the discoveries just mentioned were introduced from the west is only a conjecture; but it is a conjecture which derives considerable force from

* Lett. Ed. p. 117. The numbers 3, 4, and 5, multiplied into each other, give 60—the number of years of the Chinese cycle.
† Lett. Ed. xxvi. p. 247, ed. Paris.
‡ Mécan. Célest. vol. v. p. 246.

* Souciet, vol. ii. p. 16.
† Ibid., p. 21. ‡ Ibid., p. 24.

the circumstance, that during the whole interval between the second and thirteenth centuries after Christ, the Chinese, though continuing to observe with assiduity, made little or no progress, and certainly not one discovery. Their greatest improvements did not go beyond some trifling ameliorations in the elements of their tables. Those who feel any curiosity to examine their observations, will find in the works of Gaubil, edited by Souciet, and in the *Connoissance des Temps* for 1809, a considerable quantity of observations of solstices made with the gnomon, and of solar eclipses; with some notices of occultations, of comets, and of appulses of Jupiter to the fixed stars. Of these the most important are the observations of the gnomon; some of which have been used by La Place to determine the diminution of the obliquity of the ecliptic to the equator.

The conquest of China by Gent-Chiskhan, who brought with him men well versed in the astronomy of Ptolemy and the Arabs, gave a fresh impulse to the languid state of the Chinese astronomy; but the ameliorations then introduced belong rather to the history of the middle ages. It is time now to turn to a people whose astronomical reputation is greater, though perhaps less deserved, than that of the Chinese.

Chapter II.

The Indians.

Some learned men have been disposed to attribute an extraordinary antiquity to the cultivation of astronomy in India. This opinion, which is founded upon the elements of astronomical tables brought from India, has been supported at great length and with much ingenuity by M. Bailly, in a work professedly on this subject; which, though it may contain erroneous conclusions, must always be considered as a model of elegance in scientific composition. The tables in which the elements of the Indian astronomy are to be found have been brought into Europe at various times: the earliest known were those imported from Siam by M. de la Loubère, the French envoy, on his return from a mission to that country about the year 1687. These tables have been analysed and explained by D. Cassini (in the *Mémoires de l'Académie des Sciences*, tom. viii.), and his explanation has been adopted with some immaterial alterations by Bailly*. It appears that the epoch of these tables is the 21st of March, 638 A.D., at the moment that the sun entered the beginning of the zodiac; for the Siamese, like all the Indians, had a zodiac of twenty-seven signs or constellations, the position of which was entirely determined by the fixed stars, and which had, from the effects of precession, a progressive motion in longitude, successively occupying different situations with regard to the equinox. They had also a division of the zodiac into twelve signs, but this seems to have been merely an abstract mathematical division for purposes of calculation; and these twelve signs were by no means identified with any of the constellations. The epoch once determined, the Siamese calculate the mean motions of the sun and moon by means of two periods; the first of 800 years comprising 292207 days; the second, of 19 years corresponding to 235 lunar revolutions. The first gives us a sidereal year of $365^d\ 6^h\ 12^m\ 30^s$, about $3^m\ 24^s$ greater than the real value: in the second they appear to have taken the tropical year as equal to $365\frac{1}{4}$ days; the lunar revolution being supposed equal to $29^d\ 12^h\ 44^m\ 3^s$.

The tables of Chrisnabouram* offer little remarkable; we do not find in them the Siamese period of nineteen years, but a method of intercalation, which has the same object. The solar apogee, which the Siamese consider fixed, is here supposed moveable, though its motion is slower than it ought to be, according to our observations. The mean motions differ very considerably from those of the Europeans; but it is remarkable that the error for the sun and for the moon, in a given interval, is the same, so that the calculation of the time of an eclipse, which seems to be the principal object of all the Hindoo astronomers, is little affected by it. The epoch of these tables is fixed at sun-rise, on the 10th of March, 1491 A.D. The tables of Narsapur resemble a good deal those of Siam: they have the same period of 800 years containing 292207 days; but, instead of the second period of the Siamese Tables, they calculate the moon's motion directly by supposing that she makes 800 revolutions in 21857 days. This gives a sidereal revolution

* Astron. Indienne, chap. I.

* Sent from India by P. Duchamp. Chrisnabouram is a town of the Carnatic: Narsapur is in the vicinity of Massulipatnem: Tirvalore is near Pondicherry. The original translation of tables of Chrisnabouram into French, from Sanscrit, by Duchamp, is to be found at the end of the Astronomie Indienne of Bailly.

of $27^{d}\ 7^{h}\ 42^{m}\ 36^{s}$ considerably too great; and, indeed, the Brahmins seem to have perceived the necessity of correcting the mean motions here assigned, which they do by renewing their epoch every eighty-seven years: at least in these tables of Narsapur there are two epochs, the one in 1569, the other in 1656.

The most curious of all the Indian tables, are those brought from Tirvalore by M. Le Gentil, and analysed at length by Bailly in his Indian Astronomy*. The epoch of these tables is the year 3102 before Christ: at which time, they suppose a general conjunction on the ecliptic of the sun, moon, and planets. The object of the construction of these tables, like all those of the Indians, seems to be the calculation of eclipses; their methods have been explained at length by Bailly and Le Gentil: and the latter of these astronomers has applied their methods to an eclipse really observed by himself in India; the error he found to amount to about twenty-two minutes of time.

From a comparison of the four tables, just mentioned, Bailly† deduces that they have all one common origin: and on this head his arguments are pretty conclusive. He remarks—I. That the tables of Siam contain a reduction for the difference of meridians, which show them to have been borrowed from a place, which has about the same longitude as Benares, a town which has been always looked upon by the Hindoos with especial veneration, and which seems to have been the residence of their most learned men. The tables of Tirvalore, Chrisnabouram, and Narsapur, contain no reduction of this kind, as all these towns lie nearly on what appears to have been assumed as the first meridian.

II. The epochs of these tables are so connected by the mean motions, that from one of them all the others may be found by employing the mean motions of the tables of Chrisnabouram. This would show that the Indians have in reality only one epoch, from which the others have been deduced by calculation; and, as one of these epochs may be obtained from the other without any correction for the difference of meridians, it follows that all these tables have been originally formed for the primitive meridian.

III. There is an extraordinary resemblance in all the elements of the solar orbit, as given in these tables, though those of the lunar orbit differ. We find the same mean motion of the sun in all, the same length of the year, the same equation of the centre*. The coincidence is too striking to be fortuitous, and proves, that the tables of the sun have been taken from the same source, while those of the moon have undergone various alterations.

IV. All the Brahmins versed in astronomy agree in considering these tables as borrowed from ancient works, and principally from one called the Surya Siddhanta†: a work that, in the time of Bailly, was supposed to be guarded with great jealousy, and to be entirely shut up from all but a few learned men. This, however, turns out not to be the case: the work in question is rare, but that and many other treatises on the same subjects have been procured by Englishmen resident in India; and an interesting analysis of the Surya Siddhanta, though not so complete as might be wished, is to be found in the Asiatic Researches‡. It is evident from an inspection of the extracts given by Mr. Davis, that the solar tables of the Carnatic, which we have been discussing, are essentially borrowed from the Surya Siddhanta; in this latter the sidereal year is given at $365^{d}\ 6^{h}\ 12^{m}\ 36^{s}$: the greatest equation to the centre at $2^{\circ}\ 10'\ 32''$, the obliquity of the ecliptic at 24°. An acquaintance with the Surya Siddhanta, and the other systems of astronomy existing in Sanscrit, puts beyond a doubt, that it was the custom of the Hindoos, as we have seen it was of the Chinese, to take for epoch a fictitious

* Chap. iv. † Astron. Ind. chap. v.

* The following is a comparison of these tables, with regard to the solar orbit.

	Sidereal Year.				Greatest Equation of the Centre.		
	d	h	m	s	°	′	″
Siam	365	6	12	36	2	10	32
Chrisnabouram	365	6	12	30	2	10	32
Narsapur	365	6	12	36	2	10	32
Tirvalore	365	6	12	30	2	10	32
Surya Siddhanta	365	6	12	36	2	10	32
Modern European	365	6	9	9	1	55	26

† The term Siddhanta is applied by the Hindoos to their older systems of astronomy, in opposition to the newer. There are several Siddhantas; this of Surya (a Hindoo God, representing the Sun) is supposed to have been received by divine revelation, 2,164,930 years ago. The original, which exists in Sanscrit, (now a dead language in India,) has never been translated into any European language: it is known principally by the account given of it, and by the extracts from it, in the several memoirs in the Asiatic Researches quoted in this chapter. The same may be said of other ancient astronomical systems, particularly that of Bramagupta.

‡ Vol. ii. p. 225. London, 4to.

general conjunction of the planets obtained by calculating backwards, with the respective mean motions attributed to the several planets by the authors of the system. No doubt can remain upon this head after a perusal of the memoirs of Messrs. Bentley and Davis in the Asiatic Researches, which are founded on an examination of the original documents.

However, Bailly has devoted the greater part of his work to establishing the point that the epoch of the Indian tables in the year 3102 before Christ was not imaginary, but founded upon actual observation. The talent and research with which he has argued the question, make it worth while to give a rapid summary of the proofs he has alleged, and some remarks upon their insufficiency. His argument may be divided into two parts, I., as it regards the epochs of the tables; II., as it regards the mean motions.

I. The nature of his argument with regard to the epochs is this. The positions of the sun, moon, and planets, as well as the position of the solsticial and equinoxial colures are so accurately determined by the Indians for the time of the Calyougam*, that these positions must have been actually observed, and could not have been merely the results of calculation in later times, as the Hindoos have never possessed that high degree of science, necessary to make these calculations with any precision. But have they in reality given us these positions for the Calyougam with so much accuracy? This is a point which it is necessary to examine. Now in the first place, the Indian tables give, at the epoch, a general conjunction of the sun, moon, and planets; calculating by our modern tables, we find such a general conjunction to have been impossible. It is true that for times so distant the errors of our tables will cause an uncertainty of five or six degrees, but the calculations of Bailly himself show that the planet Venus never could have been in conjunction or indeed near it at the time specified, making every allowance for this uncertainty; and even with regard to the other planets, he is obliged to content himself with an approximation. The impossibility of this pretended conjunction appears to La Place in itself a sufficient reason to reject the pretended antiquity of the Indian tables*. Another consideration urged by the author of the Indian Astronomy is drawn from the accuracy with which he affirms the places of the sun and moon to have been determined for the epoch given. Very little stress, however, can be laid upon this argument, for, as has just been stated, we cannot be certain of the places calculated from our tables for the time of the Calyougam, to nearer than five or six degrees, and this is about the quantity by which the positions given by the Indians differ from ours. Besides, there is an uncertainty, whether the assigned longitude of the sun is the mean or true longitude. The most specious argument brought forward as to the epoch, is that founded on the pretended position of the colures. According to Bailly, the position of the equinox given in the Indian tables for the Calyougam is such, that the star called Aldebaran was in 359° 20′ of longitude: now calculating its place from the modern formulæ of precession, its longitude at the Calyougam would be 18′†: an agreement which, though it does not at first sight seem very great, was the more remarkable, because the Indians, who supposed the precession to be 54″ annually, could not have obtained it by calculating back from a modern epoch. But it turns out that this position of the colures for the Calyougam is merely a calculation of Bailly and Le Gentil, the Indian tables only giving us the longitude of the equinox 3600 years after the Calyougam‡; whence the astronomers just mentioned have deduced its position for the year 3102 B.C. The reader will see at once that this invalidates the whole of the argument urged by Bailly. Indeed the consideration of the epochs alone seems quite decisive against the pretended observations of the Calyougam. The equation of the centre of the sun, given in these tables, is much too great for the epoch in question; La Place thinks that this may be accounted for by the circumstance that in eclipses the moon's annual equation increases the sun's apparent equation of the centre, by a quantity which is very nearly equal to the difference between the equation in question as given by the Indians, and that which may be deduced from the modern tables for the period of the Calyougam. This is an ingenious ex-

* This is the name given by the Indians to their epoch 3102 B.C.

* V. Syst. du Monde, liv. v.
† Bailly, chap. v. § 30.
‡ V. Asiat. Research., vol. ii.

planation; but even should it be admitted as satisfactory, the arguments on the other side are too numerous and weighty to be much affected by it: and La Place himself, as we have before mentioned, does not hesitate to reject as fictitious the epoch of 3102 B.C. The equation of the centre of the moon presents a suspicious resemblance to that of Hipparchus: as to the elements of the orbits of the planets, Bailly himself is obliged to confess that it is only in some of them that we are to look for any precision. While the equation of the centre of Saturn agrees pretty well with theory, on the other hand the disagreement presented by other planets, and particularly by Mars and Jupiter, is very striking. But the limits of this treatise do not allow us to follow Bailly into all these details, particularly as the results are in general of a very unsatisfactory nature. In a discussion of this kind one consideration is obvious; if the Indians really observed the positions and motions of the heavenly bodies somewhere about 3000 years B.C., and if we are to take for a proof of these observations, the accordance of the elements of the sun, moon, and planets, with the values assigned for that period by modern astronomers, then we must expect to find the same degree of accordance nearly for one of these bodies as for another—allowing for the difficulty of the respective observations. No conclusion can be drawn from the agreement of some of these elements; that agreement must be general before any argument can be founded upon it.

II. The second class of arguments given by Bailly are those founded on the mean motions. But here it is necessary to make some preliminary observations, which show that little dependence can be placed on considerations of this kind in determining the age of the Indian or any other tables. The major axes and periodic times of the several primary planets are not subject to any secular inequalities; and it is very evident that they cannot in that case be used for the purpose in question. For the mean motions of the Indian tables must be well or badly determined: in the latter case it is clear that nothing can be deduced from them: in the former all that can be said is, that they were well known at the date of the formation of the tables; but as they suit any time they cannot prove the tables to belong really to one period more than another. Thus, for example, we may suppose the Indians to have framed these tables in the year 1491 A.D., and if we were to find the mean motions ever so accurate, might imagine them to have been determined by a thousand previous years of observation, which, while it would fully account for their perfection, would make the origin of astronomy in India considerably posterior to the Christian era. The planets Jupiter and Saturn are subject to a long periodic inequality of about 929 years: but it is clear that this cannot be of much assistance to us, as many of these periods are comprised between the Calyougam and the present time; and, according to La Place, the mean motions of these planets, as given in the tables now under discussion, would suit equally well the Calyougam, and the year 1491*. But the case is different with regard to the moon: the motion of this satellite is subject to a small secular acceleration, which only becomes perceptible when we compare her places at very distant periods of time. This, then, may be used as a test for trying the antiquity of Tables; but we find different and inconsistent results in the case before us, according to the particular tables we make use of. Those of Tirvalore do not give directly the moon's sidereal revolution, but Bailly has deduced it from the motions given with regard to the apogee. Now there are, in these, several periods given for the anomalistic revolution of the moon, from each of which different results may be deduced: the tables of Chrisnabouram differ again from all of these, as do likewise those of Narsapur. Nor do the two last mentioned agree with each other†. We cannot then, in this case, deduce any conclusions from the moon's mean motion, which certainly is the only one of the mean motions that affords us in general a criterion.

The theory of Bailly on this subject appears, then, totally untenable; but it may be asked, what is the real age of the Surya Siddhanta, and the Indian astronomy? and have the Brahmins

* Syst. du Monde, liv. v.

† Length of a sidereal revolution of the moon as given in the principal Indian Tables.

	d.	h.	m.	s.	
Narsapur	27	7	43	12.65	Bailly, p. 57.
Chrisnabouram	27	7	43	12.2	Ibid. p. 45.
Tirvalore	27	7	43	12.89	Ibid. 93-4-5.
	27	7	43	13.09	
	27	7	43	13.02	
	27	7	43	12.31	
Surya Siddhanta	27	7	43	13.1	As. Res. xii. 246.
Modern Astrono.	27	7	43	11.5	

borrowed their science from any other nation? In answer to the first of these questions, Mr. Bentley has published two very interesting memoirs in the sixth and eighth volumes of the Asiatic Researches. In these, his endeavour is to show, as well from internal evidence as positive historical testimony, that the age of the Surya Siddhanta may be referred to somewhere about 1000 years after the Christian era. This he concludes, from the very reasonable supposition, that whatever be the real date of the tables in question, the errors on the places of the sun, moon, and planets for that time, (as deduced from the tables,) will be less than for any other; as we cannot but imagine that any astronomer, whatever mistakes he might make in giving the positions of the heavenly bodies, for past or future times, would wish to represent faithfully the state of the heavens which he himself observed. Mr. Bentley proceeds then to find when the errors we have above-mentioned are the least in the Surya Siddhanta; and this he shows pretty clearly to have been the case about ten centuries after Christ *. If we exclude from our consideration the place of Mercury, the positions of the lunar apogee and nodes, the solar apogee, and the aphelion of Mars, as being imaginary points, and therefore more difficult to be fixed in an imperfect state of astronomy; and if we deduce the age of the Surya Siddhanta from the positions of the Moon, Venus, Mars, Jupiter, and Saturn, we shall find for this date respectively, the years A. D. 1040, 940, 1460, 924, and 994 †. Taking the mean of these five, we get A.D. 1071; if we leave out Mars, we obtain A.D. 977. From the whole of the data to be found in the Surya Siddhanta, Mr. Bentley finds A.D. 1036. The general accordance of these results seems sufficiently satisfactory ‡.

However, the Surya Siddhanta is not the oldest system of astronomy to be found among the Indians. Mr. Bentley has examined the tables of Brahma-Gupta; and, by an analysis exactly similar to that just described, has been led to fix their age about the year 536 A.D. § Indeed, it appears that Brahma-Gupta was preceded by other astronomers, and particularly by one named Aryabhatta, deserving of notice here, as having advocated the doctrine of the earth's diurnal revolution on its axis ‖. This opinion, it seems, was rejected by subsequent philosophers among the Hindoos, which will scarcely excite our surprise, when we consider that it shared the same fate in the west. Having been embraced by Philolaus, Anaximander, and Aristarchus, it was controverted by Ptolemy, and had fallen into complete oblivion, when revived by the immortal Copernicus. These doctrines of Aryabhatta render it a very interesting point to determine his age, that we may ascertain whether he borrowed this philosophical idea from the sages of Greece, or whether

* Table of the Errors in the Surya Siddhanta with respect to the places of the Planets, &c., at the undermentioned periods.

Planets, &c.	B.C. 3102.	A.C. 499.	A.C. 999.	A.C. 1499.
	° ′ ″	° ′ ″	° ′ ″	° ′ ″
Moon . .	5.52.34−	0.20.14−	0.07.39+	3.43.37+
—— apogee	30.11.25−	4.52.53−	1.21.59−	2.09.56+
—— node .	23.37.31+	3.56.06+	1.12.01+	1.32.04−
Venus . .	32.43.36−	3.33.41−	0.29.22+	4.32.25+
Mars . .	12.05.42+	2.32.42+	1.13.08+	0.06.27−
——aphelion	9.47.00+	1.30.50+	0.21.55+	0.47.60−
Jupiter . .	17.12.36−	1.48.56−	0.24.20+	2.38.36+
Saturn . .	21.25.43+	2.50.09+	0.03.33−	2.54.05−
Sun's apogee	3.15.53+	0.05.45−	0.33.45−	1.01.45−

Asiatic Researches, vol. viii. p. 289.

† Asiatic Researches, vol. vi. p. 572.

‡ However, the historical part of Mr. Bentley's argument seems open to objections, which have been urged by Mr. Colebrooke in the 12th volume of the Asiatic Researches. Mr. Bentley affirms, that an astronomer named Varaha-mihira was the author of the Surya Siddhanta, and that he is known to have lived somewhere about 1000 A.D. But it appears that it is by no means proved that Varaha was the author of the Surya Siddhanta; indeed, it is quoted by himself in a way which seems clearly to show that it was not his composition (*a*.) And another difficulty arises here, for Mr. Bentley, by reasoning entirely analogous to that just explained with regard to the Surya Siddhanta, is led to the conclusion that an ancient system of astronomy, composed by an author called Brahma-Gupta, may be referred to the year 536 A.D. (*b*) Now, Brahma-Gupta, in a work of acknowledged authenticity, quotes by name Varaha (*c*,) which circumstance corroborates the inference that Varaha could not have been the author of the Surya Siddhanta.

§ Asiatic Researches, vol. vi. p. 581. This date agrees pretty well with that attributed to Brahma-Gupta by the Hindoos themselves.—V. Asiat. Res., vol. xii. Addit.

‖ Ib. vol. xii. p. 227.

(*a*) Asiatic Researches, vol. xii. p. 221. Calcutta. (*b*) Ib. vol. vi. p. 581. (*c*) Ib. vol. xii. p. 221.

Pythagoras, who was undoubtedly well versed in the learning of the east, borrowed it himself from the Indians. But at present we have not sufficient data to decide this question, which is worthy of all the attention of Sanscrit scholars *.

The age of the tables of Brahma-Gupta, as fixed by Mr. Bentley, decides one question, by showing that the Indian astronomy was not originally borrowed from the Arabs. The tables in question were composed about 500 years after Christ. We shall see that the Arabs did not begin to cultivate astronomy till a century or two later. It is much more difficult to ascertain whether the Indians were indebted to the Greeks for any of their principal determinations; and, on this point, different opinions seem to be entertained by those who are best qualified to judge in such matters. Mr. Davis, and Delambre †, think the Hindoo methods of calculation essentially different from the Grecian; and this circumstance has been much insisted upon by Playfair. The limits of this treatise will not allow us to go into a detailed examination of these methods; we must be contented with selecting some of the most striking instances of originality.

One of these is the method given in the tables of Chrisnabouram, to find the time of the sun's continuance above the horizon, or what we call the diurnal arc, for any given day. On the day of the equinox, observe the length of the shadow of the gnomon, which is to be divided into parts, each equal to $\frac{1}{720}$th of the length of the gnomon; one-third of this measure is the number of minutes by which the day at the end of the first month after the equinox exceeds twelve hours: four-fifths of this excess is the increase of the day during the second month; and one-third of it the increase of the day during the third month. This rule involves the supposition, that the sun's declination being given, the ratio between the ascensional difference, (that is, the arc measuring the increase of the day at any place,) and the tangent of the latitude, is constant. Now this is not rigorously true, for the constant ratio in question exists between the sine of the ascensional difference and the tangent of the latitude. But between the tropics, the difference between the supposition of the Brahmins and the exact formula will be so inconsiderable as to be safely neglected. In higher latitudes, this difference will increase pretty rapidly, and soon becomes a very appreciable quantity*. It is then pretty clear that this rule of the Brahmins must have had its origin in a tropical country, and in all probability in the Indian peninsula in which it is found. The Indian methods for the calculation of eclipses which have been explained at length by Le Gentil†, and Mr. Davis.‡, are extremely curious, and bear certainly the appearance of originality. But for these we must refer to the author just mentioned; we shall only remark, in passing, that these methods show a knowledge of the celebrated property of the right-angled triangle, which tradition informs us that Pythagoras (who, it may be remarked, had travelled much in the east) first made known in Greece. The Indians demonstrate this proposition§ in a very singular way, which partakes more of the nature of algebraic reasoning than of pure geometry. Indeed, they seem to have been singularly attached to the study of algebra, in which they made great progress‖; and of which they were, very probably, the

* The length of the sidereal year, as fixed by Aryabhatta, is 365 days 6 hours 12 minutes and 30 seconds.—Asiat. Res., vol. xii. p. 249. This is the quantity adopted in most of the Indian tables. Some of them, however, make it 36, instead of 30 seconds.

† Astron. Ancienne, vol. i. p. 478.

* If we call ω the obliquity of the ecliptic, ϕ the latitude, and θ the excess of the semidiurnal arc on the longest day above 90°: then

$$\sin. \theta = \tan. \omega \tan. \phi$$

$$\text{Now } \tan. \phi = \frac{l}{h}$$

where h is the height of the gnomon, l the length of the shadow,

$$\therefore \sin. \theta = \tan. \omega \times \frac{l}{h}$$

$$\therefore \theta = \frac{l}{h} \tan. \omega + \frac{l^3}{6h^3} \tan.^3 \omega +$$

$$= 572.957. \tan. \omega \frac{l}{h}$$

in Indian minutes of time, and neglecting all the terms after the first.

$$\text{or } = 255. \frac{l}{h} \left\{ \text{supposing } \omega = 24° \right\}.$$

$$\text{The Indian rule is } 2x = 720. \frac{l}{h} \left\{ \frac{1}{3} + \frac{4}{15} + \frac{1}{9} \right\}$$

$$\text{or } x = 256. \frac{l}{h}$$

V. Bailly, p. 32; and Edin. Trans. vol. ii., p. 172.

† Mem. Acad. des Sciences for 1772, Part II. p. 221.

‡ Asiat. Research., vol. ii. p. 273. Delamb., Ast. Anc., vol. i. p. 471.

§ The 47th, 1st Book of Euclid's Elements.

‖ The Hindoos have been particularly successful in their methods for the solution of indeterminate problems; in this respect they have not been equalled in Europe till the latter half of the 18th century.

inventors. Their methods for calculating the ascensional differences prove that they were in possession of the principal theorems of spherical trigonometry*; their tables of sines, which are given for every $3\frac{3}{4}°$ throughout the quadrant, may be regarded as a mathematical curiosity†.

On the other hand, the system on which the Indians calculate the inequalities of the sun, moon, and planets, presents some remarkable coincidences with that imagined by the Greeks for the same purpose. On reference to a subsequent part of this treatise, it will be seen, that Hipparchus explained the principal inequality of the sun, and of the moon, by supposing each of these planets to revolve round the earth in a circle, the centre of which was at some distance from this last body; and thus the motion, though really uniform round the centre, appeared unequal as seen from the earth. Hipparchus also proposed another theory, which leads to the same results as that just mentioned; he supposed the sun or moon to revolve in a small circle, called the epicycle, the centre of which revolved uniformly round the earth; and he proved the virtual identity of the two hypotheses. It is remarkable enough that both these systems are made use of by the Indian astronomers; and it is equally remarkable that they appear to be ignorant of the modifications of these theories, which Ptolemy was obliged to make in the case of the moon and the planet Mercury‡. They have indeed felt the necessity of some modifications in these cases, but theirs consist in giving an oval form as well to the eccentric as to the planet's epicycle. The method of Ptolemy, which is very different, will be explained when we come to speak of the astronomers of Alexandria.

Sir W. Jones has affirmed that it is very improbable the Indians should have borrowed anything from the Greeks, as the pride of the Brahmins leads them to despise foreign nations in general, and the Greeks in particular. They have a proverb, says he, that no base creature can be lower than a Yavan; which term he explains to mean an Ionian or Greek*. But Mr. Colebrooke has quoted a very curious passage from Varaha-mihira, one of the earliest astronomers of India, who speaks with applause of the proficiency of the Yavans in astronomy†. "The Yavans," says he, "are barbarians; but this science is well established among them, and they are revered like holy sages." About the age of Varaha-mihira there seems to be some uncertainty, Mr. Bentley fixing it at about 1000 B.C.: while others make him more than five hundred years older‡; but whenever we suppose him to have lived, this acknowledgment of an acquaintance with the science of the West goes far to confirm the ideas of those who consider the Hindoo astronomy as derived from the Greek.

Chapter III.

The Chaldeans.

If we may credit Porphyry, quoted by Simplicius §, Callisthenes transmitted to Aristotle a series of observations made at Babylon during a period of 1903 years preceding the capture of that city by Alexander. This would carry back the origin of astronomy in Chaldæa to at least 2234 years before the birth of Christ. It is certainly an argument of some strength against the correctness of this statement, that Ptolemy, who has founded his theory of the moon partly upon Chaldæan observations, quotes none anterior to the year 720 B.C.; but this fact is not perhaps so decisive as it at first appears. It is impossible now to say whether or not Ptolemy had access to the whole of the observations in question; whether any had been lost in the interval—not an inconsiderable one—between himself and Callisthenes; or lastly, whether the superior accuracy of the more modern observations made him prefer them ‖. The eclipses recorded by Ptolemy are

* Delamb., Ast. Anc., vol. i. p. 470.

† These tables are remarkable as being calculated by means of second differences; they show that the Hindoos were aware that $\Delta^2 \sin. A = -(\text{chord } A)^2 \sin. A$ as well as the common theorems

$$R^2 = \sin.^2 A + \cos.^2 A$$
$$\sin. 30° = \tfrac{1}{2} R$$
$$\sin. \tfrac{1}{2} A = \left(\frac{1 - \cos. A}{2}\right)^{\frac{1}{2}}$$
$$\sin. 60° = \sqrt{\tfrac{3}{2}}\, R.$$
—V. Delamb. p. 458.

‡ Asiat. Research., vol. xi. p. 235; vol. ii. p. 249.

* As. Res., vol. ii. p. 302.

† As. Res., vol. xii. p. 245.

‡ V. Colebrooke, Asiat. Research., l. c.

§ De Cœlo, lib. ii.

‖ Another supposition, at least as likely as any of these, is, that Ptolemy selected those eclipses which he conceived to agree best with his own theories.—Delambre, Astron. Anc. Introduction, p. xxxv.

given in a rough way *, yet they are of singular interest, as with them begins, at least for the western nations, the long train of observation and discovery that has brought the science to its present perfection †.

A striking proof of the acquirements of the Chaldæans is to be found in their knowledge of the period of 6585⅓ days, in which the moon makes 223 revolutions with regard to the sun, 239 with regard to the apsides of her orbit, and 241 with regard to her nodes. This is attributed to them by Geminus ‡: Ptolemy § refers it simply to the "ancient mathematicians." The accuracy of this period is very great, and its utility no less in calculating the recurrence of eclipses. Indeed, there can be little doubt that it was by means of this period, or one very analogous to it, that they were able to predict these phenomena in the case of the moon; for, according to Diodorus Siculus ||, they did not attempt such predictions for eclipses of the sun. This is natural enough; it is sufficiently obvious to those who have any acquaintance with the science, that the calculation of the former is far more easy, the parallax not entering into it. The author just quoted also tells us that they attached great importance to the theory of the planets, which bodies they observed with care, and more particularly Saturn. In fact we find in Ptolemy several such observations ¶. Their zodiac was divided into twelve signs **: the extra-zodiacal constellations were twenty-four in number, twelve in each hemisphere. To this we may add from Herodotus ††, that to them we owe the duodecimal division of the day. This historian attributes, at the same time, to them the invention of the gnomon, and an instrument called polus. The former, we have already seen, was used in China from the earliest antiquity; of the last we have a very imperfect knowledge; it seems to have been destined to indicate the changes in the sun's meridian altitude towards the solstices. The divisions of time were measured by clepsydræ.

* V. Syntax., lib. iv. c. 5. The time is not given more nearly than within an hour—the quantity of the disk eclipsed is expressed in digits.

† Ptolemy (Syntax. xiii. 7.), speaking of astronomical observations, informs us, that the most numerous and best have been made in Chaldæa. To this we may add a curious passage of Cleomedes, ii. 6, who says, speaking of the moon being seen eclipsed, while the sun was above the horizon, that "so many eclipses of the moon having been observed and recorded, no astronomer, whether Chaldæan or Egyptian, has ever recorded one of this kind."

‡ Isagoge c. περὶ ἐξελιγμοῦ. Cf. Delambre, Astron. Ancienne, vol. i. p. 206, who disputes, upon rather inconclusive grounds, the claim of the Chaldæans to this period.

§ Lib. iv. c. 2.

|| Lib. ii. c. 31.

¶ Lib. xi. c. 7, et alibi.

** Each sign was subdivided into thirty degrees; the degree into sixty minutes.

†† Lib. ii. c. 109. What are we to say of Delambre, who assures us that no ancient author speaks of the gnomons of Chaldæa?—Astron. du Moyen Age, Discours Prélim., p. xi.

Seneca * informs us that Epigenes and Apollonius Myndius, both of whom professed to have studied under the Chaldæans, ascribed to them very different opinions on the subject of comets. According to the former they were ignorant of their nature and course; while the latter, who is called by Seneca a most scientific observer of natural phenomena, states, that they classed them with the planets, and were able to determine their motions. It is certain that very philosophical ideas were entertained on the subject of comets by the Pythagoreans, who had evidently borrowed many of their doctrines from the East. Could we admit the statement of Apollonius, few things would tend more to give us a high idea of the Chaldæan astronomy.

Chapter IV.

The Egyptians.

The Egyptians seem to have enjoyed in ancient times considerable reputation for astronomical science. It is, however, certain, that few, if any, relics of it have descended to us. It has been remarked, that the exactitude with which the Pyramids have been made to face the four cardinal points, gives us an advantageous idea of their methods of observation. However, Ptolemy and Hipparchus, who it is natural to suppose would have had access while living in the country to the Egyptian records, never quote any ancient observation made by astronomers of that nation; but, on the contrary, were forced to have recourse to the Chaldæans. On the other hand, there is some respectable testimony to prove that the Egyptians were in the habit of observing celestial phenomena in general, and eclipses in particular. Diodorus Siculus † goes so far as to say, that they were able to calculate beforehand the circumstances of these latter with much exactness. Diogenes Laertius ‡ mentions 373 solar,

* Quæst. Nat., lib. vii. c. 3.

† Lib. i. § 2.

‡ In proœmio.

and 832 lunar, eclipses observed in Egypt. The testimony of this author is in itself of no great weight, and he adds the absurd circumstance, that they had been seen in an interval of 48863 years. But it is very singular that this is the proportion of the solar to the lunar eclipses visible above a given horizon within a certain time*; and such a coincidence certainly cannot be accidental. Seneca† likewise informs us, that Conon, the contemporary of Archimedes, had collected all the eclipses of the sun preserved in Egypt. Lastly, we may remark, that Aristotle‡ mentions the Babylonians and Egyptians as having recorded a great number of credible observations. To all this is to be opposed the silence of Ptolemy, and upon this point we must refer to the remarks already made when treating of the Chaldæan astronomy.

The civil year of the Egyptians was of 365 days, but they were very early acquainted with the more accurate value, 365¼ days. This appears from the Sothiac period of 1461 years, which brought round to the same seasons their months and festivals. For this people, among their numerous singularities, had that of not wishing to connect the civil invariably with the physical year, but to suffer it to anticipate gradually, displacing thereby all the times fixed for their religious ceremonies, till at the end of the great Sothiac period they coincided once more with their original positions. One of these periods, according to Censorinus §, began in the consulship of Antoninus and Bruttius, A.D. 139. That this was not the first period of the kind, there can be little doubt ‖. The preceding one must have commenced, then, in the year 1322 B.C. Bailly¶ even, relying upon some expressions of Manetho, thinks that this was preceded by another. But, as far as the tropical year is concerned, it is necessary to observe, that the Sothiac period could not have been deduced from an actual observation of the time required for a complete restitution, for the time observed would not have been 1461 years, but 1506. It has, however,

* Between 1200 and 1300 years.
† Quæst. Nat., lib. vii. c. 3.
‡ De Cœlo, lib. ii. c. 12.
§ De die Natali, c. 21. and 18.
‖ Clemens Alexandrinus places the birth of Moses 345 years before the establishment of the Sothiac period, which must of course refer to the year 1322 B.C. This seems to receive some confirmation from a fragment of Theon, given by M. Biot, Astron. Egypt., p. 303.
¶ Astron. Anc., p. 402.

been supposed that the Egyptians had a rural year, comprising the intervals between two heliacal risings of Sirius, and that the Sothiac period must be considered as applying to this rural, and not to the tropical year. And here we meet with a very curious coincidence; for this rural year, as thus determined, had, for twenty or thirty centuries before the Christian era, very exactly the length of 365¼ days; and, consequently, the period of restitution of 1461 years would apply to it very accurately. A recent author* has disputed the fact, that such a rural year was in use among the Egyptians, before the time of Hipparchus: however, the authorities urged in its favour seem pretty satisfactory; and the coincidence above mentioned tends strongly to corroborate them. Thus it appears that the Sothiac or Canicular period had its origin when the first day of the month, Thoth, coincided with the heliacal rising of Sirius. According to Censorinus, this happened the 20th of July, A.D. 139. M. Ideler has found by calculation, that on the very same day of July, Sirius rose heliacally in the Julian years 1322 and 2782 B.C.

According to Dio Cassius†, the Egyptians were the inventors of the short period of seven days, distinguished by the names of the planets, which we call week. This period, used among all the eastern nations from time immemorial, has been called, by an eminent philosopher‡, the most ancient monument of astronomical knowledge. It is found even among the Brahmins of India with the same denominations, and the days similarly named by them and by us correspond to the same physical portions of time. The arrangement is founded upon the ancient systems of astronomy, in which the planets were placed in the following order, beginning with the most distant from the earth; Saturn, Jupiter, Mars, the Sun, Venus, Mercury, and the Moon. The day being divided into twenty-four hours, the hours were consecrated to the planets in the order just given; and each day took its name from the hour with which it began. Thus, the first hour of the first day being dedicated to Saturn, the second would be so to Jupiter, the third to Mars, and so on; then the eighth again to Saturn, the fifteenth, and the twenty-second, so that the first hour of

* Biot, Essai sur la Période Caniculaire.
† Lib. xxxvii. c. 18.
‡ La Place, Syst. du Monde, lib. v. c. 1.

the second day would belong to the Sun. Proceeding in the same way, the first hour of the third day would belong to the Moon, of the fourth to Mars, &c.

We have said, that in the ancient systems, Venus and Mercury were considered as inferior to the Sun; but upon this point there was some difference of opinion. The Egyptians seem to have perceived the real state of the case. They conceived these two planets to move round the sun, while they followed him in his annual revolution round the earth, and consequently were sometimes nearer than he to the earth, sometimes farther. This system, which is explained by several ancient authors, and among them obscurely by Cicero *, more distinctly by Vitruvius †, is positively attributed to the Egyptians by Macrobius ‡.

Some have been disposed to imagine that the constellations of the zodiac were originally invented in Egypt at a very remote period. This opinion has been advocated principally by Dupuis §, who conceives that the constellations in question had a reference to the divisions of the seasons, and to the agriculture of Egypt at the time of their invention. The sign of Cancer marks the retrogradation of the sun at the solstice; Libra, the equality of the nights and days at the equinox: the Capricorn, a climbing animal, is conceived to indicate the sun at its greatest height, or at the summer solstice; the autumnal equinox consequently falls in Aries. This system presents, certainly, some curious coincidences: thus, for example, the inundation of the Nile, which begins just after the summer solstice, would take place while the sun was in the constellations Aquarius and Pisces; and Virgo, usually represented as a woman with an ear of corn in her hand, would coincide with the time of harvest in Egypt. There is, however, one insuperable objection to this system, which is the excessive antiquity (not less than 15000 years) which it assigns to the zodiac. As this is historically inadmissible, Dupuis has modified his theory by supposing the names to have been given, not to the constellations in which the sun was, but to those diametrically opposed to him, which consequently were rising at sun-set at any given epoch. This opinion, which brings down the invention of these constellations to about 2500 years B.C., has been adopted by La Place and several distinguished philosophers *.

The scientific men who accompanied the French expedition to Egypt found in some of the temples of that country representations of the zodiac, which have given rise to much discussion in Europe. One of the most remarkable of these is on the ceiling of a portico in the temple of Denderah (the ancient Tentyra). It represents the signs of the zodiac in two rows, six in each, parallel to the axis of the temple, one to the right, the other to the left of the principal entrance; the latter all face, as if about to enter the temple, the former as if quitting it: the first of the entering signs is Aquarius, and the last Cancer: the Cancer, however, is thrown on one side out of the line, and its place filled by a head of Isis, partly plunged in the rays of the sun. It follows, from what has been said, that the first of the signs which appear to be coming out is the Lion, and the last Capricorn. Similar zodiacs are to be found in the porticoes of two temples at Esnè (Latopolis): but there the head of Isis is altogether wanting, and the bisection of the signs takes place between Virgo and Leo, instead of between Leo and Cancer. This bisection has been supposed by some to have a reference to the places of the solstices †: but the supposition is entirely arbitrary, and would give to these temples an antiquity which other circumstances by no means seem to support. Fourrier conjectures that the head of Isis, substituted in the place of Cancer, indicates that Sirius rose heliacally when the sun was in that constellation; which took place more than twenty centuries before the Christian era. M. Biot imagines that this indicates that Sirius rose with the stars of Cancer, near which the sun was at the time of the summer solstice, and refers the monument to about the year 700 B.C. There is in the interior of the temple at Denderah, another zodiac, sculptured on a ceiling, in which the signs are arranged in a circle, and here

* Somn. Scip. † Vitruv., lib. ix. c. 4.
‡ Comment. in Somn. Scip., i. c. 19.
§ Mém. sur l'Origine du Zodiaque.

* Macrobius (Somn. Scip., lib. i. c. 21.) attributes the invention of the zodiac to the Egyptians; but against this it may be urged, that Sextus Empiricus, a writer of at least equal authority, ascribes it to the Chaldæans, lib. v.

† This idea is deduced from the supposition that the constellation in which the sun was at the beginning of the year, was represented as the first in the line of the signs coming out of the temple, or appearing to lead the others: now the Egyptian rural year began at the summer solstice—thus the zodiac of Denderah would indicate that the summer solstice was in the Lion, those of Esnè in the Virgin.

again the Cancer is thrown out of its proper line, its place being occupied by a mythological figure, below which is the symbol of Isis. M. Biot has attempted to prove that this circular zodiac is a planisphere representing the appearance of the heavens at midnight on the summer solstice, about seven centuries before the Christian era.* But this opinion is exposed to many serious objections: these, however, the limits of this treatise will not allow us to enter into: we shall quit the subject with one observation. M. Champollion thinks he has decyphered among the hieroglyphics on the ceiling of the temple, the word αὐτοκρατωρ, which would seem to indicate that the sculptures in question were as recent as the Roman empire. But this by no means precludes the possibility that they may represent a more ancient sphere. That the temple itself is not of great antiquity many circumstances seem to indicate; but the question to be solved is, whether the astronomical phenomena it depicts are, or are not, to be referred to some more distant epoch, which it was intended to record?

Chapter V.

Origin of Astronomy in Greece.—Thales.—The Ionian School.—The Pythagoreans.—Meton.—The Calendar.—Eudoxus.—Pytheas.

The astronomy of Greece undoubtedly begins with Thales and the philosophers of the Ionian School, about six centuries before the Christian era. Homer, indeed, and Hesiod, the only authors anterior to this period whose works we now possess, mention some of the most remarkable constellations, though none of those composing the signs of the zodiac;† and the works of the latter author in particular show that some attention was paid in his time to the rising and setting of certain conspicuous stars. Thus he informs us that Arcturus rose heliacally sixty days after the winter solstice,‡ from which we may deduce this poet to have lived about 950 B.C., unless, indeed, which is not improbable, he copied some older calendar.* Herodotus† makes Homer and Hesiod contemporaries, and places them about 400 years before his own time, which would make them a century later than the date above assigned. But the whole question of their age is involved in much doubt and obscurity, nor is it important to the history of astronomy. We find little in these very ancient authors that throws light upon the interesting question, whether the Greek sphere was derived from that of the Egyptians, or other oriental nations. We shall see in the course of this treatise, that there can be no doubt that the zodiac was borrowed from Egypt or Chaldea, but the origin of the extra-zodiacal constellations is very uncertain. Seneca‡ attributes the division of the heavens into constellations to the Greeks; and refers this division to fourteen or fifteen centuries before Christ. An obscure author, quoted by Clemens Alexandrinus,§ ascribes the invention of the sphere to Chiron, who may be referred to the thirteenth century B.C. That the Greek sphere, whether of native or foreign origin, is as old as the time assigned to Chiron, has been attempted to be proved by a passage from Eudoxus, quoted in the commentary of Hipparchus on Aratus.|| This author, who flourished in the early part of the fourth century B.C. asserts, that there is a certain star in the celestial sphere, corresponding to the pole of the equator. Now this could not have been the polar star of our times, which was then, owing to the precession of the equinoxes, far from the pole; and upon examining that part of the heavens, there seems to be no other star that could be alluded to, except κ Draconis. About 1326 B.C. this star was within 4° of the pole, which was sufficiently near to make it appear immoveable to rude observers: and this fact has been thought to show that Eudoxus copied a sphere many centuries anterior to his own. This sphere has been supposed to be that of Chiron before alluded to; others attribute it to Musæus.¶ But not-

* Récherches sur l'Astronomie Egyptienne. Paris, 1823.

† The constellations and stars mentioned by Homer, are the Bear, the Pleiades, the Hyades, Bootes, Orion, and Arcturus. Besides these Hesiod mentions Sirius. Neither make any mention of the planets, though Homer is supposed to allude to Venus in one passage—Il. V. v. 5; others think Sirius is meant. The planets in fact do not seem to have been astronomically observed in Greece till long afterwards. "Eudoxus," says Seneca, "quinque syderum cursus primus in Græciam ab Ægypto transtulit." Quæst. Nat. vii. 3.

‡ Op. et. Dies. v. 564.

* Another remarkable proof of the accuracy of Hesiod may be found in his statement, that the Pleiades remained invisible for forty days (v. Op. et Dies. v. 385.), which has been found to be as nearly as possible the case for that epoch and latitude. V. Bailly Astron. Anc. p. 430.

† Lib. ii. c. 53.

‡ Quæst. Nat. Lib. vii. c. 25.

§ Stromat. lib. i. c. 15.

|| Lib. i.

¶ But many other claimants to the invention of the sphere might be named particularly Palamedes withstanding all that has been said upon this subject, it seems pretty evident, the positions of the stars given by Eudoxus are too discordant to admit of any conclusion being drawn from them*, and if Eudoxus copied any very ancient sphere, it must have been one of oriental origin; for in the time of Chiron (if indeed any such person ever existed) it is pretty certain that nothing was known in Greece about the existence of the ecliptic, the equinoxes, the colures, or any great circles of the sphere. The unanimous testimony of antiquity ascribes to Thales, or his immediate successors among the philosophers of the Ionian school, the invention of the zodiac, the discovery of the obliquity of the ecliptic, of the tropical revolution of the sun, and the principal circles of the celestial sphere. There is no reason for supposing that, before this time, the Greeks had advanced beyond remarking and naming a few of the most conspicuous constellations; and how little progress they had made in this, we may conjecture from the circumstance, that Thales first introduced into Greece the knowledge of the Little Bear†, by which the Phœnician pilots used to steer, while the Greeks were contented with the rough approximation to the north given by the Great Bear. For it is very remarkable that Thales, if not, as some have pretended, a Phœnician, was certainly of Phœnician extraction; and this fact, corroborated by other authors, rests on the testimony of Herodotus himself‡. Many things seem to indicate that the science of Thales was of eastern origin, and that what have been called his discoveries, were doctrines borrowed from Chaldæa or Egypt. Much stress is not to be laid upon the account of his having studied in Egypt, which rests upon the equivocal authority of his biographer Diogenes, though this is confirmed by Clemens Alexandrinus§. But the extraordinary fact of his having predicted a solar eclipse, which can scarcely be disputed, speaks volumes upon this subject. This we are told by Herodotus*, whose testimony seems above all suspicion, though he adds a singular circumstance, in which there is probably some mistake, that Thales assigned the limits of a year, within which this eclipse was to take place. It is unnecessary to remark, that could Thales have predicted an eclipse so remarkable as this, which was total in the country in which he lived—Asia Minor,—he certainly must have been able to make a much nearer approximation. But as to the fact of the prediction there can be little doubt. The eclipse is memorable in ancient history, as having separated the armies of the Lydians and Medes, at that time engaged in battle; the historian lived not more than 150 years after the event in question, and was a countryman of Thales†; but should his testimony be deemed insufficient, authority perhaps still higher may be quoted. Eudemus, an astronomer of eminence in the fourth century before Christ, composed a history of astronomical discoveries, now unfortunately lost; but Diogenes Laertius and Clemens Alexandrinus‡ both quote the authority of Eudemus for this prediction of Thales: and in farther corroboration of this, Eudemus, in a fragment preserved by Anatolius§, attributes the discovery of solar eclipses to Thales. Indeed it may be said that there is no point on which the testimony of antiquity is more decided and unvarying, than that Thales introduced into Greece the prediction of solar eclipses‖, and most probably at the same time the explanation of their real cause¶. Pliny indeed does not quite agree with Eudemus as to the date of this eclipse, which has been a subject of controversy among ancient and modern authors; but the knowledge of the exact year in which it happened is more interesting to chronologists than astronomers**.

(V. Sophocles quoted by Achilles Tatius), and Atlas (V. Diod. Sic. lib. iii. Plin. ii. 8.)

* V. Delambr. Astron. Anc. Introduct. p. 11. and vol. i. p. 122.

† See Callimachus quoted by Diogenes Laertius in Thalete, and by Achilles Tatius, Cf. Hygin. Poetic. Astronom. V. Arctus. et Theon. in Arat. That the Little Bear was discovered by the Phœnicians, is attested by Strabo (lib. i. cap. 1), and the circumstance of their navigators sailing by it, is alluded to by many authors. V. Arat. Phœnom. v. 39. Ovid. Heroid. xviii., et alibi.

‡ V. Herodot., lib. i. c. 170, Cf. Diog. Laert. in Thalete.

§ Stromat. i. 14.

* Lib. i. c. 74.

† Thales was a native of Miletus—Herodotus of Halicarnassus, both towns on the coast of Asia Minor.

‡ V. Diog. Laert. in Thal. Clem. Alex. Stromat. lib. i. c. 14.

§ V. Fabric. Bibliothec. Græc., lib. iii. c. ii. Vol. ii. p. 215. Hamburgh, 1797.

‖ Besides the authors above quoted, V. Achilles Tat. Isagog. Plin. Hist. Nat. ii. 12. Cicero de Divinatione i.

¶ This is expressly attributed to him by Plutarch de Placit. Philosoph. ii. 24.

** On this point the reader is referred to an interesting paper by Mr. Baily, in the Philosophical Transactions for 1811, where all the opinions on the subject are discussed. Mr. Baily himself refers the eclipse to the year 610 B.C.

If, then, there be no reason to doubt that Thales predicted the phenomenon in question, we can hardly fail to admit that his method was borrowed; and borrowed, in all probability, from Chaldea. For it is sufficiently clear that nothing but a very long series of observations, conducted with care and regularity, could enable any man to arrive at this knowledge; such observations, as we have no reason for supposing to have been made in Greece at these early times; while, on the other hand, we have seen that the Chaldeans were in possession of a period which enabled them to predict pretty accurately the recurrence of eclipses.

Again, if we admit that Thales explained the causes and predicted the occurrence of eclipses of the sun, we can hardly doubt that he was able to do the same with regard to those of the moon. Eudemus, indeed, according to Anatolius, attributes the discovery of the causes of the moon's light and her eclipses to Anaximenes, one of the successors of Thales. In general, we may remark, that it is very difficult to determine, amid conflicting testimony, to which of the philosophers of these times particular discoveries are to be referred, though there is a general agreement as to the doctrines taught in the Ionian and Pythagorean schools. Thus Pliny * refers the discovery of the obliquity of the ecliptic to Anaximander,—Plutarch † to Pythagoras, or Œnopides of Chios,—Eudemus to some author whom he does not name ‡, but different from all of these, and who fixed it at 24°. But Thales, who is said to have written on the length of the tropical year, and on the position of the solstices and equinoxes §, could hardly have been ignorant of the fact of the obliquity, even if he were not, as seems likely, the author of this ancient valuation. Again, the invention of the gnomon is attributed by Diogenes Laertius ‖ to Anaximander,—and by Pliny ¶ to Anaximenes,—while Herodotus, with much more probability, says it was borrowed from the Babylonians **. It is truly unfortunate, that in attempting to investigate the doctrines of these ancient philosophers, we are compelled to have recourse to authors whose ignorance of astronomy too often makes their accounts unintelligible. Thus Diogenes tells us, that Thales found the magnitude of the moon to be $\frac{1}{720}$th part of the sun; a statement clearly absurd, if meant to apply, as it evidently must, to their apparent diameters. But a passage of Apuleius * shows us the real meaning of the determination so grossly misunderstood by Diogenes. Thales, he tells us, determined the magnitude of the sun in parts of its own orbit: now, $\frac{1}{720}$th part of a great circle is 30′: the real diameter of the sun may be taken at a mean not far from 32′; so that we see the measure of Thales was a good approximation for those early times.†

We have stated above, that the constellations of the zodiac do not seem to have been known in Greece before the time of Thales. In fact, Eudemus, whose early date and astronomical knowledge make his testimony of great weight, states that they were invented by Œnopides of Chios, a Pythagorean philosopher, generally placed considerably after the time of Thales ‡. However, that they were not of Greek origin, seems highly probable, let them have been introduced into that country when they may. To establish this, it is not necessary to insist upon the zodiacs discovered in Egypt, since their antiquity has been disputed; but from the testimony of ancient authors, it is clear that the zodiac of Chaldea and Egypt was identical with that of Greece; and no doubt can remain as to which was borrowed from the other. The Syntaxis of Ptolemy establishes this identity in the case of Chaldea and Greece. In the planetary observations of the Chaldeans, quoted in that work, the place of the planet in the Chaldean zodiac is first given, and then reduced to the Greek: though the respective constellations did not quite coincide in space, yet the names are always identical, except in one instance, where it appears that the Chaldeans gave the name of the Balance

* Pliny, ii. 8. † De Placit. Phil. ii. 12.
‡ Fabric. Bibliothec. Græc. l. c.
§ Diog. Laert. in Thalete.
‖ In vita Anaximandri. ¶ II. c. 56.
** See the passage quoted above, chap. iii. With regard to Thales, we may observe, that Themistius, in a passage quoted by Fabricius, Bibliothec. Græc. vol. i. p. 239, says, that Anaximander was the first person who published any of the doctrines of Thales,—this latter having written nothing himself. This would tend to confirm the view taken in the text, that the pretended discoveries of Anaximander and Anaximenes consisted in the publication of what they had learnt from Thales.

* Florid.
† V. Montucla, vol. i. p. 106. Bailly, Astron. Anc. p. 441.
‡ Pliny refers the invention to Cleostratus of Tenedos, a philosopher rather posterior to Thales, but before the time to which Œnopides is usually referred. V. Hist. Nat. ii. 8.

to the constellation called by the Greeks the claws of the Scorpion*. From a fragment quoted by Delambre †, it seems that the Egyptians also named the claws of the Scorpion the Balance.

To this we may add the very curious circumstance, that the zodiac of India is nearly identical with that of Greece: and Humboldt ‡ has shown, in a very interesting memoir, that the twelve Indian signs are taken from among the twenty-seven lunar mansions, or constellations of the lunar zodiac, mentioned in the first and second chapters of this Treatise §.

Nothing, perhaps, is more remarkable,—and, if we refuse to admit the oriental origin of Greek science, more inexplicable,—than the circumstance of the true doctrine of the motion of the earth having been promulgated in the schools of Pythagoras ‖ and Thales. That this was the case with regard to the former, is well known; and it is generally supposed that Philolaus, the successor of Pythagoras, was the first to teach it openly¶. But there are some reasons for supposing that it had been among the doctrines professed at an earlier period by Anaximander. Eudemus, whom we have frequently had occasion to quote, affirms, in the most express terms, that this was the system of Anaximander**. If this be true, it is as probable that this philosopher merely published what he had heard from his master, as in the case of Philolaus and Pythagoras ††. Cicero ‡‡, on the authority of Theophrastus, attributes this system to Hicetas of Syracuse; and this is partly confirmed by Plutarch §§. It certainly was embraced by some very eminent men, such as Archytas of Tarentum, Timæus Locrus, and, in later times, Aristarchus of Samos*. Others, as Heraclides of Pontus, and Ecphantus, admitted, we are told, merely the earth's diurnal revolution on its axis †. One cannot help feeling some surprise, that after the true system of the world had once been promulgated,—when it had been adopted by a numerous school, and some of the most distinguished astronomers, — it should have fallen subsequently into comparative oblivion. No doubt the ancients had not the same decisive proof of the motion of the earth that we have in the aberration of the fixed stars; and the infinite distance of these bodies, which is a necessary consequence, may have staggered many of them. But the fact in question is principally to be attributed to the wide-spreading influence of the Peripatetic school, whose founder, Aristotle, had strenuously combated the Pythagorean doctrines ‡.

It is a circumstance by no means to be overlooked, that Pythagoras had travelled, according to the testimony of all his biographers, into Egypt and the East, and some say that he penetrated as far as India §. Some corroboration of this circumstance might be found in his metaphysical doctrines, evidently borrowed either from India or Egypt; but to confine ourselves merely to astronomy, we may notice opinions analogous to those known to have existed in the countries we have mentioned. We have seen the belief of the Chaldeans about comets, according to the account of Apollonius Myndius: the Pythagorean doctrine on this subject bears the greatest analogy to it. These philosophers supposed comets to be bodies as ancient as the universe, revolving round the sun, and visible only in a certain part of their orbit. Yet this sublime conception shared the fate of the system of the earth's motion. The Peripatetics were once more triumphant over truth and reason; and for eighteen centuries, it was almost universally admitted that these bodies were simply meteors engendered in the terrestrial atmosphere.

The philosophical ideas of the Pytha-

* V. Ptol. Syntax., lib. ix. c. 7.

† Astron. Anc., vol. i. p. 216.

‡ Monumens des Peuples Indigénes de l'Amérique.

§ The statements of Macrobius and Sextus Empericus, on the zodiac, have been mentioned above. They are in complete accordance with what is here advanced.

‖ This appears from Aristot. de Cœlo, l. ii. c. 13, and Plutarch in Numa, c. ii.

¶ V. Plutarch de Placit. Philos. lib. iii. cc. 13 and 17; Diog. Laert. in Philolao. Philolaus flourished about 450 B.C.

** Fabric. Bibliothec. Græc. l. c. Delambre grossly mistranslates this passage. Astron. Anc., vol. i. p. 15. It is but fair to state, that Simplicius places Anaximander among those who conceived the earth to be in the centre of the universe. De Cœlo, lib. ii.

†† Plutarch de Placit. Phil. iii. 11, certainly informs us that Thales placed the earth in the centre of the universe; his testimony, however, cannot be conclusive, when unsupported by any other.

‡‡ Quæst. Academ. iv. 39.

§§ De Placit. Phil. iii. 13.

* Archimedes in Arenario.—Plutarch de Placit. Phil. ii. 24. Plato is said by Plutarch to have regretted, in his old age, having placed the earth in the centre of the universe.—Quæst. Plat.

† Plutarch de Placit. Phil. iii. 13.

‡ De Cœlo, lib. ii.

§ Iamblichus, Diogenes, Apuleius in Floridis,

goreans about the nature of comets and the system of the world, render us not averse to admit that they taught, as is affirmed by some authors*, the plurality of worlds; that they believed the fixed stars to be so many suns, each the centre of a system similar to our own. If, on the one hand, this idea is found to be ascribed to them only by some very ignorant authors, on the other, it may be observed, that the notion itself is much too sublime to have emanated from such unphilosophical sources. What is to be deeply regretted is, that to these just and beautiful theories the Pythagoreans should have added fanciful and extravagant speculations upon numbers, harmony, and the dimensions of the celestial orbits. One of these fanciful ideas has given rise to the celebrated notion of the music of the spheres; it would seem that Pythagoras fancied he perceived an analogy between the distances of the planets and the divisions of the octave in music. But the passage in which this doctrine is explained by Pliny † appears to contain some mistake, for as he has stated it, it is incompatible with the Pythagorean system of the motion of the earth; and even putting this out of the question, with another opinion attributed to Pythagoras by Pliny himself ‡; namely, that the morning and evening stars are the same. From the way in which Pliny expresses himself on this last point, one would be led to believe that Pythagoras was aware that Venus and Mercury revolved round the sun. This doctrine we have before seen had its origin in Egypt; but on that account is, perhaps, the more likely to have formed one of the alleged discoveries of the Samian philosopher§.

The arrangement of their calendar was a subject with which the astronomers of Greece were occupied, with various success, during several centuries. The difficulties arose from the perseverance with which they attempted to conciliate the motions of the sun and moon. The month being determined by a lunar, and the year by a solar, revolution, they must soon have perceived that the former was not contained any integral number of times in the latter. Their object, then, was to find a number of years, or period, at the end of which a restitution would be effected, and the beginning of the month and the year again correspond. This problem was more difficult than they seem to have imagined; for, in the first place, the two revolutions are, strictly speaking, incommensurable; and secondly, the moon's mean motion is subject to a secular acceleration, which even if an accurate period could be found, would, in the course of time, render it inexact. However it was not impossible to find some practical solution which would be tolerably accurate for a time of no very great length, and to this object their attention was directed.

The first period of the kind alluded to was one of eight years, proposed by Cleostratus of Tenedos*. To understand its advantages and defects, it is necessary to observe that the Greek lunar year was composed of 354 days, divided into twelve months, alternately of 29 and 30 days. Cleostratus proposed, in the course of the eight years, to insert three intercalary months, of 30 days each, at the end of the third, fifth, and eighth years respectively†. He thus got a period of 2922 days, comprising 99 lunar revolutions. But, in reality, 99 lunar revolutions are performed in somewhat more than 2923 days, 12 hours; so that at the end of the period there was an error of 36 hours on the place of the moon.

Various methods were proposed to rectify this defect, but none with much success till we come to the time of Meton. This astronomer immortalized himself by the invention of a new cycle, which, taking into account the accuracy compared with the number of years contained in the period, may be considered as the most perfect ever proposed. For it is clear that it is one of the great merits of a cycle of this kind, intended

* Plutarch de Plac. Phil. ii. 15. Achilles Tatius Isagoge.

† Hist. Nat. ii. 22. Cf. Censorinum de die Natalio.

‡ ii. 8. Cf. Diog. Laert. in Pythagorâ.

§ We have studiously omitted any mention of various absurd opinions attributed to the philosophers of the Ionian and Pythagorean schools by some of the later writers of antiquity. These stories are always either at variance with what we know from more ancient and authentic sources, or unintelligible and absurd in themselves; or lastly, they arise from some palpable misunderstanding of the narrator. It would be a waste of time to discuss all the absurdities of this nature to be found in Plutarch,—an author whose inaccuracy, carelessness, and inconsistency are not yet properly appreciated. A recent historian of astronomy has drawn largely, and almost exclusively, from him in his sketch of the notions of Thales, Pythagoras, and their successors. V. Delamb., op. cit. Notions Générales.

* Censorinus, c. 18.

† V. Gemin. Isagog. c. 6.

for the purposes of civil life, to comprise as small a number of years as possible. The cycle of Meton was composed of 19 lunar years, in which seven months of 30 days were intercalated; namely, in the 3d, 6th, 8th, 11th, 14th, 17th, and 19th years. Besides this, some alteration was made in the distribution of the ordinary months: instead of having them alternately of 29 and 30 days, there were 110 only of the former, and 125 of the latter, in the period. To judge of the accuracy of the Metonian cycle, we must consider that 19 solar years comprise very nearly 6939 days, 14 hours 25 minutes; and 235 lunar revolutions comprise 6939 days, 16½ hours nearly; so that at the end of this time the moon was only about two hours behind the sun. The cycle of Meton comprising 6940 days, after one period the sun had already commenced his revolution nine hours and a half, the moon seven hours and a half*. The great accuracy and convenience of this invention procured it universal approbation; it was adopted throughout Greece, and obtained the name which it still bears of the golden number. The first cycle began in the year 432 B.C.

Callippus† about a century later proposed to remedy the slight defect of the cycle of Meton, by subtracting one day every 76 years. This was done by changing, after four periods of 19 years, one of the months of 30 days into one of 29. Callippus thus got a period of 76 years, comprising 27759 days. Now we may estimate that 940 revolutions of the moon make 27758^d, 18^h, 6^m, 76 revolutions of the sun 27758^d, 9^h, 42^m. The error on the place of the moon then was 5^h, 54^m; on the sun 14^h, 18^m‡. It was the accumulation of this error that entailed the necessity of the Gregorian reform to be explained in a subsequent part of this treatise. The first Callippic period began in the year 330 B.C.

The above-mentioned are the only periods that have been in civil usage; but Hipparchus appears to have composed one of four Callippic periods or 304 years, at the end of which he subtracted a day. By reference to what has been said, it will be seen that this will almost destroy any error on the moon's place, though it will leave one of more than a day on that of the sun. However this period never seems to have been much used even among astronomers; Ptolemy though a follower and admirer of Hipparchus, employs in preference that of Callippus.

The celebrity of Eudoxus of Cnidus, rather than his real merits, induce us to say a few words of him in this place. It is difficult to say upon what the astronomical reputation of this philosopher is founded. He was the friend and contemporary of Plato, and a distinguished geometer, but his merits, whether as an observer or a theorist in astronomy, appear to be very equivocal. He composed a description of the sphere which enjoyed great celebrity among the ancients, but it would seem from the poem of Aratus which was founded upon it, and from the commentary of Hipparchus, to have been but a rough and inaccurate production. It has been supposed by Newton and others, that Eudoxus merely copied the description of a sphere long anterior to his own time; but we rather lean to the opinion of Delambre, who seems to have shown pretty clearly that the positions on the sphere of Eudoxus are essentially inaccurate, and cannot be made consistent by a reference to any other time whatever.

Eudoxus is said to have studied thirteen years in Egypt*. Seneca informs us that he brought from that country into Greece, the theory of the five planets†; this, if true, can only allude to some very rough approximation to their motions, as we know that Hipparchus was obliged to abandon the consideration of this subject from the want of sufficient observations. We owe few thanks to Eudoxus‡ for a physico-mathematical hypothesis, which, having been adopted by the Peripatetics, spread through their influence, and became the received doctrine on these points, till the final overthrow of the school of Aristotle in the sixteenth century. He conceived that each planet had a sort of firmament composed of several concentric solid spheres, whose different motions modified each other, so as to represent the motion of the planet. Thus, in the case of the sun,

---

* Bailly, Astron. Anc. p. 225.

† Geminus, l. c. It is rather singular that Geminus does not refer the invention of the cycle of nineteen years to Meton and Euctemon, who are usually considered its inventors, but to Euctemon and Philippus.

‡ Bailly, p. 249.

* Strabo, lib. xvii. 29.

† Quæst. Nat. vii. 3.

‡ Aristot. Metaph. xi. 8. Cf. Simplicium de Cœlo, lib. ii.

there were three of these spheres, one revolving from east to west in twenty-four hours for the diurnal motion, another revolving from west to east in 365¼ days for the proper motion, and a third to represent a pretended motion of the sun in latitude. As the motion of one sphere was not supposed to have any influence on that of another, each planet was obliged to have a separate sphere for the diurnal motion common to them all. The number of spheres in the system of Eudoxus amounted to twenty-six, but, as fresh inequalities were perceived, it became necessary to augment their number very considerably, and to increase at the same time the complication of the hypothesis.

Pytheas, of Marseilles, a celebrated traveller and geographer, generally supposed to have lived about the time of Alexander the Great *, deserves particular notice, as the author of one of the most remarkable observations on record in Grecian history. Though the gnomon was certainly known in Greece, at least as early as the time of Herodotus, very few observations seem to have been made with it; very few at least have come down to us. The earliest on record is a summer solstice observed at Athens by Meton and Euctemon, in the year 432 B.C., which has been preserved by Ptolemy †. The next is the observation of Pytheas in question. Strabo informs us, that according to Hipparchus, the same ratio existed between the gnomon and the solstitial shadow at Byzantium, that Pytheas had observed at Marseilles; and in another place he informs us, that the ratio observed at Byzantium was 120 to 41⅘. Now, the fact is, that Byzantium and Marseilles are not under the same parallel; on the contrary, the latter is more than two degrees to the northward of Byzantium, and its latitude is about 46° 42′. The observation in question gives for the latitude of the place of observation 46° 48′. We see, then, that the observation was made at Marseilles, and with a care which reflects much credit on Pytheas ‡.

* 330 B.C. † Syntax. lib. iii. c. 2.

‡ A curious instance of the partiality of Delambre is to be found in his Astron. Anc. vol. i. p. 18. He says that Pytheas found Marseilles and Byzantium to be under the same parallel, and thence he concludes, that the observations of this astronomer are not to be relied upon. Now, Strabo expressly states, ii. 5., that it was Hipparchus who observed the length of the shadow at Byzantium, and found it to be the same as that observed at Marseilles by Pytheas. But this circumstance, Delambre, who wishes to exalt Hipparchus as much as possible, suppresses. It is difficult to conceive that this was done from any mistake, as he refers to de Zach, who has completely exculpated Pytheas from the charge, and shown that it ought to fall on Hipparchus.—Vide Zach, Attraction des Montagnes, vol. ii.

Chapter VI.

The School of Alexandria.—Aristarchus.— Eratosthenes. —Measure of the Earth.—Hipparchus.—Ptolemy.

We have now arrived at an important epoch in the history of Astronomy,—the foundation of the school of Alexandria. When at the division of the empire of Alexander, Egypt fell to the share of Ptolemy Lagus, and his successors, these princes, inspired with the laudable ambition of making their capital the centre of the literary and scientific world, collected round them the most distinguished philosophers of the time, founded a magnificent and extensive library, and spared no expense in the promotion of philosophical researches. Under their protection flourished a number of distinguished astronomers, who have made the school of Alexandria for ever famous. We find now, for the first time in Greece, a regularly-continued series of observations, the only real basis upon which the science could be founded. They were begun by Aristillus and Timocharis *, who, if we may judge from the observations recorded by Ptolemy, gave their attention particularly to the determination of the positions of the fixed stars. The first, however, of those great men who have reflected so much celebrity on this school, is Aristarchus, of Samos †, a distinguished advocate of the Pythagorean system of the motion of the earth, who seems, from the testimony of Archimedes, to have foreseen and answered the only serious objection to it, that arising from the non-existence of an annual parallax. He gave the only answer that could be given at present, by saying that the earth's orbit is insensible, when compared with the distance of the fixed stars.

Aristarchus is the author of an ingenious method for determining the distance from the earth to the sun, which is worth notice. Let E be the position of an observer on the earth's surface, M the centre of the moon, S that of the sun, join S M, S E, E M. It is not difficult to see, that when the line which

* About 300 B.C.

† His date is fixed by an observation of a solstice in the year 281 B.C., given in Ptolemy.

separates the dark from the enlightened side of the moon bisects her disk, the

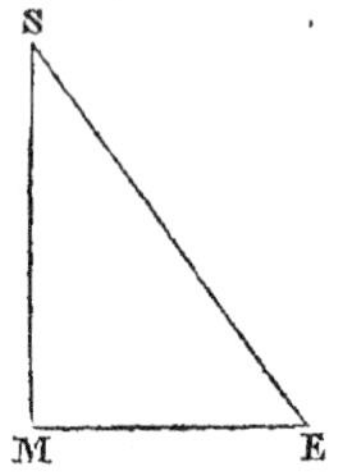

straight line E M will be perpendicular to S M. If, then, at the instant this bisection takes place, we measure the angle S E M, we shall be able to determine the ratio of E S to E M. Aristarchus found this angle S E M not less than eighty-seven degrees, whence he concluded that the sun was at least eighteen or twenty times as far from the earth as the moon. The fact is, that it is a great deal farther. The practical difficulty of the method lies in the difficulty of determining exactly the instant at which the moon is dichotomized, as it is called; still the method of Aristarchus showed much ingenuity, and his results carried the limits of the universe much farther than had been admitted before his time.

Aristarchus has attempted to determine as well the apparent diameters as the distances of the sun and moon. That of the sun, he estimated, like Thales, at thirty minutes; but he seems to have committed some mistake on that of the moon: however, the real diameter of that body he valued at rather less than one-third that of the earth, which may be considered as a near approach to the truth.

The poem called the 'The Phenomena of Aratus,' perhaps, deserves a short notice here, not on account of any intrinsic merit, so much as the celebrity it enjoyed among the ancients, having been commented upon by Eratosthenes, Hipparchus, Geminus, Achilles Tatius, and many others; and translated into Latin by Cicero, Germanicus, and Avienus. It consists principally of an account of the constellations, with their achronical and heliacal risings and settings, borrowed from the treatise of Eudoxus on the Sphere. The astronomical part has been severely but not unjustly criticised by Hipparchus.

The name of Eratosthenes has been rendered for ever memorable by the first attempt to determine the dimensions of the planet on which we live. The spherical figure of the earth had long been known in Greece: it had been taught in the Ionian and Pythagorean schools, and subsequently among the Peripatetics; in fact, it is a truth of too obvious a nature not to strike observers even in the most incipient state of the science. But, to determine the magnitude of this sphere was a problem of some difficulty, nor have we any reason to believe that it was attempted before the time of the philosopher just mentioned. Aristotle *, indeed, tells us that the mathematicians had fixed the circumference of the earth at 40000 stadii. But this seems merely to have been a rough estimation; nor is it worth discussing the value of the stadii here used, which would be very difficult to ascertain. Simplicius † tells us that Aristotle meant the surface, not the circumference, of the earth. This may be doubted; but one thing seems clear, that this commentator attached little importance to the determination of Aristotle, as he himself gives a very different one, without remarking the discrepancy as one of importance.

The method adopted by Eratosthenes is, in its principle, the same which has been used by astronomers in all subsequent measures of the same kind ‡. It consisted in determining by celestial observations the difference of latitude between two places lying under the same meridian, and then measuring the distance on the earth's surface between them. Hence he deduced the length of one degree on this surface, and multiplying it by 360°, he found the magnitude of the entire circumference. The extreme points of his arc were Syene in Upper Egypt, and Alexandria. The former place was supposed to be exactly under the tropic of Cancer, from the circumstance that on the day of the summer solstice for a space of about 300 stadii vertical bodies threw no shadow. At mid-day, then, on the summer solstice, that is, at the moment at which the sun was supposed to be vertical at Syene, Eratosthenes measured the sun's zenith distance, which gave him at once the difference of latitudes. This zenith distance he found to be $\frac{1}{50}$th part of the circumfer-

* De Cœlo, lib. ii., sub finem.
† De Cœlo, lib. ii.
‡ V. Cleomed. Meteor., lib. i., c. 10.

ence, or 7° 12′*. The distance between Syene and Alexandria he estimated at 5000 stadii: he found thus, for the length of the circumference, 250,000 stadii. To judge of the accuracy of this measure we ought to know what was the value of the stadii employed, about which there is great uncertainty; but it is at once obvious that there are several sources of inaccuracy. Syene, we know, is not under the meridian of Alexandria, but nearly three degrees to the eastward of it; it is also about 50′ to the north of the tropic; lastly, the distance between this place and Alexandria seems to have been estimated, not measured.

About two centuries later, Posidonius made an attempt to verify the measure of Eratosthenes †. He observed, that in the island of Rhodes the star Canopus just grazed the horizon, while its meridian altitude at Alexandria was 7¼ degrees. The distance of the two places he estimated at 5000 stadii; hence he got the length of the circumference, 240,000 stadii. But this measure is perhaps still more inaccurate than the former: the distance being across the sea, could only be most roughly estimated; and there is also more than a degree of difference in the longitudes of the extreme points.

But to return to Eratosthenes: he seems to have observed the winter as well as the summer solstice at Alexandria, for we possess a valuable determination of the obliquity of the ecliptic by him ‡: he is said to have fixed the angle between the tropics at $\frac{11}{83}$ parts of the circumference, or 47° 42′ 27″, whence we get for the obliquity of the ecliptic 23° 51′ 13″; the theory of universal gravitation would make it about 7′ less, which, under all the circumstances, is an inconsiderable difference.

Among the distinguished men produced by the School of Alexandria, Hipparchus § stands pre-eminent. He has been called the Father of Astronomy; and it is unquestionable that by his labours were laid the foundations of the science. Far surpassing his predecessors, he has been equalled by few of his successors, perhaps by none except Kepler and Bradley. Of all his writings only one, and that one of the least importance, has descended to posterity; a commentary on the Phenomena of Aratus. This astronomical poem is in fact a description of the sphere, the materials of which Aratus seems to have taken from Eudoxus; the positions of the stars are given in a very rough and often in a very inaccurate manner, and scarcely deserve a commentary from an astronomer so eminent as Hipparchus. This commentary seems to have been written when Hipparchus was a young man; at all events before he had discovered the general motion of the stars in longitude, to which it contains no allusion. The most interesting fact that can be elicited from it is, that Hipparchus was then in possession of a method for the resolution of spherical triangles.* As we find no traces of the science of spherical trigonometry in any preceding author, Delambre concludes, and apparently with reason, that Hipparchus was the inventor of it. This is certainly not the least of the obligations we owe him; for it is evident that astronomy could make little progress without the assistance of trigonometry. But though the works of Hipparchus, with this exception, are lost, we are able to ascertain pretty exactly the extent and nature of his discoveries from the great Syntaxis of Ptolemy. We see there that the foundations of nearly all the theories developed by Ptolemy, were laid by Hipparchus: the additions made by the former will be examined in a subsequent part of this treatise.

The astronomers of Greece for several centuries had supposed the exact length of the solar year to be 365 days and a quarter. Hipparchus by comparing one of his own observations of the summer solstice, with one made 145 years previously by Aristarchus of Samos, discovered this to be too great. He found that the solstice arrived 12 hours sooner at the end of these 145 years than it ought to have done, on the supposition of the solar year being 365¼ days; 12 hours divided by 145 gave him the diminution to be made on the length of the year. In this way he found for the length of the tropical year 365 days 5 hours 55 minutes 12 seconds.

The sun appearing to move in a circle round the earth, it was natural to suppose that his motion in the ecliptic was uniform; and such was probably the

* To this must, in all probability, be added 15′ for the sun's semidiameter: the Greeks generally seem to have neglected this correction in observing with the gnomon.

† Cleomed. cap. cit.

‡ V. Ptol. Syntax.

§ Born at Nicæa in Bithynia: flourished about 150 B.C.

* V. Delambre, Astron. Anc. Vol. i. p. 142.

opinion of the early astronomers of Greece. However when they began to make observations with the gnomon, they could not help perceiving a considerable difference between the intervals of the equinoxes and solstices; intervals which must be equal were the motion of the sun round the earth uniform. Hipparchus undertook to investigate this point. He observed that the interval between the vernal equinox and the summer solstice was 94½ days; between the summer solstice and the autumnal equinox 92½.* Thus the sun took 187 days to describe the northern half of the ecliptic, and only 178¼ for the southern half; indicating a considerable increase of velocity during the latter. To explain this irregularity, Hipparchus supposed the sun to move round the earth in an excentric circle; that is in a circle, whose centre did not coincide with that of the earth. It is clear, that in this case the sun, though moving uniformly in its orbit, would appear to a spectator at the earth to move with an unequal velocity, on account of the variation of its distance. The question was to determine the quantity of this excentricity, that is to say, the distance of the earth from the centre of the solar orbit; and the position of the apogee and perigee, or of the points of greatest and least

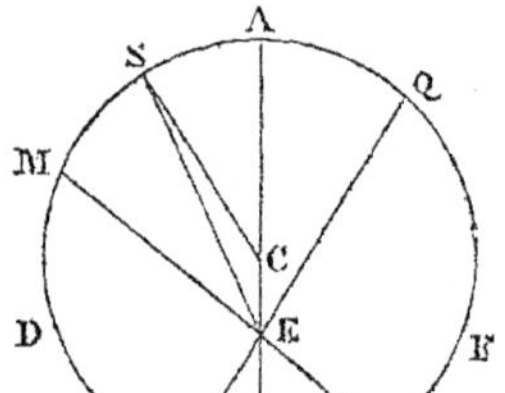

distance. Let A D B F represent the circle in which the sun is supposed to revolve; let the centre of this circle be at C, and the earth at E: according to Hipparchus the sun revolves with an uniform motion round C: it is evident that, seen from E, his motion will appear unequal: it will be fastest at the point B or the perigee; slowest at A, the apogee: let M N be the line joining the sun's places, at the two solstices; P Q at the equinoxes: the object of Hipparchus was to ascertain the ratio of E C to B C, and the arc Q A which determines the position of the apogee. By combining his observations of the equinoxes and solstices, he found the excentricity equal to $\frac{1}{24}$th part of the radius, and the longitude of the apogee, or the arc Q A, equal to 65° 30′. This value of the excentricity is, however, too great by about one-sixth. The excentricity and place of the apogee being once known, it was easy to construct Tables which should give the sun's position at any time. For, suppose the sun to be at S, then as he is supposed to revolve uniformly round C, we can find from the time taken to describe the arc A S, the value of the angle A C S, and therefore S C E; and in the triangle S C E, C E, and C S are known, whence we may find C S E, which is the difference between the angles A C S and A E S, or between the mean and true anomaly. This difference is called the equation of the centre. From what has been just said, we may see how Hipparchus calculated the values of the equation of the centre corresponding to successive values of A S, or the angle A C S, in his solar Tables. We must recollect that the instant of the sun's passage through the equinox at Q may always be supposed known: the arc Q S is proportional to the time elapsed since the equinox, and is soon found: Q A is known: hence we find A S, and looking into the Tables, find the corresponding equation of the centre. This gives us A E S, and consequently Q E S, or the sun's apparent longitude for any given time.

From the theory of the sun Hipparchus proceeded to that of the moon. By comparing some ancient eclipses with those observed by himself, and dividing the interval of time by the number of revolutions, he obtained the value of a synodic revolution of the moon. By methods similar to those employed for the sun, he determined the excentricity of the lunar orbit, and its inclination to the ecliptic, which latter he fixed at 5°. Finally, he is said to have measured the motions of the lunar apogee and node. With these data he calculated the first Tables of the sun and moon of which history makes mention. This alone would have secured for him the gratitude and admiration of posterity. The want of observations, and perhaps the difficulty of their theory in his system, prevented

* Ptol. Syntax. Lib. III.

him from attempting a similar undertaking with regard to the planets.

But the most important, perhaps, of all the services rendered to astronomy by Hipparchus, was the formation of a catalogue of the fixed stars. If we consider the boldness of the attempt, the labour of the execution, and the importance of the result, the author of it seems not undeserving the enthusiastic praises of Pliny *. Such a catalogue is, in fact, the foundation of all astronomy. The fixed stars are so many standard points to which the celestial motions are referred, and the determination of their relative distances is of the utmost importance. By comparing their positions at distant periods, we may detect those small variations which require centuries to become sensible; and there is every reason to believe that, if we possessed a really accurate catalogue of twenty or thirty centuries back, we should be in possession of many valuable discoveries, which perhaps are destined to lie hid for ages. It was, indeed, in this way that Hipparchus was led to his great discovery of the precession of the equinoxes. On comparing his own observations with those of Aristillus and Timocharis, made 150 years previously, he perceived that all the fixed stars, while they retained their latitudes sensibly unaltered, had advanced about two degrees in longitude; or what comes to the same, the equinoctial points appeared to have retrograded along the ecliptic by the same quantity. It was reserved for Newton to explain the causes of this singular phenomenon.

Such is a brief account of the astronomical discoveries of Hipparchus: we have already seen that he was the inventor of trigonometry; it also appears that he was the first who suggested the method of fixing the positions of places on the earth's surface by their longitudes and latitudes, and that he proposed to determine the former by means of lunar eclipses; a method excellent in its principle, though now abandoned on account of some practical objections.

As nothing connected with astronomy seems to have escaped the sagacity of Hipparchus, he did not overlook the correction of the Calendar. We have seen that the period of Callippus was far from exact: according to the calculations of Hipparchus, the error at the end of a period was about one-fourth of a day. He proposed to quadruple the period of Callippus, and then to subtract a day. This new period brought the moon again to the same place pretty exactly: the error on the sun's motion was about a day and a quarter, which is one-fourth of the error of Callippus in the same time.

We have some reason to be surprised that the discoveries of Hipparchus were not followed up by succeeding astronomers. One might have imagined that such brilliant success would have stimulated others to the further development of the science; but, extraordinary as it may appear, history records not one astronomer of note in the three centuries between Hipparchus and Ptolemy. The attempt made by Posidonius to measure a degree of the meridian has been already noticed: a few authors on spherical trigonometry flourished in this interval, among whom may be distinguished Theodosius and Menelaus; but astronomy itself seems to have made no progress till the time of Ptolemy. This eminent and laborious philosopher felt the necessity of uniting all the scattered materials existing in the works of Hipparchus and others, which, combined with his own discoveries, formed, as far as the knowledge of the time allowed, a complete system of astronomy: by so doing he rendered a distinguished service to science; and the publication of his μαθηματίκη σύνταξις forms an important epoch. This work, which has fortunately survived the barbarism of the middle ages, formed the basis of all the astronomy of the Arabians, and for a considerable time that of modern Europe. Its importance requires here a concise analysis.

Ptolemy begins his work with a discussion of the relative positions of the earth, sun, and planets. We have already seen that the Greek astronomers were divided on the subject of the earth's motion. Though many distinguished philosophers held the opinions of Pythagoras, the majority seem to have embraced the opposite doctrines. Ptolemy followed these latter, and, unfortunately for him, his name has become attached to a system now universally admitted to be erroneous. It is true that the ancients wanted some decisive and convincing proofs of the earth's motion, which we possess; but though much has been said to excuse Ptolemy, his justification remains very incomplete. The arguments that he urges

* Hipparchus nunquam satis laudatus, ut quo nemo magis comprobaverit cognationem cum homine syderum, animasque nostras partem esse cœli, ... ausus rem etiam Deo improbam, annumerare posteris stellas.—Hist. Nat. ii. 26.

against the earth's motion, such as that in this case the poles would not be immoveable points on the celestial sphere, that the fixed stars would not always preserve the same apparent distances from one another, and other objections of a similar kind, are all obviated by the single remark made by Aristarchus, four centuries previously, that the earth's orbit was a point in comparison with the distance of the fixed stars. On the other hand, the motions of the planets, so complicated and almost inexplicable in the one hypothesis, are accounted for so simply in the system of Pythagoras, that one cannot but feel astonished that Ptolemy should have felt so little hesitation in rejecting it. "The same reasons," says he, "which show that the earth is a point in magnitude compared with the heavens, will show the impossibility of its having a motion of translation:" and the only argument he combats at any length, is that which appears to have been urged by some Pythagoreans, that the earth being spherical and unsupported, could not remain at rest in the centre of the heavenly motions. Having discussed this point, with a singular mixture of truth and error, he adds these remarkable words: "But if there were any motion of the earth common to it and all other heavy bodies, it would certainly precede them all by the excess of its mass, being so great; and animals and a certain portion of heavy bodies would be left behind, riding upon the air, and the earth itself would very soon be completely carried out of the heavens. But such things are most ridiculous, even only to imagine." This passage is remarkable, because it shows how little the Greeks had studied natural and experimental philosophy, and how falsely their geometers could reason on purely physical subjects. A heavy body in vacuo does not, as Ptolemy supposes, move faster than a lighter one, as may be verified by direct experiment; yet this he clearly considered a self-evident truth, and founded on it arguments which must be classed among the weakest ever urged against the Pythagorean system of the world.

After rejecting the motion of translation assigned by some to the earth, he proceeds to examine the probability of its diurnal motion on its axis. This system he confesses simplifies very much the appearances of the heavens; but it appears to him equally ridiculous with the former; as in this case, the earth revolving with great rapidity from west to east, would leave behind it the clouds, birds flying in the air, and, generally, all objects suspended in the atmosphere. A stone thrown to the east would not advance, the earth constantly preceding it by the excess of its velocity. These objections are all founded on an ignorance of the principles of mechanics, and seem to be on a par with those urged by some against the roundness of the earth, and the possibility of the existence of Antipodes; arguments which he has himself successfullyrefuted.

The earth then, according to Ptolemy, was fixed and motionless in the centre of the heavens; he supposed the different planets to revolve round it, arranged in the following order, according to their distances: first, the Moon, then Mercury, Venus, the Sun, Mars, Jupiter, Saturn, and, lastly, the sphere of the fixed stars. With regard to Venus and Mercury, Ptolemy remarks, that some astronomers had placed them beyond the sun, while others made them nearer: the most ancient writers had adopted the latter opinion, which had been rejected by subsequent authors, because these two planets had never been seen on the sun's disk. This reason Ptolemy rightly rejects as insufficient; for such passages over the sun's disk would not happen, unless the planes of the orbits coincided with the ecliptic, or else the nodes happened to coincide nearly with the sun's place at the time of inferior conjunction. He does not seem to be aware that these passages or transits really do take place; and sufficiently often in the case of Mercury, though but rarely in that of Venus. But the difficulty of observing these phenomena renders it by no means extraordinary that they should not have been noticed, though it might have taught caution to those who affirmed positively their non-existence. It is much more remarkable that Ptolemy should not have perceived that it was possible to conciliate the two hypotheses in question, by making these two planets revolve round the sun; in which case it is clear they would be sometimes more distant, and sometimes nearer, than that body. And this inadvertence is the more singular, as the doctrine just mentioned is said to have been maintained by the ancient Egyptians. It seems probable that the systematic ideas of Ptolemy made him unwilling to place the sun in the centre of any of the heavenly motions; or he might have been repugnant to consider any of the planets merely as satellites or secondaries, which, in his system, Venus and Mercury would thus have become.

In considering the theory of the sun, Ptolemy adopted without alteration the elements of Hipparchus. But to the theory of the moon he made several important additions. We have seen that Hipparchus explained the irregularity of the sun's motion by the hypothesis of an excentric circle. There is another way, however, of explaining this irregularity, by the hypothesis of an epicycle. In this case the planet was supposed to move in a small circle, called the epicycle, the centre of which revolved uniformly round the earth: in the case of the sun, this epicycle had for its radius the observed excentricity of the orbit; and the sun's motion in it was such, that during the interval between the apogee and perigee, that planet had approached the earth by exactly the diameter of its epicycle.*

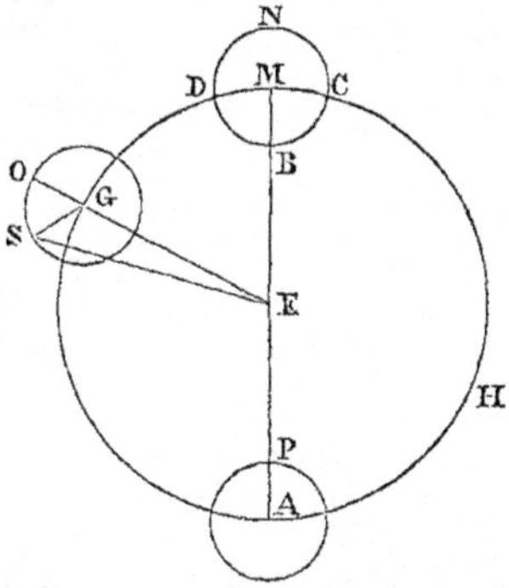

Thus let E be the earth; M, which is the centre of the epicycle D B C N, represents the sun's mean place, and describes uniformly the circumference of the deferent M G A H; while the real sun describes the circumference of the epicycle; at the apogee the sun is in N; the true and mean places coincide, and the distance of the sun from the earth is EM+MN; MN being equal to the excentricity. At the perigee S is in P; the true and mean places again coincide, and the distance is EA − AP = EM − MN. In any intermediate position the true and mean places will differ by the angle GES, and the distance will take every value between the limits EA ± MN. This angle GES, which in fact is the equation of the centre, may easily be calculated for any given value of the arc MG. For EG, and GS the radii of the deferent and epicycle are known, and the angle OGS is equal to MEG: the motion in the epicycle being supposed in a contrary direction to that in the deferent. This hypothesis of the epicycle coincides with that of the excentric, when the radius of the epicycle in the one case is equal to the excentricity in the other.

But as the motion of the moon is much more complicated than that of the sun, it was necessary to have recourse to a combination of excentrics and epicycles. Hipparchus had discovered in the moon's motion an inequality similar to that of the sun, and depending on the same cause, the excentricity of its orbit: Ptolemy detected another depending on the angular distance between the moon and sun. This inequality, usually called the evection, is greatest in the quadratures, and least in the syzygies; but its magnitude also depends on the combination of the places of the lunar apsides with those of the conjunctions. When the conjunctions happen in the moon's apogee, the inequality we are speaking of becomes the greatest possible in quadratures, and amounts to about 2° 40′. It is then negative in the first two quarters (that is to say, the moon is behind her calculated place), and positive in the last two. When the conjunction takes place in the perigee, the inequality in the quadratures is also at its maximum; but it is positive in the first two quarters, and negative in the latter two. In intermediate positions of the lunar apsides, the inequality diminishes; when they are in quadratures, it is reduced to nothing. Finally it is negative in the first two quarters, and positive in the last two, or the converse, according as the conjunctions happen in the first or second quadrant of a circle, counting from either of the apsides in the direction of the moon's motion. The detection of the law existing between these complicated phenomena reflects great credit on the sagacity of Ptolemy.

To explain the first inequality of the moon, that depending on the excentricity of its orbit, he imagined an epicycle carried on an excentric; an hypothesis which is the same as that of a simple excentric, if the excentricity and radius of the epicycle together are equal to the excentricity of the simple excentric,

* There is a slight mistake in the figure given above: E S should be drawn on the other side of E G, so that the point S should fall within the angle M E G: otherwise the point S would not appear to revolve in the epicycle with a contrary direction to that of G in the deferent.

To represent the second inequality, which depends upon the elongation and the position of the apsides, he supposed that the centre of the deferent revolved round the earth, with a motion equal to that of the centre of the epicycle, but in a contrary direction. Thus the centre of the epicycle would always coincide in quadratures with the perigee, and in syzigies with the apogee.

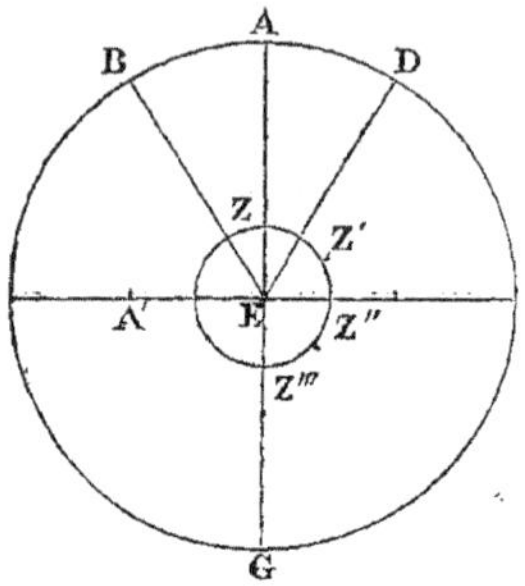

Let ABGD be a circle homocentric with the ecliptic, let E be the place of the earth's centre, AEG is the diameter. Suppose the moon, when in conjunction, to be in apogee, and let her place at that time be A; let the centre of the excentric be Z. Now if the moon in one day revolve through the arc AB, the apogee in the same time will move through the arc AD, equal to AB, and Z will have moved to Z′. In a quarter of a month, the points D and B will be diametrically opposed, and the centre Z will be at Z″; and the centre of the epicycle will be at A′, its nearest point to the earth. The inequality will be at a maximum, and its general effect will be to augment the first inequality, by making the radius of the epicycle appear larger: consequently in the first two quarters the moon will be retarded. At the end of half a month, D and B will be in conjunction at G; the inequality will vanish. After this it will augment again gradually till the quadrature, and then diminish till conjunction; but in these two last quarters the moon will be accelerated. Were the moon in perigee in conjunction, the same phenomena would take place, with the difference, that the acceleration would be in the first two quarters, and the retardation in the two last. If the moon's perigee and apogee were in quadratures, the inequality would altogether vanish; as at the end of each quarter, the centre of the epicycle would always be ninety degrees from the apogee. This hypothesis of Ptolemy represents pretty well the greatest of the moon's inequalities; and it certainly was a very ingenious effort for the time; but it had several defects, the principal of which was, that in consequence of the proportion that Ptolemy was obliged to establish between the excentricity of the moveable orbit or deferent, and the radius of the epicycle, the moon's distance from the earth in quadratures would sometimes be only half of what it is in syzigies, which is entirely contradicted by observation; the variations of the moon's distance are comprised within limits comparatively very small.

From the theory of the moon, Ptolemy proceeded to that of the planets, which it appears that Hipparchus had not ventured to touch, deterred in all probability by the apparent complication of their motions. Ptolemy, however, attempted to represent them, by a combination of epicycles and excentrics. For the superior planets, he supposed the centre of the epicycle to make a revolution on its deferent in the time of a mean revolution of the planet, while this latter revolved in its epicycle in such a way, that it was always at the lowest point of the epicycle at the instant of mean opposition with the sun. The deferent itself was an excentric. It is evident that by determining properly the magnitudes of the epicycles, he could represent all the phenomena observed. For when the planet was in the superior part of its epicycle, its motion was direct; when in the inferior part, it moved in a contrary direction to that of the centre of the epicycle, and its motion, seen from the earth, would appear direct, stationary, or retrograde, according as the motion in the epicycle was less rapid, equal to, or greater than that of the centre on the deferent. We see too that each retrogradation was preceded and followed by a station, and that the place of the opposition sensibly bisected the arc of retrogradation. Finally, the excentricity of the deferent explained the inequality of the intervals between the oppositions and of the arcs of retrogradation. But even this was not sufficient to satisfy all the phenomena observed; Ptolemy was compelled to make the centre of the epicycle revolve with a motion that was uniform, not round the centre of the excentric it described, but round a point as far beyond this centre in one direction, as the earth was from it in the opposite direction: thus virtually abandoning the perfect regularity which was long thought necessary in the orbits of the heavenly bodies.

For the inferior planets the same hypothesis of an excentric and epicycle was employed. In this case the centre of the epicycle always coincided with the mean place of the sun, while the planet described its circumference with a velocity proportional to the time employed in going from one point of greatest digression to another. As the ellipse of Mercury is much more sensibly excentric than that of the larger planets, Ptolemy found the hypotheses which satisfied the others insufficient in this case. He was compelled to suppose that the point, which was the centre of the uniform motion called the centre of the equant, instead of remaining fixed, revolved in a small circle round the centre of the excentric; the radius of this circle being equal to the distance between these two centres, and the direction of the motion against the order of the signs. But it is impossible, in this place, to follow him into all the artifices he was forced to have recourse to in explaining the irregularities of the planetary motions.

The extreme complication of this system arose in a great measure from the law he had imposed upon himself of admitting none but circular motions in the heavens: "uniform and circular motions," says he, "belonging by their nature to celestial bodies." (Syntax. lib. ix. c. ii.) That astronomers should have attempted to represent all the celestial motions by circles, was natural enough in the infancy of the science; and as long as the apparent inequalities could be represented by a combination of these circles, they were justified in so doing: but it is lamentable to observe that men of talent could mistake gratuitous and arbitrary assumptions of their own for laws of nature;—these metaphysical fancies, principally borrowed from Aristotle, about the perfection and incorruptibility of circular motion, long retarded the progress of science.

But though the Aristotelian physics of Ptolemy form a strange contrast with the geometrical knowledge displayed in his work, it would be unfair to charge him with having admitted the monstrous doctrine of solid transparent spheres, revolving the one within the other, and each carrying a planet attached to it, which was promulgated by Eudoxus. To these he makes no allusion; and it is but justice to him to suppose that he himself considered his system of deferents and epicycles merely as a means of determining mathematically the positions of the heavenly bodies for any given time.

Ptolemy was unsuccessful in his researches on the quantity of precession. Hipparchus had supposed it to be about one degree in 75 years: Ptolemy, undertaking to correct this determination, went much wider from the truth; he fixed it at one degree in 100 years, whereas the real value is one degree in 72 years. But there is a heavier charge against him: that of having appropriated and published as his own the catalogue of fixed stars, formed by Hipparchus. This seems to be but too well proved. He states the quantity of precession in the 265 years between himself and Hipparchus at $2^\circ\ 40'$: this alone would show that he had not observed, as he would have found it considerably more; but if we subtract from all his longitudes $2^\circ\ 40'$, the precession he supposed for 265 years, we get exactly the longitudes, such as they were in the time of Hipparchus, and such, in fact, as that astronomer seems to have fixed them, judging from the positions given in the commentary on Aratus.

As all astronomy must be founded on observation, Ptolemy has not neglected to describe the instruments used for that purpose at Alexandria. To determine the sun's altitude, the Eastern nations had long been in the habit of measuring the shadow of a vertical gnomon; and, if a few simple and obvious precautions be attended to, this method may give very accurate results. The Greeks learned the use of the gnomon from the Chaldeans at a very early period; and we have seen that it was employed by Meton, Pytheas, and others. But at Alexandria it seems to have been but little used; the astronomers of that place substituting for it armillary spheres of different kinds. To observe the sun's passage through the equinox they used two circles, firmly attached to each other, and placed one in the plane of the meridian, the other in that of the equator: at the moment of the equinox this latter was not illuminated by the sun on either side. For the solstice they used two concentric circles in the plane of the meridian, the one revolving within the other, and carrying two small prisms at right angles to the limb, and fixed at points diametrically opposed on the circle. To observe the sun's meridian altitude with this, the inner circle was turned till the shadow of one prism completely covered the other: the shadow of this second fell on the graduated limb of the outer circle, and the middle of it being marked, gave the altitude of the sun's centre. For this solstitial circle Ptolemy substituted a quadrant, on which the observation was made in a manner very similar: but his most important invention was that of the parallactic rulers. These rulers formed an isosceles triangle, susceptible of being opened to any angle at the vertex; one of the equal sides was always vertical, the other being pointed on the star; the observer read off on the graduation of the base the length of the chord: a table of chords gave him the value of the angle at the vertex, that is, of the zenith distance.

The construction of the astrolabium, with which the longitudes and latitudes of the planets or fixed stars were observed, was rather more complicated than that of the solstitial or equatorial armillæ. They carried circles representing the equator, the ecliptic, the meridian, &c., and placed respectively in the planes of the celestial great circles they represented. Two other circles, moveable on the poles of the ecliptic, were made to pass through two stars: the observer then read off on the graduation of the ecliptic and circles in question respectively, the latitudes of the two stars and their difference in longitude.

As the Greeks had no means of measuring time with any accuracy, they were obliged, when they wished to compare the place of the sun with that of the fixed stars, to measure in the daytime the difference of longitude between the sun and moon, and at night that between the moon and a fixed star. The moon's rapid and variable proper motion necessarily rendering this method very inexact, the Arabs improved it considerably by substituting the planet Venus for the moon. A still greater improvement will be noticed when we consider the observations of El-Batani.

Ptolemy was the author of a most important discovery not recorded in the Syntaxis, the effect of refraction in augmenting the apparent altitudes of the heavenly bodies. This is clearly shewn in his Optics*, where he investigates the theory of refraction in general. He was aware of the existence of a certain constant relation between the angle of incidence and that of refraction, and made several experiments to determine the value of the latter when a ray passes from air into water. Though he perceived clearly the nature of the effect produced on the altitudes of the stars, and that it diminished with the zenith distance, yet he declared himself unable to determine the absolute quantity of refraction, from not knowing the height of the terrestrial atmosphere. However, this treatise is extremely remarkable, and one of those that reflect the greatest honour on its author. To point out the existence of refraction, even without measuring it, was to render an important service to astronomy; to which we must add, that this is the only work of the ancients in which there is anything resembling the experimental philosophy of the moderns. We also find here an ingenious explanation of the optical illusion which makes the disks of the sun and moon apparently much larger when near the horizon; and this explanation is the one generally received at present, though there still seems to be some doubt on the subject.

* Vid. Delambre, Astr. Anc., vol. ii. This work of Ptolemy, though known to the Arabs and to Roger Bacon, was for a long time lost in Europe. A Latin translation of it was fortunately discovered by La Place in the Royal Library at Paris.

Chapter VII.

Astronomy of the Arabs.—The Persians.—The Chinese.

With the Syntaxis we take our leave of the astronomy of the Greeks. The interval between the publication of this work and the conquest of Egypt and Syria by the Arabs did not produce a single astronomer; for we cannot give that name to one or two commentators on Ptolemy, of whom Theon is the most generally known. But when the Arabs had firmly established themselves in the East, they began to cultivate all the branches of mathematical science, and astronomy in particular, with extraordinary zeal. This revolution in the character of the Arabs, the beginning of which dates from the Caliphs, El-Mansour and Haroun-el-Reschid, at the end of the eighth century after Christ, was finally accomplished under El-Mamoun, who reigned in the beginning of the ninth. Ibu Jounis has recorded several observations made by the astronomers of this prince, the most interesting of which are those instituted to determine the obliquity of the ecliptic. This was found by some 23° 33′, by others 23° 33′ 52″, which is exact within 3½ minutes, and more correct than any of the determinations made by the Greeks. Justly dissatisfied with the rough attempts of the Greek astronomers to measure an arc of the meridian, El-Mamoun ordered his astronomers to proceed to a new measurement. The method they followed is sufficiently simple. Having chosen a large plain in Mesopotamia, they divided themselves into two parties; then, starting from a given point, each party measured in a right line an arc of one degree, the one towards the north, the other towards the south. The former found for the length of a degree fifty-six miles, the latter fifty-six and two-thirds; the mile being equal to 4000 cubits. But here arises the question, what was the length of these cubits? Unfortunately this is not easy to decide. Two Arabian authors agree, that the cubit employed was the black cubit of twenty-seven inches; but one says that the inch was determined by six grains of barley placed in contact sideways; the other makes it equivalent to five similar grains*. The latter seems to agree better with the real length of the degree; but the error is still very considerable, being between three and four miles in excess. But if we suppose the cubit employed to be the royal cubit of twenty-four similar inches, the length of the degree will then be brought within about a third of a mile of its real value.

The two centuries immediately following the reign of El-Mamoun were extremely fertile in astronomers, and particularly in observers; forming thus an advantageous contrast with the Greeks, who seem, with very few exceptions, to have had little taste for observation and experiment in any of the sciences. In this respect the Arabs effected a complete reform in astronomy. They have left behind them an immense mass of recorded observations, of which the greater part has never been printed; and which might be of great service to astronomy, did not the superiority of our instruments render the modern observations so much more accurate, as to compensate for the smaller interval of time existing between them.

The most distinguished of the Arabian astronomers is Albategnius or El-Batani, who rectified, in many points, the determinations of Ptolemy, and added the important discovery of the motion of the solar apogee*. Ptolemy had fixed the precession at one degree in one hundred years, instead of seventy-two, the real value: El-Batani corrected this mistake, but made it, on the other hand, a little too rapid; namely, one degree in sixty-six years. Similarly he made the length of the solar year about two minutes and a half too small; but it is just to remark, that the errors of El-Batani proceed from the confidence he placed in the observations of Ptolemy,—observations which, as we have seen before, appear to be fictitious†; had the Arabian astronomer compared his observations directly with those of Hipparchus, he would have approached much nearer to the truth. The excentricity of the solar orbit was determined by him with great accuracy, the equation of the centre fixed at 1° 58′, and the obliquity of the ecliptic at 23° 35′. The observations used to determine these quantities seem to have been made with great care, and are much superior to any recorded by the Greeks. El-Batani, who gives in his writings many proofs of a sound judgment, rejects, with reason, a pretended motion of the fixed stars, by which they appeared to oscillate about a certain point, their motion in longitude becoming sometimes direct, and sometimes retrograde. To explain this pretended motion, which was called trepidation, the equinoctial points were supposed to revolve in a circle of 4° 18′ 43″ radius round their mean places, which retrograded along the ecliptic, according to the laws of precession. This, at least, is the way in which the theory was represented subsequently. El-Batani merely states that the stars were supposed to move directly through 8°, then to retrograde through the same arc. The Arabian astronomer, while refuting this theory, attributes it distinctly to Ptolemy. It is remarkable that none of the extant works of Ptolemy make the slightest allusion to trepidation; the first mention

* According to Thevenot, 144 grains of Oriental barley, placed side by side, are exactly equal to one foot and a half of the old French measure.

* This is not expressly stated by El-Batani, but it is an evident consequence, from his discovery, that the apogee, which Ptolemy found to be in 65° 30′, was now in 82° 17′. This gives an annual motion of 79″. Now the Arabian astronomer allowed 54″ for the annual effects of precession; there would remain, therefore, about 25″ for the annual proper motion of the apogee.

† The equinox taken by El-Batani to compare with his own observations, is recorded by Ptolemy with the mistake of a whole day on its date. On the observations of Ptolemy in general, Halley has expressed a severe but just opinion. V. Delamb. Ast. du Moyen Age, pp. 61, 62. Delambre is of opinion that Ptolemy never observed at all. For his catalogue of fixed stars, see what has been said above, p. 31.

of it is found in Theon *, whose commentary on the Almagest has been noticed above. It is very unjustly that Thebit ben Corah, an Arabian astronomer, has been considered as the inventor of trepidation; we may see from Theon, that it was a fancy of the Greeks, probably anterior to Ptolemy, though never noticed by that author in the Almagest. However, Thebit adopted it, and even wrote a treatise purposely to establish it; at least if he be really the author of the work on the eighth sphere generally attributed to him. But there are some reasons for doubting of this; as Ibn Jounis has preserved an original letter of Thebit, in which he expresses himself as far from convinced of the existence of trepidation †.

The limits of this treatise do not allow us to enter into an examination at length of the writings of the numerous Arabian astronomers to be found in the libraries of Europe, nor would a bare catalogue of names offer any interest for the reader. It will be enough to notice shortly the tables of Ibn Jounis and Arsachel. The former, an Egyptian astronomer of great merit, has left behind him a considerable mass of observations, and a treatise on astronomy, in which are some remarkable improvements in trigonometrical calculation. His tables are, in fact, those of Ptolemy, with many ameliorations in the constants and epochs. It appears rather singular, that though coming after El-Batani, he does not admit any other motion for the solar apogee than that of precession: we have seen that the observations of El-Batani indicated very clearly an annual proper motion. The tables of Arsachel, like all the Arabian tables, are, in substance, those of Ptolemy; in the numerical determinations they seem inferior to those of El-Batani; nor would they deserve mention here, were they not supposed to have been of great assistance to the composers of the famous Alphonsine tables.

The instruments of the Arabs were essentially the same as those of the Greeks; the gnomon, various kinds of armillary spheres, and a sort of mural quadrant. But they added (and we owe this apparently to Ibn Jounis), the method of determining the time by observing the absolute altitude of a fixed star or planet *. This was probably the best method that could be employed before the invention of pendulum clocks.

The science of trigonometry is necessarily and inseparably connected with astronomy. The Arabs, who cultivated the latter so zealously, made considerable additions to the former. The most important of these was the substitution of the sine instead of the chord of the double arc employed by the Greeks. We owe this very important amelioration to El-Batani. It enabled him to simplify very much the solutions of several cases of oblique angled spherical triangles; particularly that in which the two sides and the included angle are given to find the third side, or either of the remaining angles. Both El-Batani and Ibn Jounis make use of tangents and cotangents in their treatises on dialling; and even give tables of these quantities; but it was reserved for Aboul Wéfa, an astronomer of Bagdad, of the eleventh century after Christ, to introduce them into trigonometry. This was a second important improvement. The same author is also the first who treats of secants and cosecants; but the Arabs do not seem to have been aware of the advantage of introducing the cosine into trigonometry, till a century later, when Geber, a Mahometan Spaniard, gave for the first time a formula into which it enters. Ibn Jounis was the first author who made use in trigonometrical problems of the tangents, cosines, and secants of subsidiary arcs. This elegant method, which in many cases simplifies extremely numerical calculations, seems to have been unknown to the mathematicians of Europe till the middle of the eighteenth century, when it was reinvented by Simson.

The Tahtar conquerors who succeeded the Arabs in the east, seem to have been as zealous as the caliphs in their attachment to astronomy. The grandson of Gent-Chis Khan founded in Persia an observatory, which he fitted up with the best instruments of the time, and all the most valuable works on astronomy existing in the east. Under his protection the astronomer Nasseer-ad-Deen published tables, which are supposed to have been entirely borrowed from Ptolemy. These are not the only tables existing in Persia:

* *κανονες προχειροι.* V. Delambre, Astr. Anc., vol. ii. p. 625. Indeed, Theon says distinctly, that Ptolemy did not admit these alterations in the precession of the fixed stars; *ὅπερ Πτολεμαίῳ ου δόκει.*

† V. Delambre, Astr. du Moyen Age, Ibn Jounis.

* Delambre, Astr. du Moyen Age, p. 78.

Chrysoccas, a Greek physician, has translated others brought from Persia by Chioniades, which seem to have been composed in the eleventh century. They offer, however, little interest, being evidently borrowed from the Greek.

The descendants of Timour were as much attached to astronomy as those of Gent-Chis. Ulugh Beg, grandson of Timour, and sovereign of Samarcand, devoted himself with extraordinary zeal to the cultivation of this science. Having erected an immense observatory, and procured the assistance of a number of mathematicians, he published a collection of tables and a catalogue of the fixed stars, which acquired, and still continue to enjoy a great reputation in the east. As far as we can judge this reputation seems well deserved; but we only know that part of the tables containing the motions of the sun, and the catalogue of the stars,—the rest has never been translated, or at least never published. The exactitude of the solar tables is very creditable to Ulugh Beg, and shows that his observations, which are said to have been made with the gnomon, were very good. The epoch of these tables is the 4th of July, 1433, A.D. The epoch of the catalogue is the year 841 of the Hegira, or 1447, A.D. The stars were observed with a quadrant of enormous dimensions; but though superior in accuracy to the Greek catalogues, the errors in longitude sometimes amount to half a degree*.

We have noticed the protection given to astronomy by the Tahtar princes in Persia and Bokhara: equal favour was shown to the science by the successors of Gent-Chis Khan on the throne of China. Though the observations of the Chinese go back to the earliest antiquity, it is not the less certain that their knowledge was extremely limited and confined to the most elementary parts of astronomy. But when the Tahtar conquest brought them into contact with the nations of Western Asia, a very sensible amelioration took place. The thirteenth century may be considered as the most brilliant epoch of Chinese astronomy. Cocheou-King who had been appointed by Kubla-Khan, the descendant of Gent-chis, to the presidency of the tribunal of mathematics, undertook to rectify from his own observations the principal elements. This he effected with considerable success. His observations of the sun were made with a gnomon of forty feet, and appear very accurate. He fixed the length of the solar year at $365^d\ 5^h\ 49^m\ 12^s$; and the obliquity of the ecliptic at $23^\circ\ 33'\ 39''$. The date of his tables is A.D. 1280. It is probable that at this time the Chinese astronomy borrowed a good deal from the Arabs. We now hear, for the first time in China, of spherical trigonometry; and the invention of it is attributed to Cocheou-King; but there is every appearance that he learnt it from the astronomers of the west; for it is known that under Kubla-Khan Persia and China were in frequent communication*. It is extraordinary that subsequently to Cocheou-King the mathematicians of China should have degenerated to such a point, that at the arrival of the Jesuits the president of the mathematical tribunal was unable to solve a plane right angled triangle†. This, indeed, is the more singular, since we are told that before the Christian era the Chinese could calculate the lengths of the shadow of the gnomon, and even had methods for the prediction of eclipses.

Chapter VIII.

Astronomy of the Middle Ages.

It is impossible to pay any attention to the history of the Romans without perceiving that in that nation there prevailed at all times a singular indisposition to the pursuit of mathematical and physical science. The poets and orators of Greece, and her metaphysicians were studied with ardour in Italy, but her geometers and astronomers were totally neglected; and it appears that these sciences, so highly estimated in one country, were thought in the other to be beneath the notice of a man of good birth and liberal education. This difference, so little creditable to his countrymen, is remarked by Cicero; nor does the Roman character seem to have changed in this respect in subsequent ages. The extent to which astronomy was neglected is evident from the circumstance, that the difference between the beginning of the civil and of the solar year, amounted in the time of Julius Cæsar to three months. During the whole existence of the republic we hear but of one Roman who attained any emi-

* V. Delamb., Ast. du Moyen Age, p. 207.

* V. Montucla, vol. i., p. 463.

† Delamb., Ast. Anc., vol. i. p. 361.

nence in astronomical studies. C. Sulpitius Gallus is mentioned by Cicero as an indefatigable calculator of eclipses*; and he is known to have predicted an eclipse of the moon on the night preceding a decisive battle between the Romans and the king of Macedonia†. On this occasion his science may be said to have rendered his countrymen an essential service; for, from the well-known superstition of the ancients, the Roman soldiery would have been much terrified by what was supposed to be an unfavourable omen. At the celebrated siege of Syracuse an unexpected eclipse of the moon deterred the Athenian commanders from commencing their march at the proper time, and caused eventually the destruction of a fine army‡. Besides Sulpitius Gallus, we find the names of one or two Romans, who seem to have written on astronomical subjects (among whom is Varro); but their works are lost, and we have no means of judging of the extent of their scientific acquirements. Perhaps Cicero himself ought to be quoted here, as he has translated into Latin verse the Phenomena of Aratus, a poem which has been already noticed. One of the books of this translation is still extant, containing a description of the celestial sphere, and in particular of the constellations of the zodiac. Nothing is to be found in this poem beyond the most elementary doctrines of astronomy: the same may be said of the astronomical poem of Manilius, the date of which is a little later.

Of the last-mentioned poem a small part is devoted to the description of the sphere; by far the greater part being occupied with astrological precepts. This superstition, the influence of which has been so long and so widely felt, and the traces of which are not yet extinct in Europe, is in all probability of Chaldæan origin. It seems to have been unknown in Greece before the conquests of Alexander; and, indeed, it is but justice to the astronomers of that country, to say that in general no traces of belief in astrology are to be found in their writings. It is true that there exists an astrological treatise attributed to Ptolemy, but it may be doubted whether it is really one of his productions. His great and very complete work the μεγαλη Συνταξις contains nothing which could lead us to believe that he was infected with these ideas. At Rome, under Augustus, and the succeeding emperors, the city was inundated with Syrians*, Chaldæans, and other natives of the East; and numerous passages in ancient authors show that a belief in judicial astrology was extremely prevalent. From this time, down to the seventeenth century, the whole of the civilized world in the West as well as the East was enslaved by this childish superstition, against which philosophy and religion seem to have combated in vain. It is humiliating to know that some of the great restorers of astronomy in Europe, were not superior to the follies of their age; and that even the bold and acute Kepler could dispute on the best manner of drawing up a scheme of nativity. Our posterity will, perhaps, be equally astonished to think that in the nineteenth century, in the country of Newton and Bradley, almanacks in general circulation should contain predictions founded on the aspect of the planets, and all the mummery of an art, the existence of which is a disgrace to any nation possessing claims to civilization, or even common sense. It is difficult to imagine that there can still be persons who believe that the distances between Jupiter and Saturn can influence the fate of empires, or that on the sign of the ecliptic which is just rising, at the moment of a man's birth, depend the principal events of his life†. But vain and futile as is this pretended art, it has exercised too great an influence on the cultivation of astronomy to be passed over here without an allusion. There can be no doubt that an attachment to astrology was at least one of the reasons that induced the Arabs to study with so much ardour the phenomena of the heavens; and it is quite certain that the encouragement given at the revival of letters by many European princes to astronomy, was entirely owing to their wish to read the future by means of the stars.

But to return to the Romans. Their calendar, in the time of the republic, fell as has been already noticed, into great confusion. This, Julius Cæsar‡, who

* Cato Major de Senectute, c. 49. V. Also Pliny, ii., 12.
† V. Liv. Hist., xliv., c. 37.
‡ Thucydid., vii., c. 50.

* Jampridem Syrus in Tiberim defluxit Orontes. —Juv. Sat. iii.
† This sign was called the horoscope; a term familiar to most persons, though perhaps at this day few are acquainted with its exact meaning.
‡ It appears from Pliny, vii., 25, that Cæsar was himself an observer, and had composed some

was a man of extraordinary knowledge and universal genius, endeavoured to correct, with the assistance of a Greek astronomer, named Sosigenes. At this time, Hipparchus had already shown that the length of the solar year was something less than 365 days and a quarter, and Sosigenes could not have been ignorant of this fact: but, in all probability, he considered the difference too small to be worth taking into account. Be the reasons what they may, he proposed a method of intercalation which, to be quite rigorous, supposes the solar year to be 365 days and a quarter; namely, he proposed to make the civil year of 365 days, and every fourth year to insert an additional day. The year in which this intercalation took place was called bissextile, because the additional day inserted, bore the designation of *bis sexto calendas Martii;* the day named *sext. calend. Mart.* (corresponding to our 28th of February) being repeated. It is easy to see that this intercalation, and in general the Roman form of year is the same as that now in use; but it has been found necessary to introduce some modifications, the nature of which will be explained in a subsequent part of this treatise. But without going further at present, it may be noticed that the tropical year, not having the value supposed by this method, but being in fact, at the time of Cæsar, about ten minutes shorter, at the end of about a century and a half, the beginning of the civil year would be a day behind that of the astronomical; and this difference would, of course, go on increasing. We shall see that this was in reality what took place, and that ultimately the error became pretty considerable.

The great division of the Roman empire under the sons of Theodosius, probably exercised considerable influence on the fate of letters and science. The Greek language, which for a long time had been familiar to the Romans, ceased to be cultivated in the West; and this circumstance is of some importance in the history of astronomy; for we have seen that this science was almost completely neglected in Italy: those who wished to cultivate it were necessarily obliged to have recourse to Greek authors. But, during the middle ages, the little learning that existed in Western Europe was confined entirely to a knowledge of the Latin language: the works of Ptolemy were a sealed book for the few learned men (if they may deserve the name) of those days. However, the conquests of the Arabs having brought them into immediate contact with the Gothic nations in Spain, they began to impart to their neighbours some of their zeal for science; and the works of Euclid and Ptolemy became known in Europe, through the means of Arabic translations. Gerbert, a man of great talent and very superior to the age in which he lived, found himself obliged to attend the Moorish universities in Spain, to acquire some knowledge of mathematics and astronomy, which it was impossible at that time to study in any Christian country. On his return to France, he composed several treatises, which show a knowledge of geometry remarkable for that time, and a considerable familiarity with the works of Euclid and Archimedes.

At a later period the example of Gerbert was imitated by an English monk named Adhelard, who for the sake of acquiring scientific knowledge, travelled in Spain and Egypt, and having become acquainted with Arabic, turned it to account, by translating from that language into Latin the Elements of Euclid. This appears to be the first translation of Euclid executed in the west; but it seems to have been little known, and has remained in manuscript till the present day. The translation of Euclid by Campanus, who lived about a century later, was the first that acquired any popularity; and all the early editions were printed from his text, which was like that of Adelhard, a translation from the Arabic, a language apparently known to several of the literati of that day, while they still remained in profound ignorance of Greek.

In the thirteenth century many symptoms of a reviving love for letters and science began to show themselves; and knowledge which had been hitherto confined in monasteries, now spread into cities and courts. The barbarous monarchs of the middle ages were succeeded by princes who cultivated and protected letters; and though it is to be feared that the encouragement given by some of them to astronomy, was founded on a superstitious belief in the influence of the stars, we may view with some indulgence a weakness which led to such beneficial effects. The emperor Frederick the Second, a prince distinguished as well by a very cultivated mind, as a

works on astronomical subjects which are now lost.

generous protection of learned men, deserves particular mention here, as to his encouragement we owe the first translation into Latin of the Almagest of Ptolemy. This translation it is to be noticed was made from the Arabic, (Greek being still unknown in the west,) and consequently was a good deal disfigured; the Mahometan writers by no means piquing themselves on a scrupulous adherence to the text of the authors they translate: but still the service thus rendered was important; and no doubt had a favourable influence on the progress of science. The writer* who had executed the translation of the Almagest, added a translation of the commentary of Geber on the same work, and of the treatise of Alhazen on twilight. But in this latter respect he seems to have been anticipated by Vitellion, a Pole, whose voluminous treatise on optics is little more than a translation of Alhazen.

To Alphonso X., king of Castile, astronomy owes still more than to Frederick; and his reign will always form a memorable epoch in the annals of that science. His situation in a country bordering on the Arabs, who then occupied the south of the Spanish Peninsula, was very favourable for collecting about him able astronomers, who were then only to be found at the Moorish universities. Of this he profited to draw to his court a number of learned men, whom he employed for four years in constructing new and complete tables; those of Ptolemy having become in the lapse of time quite insufficient. These tables, generally called the Alphonsine, cost the prince an immense sum; and if they fell short of the degree of perfection that might have been expected, it was not from any want of munificence and zeal. Their principal defect was the introduction of an inequality in the motion of the fixed stars in longitude, by which this motion appeared to be sometimes accelerated and sometimes retarded: as the equinoctial point was supposed to describe the circumference of a small circle, the centre of which moved along the ecliptic according to the ordinary law of precession. This pretended inequality, known by the name of the trepidation of the fixed stars, or the motion of the eighth sphere, has been already noticed in speaking of the Arabs. It seems to have originated in Greece; but it was certainly not admitted by Ptolemy, or the most judicious of the oriental astronomers. The introduction of it into the Alphonsine tables seems owing to the Jews, who had a large share in their formation; at least this may be conjectured from the numbers 7000 and 49,000 years, in which the small circle above-mentioned and the ecliptic were respectively described. These are cabalistic numbers for which the Jews from fanciful ideas felt great veneration; and, indeed, the astronomers of that nation seem to have been singularly attached to the doctrine of trepidation, since the time of Thebit-ben-Corah, unjustly accused of being the inventor of the system. But it reflects little credit on the Alphonsine astronomers, to have admitted an inequality founded on no observations, and rejected by every author of eminence.

It seems but reasonable to suppose that Alphonso, who protected so zealously astronomy, was himself versed in the science. Of this we have no direct proof, but there is on record a saying of his, which has been accused of impiety, though it would be fairer to regard it as an expression of the disgust caused to a sound judgment, by the complication of the Ptolemaic system. "Had the Deity," said Alphonso, "consulted me at the creation of the universe, I could have given him some good advice." If this exclamation may be justly blamed as irreverent in its expression, it certainly conveys a condemnation of the theories of his time, and the monstrous combination of—

Cycle and epicycle, orb on orb,

then generally received as the system of the world.

On comparing the Alphonsine tables with those of Ptolemy, we see how little progress astronomy had made in eleven centuries. Some ameliorations in the elements of the syntaxis, are perhaps more than counterbalanced by the introduction of the imaginary inequality called trepidation; which continued to disfigure the best tables as late as the time of the celebrated Copernicus. Nor to judge from the slowness with which physical science recovered from its long torpor, could any one have guessed at the rapid progress it was about to make in the sixteenth and subsequent centuries. The real restoration of astronomy in Europe can scarcely be placed earlier than two hundred years subsequent to the publication of the Alphonsine tables. At this time Pur-

* Gerard of Cremona.

bach, professor of astronomy at Vienna, undertook to ameliorate the hypotheses and tables then existing. He felt the necessity of beginning by making an accurate translation of Ptolemy, the want of which was very sensibly felt, though in addition to numerous translations from the Arabic, George of Trebizond had published one from the original Greek. This, however, was very faulty; nor was it in the power of Purbach to remedy, the defect, as he was ignorant both of the Greek and Arabic languages. His anxiety to acquire the former of these, induced him to accept an invitation made by Cardinal Bessarion, to visit Rome; but death prevented the accomplishment of this design, and the translation of Ptolemy was left to be performed by his pupil Regiomontanus.

Purbach's theory of the planets is in all essential points the same as that of Ptolemy; but he has unfortunately introduced into his tables the imaginary trepidation of the fixed stars we have just spoken of. The most considerable service that he has rendered to science, was by calculating a table of trigonometrical sines, from ten to ten minutes throughout the quadrant, for a radius of 600,000 parts. Ptolemy had employed a sexagesimal division of the radius, which was singularly inconvenient for arithmetical computations; the decimal division introduced by Purbach was an important improvement.

The design entertained by Purbach of making a new translation of Ptolemy's great work, was executed by his pupil and successor John Müller of Königsberg, commonly called Regiomontanus*. He added to this a commentary, containing a number of problems, likely to be useful in astronomical calculations: and made very considerable ameliorations in the solution of plane and spherical triangles. Regiomontanus enjoyed for a long time the reputation of having been the first to introduce tangents into trigonometry. But we have seen that this had been previously done by the Arabs: however, he was certainly the first in Europe to calculate a table of tangents, which he did for every degree of the quadrant. Nor is this the only benefit of the kind that we owe him: he extended the table of sines calculated by Purbach for every ten minutes, to every minute of the quadrant, for a radius of 1,000,000 parts; which was a second important improvement added to that of his predecessor. Indeed, the decimal division of the radius has been found so convenient, that it remains unaltered to the present day. It is, perhaps, to be regretted that he did not extend the decimal division to the quadrantal arc itself: such a reform, which, however useful, it has been found impossible to effectuate in the present day, might have been practicable in the infancy of astronomy. The great reputation of Regiomontanus induced Pope Sixtus IV. to request his assistance in the reformation of the calendar, an operation of which we shall say more shortly. For this purpose he proceeded to Rome, but had not been long engaged in the prosecution of this important work, when he was carried off either by an epidemic distemper, or as some have said, by poison, administered by the sons of George of Trebizond, to avenge the criticisms on the translation of Ptolemy, executed by their father.

The labours of Purbach and Regiomontanus form the link between modern astronomy and that of the middle ages. The end of the fifteenth century was not distinguished by any great discoveries; yet it is easy to see in the rise of a spirit of inquiry and investigation, the dawn of that light which was about to illuminate Europe with so much brilliancy. No doubt the progress of this spirit was at first slow and uncertain; but it may fairly be traced back as far as those who felt the necessity of establishing their theories upon observation alone, and who aspired to something beyond commenting Ptolemy. Among those who contributed most by assiduous observation was Bernard Walther, a rich citizen of Nuremberg; one of the earliest and certainly of the most zealous of modern astronomers. His observations, interesting on many accounts, are particularly remarkable for having been made with clocks regulated by wheels; which seem to have answered tolerably well for the division of time. We have seen what were the various and unsatisfactory contrivances of the Greeks and Arabs to accomplish this important point. Walther was one of the first, it appears, of the moderns who recognized the existence of refraction; of which it is true he formed a very incomplete idea; supposing it to exist only near the horizon. Alhazen and Vitellion had both treated of

* Regiomontanus means 'of Königsberg,' or 'King's Hill.'

this subject, but Walther affirms that his discovery was made previously to becoming acquainted with the works of either of these philosophers.

Chapter IX.

Copernicus.—Tycho.

The limits of this treatise will not allow us to notice various meritorious astronomers of the early part of the sixteenth century, whose names are now little known, and whose writings can offer little interest. We shall proceed at once to a work which was destined to change for ever the face of astronomical science—the Revolutions of Copernicus*. Struck with the complication of the Ptolemaic theory, and the weakness of the arguments by which it was supported, this great man had passed nearly forty years of his life in meditating on the true system of the world. He was led irresistibly to the conclusion that the system taught in Greece by the Pythagoreans, of the earth's motion on its own axis, and round the sun, was the only one consistent with the observed phenomena and the simplicity of nature. But to make this very interesting and important question fully understood, it will be better to refer to the opinions of preceding philosophers on the subject.

From a very early age in Greece, it had generally been recognized that the earth was of a spherical figure, and upon this point there was not, and indeed there scarcely could be, any difference of opinion. But this fact once conceded, the question naturally arose, was it suspended motionless in the universe, the centre of the heavenly motions, or did it of necessity, as some argued from the supposed impossibility of its remaining unsupported, revolve round another body? It is singular enough, that the philosophers who first taught in Greece the rudiments of science, generally advocated what we now know unquestionably to be the real system of the world, the revolution of the earth round the sun. This was taught by the followers of Pythagoras, and by those of Thales: and particularly by Philolaus and Anaximander. If we consider how contrary this theory apparently is to the evidence of the senses, and how unlikely to be one of the first truths discovered in the infancy of science,—if we consider the prediction of an eclipse by Thales (a most extraordinary fact for this early age), his Phœnician origin, the travels of Pythagoras in the east,—if we compare these circumstances with the doctrines of Aryabhatta in India,—little doubt will remain that the motion of the earth was borrowed, with other truths, from one of the oriental nations. Unfortunately we know not much of the arguments by which these doctrines were supported in Greece; though we possess in sufficient detail the nugatory reasons urged against them by Aristotle, and in later times by Ptolemy*.

However, it was not difficult for Copernicus to conceive all the reasons which led the Pythagoreans to the doctrine of the earth's motion; and these he has expounded with singular judgment. It would be too much to expect him to be always superior to the erroneous physical ideas of his age; but his argument is in general equally distinguished by sound sense and moderation. He begins by remarking that if we suppose the distance from the earth to the fixed stars to be infinitely great, compared with its distance from the centre of the universe; but, on the contrary, this latter distance to be very considerable when compared with the orbits of the planets; all the phenomena may be just as well explained by supposing the earth to revolve on its axis from west to east in twenty-four hours, and to have besides this a motion of translation in the heavens; as by supposing the earth to be immoveable while the fixed stars and planets revolve round it in their different spheres†. That the earth itself was a point compared with the distance of the heavens (meaning the fixed stars) was a point conceded on all sides; but as Copernicus very well remarks, it by no means follows from this that the earth is at rest in the centre of the universe: on the contrary, it seems the more extraordinary that such a vast circumference should revolve in twenty-four hours, rather than this infinitesimally small part of it, the earth‡.

He then proceeds to consider the reasons urged by ancient philosophers against the earth's motion. The first argument he combats is a very futile and fanciful one urged by Aristotle. The earth according to him was the heaviest of the elements, and all heavy bodies tended to

* A native of Thorn in Polish Prussia, was born in the year 1473; died in 1543.

* Plutarch (Quæst. Plat.) tells us that the system of the earth's motion, which was proposed as an hypothesis by Aristarchus, was proved by Seleucus; but of the nature of this proof we are in ignorance.

† Revolutionum, lib. i. cap. 5.

‡ Revol. lib. i. cap. 6.

its centre. The latter part of this affirmation we should have in these days little difficulty in admitting, but the inference deduced was rather extraordinary. "So much the more then," said Aristotle, "will the whole earth rest in the centre, and that which receives all heavy bodies falling on it, remain immoveable by its own weight." To this he adds an argument still more fanciful. "All simple motion must be rectilinear or circular; to a centre, from a centre, or round a centre. It suits earth and water, which are heavy bodies, to tend downwards; air, and fire, which are light, to rise upwards: it seems reasonable to give these four elements rectilinear motion, but to the celestial bodies, a circular motion." It must be confessed that the objections urged by Ptolemy, though sufficiently trivial, were a little more rational. This latter author objected to the diurnal revolution of the earth, that from its extreme rapidity it would overcome the force of gravity, and everything on the earth's surface be scattered and dissipated into space. The reply of Copernicus is not completely satisfactory; he might have said that such effects would not necessarily take place, unless the velocity of rotation were sufficiently great to counteract the force of gravity: but he replied, in a style too consonant to that of his adversaries, that the motion was natural and not violent; that natural motions have not the same effects as violent ones; the latter tending to dissolution, the former to conservation*. He adds, however, much more reasonably, that if Ptolemy's argument be worth anything, it will apply with still greater force to the celestial sphere, which must revolve with a velocity infinitely greater, and consequently be exposed in an infinitely greater degree to this dispersion. "Why, then, do we hesitate," he exclaims, "to give to the earth the mobility suitable to its form, rather than that the universe, whose bounds we do not and cannot know, should revolve? why should we not confess that the diurnal revolution is apparent only in the heavens and real in the earth? Thus Eneas in Virgil exclaims—

Provehimur portu, terræque urbesque recedunt.

Since while the ship glides tranquilly along, all external objects appear to the sailors to move in proportion as their vessel moves, and they alone and what is with them, seem to be at rest." It were to be wished that Copernicus had always contented himself with reasoning as soundly; but we have seen that he frequently combats the Aristotelians with arguments frivolous and futile as their own objections. The most illustrious sages have shown that on some weak point they were as fallible as their brethren, and Copernicus has not avoided paying this tribute to mortal nature. But we must not class with such errors his speculations on the existence of several centres of gravity in the universe. The Aristotelians, observing that heavy bodies on the earth's surface tended to its centre, hastily concluded that this point was the centre of gravity of the universe. But this, Copernicus remarks, is very doubtful. Gravity, according to him, is nothing but the tendency of parts to draw together and coalesce in the form of a globe*. "Now it is probable that such a tendency exists in the sun, moon, and other heavenly bodies; but this does not hinder them from describing their respective orbits. If, then, the earth have other motions, these must be the same as we appear to observe in other bodies; and if we change the solar orbit into a terrestrial one, the risings and settings of the signs and fixed stars, in the evening and morning, will appear the same: the retrogradations, and precessions of the planets will be no longer their real motions, but appearances borrowed from that of the earth: the sun, lastly, will be in the centre of the universe, as the order in which these phenomena succeed each other, and the harmony of the whole world, sufficiently show."

After having combated the opinions of preceding philosophers, with regard to the immobility of the earth, Copernicus proceeds to explain his own system; which placed the sun in the centre of the world, the planets revolving round it in the following order, beginning with the nearest: Mercury, Venus, the Earth, Mars, Jupiter, and Saturn. The Moon revolved in a circle, which had the Earth for its centre, and consequently participated in the annual motion of that body. As the Copernican system is now explained in all treatises on astronomy, we shall not enter into details respecting it, but shall merely notice one or two circumstances connected with it, which are not so generally known. Ever since the time of Aristotle it had been

* Revol. i. 8.

* Revol. i. 9.

received that all the heavenly motions were circular. This doctrine was founded on some very false metaphysical notions, about the excellence and incorruptibility of circular motion, but it led astray Ptolemy, and, we must add with regret, Copernicus; indeed, the latter pushed these ideas so far as to blame Ptolemy for having admitted, for the minor planets, a motion which was not uniform round the centre of the circle, but round a point at a certain distance from it. His reasoning on this subject is completely Aristotelian. "It is impossible that a single celestial body can move unequally in one orbit; for that must happen, either through the inconstancy of the moving power, whether it be extraneous, or belonging to its intimate nature; or through a disparity in the body revolving. But both of these suppositions are repugnant to our understandings*." However, it was necessary to have recourse to some hypothesis for explaining the evident excentricity of the planetary orbits. Copernicus employed the ancient hypothesis of an epicycle for this purpose; and this, perhaps, was the best that could be adopted before the discovery of the real form of these orbits by Kepler. But he had the advantage, in his system, of being obliged to introduce epicycles to account for the real inequalities only of the planets, while Ptolemy was compelled to combine with these numerous others, to explain their stations and retrogradations.

In speaking of Hipparchus, we have noticed the discovery he made of an apparent retrogradation of the equinoctial points. Copernicus pointed out that this phenomenon was the effect of a libration of the earth's axis, which did not remain parallel to itself, but had a slow retrograde conical motion, the cone in question having its vertex at the earth's centre. He seems to have imagined that the motion of translation would derange the parallelism of the earth's axis, and that it was necessary to give this latter a retrograde conical motion of the nature described, and in quantity such as nearly to counteract the effect of the libration from the annual motion. But he did not suppose this to be exactly the case: the retrograde motion of the axis was made to surpass a little the other, and this excess was supposed to produce the phenomena of precession. In this respect the system of Copernicus was unnecessarily complicated: there is nothing in the motion round the sun to derange the earth's axis, which always remains very nearly parallel to itself: consequently, the two counteracting motions of Copernicus should be suppressed. Delambre, Astron. Moderne, vol. i., p. 95, affirms that Kepler was the first to point out the propriety of this suppression; but the fact is that it was most clearly indicated by Rothmann, astronomer to the Landgrave of Hesse, before Kepler. In a letter to Tycho Brahé, he remarks "there is no occasion for the triple motion of the earth; the annual and diurnal motions suffice. . . . . . The axis of the earth is so carried round in its annual motion, that it always remains pointed in a parallel direction to the same part of the universe; and on account of the evanescence of the terrestrial orbit, compared with the immensity of the sphere, it always remains directed exactly to the same point." This letter, printed in Tycho's Epistolæ, lib. i., p. 184, is dated in the year 1590. Kepler's earliest work was printed 1596; his Epitome of the Copernican Astronomy, in 1618.

Copernicus has also fallen into the mistake of admitting an inequality in the precession of the equinoxes, analogous to that already spoken of under the name of trepidation, which existed only in the imaginations of certain authors of the middle ages; having been passed over in silence by Ptolemy, and distinctly rejected by El-Batani.

The illustrious author of the Revolutions was well aware that his system of the world, as well from its novelty, as from the intellectual monopoly then exercised by the followers of Aristotle, was likely to meet with great opposition; and he seems to have been anxious to present it in a form as little offensive as possible. But he does not appear to have anticipated the outcry that would be made against him upon what were called religious grounds. To such objections he alludes briefly and contemptuously; and it is somewhat singular that he not only dedicates his work to Pope Paul the Third, but mentions that he was principally induced to publish it by the persuasions of his friends, Schonberg, Cardinal of Capua, and Gisias, Bishop of Culm. Apparently, these prelates suspected as little as himself that any charge of impiety could be extracted from an astronomical theory.

* Revol. i. 4.

The doctrines of Copernicus found, at first, few partizans; but these were all men of great scientific merit, and among the first astronomers of their day. Such were Rhæticus, who has written a commentary on the Revolutions, and to whom we owe the valuable Opus Palatinum *, and Erasmus Rheinold, author of the Prutenic Tables, which may be considered as an amelioration of those of Copernicus, and which enjoyed for some time considerable reputation. The commentary of Rhæticus informs us of a curious fact; that it was the observation of the orbit of Mars, and of the very great difference between his apparent diameters at different times, which first led Copernicus to embrace the system of the Pythagoreans; we shall see that the same planet led Kepler to the discovery of one of the most important facts connected with our system. Rheinold is remarkable, as having taught that the orbit of Mercury was elliptic; and in his theory of the moon he made her epicycle to revolve on an elliptic orbit, thereby partially anticipating the great discovery of Kepler, to which we have just alluded. To Rheinold and Rhæticus we may add the names of Rothmann, astronomer to the Landgrave of Hesse, and Mæstlin, the instructor of Kepler. The Landgrave of Hesse is a remarkable instance of a sovereign prince animated with such an ardour for science, that for many years he devoted himself to assiduous observation, and produced a catalogue of the fixed stars, which was enabled to bear a comparison with that of Tycho Brahé. He was assisted by Rothmann, just mentioned, and Justus Byrgius, a mathematician of considerable eminence, and well versed in the construction of astronomical instruments. Mæstlin, one of the few partisans in those days of the system of Copernicus, is known, not merely as the preceptor of the illustrious Kepler, but as having been the first to explain the real cause of the light seen on that part of the moon's disk which is not directly illuminated by the sun. The opinion previously entertained upon that subject had been, that this was produced by a proper light belonging to the moon itself †; but Mæstlin attributed it, and with reason, to the reflection of the solar light from the illuminated part of the earth's surface. This explanation is so simple and so natural, that it has been universally admitted by astronomers ever since. It is said, however, that the celebrated painter, Lionardo da Vinci, had made this remark before Mæstlin *.

The middle of the sixteenth century was rendered memorable by the publication of the immortal work of Copernicus; the close of it was adorned by the labours of Tycho Brahé †. The former had passed his life in meditation on the sublimest truths of astronomy, the latter devoted his time and fortune to diligent observation of the heavens. He was rewarded by a number of brilliant discoveries, which have secured for him a fame equal to that of his most distinguished predecessors. Fortunately for science, he found a protector in Frederic, king of Denmark, who granted him the island of Huene in the Baltic, and assisted him in building a splendid observatory, furnished with instruments superior to any that had yet been constructed. Here he continued for twenty years; and in that time collected a mass of observations, which were of the greatest use to succeeding astronomers, as well as to himself, in the reformation of the sciences. But, at the death of Frederic, the enemies of Tycho induced the minister Walchendorp, to withdraw the donations of the late king, and the assistance he had been in the habit of receiving. These, and other circumstances, induced the illustrious astronomer to withdraw in disgust into Germany, where he met with Kepler, who, of all men in Europe at that time, was perhaps the best able to make a good use of his extensive and valuable observations.

We have seen that the Greeks, in comparing the position of the fixed stars with that of the sun, made use of the moon; they determined in the day-time the distance of these two bodies, and at night the difference of position between the moon and a given fixed star. The rapid proper motion of the moon made this method very inaccurate: Tycho improved it materially by substituting for the moon the planet Venus ‡, whose proper motion is at once much smaller and more uniform, and from its great brightness is frequently visible for some

* An extremely extensive table of sines, tangents, &c.

† Others said that it was produced by the light of Venus. V. Life of Galilee, p. 34.

* Life of Galileo, p. 33.

† Born in 1546 at Knudstrup in Denmark—died in 1601, at Prague in Bohemia.

‡ It appears, however, that the Landgrave had sometimes used Venus in a similar way: we have no means of judging to which the priority is due.

time while the sun is yet above the horizon. Having once determined in this way the places of a few principal fixed stars, the others were referred to them by measuring their angular distances to two fixed stars, one to the east, and one to the west, and by determining their meridian altitudes. It was easy from these data, to calculate the right ascensions; and, as the meridian observation gave the declination, to determine the longitudes and latitudes. It appears that Walther was the author of the method of determining a star's place by observing its distances to other known fixed stars. These contrivances would have been unnecessary had astronomers possessed any means of measuring time accurately. The Greeks used for this purpose water-clocks, which were necessarily very imperfect: Tycho had substituted for water, mercury, and this not answering, he had made several attempts, as Walther and the Landgrave of Hesse had done before him, to measure time by means of clocks moved with wheels, but had not been able to construct any which gave satisfaction. This difficulty was not surmounted before the time of Huyghens. Tycho was, however, enabled to form a catalogue of the fixed stars, surpassing considerably in accuracy those of Ptolemy and the Arabs. Indeed, he flattered himself that the errors never exceeded a minute; but in this he seems to pretend to a greater degree of accuracy than his instruments were susceptible of.

Hitherto astronomers had always determined the latitude of the place, where they made their observations, by observing the zenith distances of the sun, at the summer and winter solstices. Half the sum of these quantities was the latitude required. Tycho invented another method, much more convenient, as the latitude could by it be found in twelve hours instead of six months, during which it was necessary to wait in the old method. He observed the zenith distances of some circumpolar star on the meridian when above and below the pole: half the sum of these was the colatitude of the place. On comparing the latitude thus found with that determined by observations of the solstices, he found a difference of four minutes, which he rightly imputed to the effects of refraction. This led him to form from observation a table of refractions, which though necessarily imperfect secures for him the glory of having been the first who introduced this important correction. He made the horizontal refraction 34′, which agrees very well with modern determinations; but he was mistaken in supposing that refraction does not exist above 45° of altitude. This mistake in observation led him to another in theory. In the beginning he had formed a just conception of the causes of refraction; he attributed it to a difference in density between the atmosphere and the ethereal matter which he supposed to pervade the planetary regions. On this subject he had a warm discussion with Rothmann, who attributed refraction to vapours arising from the earth, and rendering more dense the lower regions of the atmosphere*. The argument of Rothmann against the theory of the Danish astronomer was indeed irrefragable, had the facts been as admitted on both sides. If refraction were owing to the causes assigned by Tycho, he contended that it would extend to the zenith; and though Tycho at one time admitted that this might be so†, he seems ultimately to have been convinced of the contrary, as in the Progymnasmata ‡ he adopts the explanation of Rothmann.

The year 1572 is memorable in the annals of astronomy for the appearance of a new star of extraordinary brilliancy in the constellation Cassiopeia. It appeared on a sudden with a light greater than that of any of the fixed stars, or even Jupiter, and nearly equal to that of Venus when brightest. But it did not long shine with this degree of splendour: it was first seen by Tycho on the 11th of November, and by the month of January, its light was less than that of Jupiter; in February and March it was comparable to the fixed stars of the first magnitude; during April and May to those of the second, and so it went on diminishing till it finally disappeared in March 1574. A phenomenon so extraordinary could not fail to fix the attention of all astronomers. It furnished Tycho Brahé with matter for a considerable treatise, in which he has compared and discussed the various observations made on it in different parts of Europe. From all of these it evidently resulted that the star in question had no sensible parallax, and consequently was infinitely beyond the planetary regions. It has been said that this phenomenon, rare as it is, was not altogether unprecedented in

* V. Epist. Astron. p. 81.
† Epist. Astron. p. 108.
‡ De Stellâ. novâ. p. 91.

the history of the world. Pliny narrates that a similar circumstance suggested to Hipparchus the idea of forming a catalogue of the fixed stars. The fact is certainly possible; but if it be true, it seems singular that the star should not be mentioned in the catalogue of that astronomer. Some authors have quoted on this subject the tradition with regard to the constellation of the Pleiades; which are said to have been originally seven in number, whereas six only are usually perceived by the naked eye; but this, even if well established, would merely prove a diminution in brightness of one of the stars forming that constellation.

Tycho was too good an astronomer to admit the pretended inequality in the precession of the equinoxes, which had been introduced into the tables of Alphonso and Copernicus. This invention of a barbarous age was now finally discarded; nor would it have stood so long, were it not for the undeserved confidence some astronomers placed in the position of the fixed stars as given by Ptolemy. But Tycho and his correspondent Rothmann were satisfied that Ptolemy had only borrowed the catalogue of Hipparchus, reducing it to his own times, by the value he assigned to Precession *; an opinion which has generally been adopted by succeeding philosophers. In comparing the positions given by Hipparchus, with those of his own catalogue of the fixed stars, Tycho Brahé was led to the important discovery that the inclination of the earth's equator to the plane of the ecliptic is not constant, but subject to a very slow diminution. As the diminution in question is very gradual, the existence of it was for a long time disputed; but modern observations have established it beyond dispute, and La Grange has shown that it is a consequence of the theory of universal gravitation.

We have seen that Ptolemy attributed to the moon two inequalities, the one depending on the excentricity of her orbit, the other on the position of the line of the apsides with regard to that of the syzigies. Tycho discovered a third, called the variation: this is greatest in the octants, that is to say, at 45° of elongation from the sun (at which time it amounts to more than 40′), while the evection discovered by Ptolemy is greatest in the quadratures. This discovery was an important improvement in the lunar theory; it is not the only one that we owe to the same illustrious author. Hipparchus had shown that the lunar orbit was inclined to the ecliptic at an angle of nearly 5°; but Tycho proved that this inclination is not constant, it varies nearly 20′; it is at its maximum when the moon is in quadratures, and at its minimum when in syzigies*. The third discovery of Tycho on this subject was that the motion of the lunar nodes, is not, as had been supposed, uniform, but variable, at one time appearing to advance, at another to retrograde.

It has been shown that the Chaldæans, according to Apollonius Myndius, believed comets to be bodies of the same nature as the planets; and this opinion is warmly embraced by Seneca. But it never seems to have enjoyed much favour among astronomers before the latter part of the fifteenth century. The best philosophers entertained the most inadequate ideas upon this point, believing them to be not merely sublunary bodies, but even within the terrestrial atmosphere; and to these false notions they added others still more absurd about the generation of these bodies in the upper regions of the air. Tycho was the first to overthrow these prejudices; and his labours on this point form one of his most solid titles to glory. Having observed carefully the comet of the year 1577 through the whole visible part of its orbit, he established beyond a doubt, that it had no sensible parallax; whence he deduced two conclusions of great importance, and quite fatal to the established theories on the subject. The first, that comets move far beyond the orbit of the moon; the second that the heavens are not formed, as was then supposed, of solid transparent spheres, since they are traversed by comets in every direction. The Aristotelians were reluctant to concede two points so opposed to their doctrines; and Tycho was violently attacked upon this subject by a Scotchman named Craig; but his adversaries, totally unable to meet his reasons, had recourse to personalities, which could not shake the facts laid down by the Danish astronomer. The debate was renewed in the time of Kepler, but the question was decided in the minds of all really scientific men by the discussion of Tycho.

It being once established that the cometary orbits were of a magnitude comparable to those of the planets, the

* Epist. Astron. p. 87.

* The inequality of the inclination depends also on the position of the moon's nodes; but this Tycho does not seem to have perceived.

question naturally suggested itself, what was the nature of these orbits? "Mæstlin, a zealous Copernican, supposed comets to revolve in a circle round the sun; and he explained the inequalities of their motion, by introducing an epicycle on which he made them move; but Tycho who did not admit the earth's motion round the sun, proposed to explain their irregularities by combining their own circular motion round the sun, with the revolution round the earth that he attributed to that body.

It is melancholy, after relating so many brilliant discoveries, to have to record that Tycho rejected the Copernican system of the world. Whether he was influenced by a wrong interpretation of some passages of Scripture, or the desire to attach his name to a new theory of the universe, or was really persuaded that the arrangement proposed by Copernicus was physically untenable, it is not easy to decide. It is certain that he generally speaks of this great astronomer, in terms, not of respect merely, but admiration: and he refutes himself the arguments urged against the diurnal motion of the earth by Ptolemy and others, remarking that the atmosphere would in this case revolve with the earth, nor would those absurd consequences result which had been supposed*. Yet led by other reasons of no greater weight, he rejected this diurnal motion: and placed, with Aristotle and Ptolemy, the earth at rest in the centre of the universe. He supposed the sun to revolve round the earth; but his system differed from that of Ptolemy in this, that he made the five minor planets to revolve round the former, which carried them along with it in its annual course round the earth. It is not to be denied that, mathematically speaking, this system satisfies the phenomena observed; but it is so immeasurably inferior to that of Copernicus in simplicity, that it appears very extraordinary, it should have been imagined after the other was known. Holding, as the system of Tycho does, an intermediate place between those of Ptolemy and Copernicus, one would have expected that it should have been the first attempt to simplify the complication of the former. Indeed we may easily pass from the first to the second of these, by making, for the inferior planets, the deferent coincide with the annual orbit of the sun, and the epicycle with the planet's proper orbit; and in the case of the superior planets, the converse.

It cannot but be interesting to know the reasons which the Danish philosopher urged in defence of his own theory against the adherents of Copernicus. We shall not notice the arguments drawn from the Scriptures, because it is now generally admitted, as indeed was asserted by the Copernicans of those days, that we are not to look in them for strictness of scientific expression on such subjects, since they naturally use only such language as would be intelligible to those to whom they were addressed*. It can scarcely be supposed that the Pope, to whom Copernicus dedicated his work, or the prelates who exhorted him to publish it, imagined that his doctrines contained anything contrary to the Scriptures. It was in later times that the cry of impiety was set up; and it was pretty evident that this clamour arose from the offended pride and prejudices of the Aristotelians, determined to punish as heresy what they could not refute as false philosophy. Tycho Brahé, indeed, was not an Aristotelian, and it is probable that he was sincere when he quoted certain phrases used by Moses, as hostile to the opinion of the earth's motion; but it is lamentable to see such a man leading the way in an opposition of this kind. He would probably have repented of it, could he have foreseen the consequences to which it led in the case of Galileo.

The great argument used by this eminent man against the earth's diurnal motion was this: if a stone be suffered to fall from the top of a high tower, it would not, as we see it does, fall at the foot of that tower; the earth's velocity of rotation being so great, that during the few seconds it took to fall, the tower itself would have passed through an arc of several hundred feet, and the stone be left far behind ere it touched the ground. To this it was answered by his correspondent Rothmann, that every body on the earth's surface partakes of the earth's motion†; that consequently the motion of a falling body is compounded of a rectilinear and circular motion; the former tending to the centre of the earth; the latter in the circumference of the circle described by the point from which it falls; and if the velocity with which this point revolves be so great, what then must be

* Epist. Astron. p. 74.

* V. Rothmann in Epist. Astron. p. 130.

† Epist. Astron. p. 185.

that of the sphere of the fixed stars, if we, taking the other side, suppose the earth to be at rest.

The arguments of Tycho against the annual motion of the earth were much more weighty: the most important of them was drawn from the fact of the fixed stars having no annual parallax. If, as Copernicus teaches, the earth revolves round the sun in an orbit nearly circular, her places, at intervals of nearly six months, will be distant from each other by the whole diameter of the orbit. If then we suppose lines drawn from two places diametrically opposite to the nearest fixed star, these two lines will form an angle (called the annual parallax), which must be appreciable, unless the distance of the star is so great, that, compared with it, the diameter of the earth's orbit is insensible. But this diameter is in itself of such immense length, that, according to Tycho, the supposition just mentioned would be preposterous; and as it is established by observation that the parallax of the orbit is insensible, we must conclude that the earth does not move round the sun. This is the most specious of all the arguments urged against the Copernican system. We have now incontestable proofs of the earth's motion, and we know, astonishing as the fact may appear, that the distance of the fixed stars is infinitely great when compared, not merely with the diameter of the earth itself, but even with that of its orbit. But it is not surprising that in earlier ages men should have been reluctant to admit such a conclusion; however as there is no absurdity in it, they were not justified on this single ground in rejecting a system so simple and beautiful as that of Copernicus. Tycho Brahé has urged another argument, ingenious enough though founded on a mistake in facts. The imperfection of the instruments and methods of observations used in those times, led him to ascribe a sensible, though small, apparent diameter to some of the fixed stars. If then, he argued, the fixed stars be as distant as Copernicus supposes, in order to subtend a visible angle, the diameters, even of stars of the third magnitude, must be greater than that of the annual orbit. This reason, even if true, would not be conclusive; as we are in absolute ignorance of the real magnitudes of the fixed stars; nor can we take upon us to affirm that they may not extend even beyond the limits here mentioned: but the fact is, that, seen in the best telescopes, they have no appreciable diameter, appearing simply as highly luminous points.

Chapter X.

Reformation of the Calendar—Kepler—Laws of the Planetary Orbits.

When Julius Cæsar, and his adviser Sosigenes, determined that in the Roman calendar every fourth year should be bissextile, they seem to have supposed the year to have been composed of exactly 365 days, 6 hours. Yet Hipparchus had proved, a century before, that this value was about five minutes too great; and even his determination was considerably in excess, as the real length was not more than 365 days 5 hours, 48 minutes, 45 seconds; and consequently the excess of the Julian above the tropical year amounted to more than 11 minutes. At the end of a century the difference had accumulated to more than 18 hours, and at the end of fifteen centuries to nearly 11 days, by which the seasons had moved from the places in which they were originally fixed. Thus the vernal equinox, which at the institution of the calendar by Julius Cæsar, fell on the 21st of March, had retrograded to the 10th: and if no steps had been taken to correct this, in the course of time, spring would have commenced in December, summer in March, autumn in June, and winter in September. In process of time, the equinox, having passed successively through all the intermediate months, would have returned again to that of March. But before this took place a great many centuries would have elapsed; and even supposing the derangement to have been much more rapid, it may be questioned whether it would have caused any practical inconvenience. The ancient Egyptians knew that the solar year comprised about 365 days and a quarter, yet they made their civil year of 365 days only; the consequence of which was that each month corresponded in succession to different seasons, a complete restitution being effected in about 1461 years. It is probable that the principal motive which induced many well-informed men in the sixteenth century to urge a reformation of the calendar, was a desire to fix in a more correct way the day on which the festival of Easter ought to be celebrated. This celebration having been connected with the equinox by a decree of the council of Nice, it became of importance to the Church, to fix definitively the place of the equinox in the calendar. A proof that religious and not civil considerations led to the reform may be found in the extent of the changes effected. It may be said that a certain degree of inconvenience would result to the public, from having a moveable instead of a fixed year; and though, in the Julian system, the anticipation of the seasons is so slow that the inconvenience must be nearly inappreciable; yet there could have been no objection to fixing permanently the different seasons, as it might have been effected without difficulty by adopting a different intercalation for the future. But on the other hand, the time at which the year shall be made to begin is entirely arbitrary, and in practice a matter of perfect indifference. In the age of Cæsar, the year began a few days after the winter solstice, and the vernal equinox fell on the 21st of March. In the age of Pope Gregory XIII., this equinox fell on the 10th: here it might have been fixed for the future without any inconvenience; but the Pope and his astronomers took the very unnecessary step of suppressing altogether eleven days in the year 1582, in order to bring the equinox to the 21st. This uncalled for measure had the inconvenience of introducing into Europe two styles, or modes of reckoning dates, as the new calendar was for a long time rejected by the Protestant states of Europe, and to this day has not been received in the empire of Russia. In the north of Germany it was not admitted till the year 1699, nor in England till 1751; 169 years after its publication by Gregory at Rome.

The equinox being once brought to the 21st of March, the object of those who effected the reform was to keep it as nearly as possible to that day by a proper system of intercalation, and to effect this the Julian calendar was modified in the following manner. It was arranged that, for the future, in the year concluding every century, which ought in the Julian system to be bissextile, the intercalary day should be suppressed; but that this should only be done for each three successive centuries, and not in the fourth. Thus the years 1700, 1800, 1900, are not bissextile, but the year 2000 is so: again, in 2100, 2200, 2300, the intercalary day is suppressed, but re-established in 2400. La Place has remarked, that to give this intercalation all the accuracy of which it is susceptible, it would be necessary, at the end of 4000 years, to render the secular year common instead of bissextile: that is to say, in this interval to intercalate only 969 instead of 970 times*.

It is a singular fact, that the Persians have been for several centuries in possession of a calendar constructed on much more scientific principles, than Europe, with her superior knowledge, can boast of. It has been stated by La Place, Montucla, and Bailly, that the Persian intercalation consisted in inserting eight days in thirty-three years. This, if true, would at once be a much more accurate and simple method than the Gregorian; but the fact is, that the Persians combine two periods, each of considerable accuracy, the one erring a little in excess, the other in defect†. The first period is one of twenty-nine years, in which they intercalate seven days: this is followed by four successive periods of thirty-three years, in each of which they intercalate eight times: forming a whole period of 161 years, which includes thirty-nine intercalary days. To show the extreme accuracy of this method, it is only necessary to remark, that it supposes the length of the year to be $365^d\ 5^h\ 48^m\ 49^s\ 1875$, the real length being $365^d\ 5^h\ 48^m\ 49^s.7$, the difference is less than a second; while in the Gregorian calendar it amounts to more than twenty-one seconds. The first year of the Persian æra began with the vernal equinox (A.D. 1070); the astronomers of that country have very wisely avoided subjecting themselves to the unnecessary and embarrassing condition, that the equinox should always coincide with the first day of the year. However in their system it never can be far from it, while in the Gregorian, the real equinox which ought to fall on the 21st of March may sometimes fall on the 19th. There

---

* In order to judge of the accuracy of the Gregorian intercalation, we must recollect that the length of the tropical year is 365.2422414 days: resolving the decimal part into converging fractions, we get, successively—

$$\frac{1}{4},\ \frac{7}{29},\ \frac{8}{33},\ \frac{39}{161},\ \left(\frac{47}{194}\right),\ \left(\frac{86}{355}\right),\ \left(\frac{125}{516}\right),$$

The Gregorian calendar intercalates 97 days in 400 years: the fraction $\frac{97}{400}$ does not appear among the convergents, and consequently the calculation is not as accurate as it might be: it supposes the length of the year equal to 365d. 2425, or too great by 0.0002586, that is, rather more than twenty-one seconds: this excess in 4000 years would amount to little more than a day.

† This was first made known in a note from Sédillot to Delambre, printed in the Astron. Moderne, vol. i. p. 81.

can then be little doubt that the Persian system is the most elegant and scientific of any that has hitherto been used: the principal objection to it is that the intercalations cannot follow a law so simple as those in the Gregorian Calendar; on the other hand it surpasses this latter in accuracy; as it does that adopted during the revolution in France, by being freed from the extreme complication, consequent on making the beginning of the year invariably coincide with the equinox*.

When Tycho Brahé retired in disgust from his native country to Bohemia, a fortunate chance caused him to fall in with a young man of a genius the most opposite to his own, but also perhaps, on that very account, the best qualified to deduce novel and important truths from the observations he had been so long accumulating. Kepler†, at a very early age, had distinguished himself by a work on the distances of the planets from the sun, full, it must be confessed, of fanciful and erroneous views, but still, in the opinion of Tycho, bearing the stamp of genius. The objects of this singular treatise, called the Mysterium Cosmographicum, was to point out a supposed relation between the magnitudes of the orbits of the five principal planets, and the five regular solids of geometry. His theory with regard to these was the following:—"Round the orbit of the earth circumscribe a dodecahedron,—the circle comprising it will be that of Mars. Round Mars circumscribe a tetrahedron, the circle comprising it will be that of Jupiter. Round Jupiter circumscribe a cube—the circle comprising it will be that of Saturn: now within the earth inscribe an icosahedron, —the inscribed circle will be that of Venus: in Venus inscribe an octahedron —the circle inscribed in it will be that of Mercury‡." It is scarcely necessary to observe that these proportions are altogether fanciful, and indeed incompatible with the very laws discovered subsequently by Kepler himself, but they are interesting, as these speculations, apparently so chimerical, led to one of the most brilliant discoveries ever made in astronomical science. It does credit to the sagacity of Tycho, himself eminently a practical man, and little disposed to admire such subtleties, that he seems to have perceived at once all the force of the genius thus misdirected; and did not rest till he succeeded in fixing the young author near his own person. Indeed the protection he extended to Kepler is the more remarkable, as the latter had been from his youth an ardent partisan of the Copernican theory, which Tycho pertinaciously, though vainly, endeavoured to supplant. His obstinacy on this point was carried so far, that he is said to have requested of Kepler on his death bed, whatever might be his real sentiments, to adopt in his published works the Tychonian system of the world. It was not to be expected that such an injunction should be complied with; and the person to whom the dying astronomer addressed this extraordinary request, was of all men least likely to sacrifice truth and reason at the shrine of authority.

It had been universally admitted, from Aristotle down to Copernicus and Tycho, that the orbits of the planets were circular, or rather formed by a combination of circles, and a variety of weak and vague reasons were adduced to account for the supposed fact. However, the hypothesis was natural, and its deviations from the truth were not much greater than the errors of the ancient observations: its inadequacy could not be detected, till a mass of more accurate observations had been collected, and then submitted to discussion by a philosopher of quick conception and unfettered judgment. These conditions were eminently satisfied in Kepler, and the observations of Tycho supplied him with the most valuable data. It was then found that the planets, supposed by Copernicus to revolve round the sun in circles, move in ellipses, of which that body occupies one of the foci; and their motion is such, that straight lines being drawn from the focus in question to any two points of the orbit, the area thus intercepted is proportional to the time employed by the planet in passing from one of these points to the other. To these two most important

* The two fractions $\frac{7}{29}$ and $\frac{8}{33}$ are very near approximate values of the part of a day, by which the year exceeds 365. The former of these is a little too small, the latter rather too large; if then we make these two periods alternate, the errors being in contrary directions will, to a certain point, compensate each other; however, as $\frac{8}{33}$ lies nearer to the real value of the year than $\frac{7}{29}$, the former period must occur oftener than the latter to make the compensation exact. We have seen in the text, that one of the periods $\frac{7}{29}$, was followed by four of $\frac{8}{33}$ making on the whole $\frac{7\;4\times 8}{29+4\times 33}=\frac{39}{161}$. See the note to page 48.

† Born at Wiel in the duchy of Würtemberg, 1571, died at Ratisbon in 1630.

‡ Myst. Cosmograph. p. 10. Frankfort, 1621.

theorems a third was subsequently added, that the squares of the times of revolution of two planets are to each other as the cubes of the greater axes of their respective ellipses. These are the three laws which have immortalized the name of Kepler, and effected a revolution in astronomy. Beautiful and important as they are in themselves, it is impossible to appreciate their full value, without a knowledge of the sublime theory founded on them by Newton; nor can we appreciate, as we ought, the genius of their inventor, if we forget the immense strength of the prejudices that he dared to break through. Perhaps his contemporaries alone could properly estimate the courage required to discard the circular motions, which even that great reformer, Copernicus, considered solely admissible in the heavens; and to introduce squares and cubes into the proportions of revolution and distance: but they could not foresee the results to which these discoveries have led; and it may safely be affirmed that it is only since the publication of the Principia, that they have met with their deserved tribute of admiration. It is remarkable enough, that, whether from this cause, or from the fanciful speculations with which truth in Kepler's works is so often accompanied and obscured, his distinguished contemporary Galileo does not seem to have perceived the importance of the three famous laws; though his own knowledge qualified him better than any one to appreciate them justly, and they would have furnished him with some strong arguments in favour of the Copernican system.

The way in which Kepler was led to these great discoveries is one of the most interesting objects of study in the whole range of physical science. Supposing the planetary orbits circular, it is generally possible, by a proper combination of epicycles, to represent the various inequalities of their motions; but it is extremely difficult, in this hypothesis, to satisfy, at once, the variations of velocity and those of distance. Thus we have seen that Ptolemy, in representing the evection of the moon, was led to a theory which would give the variations of distance considerably greater than they really are; a fact which he seems to have overlooked, but which could not long escape the notice of astronomers. Mars, whose apparent diameter varies from 4″ to 18″, and whose extreme distances from the earth are ½ and 2½ parts of the mean radius of the ecliptic, was one of the heavenly bodies, whose motion and distances were most difficult of conciliation. Tycho had already remarked that the annual orbit of Copernicus, or the epicycle of Ptolemy, was not always of the same magnitude with regard to the excentric; but that it produced a sensible alteration in the three superior planets, and that for Mars, the difference amounted to 1° 45′. This, however, is exaggerated, as the difference in question does not exceed 38′; but the statement of Tycho, combined with other observations, led Kepler to the first step in the progress of his discoveries. We have seen that Ptolemy had been induced to bisect the excentricity of the orbits of the planets, by placing the centre of equal motion as far from the centre of equal distances in one direction, as the earth was from it in that opposite. Kepler, who, following Copernicus, considered the earth itself as a planet, was led to extend this system to the terrestrial orbit, and by this he was enabled to represent better than had hitherto been done, what may be called the optical part of these inequalities: that is to say, the part depending on the motion being seen from a point not coincident with the centre of the circular orbit.

The next step was to represent the real inequality of the planet's motion more accurately than had hitherto been done by epicycles and excentrics. Kepler began by supposing the times occupied by the earth in equal parts of the excentric to be to each other in the proportion of the distances of those parts; and hence that the periodic time was to the sum of these distances as any given time to the sum of the distances to the corresponding arc*. This principle is only approximately true, and the calculation of the sum of the distances was involved in much difficulty. While meditating on the best method of surmounting these, Kepler was led to substitute, for the sum of the distances, the areas comprised between two positions of the radius vector; and he thus arrived at the first of his great laws, that the radius vector of a planet describes areas proportional to the times. The reasoning by which he was led to this substi-

* De Mot. Stellæ Martis. p. 192. Kepler was led to this hypothesis from observing that the law in question might be deduced as a consequence of the Ptolemaic theory when the planet was near its apogee or perigee. V. Op. Cit. p. 165.

tution of the arcs for the areas is curious enough: let AB be the line of the

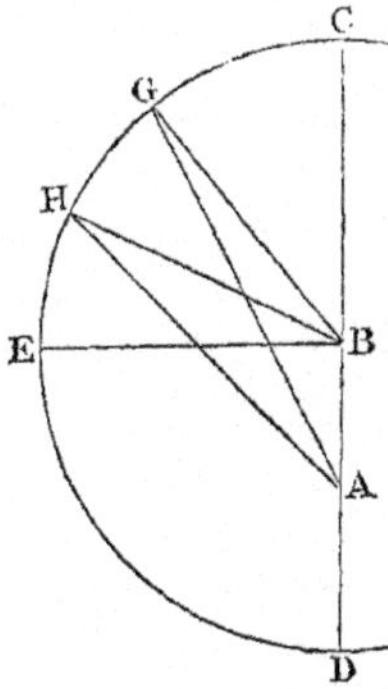

apsides, A the sun; B the centre of the excentric CD; let the semicircle CD be divided into any number of equal parts at G, H;.... join BG, BH,.... AG, AH:.... the areas of the sectors CBG, GBH.... will be equal, and their sum forms the area of the semicircle, as the sum of the arcs forms the semicircumference; the total area will then be to the partial area as the total arc to the partial arc. We may then substitute the areas for the arcs in question, which represent the degrees of excentric anomaly. But, argues Kepler, this same surface, which is the sum of all the equal radii BC, BG, BH, is also the sum of the distances AC, AG, AH, and therefore the sum of these may be substituted for those of the circular radii, or the areas proportional to them. Hence, then, the area CED is to the time of describing 180°, as any area CAH, to the time of describing CH*. Kepler must have been aware that the reasoning he has followed was altogether unsatisfactory, though the theorem at which he arrived was accurate, with two small exceptions, which he has himself remarked; the first, that the orbit of the planet is not, as here supposed, circular; the second, that the plane of the orbit is that in which this proportionality of the areas takes place, and not the plane of the ecliptic. This last source of error would, however, be very small in the case of Mars, as the inclination of the two planes just mentioned is only 1° 51′ 6″†.

Kepler had already determined, with sufficient accuracy, the mean distance, the excentricity, and the longitude of the apogee of Mars; on calculating, then, from these data, and from the law of the areas which he had found for the earth, the place of the planet, he was surprised to find it differ from the place given by observation by about eight minutes. The accuracy of the observations of Tycho with which this comparison was made was such as to leave no doubt that the error lay on the side of the theory, and Kepler seems now to have suspected, for the first time, that the orbit was not circular*. This supposition was soon verified, for on calculating three places, one about the aphelion, the others near the mean distances, they were all found too great; the first by 350, the two latter by 783 and 789 parts†. It was evident then, that the orbit was not circular but oval; and the author expressly remarks, that when he uses the term oval, he does not mean elliptical in the proper sense of the word, but resembling an egg in form, by having a greater curvature at one end of the axis major than the other. This, indeed, the observations seemed to indicate; the diminution of the mean distances proved the curve to be flattened in the middle, and that of the aphelion distance seemed to indicate a similar flattening, though in a less degree, towards the point‡.

The extreme candour with which Kepler details the steps by which he was led to his discoveries, and the errors into which he fell before arriving at the truth, render his work on the motions of Mars eminently interesting and instructive. It were to be wished that, in this respect, his example had been followed by some of the great philosophers who succeeded him. He confesses that for a long time he remained in the conviction that the orbit of Mars was really oval, and not elliptical; and he has detailed various attempts to deduce its positions in this hypothesis. But the calculations involved considerable difficulties, and though for some time Kepler flattered

* De Mot. Stellæ Martis, p. 193.

† In the figure above given, the area CAH will measure the time, or what astronomers call the mean anomaly; the area CBH will measure the excentric anomaly; and the area BHA will represent the excess of the former above the latter.

* The ellipse of Mars being much more excentric than that of the earth, the error of the circular orbit, which was inappreciable in the latter, became sensible in the former.

† The mean radius of the orbit being 100000. V. Op. Cit. p. 213.

‡ The notion of an oval was not altogether new; it had been thrown out by Purbach in the case of Mercury, whose path he considered to be oval, and more flattened at the apogee than the perigee. V. Purbach, Theoric. nov. de Mercurio, Coroll. sext.

himself that he had succeeded in representing pretty accurately the motions of Mars, he was not long in finding that his attempts had in reality completely failed. The terms in which he announces this discovery are a curious sample of his very peculiar and amusing style. "While in this way I triumph over the motions of Mars, and form for him, as one completely conquered, the prison of tables, and the chains of the equations of the excentric, intelligence is brought me from different places that the victory is all vain, and the war as fierce as ever. For the captive enemy has broken through all the bonds of the equations, and the dungeons of the tables.... And the fugitive would have effected his junction with the rebels, and driven me to despair, had I not suddenly brought up a new reserve of physical considerations, the first having been beaten and scattered*." These new physical considerations consisted in attributing to the planet a libration in the diameter of its epicycle; a supposition which, in reality, coincided with that of an elliptic orbit, though at first he did not perceive the identity of the two hypotheses. Indeed he himself compares Nature to the Galatea of Virgil, who, the nearer she is approached, is the more petulant in her sportive attempts to escape; and while she invites the philosopher to seize her, seems to enjoy his errors and fruitless efforts†. But at last perceiving clearly the real nature of the curve described, he announced, formally, that it was an ellipse. Thus was finally overthrown the prejudice regarding circular motions, which had hitherto retarded the progress of astronomy, and the way opened to the great physical discoveries of Newton, and his successors.

It is worth while to remark that Kepler was, in a great measure, led to the discovery of the ellipse by his perseverance in attempting to represent as well the variations of distance as those of longitude and latitude; the former of which had been much neglected by Ptolemy, and subsequent astronomers. The discovery was ultimately and immediately due to a singular accident. The error of his calculated distance at the mean point of the orbit was .00429, the radius being equal to unity. Happening to consider the greatest optical inequality of Mars, which was 5° 18′, he found .00429 to be exactly the excess of the secant of this angle above the radius*.

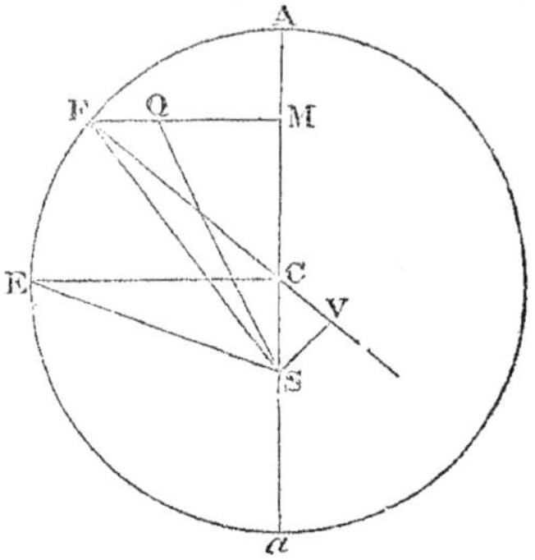

In the figure let A *a* be the line of the apsides, C the centre of the excentric, S the place of the sun, CE perpendicular to A *a*, the angle CES will be the greatest optical inequality. The excess of the calculated over the real distance was that of SE over CE, or of the secant of CES over the radius. This single accidental remark induced Kepler to substitute every where for any distance SF, FV found by drawing SV perpendicular on FC produced, since clearly the excess of SF above FV, will be that of the secant of optical equation above the radius. The next point was to know in what direction from S the line equal to FV should be drawn; and, finally, after many trials, he found that it should be taken so as to meet FM perpendicular to A *a*, as at Q, and he had the gratification to find that the curve traced by the several points Q was an ellipse.

The third great law of Kepler, that the squares of the periodic times are as the cubes of the mean distances, was first enounced by him in the Epitome of the Copernican Astronomy†. In the 20th chapter of the Mysterium Cosmographicum, he had attempted to investigate, from physical considerations, the proportion that existed between the distances and the periodic times; for of the existence of some law he made no doubt; and he quotes Aristotle in support of his opinion‡. The idea of the latter on this point was, that an equality of motion resided in the moving bodies themselves, the inequality of their returns to the same place depended on the orbits: thus, that the velocity of Saturn in its orbit was the same as that

* De Mot. Stellæ Martis, p. 246.
† V. de Stell. Mart., p. 283.
Malo me Galatea petit, lasciva puella,
Et fugit ad salices, et se cupit ante videri.—
Virg. Eclog.

* P. 167. † Lib. iv. c. 4.
‡ De Cœlo, lib. ii. c. 10.

of the moon, but the orbit being so much larger, it was much longer in returning to the same point of the heavens. Kepler, perceiving the inaccuracy of this principle, began to investigate the subject in a different manner. He supposed the sun to be a centre of attractive force, which diminished as the distance from that body increased: again, the magnitude of the orbit increased as the distance: from these two causes combined, he concluded, that the periodic time varied inversely as the square of the distances. But in the Epitome he modified his theory, by introducing the consideration of the mass and of the volume of the planet. His reasoning on this head is very curious, "There are four causes on which the length of the periodic time depends:—1, the length of the path described; 2. the mass of the body to be transported; 3. the strength of the moving force; 4. the volume under a given mass of the body to be moved." According to Kepler there is an exact compensation between these two last causes, so that the orbit in reality only depends upon the two first. Now, he conceived that he had established a law showing that the masses of the planets varied as the square roots of the distances: and the circular paths of the planets were certainly in the simple proportion of these distances; compounding, then, these two ratios, he found the squares of the periodic times to vary as the cubes of the distances.

It is equally extraordinary and discouraging to know that reasoning so vague, founded on arbitrary and erroneous assumptions, should have led to one of the most brilliant discoveries on record. This success may well inspire with dismay those who are accustomed to consider experiment and rigorous induction as the only means to interrogate nature with success. But it is to be remarked, that Kepler united to a very lively and ardent imagination considerable mathematical knowledge, and great perseverance in calculation; that his hypotheses, though founded on the most arbitrary assumptions, were always rigorously compared with observation, and rejected without hesitation when found to contradict it. His genius, at once fertile and persevering, led him ultimately, with great labour and after many unsuccessful trials, to brilliant results: but the route which he followed is not likely to be again attempted by his successors.

It has been a matter of just surprise, that a man of such acute conceptions should have suffered the analogy between the planetary and cometary orbits to escape him: and even opposed the system of Tycho, who had supposed the latter to move nearly in a circle round the earth. Kepler, misled by false ideas with regard to the physical constitution of these bodies, which he considered to be of very transitory nature and short duration, attempted to calculate their orbits on the supposition of their being rectilinear; and in favour of this last assertion, he urged a reason which turned out to be singularly inapplicable,—namely, that these bodies never returned: it so happened, that the very comet whose orbit he was considering, was the famous comet of Halley, which is well known to have visited the earth repeatedly.

An extraordinary phenomenon, which had occurred in the time of Tycho Brahé, was renewed in that of Kepler. In the year 1604 there appeared suddenly in the constellation of Serpentarius, a star of the first magnitude, which, after having lasted upwards of a year, disappeared, and has never been since seen. When brightest, it surpassed the fixed stars of the first magnitude, and was even compared by some to Venus: the intensity of its light diminished gradually to its extinction; the colour, which was at first yellow, became gradually purple, and then red: in all these respects, and particularly in having no sensible parallax, as well as in its sudden apparition and slow extinction, it presented a great analogy with the star of the year 1572. Of the causes of such extraordinary phenomena we remain in profound ignorance, nor is it easy to decide upon the relative probability of different suppositions: the idea of Kepler was, that they proceeded from a vast combustion, and this appears to derive some corroboration from the variations in the colour.

We cannot quit Kepler without noticing his labours to improve the Theory of Astronomical Refractions. We have seen the mistake of Tycho Brahé upon this subject, who, with a right notion as to the physical causes of the phenomenon, imagined that refraction did not exist beyond 45° of altitude. Kepler clearly saw that it must reach to the zenith; and he attempted to determine, from experiment, the law of its diminution from the horizon upwards. The law at which he arrived gave him results approximating very nearly to the truth, except for low altitudes*. On the causes of refraction he reasons very justly: and his treatise on this subject contains a number of curious and important remarks, of which we shall only quote here the observation that the air is a substance possessing weight†; that the refractions will vary with the temperatures, and that during a lunar eclipse, the moon's surface is still illuminated by a certain portion of rays refracted by the atmosphere of the earth ‡.

CHAPTER XI.

Galileo—Copernican System.

THE discoveries of the celebrated Galileo § were nearly contemporaneous with those of Kepler; though perhaps of less real importance, they were of a nature to be more generally appreciated, and contributed unquestionably in an eminent degree to the overthrow of the Aristotelian philosophy, and the definitive triumph of the Copernican system of the world. In judging of the character of Galileo, it is fair to recollect that his astronomical discoveries form but a small part of his claims to distinction; but as it is with these alone that we have anything to do in this place, we must refer the reader for a complete account of the labours of this eminent philosopher to the Life of Galileo, in the series of these Treatises. At an early period of life he had commenced his scientific career by some important mechanical inventions, and held the mathematical chair successively in two of the most distinguished universities of Italy—Pisa and Padua. It appears, that, while resident at the latter place, he heard of the discovery made in Holland of an instrument by which distant objects were apparently brought nearer to the eye. Some enemies of Galileo have pretended that he had actually seen one of the Dutch instruments before constructing any of his own, but this assertion is not sufficiently established; and there seems no good reason to doubt the account given by the philosopher himself, who tells us, that having heard of the effects of the telescopes made in Holland, but knowing nothing of their construction, he was led to meditate upon the combination of glasses required, and thus in reality re-invented in Italy, and unquestionably in a more perfect form, the instrument to which some Dutch spectacle-makers had been led by accident. There can be no doubt that the instruments of Galileo were very much superior to those made in the north: and indeed there is some reason for believing that they were constructed on principles widely different: it would appear that the latter were composed of two convex lenses distant by the sum of their focal lengths (a combination similar to that of the modern astronomical telescope); while there is no doubt that the telescope of Galileo, like our opera glasses, was formed by a plano-convex and plano-concave lens (the latter being nearest to the eye) placed at a distance from each other equal to the difference of the focal lengths*.—The former of these combinations represents objects in an inverted position, which is not the case with the latter: some accounts represent the early Dutch telescopes as inverting the objects seen through them, which fact, if sufficiently well established, would decide the question as to the originality of Galileo's invention †.

The use of single lenses, in magnifying objects placed near them, had been known for a long time. Spectacles had been in use for three centuries previous to the age of Galileo; but the magnifying distant objects by a combination of lenses seems to have been entirely the result of accident. It is in vain that some authors have attempted to trace this invention to Baptista Porta and Roger Bacon: there can be no question that the first telescope ever made was constructed at Middleburg, in Holland, by an optician, who having accidentally regarded a distant object through two lenses at some distance from each other, was struck with the magnifying effect produced. It is equally certain, that Galileo immediately began to construct telescopes of a quality much superior, and, as we have seen to be very pro-

---

* The law of Kepler was $r = \frac{\text{constant } (z+r,)}{\cos. (z+r)}$ when r is the refraction and z the zenith distance; had he put in the numerator, sin. $(z+r)$ for $(z+r)$, he would have got

$$r = \text{constant} \frac{\sin. (z+r)}{\cos. (z+r)} = \text{constant } \tan. (z+r),$$

which is the real law. But the error of his formula was only 26″ at 80° zenith distance. V. Delamb. Astron. Moderne, vol. i.

† Paralipom. ad Vitellion, p. 128.

‡ P. 271.

§ Born at Pisa in 1564, died at Arcetri in Tuscany in 1642.

* Nuncius Sidereus, p. 11, edition of 1682.

† V. Life of Galileo, p. 23.

bable, with a different combination of lenses. He has always been considered by posterity as one inventor of the telescope; and there is no ground to disturb his title to that honour *.

Such an instrument, in the hands of such a man, was not likely to remain long useless. One of the first objects to which it was directed was the planet Jupiter, which, to the astonishment of the observer, appeared to be accompanied by four small planetary bodies or satellites revolving round him, as our moon does round the earth †. The smallness of these bodies renders them invisible to the naked eye; but with a telescope of moderate power, they are very easily discernible. It is an extraordinary proof of the extent to which prejudice may be carried, that in a matter which admitted of being decided by a direct appeal to the senses, numbers of men should have been found who persisted in contesting the existence of these secondary planets, because it contradicted the received ideas with regard to the number of bodies in our system, and furnished an additional probability in favour of the doctrines of Copernicus. Some declared that they had examined Jupiter carefully with the telescope without noticing any such appearances as Galileo described, and attributed his observations to an optical illusion. Others went so far as to say that the telescope represented terrestrial objects correctly, but could not be relied upon for the heavens; while the climax of bigotry was exemplified in a professor of Padua, who refused to look at Jupiter through Galileo's instrument, lest he should be constrained to acknowledge what he had predetermined never to admit. Kepler, who had supposed the number of the planets to correspond to the regular solids of geometry, had the candour to admit, without dispute, facts that overturned his theory; and, with a pliancy of genius and fertility of invention very characteristic of him, immediately formed a new system, supported by reasons just as good as those he had alleged for his first. However, the observations of Galileo were so easily susceptible of verification, that they ultimately extorted consent from the most reluctant; and a contest of another kind arose with some who pretended to a prior, or, at least, a contemporaneous discovery of them. Simon Mayer, more usually called Marius, in a work entitled the *Mundus Jovialis*, lays claim to having perceived them about the same time as Galileo; but, even admitting the statement there given, the latter appears to have the priority *.

Ptolemy had proposed to employ lunar eclipses for determining the differences of terrestrial longitude. The physical instant at which the moon enters the earth's shadow being the same for all places, the difference of the times counted by each observer when this phenomenon takes places, gives directly the difference of longitudes. But in practice, this method is subject to a considerable defect, arising from the difficulty of determining with accuracy the instant of the beginning or end of the eclipse. The cone of shadow projected by the earth is surrounded by a penumbra of very sensible magnitude, which causes the satellite to lose a good deal of light before and after the real eclipse, and renders the phenomenon more gradual than is consistent with the purpose to which Ptolemy wished to apply it. The satellites of Jupiter suffer eclipses exactly analogous to those of our moon, and much more frequent, as not only is the nnmber of these secondaries greater, but their revolutions round their primary much more rapid. The synodic revolution of the first satellite is only about forty-two hours; and that of the fourth, which is much the slowest, does not exceed seventeen days. The rapid motion of these small bodies, and the frequent recurrence of their eclipses, render them much more suited for the determination of the longitude than lunar eclipses; and Galileo did not fail to point out this application of his discovery. In every method for this purpose, it is evidently necessary that the observer should know exactly the time of the observation counted on his own meridian. Galileo first suggested the application of the pendulum to the measurement of time, for the purpose just mentioned; but it must be confessed that his instruments were too rude to be of any practical utility.

Jupiter was not the only planet which, on being examined with the telescope,

* For an interesting discussion on the invention of the telescope, see Journal of the Royal Institution, No. II. p. 319.

† This was announced in a small pamphlet by Galileo, called the Nuncius Sidereus.

* According to Marius, his first observation was made on the 8th of January, 1610,—the first of Galileo on the 7th. He only quotes one observation in his whole book, which was not published till two years after the Sidereus Nuncius of Galileo.—Cf. Delamb. Astron. Moderne, vol. i. pp. 695, 702, 703.

presented new and unlooked-for appearances. Saturn seemed, to Galileo, whose instrument, it must be recollected, was of a very inferior description to those constructed in more recent times, to present some very extraordinary and unaccountable phenomena. At first this planet seemed accompanied by a satellite on each side; but on a more narrow examination, it appeared that these appendages were connected with the body of the planet, thus giving it an oblong or flattened form, compared by the discoverer himself to that of an olive *. But following assiduously the appearances of the planet for some years, he was astonished to see the appendages mentioned totally disappear, remain invisible for some time, and then again slowly resume their former shape. From the imperfection of his telescope, he was at a loss how to explain these singular phenomena, the real nature of which were first ascertained by Huyghens.

On contemplating Venus, it was not without much gratification that the illustrious explorer of the heavens observed her to present phases similar to those of the moon; thereby establishing a fact which was strongly corroborative of the theory of Copernicus, and utterly subversive of the Aristotelian notions on the subject, as it proved that she shone merely by the reflection of light from the sun. Phases of a similar kind, but less extent, were also remarked in Mars. It is scarcely necessary to observe, that the more distant the planet, the less sensible these appearances become; it is not therefore surprising that they should be inappreciable for Jupiter and Saturn. Before Copernicus, it had been generally admitted that all the planets shone by light of their own. The observation of the phases of Venus and Mars, and that subsequently made by Horrox of the transit of Venus over the sun's disc, were a complete refutation of this groundless notion.

It seemed as if every discovery of Galileo were destined to be fatal to some favourite idea of the Peripatetics, and to afford some new confirmation of the analogy existing between the earth and planets. The moon's surface was found by him to resemble in structure that of the earth; to be covered with inequalities similar to our mountains and valleys, and even exceeding them in magnitude. The sun itself, so far from possessing that perfection which many were willing to ascribe to the least of the heavenly bodies, was seen by Galileo to be covered in certain places with dark spots of irregular form and variable magnitude. They were of short and uncertain duration, but lasted long enough to establish the important fact that the sun revolved on an axis inclined to the ecliptic, the period of his rotation being nearly a lunar month. This rotation of the sun had been before conjectured by Kepler, from some fanciful ideas as to the physical causes of the planetary revolutions, and without any attempt to support the notion by actual observation. It is to be remarked, that Kepler had, without being aware of it, actually observed a spot on the sun's disc, which he mistook for a transit of Mercury, and announced as such to the world. It is much to his credit that, on hearing Galileo's discovery, he readily confessed his error. A German Jesuit, named Scheiner, who had observed these solar spots, advocated the idea of their being small planets revolving round the sun *. This led him into a controversy with Galileo, who established beyond doubt that they were really adherent to the sun's disc, and participated in the common motion of rotation. It is only within the last few years that it has been discovered that Galileo had been anticipated in his discovery of the solar spots by the celebrated English mathematician Harriott †. The observations of the latter have never been published; but it appears, from what has been made known, that he first observed them on the 8th December, 1610, while the discovery of Galileo dates from March, 1611. It is hardly necessary to observe, that the observations of Harriott could not have been known to Galileo; indeed they have remained buried in profound obscurity up to the early part of the present century, when they were found among the manuscripts of Harriott by a distinguished foreign astronomer, then on a visit to this country. The reputation of the great Italian cannot be affected by such

* See a letter of Galileo, inserted by Kepler in the introduction to his Dioptrics.

* Not content with having thrown out this idea, which he was subsequently obliged to abandon, Scheiner disputed with Galileo the discovery of the solar spots. Some controversy ensued, in which Galileo had decidedly the advantage.—See the Life of Galileo, chap. x.: and Delamb. Ast. Mod. vol. i. p. 681; Scheiner's Rosa Ursina, Bracciano, 1626-30. Galileo de Maculis Solaribus tres Epistolæ.

† Edinburgh Philosophical Journal, vol. vi p. 317.

a discovery; his observations, if not the first, were probably the most exact, and certainly the most ably followed up. From them he deduced the fact of the rotation of the sun, and determined the period in which it was accomplished to be nearly a lunar month,—phenomena of greater importance than the mere knowledge of the existence of the maculæ on its surface.

The name of Galileo is indissolubly associated with the Copernican system of the world. The zeal and talent with which he advocated this system were decisive of its final triumph, and the persecutions he underwent in this cause have endeared his memory to posterity. While yet a very young man, and a student at the University of Pisa, he became a convert to the doctrines of Copernicus, which were then gaining ground among the astronomers of the north. Of these doctrines, which his own discoveries so beautifully corroborated, he was through life a most ardent and enlightened advocate; and through his influence and exertions they began to spread rapidly throughout Italy and all Europe. The Peripatetics now seriously took the alarm: while these opinions were confined to a few scientific men, they had looked on with indifference; but when they became popular, they felt that the empire of Aristotle was shaken to its foundations. Overwhelmed by the facts and reasonings of their adversaries, to which, in general, they were able to oppose nothing but the dicta of their great master, they had recourse to religion, and declared the opinion of the earth's motion to be not only unphilosophical, but heretical! It was not difficult for the Copernicans to expose the absurdity of this assertion; but their enemies had the ear of the ecclesiastical authorities, and a decree was procured from the Court of Rome in the year 1616, prohibiting the Revolutions of Copernicus, the Epitome of Kepler, and other works of a similar tendency: at the same time Galileo received an intimation to desist for the future from teaching and advocating the theory contained in them. But, notwithstanding the evident danger to be incurred by violating such a prohibition, the truth was too dear to Galileo to be abandoned without a struggle; and after the lapse of some years, he published his famous Dialogues on the Ptolemaic and Copernican systems, a work which gave the last fatal blow to the adherents of Aristotle. Dreading, however, the censures of the Church, he thought it necessary to use some precautions in the publication. Without affirming anything in his own person, he makes the interlocutors on each side adduce the best arguments they could urge: it may easily be conjectured which is victorious, but this Galileo leaves to the sagacity of the reader to perceive. In his preface he even pretends that he composed the work for the sake of satisfying the ultramontane philosophers, that the condemnation of Copernicus at Rome had not its origin in ignorance, but was pronounced by those who had well weighed the arguments on either side, and were competent to decide. But this veil of irony was too flimsy not to be seen through at once; nor could it protect him from the vengeance of the Inquisition, which was prompt and severe. He was summoned to Rome, brought before the tribunal of the Holy Office, condemned to imprisonment during pleasure, and forced to sign a solemn abjuration of the doctrine of the earth's motion; a doctrine declared by the inquisitors to be false in philosophy and heretical in religion. The imprisonment was subsequently commuted for relegation to his villa at Arcetri, where he was compelled to pass the remainder of his life in strict seclusion. Indeed, there is every reason to suppose, that, had it not been for the protection of the Grand Duke of Tuscany, he would have been more hardly dealt with, and in all likelihood never quitted the dungeons of the Inquisition *. It is difficult to conceive what the inquisitors could have expected to gain from a recantation extorted by intimidation—perhaps even by actual torture. "All Europe," says La Place, "was revolted at the sight of an old † man, rendered illustrious by a long life dedicated to the study of Nature, forced to abjure on his knees, and against the testimony of his conscience, the truths he had incontestably proved." In spite of decrees and prohibitions, the reasonings of Galileo were generally felt to be of overwhelming force, and the court of Rome shook its own authority much more than the Copernican system by its ignorant and tyrannical censures.

* For the particulars of this scandalous trial, and the abjuration of Galileo, the reader is referred to the Life of Galileo, c. 13. The date of this event was 1633.

† Galileo was at this time seventy years of age.

As we shall not have occasion to revert again to the great dispute about the motion of the earth, we shall now proceed to make a few remarks on the nature of the discussion, and the mode in which it was carried on. Galileo has remarked in his Dialogues, that the arguments on either side might be classed under two heads, those purely astronomical, and those derived from physical and mechanical considerations. But among the latter it will be necessary to distinguish those really physical from others which might more properly be called metaphysical, being merely founded on fanciful notions of Aristotle and his followers, as to the essence and nature of certain objects. A very numerous class of objections to the motion of the earth may be referred to the idea of Aristotle, that all celestial bodies were of a totally different nature from the terrestrial, being essentially perfect and incorruptible, while the latter were imperfect and liable to decay. This idea, which was a fundamental one in Aristotle's natural philosophy, made the Peripatetics utterly averse to admit that the earth was merely one of a system of planets revolving round the sun, and that there was no essential difference between them, either in their motions or physical constitution. The discoveries of Galileo, however, had eminently contributed to establish that such was the fact: he had shown that Jupiter was the centre of a system of satellites, revolving round him as our moon does round the earth; that the planets shone merely by light reflected from the sun; that the moon's surface presented inequalities exactly similar to those of the earth; that neither it nor the planets were the smooth, round, spontaneously luminous bodies they were asserted to be, but bore every analogy to the globe we inhabit. He proceeded in his Dialogues to overthrow the pretended perfection and immutability of the heavens, by referring to the apparition of a new fixed star in the year 1572, and to the spots on the surface of the sun, which he had discovered to form and disappear, indicating continual and extensive changes on the surface of that body. Tycho Brahé had already shown that comets were not meteors engendered in the atmosphere, but bodies revolving round the sun, and the direction of their orbits disproved the existence of the solid transparent spheres which had been imagined to explain the motions of the sun and planets, as they must have traversed these pretended spheres in every direction. The physical theories of the Aristotelians were thus completely overthrown: the epicycles of Ptolemy might be considered as a purely mathematical hypothesis for the purposes of calculation; but, unfortunately for his reputation, this author has urged some mechanical objections to the motion of the earth, which were much insisted upon in later times: and these we shall now proceed to consider.

The real mechanical objections to the doctrine of the rotation of the earth on its axis may nearly all be reduced to one head, though they have been presented in many different forms. Thus it was argued, that, did the earth revolve on its axis with the velocity which must be attributed to it in that hypothesis, stones thrown up into the air would not descend to the same place; a bird on the wing would find the ground over which it hovered pass away from under it; projectiles would range differently, according to the direction in which they were fired, and so on. The answer to all these, and innumerable objections of the kind which have been urged, is the same. Bodies on the earth's surface, as well as the atmosphere which surrounds it, participate in the common motion of rotation: the motion of a projectile in space is determined not merely by the immediate impulse, but also by the motion it had already acquired; though, as this is common to us, and all the points with which we compare it, we perceive only the relative motion. This point was well illustrated by Galileo *. Suppose, said he, a painter on board a vessel going from Venice to Alexandria were to begin to draw on quitting the port, and continue this till he arrived at Alexandria, it is clear that the point of his pencil will have traced out a very long line on the earth's surface; yet the peculiar work on which he was engaged, a landscape, for example, would be exactly the same as if the vessel had remained on the same spot, it being merely the relative motion of the pen that determined the outline of the figures on the paper. Another good illustration of the same point is this:—Suppose that, in a vessel under sail, a heavy body is suffered to fall from the top of the mast to the deck, it will strike the deck

* Opere di Galileo, vol. iv. p. 132. Padua, 1744.

exactly at the foot of the mast: (and not behind it, as many had affirmed in the teeth alike of reason and direct experiment;) for the same reason that the heavy body, participating originally in the common motion of the vessel, strikes the deck at the foot of the mast, will a similar body, suffered to fall from the top of a high tower on the earth's surface, strike the ground at the foot of the tower, and not at a distance from it*. This, indeed, is only true approximately; for, if we come to great accuracy, the summit of the tower revolves rather faster than the base, and the heavy body will deviate a little in its fall, but by so small a quantity as hardly to be perceptible in the most accurate experiments.

The physical difficulties being once disposed of, it only remained to compare the Ptolemaic and Copernican systems in an astronomical point of view. And here the advantage of the latter was palpable and immense. The annual motion of the earth suppressed at once the epicycles of the superior and the deferents of the inferior planets: it pointed out the cause of their stations and retrogradations, phenomena otherwise totally inexplicable. The diurnal motion introduced a still greater simplification, by getting rid of the monstrous revolution of the *primum mobile*, by which not merely the sun, a body enormously larger than the earth, and at a distance of more than ninety millions of miles with the planets, but even the fixed stars, whose distance is so great as almost to surpass our powers of conception and expression, revolved round an atom like the earth, all exactly in the same period of twenty-four hours. Indeed, the beauty of the Copernican system extorted a tribute of warm admiration from its very adversaries. The Jesuit, Riccioli, who wrote expressly to overthrow it, exclaims†, "Never can we sufficiently admire the genius and sagacity of Copernicus, who, by three motions of a globule like the earth, has explained what astronomers have never been able to represent without an absurd complication of machinery; and who, disengaging the fixed stars from their rapid diurnal motion, so difficult to reconcile with their general motion round the Poles of the Ecliptic, has happily explained the stations and retrogradations of the planets, and the precession of the equinoxes; who has destroyed three enormous spheres; who, lastly, like Hercules, has been able to sustain alone a weight that has so often crushed an Atlas." We have quoted this passage to show that the real difficulties in the admission of the system of Copernicus lay merely in the erroneous physical notions of the time, in the authority of Aristotle, and in religious prejudices. Indeed, but one argument has ever been urged against it, which would have any weight in the eyes of an astronomer: this is the enormous distance at which it places the fixed stars from the earth. For since the earth describes round the sun an orbit nearly circular, its positions at the interval of half a year will be distant the whole diameter of the orbit, or nearly 190,000,000 of miles; yet this immense distance produces no change in the apparent places of the fixed stars. In other words, a straight line of that length subtends no sensible angle at the distance of the nearest fixed star. Though there is no physical impossibility in this, it was felt for a long time as a difficulty in the Copernican system, and great exertions have been made to discover some annual parallax in the fixed stars, but, up to this time, without much success. It is curious enough, that while prosecuting researches of this kind, Bradley was led to the discovery of aberration, which furnishes us with a direct proof of the earth's annual motion. There is, then, no longer any hesitation in admitting the immense distance we have alluded to; however much it may surpass any magnitude of which we can form a conception.

* Galileo, Opere, vol. iv. p. 112.
† Almagest. Novum, vol. ii. p. 309.

CHAPTER XII.

Measure of the Earth—Snell—Norwood —Transits of Venus and Mercury— Gassendi--Horrox--Diminution of the obliquity of the Ecliptic—Wendelein — Morin— Longitudes— Hevelius— Libration of the Moon—Comets— Mouton.

THE attempt of Eratosthenes to measure the earth was repeated in the beginning of the seventeenth century by Willebrord Snell*, a distinguished professor of mathematics at Leyden. This, however, was not quite the first experi-

* Snell is illustrious as the discoverer of the law, that when a ray of light passes from one medium into another, the sine of the angle of incidence is to that of refraction in a constant ratio, depending upon the media.

ment of the kind in modern times. Fernel, a French physician, had, in the year 1528, measured the length of a degree between Paris and Amiens, by counting the number of turns made by a carriage wheel in passing along the high road: he found, for the length of the degree, 55746 toises of Paris: and it is remarkable that proceeding so roughly he should have obtained a result, which comes nearly within 300 toises of the truth. The method of Snell was much more scientific, and, in fact, is the same with that which has been pursued in similar researches ever since. Having measured in the plains near Leyden, a straight line or base of considerable extent *, he proceeded to observe at each end of it the angle made with a given signal. In the triangle thus formed, he calculated the remaining sides; one of which being taken for a fresh base, and connected with a fresh signal, formed a new triangle, and proceeding in a similar way, he was enabled to connect the two extremities of an extensive arc by a series of triangles. But as these two extremities did not lie exactly under the same meridian, it became necessary to determine the arcs of the meridian intercepted between the vertices of the respective triangles. To effect this, it was necessary, at the extremity of the arc, to measure the azimuth or deviation from the meridian of one of the sides of the first or last triangle. The sum of the arcs of the meridian thus found gave Snell the whole arc sought. The arc between Alcmaer and Bergen-op-Zoom, he found to be 34018 perches, each perch containing twelve Rhenish feet; the difference of latitude was $1° 11' 30''$: and hence, he concluded the length of the degree to be 28473 perches. He also observed the latitude of Leyden, the situation of which is intermediate between Alcmaer and Bergen-op-Zoom, and found by this operation 28510 perches, whence, taking a mean, he estimated the degree at about 28500 perches, or 331056 English feet †.

This value is not so exact as might have been expected from the goodness of the method, the real value of the degree being about 364870 feet. It appears that the principal error was in the determination of the azimuth, a delicate and difficult operation. It is, however, but justice to the memory of Snell, to observe, that he himself was aware of his error, and had repeated the whole operation, but was cut off by death before he could publish the corrections he had obtained. It is easy to see how superior his method was to that of Eratosthenes. In the first place, the latitude was actually observed at the two extremities of the arc; but the principal advantage was in the terrestrial measurements. Eratosthenes not merely neglected the reduction to the meridian, but, instead of determining the arc between Syene and Alexandria by any mathematical process, took the itinerary distance as given by the royal surveyors.

Soon after this measurement, a similar operation was undertaken in the same country by Blaeu, a countryman of Snell: it appears to have been executed with great accuracy; but the details have never been published. In the year 1635, Richard Norwood, having observed the difference of latitude between York and London, proceeded to measure the distance between these two places along the high road, partly with a chain, partly by stepping, and making the best allowance he could for the windings and inequalities of the road. This he did by observing the deviations from the meridian with a compass, and measuring the inclination of the various declivities. It must be regarded as a most curious instance of the tendency of small errors to compensate each other in a great arc, that in this way he should have found a value of the degree a great deal more accurate than that of Snell *. It is now hardly necessary to allude to the operation of the Jesuits, Riccioli and Grimaldi, in Italy, which was similar in method to that of Snell, but equally unsuccessful in the execution: their error was upwards of 31000 feet in excess.

In the system of Copernicus, or indeed that of Tycho Brahé, the inferior planets, Mercury and Venus, revolving round the sun in planes very slightly inclined to the ecliptic, must occasionally pass

* In the year 1617.

† See his work called Eratosthenes Batavus. Leyden, 1617, lib. ii. c. ix. p. 197.

* The error of Norwood was not much more than 2000 feet. He made the length of the degree 36716 English feet; it is in reality nearly 364870. For an account of this measure, see his Seaman's Practice. London, 1659, c. 2. Some idea of his manner of proceeding may be obtained from the following passage:—'For I did often observe a mile or two before me, some mark in the highway, noting the degree, and measuring to it in the way, neglecting to observe the intermediate swervings of the way, sometimes three or four degrees towards the right hand, sometimes as much to the left, but making such allowance for that and for the unevennesse as I judged sufficient.'—p. 35

between the sun and the earth. At this time, the planet seen from the earth will appear to traverse the sun's disc; and these phenomena are called transits of Venus or Mercury over the sun. It is easy to conceive their importance in astronomy. They are a decisive confirmation of the fact, that the inferior planets revolve round the sun; and they are of the greatest use in determining the elements of the orbits of these planets. It is evident, that at this time they must be very near their nodes: it is easy, when they are visible on the disc, to measure their distance from the sun's centre, to determine the instant of their appearance and disappearance, and hence to conclude the longitude of the node, and the inclination of the orbit to the ecliptic. These considerations induced Kepler to call the attention of astronomers to the transit of Mercury, announced for the 7th of November, 1631. Kepler had once imagined that he had observed such a transit on the 28th of May, 1607; but he recognized afterwards his error, and confessed that he had mistaken a solar spot for the planet. However, the transit of 1631 was certainly observed by several astronomers; but the only one whose observation is on record is the celebrated Gassendi *. To his great surprise, he found the apparent diameter of Mercury, which had usually been estimated at 2′, not to exceed 20″: he immediately conjectured that the apparent diameter of Venus would not be found to exceed much 1′; which was verified at the transit of 1639, when Horrox found it about 1′ 10″. This was the first transit of Mercury ever observed: these phenomena are of the more consequence in determining the orbit of the planet, since, from its proximity to the sun, it is always difficult to observe it with advantage by the ordinary methods. Those of Venus are still more rare and more important. One was announced for the 6th December, 1631; but it was invisible in Europe, Venus having traversed the sun's disc during the night. It is remarkable, that the next transit, in the year 1639, though visible, was very nearly missed, and only observed, it may be said, through an accident. The extreme importance of this observation may be collected from the fact that no other transit of Venus occurred before the year 1761, when the different governments of Europe sent astronomers to all parts of the world to make the observation.

We have said that the observation of the transit of Venus was owing to an accident. There lived at that time in Lancashire a young astronomer of the name of Horrox, who, though carried off by a premature death at the age of twenty-three, has left behind him proofs of enthusiasm and genius for science altogether uncommon. Kepler, from some error in his tables, had announced, that, after the transit of 1631, no similar phenomenon would occur for more than a century. It so happened, however, that Horrox had been in the habit, when he commenced his astronomical pursuits, of consulting the tables of Lansberg, though he had subsequently abandoned them for those of Kepler; and also, that Lansberg's tables, though in general very defective, yet, by a compensation of errors, were right in announcing a transit of Venus for the year 1639. Struck by this announcement, Horrox was induced to examine more minutely Kepler's tables; and he perceived that, after making certain necessary corrections, they made the transit visible, and fixed it for the 4th December, (N. S.) 1639. He gave notice of this to his friend and correspondent, Crabtree; and both had the pleasure, on the day mentioned, of seeing Venus on the sun's disc, though circumstances prevented the observation of the beginning and end of the phenomenon. The observation, however, was of the greatest use in fixing the elements of the orbit; and the two friends had the satisfaction of being the only individuals in Europe who had witnessed such a spectacle. They did not live long to enjoy the distinction. Both died at a very early age *.

It was about this time that the idea of a variation in the obliquity of the ecliptic was first entertained. It seems to have originated with Godfrey Wendelein, a Dutch astronomer of some merit. He induced Gassendi and Peyresc to repeat, at Marseilles, the observation of Pytheas on the sun's altitude at the summer solstice. Having em-

* Gassendi's observation is to be found in a treatise composed expressly on this subject, called De Mercurio in Sole viso. Hague, 1662.

* The observation of Horrox is detailed in a small treatise composed by him, called Venus in Sole visa, first printed as a supplement to the Mercurius in Sole visus of Hevelius, Dantzic, 1662.

ployed for this purpose a gnomon of 52 feet, they found the ratio of this altitude to the solstitial shadow to be $120:42\frac{3}{5}$. Pytheas had found $120:41\frac{4}{5}$: the comparison indicated clearly the diminution suspected; but astronomers do not seem to have placed much reliance on this observation of Gassendi, or indeed that of Pytheas, with which he compared it *. We have seen, however, that the latter was made with considerable care, and the doubts which have been thrown upon it are very unjust: it were to be wished that Gassendi had observed with as much success.

Wendelein deserves notice here for other services rendered to astronomy. He ascertained that the law of the periodic times discovered by Kepler to exist among the planets, was also true for the system of the satellites of Jupiter: the squares of the times of their revolutions round the primary planet, being as the cubes of the major axes of their orbits. We now know this to be a necessary consequence of the theory of universal gravitation: at that time it furnished a curious and not an undesirable corroboration of this remarkable law. Wendelein has also the merit of having corrected, materially, the value of the sun's parallax, then estimated at 3′, though it does not in reality exceed 9″. Aristarchus had placed the sun nineteen times as far from us as the moon: Wendelein, by following the same method, with the assistance of the telescope, was able to make a much more accurate determination: he placed the sun 229 times farther from us than the moon; which gives a parallax of 15″, supposing the moon's distance to be sixty semi-diameters of the earth.

The discoveries of India and America had given such an impulse to navigation, that the want of a method for determining the longitude at sea began to be severely felt. Ptolemy had proposed eclipses of the moon for the determination of terrestrial longitudes: the suggestion of Galileo to employ the eclipses of Jupiter's satellites was much more important; but the observation is made with difficulty at sea; and had it been otherwise, the satellites could not have been thus used without accurate tables. The object of the seaman being to discover his longitude at once, he cannot of course wait to compare his observation with one made under another meridian. Jean Baptiste Morin was the first to propose a method analogous to that of lunar distances, now in general use. He proposed to observe the altitudes of a fixed star and the moon, and at the same time their apparent distance; hence to obtain the true distance, and from this and the altitude of the moon to find her right ascension at the instant of observation *. It is always possible to find by the tables, the moment when the moon has a given right ascension under a known meridian; and the difference between these times gives at once the difference of the meridians or the terrestrial longitude. This method is quite good in theory, and, indeed, bears considerable resemblance to that now in use; but in practice it was inapplicable without tolerably exact tables of the moon; and upon this ground the commissioners appointed by Cardinal Richelieu to examine the merits of the invention, condemned it. It has been said by some authors that Morin was hardly used by them; yet if we consider that the reward offered by the government was evidently intended for some method of discovering the longitude directly applicable to navigation, which this method could not be made, without the construction of good lunar tables; that in fact the real difficulty of the problem lay in the formation of these tables, which required a much more perfect state of astronomy than existed in the seventeenth century; that the mere idea of employing the place of the moon for the object in question, was not difficult to hit upon, nor in fact original †; we shall probably think the commissioners had sufficient ground for the opinion they pronounced. Nor is the character of Morin such as to excite much sympathy for his alleged ill-treatment. Not content with being an infatuated follower of judicial astrology, he had waged an implacable war against the doctrines and followers of Copernicus; upon whom he had heaped the

* For the observation of Gassendi, see his Proportio Gnomonis ad Umbram Solstitialem Massiliæ Observata. Hague, 1662.

* Longitud. Scient. p. 71. See also p. 49. Paris, 1640.

† It had been proposed by Gemma Frisius, Usus Globi Astronomici, cc. 17 and 18: no doubt the method of Gemma Frisius was so inaccurate as to be useless in practice. He employed a globe and compass instead of trigonometrical calculation, and neglected altogether the moon's parallax: Morin might justly claim the merit of having corrected the method, and made it applicable, provided good tables could be found: but here was, as we have said, the real difficulty.

most unsparing and unmeasured abuse *.

It has been remarked by Montucla that none of the contemporaries of Kepler seem to have understood the importance, or, indeed, the real nature of the laws discovered by that philosopher. They saw in them nothing but the motion of the planets in ellipses, of which the sun occupied one of the foci. The important law of the areas was either overlooked or disbelieved. It may justly excite our astonishment, that long after the publication of Kepler's work on Mars, Bouillaud, a French astronomer of great erudition, should have proposed to represent the celestial motions by another hypothesis †. He conceived each planet to move in an ellipse, (one of the foci of which was occupied by the sun,) adapted to an oblique cone, in such a manner that the axis of the cone passed through the other focus. The planet was supposed to move in this ellipse, so that the times were proportional to the angles described round the axis of the cone. To an eye placed in the summit of the cone the planet will appear to move uniformly in the circumference of a circle. We see here the last remaining traces of the ancient prejudice in favour of circular motions. Seth-Ward, Bishop of Salisbury and professor at Oxford, modified the idea of Bouillaud. He made the planet revolve in an ellipse similar to that of Kepler, but in such a way that the times were proportional to the angles described round the focus not occupied by the sun ‡. This has been called the simple elliptic hypothesis. It offers advantages for the calculation of the true anomaly from the mean, which in Kepler's theory was sufficiently difficult; but it is not the law of nature, and has been universally abandoned. However, it enjoyed for some time considerable vogue: being subsequently discovered to give very erroneous results when the excentricity was considerable, it was again modified by Mercator §: he divided the distance between the foci of the ellipse in extreme and mean ratio, so that the point of section fell between the centre, and the focus not occupied by the sun. Cassini went so far as to reject altogether the common ellipse, and substitute an analogous curve, where instead of the sum of the distances from the foci to a given point, it is their product which is constant. Indeed it may be said that it was not till Newton had shown that the laws of Kepler were deductions from one general principle which governs all the planetary motions, that they seem to have been universally appreciated as the real expression of the phenomena.

John Hevel or Hevelke *, commonly called Hevelius, (a senator of Dantzick), was the most assiduous practical astronomer that Europe had seen since the death of Tycho Brahé. The revolution in the art of observing, caused by the adaptation of telescopes to instruments for measuring angular distance, which occurred towards the latter end of his career, rendered his labours less valuable than they would have otherwise been; and we must condemn the obstinacy with which he constantly refused to adopt this improvement. Yet the industry with which he has executed some laborious undertakings, such for example as an accurate delineation of the moon's surface, deserves our gratitude †. While observing this satellite, he was led to the discovery of a phenomenon which had escaped the sagacity of Galileo. This great man had remarked, that though the moon always presents very nearly the same face to the earth, yet this law is subject to some small variations, which depend upon the following causes:—In the first place, the moon's parallax being a very sensible quantity, a spectator supposed to be at the centre of the earth, and one on the surface, would not see exactly the same disc. When the moon is near the horizon she is depressed by parallax; the eye, being above the centre of the earth, perceives spots and mountains on the upper edge, which disappear as she rises towards the zenith, where the parallax vanishes, and other spots on the lower edge come into view. As she descends again towards the horizon, the latter vanish, and the former reappear. This is called the diurnal libration: there is another libration in latitude, the nature of which is easy to be conceived, which arises from the moon not moving in the plane of the ecliptic: when she is some degrees above

* See the article Morin in Delambre, Astron. Moderne, vol. ii.
† Astron. Philolaica, lib. 1. c. 17.
‡ Astronom. Geometrica, London, 1656.
§ Astronom. Sect. 2. London, 1675.

* Born, 1611; died in 1687.
† The names given to the spots on the moon in modern maps, are not those of Hevelius, but Riccioli, whose idea of naming them from great astronomers is sufficiently happy.

this plane the disc that we see, does not comprise exactly the same portion of her real surface, with the disc seen when she is below it. These phenomena have been designated by the name of libration: the former is called the diurnal libration; the latter, the libration in latitude. The observations of Hevelius led him to the discovery of a libration in longitude*. Its cause is this;—The moon revolves unequally round the earth, while her motion of rotation on her axis is uniform. The two motions being accomplished exactly in the same time, it follows that the same part of her surface is always presented to the centre of the orbit, but not to the earth: the disc seen from these two points coincides only when the satellite is in the line of her apsides. All these appearances, of course, are merely optical; theory indicates a real libration in longitude, but this has not been sensible in any observations hitherto made.

Hevelius deserves further notice here as being the first astronomer who had a correct notion of the nature of the orbit described by comets. Tycho Brahé had demonstrated that these bodies existed beyond the limits of our atmosphere: the comet of 1577 he had shown to be three times farther from the earth than the moon. Hevelius showed that the comet of 1664 was five times farther from us than this satellite. Tycho supposed them to move in a circle: Hevelius, guided by analogies partly false and partly true, supposed them to move in a parabola, which we now know to be the real law†. He was unwilling to believe that bodies of a transitory nature, such as he conceived comets to be, could revolve in a circle: a motion in that or any closed curve belonging only to such as had a periodic rotation and an eternal duration. He rather, therefore, leaned to the opinion of Kepler, who supposed the orbits to be rectilinear; but observing that all projectiles on the earth describe a parabola, he was led by rather vague but fortunate reasoning to the conclusion, that such, also, was the curve described by comets. As the path of a projectile is determined by the force of impulsion and gravity combined, so was that of the comet, by the impulsion with which, in his theory, they were thrown out of the atmosphere of the planets in which they were generated, and by an attractive force directed to the sun. These notions of Hevelius approximate curiously to the Newtonian theory, which we shall subsequently have occasion to develope.

Gabriel Mouton, of Lyons, an astronomer of considerable merit, but little known, flourished about the year 1660. He has the merit of having introduced into astronomy the method of determining, by interpolation, the place of a planet at some instant, intermediate to other instants, for which its place is given in the tables. It is unnecessary to make any remarks upon the extreme importance of this method, which is of daily and hourly use in almost all astronomical calculations. But this is not our only obligation to Mouton: he was the first observer who used the pendulum to determine differences of right ascension. Hevelius, indeed, had employed a pendulum in some observations a few years before; but we do not know how he measured the number of oscillations made in a given time. Galileo had discovered, many years before, that the oscillations of a pendulum in a small circular arc are sensibly isochronous; and Huyghens had already applied the invention to clocks. But those clocks were unknown to Mouton, who employed a simple pendulum: he determined the number of vibrations in a given time, by counting how many took place, while the sun traversed a known arc of the equator. Having ascertained this fundamental point, he proceeded to observe the time taken by the sun's diameter to pass the meridian. He thus found the value of this diameter, when the earth is in its aphelion, about 31′ 31″, a value as accurate as any that can be assigned.

Chapter XIII.

Origin of Scientific Societies—National Observatories — Application of the Telescope to the Quadrant—Gascoyne, Auzout, Picard — Invention of the Micrometer — Pendulum Clocks — Huyghens — Transit Instrument — Römer.

We have now arrived at an important epoch of our history. The latter half of the seventeenth century witnessed a complete revolution in astronomical science. This revolution was owing, in some measure, to the foundation of scientific societies and national observatories, but principally to the great and important improvements in the methods of obser-

---

* De Motu Lunæ Libratorio; Gedani, 1654.
† Cometographia; Gedani, 1668; lib. ix. p. 683.

vation. The utility of scientific societies is sufficiently evident; but, perhaps, it was still greater at the time of which we are now speaking, than in the present day. Without the constant communication, and the periodical journals that apprise us now, not merely of every interesting discovery, but of the slightest improvements in scientific methods, the philosophers of the seventeenth century lived in a state of isolation, which must have materially retarded the general progress of knowledge. The remarkable observation of Horrox on the transit of Venus was not published till thirty years after it was made. In the meanwhile, the important conclusions to be drawn from it were lost to the world. The observations and discoveries of Gascoyne (see page 66) were still longer buried in oblivion.

The foundation of the Royal Society of London, the oldest existing scientific corporation, has been referred to the meetings of several distinguished lovers of natural philosophy, who, being attached to the party of the king, had retired to Oxford during the reign of Cromwell. On the restoration of Charles II. they adjourned to London; and, having increased their number, and being much favoured by the monarch, who had some taste for chemical pursuits, they obtained a charter, and were regularly incorporated in the year 1662*. Scientific societies, something similar, had existed previously in Italy: the first had been formed by the Marchese Frederico Cesi, as early as the year 1611, under the name of the Lyncean Society: of this the celebrated Galileo was a member; but it seems to have been dispersed at the death of its founder in 1632†. It was succeeded, in 1657, by the Accademia del Cimento, of Florence, which owed its origin to the distinguished pupils of Galileo, Viviani and Torricelli; and enjoyed but a short though brilliant career of about ten years. The date of the foundation of the Royal Academy of Sciences at Paris is posterior only by four years to that of the Royal Society. It met for the first time in the year 1666‡.

Nearly contemporaneous with this event was the foundation of a Royal Observatory at Paris. There can be no doubt that this was an invaluable benefit to astronomy. The foundations of this science are an accurate knowledge of the distances of the fixed stars from each other, and of the varying positions of the sun and planets. The latter, in particular, can only be determined by long series of observations, carried on steadily and without interruption, from year to year, and even from century to century. Without such observations, the profoundest analysis could not have constructed those tables which have now been brought to incredible accuracy; by which the astronomer can predict, with unerring certainty, the place occupied on a given moment of a future year by the most refractory planet, and the navigator steer with confidence and security across the trackless breadth of the Atlantic. The Royal Observatory of Greenwich, which has rendered greater services to astronomy than any similar institution without exception, was founded in the year 1675.

But these observatories could not have rendered such eminent services, without great improvements in the instruments employed. One of the most important of these improvements was the adaptation of telescopes to instruments for measuring angular distance. It is easy to conceive the advantages gained by this addition. In pointing upon any heavenly body the moveable radius, the extremity of which measures on the limb the altitude or distance required, the exactitude of the observation evidently depends upon the magnitude of the angle under which we see a given portion of space. The observer then who is employing the telescope, will make his observation with an accuracy proportional to the magnifying power of the instrument. This circumstance is so very obvious, that it must always remain a matter of surprise that the invention was so long deferred. Some doubts have been raised about the person to whom it is to be ascribed. It seems pretty certain that it was used in England about the year 1640 by Gascoyne, an astronomer of genius, the contemporary and friend of Horrox and Crabtree, and carried off like them by a premature death*. It is equally certain that the observations and discoveries of Gascoyne remained unpublished, and

* The most distinguished among the original members of the Royal Society were Boyle, Hooke, Wren, Wallis, Ward, Lord Brouncker, &c.

† For an account of the Lyncean Society, see the Life of Galileo, c. ix.

‡ Among its members were Auzout, Picard, Roberval, Richer, &c., and three distinguished foreigners, invited to settle in France by Louis XIV. Cassini, Huyghens, and Römer.

* He fell at the battle of Marston Moor, aged twenty-four.

unknown to the scientific world, when the same happy improvement was introduced in France by Picard and Auzout, about the year 1667*.

But the most remarkable circumstance is, that the idea of adapting the telescope to the sextant was published as early as the year 1634 by Morin, who made it one of the proposed improvements on which he rested his claim to the discovery of the longitude †. Yet this proposition seems to have attracted no attention at the time, and to have been completely overlooked by subsequent authors. Delambre we believe to have been the first to notice it. It is important to remark, that Morin never actually practised the method he proposed; while we shall see that Gascoyne, notwithstanding Delambre's assertion to the contrary, really did use the telescope applied to the quadrant. Morin went so far as this; namely, to attach a telescope to the alhidade of what he calls a planisphere, an instrument, he says, which performs all the offices of an equatorial armilla ‡; but, and this is the important circumstance, he did not employ it to measure angular distance, merely to follow a star when in the field of view. The telescope of Morin, thus adapted to his planisphere, had apparently some analogy with the equatorially mounted telescope of modern times. Now, Gascoyne, it appears, had actually used the telescope in the observation of altitudes, &c. But as this has been virtually denied by Delambre §, who even will not allow that he ever had the intention of applying a telescope to his sextant, we think it necessary to put before the reader the original passages from the correspondence between Gascoyne and Crabtree, from the *Philosophical Transactions* for 1717, No. 352, III. In a letter from the former to the latter, dated 25th January, 1640-1: after speaking of his micrometer, which was used in a telescope, he says, "or if here a hair be set, that it appear perfectly through the glass, you may use it in a quadrant, for the finding of the altitude of the least star visible by the perspective wherein it is. If the night be so dark, that the hair or the pointers of the scale be not to be seen, I place a candle in a lanthorn, so as it cast light sufficient into the glass, which I find very helpful when the moon appeareth not, or it is not otherwise light enough." Again, in another letter, dated Christmas eve, 1641: "I am fitting my sextant for all manner of observations, by two perspicills with threads. And also I am consulting my workmen about the making of wheels, like β, γ, δ, ϵ of diagram 3" (this diagram is lost,) "to use two glasses like a sector. If I once have my tools in readiness to my desire, I shall use them every night. I *have fitted* my sextant by the help of the *cane, two glasses in it*, and a thread, so as to be a pleasant instrument, could wood and a country joiner or workmen please me." Again, in a letter from Crabtree to Gascoyne, 30th October, 1640: "Something I am sure you were telling me concerning a way of observing the places of the planets by your glasses. . . . . . . One means, I think you told me, was by a single glass in a cane, upon the index of your sextant, by which, as I remember, you find the exact point of the sun's rays; but the way how I have quite forgotten, and much desire. . . . . . I cannot conceal how much I am transported beyond myself with the remembrance (of that little I do remember) of those admirable inventions which you showed me when I was with you. I should not have believed the world could have afforded such exquisite rarities, and I know not how to stint my longing desires without some further taste of these selected dainties. Happier had I been had I never known there had been such secrets, than to know no more but only that there are such. Of all desires, the desire of knowledge is most vehement, most impatient; and of all kinds of knowledge, this of the mathematics affects the mind with the most intense agitations. I doubt not but you can experimentally witness the truth hereof, and one time or other have been no stranger to such thoughts as mine. And, therefore, although modesty would forbid me to request any thing (until you give me leave) but what you please voluntarily to impart, yet the vehemence of my desire forceth me to let you know how much I desire, and how highly I should prize anything you would be pleased to communicate to me in those optick practices. Could I purchase it with travel or procure it with gold, I would not long be without a telescope for observing small angles in the heavens; nor want the use of your other

* Delisle informed Lalande that it was suggested by Roberval. V. Lalande, Astron., art. 561.
† Long. Scient., p. 56. ‡ Ibid., p. 210.
§ Astron. Mod., vol. ii. p. 589.

device of a glass in a cane upon the moveable ruler of your sextant, as I remember, for helping to the exact point of the sun's rays." These passages will probably leave no doubt in the mind of the reader, that Gascoyne had constructed and used sextants fitted with telescopes for the purpose of measuring angular distance; and if he did so, he was unquestionably the first. That his discoveries were not published, is sufficiently explained by his early death on the field of battle, and the political troubles that followed. We wish to vindicate his memory without diminishing the well-earned reputation of Picard and Auzout, who could have known nothing of Gascoyne and his inventions.

It may be necessary to observe, that in the quadrants used by the older astronomers, by Tycho and Hevelius, the moveable radius carried, at each extremity, flat pieces of metal projecting perpendicularly to the plane of the limb, and facing towards each other *. Each of these was pierced with a circular aperture; a very small one in that near the eye, and one something larger in the other †. In making an observation, the radius was moved till the star was seen through both holes, and as nearly as possible in the centre of the larger. Of course measures were taken to ascertain that the line joining the centres of the two circular apertures mentioned, was parallel to the plane of the instrument, as is done now with regard to the optical axis of the telescope. There can be no doubt that observations might be made with some accuracy in this way; under favourable circumstances and with a good observer, the error might be not much above a minute: at least the difference was always less than this between the observations of Hevelius, and those made with the assistance of the telescope by Halley, for the express purpose of comparison.

The form of these ancient instruments was of the simplest kind; they consisted merely of a segment of a circle, generally a sextant or a quadrant, of several feet radius, solidly constructed in metal; a radius, moveable on the centre of the circle, carried the pinnules, and traced out with its extremity, on the divided limb, the arc it was wished to measure. Were it intended to observe the altitude of a star, the instrument, which was moveable on its support, was placed in a vertical plane; a plumb-line suspended from the centre, showed whether the zero point of the arc was exactly below it. When this was ascertained, nothing further was to be done than to point the moveable radius on the star, and read off on the graduated limb the altitude. If it were wished to measure the distance between two stars, two pinnules, one at the centre and one at the zero point, were substituted for the plumb-line; the instrument was placed in the plane of the two stars; one observer pointed upon one star through the fixed pinnules, the other through the pinnules of the moveable radius upon the other *. Some ingenious methods had been introduced for subdividing, with accuracy, the graduated limb of the instrument. Early in the sixteenth century, Pedro Nuñez or Nonius, had proposed the following method of subdividing the limb of an astrolabium:—On the surface of the instrument, supposed to be a perfect plane, he traced 44 concentric quadrants; the external one was divided into 90 parts, the next into 89, the next into 88, and so on,—the innermost consequently into 46. The altitude of any object having been observed, could only be read off on the outermost quadrant to degrees: in order to obtain fractions of a degree, he proceeded thus:—The moveable radius is pretty certain to intercept some one of the concentric quadrants at a complete division; read off the number of parts intercepted on that quadrant; multiply this by 90, and divide by the number of parts in the whole quadrant; this will give you the altitude to fractions of a degree †. Thus, if the radius intercept exactly 23 parts on the quadrant divided into 87, the angle measured will be 22°.793; the fraction is easily converted into minutes, or any other parts of a degree wished. This method is ingenious, but offers in practice several inconveniences, which caused it to be soon given up for another, the author of which is not known: it was certainly employed by Tycho Brahé ‡, but seems to have been invented some time before

* These were usually called pinnules.

† Sometimes the pinnules were slit longitudinally. V. Hevel., Machina Cœlestis, c. 14.

* A good description of the instruments used in the sixteenth and first half of the seventeenth century, may be found in the Machina Cœlestis of Hevelius. Gedani, 1673.

† Nonius de Crepusculis. Basle, 1592. p. 392.

‡ De Cometâ Anni 1577, p. 22. Frankfort, 1648. For an account of the different persons to whom the division by transversals has been attributed, see Lalande, Astronomie, Art. 2336, *et seq.*

it was adopted by him. It consisted in drawing parallel transverse lines, from each of the points of division on the limb of the quadrant at any given angle, and dividing each transversal into the same number of equal parts; then, by observing the point at which the alhidade or moveable radius cuts any one of these, it was easy to determine the corresponding fractional part of the space between two divisions on the limb. If, for example, the limb were divided from degree to degree, and the transverse lines into thirty parts, we might thus read off the arc to two minutes. This method has been finally supplanted by the Vernier, an instrument first made known to the public by Pierre Vernier, the inventor, in the year 1631. It is unnecessary to explain the construction of an object in such general use; we shall merely remark, that the original Vernier differed slightly from that now employed, in having the divisions counted in a contrary direction to those of the limb, and also, from having the number of divisions less by one than the corresponding number on the limb, instead of exceeding them by one, as they are now usually constructed*.

The invention of the micrometer is perhaps not inferior in importance to the application of the telescope to the quadrant. This instrument was first constructed by the ingenious and unfortunate Gascoyne. But though there can be no doubt about this, yet as it remained unknown till after the rediscovery of the same instrument on the Continent, we shall defer speaking of Gascoyne till we have considered what was done by Huyghens† and Auzout. The former of these philosophers led the way by observing, that anything placed in the interior of a telescope, at the focus of the object-glass and eye-glass, was magnified in the same way as the distant body on which the instrument was pointed‡. At this focus, then, he inserted into the tube a circular ring, traversed by a slip of metal of unequal breadth. The diameter of the ring was found by observing how long a star took to cross the field; then, with a compass, the different breadths of the slip of metal were compared with the diameter of the ring. Knowing these pretty exactly, it was easy to find to which of them the diameter of any given planet was equal. Very soon afterwards this method was modified by Malvasia in Italy, who, in the place of the ring and slip of metal, substituted a number of fine threads, cutting each other at right angles, and thus subdividing the field into a number of small squares: the ratio of these squares to the field being known, it was easy to estimate the apparent magnitude of any object by the number of these subdivisions that it occupied. In this form of the instrument Auzout made an inestimable improvement, by suppressing all the threads but two, in a parallel direction,—one fixed, the other moveable, and susceptible of being placed by a screw at any distance, but always remaining parallel to the first. If it were required to measure the apparent diameter of a planet, one limb was brought into contact with the fixed thread, and the moveable thread placed so as just to touch the other limb; the distance between these two threads might always be compared to the diameter of the field, as in the method of Huyghens; or as Auzout himself remarks, by a knowledge of the focal length of the telescope, and the distance of the threads, we might get at once the apparent angle under which the body is seen*.

This invention making considerable noise in Europe, Mr. Townley inserted in the Philosophical Transactions for 1667†, some account of a micrometer by Gascoyne, which he explained more fully in a subsequent number‡. In all the essential parts, this is clearly identical with that of Auzout; and it is also to be remarked, that this instrument had not only been actually constructed, but also used by Gascoyne for measuring the diameters of the moon, Jupiter, Venus, and Mars§. There can be no doubt, as we have before stated, that the Continental astronomers knew nothing of what had been done in England: in fact, the Royal Society of London was in equal ignorance before the publication of Mr. Townley's Memoir. So far had

* The instrument of which we are speaking has been often, though erroneously, called Nonius; this last name has still more frequently, and quite as mistakenly, been given to the division by transverse lines; the real Nonius is the division by concentric arcs of circles.

† Christian Huyghens was born at the Hague in 1629,—died in 1695.

‡ It is necessary to recollect, that the astronomical telescope is essentially composed of two convex lenses, distant by the sum of their focal lengths. This combination was originally imagined, though never executed, by Kepler. *Dioptria*, Prob. 86.

* Mem. de l'Academie des Sciences, vol. vii. p. 118.

† No. 25, p. 457. ‡ No. 29.

§ Philosoph. Transac. for 1753, p. 190. Flamstead Historia Cœlestis, vol. i. p. 5.

Gascoyne anticipated the French discoveries, that he had even suggested the application of the micrometer threads to the telescope of the quadrant, an improvement usually referred to Louville, who proposed it towards the beginning of the eighteenth century.

In the course of this work we have seen the various contrivances to which the older astronomers were forced to have recourse, from the want of any exact method of measuring time. Before Huyghens no clocks were constructed whose rate was sufficiently regular to make them serve for astronomical purposes. It would appear that the Landgrave of Hesse and Tycho made fruitless attempts to procure clocks upon which they could rely, and that they subsequently abandoned the idea in despair*. They determined the difference of positions at night by taking the distances to two known fixed stars: if it were wished to refer them to the sun, this was done by means of Venus, as explained in a preceding chapter. Galileo had first remarked that when a pendulum vibrates in a small circular arc, the oscillations are isochronous. He at once conceived the possibility of applying this property to the measurement of time, and the improvement of astronomical observations. But to the use of the simple pendulum in this way, there were two principal objections. First, that the pendulum kept describing smaller and smaller arcs, and, finally, came to rest: secondly, that it was necessary to count every vibration, a process manifestly impossible if the time were considerable, and always very troublesome and inaccurate. This last difficulty Galileo proposed to remedy by a contrivance of this kind. The pendulum was to carry a delicate needle at right angles to its axis, which, in passing, strikes a rod fixed at one end and with the other resting on the teeth of a light horizontal wheel. The rod being forced against the perpendicular side of a tooth moves it, (and consequently the whole wheel,) but in returning it slips over the oblique side and falls at the foot of the following tooth, so that the motion is always in one direction. The number of teeth passed will then show the number of vibrations that have elapsed. It is easy to see that an index might be attached to the axis of the tooth wheel, like the second hand of a clock, or various other contrivances might be conceived to be adapted to it for the same purpose. The second difficulty it must be confessed Galileo never fairly overcame, as he talks of having an assistant to give the pendulum a smart push from time to time, a method, the inconvenience of which it is unnecessary to characterize.

In the invention of Huyghens this was completely and elegantly mastered. To understand how it was effected it is necessary to call to mind the manner in which in previous times the motion of clocks was regulated. The weight which was the prime mover of the system was connected with a vertical toothed wheel: the balance was a vertical arbor with two pallets*, carrying a cross bar with weights at top: as the wheel revolves, imagine the upper pallet to become engaged in its teeth, the balance is forced to revolve till the opposite pallet becomes engaged below against the perpendicular side of a tooth; the motion of the balance is then suddenly checked, and it is forced to revolve in the opposite direction till stopped again by the upper pallet. At each of these turns of the balance the weight is brought to rest: consequently instead of descending with continually accelerated velocity, it moves nearly uniformly; and by the common mechanism used in these cases, the clock may be made to indicate the number of oscillations of the balance, which evidently must always be of nearly equal duration. However, as we have before mentioned, the accuracy of these clocks was not sufficiently great for astronomy. Huyghens had the happy idea of substituting for the old balance, a pendulum, the oscillations of which being necessarily isochronous, the beats of the clock became strictly equidistant, and its motion uniform. To understand how Huyghens adapted the pendulum to the common clock, we must suppose the vertical arbor and pallets to remain as before, but the cross bar at top to be taken away and replaced by an horizontal pinion wheel. The pendulum a little below its point of suspension passes through a horizontal axis connected with a vertical rack working into the pinion we have mentioned. The oscillations of the pendulum cause the axis and rack wheel to vibrate with them, and by means of the pinion the pallets

* De Stellâ Novâ. Anni 1572, c. 2. p. 109.

* These two pallets are placed so as to engage, one with the upper, the other with the lower extremity of the toothed wheel.

to engage alternately in the toothed wheel as before*.

Some time after the death of Galileo, a very unfounded claim was raised on his behalf in Italy, to the invention of pendulum clocks. Count Lorenzo Magalotti, secretary to the Accademia del Cimento, and afterwards Tuscan minister at Vienna, distinctly attributed this improvement to Galileo and his son Vincenzo. But the letters of Galileo himself are sufficient to disprove the assertion; it is quite clear from them what his clock was, and how much it differed from the pendulum clock of Huyghens: of this the reader may judge from the description we have given of both. Galileo had proposed to employ the pendulum for the measurement of time, this is undoubted: it seems quite as certain that he never intended to substitute it for the balance of an ordinary clock. It is in this substitution that the merit of Huyghens consists†.

The transit instrument was the last, and certainly not the least important of those discoveries which, towards the end of the seventeenth century, effected a complete revolution in practical astronomy. It is not our intention here to explain the construction and uses of an object so well known; we shall merely remark, that this and the clock may be said to be the capital instruments of an observatory. Indeed, with these, an astronomer might almost dispense with any other. The transit placed in the plane of the meridian gives him his time; in the prime vertical, his latitude. In the former plane he can measure the differences of right ascension; in the latter the declinations of stars‡. Olaus Römer, rendered illustrious by other discoveries, was the inventor of this invaluable instrument§. Picard had been the first to see the advantages of observations made exclusively in the meridian; and to propose to determine differences of right ascension by the intervals between the transits in a telescope attached to a mural quadrant in that plane. In this way the right ascension and declination might be found by a single observation. He had previously ascertained the possibility of seeing the fixed stars and planets in the day-time in the field of a telescope, which indeed had been done before by Morin; Morin's discovery, however, had been completely forgotten in the interval, and Picard appears to have known nothing of it. But from some causes, which we are not now able to appreciate, the latter was unable to realize his intentions, and the mural quadrant was first employed in the observatory at Paris by his successor La Hire*. In the mean time Picard employed a telescope fixed in the plane of the meridian, "lunette murale," for the observation of altitudes; of the construction of this instrument we are ignorant, it seems to have had some kind of graduation attached to it, as he was able to measure altitudes between 56° and 61°. Römer the friend and assistant of Picard, in attempting, some years afterwards at Copenhagen, to improve the mural quadrant, fell upon a construction analagous to that of the transit instrument. At right angles to an horizontal metallic cylindrical axis of five feet long, and an inch and a half thick, he fixed a telescope, (revolving in the plane of the meridian,) not however in the middle, but nearly at one end of the axis; the extremities of the latter rested on firm supports let into the wall, resembling the modern Y's. This axis carried besides the telescope, and towards the other extremity, an arm projecting at right angles, the extremity of which carried a microscope, and corresponded to the segment of a circular arc firmly attached to the wall; thus affording the opportunity of measuring altitudes, as well as differences of right ascension. The field was traversed by three horizontal, and ten vertical threads; but, in observing transits, three only were used: the equatorial interval of these was 34 seconds. The threads were illuminated by a concave speculum, just behind the object glass, perforated in the centre, with the concavity towards the eye, which reflected

* The construction explained by Huyghens in the Horologium Oscillatorium is an improvement upon that given in the text. The horizontal axis attached to the pendulum carries two pallets which engage alternately the teeth of a horizontal crown wheel, carried on the vertical arbor of which we have spoken: the pendulum is suspended by a thread of some inches length, which winds and unwinds alternately along two cycloidal cheeks.

† The reader who feels interested in this discussion is referred to an article by Van Swinden in the Edinburgh Philosophical Journal, vol. vi. p. 197. The question is also well stated in the life of Galileo, chap. 18. The invention of Huyghens took place in the year 1657, V. Horolog. Oscill. p. 1.

‡ We do not mean to say that this would be a good method of determining the declinations of stars; we only state the possibility of employing it. It is evident that it would be impracticable for stars that had not some northern declination.

§ Römer was a Dane. He was born in 1644, died in 1710 at Copenhagen: he was a member of the French Academy of Sciences, and for some time was attached to the observatory at Paris.

* In the year 1682.

the light from a lantern over the centre of the instrument, the upper part of the tube being cut away to give a free passage to the rays. The microscope was a sort of vernier, as it carried in the field eleven divisions by threads just corresponding to ten divisions on the limb. The construction of this instrument was afterwards improved upon by Römer at his observatory in the country. The axis of this second instrument was composed of two equal cones joined by their bases; at right angles to the axis, and at one extremity of it was a circle of five feet diameter, carrying a telescope of about the same length: the graduation of this circle was read by two microscopes, fixed to the supporting pillar, and the instrument of course was placed so as to move nearly in the plane of the meridian. In this form, the instrument bore some resemblance to the circles lately constructed by Reichenbach; its principal difference from them consists in the telescope not being placed in the middle of the axis. The adjustments were made in the manner that has been followed ever since. The deviation of the optical axis was corrected by reversing the instrument; the deviation from the meridian, ascertained by transits of α Lyræ above and below the Pole; the verticality of the circle was observed with the plumb-line; and hence the horizontality of the axis established. All these adjustments having been completed, Römer, in order to try the goodness of his instrument, proceeded to make a series of observations, during three consecutive days, the excellence of which was the most satisfactory proof of his success. It could not be expected that the altitudes should be given by this instrument as exactly as by the mural quadrant: but the agreement of the transits is really very remarkable, the mean error not much exceeding half a second of time *.

Chapter XIV.

Saturn's Ring and Satellites—Huyghens—Cassini—Rotation of Jupiter, Mars, and Venus—Refractions—Parallax of the Sun—Zodiacal Light—Libration of the Moon—Römer—Successive Transmission of Light.

The telescope in the form originally invented by Galileo was not susceptible of being carried to any great perfection. It would hardly be possible to use with this construction a greater magnifying power than from 30 to 40 times, which was the power employed by the original discoverer. Kepler seems first to have suggested the combination of two convex lenses, which constitutes the principle of the astronomical telescope: but there is no reason for believing that he ever executed an instrument of this description. However, the idea was soon realized by other astronomers; and a very great improvement in the power of the telescope was the consequence. A Neapolitan, named Fontana, laid claim to the honour of having first made an instrument on these principles in the year 1608; others have attributed it to Father de Rheita; but Montucla is of opinion that Scheiner, the author of the Rosa Ursina, was the first who reduced to practice the theory of Kepler *. But, on executing these telescopes on a large scale as was done by Campani and others towards the latter end of the seventeenth century, it was found necessary, in obtaining any considerable aperture, to give the object-glass such an enormous focal length, as to make the instrument very heavy and unmanageable. To overcome this difficulty, it was proposed by Huyghens and others, to suppress altogether the tube of the telescope: the object glass being attached to some solid support, the observer was to place his eye in the focus, and there magnify the image with a lens or combination of lenses, according to the usual method †. In this way Huyghens constructed telescopes upwards of a hundred feet long. The difficulties of managing such an instrument are easy to conceive, and appear to us at the present day almost insurmountable; yet that they were overcome by patience and ingenuity is sufficiently shown by the remarkable discoveries of himself and Cassini. The appearances exhibited by Saturn in the telescope had been inexplicable to Galileo and succeeding observers. A variety of conjectures had been thrown out on the subject. Roberval imagined the equator of Saturn to be surrounded by a mass of vapours which reflected to us the solar light: Cassini that the planet

* For a very detailed account of the instruments and observations of Römer, see the work of his pupil and assistant, Horrebow, entitled 'Basis Astronomiæ,' Copenhagen, 1735: a treatise very interesting on many accounts to the astronomical reader.

* Vol. iii. p. 234. The Rosa Ursina of Scheiner was published in 1650.

† It is to be understood, that though there was no tube, there was machinery, by which the object glass could be turned so as to point on any object required.

was surrounded by a string of satellites very close to each other. Huyghens having observed Saturn carefully with a good telescope, perceived the real nature of the appearances. He announced that Saturn was surrounded by a slender plane ring, nowhere adhering to its surface, and inclined to the plane of the ecliptic. This inclination Huyghens supposed to be about 23°: modern observers have made it about 28°: the distance between the ring and the planet he estimated to be equal to the breadth of the ring, which is about one-third of the diameter of the planet. Whenever the plane of the ring produced passes either through the sun or through the earth, the ring will disappear, being too thin to reflect any sensible quantity of light from its edge. But in other positions, particularly when the earth is 90° from the nodes of the ring, it becomes extremely distinct; the blue of the sky, and even occasionally small stars, being beautifully visible between it and the planet. The great improvement of telescopes in modern times has enabled astronomers to add some facts to those discovered by Huyghens. It has been found that the ring is double, being in fact composed of two concentric rings, very discernible in a good telescope: the breadth of the exterior being rather less than that of the interior. It has also been ascertained that the ring revolves round an axis perpendicular to its plane, and passing through the centre of the planet, in about ten hours and a half. It is sufficiently remarkable that this is the period in which a satellite, assumed to be at a mean distance equal to the mean distance of the ring, would revolve round the primary planet, according to the third law of Kepler.

The attention with which Huyghens watched the system of Saturn was rewarded by a second discovery; that of a satellite, which revolved round the planet, in the plane of the ring, in the space of about sixteen days. The fanciful notions in which Kepler and others had indulged so much were not yet quite exploded, and Huyghens ventured to predict that there could be no more satellites, as the number known was just equal to that of the primary planets. This was an unfortunate prediction: for not more than three years after it was made, Cassini discovered a second, and subsequently three more satellites, thereby carrying the whole number to five*: nor are these all, for in later times Herschel has added two. The satellite of Huyghens is the sixth from the planet, and the most considerable of any. With the exception of the seventh they all appear to lie in the plane of the ring. Their respective times of revolution are roughly twenty-three, thirty-three, forty-five, and sixty-five hours: four and a half, sixteen, and seventy-nine days†.

Cassini‡ of whom we have just been speaking was one of the greatest observers that Europe has produced. To him we owe a quantity of interesting and important facts regarding the system of which we form a part. By the assiduous observation of spots on the surface of Jupiter, he ascertained that this planet revolves on an axis nearly perpendicular to its orbit in about ten hours. In a similar way he ascertained that Mars revolved in about twenty-four hours on an axis, like that of Jupiter, nearly perpendicular to its orbit§. The observations of spots on Venus were much more difficult, for when this planet is brightest she is so near the sun as to make it almost impossible to observe her; and when she is farthest from that body her disk is dichotomized. However, Cassini succeeded in discovering that the rotation took place nearly in twenty-four hours round an axis very much inclined to the ecliptic. It is to him likewise that we owe the first observation of the division in Saturn's ring; and of the flattened form of Jupiter's disk; he determined this to be an ellipse, the major axis of which corresponded to the equator, and was to the minor axis in the ratio of 15 to 14‖. We shall see the importance of this observation, when we come to consider the figure of the earth. The importance of the eclipses of Jupiter's satellites for finding terrestrial longitudes

* Hist. de l'Acad. des Sciences, tom. i. pp. 150, 159, and 415.

† The discoveries of Huyghens with regard to Saturn were made in the year 1654; he employed a telescope of twenty-three feet in length, with an aperture of four inches. They are found in the Systema Saturnium, Hague, 1659.

‡ He was born in the county of Nice in 1625, and naturalized in France; he died at Paris in 1712.

§ It is really inclined between 20° and 30° to it. The English astronomer Hook has a claim to some share of the discovery of the rotation of Jupiter and Mars. His observations were nearly contemporaneous with those of Cassini, but much less numerous and complete; the results at which Hook arrived with regard to the times of rotation were mere approximations. V. Philosoph. Transact. Nos. i, viii. and xiv. For the discoveries of Cassini on the rotation of Jupiter, Venus, and Mars, see Elémens d'Astronomie, vol. i. pp. 402, 457, 511.

‖ The ratio now admitted is 177 to 167.

induced Cassini to study this elegant system with peculiar attention. He determined the times of their revolutions: he shewed that all moved in the same plane inclined about 3° to the orbit of the primary planet, and fixed the longitude of the intersection of these planes. He succeeded by assiduous observation in mastering all the difficulties of their motions, and finally published exact tables of their motions and eclipses: a present equally valuable to astronomers and geographers *.

One of the most useful of the labours of this eminent astronomer was his great improvement of the existing tables of refractions. Many erroneous theories existed at that time upon the extent and laws of this phenomenon. Most astronomers supposed that the refractions did not extend beyond 45° of altitude: that the absolute quantity was different for different celestial bodies. These ideas had been combated by Kepler, but the scientific world was still undecided. The table of refractions given by Tycho was purely empirical: that of Kepler was, as we have seen, founded upon an erroneous law. The method adopted by the former was to observe a great quantity of different altitudes of the sun and fixed stars, and to compare these with the altitudes calculated from a knowledge of their true places. The differences gave the effect of the refraction for each degree of altitude. But the instruments of Tycho were as we have seen extremely imperfect; and with them he was quite unable to appreciate the small refractions within 45° of zenith distance. The important discovery of Willebrord Snell, that when a ray of light passes from one medium to another, the sine of the angle of incidence bears a constant ratio to the sine of the angle of the refraction, enabled Cassini to construct tables on a more scientific principle than had hitherto been followed. To determine the constants of the problem, he took the refractions found by observation at two given altitudes, namely at the horizon, and at 80° of zenith distance. The former he found 32′ 20″ the latter 5′ 28″. Knowing then the law and the absolute values for these two altitudes he was enabled to determine the quantity of refraction for any other degree of zenith distance. It is easy to deduce from the theorem of Snell, that a ray of light entering our atmosphere is so inflected, that the whole deviation observed is proportional to the odd powers of the tangent of the zenith distance of the star from which it proceeds. If we only take into account the two first terms of the series, two observations will determine their two coefficients, and then every thing is known in the formula*. The form of the series shows that the refraction continues to the zenith.

The accurate knowledge of refraction led Cassini to correct materially the generally admitted value of the solar parallax. This, in conformity with the ideas of Kepler, was estimated by some at one minute, by others at twelve seconds, a quantity almost inappreciable in the state of astronomy at that time. Cassini having satisfied himself that the refraction was the same for the sun and the fixed stars, and perceiving that a parallax of one minute could not be made to accord with his tables of refraction, adopted the opinion of those who made it nearly insensible. But not satisfied unless he could ascertain its exact magnitude, however small, he proposed a method of considerable ingenuity for its determination. The principle of it is this: if we know the parallax and apparent semidiameter of any one planet, its distance and magnitude may be easily deduced; and knowing this distance, that of any other may be found by the third law of Kepler. Cassini proposed then to begin by determining the parallax of Mars. Mathematically speaking, this may be done by a single observer without quitting the same spot, by observing the planet when on the meridian, and also when but little above the horizon. But there are practical difficulties in this, arising from the uncertainty and magnitude of the refraction at low altitudes, and the smallness of the parallax sought. A better way is for two observers, under very different latitudes, to determine simultaneously, the position of Mars; for example, by comparing his place with that of a known fixed star. To effect this, Richer, a member of the Academy of Sciences, was sent to Cayenne †, but owing to the imperfect means of observation then existing, all he could ascertain was, that the parallax of Mars was insensible. This, however, was something gained, as giving a limit which the parallax of the sun could not exceed.

* Paris, 1693.

* This supposes the latitude known, but as a knowledge of the latitude involves that of the refraction, the problem is really indeterminate.

† Anno 1671. V. Hist. de l'Acad. des Sciences, tome i., p. 168.

For if that of Mars had amounted to much above a quarter of a minute, it would have been perceptible; estimating it, then, at 25″, it followed that the solar parallax could not exceed 10″. This was a great step made in the knowledge of our system, and showed that the sun was at the distance of at least 22326 semidiameters of the earth removed from our planet. It is curious to trace the progress of our knowledge on this interesting point. Aristarchus began by proving that the sun was nineteen times more distant from us than the moon: this determination seems to have been admitted by Ptolemy and Tycho Brahé, who made the sun's parallax three minutes: Kepler reduced it to one minute, and Cassini to 10″: astronomers of the present day only admit between 8″ and 9″, corresponding to a distance of about ninety-five millions of miles. The comparison of the apparent semidiameter shows the radius of the sun to be more than a hundred times that of the earth. Were the centre of the former supposed to coincide with the centre of the latter, the sun's external circumference would reach half as far again as the orbit of the moon. We shall perhaps form the best idea of the relative magnitudes and distance of these two bodies, by considering that if the former be represented by a globe of nine inches diameter, the latter must be a globule of one-twelfth of an inch in diameter, revolving round the former at the distance of seventy-five feet.

There is a curious phenomenon connected with the sun, first noticed by Cassini in the year 1683*. It consists in a pale light, very much resembling in appearance the tail of a comet, which is occasionally visible towards the time of the equinox, extending a considerable way along the zodiac, whence it has received the name of zodiacal light. The smallest stars are visible through it; its form may be compared to that of a cone whose base is the diameter of the sun. It does not always appear with equal brightness; in many years it is altogether invisible; and it would seem that it disappears when the spots on the solar surface are least numerous. The zodiacal light is not rigorously parallel to the ecliptic, but rather to that of the solar equator, which is somewhat inclined to the former. Putting all these circumstances together, Cassini was led to imagine that this phenomenon indicated the existence of a solar atmosphere of very rare and luminous matter and immense extent, since it reaches far beyond the orbit of Venus, and even perhaps as far as the earth. There are, however, difficulties in this explanation, which have caused it to be rejected by La Place; the subject is one of those regarding which we must be compelled to confess our ignorance.

Finally, Cassini has the honour of having completed the theory of the moon's libration. Galileo had explained the diurnal libration, and that in latitude. Hevelius had led the way in explaining the libration in longitude, by observing that the moon always presented the same face to the centre of her orbit, and not to the earth, which is in the focus of the ellipse. This Newton showed to be equivalent with supposing her rotation on her axis to be uniform, while her motion round the earth was unequal. Lastly, Cassini showed that the axis of rotation was not perpendicular to the ecliptic, but slightly inclined, and that the nodes of the lunar equator always coincided with those of the orbit. This explained satisfactorily what had been before observed, that the period of the inequalities of the libration coincided with the revolution of the nodes of the orbit.

This long list of discoveries sufficiently shews Cassini to have been one of the greatest observers who have ever existed: perhaps in this respect he is second only to Herschel. Yet we must confess with Delambre, that he has been in France the object of exaggerated eulogium; and several astronomers, whose labours were of a nature less generally appreciated, have in reality done more for the advancement of science. The observer who has devoted himself to forming a catalogue of the fixed stars, or the correction of the solar, lunar, and planetary tables, if he succeeds in his object, has rendered a more essential service than he who has added to the number of satellites of Saturn, or taught us the rotation of the planets on their axes. Yet the latter will enjoy a popular reputation, for which the former must not hope, contented to be properly appreciated by the scientific few.

We must now revert to Römer, whose invention of the transit instrument we have before noticed. But it is by a discovery of higher importance that his name is immortalized. It required a genius of no common order to detect the

* Mém. de l'Acad. tom., viii., p. 122.

interval of time occupied by light in traversing the planetary orbits; or even to conceive the possibility that its transmission was not instantaneous. This bold idea was not the result of a mere fanciful speculation, like those which so often led Kepler to stumble upon a sublime truth, and so often to plunge into the wildest errors; it was founded upon a series of careful and systematic observations: and Römer deserves equal credit for his assiduity in following the phenomena, and his acuteness in generalizing from them. Careful observation of the eclipses of Jupiter's satellites, and more particularly of the first, led him to remark, that sometimes these eclipses occurred from ten to twelve minutes later than was indicated in the tables of Cassini, while at other times no such retardation was perceptible. The quantity of this retardation was found to be constant for the same time, and to be the same for all the four satellites. It could not, then, depend upon the inequalities of the motions of these satellites; nor even upon that of Jupiter, for these would affect differently the different satellites. The cause then being extraneous to Jupiter and his system, Römer endeavoured to ascertain upon what other circumstances the fact could depend. In this he was led to the fortunate remark that the retardation was the greatest when Jupiter was farthest from the earth, and that its period varied with the distance of the planet. We can thus see the train of ideas that guided him to the theory of the successive propagation of light; for what other cause could make the eclipses occur later as Jupiter receded from us? The eclipse is determined by the satellite entering the cone of shadow cast by the primary planet. The position of this cone and of the satellite at a given moment evidently cannot depend upon the distance to the earth: there is, then, nothing left for us but to suppose that we see them later when we are farther removed; or that light is not transmitted instantaneously, but successively. The time taken by it to traverse the diameter of the earth's orbit was estimated by Römer first at eleven and then at fourteen minutes: modern astronomers have fixed it at sixteen minutes and a quarter. It is remarkable that an explanation so simple and beautiful of a phenomenon at first sight so perplexing should not have extorted at once general assent and eulogy. But this was far from being the case: it was not admitted even by the great Cassini; it was spoken of by the acute Fontenelle as a seductive error: and though the truth ultimately overcame all opposition, it remained sterile till the beautiful application of it made by the immortal Bradley*.

CHAPTER XV.

Picard—Measure of a Degree—Length of the Pendulum—Richer—Huyghens—Figure of the Earth—Flamstead—Halley—Transit of Venus—Theory of the Moon—Orbits of Comets.

THE attempts to measure a degree by Snellius and Riccioli, though conducted on scientific principles, had left much obscurity on the interesting question of the magnitude of the earth. The subject was well worthy of the attention of the French Academy of Sciences, and they determined to repeat the measurement, with all the accuracy which the recent improvements in the art of observation enabled them to aspire to. The person selected to conduct the operations was Picard, of whom mention has already been made in speaking of the micrometer, and the adaptation of telescopes to astronomical instruments. The care and skill with which he executed this very important duty, merits for him a high place among the astronomers of the seventeenth century.

The principle of this measure was, it is hardly necessary to say, exactly the same as that of Snellius; to connect two points by a series of triangles, and thus ascertaining the length of the arc of a meridian intercepted between them, to compare it with the difference of latitudes found from celestial observations. The extreme points of the arc of Picard were Malvoisine in the vicinity of Paris, and Sourdon near Amiens: the base was measured between Ville-Juif and Juvisy, and found to amount to 5663 French toises. In order to verify the observations, another base, of 3902 toises, was measured near Sourdon, and its length found to agree with that deduced from calculation. The difference of latitudes was determined by observing the zenith distances of δ Cassiopeiæ: this was found to be 1° 11′ 57″. The arc of the

* Römer's discovery was announced to the Academy of Sciences in 1676. V. Hist. de l'Acad., tom. i. p. 213.

terrestrial meridian between the two was 68430 toises, giving for the length of the degree 57064 toises. But the system of triangles having been subsequently extended to Amiens, and the latitude observed there, the length of the degree was found to be 57057; and by a mean between this and the former, 57060 toises*. If the arc of Snellius was the first that was measured upon scientific principles, that of Picard was the first that was executed with sufficient accuracy in the details, to bear a comparison with modern determinations of the same kind. It involved however two small errors, which fortunately were of such a nature as to compensate each other. Though the base had been measured twice over, and the difference between the two measurements only amounted to two feet; yet it was found by Clairault and Maupertuis, who verified the operations of Picard in the next century, that an error of six toises had been committed on this length. Again, in determining the latitudes, no allowance was made for the aberration of the fixed stars, a phenomenon at that time unknown; but this error fortunately acted in a contrary direction to the former. This operation of Picard is interesting, among other reasons, for having been the first in which instruments furnished with telescopes were employed. The terrestrial angles were measured with a quadrant of 38 inches radius; the zenith distances with a sextant of ten feet radius, both fitted with telescopes†. A novel and excellent part of the operation of Picard, consisted in comparing the length of the toise he made use of, with that of the pendulum beating seconds of time at Paris. The length of this pendulum was found to be 36 inches 8½ lines, the toise being supposed to contain six feet. The determination of the length of the seconds pendulum is so very delicate an operation, that perhaps no great reliance is to be placed upon the measure of Picard, but the idea is in the highest degree philosophical, and has received a splendid extension in later times. Mouton first proposed the length of the pendulum as a standard measure, but he complicated the idea so much by connecting it with the erroneous measure of a degree taken from Riccioli, as to make it impracticable. The advantage of an invariable physical standard, susceptible of being verified, or recovered if lost, is so great as to have engaged the attention of the most enlightened governments in Europe for some time past: and no standard seems, on the whole, better to satisfy the conditions required than the seconds pendulum. But on this point we shall have more to say in a subsequent part of this treatise.

In the preceding chapter it was mentioned that Richer had been sent to Cayenne to observe the parallax of Mars. While engaged with these observations, an unexpected phenomenon presented itself to his notice*. The pendulum that he used made, in the course of twenty-four hours, 148 vibrations less than it had done at Paris. Yet its length remained unaltered; and it was found necessary to shorten it, at least one line and a quarter, to make it beat seconds, as it had done at Paris. Picard had already perceived that a difference of temperature, by causing the metal to expand or contract, would affect the duration of an oscillation; but it was not possible to explain from this cause the phenomenon observed by Richer. For the change of temperature in this case could not at the utmost have affected the pendulum more than the third of a line. It was impossible then to assign any cause for the fact, which observations, carried on during ten months, had indisputably established, except the diminution of gravity towards the equator. The cause of this diminution was perceived by Newton and by Huyghens. We shall defer giving any account of the explanation of the former till we come to consider his theory of universal gravitation. The reasoning of the latter was founded on his own theorems regarding centrifugal force. Without entering into geometrical details, the nature of his theory may be thus briefly conceived:—Supposing the earth spherical, and to have a motion of rotation on its axis, the centrifugal forces which act on different points of the surface, will be different, according to the velocity with which they revolve in their different circles. A body at the equator revolves with the greatest velocity, and the centrifugal force soliciting it is the greatest; at the latitude 45°, both the velocity and the centrifugal force are much less; at

* Equal to 122943 English yards.

† For Picard's account of his operations, see Mémoires de l'Acad. des Sciences, tom. viii.

* Hist. de l'Acad., tom. i. p. 176.

the Pole the body is immoveable, and the force vanishes. Also the centrifugal force acts in the plane of the circle in which the body revolves, gravity in the direction of the centre of the sphere; the compound force resulting from these two will then not tend to the centre, except under the equator or the pole. The direction of the compound force must be perpendicular to the earth's surface at every given point; for the fall of heavy bodies must take place in this direction, and the surfaces of fluids at rest be perpendicular to it. But since this force does not tend to the surface of the sphere, and yet the earth's surface must be perpendicular to it, the earth cannot be a sphere. Under the equator the whole centrifugal force goes to diminish gravity, and this force is, at the same time, at its maximum; at any other place it is only a certain resolved part of the force in question, which counteracts gravity, and this force also diminishes towards the Pole. We see then the reason of the diminution of gravity from the Pole to the equator: a diminution which Huyghens from theory estimated at $\frac{1}{385}$th part of the whole. He proceeded to determine from theory the ratio of the equatorial to the polar semi-axis*. Following the example of Newton, he supposed two cylindrical canals, reaching from the centre of the earth, the one to the pole, the other to the equator. These canals he supposed filled with water; and it is evident that they must be in a state of equilibrium. The quantity by which the weight of the equatorial branch is diminished, is equal to half the product of the length of this branch into the centrifugal force. The whole weight of this branch then will equal the product of the absolute gravity into its length, minus half the product of this length into the centrifugal force. This must equal the weight of the polar branch, which is represented by the product of its length into the absolute gravity. Equating these two expressions, we get the ratio of the length of the equatorial to the polar branch, or which is the same, of the length of the equatorial to the polar axis of the earth. This ratio is that of the absolute gravity, to the same diminished by half the centrifugal force. But Huyghens had found before the centrifugal force at the equator equal to $\frac{1}{289}$th part of gravity: the half would be $\frac{1}{578}$. The ratio in question then which is the ratio of the two semi-axes, is 578 : 577. Though subsequent observation has fully established the spheroidal figure of the earth, it has made the difference of the axes much greater. Yet these early attempts to determine theoretically this important question are too interesting to be passed over without notice.

The foundation of the Royal Observatory at Greenwich, in the year 1675, has been already mentioned. The first person upon whom the office of astronomer at this observatory was conferred was John Flamstead, who had already distinguished himself by two small but excellent Treatises on the Equation of Time, and the Lunar Theory, which were published with the posthumous works of Horrox. It may justly be a matter of surprise, that the subject of the equation of time should not have been completely understood by the astronomers of Modern Europe, long before the work of Flamstead appeared. Yet such seems not to have been the case. Kepler had perceived, as indeed Hipparchus had done before him, that this equation was composed of two parts, the one resulting from the reduction to the equator of the sun's path in the ecliptic, the other from the inequality of the sun's motion in his orbit: but to these he had added a third, which had no real existence,—a pretended acceleration of the earth's rotation on its axis. Tycho had only admitted the first part of the equation; and the opinions of scientific men were still divided, when the treatise of Flamstead first put the question in its proper light. The numerous lunar inequalities have, in all times, given much trouble to astronomers. Besides the inequality depending upon the eccentricity of the orbit, Ptolemy had discovered another, the Evection, which appeared to be connected with the eccentricity, and which he attempted to represent in the way that we have seen in a preceding part of this treatise. Horrox, with great sagacity, perceived that it depended upon a periodical change of the eccentricity of the lunar orbit. To represent this, he supposed the centre of the lunar ellipse to revolve in a small circle, the earth remaining always in the focus. The least distance of the centre from the earth gave the first inequality alone, the greatest gave this inequality increased by the whole evection. In detecting the connection between these two inequalities,

* De vi Gravitatis, 1690.

Horrox showed great sagacity*; and Flamstead adopted the theory in his own work on the subject. But these geometrical hypotheses have ceased to possess any interest: for those who wish to construct lunar tables, it is sufficient to determine empirically a certain sine or cosine which represents the variations of any given inequality, and its maximum value. Hence its value, at any given moment, is found with the greatest facility. Kepler seems to have given the first idea of this method in the lunar theory, by remarking that the variation discovered by Tycho, was proportional to twice the distance from the moon to the sun.

When promoted to the place of Astronomer Royal, which he filled for many years, Flamstead devoted himself to observation with most meritorious zeal and constancy. The fruit of his long labours was ultimately published in three folio volumes, under the name of "Historia Cœlestis Britannica," containing a great mass of observations, and an extensive and accurate catalogue of the fixed stars. The Greenwich observations begin in the year 1676: at this time, Flamstead followed the method of Tycho and Hevelius; that of determining with a sextant the distances of the observed body from two given fixed stars; the only difference between his instrument, and that of the astronomers just mentioned, being that it was furnished with a telescope. But seven years afterwards he abandoned this system for that introduced by Picard, namely, the observing the transits and meridian altitudes simultaneously by means of a mural quadrant placed in the plane of the meridian. In this way the declinations of the stars may be obtained with considerable accuracy; but the right ascensions must be rather uncertain, from the difficulty of adjusting the plane of the instrument exactly to the meridian, or, at all events, of estimating the deviation. It is remarkable that the transit instrument did not come into use, at the observatories of Greenwich and Paris, till fifty years after its first invention by Römer. Bradley and La Caille have the merit of its introduction. It is still more curious that Halley, the successor of Flamstead, adopted it for a short time, and then abandoned it again, to confine himself to the mural quadrant. In the transits observed by Flamstead, with his mural instrument (and this may well excite our surprise), he employed only one wire: the time is generally given to whole seconds, very rarely to the half. From all these circumstances, these observations are now of little use to science: a tenth part of this mass made with the same care and skill as those contained in the triduum of Römer* would be a valuable treasure. Yet we must not undervalue the merits of this first Greenwich Catalogue: it was an important addition to the astronomical knowledge of the age; and its merits are sufficient, in the opinion of the best judges, to place its author by the side of Tycho among the most illustrious of the benefactors to science.

To accompany his catalogue, Flamstead undertook to construct a celestial Atlas on a large scale. This, indeed, had been done before by Bayer, of Augsburg, but with a very inferior degree of accuracy: the only happy idea in his atlas was that of distinguishing the different stars of a constellation by the letters of the Greek alphabet. The constellations, by a curious mistake, he had drawn as seen from the outside of the sphere. This was rectified in the charts of Flamstead, who invented a new method of projection for laying down the arcs of the meridian and the parallels. This method consisted in developing the arcs of parallels in parallel straight lines, perpendicular to a given meridian, also represented by a straight line: the degrees of right ascension measured on these parallels diminish as the cosine of the declination.

The successor of Flamstead, at the Greenwich Observatory, was Edmund Halley, one of the greatest names in the whole history of ancient and modern discovery. His brilliant career commenced at an unusually early age. At twenty he undertook a voyage to the southern hemisphere, for the purpose of forming a catalogue of those stars which were too remote from the North Pole to be visible to European observers. For the execution of this project he selected the island of St. Helena, a choice which turned out singularly unfortunate, as the climate of that island, which had been represented to him as well adapted to observation, proved cloudy and unfavourable in the extreme. During a year's residence (1676-7) he was not able to determine the places of more

* Horrox, Lunæ Theoria, Nova Opera Posthuma. London, 1673.

* V. Page 71.

than 360 stars, though he neglected the planets altogether, and, for the sake of greater expedition, merely observed the distances to stars known by the catalogue of Tycho. Yet so little was previously known of the southern hemisphere, that Halley's catalogue, though imperfect, was received with applause and gratitude in Europe; and its author was complimented by Flamstead, with the name of the southern Tycho. The survey of a part of the heavens so little known, gave Halley an opportunity of paying a compliment to Charles II., by forming a new constellation, to which he gave the name of *Robur Carolinum*, from the well-known oak in which that prince took refuge after the battle of Worcester. The patronage extended by Charles to the sciences, and particularly to astronomy, made him more deserving of the honour than some other sovereigns to whom similar compliments have been paid.

An important observation was made by Halley during his residence at St. Helena; that of a transit of Mercury over the sun's disc. It is not, however, in this case the observation itself which was of so much importance, as the idea which it suggested to Halley of employing these transits to determine the sun's parallax. Practically speaking, the transits of Mercury, though recommended by the illustrious proposer of the method, are not applicable to the purpose mentioned; but the transits of Venus offer the best means with which we are acquainted for this determination. This Halley has the credit of having perceived distinctly, and having most earnestly recommended to the notice of astronomers. But these transits of Venus are phenomena of excessive rarity: Halley could entertain no hopes of surviving to witness in person the next, which was announced for the year 1761. He addressed then to future astronomers a most solemn and affecting exhortation, not to suffer so precious an occasion to pass unprofited, but to unite all their efforts to deduce from this observation one of the most important elements of our system. His exhortation was not addressed in vain: we shall see in a subsequent chapter how successfully the idea was carried into execution.

It may be worth while, considering the extreme importance of the method proposed by Halley, to say a few words as to its nature. When a planet like Venus passes over the sun, it appears to describe a chord of the solar disc, greater or smaller, according to the latitude of the planet at the time. The extent of this chord is measured by the time the planet occupies in describing it. If there were no sensible parallax of either body, the chord described would be the same for the whole earth. But this is not the case; and the difference observed by two persons in favourable situations may amount to twenty-five minutes or more; and an error of ten seconds of time on the duration of the transit would not amount to more than the fifteenth of a second on the sun's parallax. It is easy to conceive the cause of this difference in the duration of the transit. Parallax always acting in a vertical, depresses the sun's disc, but it depresses Venus more, her parallax being much larger; and thus the chord described is diminished or increased, according to circumstances. Now, for an observer who at the time of the observation has the sun near his zenith, this effect of parallax will be insensible; while for another who has it near the horizon it will be very great: the difference between the duration of the transits observed by these two, will evidently be entirely due to the difference of the parallaxes of the sun and Venus; for, were these parallaxes equal, the duration of the transit could not be affected by it, both planets being equally depressed. We might choose two stations still more favourable for determining this difference, by placing the observers in different hemispheres. It is easy to see that, in this case, the parallaxes would act in different directions, the one bringing the planet nearer to the northern, the other to the southern limb of the sun, and in this the whole effect will be double of that which would take place in our first supposition. By a comparison then of observations thus made, we determine with much accuracy the difference of the parallaxes of the sun and Venus: from their periodic times we conclude the ratio of these parallaxes, and thus can find from two equations of the simplest form the absolute values of the parallaxes themselves. But the parallax of any one planet being known, those of all the rest may be deduced by the third law of Kepler. From this one phenomenon then of the transit of Venus, we may conclude the dimensions of all the planetary orbits; and though the whole parallax of the sun does not exceed nine seconds, it may be found, by these means within a tenth of a second. We may be allowed to reflect with some pride upon this triumph of human ingenuity *.

The theory of the moon attracted in a particular degree the attention of Halley; and he introduced into it several minor improvements, into the details of which it is unnecessary to enter. But we must not pass over without notice an ingenious suggestion for the amelioration of the lunar tables. The principal inequalities, both in longitude and latitude, depend upon the position of the moon with regard to its apogee, its node, and the sun. If, then, we could find any period at the end of which the moon returned to the same place with regard to all of these, it is evident that the inequalities would be renewed in the same order, and it would be easy to predict their recurrence, had they been observed during the period. Now, it has been seen that the Chaldæans had discovered such a period, comprising 223 lunar months; to which they gave, it is said, the name of Saros †. It is not to be understood that this period is rigorously and mathematically exact, yet the error is small, and, at all events, would not affect sensibly the magnitude of the equations. When Halley succeeded to the post of astronomer royal, in the year 1720, he began to execute his plan of observations on the moon throughout the duration of a saros, and he succeeded in carrying them through half this period. Lemonnier at Paris subsequently completed a whole period: but the progress made latterly in physical astronomy, has rendered these empirical methods almost useless. It is to Halley that we owe the first suspicion of a very important fact regarding the moon's motions. It had been hitherto the universally received doctrine, that all the planets, without exception, were subject to such inequalities only as are renewed within a certain space of time, and which, on that account, have been called periodic inequalities. The mean motion is determined by a comparison of the planet's places at very distant times, embracing a great number of the periods within which the inequalities are renewed, so that the result obtained is quite independent of these inequalities. No astronomer had ever ventured to doubt the uniformity of these mean motions; and, in fact, we now know that the major axes, and, consequently, the periodic times of the primary planets, are not subject to any but the periodical inequalities of which we have spoken. But this is not the case with the moon. The mean motion of that satellite is continually accelerated; and as this acceleration, though, mathematically speaking, it has a limit, yet will continue for perhaps many thousands of centuries, it is called the secular acceleration, to distinguish it from those inequalities which have a period that falls within the limits of observation. The cause of this phenomenon was discovered by La Place: the fact itself was first suspected by Halley. The reality of it was disputed for a long time, but it is now incontestably established. The mean motion of the moon, as determined by modern observations, exclusively, is between three and four minutes more rapid than that found by comparing modern observations with the eclipses observed at Babylon 700 years before Christ; and this result is fully confirmed by two eclipses of the sun, and an eclipse of the moon observed at Cairo by Ibn Jounis, towards the end of the tenth century.

It has been mentioned in a preceding chapter, that Hevelius had remarked that comets do not move as supposed by Kepler in right lines, but in orbits slightly curvilinear, and bearing considerable analogy to the parabola. But though Hevelius completely proved the deflection from a straight line, he did not prove the orbit to be parabolic, nor did he explain himself very clearly upon the subject of the focus being occupied by the sun. Delambre * has quoted a passage from the correspondence of Borelli, which shews this astronomer to have perceived the analogy between cometary orbits and the parabola, previously to the publication of the Cometographia of Hevelius. But the honour of having clearly proved this important point belongs to Dörfel, a Saxon, who, in a small treatise published in 1681, demonstrated that the great comet of the preceding year had described a parabola of which the sun occupied the focus. Without any knowledge of the work of Dörfel, Newton had been led to the conclusion, that comets describe

* Halley developed his idea, first proposed in the Catalogus Stellarum Australium, in the Philosophical Transactions for 1691 and 1716.

† V. Philos. Transact. for 1691, p. 535.

* Astron. Moderne, tome ii. p. 334.

round the sun very excentric ellipses, the visible arcs of which cannot be distinguished from a parabola. Halley, embracing with ardour this idea, and anxious to verify it, undertook the laborious task of calculating the elements of all the comets known, with a view to their recognition, should they at any time reappear. Having formed for this purpose a catalogue of twenty-four comets, he was struck with the resemblance between those of the years 1531, 1607, and 1682. The difference, however, of the inclinations of the orbits, and of the periods, made him doubt whether this could be the same comet that had reappeared three times at intervals of between seventy-five and seventy-six years; and the observations of Apianus and Kepler on the two first occasions were not sufficiently exact to decide the question. On further consideration Halley perceived in his catalogue three other comets, which returned in the same order and at the same intervals, namely, in 1305, 1380, and 1456. This gave him more confidence, and having re-examined and recalculated the elements, he ventured to affirm all these to be reappearances of the same comet, and to predict its return in the year 1758. This prediction has been completely verified, and its fulfilment forms one of the proudest triumphs of modern astronomy, while it has rendered the name of Halley immortal. But on this interesting point more will be said in a subsequent chapter*.

Chapter XVI.

Origin of Physical Astronomy.—Kepler—Descartes—Newton.

The causes which determine the motions of the heavenly bodies have from the earliest times been an object of speculation to mankind. For a long time these motions were explained by supposing the planets fixed to solid transparent spheres, the one revolving within the other, and having the earth for their common centre. But when this rude and barbarous hypothesis was overthrown by increasing knowledge—when it was shown that the sun was the centre of the planetary motions—that the planets themselves were bodies in all respects comparable to the earth on which we live—thinking men were led to reflect upon the cause of those motions round a common centre, and to compare it with those forces which may be observed at the earth's surface. Aristotle had chosen to assert that heavy bodies were attracted to the centre of the earth by a property peculiar to that point as the centre of the universe; and so completely was this power, according to him, inherent in that point, that heavy bodies he affirmed would equally tend to it were the earth removed. This fallacy was combated by Kepler, who was one of the first to entertain sound notions on the subject of gravity. He contended that a mathematical point, whether the centre of the universe or not, could have no power to attract heavy bodies to it; neither were they urged to this point from a desire to avoid the extremities of the universe, the distance of which is infinite, compared with the distance to the centre; nor by the revolution of the primum mobile, which would carry away the earth itself. Kepler then proceeds to explain the real nature of gravity. He begins by establishing that every particle of matter has a tendency to remain in repose, as long as it is beyond the sphere of influence of a cognate particle. Gravity is a tendency to conjunction between cognate particles, similar to the force of magnetism. But the earth will attract a stone much more than the stone attracts the earth. Heavy bodies are not carried to the centre of the world, in its quality of centre of the world, but as a centre of a cognate round body, namely the earth. Wherever the earth is transported heavy bodies will always tend to its centre. If the earth were not round, these bodies would not tend to its centre but to different points. Two stones placed in any part of the world near each other, and beyond the influence of a cognate body, would come together in an intermediate point, each moving through a space proportional to the comparative mass of the other. If the moon and the earth were not retained in their orbits by an animal force or some equivalent, the earth would ascend to the moon one fifty-fourth part of the distance, the moon would fall to the earth through the other fifty-three parts, supposing both to have the same density*.

The accuracy of these notions will probably surprise those who have imbibed the vulgar error, that the idea of

* The discoveries of Halley about comets are to be found in his Synopsis Astronomiæ Cometicæ, appended to his tables, London, 1749.

* Consult the Life of Kepler, chap. v.

a general attractive force between the different bodies of the solar system and all particles of matter originated with Newton. Such ideas had been prevalent for half a century before the publication of the Principia ; even the real law according to which the attractive force diminishes as the distance increases had been surmised by more than one individual. To those who know in what Newton's discoveries really consist, these statements will not appear to be any disparagement of that extraordinary man. He has created a science where others have conjectured.

But to return to Kepler, who proceeds in the same remarkable strain. The sphere of attraction of the moon extends to the earth, where it causes the tides ; much more then will the sphere of attraction of the earth extend to the moon, and even much farther. In other places Kepler recognizes the mutual attraction of the sun and planets, and even proceeds to conjecture that the attraction of the former may cause the irregularities in the moon's motion. But it must be confessed that Kepler could not explain the cause of the elliptic orbits of the planets, and he was obliged to have recourse to the singular idea that each planet was the seat of an intelligent principle, bearing the same relation to its material substance, that the mind of a man does to his body. The science of dynamics was not yet sufficiently advanced for him to see, that a tangential impulse, combined with gravity, would explain the revolutions without such a monstrous supposition.

The ideas of Galileo, though not so distinctly expressed, seem to have resembled those of Kepler. Galileo certainly recognized the mutual attraction of the particles of matter; and asserted that a certain number meeting together would have a tendency to unite in a spherical form. He also evidently saw that gravity, at the earth's surface, is the resultant of the attractions of the particles of which the earth is composed*. The idea of a general attractive force between the sun and planets being admitted, and the identity of this force with gravity having been clearly stated by Kepler, it was natural that the next object of speculation should be the law according to which this force decreased as the distance increased. Kepler seems to have imagined that the force varied inversely as the distance. Bouillaud has the honour of having first conjectured what has subsequently been proved to be the real law of nature, that the attractive force varied inversely as the square of the distance*. This, however, was a mere conjecture on the part of Bouillaud, and it was quite out of his power to offer any proof of it. However, the same idea was entertained by several men of eminence. In England it appears to have been adopted by Wren, Halley, and Hooke†; and the latter even boasted among his friends of being able to demonstrate it mathematically. However he was unable to produce this demonstration when called upon to do so ; to which we may add, that in a very clear and remarkable statement of the nature of the mutual attractive force between the celestial bodies and all particles of matter, he does not venture to state the law of its variation, a sufficient proof that, at least, he was unable to offer any satisfactory demonstration of its existence.

The physico-astronomical system of Descartes enjoyed such great popularity during the seventeenth century, that it is necessary to give some account of it in this place. It consisted in supposing the different bodies of the solar system to float in a fluid in motion round the sun, and forming a vortex of which that body occupied the centre‡. The different parts of this vortex moved with unequal velocities, and were of various densities, each planet floating in a stratum of a density equal to its own. In this great vortex there were again minor vortices, in which the satellites were carried round their primary planets. The mechanical and physical objections to this theory are so numerous and so completely insoluble, that its great success may well astonish us. That it should have been defended by some of the ablest mathematicians in Europe even after the appearance of Newton's *Principia*, is still more surprising. To mention a few only of the most striking difficulties, we may remark that it has been shown by D'Alembert that a vortex, whether spherical or cylindrical, could not exist unless all its strata revolved in the same time. The third law of Kepler being in

* Dialogo primo.

* Astronom. Philolaic. Lib. i., p. 23.

† V. Newton. Princip. Mathemat. Lib. i., Prop. 4, Scholium.

‡ The system of Descartes may be found as laid down by its author in the third part of his Principia Philosophiæ. Amsterdam, 1677.

direct opposition to this, proves that the planets are not carried round in a vortex. Again, in the system of Descartes the vortex is composed of strata of very different densities, superposed irregularly, so that in many cases a lighter stratum is lower or nearer the centre than one specifically heavier. Such an arrangement is totally irreconcilable with the laws of hydro-dynamics. To this we may add that the very existence of a spherical vortex seems, from several considerations, to be impossible. Were it worth while to go farther into the examination of a theory now totally exploded, several other arguments might be adduced. The various inclinations of the planetary orbits, their elliptical form, are all difficulties, which no ingenuity of the Cartesians was able to overcome. Yet, notwithstanding this, the theory of vortices reigned paramount in Europe during the greater part of a century. Its duration in England was shorter than on the Continent; the Newtonian theory having been received with universal approbation in the former country from the first moment of its publication. But on the Continent, where Cartesianism seems to have taken a firmer hold on men's minds, it continued to flourish for twenty or thirty years longer; and, indeed, cannot be said to have died away entirely before the middle of the eighteenth century.

It was reserved for the immortal Newton to lay the foundations of the science of physical astronomy, by applying the laws of dynamics to the motions of the celestial bodies, by demonstrating the identity of gravity with the attractive forces of the sun and planets, and by showing all the inequalities and irregularities of their motions to be consequences of the same principle that determined their elliptic orbits. At a very early age, Newton had been led to reflect upon the probable identity of gravity with the force by which the moon is retained in its orbit round the earth. There is a popular story current about his having been led to these reflections by observing the fall of an apple; but the fact is, that the relation of these forces had long been a subject of meditation to every philosophical mind. The difference between Newton and all his predecessors is, that he undertook at once to submit his conjectures to the test of calculation. It was natural to suppose, that if the force which at the surface of our planet determines the fall of heavy bodies to the centre of the earth be identical with that which retains the moon in her orbit, that it will act with less energy at that increased distance. But in order to compare the former effect with the latter, it was necessary to make some supposition as to the law of decrease. Newton adopted the hypothesis that the force varied inversely as the square of the distance. This, as we have seen, had been the supposition of Bouillaud and others. The distance from the centre of the earth to the moon being nearly sixty semidiameters of the former, gravity at the distance of the moon would be about 3600 times less than at the surface of the earth. If gravity then were the force which retained the moon in its orbit, that body, abandoned to the action of this force alone, during a small given time, ought to fall through a space 3600 times less than that through which a heavy body near the surface of the earth would fall in the same time. The former of these spaces is measured by the versed sine of the arc described by the moon in the given time. To calculate this versed sine, we must know the dimensions of the lunar orbit; or if we suppose the moon's distance to comprise sixty semidiameters of the earth, we must know the length of this semidiameter.

Unfortunately when Newton first undertook these calculations, the dimensions of the earth were not known with any accuracy; and having assumed a wrong value of the degree, he obtained results which appeared to overthrow his hypothesis. But Picard having in the mean time determined the length of the degree with considerable care, Newton resumed the consideration of the subject some years afterwards. His data being corrected, he found that the versed sine of the arc described by the moon in one minute, was equal to the space through which a heavy body at the earth's surface falls in one second. Consequently, the space through which the latter would fall in one minute, would be three thousand six hundred times greater than that through which the moon would fall in the same time. But this is exactly the proportion which ought to exist, supposing the motions in question to be determined by a force varying inversely as the square of the distance. Thus was established beyond contradiction the nature of the attractive force that determined the moon's revolutions.

The next step in this inquiry was to investigate the analogy between the forces by which the primary planets are retained in their orbits. In the execution of this, Newton began by showing that any body revolving round a point either immovable or uniformly progressive in a right line, and moving in such a way that its radius vector describes round the point areas proportional to the time, is urged by a centripetal force tending to that point. But the first law of Kepler established that the planets did revolve in this way; and Newton therefore concluded that they were acted upon by a centripetal force. The second law of Kepler was, that they moved in ellipses, of which the sun occupied one of the foci. This Newton showed could only be the case if the centripetal force varied inversely as the square of the distance from the sun. The third law of Kepler completed the proof of the identity of this attractive force for all the planets. It may be shown to be a consequence of it, that the intensity of the force in question is the same for all the planets at the same distance; as, on the surface of the earth, gravity gives the same motion to all bodies placed at the same distance from the centre. What then had been vaguely surmised by Kepler, Boulliaud, Hooke, and others, was proved, with all the certainty of mathematical demonstration, to be the law of nature. It may be necessary to explain here, for the sake of those who are not familiar with dynamical reasoning, that the elliptic orbit of a planet results from the combination of gravity with a primitive impulse in the direction of a tangent to the curve. Without gravity, the effect of the original impulse would have been to make the body move uniformly forwards in a straight line: its continual deflection, from a straight line towards the sun, indicates the existence of an accelerating force in the latter, and the form of the planetary orbit enables us to conclude the law of the force.

The comparison of gravity with the force determining the lunar revolutions, shows clearly, that in calculating the attractive forces, we must measure the distances from the centres of the sun, or the respective planets. This is demonstrated in the case of the earth and moon; and the attraction of the earth being evidently identical with that of the sun and the planets, we conclude the same with regard to them. But this property of attraction does not merely belong to the masses of the celestial bodies, but to each of their particles. It happens however, that the attraction of a sphere upon a point without it is the same as if all the matter of the sphere were united at its centre, provided the sphere be homogeneous, or at least composed of concentrical homogeneous layers. This proposition, which is due to Newton, explains the fact just stated with regard to the distances being measured from the respective centres of the heavenly bodies.

It being admitted that gravity is a property belonging to all particles of matter, it follows as a consequence that if the sun attracts a planet, the planet again will attract the sun. Supposing both these bodies spherical, we may imagine the mass of each united at its centre; and the attractions will be proportional to these masses. Strictly speaking then, the centre of the sun will not be immovable, while that of the planet revolves round it, but both will revolve round their common centre of gravity. However, the mass of the sun being so enormously greater than that of the largest planet, the latter may be neglected for all practical purposes in our calculations. Yet considerable disturbances arise in the motion of the planets from their mutual attractions. Thus the moon, while it revolves in an ellipse round the earth, is exposed to the attraction of the sun, causing a number of inequalities in its motions; thus too, the earth is troubled by the action of the neighbouring planets, each of which, in its turn, is similarly affected by those near it. The development of these perturbations, as they are called, forms a most important part of the theory of universal gravitation; a theory that shows us not merely the causes and connexion of the complicated and seemingly inexplicable inequalities already detected by astronomers, but has indicated the existence of many, which, but for its assistance, would be still unknown.

The inequalities of the moon being the most numerous and the most evident, engaged more particularly the attention of Newton. To form some idea of his method of investigation, let us begin by supposing the moon to move in the plane of the ecliptic. The forces acting upon her are the attraction of the earth, directed to the centre of the latter body, and that of the sun, the tendency of which in conjunction and opposition will be to diminish her gravity to the earth: for, in the former case, the moon is more strongly attracted by the sun than the earth; in the latter, the earth more strongly than the moon: the effect, in both cases, is the same, and equals twice the product of the sun's mass into the radius of the lunar orbit, the whole divided by the cube of the distance from the earth to the sun. In quadratures, the action of the sun on the earth, decomposed in the direction of the lunar radius vector, tends to increase the moon's gravity towards the earth, but the whole effect will be only half of that which takes place in a contrary direction at the syzygies. From all the actions of the sun on the moon, in the course of a synodical revolution, results a mean force diminishing the moon's gravity, and equal to half the product of the sun's mass into the lunar radius vector divided by the cube of the distance from the sun to the earth. Now, Newton proved that supposing one body to revolve round another which attracted it with a force varying inversely as the square of the distance, if we diminish the attractive force, by a quantity proportional to the distance, the orbit described will still be an ellipse, but its apsides will have a progressive motion; such in fact, as is actually observed, the lunar perigee having, as is well known, a rapid direct motion.

By the diminution of its gravity to the earth, the moon is maintained at a greater distance than would otherwise be the case: the sector described by its radius vector in a given time round the earth is not changed, since the force which produces it is in the direction of this radius. But the velocity and angular motion are diminished; as must evidently be the case, since the sector remains constant while the radius vector is increased. The preceding considerations show also that the radius vector varies sensibly, in the course of a lunar revolution, from the effect of the sun's attraction. The same is the case with regard to the excentricity. For the augmentation of the force in quadratures and its diminution in syzygies gives the orbit a greater curvature in the former and dilates it in the latter case. The excentricity then will alternately augment and decrease in each lunation: and, by following the same reasoning, it might be shown that the excentricity is greatest when the line of the apsides coincides with the line of the syzygies; thus indicating the cause of the inequality called the evection.

There is also an annual variation of the lunar excentricity, resulting from the excentricity of the earth's orbit round the sun. The disturbing force, it has been seen, varies inversely as the cube of the distance from the sun to the earth, and this distance not being always the same, the lunar orbit participates in its variations. When the earth is in its perihelion the disturbing force is greatest, and the orbit of the moon most dilated: the contrary takes place at the aphelion. Hence arises an inequality, which is called the annual equation, as it has for its period an anomalistic year.

Hitherto we have only considered the part of the disturbing force decomposed in the direction of the lunar radius vector: the other part tends to increase the velocity of the moon in its orbit or the area described in a given time. The increment of the area is proportional to the sine of twice the angular distance from the moon to the sun. This in fact is the inequality discovered by Tycho Brahé, which is greatest in the octants, and to which the name of variation has been given.

Let us now proceed to consider the effects of the disturbing force resulting from the inclination of the lunar orbit to the ecliptic. The most important of these is the retrograde motion of the nodes. To conceive the cause of this, let us suppose, as before, the solar attraction decomposed in two directions, the one parallel to the line joining the earth and the sun, the other in the direction of the lunar radius vector. The first will be entirely in the plane of the ecliptic, and therefore can only affect the longitude of the planet; but the second will be inclined to this plane and may again be decomposed into two, one parallel to the ecliptic, the other perpendicular to it. The effect of the former will fall entirely upon the longitudes: it is then only the latter which is now to be considered. The constant tendency of this force is to bring the moon nearer to the ecliptic, and consequently to accelerate its return to this plane. The effect will be to make the moon, at each semi-revolution, meet the plane sooner than it otherwise would have done, and at a point less advanced on the orbit. The node then will appear to have retrograded a certain quantity; and the phenomena will be just the same as if, while the moon moved in the plane of its orbit, the orbit itself had a retrograde motion. The retrogradation in question will be continued, but it will not be always equal. When the nodes are in the syzygies, the force perpendicular to the orbit will vanish, since the plane of the orbit will pass through the sun, and consequently the retrogradation will cease with it. But the moment the sun has passed the line of the nodes, the force will begin to act, and will reach its maximum when the nodes are in quadratures.

Such is a short sketch of the way in which Newton laid the foundations of the theory of the moon; a theory that from its complication has at all times occupied a large share of the labours of the mathematician and the astronomer. His numerical determinations must be considered as rough approximations; and even his methods of investigation can now be studied only as matters of scientific curiosity. The exact determination of the arguments and coefficients of the numerous lunar inequalities, requires all the resources of modern analysis, and all the accuracy of modern observations. The geometrical methods of Newton are inadequate to the complete solution of this complicated problem. Yet to mathematicians they will probably always prove an object of interest; while the sublimity of the results will be an object of unqualified admiration to all.

Observation alone might teach us the magnitudes and distances of the sun and planets. But the Newtonian theory gives us the means of determining an element that might have seemed for ever inaccessible, namely, their masses. The whole attraction of the sun or any planet in this theory is proportional to its mass. If we were able to measure the weight of bodies at the surface of the sun, we might, by comparing with this their weight at the surface of the earth, ascertain the ratio of the mass of the sun to the mass of the earth. But such an experiment is clearly impracticable. We have, however, another in our power which is equivalent to it. If we can determine the force with which the sun attracts a given planet, Mercury for example, we may ascertain the force with which it attracts a body at its own surface. For these forces are to one another inversely as the squares of the distances from the sun's centre. But the force with which Mercury is attracted may be determined in the same way that the attraction of the earth on the moon was found. It is measured by the versed sine of the arc, described in a given time, as, for example, a second. In this method, the mass of Mercury being quite insensible, compared with that of the sun, is neglected. In the same way we may compare the masses of Jupiter and Saturn with that of the sun: for each of these planets is accompanied by satellites, the revolution of any one of which may be employed as that of Mercury, as has just been shown. Venus and Mars being without satellites, their masses must be determined by the perturbations they cause in the motions of the neighbouring planets, as, for example, the earth. To find the mass of Mercury, we are forced to make an hypothesis somewhat precarious; namely, that the densities of this planet and the earth are inversely as their mean distances from the sun. Newton attempted to determine the mass of the moon from certain phenomena of the tides, of which more will be said presently. Other and probably better methods have been since proposed. One of them is founded on a comparison of the quantity of the nutation of the earth's axis, and the precession of the equinoxes. The former phenomenon results solely from the moon's action, the latter from that of the sun and moon combined. In this way the lunar action is found to be about double that of the sun. Each of these actions is proportional to the mass of the body, divided by the cube of its distance to the earth. Hence we obtain for the ratio of the masses in question $\frac{1}{68.5}$. The theory of centrifugal force, due to Huyghens, had made it evident that the rotation of the earth on its axis was incompatible with the spherical form, supposing the earth to have been originally fluid. This subject was investigated by Newton, and considered by him in a more general and satisfactory way. In determining the ratio of the polar to the equatorial axis, Huyghens had supposed every particle of matter in the earth to have the same weight, at whatever distance it is from the centre. This is quite incompatible with the Newtonian theory, that the whole attraction of the earth is the resultant of the attractions of all the particles of which it is composed: a doctrine which, in fact, Huyghens never admitted, though he allowed the Newtonian law of gravity to exist between the different planets. Newton, on the other hand, demonstrated that a body placed within a spherical stratum of uniform thickness is equally attracted on all sides; and that the same is the case with an elliptical stratum, whose interior and exterior surfaces are similar and similarly situated. Supposing the planets to be homogeneous spheres, gravity in their interior will vary directly as the distance from the centre; for, taking a point anywhere, the stratum exterior to it does not contribute to its gravity, which is, in consequence, only produced by the attraction of a sphere, whose radius is equal to the distance from the centre. This attraction is proportional to the mass of the sphere divided by the square of the radius, and the mass of the sphere is proportional to the cube of the radius; whence we see that the attraction varies as the radius. But if the planets be denser towards their centres, gravity will no longer diminish in the same ratio.

It is very easy to see that this theory gives a greater flattening at the poles than that of Huyghens; the particles of the equatorial column weighing less in succession as they approach the centre, will yield more to the centrifugal force acting upon them. The determination of the axes is much more difficult in this hypothesis. Newton contented himself with supposing, that an elliptic figure would satisfy the conditions of equilibrium of a fluid mass in rotation, whose respective particles attracted each other mutually as the inverse squares of the distances. Supposing, at the same time, the earth homogeneous, he found that the two axes of the planet would be in the ratio of 229 : 230. This supposition of Newton, that the elliptic figure satisfies the conditions of equilibrium of an homogeneous fluid mass, was subsequently demonstrated by Maclaurin. But repeated measurements having shown that the ratio assigned is too great, it follows as a consequence that the earth cannot be homogeneous.

The connexion between the phenomena of tides and the diurnal revolution of the moon had not escaped the notice of the philosophers of antiquity. Pliny*, among others, distinctly observes that the cause of the tides is in the sun and moon, and refers them in particular to the attraction of the latter. In modern times Descartes proposed an explanation of these causes, which bears some analogy to that of an ancient philosopher, named Seleucus*. He imagined that the moon, passing through the zenith of a spot covered by the ocean, pressed upon that spot, and forcibly lowering the surface of the water immediately beneath it, forced this surface to rise in compensation along the distant coasts, attaining its maximum of elevation at 90 degrees from the spot in question. In any given place, then, the waters will rise and fall alternately at intervals of six hours, and the moon's passage over the meridian retarding daily three quarters of an hour, the tides will be retarded by the same portion of time, which is consistent with observation. But this explanation is quite untenable even in the Cartesian system of vortices. How can the moon, suspended in a vortex of the same density with itself press upon the surface of the ocean? The pressure will be uniform all round the earth, and will depend upon the density of the vortex in question: it can make no difference whether any given part of this is occupied by the moon or by a fluid sphere of equal specific gravity. Besides, Descartes does not explain by these means how it happens that while it is low water under any one meridian it is also low water under the meridian removed from it 180 degrees. The pressure of the moon seems inadequate to explain this.

The celebrated Galileo imagined that the phenomena of tides were produced by the rotation of the earth on its axis, and he seems to have imagined that we might find in them a proof of the Copernican theory. When a given point on the earth's surface has arrived under a meridian exactly opposed to the sun, the motions of rotation and translation take place in the same direction; when it is in conjunction with the sun these two motions take place in contrary directions. Galileo imagined that, from this circumstance, a particle of fluid would be accelerated in the first case, and retarded in the second; and that the whole effect on the ocean would be to form two waves, the axis of the first of which would be in the direction of the meridian passing through the sun, the other exactly opposite to it†. But this is founded

* II. 97.

* Plutarch. de Placit. Philosoph. III. 17.
† Dialogo quarto.

on false physical notions. The waters of the ocean we now know from hydrostatical principles would not be affected at all by the causes here assigned: they participate in the uniform rotation of the globe. In addition to this, the hypothesis of Galileo has the disadvantage of making the tides depend entirely upon the sun, whereas their connexion with the moon is so evident as to have been recognized from the earliest times. It will hardly be credited, that, to obviate this last objection, Baliani adopted a system which made the earth revolve as a satellite round the moon. By this means, following the principles of Galileo, he explained the two high tides depending upon the moon's passing the meridian above or below the horizon; and the ebb when she is rising or setting. But such a theory is quite undeserving of a serious refutation.

Though Galileo treated the theory of attraction by the moon as absurd, yet it was generally embraced by modern philosophers, even before it was completely developed by Newton. Kepler says that the sphere of attractive virtue which is in the moon extends as far as the earth, and entices up the waters. The same, nearly, was said by the Jesuits of Coimbra, and by Antonio de Dominis, Archbishop of Spalatro, and had been stated before them, though in rather a vague and fanciful way, by Gilbert*. The difficulty experienced by these writers was to explain the cause of the inferior tide: this was first done in a satisfactory way by Newton. Just in the same way that, in treating of the inequalities of the moon, we have seen that, when this satellite is in conjunction with the sun, she is more strongly attracted by the latter than the earth, and, when in opposition, the earth more strongly than the moon; the effect produced being in each case exactly the same, namely, to increase the distance from the moon to the centre of the earth: so may we reason with regard to the surface of the ocean and the centre of the earth. That portion of the water which has the moon on the meridian and above the horizon, will be more strongly attracted than the earth's centre; and again, that centre more strongly than the portion under a meridian distant half a circumference from the first. Towards each of these meridians then the water will rise, and its whole surface will form a spheroid whose major axis is directed to the moon. If the earth were at rest, this axis, always directed to the moon, would traverse the earth's circumference in the period of a lunar revolution.

The rotation of the earth, however, causes a diurnal motion on the axis, corresponding with the tides observed in twenty-four hours. The inertia, however, of the waters of the ocean does not permit them to move with quite as much rapidity as the moon; and the major axis of the spheroid mentioned will remain somewhat behind the direction of the moon's place. The exact quantity of this retardation at any given port will depend upon local circumstances, and can only be ascertained by observation.

The sun acts upon the ocean exactly in the same way as the moon, but its effect is much less, in consequence of its distance. There will then be a solar as well as a lunar tide twice in every twenty-four hours. In syzygies the two tides will coincide, and the whole rise will be the greatest; in quadratures, they will counteract each other, and the tides will be at their minimum. The first have been called spring, the latter neap tides. In general, the force of the sum being much less than that of the moon, the only way in which the solar tides will be perceptible is by their effect in augmenting or diminishing those owing to the moon. These phenomena were employed by Newton for finding the ratio of the masses of the sun and moon; for at spring-tides we observe the effect of the sun, and at the neap-tides of the difference of their forces. These effects depend upon the distance and mass of each body: now their distances are known,—hence, then, the ratio of their masses may be found. It is, however, necessary to take into account several circumstances which must clearly affect the magnitude of the tides, such as the declination of the sun and moon, the distance of each from the perigee of its respective orbit, &c. Newton found that the ratio of the forces was about $4\frac{1}{2} : 1$. More extensive observations have made it nearly $3 : 1$.

If the earth were a perfect sphere, or if, being of a spheroidal form, the axis of rotation were perpendicular to the plane of the ecliptic, the motion of the equinoctial points, called precession, would not take place; nor would the action of the sun tend in any degree to produce an alteration in the motions of the earth:

* Consult the Life of Galileo, p. 71.

for in either case, the action upon the two parts of the latter, above and below the place of the ecliptic, would be equal, and would act in contrary ways to produce any change of position in the axis of rotation. But this is no longer the case when the earth is spheroidal, and the axis inclined to the ecliptic. To examine what takes place in this case, Newton considered the earth as a sphere surrounded by a meniscus of matter, whose greatest thickness is at the equator, and he compared each particle of this meniscus to a small satellite revolving round the earth in twenty-four hours. To each of these satellites we may apply what has been already said relative to the orbit of the moon; and for the same reasons the sun will cause the nodes of each of these particles to retrograde, at each diurnal revolution, as the nodes of the moon do at each lunation. From the sum of all these small motions will result a daily retrogradation of the points of intersection of the equator and the ecliptic—that is to say, of the equinoxes; and though it is extremely small, yet, as it is daily repeated in the same direction, the effect at the end of a year will be very sensible.

It must not, however, be supposed, that the whole precession, or even the greater part of it, results from the action of the sun. The attraction of the moon produces an effect exactly similar in its nature to the former, and much more considerable. It depends upon the spheroidal figure of the earth, and the inclination of its axis to the plane of the lunar orbit. The lunar orbit, however, being but slightly inclined to the ecliptic, the retrogradation of the intersections of the equator and this orbit will not differ materially from that of the intersections of the equator and the ecliptic. The joint effect of the sun and moon will produce a precession of about 50″ annually, and of this, about 33″ are due to the latter alone.

The explanation of Newton indicates the true cause of the great phenomenon of the precession of the equinoxes, so long involved in mystery, and is fully worthy of the sagacity of its illustrious inventor. But in his day analysis and mechanics were not sufficiently advanced, to admit an exact investigation of the various motions of the earth's axis produced by the attractions of the sun, moon, and planets. The glory of Newton consists in having shown that not merely the elliptic orbits of the planets, and the law of their periodic times, but all the irregularities of their motions were consequences of one great principle, and might be deduced from it by rigorous calculation. In the complete execution of these calculations, it has been found necessary to have recourse to methods differing considerably from those of the inventor; but the most distinguished of his successors have cheerfully recognized in him the first of mathematicians and philosophers, the father of physical astronomy*.

Chapter XVII.

Bradley.—Aberration of the Fixed Stars —Nutation.

From the time that the Copernican hypothesis was generally received, the attention of astronomers was more and more drawn to the parallax of the fixed stars. It was well known that this parallax, if it existed, must be very small; but it seemed reasonable to suppose, that a line of such magnitude as the diameter of the earth's orbit would subtend an angle of at least a few seconds at the distance of the nearest fixed star. Though the non-existence of any parallax was not considered sufficient to shake the conclusive arguments urged in favour of the earth's annual motion, yet it was felt that, were it once clearly established, however small the quantity, the question would be decided beyond all controversy. With this view, Galileo proposed a method of observation, which he never seems to have realised, but which has been re-invented in an improved form in modern times, and successfully used by M. D'Assas de Montdardier. This method consists in fixing invariably a telescope, and at a considerable distance from it, in the field of view, an opaque object; as, for example, an horizontal beam. In this way artificial occultations are produced by the disappearance of the star behind the object; and should the latter pass at different altitudes at different times of the year, it will at one time suffer an occultation which it will not at another,

* Newton was born at Woolthorpe, in Lincolnshire, in the year 1642: he died in London in 1727. The first edition of his famous work, the Principia Mathematica, was published in 1687. It is said that he had been in possession of his most remarkable discoveries for many years.

or it may be seen to pass at one time above, at another below the beam: the fact, then, of a change of position, even though very small, may be thus ascertained. It is better, however, as has been done by M. D'Assas, to make the object triangular, and suffering it to extend across the whole field, to measure the time between the apparent immersion and emersion; this interval evidently depending upon the altitude of the star. By observations of this kind, M. D'Assas seems to have established satisfactorily an absolute parallax of 2″ for the star Keid*, 1″.43 for Rigel, and 1″.24 for Sirius†.

To go back to the year 1660, we find at that time Hooke engaged in researches of the same nature. With a telescope of thirty-six feet, placed vertically, he observed during several years the bright star in the Dragon's head‡ passing very near his zenith: he found its altitude generally less near the winter solstice than in the summer. This he concluded to be the effect of parallax; and published his results as a proof of the motion of the earth. Following the same ideas, Flamstead, with a mural quadrant, made a series of observations on the altitude of α polaris, in which he found variations of the same kind. These variations, indeed, had been previously noticed by Picard, when sent to determine the latitude of Tycho's observatory at Urainburg. But it was shown by several astronomers that these could not be the result of any annual parallax; since this would not tend to make the zenith distance of this star greater in winter than in summer, but greater at the vernal than at the autumnal equinox.

Flamstead, having been unsuccessful in his attempt to discover the parallax by the variations in declination, Römer conceived that the variations in right ascension would be better adapted for the purpose. For many years he prosecuted observations with this view, which were continued and published by his pupil Horrebow. The results were not very satisfactory: all that Römer ventured to affirm was, that the sum of the parallaxes of Sirius and α Lyræ was between a minute and a minute and a half in space§: Horrebow, from his own observations on several stars, found for a mean about 15″. But though he announced his discovery with much confidence in the *Copernicus Triumphans*, we know that his determination is considerably too great, and his means of observation were apparently inadequate to the solution of so nice a problem. About the same time, Cassini having observed the variations of altitude in Sirius, conceived that he had been able to distinguish the effects of a parallax. But it is unnecessary to enumerate at any length the attempts of a similar kind before the time of Bradley, as the small motions observed involved the effects of aberration and nutation, phenomena which were then unknown.

In the year 1725, two English astronomers, Molyneux and Bradley, began to observe with a zenith sector, the star γ Draconis, which was selected from its passing very near the zenith of their observatory at Kew*. Observations made on the 3d, 5th, 11th, and 12th of December, not having indicated any material difference in the place of the star, they were about to be discontinued for a time, when Bradley, having from curiosity repeated his observation on the 17th, perceived that the star passed rather more to the southward than before. This was at first attributed to the uncertainty of the observations; but it was confirmed by the star's passage, on the 19th, still farther to the southward. These appearances surprised Bradley and Molyneux, since they could not be the result of annual parallax; and at first they imagined that they might arise from some change of place in the instrument. But this idea was soon excluded by the regularity of the motion observed. At the beginning of March, 1726, the star was found 20″ more southerly than at the time of the first observation; and this was the limit of its progress in this direction: by the middle of April it began to return to the north; and about the beginning of June had returned to the same zenith distance that it had in December. It then proceeded to the northward for about 20″; and in September began to return to the south, reaching in December the place it occupied twelve months before, allowing for the precession of the equinoxes.

The order of these changes excluding the idea of an annual parallax, Bradley

* 29. Eridani.
† Connoissance des Tems, 1831. Additions, p. 134.
‡ Philosoph. Transact., No. 101.
§ Horrebow, Basis Astronomiæ

* V. Philos. Transact. Vol. xxxv., p. 637.

at first supposed they might be caused by a nutation of the earth's axis. But this idea was rejected, as a small star, with nearly the same declination as γ Draconis, but differing from it in right ascension about 180 degrees, experienced in the same time a change of declination, not amounting to more than half of the former: had the earth's axis moved, the alteration in both stars from this cause must have been nearly equal. In order to investigate the subject more completely, Bradley erected at Wanstead a sector of $12\frac{1}{2}$ feet radius, with an arc of about $6\frac{1}{4}^{\circ}$ on each side of the zenith; and with this he examined carefully during a whole year the motions of γ Draconis, and several other stars. He soon remarked that only those stars situated near the solstitial colure had their greatest digressions when the sun was in the equinox; but that these digressions took place for all when they passed the zenith at six in the morning or evening; and again, that each was farthest north when it passed at six in the evening, and farthest south when it passed at six in the morning.

After various conjectures the happy idea occurred to Bradley, that all these phenomena proceeded from the progressive motion of light, and the annual motion of the earth in its orbit. He considered this matter in the following way. Let CA be a ray of light falling perpendicularly on the line BD; then

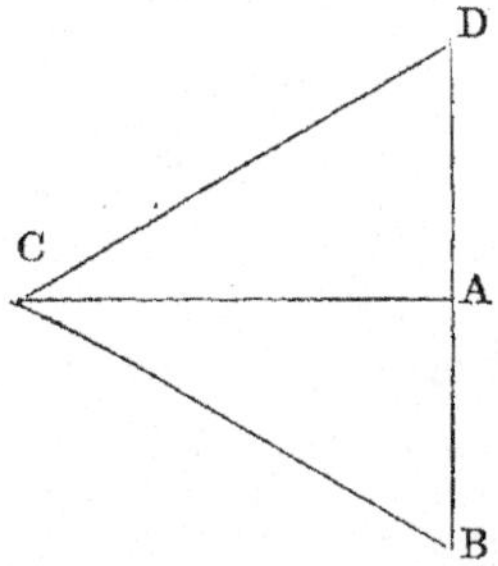

if the eye is at rest at A, the object must appear in the direction AC, whether light be propagated in time or instantaneously. But if the eye be moving from B towards A, and light be propagated in time with a velocity that is to the velocity of the eye as CA to BA, then light moving from C to A, while the eye moves from B to A, that particle of it which will be discerned when the eye comes to A, is at C when the eye is at B. Suppose BC a tube of such a diameter as to admit but one particle of light: then the particle of light thrown off from C, by which the object must be seen when the eye arrives at A, will pass down the tube BC if it be inclined to BD, at the angle DBC, and accompany the eye in its motion from B to A; and it will not pass down the tube if it have any other inclination to BD. Similarly if the eye moved the contrary way, with the same velocity, from D to A, the tube must be inclined at the angle BDC. Although then the line drawn to the real place of an object be perpendicular to the line in which the body is moving, the line drawn to the apparent place will not be so; since it must depend upon the direction of the tube; and the difference between the true and apparent places will depend, *cæteris paribus*, upon the proportion between the velocity of light and that of the eye.

If the earth revolve round the sun annually, and the velocity of light were to the velocity of the earth's motion in its orbit (supposed circular) as 1000 to one, a star really placed in the pole of the ecliptic, would, to an eye carried along with the earth, seem to describe a circle round that pole of $3\frac{1}{2}'$ radius. If, on the other, hand the greatest alteration of this star be known, we may deduce the proportion between the velocity of light and that of the earth in its orbit; and from this we may find the difference between the true and apparent place of any other star for any time; or conversely, given this difference, we may find the proportion between the two velocities.

A star neither in the pole nor in the plane of the ecliptic will appear to describe about its true place an ellipse, whose axis major is parallel to the ecliptic, and equal to the diameter of the little circle described by the star in the pole, while the axis minor will be to the axis major as the sine of the star's latitude to the radius. Bradley proceeds to show how the semi-axis major of the ellipse may be deduced from the differences of declination; and, applying his formulæ to the preceding observations on γ Draconis, he found for the diameter of the small circle at the pole 40″.4. One half of this is the angle ACB in the figure: hence we may conclude that

AC is to AB as 10210 to 1; and it follows that light moves from the sun to the earth in 8′ 12″ of time. The result given by γ Draconis was confirmed within a second, by various other stars; and it agrees very well with observations made in later times on the eclipses of Jupiter's satellites.

To these apparent changes in the places of the fixed stars the name of *aberration* has been given; and the discovery ranks in importance next only to the laws of Kepler, and the precession of the equinoxes. But this is not the only triumph of the English Hipparchus, as Bradley has been justly called. Scarcely had he determined the laws of the motions produced by the successive propagation of light, than his attention was excited by a new phenomenon *. Continuing his observations with the sector, he found a greater apparent change of declination in stars near the equinoctial colure, than could arise from a precession of 50″ in a year; and at the same time an effect of exactly an opposite nature in stars near the solstitial colure. From the year 1727 to 1732 the changes in question amounted for some of the stars to 9″ or 10″. It being evident that an alteration of the quantity of the precession would not explain these changes, Bradley reverted to the idea which had occurred to him first, when considering the causes of the aberration, namely, a nutation of the earth's axis. This he found confirmed; for, while γ Draconis appeared to have moved northward, the small star, 35 Camelopardali, which was almost opposite to it in right ascension, seemed to have gone as much towards the south; and the same thing was observed in a similar comparison of other stars. These motions were found to be in connexion with the position of the nodes of the lunar orbit. Thus γ Draconis, in the course of nine years (from 1727 to 1736), or half a revolution of the node, moved 10″ to the north: and at the end of the nine succeeding years had returned exactly to its original position in 1727. Two things were thus ascertained; that the motions in question resulted from a change of position in the earth's axis, and that these changes were connected with the places of the nodes of the moon's orbit. It remained to discover the cause of the connexion between these variations.

'I suspected,' says Bradley, 'that the moon's action upon the equatorial parts of the earth might produce these effects. For if the precession of the equinox be, according to Sir Isaac Newton's principles, caused by the action of the sun and moon upon those parts, the plane of the moon's orbit being at one time above ten degrees more inclined to the equator than at another, it was reasonable to conclude, that the part of the whole annual precession, which arises from her action, would in different years be varied in its quantity; whereas the plane of the ecliptic, wherein the sun appears, keeping nearly always the same inclination to the equator, that part of the precession which is owing to the sun's action, may be the same every year; and from hence it would follow that, although the mean annual precession, proceeding from the joint actions of the sun and moon were 50″, yet the apparent annual precession might sometimes exceed, and sometimes fall short, of that mean quantity, according to the various situations of the nodes of the moon's orbit.' It is difficult to give a more lucid account of the causes of nutation than in these few words of Bradley. The phenomena both of precession and nutation, considered as consequences of the solar and lunar attractions upon the spheroid of the earth, were subjected to rigorous investigation by D'Alembert, and the comparison of his results with observation, placed beyond dispute the correctness of the explanation given by Bradley.

Machin suggested an elegant construction for representing the various effects of nutation. The precession of the equinoxes may be conceived to be caused by a slow retrograde conical motion of the axis of the equator round that of the ecliptic, the angle of the two remaining invariable, so that the pole of the equator describes the circumference of a small circle perpendicular to the former axis. This at least would represent the phenomena of precession, were it uniform; as it may easily be seen that the equinoxes in this case would have an uniform retrograde motion through the circumference. We may consider the small circle above-mentioned as the locus of the mean pole of the equator, round which the real pole oscillates in consequence of the nuta-

* V. Philosoph. Transact. for 1748, p. 1.

tion. Machin supposed the real pole to describe round the mean, in the time of a revolution of the lunar nodes through the ecliptic, a small circle of 18″ diameter. It would be more accurate to substitute for this small circle an ellipse, whose major axis occupies an arc of 18″, on the circle of latitude passing through the poles of the equator and the ecliptic, to which it is a tangent; the minor axis being a tangent to the circle on which the pole of the equator moves parallel to the ecliptic, and occupying an arc of 13″ of a great circle of the sphere.

Chapter XVIII.

The Figure of the Earth.—Arc of the Meridian and Parallel measured by Cassini in France—Measure of a Degree in Lapland—Verification of the Arc of Cassini—Measure of a Degree in Peru—Theoretical Investigations—Maclaurin—Clairault—Measure of an Arc of Parallel by Cassini—Degrees of La Caille, Boscovich, Beccaria, Mason, and Dixon.

Scarcely had Picard completed his measure of a degree, before the Academy of Sciences resolved on executing a more extensive and important operation. It was proposed to prolong the arc of the meridian measured by him in each direction, and to make this the basis of a new and accurate map of France. This project, first entertained in the year 1671, was partially carried into execution in 1679 by Picard and La Hire. From the year 1680 to 1683 the prolongation of the meridian was carried on with activity by La Hire and Cassini, when the death of the minister Colbert suddenly interrupted their proceedings. The measure of the meridian was not resumed till 1701, when Dominique Cassini, assisted by his son, prolonged it to the southernmost extremity of France, determining, geometrically, the length of an arc of more than six degrees; he concluded the mean length of a degree to be 57097 Parisian toises. It still remained to measure the arc intercepted between Paris and the northern extremity of the kingdom. Though this had been partially done by La Hire, it was found necessary to repeat the whole operation from the extremity of Picard's arc. This was effected in 1718 by Jacques Cassini, who prolonged the meridian to Dunkirk; he obtained for the mean length of a degree 56960 toises.

These measures indicated a decrease in the length of the degree from the equator to the poles; but it is rather remarkable that Cassini and his coadjutors did not at first draw the legitimate inference from them. For if the earth be flattened towards the poles, it is not difficult to see that the length of the degree must increase as we recede from the equator. The less the curvature the greater will be the radius of the osculating circle to the meridian at any given point. In comparing two degrees measured at different points, we may suppose each to coincide with a degree of the respective osculating circle: these degrees then will be as the radii of curvature, and consequently will increase towards the poles. Now the degrees of Cassini diminished in this direction, yet he supposed them to indicate a flattening at the poles, consistent with the theories then generally received of Newton and Huyghens. This mistake of Cassini however, was soon rectified in the Academy of Sciences; and he was reduced to the dilemma of making the earth a spheroid, elongated not flattened towards the poles, or of giving up the accuracy of his own admeasurements. He embraced without hesitation the former alternative, in which he was supported by La Hire, Maraldi and others of his coadjutors. The mathematicians and philosophers of the Academy could not however, be brought to admit a conclusion, which appeared to them repugnant to the laws of mechanics and the theory of gravitation; and they preferred very reasonably to suppose some error in the observations, rather than concede consequences so much at variance with all physical considerations. On the other hand Cassini and his partisans stood firm; and to solve the difficulty, it was determined in the year 1773 to measure an arc perpendicular to the meridian of Paris, and thence to conclude the length of a degree of the parallel in that latitude. By a sort of fatality this arc, when measured and compared with a degree of the meridian, conspired to make the earth elongated, not flattened towards the poles, and to corroborate the system and measures of Cassini. It was however urged on the other side that this measure of a perpendicular was inadequate to throw any additional light on so delicate a question.

This in fact is quite true: the determination of a degree of a parallel is a difficult and delicate operation; and it is certain that the arc of Cassini deserves no kind of consideration. It is sufficient to observe, that the difference of longitude between the extremities, was determined by eclipses of Jupiter's satellites, and many of these too, eclipses observed formerly by Picard and La Hire, before pendulum clocks were in use, and when observers never could have been certain of their time to several seconds. Even allowing the observer to have known his time accurately, it is not perhaps too much to allow half-a-minute for the possible error of observation of the eclipse of a satellite; an error corresponding on the parallel between the meridian of Paris and Brittany to 5000 Parisian toises.

In order to put an end to these doubts and differences of opinion upon so interesting a subject, the French Academy, with the assistance and countenance of the king and his ministers, resolved upon sending out two scientific expeditions to measure degrees of the meridian, the one under the equator, the other as near to the pole as circumstances would admit. An inspection of the formulæ for determining the ellipticity of the meridian, from the comparison of two degrees of latitude, shows that the most favourable combination to be employed is that, where the middle point of one of these degrees is at the equator, and that of the other at the pole. The last condition evidently cannot be satisfied, but it is desirable to approximate as much as possible to the combination here indicated. For these reasons Bouguer, La Condaminè and Godin, were sent to Peru in 1735, while, the following year, Maupertuis, Camus, Clairault, and Lemonier, were despatched to Tornea in Swedish Lapland. This latter detachment completed their operations several years before the return of the Peruvian expedition, which had to contend against a long series of physical and moral difficulties. Not that the task of Maupertuis and his companions was by any means an easy one. Their original intention had been to make use of the islands in the Gulf of Bothnia for their stations; but this being found impracticable, they repaired to the valley of the river Tornea in Lapland. The southern extremity of their arc was near the town of Tornea; the northern, at a place called Kittis. Their base was measured on the frozen surface of the river, about the middle of the arc; and two independent measures, by two sets of observers, differed only four inches: its length was 7406 toises, 5 feet. The terrestrial angles were observed with a quadrant of two feet; the azimuth at Kittis was found by observing the sun's transit over the vertical of the next signal. The distance in the direction of the meridian, between the extremities of the arc, was determined by the method of parallels and perpendiculars, and found equal to 55023.4 toises. The latitudes were observed with a zenith sector of nine feet radius, constructed by Graham: the same stars, namely α and δ Draconis, were observed at each extremity. The difference of latitudes was thus found to be 57′. 29″. 6: and hence an arc of one degree = 57422 toises.

These operations were carried on under circumstances of great difficulty, arising from the nature of the country, and the severity of the climate. Yet so much care was taken in its execution, that it seems fully deserving of confidence. However, doubts having been thrown upon it, the arc was remeasured and extended in both directions by Swanberg, in the years 1801, 1802, and 1803. In both the geodetical and astronomical part he employed exclusively the repeating circle. The length of his arc was 92778 toises; the difference of latitudes 1° 27′ 19″.5: and hence the arc of 1° = 57196 toises. The difference between this and the determination of Maupertuis is very considerable; but it is not easy to point out its origin. The geodetic measures agree very well as far as they go together: the latitude of Tornea, as determined by Maupertuis, agrees very well with that found by Swanberg; unfortunately, the latter has not verified the latitude of Kittis. It is at all events the opinion of several able modern astronomers, that the arc of Maupertuis deserves full confidence. It has been fully discussed by M. Rosenberger, who deduces from it for the length of 1°, 57405 toises *.

The comparison of the degree of Lapland with those measured in France, established manifestly the fact of the compression of the earth towards its poles. Yet Cassini and others, who

* V. Schumacher's Astronomische Nachrichten, Nos. 121 and 122. For the original observations, consult Maupertuis, Figure de la Terre.—Amsterdam, 1738. 12mo.

had made it a prolate spheroid, attaching an undue confidence to the accuracy of their measures, refused pertinaciously to admit the oblate figure; and this scientific controversy was carried on for some time, not without considerable acrimony. At length, in the year 1739, Cassini de Thury, grandson of Dominique, assisted by La Caille, undertook to verify the arc of the meridian measured in France. In the course of this verification, which was conducted with very great care, it was discovered that a considerable error existed on the base measured by Picard, which had contributed to the false conclusions of Cassini regarding the figure of the earth. The operations now carried on were executed with all the precautions desirable: six bases for verification were measured in different parts of the kingdom. The latitudes were found with a zenith sector of six feet radius, and the same stars observed at each station*. The results fully confirmed the flattening of the meridians towards the poles; and from this time the figure of an oblate spheroid was conceded to the earth by the common consent of mathematicians and astronomers.

In the meantime Bouguer, La Condamine, and Godin, were carrying on their measure in Peru, under extraordinary difficulties. These difficulties were caused partly by the localities, and partly by the ill-will and indolence of the people of the country. The place selected for their measure was in an elevated valley, between the two principal chains of the Andes. The lowest point of their arc was at an elevation of a mile and half above the level of the sea; and in some instances the heights of two neighbouring signals differed more than a mile. Encamped upon lofty mountains, they had to struggle against storms, cold, and privations of every description, while the invincible indifference of the Indians they were forced to employ was not to be shaken by the fear of punishment or the hope of reward. Yet by patience and ingenuity they overcame all obstacles, and executed with great accuracy one of the most important operations of this nature ever undertaken. To accomplish this however took them nine years, of which three were occupied in the determination of the latitudes alone.

* For an account of these measures see Cassini de Thury, Méridienne de l'Observatoire Royal Vérifiée. Paris, 1744. 4to.

The northern limit of the arc measured by Bouguer and his companions was at a place called Tarqui, $2\frac{1}{2}'$ north of the equator: the southern extremity at Cotchesqui, in south latitude, $3° 4\frac{1}{2}'$. A base of 6272 toises was measured in the neighbourhood of Quito, near the southern extremity, and a base of verification of 5259 near Tarqui. These bases were measured with wooden rods, compared with a standard iron toise, the expansion of which, for 1° of temperature, had been elegantly determined by the observation of its vibrations in a given time, when swung as a pendulum under different temperatures. In calculating the arc of meridian from the triangles, 33 in number, the method of parallels and perpendiculars was used. The azimuths were found by observations of the sun near the horizon. Proper reductions were made for the difference of elevation between the extremities of the base, as also between the different signals. Finally, the distance of the parallels, passing through the extremities of the arc, was found by Bouguer, 176940; by La Condamine 176950 toises. The latitudes were observed with a zenith sector of 12 feet radius, the limb of which was subsequently redivided by the academicians themselves by a new and ingenious method. Perplexed for some time by the discrepancies of their observations, (for they were not yet acquainted with the inequality of nutation,) they at last resolved to make simultaneous observations of the same stars at the two observatories. The star principally employed for this purpose was ϵ Orionis. In this way the difference of latitudes was found to be $3° 7' 1''$; and they concluded the length of 1°, the one = 56746, the other = 56749 toises. Delambre, who has recalculated this degree, finds 56737. These trifling differences are not of much moment in determining the quantity of the earth's compression, and of none at all in proving the fact of the flattening towards the poles. This, indeed, had been established by the measures of Maupertuis and Cassini, before the return of the academicians from Peru. The arc of Peru afforded a satisfactory confirmation of it, and furnished valuable data for determining the excentricity of the elliptic meridian. It was found however, that the three measures of France, Lapland, and Peru, gave, when compared two and two together, different results for this quantity; and the numerous arcs since measured have offered similar anomalies. With the cause of these we are not fully acquainted; but in some instances, at least, we may attribute them to local disturbances, such, for example, as the attraction of a considerable chain of mountains *.

While these important operations were being carried on in Peru, considerable advances were made in the theoretical branch of this interesting investigation. In the year 1740, Maclaurin showed that the oblate spheroid was a form of equilibrium, and gave an equation by which the ellipticity can be found when the proportion of the centrifugal force at the equator to gravity is known, supposing the earth to be an homogeneous fluid. Maclaurin showed that when the centrifugal force is small, the ellipticity is expressed by five-fourths of the ratio at the equator of this force to gravity, that is in the case of the earth $\frac{5}{4} \times \frac{1}{289} = \frac{1}{231}$; and that gravity at the pole exceeds the gravity at the equator in the same proportion; and that the increase of gravity is as the square of the sine of the latitude. In 1743, Clairault published his Figure of the Earth, in which he considered the subject in a more general way, discovering the form of equilibrium on the following suppositions: First, that the fluid is homogeneous, with a spheroidal nucleus of different density. Secondly, that the whole mass is fluid and heterogeneous. The form of possible equilibrium in these cases is approximately an elliptic spheroid: the ellipticity varies with the law of density, and other circumstances, but in all cases the following theorems are true. The increase of gravity is as the square of the sine of the latitude; and, secondly, the sum of the ellipticity, and of the ratio of the whole increase of gravity, to the equatorial gravity, is $\frac{5}{2} \times$ the ratio of the centrifugal force at the equator, to the force of gravity there. This last important theorem has been distinguished by the name of its inventor.

It had been the intention of Godin and La Condamine to measure a degree of longitude at the equator; and they even wished to have commenced their operations with this. But the idea was strenuously combated by Bouguer, who procured an order from the minister to begin with the measure of the meridian; and owing to the time occupied by this latter, the former proposition was never carried into effect. Cassini de Thury, while occupied with the verification of the meridian, undertook to measure a degree of a parallel in the south of France. The extremities of the arc were Mont St. Victoire, and a station near Cette. The explosion of gunpowder at a post intermediate to these was observed, and from the difference of apparent times, the difference of longitude was concluded to be equal to 1° 53′ 19″; the length of the arc of the parallel between the two meridians was 78599.6 toises †. The comparison of this with the arc of the meridian between Perpignan and Rhodéz, gave an ellipticity of $\frac{1}{168}$. A comparison of the Peruvian and French arcs of the meridian had given an ellipticity of $\frac{1}{314}$; that of the Peruvian and Swedish arcs, $\frac{1}{213}$; and as we proceed, we shall find in other measures discrepancies equally considerable. These difficulties weighed so strongly with Bouguer, that he was inclined to reject the ellipse altogether, and to suppose that the increase of the length of the degree was proportional to the fourth power of the sine of the latitude, and that the proportion of the axis was as 178 : 179. But Laplace has shown that such a figure is incompatible with the observed law of the decrease of gravity, from the equator to Pello in Lapland.

The anomaly offered by a degree of the meridian measured in 1752, near the Cape of Good Hope, by La Caille, was still more remarkable than any of the preceding. This astronomer, who enjoys the highest reputation as an observer, took advantage of a residence at the Cape, undertaken for the purpose of observing the southern stars, to measure a degree of latitude in the southern hemisphere. All the usual precautions were taken, the instruments appear to have been good, La Caille himself was one of the ablest astronomers of the eighteenth century; yet according to this measure, a degree in the southern hemisphere, whose mean latitude is 33° 20′, is equal to a degree in the northern hemisphere, whose latitude is 45°. The amplitude of the arc, deduced from observation, was 1° 13′ 17″.5; the length 69669.1 toises; and hence 1° = 57034.4

* For a complete account of the Peruvian degree, see Bouguer, Figure de la Terre. Paris, 1749. 4to.

† For an account of this measure, see the Méridienne Vérifiée, p. 96.

toises. So surprised was La Caille at this result, that he remeasured the base, and recomputed all the triangles, still obtaining the same result. It did not appear to him that the observations of the terrestrial angles or celestial zenith distances were open to any doubt.

The limits of this treatise compel us to pass rapidly over two degrees measured in Italy, the one near Rimini, by Boscovich and Le Maire *, the other in Piedmont, by Beccaria. Astronomers never seem to have placed much confidence in the latter. The same arc has been remeasured a few years ago by Plana and Carlini, who found an error of 13″ in the amplitude. The former degree (that of Boscovich) has been criticised, but it has found able defenders. This is not the place for entering into discussions on the details of such operations; nor can we even pretend to give an account of the various experiments of this kind, instituted in different countries, and at different times. But we cannot pass over in silence the arc measured in Pennsylvania by Mason and Dixon, in the year 1764, from the remarkable circumstances connected with it. This operation differs from all others in this respect, that no triangles were used, but the whole line (nearly 100 miles) measured with rods. By means of a clock, the transit instrument was directed upon certain stars at the moment of their culminating, and then used to place a mark in the direction of the meridian. Thus the line might be prolonged in that direction, or if it deviated from it, the azimuth might be measured and allowed for. The latitudes were observed with a six-feet sector, which was reversed at each station. The amplitude of the arc was found 1° 28′44″.99; the length, 538079 feet: hence the degree of a meridian, whose mean latitude is 39° 12′, is equal to 60628.7 English fathoms, or 56888.3 French toises.

* In the year 1750.

Chapter XIX.

Improvement in the art of Observing.—Bradley—La Caille—Mayer.—Equations of condition.—Return of the comet of Halley.—Periodic Comets subsequently discovered.—Transits of Venus in the years 1761 *and* 1769.

The middle of the eighteenth century forms a very important epoch in the history of astronomy. It owes this character generally to the great improvements in the art of observing then introduced, and more particularly to the labours of Bradley, La Caille, and Mayer. It was now that the transit instrument and mural quadrant began to be extensively employed; and from this time we can date the formation of good catalogues, and collections of really trustworthy observations. For scientific purposes, observations made before the year 1750 are rarely if ever used. It may seem surprising that the transit instrument invented by Römer, and the mural quadrant by Picard, should not have been more profitably employed before this time. When Bradley succeeded to the place of Astronomer Royal in 1742, he found a transit instrument of five feet, which had been erected, but very little used, by Halley. Similar to the original instrument of Römer, the telescope of this transit was 26 inches nearer one extremity of the axis of rotation than the other; and in other respects its construction was objectionable. Bradley procured a new transit instrument of eight feet, with an axis of 4½ feet, and an additional mural quadrant of 8 feet radius *. The zenith sector, with which he had discovered the aberration and nutation at Wanstead, was so placed, as to be used for verifying the line of collimation of the two mural quadrants. With these instruments, and an excellent clock, Bradley made, during twelve years, (from 1750 to 1762,) one of the most extensive and valuable series of observations ever executed by an individual. They have been discussed at great length by M. Bessel, who has deduced from them a number of important conclusions, with regard to the places of the fixed stars, to refraction, and other objects of equal interest†.

The labours of La Caille in France were contemporaneous with those of Bradley in England, and much of the same nature. We have already had occasion to allude to his voyage to the Cape of Good Hope. This was undertaken partly to determine more accurately the parallax of the sun, and partly

* The object-glass of this transit was not achromatic. The eye-piece magnified fifty times. There were five parallel wires in the field.

† V. Fundamenta Astronomiæ. Königsberg. 1818. Folio. According to Bessel, Bradley's errors in right ascension are under 1″; those of declination under 4″ of space.

to form a catalogue of the stars in the southern hemisphere, that of Halley being altogether insufficient. In the latter object La Caille succeeded completely; having within ten months observed ten thousand stars: in the former he was not quite so fortunate; in fact, the transits of Venus in 1761 and 1769, have now superseded altogether other methods; and the attempt to conclude the solar parallax from observations, such as those of La Caille, will not probably be again resumed. On his return to Paris, he devoted himself assiduously for the rest of his life, to observation and calculation. The perfection of the theory of the Sun, researches on the law and magnitude of refraction, and a catalogue of four hundred of the principal stars, are the most remarkable productions of this distinguished astronomer. It is a curious illustration of what has been said above regarding the transit instrument, that La Caille, in the middle of the eighteenth century, employed the method of corresponding altitudes, for determining the right ascensions of these stars.

Tobias Mayer deserves, on many accounts, the honour of being associated with Bradley and La Caille, among the greatest astronomers of the last century. As we shall have occasion to notice him again, we shall now merely allude to his important researches on the theory and laws of refraction, his Tables of the Sun and Moon, and his catalogue of Zodiacal stars. The works of Mayer have not been so numerous as those of many others, but for their importance, and the talent shown in their execution, they claim the very first place. While yet quite a young man, he had given, in a Memoir on the Libration of the Moon, the first example of the use of equations of condition in determining simultaneously the corrections to be applied to all the quantities, which enter into the analytical expressions for the longitude and latitude of a planet. This method, which has now become of daily, and indeed it may be said indispensable, use to astronomers, enables them to employ thousands of observations for correcting the elements of their tables; and consequently has given a precision to the latter, which has produced most important results for astronomy, navigation, and geography.

A considerable discussion has arisen upon the point to whom we are indebted for the first idea of the motion of comets in a parabola or an ellipse: for the former curve may always, and in this instance more especially, be considered a particular case of the latter. The facts of the case, all prejudice being put aside, appear to be these. From the earliest times there were some philosophers who conceived that comets were analogous in their constitutions and motions to the planets of our system. This has been sufficiently shown in a former part of this treatise. It has been seen too, that in modern times there were found those who, unbiassed by Aristotelian prejudices, maintained the truth of this analogy. As the Peripatetic philosophy in general declined, the number of those who conceived comets not to be meteors, but permanent bodies, like the planets, increased. And very naturally, as long as the planetary orbits were supposed circular, such persons would suppose comets to describe circular paths. But as soon as it was known that the planetary orbits were elliptic, they would ascribe the same form to the orbits of comets. This indeed is not altogether hypothesis: we know that Tycho Brahé, who maintained that the earth was the centre of the celestial motions, supposed every comet to move in an arc of a great circle of the sphere; and when the doctrines of Copernicus had triumphed, and those of Kepler were beginning to obtain credit, we find the Earl of Northumberland remarking, in a letter to Harriott, when speaking of Kepler,—"I am much in love with . . . . . his elliptical iter planetarum,—for methinks it shewes a way to the solving of the unknown walkes of comets *." Thus, too, Borelli, in 1665, had conjectured the elliptic form of these orbits;—as has already been stated. Kepler had not recognized the analogy perceived by so many of his successors, and had contended that the trajectory of a comet was a right line. Hevelius followed the same ideas, but we must allow him the credit of having remarked, and indeed proved, that the rectilinear orbit was not rigorously accurate, but that there was a sensible deflection from it, so that the trajectory in fact approximated to the nature of a parabola. But Hevelius by

* De Zach. Correspond. Astronom. vol. vii. p. 113. We do not know the date of this remarkable letter; it must be some time between 1606 and 1618; as it was written while the Earl was a prisoner in the Tower, to which he was sent in 1606, and where he remained till 1621; Harriott died in India.

no means admitted the permanent nature and return of comets; nor did he suppose as we do now, the visible arc to be that of an ellipse, which from its great excentricity is sensibly identical with a parabola. The parabola, according to him, was the result of the deflection from a straight line, produced by the attraction of the sun: it does not appear clearly whether or not he meant to place the sun in the focus of the parabola. It has been supposed by nearly all the astronomers who have discussed this point in later times, that the honour of having first established the parabolic form of the orbit, the sun occupying the focus, belongs to Dörfel, minister of Plauen, in Upper Saxony, who showed that this was the case for the famous comet of 1680, and extended the proposition by induction to all comets*. But it has been said by some, that he was preceded in this discovery by Frederic Madeweis, an astronomer of Berlin, in a work on the same comet. This work of Madeweis is only known (as far as we are aware) by a notice of Ries, professor at Tubingen, who has stated no particulars, and with whom we are only acquainted from a quotation given by de Zach†.

Finally, the question was taken up by Newton, who taught that comets like the planets revolved in ellipses round the sun, but that their ellipses being, unlike those of the planets, very excentric, the arc visible to us was sensibly identical with a parabola; and he showed how to calculate the elements of a comet's orbit from three observations in this hypothesis. The subject was taken up with ardour by Halley; and it has been seen that he ascertained the identity of the comet of 1682, with those of the years 1607 and 1532. So satisfied was he of their identity, that he did not hesitate to predict its return about the year 1758; and conjured posterity to remember if his prediction was verified, that the discovery was of English origin. It may easily be supposed, that as the year 1758 drew near, great curiosity was excited, to see whether the prediction would be accomplished. The greatest difficulty of Halley in predicting the return, had been in the inequality of the two preceding periods; and though, with his usual genius, he perceived that the cause of this was in the planetary attractions, he was unable to calculate in any satisfactory way their effect upon the time of the next apparation. But the great progress of mathematical science in the interval, enabled Clairault to resume with success the calculation of those perturbations. Having by profound and ingenious calculations determined the disturbing effects of Jupiter and Saturn on the comet at its last reappearance, he found for a definitive result, that the new period would exceed the preceding by about 618 days, and he fixed for the time of its next passage through the perihelion 4th of April, 1759*. The realization of this prophecy was one of the most remarkable events in the annals of astronomical science, and one of the proudest triumphs of the Newtonian theory. It has immortalized Halley, whose name has been justly attached to this memorable comet, and it conferred additional honour on the already illustrious Clairault. It is singular that, though several able astronomers were on the look-out for it in France, it was first perceived by a Saxon peasant named Palitsch, on the 25th of December, 1758. It was soon recognized for the comet of 1682 in Germany; yet the news of its discovery did not reach Paris till long afterwards. In the mean time it had been discovered on the 18th of January by Messier, who continued to observe it in secret according to the orders of his patron De l'Isle, till the 14th of February, when the matter could no longer be concealed. But the comet had then approached too near the sun to be seen; and the French astronomers were obliged to content themselves with observing it in the second branch of its orbit. Such disgraceful conduct as that of De l'Isle is fortunately of rare occurrence among men of science.

For a long time this was the only comet certainly known to return, and whose periodic time had been fixed. Analogy indeed led men to suppose, that if one comet certainly revolved in an ellipse, and revisited our system at regular intervals, that such also would be the case with others; but this was a point very difficult to verify. For in most of the ellipses hitherto calculated

* The work of Dörfel is excessively rare; neither Pingré nor Delambre had ever seen it; it was published in 1681; a short account of it by Mathieu, abridged from Burckhardt, may be found in the Histoire de l'Astronomie, au 18me. siècle, p. 671.

† Correspond. Astron., vol. vii. p. 138.

* Clairault himself did not pretend to predict this passage within a month, from the unavoidable uncertainty of several elements of his calculation. The comet passed its perihelion on the 13th of March.

for various comets, the periodic times have been so considerable, that future generations alone can witness their reappearance. This however was not the case of the first comet of 1770, for which Lexell found an ellipse, with a periodic time of five years and seven months; yet from causes not sufficiently known it has never since reappeared. But the present century has witnessed the singular and interesting discovery of two comets returning at short periods; the one accomplishing its revolution in about three years and four months, the other in rather more than six years and a half. The orbits of both of these remarkable bodies lie within that of Jupiter, so that in every sense of the word they form a part of our system, differing from the planets only in the great excentricity of their ellipses. The former of these was recognized as a periodic comet for the first time in the year 1819. M. Arago then remarked the great analogy of its elements with those of a similar body seen in 1805; and Dr. Olbers pointed out its identity with the comets of 1759 and 1789. M. Encke found the orbit to be sensibly elliptic, and that it had performed four complete revolutions between 1805 and 1819. He was enabled to predict its reappearance for the year 1824, which prediction was exactly verified, and it has been observed again in its subsequent approaches to the sun, so that the elements of its orbit are known with as much accuracy as those of many of the planets. The latter of the two periodic comets was first discovered by M. von Biela, in Bohemia, in the year 1825: it was soon recognized as identical with the comets of 1772 and 1806; and its elliptic orbit and perturbations have been calculated with considerable precision. From the increased attention now paid to the observation of these small but interesting bodies, we may confidently anticipate results of a very remarkable nature.

The method of determining the sun's parallax from observations made on the transit of Venus, has been explained in a preceding chapter, and mention has been made of the pathetic address of Halley to future astronomers, conjuring them not to suffer the transits of 1761 and 1769, to pass without the accomplishment of this purpose. The ideas of Halley have been executed to their fullest extent. The object was thought of sufficient importance to be taken up by several of the European governments, who sent, at their expense, astronomers to various parts of the globe favourably situated for the observation. The French government sent Le Gentil to Pondicherry, La Chappe to Tobolsk, and Pingré to the isle of Rodriguez in the Indian ocean, not far from the Mauritius. Great Britain despatched Mason to the Cape of Good Hope, and Maskelyne to St. Helena. Cloudy weather prevented Le Gentil from making any observation: the same cause hindered Maskelyne from observing the end, and Pingré the beginning of the transit; and unfortunately the observations at the Cape, and the island of Rodriguez, when compared with the northern observations, gave different results for the sun's parallax. We now know that the observation at the Cape was correct; but this was at that time uncertain, and the doubts still hanging over the question induced the different governments to make still greater exertions for the observation of the transit of 1769. France sent the Abbé Chappe to California; England, Dymond and Wales to Hudson's Bay, Call to Madras, and Green to the island of Otaheite, in the South Sea: several Russian observers were stationed at various points of Siberia, and the Russian empire: the king of Denmark sent Father Hell to the island of Wardhus, near the North Cape; and the king of Sweden, Planmann to Caianebourg in Finland. At the stations of Hudson's Bay, California, Otaheite, and Wardhus, the observations were completely successful; and this important question was now finally decided. After a great deal of discussion by the most eminent astronomers, the parallax of the sun has now been fixed at $8''.6$, and we are pretty sure that the error does not exceed the tenth part of a second.

Chapter XX.

Determination of Terrestrial Longitudes.—Method of Lunar Distances.—Improvement of the Tables.—Clairault.—Mayer.—La Place.—Reflecting Instruments.—Repeating Circle.—Mayer.—Borda.—Chronometers.—Harrison.

The determination of terrestrial longitudes by the observation of celestial phenomena, is the most important application of astronomy to the purposes of civil life. The problem is, in fact, to find the time reckoned at a given instant at Greenwich, corresponding to the time counted at the same instant by the observer under any other meridian. All the methods, of any practical utility at sea, hitherto proposed, may be referred to one of two classes: those depending on observations of the place of the moon; and those which depend on the transport of time from the first meridian *. The former are founded upon this principle: the true place of the moon's centre, that is, her place corrected for parallax and refraction, is the same at a given instant for every meridian. If, then, the observer can at any moment find that place, he may calculate from the lunar tables the Greenwich time corresponding to that position of the moon's centre. The reason why the moon is selected for this purpose is, the great rapidity of her motion, compared to that of any of the other planets. For if we take the sun, which moves so much slower, and if we allow his place to be determined with an uncertainty only of a quarter of a minute in space (a supposition in reality much too favourable), this will leave an uncertainty of nearly six minutes in time on the longitude. But if we take a body which, like the moon, moves thirteen times faster, the uncertainty on the longitude with the same accuracy of observation, will be thirteen times less. This advantage is so obvious, that the method of finding the longitude by observation of the moon's place had occurred to several astronomers as early as the sixteenth century. When speaking of Morin, we have seen that it had been enounced by Gemma Frisius, Nonius, and others; but without good lunar tables, and good means of observation, the idea could have no practical utility. On this account, the commissioners appointed to report on Morin's claims decided with good reason, that he could claim no reward, the real difficulties of the question lying in the circumstances just mentioned. And these difficulties were not surmounted till the middle of the eighteenth century. At that time, the great progress of physical astronomy, by indicating *à priori* the form of all the inequalities in the lunar theory, and the increased mass of good observations, enabled Mayer, who was at once a profound mathematician, a patient calculator, and a good practical astronomer, to form lunar tables of sufficient accuracy to be employed for the purposes in question *. For this valuable service, the widow of Mayer received a recompense of 3000*l.* from the British government; nor does this reward appear at all excessive. Still, by this, only one part of the difficulty was removed; there remained that of the observation. Gemma Frisius had proposed the plan now adopted, of observing the distance of the moon from a known fixed star. But the difficulty of this observation. before the invention of reflecting instruments, caused Lemonnier and Bouguer to suggest other methods. That of the latter consisted in determining the time of the moon's transit over the meridian by the observation of corresponding altitudes; but this is objectionable from the corrections required for the irregularity of the moon's motion, and the motion of the vessel in the interval, besides other practical inconveniences which are sufficiently obvious. Lemonnier proposed to find the moon's place by absolute altitudes; this does not seem to be by any means a convenient method, since it supposes us to know the moon's declination, which itself depends upon the longitude to be found; and it can only lead to an accurate result by a series of trials. The invention of Hadley's quadrant permitting the observation of distance to be made with convenience and accuracy, La Caille with great justice strongly advocated this method in preference to all others. The same opinion was expressed by Mayer, and warmly taken up by Maskelyne, to whom the definitive success of this important method is in a great measure to be attributed. It cannot be expected that navigators should be able to spare time for the laborious task of computing from the lunar tables the distances to the fixed stars that they may have occasion to employ. To obviate the necessity of this, Maskelyne induced the British Board of Longitude to publish annually a Nautical Ephemeris, containing the places of the sun and moon for every twelve hours, and the distances of the latter from the fixed stars. These latter are now published for every three

* The method proposed by Halley, of finding the longitude at sea from observations of the variation of the magnetic needle, is impracticable, from the irregularities in the lines of equal variation, and the constant changes in their position.

* The Tables of Mayer were published by Maskelyne, London, 1770. 4to.

hours, so that by an easy interpolation, the distance for any intermediate instant may be found.

The intimate connexion between the problem of the longitude and the theory of the moon, induces us to say a few words in this place upon the ameliorations introduced into the latter since the time of Newton. Clairault was the first to investigate, in a general and scientific way, the problem of the three bodies, as it has been called: viz. to find the motions of the moon, exposed simultaneously to the attractions of the earth and sun, and attracting them in return. The problem is simplified by the mass of the sun being incomparably greater than the other two, and consequently nearly at rest, relatively to them; and its distance so great, that it may be supposed the same for both. With these data, the question is to determine the irregularities produced in the elliptic motion of the moon round the earth,—assuming the Newtonian law of gravitation and a knowledge of the masses. In his first essays, Clairault found for the movement of the lunar perigee only half the quantity given by observation; and this fact was gladly seized by the enemies of the Newtonian system to throw discredit upon the whole theory of gravitation. But their triumph did not last long; for Clairault himself, upon reviewing his calculations, perceived that he had neglected a quantity which, when taken into account, reconciled the motion of the apogee with the Newtonian law. For many years subsequent to this, the lunar theory was the subject of the labours of the first mathematicians of the age, more particularly of D'Alembert, Clairault, and Euler. It is upon the theory of the latter that the tables of Mayer are principally founded: at the same time they owe their excellence in great part to the care taken by Mayer to establish the coefficients of the inequalities by the comparison of numerous observations; nor did he hesitate to introduce empirical equations, where observations appeared to require them. The most remarkable of these was one for the secular acceleration of the moon, first suspected by Halley, and now fully established. The cause of this acceleration was long a subject of doubt and abstruse speculation to the first mathematicians of Europe. At one time it was attributed to the resistance of the ether; at another, to the successive propagation of gravity; but these and other reasons were found upon examination to be insufficient. At last, La Place discovered that it was owing to the secular diminution in the excentricity of the terrestrial obit. It had been shown by La Place and La Grange, that amid all the changes produced by the mutual action of the planets on each other, the major axes of their orbits remain invariable; but the excentricities, the inclinations to a fixed plane, the positions of the nodes and perihelia vary slowly. These variations are comprised within limits not very extended, but their progress is so slow as only to be perceptible at the end of several centuries. Of this nature is the gradual diminution of the obliquity of the ecliptic, long disputed, and now known to be a consequence of the law of gravitation. Such, too, is the secular diminution of the excentricity of the earth's orbit, of which we have been speaking.

We owe to La Place the determination and explanation of another inequality which had for some time embarrassed the calculators of lunar tables. The comparison of different epochs showed an acceleration at one time, and a retardation at another; but the period being evidently long, and the inequality perhaps dependent upon many elements, there was little hope of detecting the law by observation. But a new examination of the theory of the moon on the principles of gravitation, showed to La Place the existence of an inequality hitherto unnoticed, of a period of 184 years, and rather a complicated form, which explained all the discrepancies observed. This is a good example of the advantages derived to our tables from physical astronomy. Indeed, to such perfection has this science been carried in the case of the moon, that M. Damoiseau has lately published tables, founded exclusively on theory, which are at least equal, if not superior, to any we possess.

But to return to the subject of the longitude. The method of lunar distances could never have come into general use without the invention of reflecting instruments. The construction of Hadley's quadrant is explained in most optical treatises. We shall only remark, that it is essentially composed of two mirrors, both in planes perpendicular to that of the instrument. The one is fixed, the other is moveable, being attached to the centre of the moveable radius, the extremity of which indicates on the graduated limb the inclination of the two mirrors. The fixed mirror, which is perpendicular to the line of sight, only intercepts half the field of view; so that at the same instant, and in the same field, the observer may see the horizon by direct vision, and the image of the sun after two reflections. He then reads off on the limb the inclination of the two mirrors. By a well-known optical theorem, this is equal to half the angle between the two objects, or in this case, the altitude of the sun above the horizon. We owe to the inventive genius of Mayer the idea of a most important improvement in these instruments *. This idea, which has been more lately revived by Borda, consisted in substituting for the quadrant or sextant a complete circle, and making both mirrors moveable upon it, the line of sight always remaining perpendicular to the small mirror. After making an observation, by moving the small mirror till the coincidence of the images is established, the instrument is reversed, so as to face east instead of west, or the converse. The great mirror is moved till the coincidence is again made, and it may be seen that it must have moved through an arc double of the angular distance to be measured †. Instead of reading off this arc, leave the alhidade of the great mirror fixed, reverse the instrument, make another observation by moving the small mirror. We have now arrived exactly to the position from which we set out, only that the alhidade of the great mirror has advanced on the limb an arc double of the distance to be measured. Recommence all the operations described; at the end of them this alhidade will have described an arc quadruple of the angular distance, and so on. We may obtain then any even multiple of the arc sought to be measured; and we need only read off once at the beginning and once at the end of our observations. In this way the errors of reading off may be indefinitely diminished, being divided by the number of observations. Any error from defective centering is obviated by measuring successively all round the circle; and we are rendered independent of the knowledge of the point of the graduation corresponding to the position of the mirrors when parallel. Such are the principal advantages of the reflecting repeating circle; to which we may add, in taking sets of lunar distances, the eye is not fatigued, by the application of light at each observation to read off the arc.

* See the Discourse prefixed to Mayer's Tables, and called Methodus Longitudinum Promota, p. xxi., and the accompanying plate: circles on this plan were actually constructed, but fell into disuse till again brought before the public notice by Borda.

† It is to be observed, that these instruments are so graduated as to give at once twice the inclination of the mirrors, or the real distance between the objects observed.

The first idea of reflecting instruments is certainly due to Hooke, who proposed by a sextant, with a single mirror attached to the moveable radius, to obviate the necessity of having two observers in measuring the distances of stars, as practised by Hevelius: to measure altitudes he was obliged to attach a level to the sextant*. It is evident that the instrument in this form was not applicable to the purposes of navigators; and we may consider the construction given by Hadley as a new invention of the most important character. It is not easy to understand why so much pains have been taken to deprive Hadley of the merit of this invention: Newton, it is said, had previously invented a reflecting instrument with two mirrors; at all events this was not known to Hadley, whose discovery was published in 1731 †, that of Newton in the year 1742. Lalande ‡ mentions Godfrey of Philadelphia, as having anticipated Hadley; he quotes the Philosophical Transactions for 1734, and the first volume of the Philadelphia Transactions. The latter we have not seen, the former contains the account of an instrument, bearing no analogy to Hadley's quadrant, but it alludes to the discovery of a reflecting instrument by the author §. Be the invention of Godfrey what it may, it was not known in England till after the publication of Hadley's paper in the Philosophical Transactions.

The method of determining the longitude by the transport of time is one which

* See Animadversions on the first part of the Machina Cœlestis of Hevelius. London, 1674, 4to.

† See Philosoph. Transact. for that year; No. 420, p. 147: there is no reason to doubt that Newton had invented an instrument similar in principle to Hadley's.—Philos. Transact. No. 465. It is doubtful whether it was ever executed.

‡ Montucla. Hist. des Mathemat., vol. iii. p. 522. See also Wales and Bayley. Astronomical observations made in Cook's voyage.—Introduction, p. 30.

§ Philos. Transactions, vol. xxxviii. p. 441.

depends for its excellence entirely upon the perfection of the mechanical means used. Huyghens had no sooner discovered the application of the pendulum to the balance of clocks, than he attempted to apply his invention to the use of navigation. But none of his ingenious endeavours were successful; though for some time time he flattered himself that he had overcome the difficulties offered by the violent and irregular motions of the vessel. He finally turned his ideas into a new channel, and fell upon a discovery of perhaps equal importance with that of the pendulum clock. This was the isochronism of spiral steel springs, when used as a balance in watches; the first great step made towards the finding the longitude at sea by chronometrical observations. The priority in this discovery was contested by Hooke. It would appear that watches upon this construction were executed under the direction of Huyghens in the year 1674: it is stated upon the authority of Leibnitz, that about this time a dispute as to the right of the invention, arose between Huyghens and the Abbé Hautfeuille, and was decided in favour of the former in a court of law *. There can be little or no doubt that at the time of which we are speaking, watches on this principle were constructed in France. In a postscript to a small treatise on Helioscopes, published 1676, Hooke refers the date of his own invention to seventeen years back. But such testimony unsupported is insufficient. His theory of springs was not published till 1678 †. He alludes also in a work called Lampas, published 1677, to a new contrivance for the balance of watches, which he says he showed to the Royal Society ten or twelve years before ‡; his explanation is very concise and obscure; the contrivance seems to have been a combination of the centrifugal pendulum and spiral spring. If we consider the avidity shown by Hooke to claim every discovery of importance made in his time, we must look with distrust upon his somewhat dubious pretensions, in the case of the spiral spring.

In the year 1714 the British parliament passed an act, holding out a great recompense to those who should contribute to the discovery of the longitude at sea,—namely, 10,000*l.* if the longitude were found within a degree; 15,000*l.* if within forty minutes; and 20,000*l.* if within half a degree. These munificent rewards produced the desired effect. John Harrison, a man of humble origin but great genius, devoted a long and laborious life to the construction of clocks and watches for navigation. So numerous and important were his improvements, that we may with justice consider him as the inventor of the marine chronometer. It will, perhaps, give the best idea of his merits, to say that he has done as much for this instrument, as James Watt for the steam-engine. It is impossible in this place to give any idea of the numerous improvements introduced by him, without entering into details incompatible with the nature of this treatise. It is enough to observe, that in 1749 he obtained the Copley medal of the Royal Society; and in 1761 obtained from the Commissioners of the Board of Longitude a trial of his chronometer by a voyage to Jamaica. It was found at the end of sixty-one days that his chronometer gave the longitude of Port Royal within five seconds of time; and on the return to England, after an absence of 161 days, the whole variation was only $1^m\ 5^s$. It was evident that the conditions of the Act of Parliament were satisfied, and Harrison received a payment, on account, of 5000*l.*, but the Commissioners decided that the remainder of the reward should not be paid till a second trial had taken place. This was executed in 1764, and crowned with complete success; it was decided unanimously by the Board of Longitude, that the longitude of Barbadoes had been determined within the limits prescribed by the act; 5000*l.* was immediately granted to him; and 10,000*l.* more when he had explained to commissioners, appointed for the purpose, the details of his construction. This took place in 1765. From this date the use of chronometers began rapidly to spread among navigators; and if the method of lunar distances be still more extensively used, it is perhaps to be attributed to the expense of the former instrument. At all events it is of great importance to have two independent methods for the longitude, the one acting as a salutary check upon any accidental errors in the other.

* V. Lalande in Montucla, vol. iv. p. 550.

† Lectures de Potentia restitutivâ. London, 1678. 4to.

‡ Lampas, London, 1677. 4to. London, p. 43.

Chapter XXI.

Attraction of Mountains.—Bouguer.—Maskelyne—de Zach.—Experiments of Cavendish.—Measures of the Meridian in England and India.—Arcs of parallel.—French measure of the Meridian between Dunkirk and Barcelona.—Determination of the Mètre.—Various measures of the length of the seconds' pendulum.

Many of the operations for measuring arcs of the meridian, having been carried on in mountainous countries, it became reasonable to suppose, that a part, at least, of the anomalies observed, were due to the attractions of the mountains themselves. Bouguer and La Condamine endeavoured to put this to the test of experiment in Peru. They attempted to measure the attraction of the mountain called Chimboraço, the most considerable of that province. It is easy to conceive how this can be effected. In a station north of the mountain, the attraction of the mountain causes the plumb-line of the sector to deviate from the real vertical to the south, and the converse, in a station to the south. The apparent latitudes found at each station will differ from the real; the difference between these latitudes will be augmented by the sum of the attractions of the mountain at the two stations. If then we determine by other means, such as trigonometrical measures, the difference of latitudes, and compare it with that obtained astronomically, the difference of the results will give the sum of the attractions. The attempt made by Bouguer was not very satisfactory in its results; he found for the attraction of the mountain rather more than 7″, being about half of what he had expected; but his stations were not favourably placed, being both to the south of the mountain, in which case the difference of the attraction of the two stations is observed, instead of their sum*.

The experiment was repeated under more favourable circumstances by Maskelyne in 1774 on the mountain of Schehallien in Perthshire, elevated about 2000 feet above the level of the sea†. The meridional distance of two stations, the one to the north, the other to the south of the mountain, was found to be 4364.4 feet, corresponding to an amplitude of 42.″9. This amplitude, determined astronomically by the zenith sector from very numerous observations, was found 54.″6: giving 11.″7 for the sum of the attractions of the mountain at the two stations. The density of the mountain having been determined by a careful survey, the mean density of the earth was concluded to be about 1.8 times the density of Schehallien, or five time the density of water.

More recently* the Baron de Zach has attempted to measure the attraction of Mont Mimet near Marseilles. Both his stations were to the south of the mountain; the one close to its foot, the other on the island of Planier. The latitudes were observed with a repeating circle; he found for the whole effect of attraction 1″.98; but it may be questioned, whether, with the instruments used, it was possible to be certain of so small a quantity †. In the mean time, in the year 1798, Cavendish had instituted a very beautiful series of experiments for finding the earth's density ‡. By means of a delicate instrument, called the balance of torsion, the attraction of a leaden sphere, eight inches in diameter, was made sensible, and the effects of this attraction compared with that of the earth. The result found for the mean density of the latter agreed very well with that found from Schehallien.

In the year 1787, the governments of France and England in conjunction, resolved to connect the two observatories of Greenwich and Paris, by a series of triangles. The cause of this was the uncertainty that still prevailed upon the exact difference of longitude between these places. On the French side the measure was conducted by Cassini IV., Legendre, and Méchain; on the English side under the superintendence of General Roy. The latter officer founded his chain of triangles on a base measured with great care on Hounslow Heath, in which he used glass rods twenty feet long. These were afterwards abandoned, and the base of verification on Romney Marsh was measured with steel chains. Such was the care with which these operations were conducted, that the base

* V. Bouguer. Figure de la Terre, p. 379.

† Philos. Transact. for 1775, p. 500; and also for 1811, p. 346.

* In 1810. His operations are given in detail in his work on the Attraction des Montagnes.

† V. Attraction des Montagnes. Avignon, 1814. 2 vols. 8vo.

‡ Philos. Transact. for 1798, p. 469.

of verification, as computed from the original base on Hounslow Heath, differed from that actually measured by only two feet, in a total of 28533, though the connecting chain of triangles extended over a space of eighty miles. These measures are interesting as having formed the commencement of the great trigonometrical survey, extended over Great Britain and Ireland, which is at this moment in the progress of execution. In the course of this excellent survey the terrestrial angles were measured with the theodolite of Ramsden, three feet in diameter; the latitudes with a zenith sector of Ramsden, of eight feet radius; both of them superb instruments of their kind. In spite however of the excellence of the instruments, and the unquestioned accuracy of the observers, the degrees of the meridian in England offer anomalies analogous to those previously noticed, and which, it cannot be doubted, arise from local attractions, or from real irregularities of the earth's surface *.

A very extensive arc has been measured in the Indian peninsula with instruments similar to those used in Great Britain. The whole amplitude of this arc, as determined by Colonel Lambton, is near ten degrees; and it has been subsequently prolonged six degrees to the northward, by Captain Everest. Very competent judges have pronounced this arc, considering its extent, and the care used in every part of the operations, to be superior to all other measures of the same kind hitherto executed †.

The astronomers who conducted the survey of Great Britain, took the opportunity of measuring a parallel of latitude between Beachey Head and Dunnose. On this occasion the difference of longitudes was determined by a new method. The two signals being visible from each other, the azimuths were reciprocally observed, the latitude of each station being deduced from the triangles of the survey. This arc is interesting, from the novelty of the means employed, and the rarity of measures of this description. The only other measure of a degree of longitude offering any interest, except perhaps, a degree measured by Colonel Lambton in India, is one recently executed on an extensive scale in France and Italy. It extends from Padua on the east, to Marennes near Bourdeaux, on the west. These two points had been joined trigonometrically, in the course of very extensive surveys made by the French and Austrian engineers. They were joined astronomically by fire-signals, made on the Monte Baldo near Verona, Rochemelon on the Mount Cenis, Mount Tabor in Savoy, Pierre sur Autre in Auvergne, and other stations in France,—these signals being observed at Padua, Milan, Turin, Geneva, La Jonchére, and several points to the westward, as far as Marennes. This arc is very remarkable from its great extent, and for having been carried completely across the vast chain of Alps which separates France from Italy. It is intended to carry it still farther to the eastward, to Orsova, on the frontiers of Hungary and Wallachia, embracing a total arc of 24° in longitude *.

One of the most celebrated determinations of an arc of meridian yet undertaken, was that commenced at the beginning of the French Revolution, for the purpose of finding the length of a quadrant of the meridian, the ten-millionth part of which it was proposed to take as the standard for a new measure, called the métre. This celebrated undertaking was executed by the astronomers, Delambre and Méchain. The extremities of the arc were to the north, Dunkirk, to the south, the town of Barcelona, in Spain: but the arc was subsequently extended by Biot and Arago to the island of Formentera near Minorca. In this measure the repeating circle, which had been previously tried for the first time at the junction of the observatories of Greenwich and Paris, was used exclusively for both the geodetic and astronomical angles. The value of this elegant and ingenious instrument appears to have been overrated by the French astronomers; and it seems to be thought at present that the latitudes at the extreme points of their arc are not so certain as could be wished. This is more particularly the case with Formentera, where the latitude was found with a repeating circle, having a fixed level attached to it; a construction

* A detailed account of the English Trigonometrical Survey has been published in three 4to vols. London.

† An account of Colonel Lambton's arc may be found in the 8th, 10th, 12th, and 13th volumes of the Asiatic Researches. See also Philosophical Transactions for 1818, p. 486.

* A detailed account of the operations in Italy and Savoy may be found in the "Opérations Geodesiques et Astronomiques pour Mesurer un Arc de Parallèle Moyen." 4to. Milan, 1825.

justly esteemed to be very objectionable. It has also been ascertained that the best repeating circles are subject to errors of a very perplexing kind, which can only be eliminated by observing stars to the north and south of the zenith. This precaution was not always taken in the French survey. However, this French arc upon the whole, considering its extent, and the great talents and celebrity of the astronomers by whom it was executed, is one of the most important measures that we possess *.

It has been mentioned that the French survey was undertaken for the purpose of fixing a new standard measure. The French savans, struck with the disadvantages arising from the variety of measures in Europe, and from the inconvenience of not being able to determine the exact length of several of the measures employed in former geodetic operations, conceived the beginning of the revolution to offer a favourable opportunity for introducing a new universal measure, which being founded on some invariable physical standard, could always be recovered, should its exact magnitude, in the course of time, become involved in any doubt. The idea was not new, for Picard had proposed, long before, the length of the seconds' pendulum as the standard for an universal measure; and it is perhaps to be regretted that this idea was not followed at the revolution. But notwithstanding the practical facilities of this determination, it was rejected by the French savans, as involving an arbitrary and heterogeneous element, the division of time; to which rather fanciful objection they added the variation of the length in the pendulum at different latitudes. But this might have been obviated by taking, for the standard, the length of the pendulum at the level of the sea, under the equator, or under the parallel of 45°; and it is difficult to see how the vanity of any nation could have been offended by such a choice. However, it was determined to take the quadrant of a terrestrial meridian, which, supposing a spheroid of revolution, would be the same for all countries, and to divide it into ten million parts. One of these parts was taken for the unit of length, and called the métre. As it was impracticable to measure the whole quadrant under any meridian, they were obliged to calculate the ellipticity and dimensions of the earth, and the length of the quadrant, from a comparison of the arc between Dunkirk and Barcelona, with that measured by Bouguer and La Condamine in Peru. They adopted for the ellipticity $\frac{1}{334}$, and the métre was fixed at 443.296 lines of the old measure. This ellipticity agrees pretty well with that found by Laplace, from a very curious method. The oblateness of the earth's figure causes an inequality in the moon's motion in latitude, proportional to the sine of the moon's true longitude: the coefficient of this inequality depends upon the ratio of the earth's axes. Now this coefficient can be determined by observation, and hence the ratio in question may be obtained. In this way La Place found for the ellipticity $\frac{1}{304}$; another inequality in longitude gave him $\frac{1}{305}$.

It is to be feared that the comprehensive views of the French philosophers have not been answered. The métre, as far as we know, has not been adopted in other countries. In many continental works of science the old French toises and feet are still used; and as these denominations are now given in France to measures comprising two métres, and one-third of a métre respectively, the confusion is greater than ever. Perhaps the French would have acted more wisely in contenting themselves with doing what has since been adopted in England, determining accurately the ratio of the length of the second's pendulum to the foot in common use. As the latter may be supposed invariable, the former may be always recovered if lost. The introduction of one uniform measure of length throughout Europe is indeed much to be wished: but the difficulties in the way of such a consummation are, we fear, too great to be surmounted.

The theorems of Clairault, given in Chap. XVIII. enable us to determine the ellipticity of the earth, from observations on the length of the second's pendulum, at different latitudes. According to those theorems, in going from the equator to the poles, gravity increases as the square of the sine of the latitude; and the sum of the ellipticity of the earth, and of the ratio of the whole increase of gravity to the equatorial gravity, is

* A complete account of all the operations is to be found in the work called "Base du Systême Métrique." 4to. Paris, 1806. And in the additional volume of observations by Biot and Arago. 1821.

equal to $\frac{5}{2}$ × ratio of the centrifugal force at the equator to the equatorial gravity, or to $\frac{5}{2} \times \frac{1}{289}$. This important application gave a new interest to pendulum experiments, which in later times have become very numerous. These observations demand so much nicety of execution, and so great an attention to minute details, that the early observations, to which allusion has been made in a former part of this treatise, (those of Picard, Richer, &c.) can now only be looked upon as matters of curiosity. The first good observations of the pendulum to be met with, are those made in Jamaica, of which an account has been given by Bradley in the Philosophical Transactions for 1734; and to these we may add experiments made by Mairan, at Paris, in 1735. Bouguer in Peru, Maupertuis in Lapland, La Caille at the Cape of Good Hope, did not omit, when occupied with the measure of degrees, to determine at the same time the lengths of the second's pendulum. But it is particularly within the last forty years, that measures of this kind have become numerous and exact. At the time that the measure of the meridian was undertaken in France, Borda was desired to determine the length of the pendulum beating seconds at Paris. He conducted with great skill this delicate task; and this may be considered the first measure executed with all the perfection required by the actual state of science *. This experiment has subsequently been repeated without any material difference in the results, by M. Biot, and by Captain Sabine. The former of these philosophers has made an extensive series of observations at different points of the French and English arc, from Unst, among the Shetland islands, to Formentera in the south, and at different stations on the parallel of 45°. In the French observations in general, the measure has been made by noting the number of oscillations made in a certain time by the pendulum, at a given station, and then actually measuring its length. The English observers have of late years usually employed a pendulum of invariable length. The number of oscillations which this pendulum makes in a certain time, at a given place, as, for example, London, having been ascertained, it is transported to another station, and a similar observation made there. From the comparison of these, the variation in the force of gravity is easily deduced. In this method, the necessity of actually measuring the length of the pendulum employed is superseded; and this is no slight advantage, from the great difficulty of the operation.

* V. Base du Syst. Métrique, vol. iii., p. 337.

The absolute length of the second's pendulum at London was determined in a peculiarly ingenious and elegant way by Captain Kater *. To Captain Sabine we owe a series of most extensive and important observations, made over a great part of the northern hemisphere, on the coasts of Europe, Africa, and America, from 12° of south to 80° of north latitude †.

It is impossible in this place to give even a cursory account of the numerous and valuable experiments made by various observers in our own times. But we must observe, that the various measures of the pendulum, like those of the degrees, are far from conspiring to give the same ellipticity to the earth. Local causes have doubtless something to do with these irregularities. But it would appear also that it is necessary to take into account a term involving the product of the squares of the sine and cosine of the latitude ‡. Theory shows that such a term ought to exist, but the coefficient had not hitherto been supposed to be of sensible magnitude. One fact is remarkable, that the pendulum experiments in general indicate a greater compression than the measure of degrees. The former is stated by Captain Sabine at $\frac{1}{288}$ §, by others at $\frac{1}{285}$.

Chapter XXII.

Discovery of the Planet Uranus.—Law prevailing between the mean distances of the planets.—Discovery of Ceres.—Of Pallas.—Conjectures of Olbers.—Discovery of Juno and of Vesta.—Systems of double and multiple stars.—Supposed motion of the sun in space.—Nebulæ.—Hypothesis of La Place respecting the origin of the Solar System.

In the year 1781, a curious and altogether unexpected discovery was made

* Philosoph. Transact. for 1818, p. 33.

† An Account of Experiments to determine the Figure of the Earth. 4to. London, 1825.

‡ Encyclopædia Metropolitana. Article, Figure of the Earth, p. 231.

§ Experiments, p. 352.

with regard to the system, of which the earth forms part. The five planets, Mercury, Venus, Mars, Jupiter, and Saturn, had been known from time immemorial; nor does it seem to have been suspected that there existed in the solar system other bodies analogous to them. It is to a fortunate accident that we owe the extension of our knowledge on this subject. On the 13th of March, of the year abovementioned, William Herschel, an able but self-taught optician and astronomer, having directed his telescope upon some small stars in the constellation Gemini perceived in one of them some peculiarities, which induced him to examine it more closely. The excellence of his telescope enabling him to apply a magnifying power of several hundred times, he remarked that the apparent diameter of the body in question increased sensibly, while that of the small fixed stars near it remained unaltered*. This and other circumstances induced him to suspect that it might be a comet, and his conjecture seemed to be confirmed when two days afterwards it was found to have changed its place. Being then satisfied that the body he observed was not a fixed star, Herschel communicated his discovery to Maskelyne the astronomer royal, by whom it was made known to the continental astronomers, and observations were made upon it at Paris, Milan, Berlin, Stockholm, and other places. No doubt being entertained of its cometary nature, several astronomers, particularly Méchain, Boscovich, La Place, Lexell, and president de Saron, attempted to calculate its orbit in the usual parabolic and elliptic hypotheses. It was found however impossible to obtain elements which would represent tolerably any sensible portion of the orbit, which arose principally from the distance being assumed a great deal too small. De Saron was the first to perceive that it was necessary to place his body much farther than an ordinary comet in the visible part of its orbit: he supposed it to be twelve times farther from us than the sun, and on this supposition the calculations agreed much better with observation. After some time it was perceived that the distance did not sensibly vary, that the orbit in fact was circular, and the newly-discovered body a real planet revolving round the sun, at a distance about double that of Saturn, in a period of eighty-four years. The name given by Herschel himself to the new planet was the Georgium Sidus, but this does not seem to have been very favourably received; Lalande proposed to give it the name of its discoverer; but that which appears to have finally prevailed is the name of Uranus, applied from mythological analogies. In the Greek fables Uranus, the oldest of the gods, was the father of Saturn, as the latter was of Jupiter: in the solar system the new planet, the most distant of all, lies beyond Saturn, as the latter does beyond Jupiter. We owe this elegant denomination to the German astronomers. It was first used in the tables of Wurm, and was subsequently adopted by the French Board of Longitude. The discovery of the planet Uranus afforded an altogether unexpected confirmation of a curious empirical law discovered by Professor Bode of Berlin to prevail between the distances of the various planets from the sun, which may be thus represented. Suppose the distance from Saturn to the sun to be divided into 100 parts: then the mean distance of the other planets, respectively, will be,

Mercury	4
Venus	$4+3=7$
The Earth	$4+2.3=10$
Mars	$4+2.2.3=16$
. . . .	
Jupiter	$4+2.2.2.3=52$
Saturn	$4+2.2.2.2.3=100$
Uranus	$4+2.2.2.2.2.3=196$.

This law was remarked by Bode in 1772, consequently some years before the discovery of Uranus, and the distance of that planet being found to correspond with it, afforded a very remarkable confirmation of its truth. But it will be perceived that there is a deficiency of one term between Mars and Jupiter. This led several German astronomers to suspect the existence of a planet at the distance required by Bode's law, which from its smallness had hitherto escaped observation; and de Zach had actually computed from analogy what should be the elements of the orbit of the supposed missing planet. These surmises were destined to receive a curious confirmation. On the 1st of January, 1801, Piazzi, astronomer royal at Palermo, being occupied with a revision of the heavens for the purpose of forming a new and extensive catalogue

* Philosoph. Transact. for 1781, p. 492.

of the fixed stars, perceived in the shoulder of the Bull a small body which from its change of place in the course of a few days he at first supposed to be a comet; and as such he announced it to Bode, who found directly from the observations communicated to him, that the orbit was circular, and that the body in question must be ranked among the planets of our system. Piazzi himself had recognized this fact in the mean time, and Oriani, de Zach, and other astronomers came to the same conclusions. The elements calculated from the observations by the former, presented the most remarkable resemblance with those of the imaginary planet previously calculated by de Zach. Thus de Zach had given for the mean distance 2.82 and for the periodic time four years, nine months: Bode found from observation 2.75 for the distance, and for the periodic time four years, nine months as before. This remarkable coincidence, occurring after the discovery of Uranus had already afforded an unexpected corroboration to the law of Bode, has made the latter, though entirely empirical, and not to be referred to any known cause, extremely deserving the notice of astronomers. The new planet, which was supposed to complete our system, received from its discoverer the name of Ceres*.

After this discovery there was no reason to anticipate any others of a similar nature. However on the 28th of March, 1802, Dr. Olbers of Bremen observed in the constellation Virgo a moveable star, bearing a great resemblance in light and form to Ceres, and not easily to be distinguished from a star of the seventh magnitude. Though the external appearance of this body was by no means similar to that of a comet, yet the impossibility of representing its orbit by the arc of a circle, induced Olbers to suspend his opinion as to its nature, giving it in the interim the name of Pallas.† But when the observations had been continued for about three weeks, Gauss undertook to determine the orbit, without any previous hypothesis as to its form, and found that Pallas was a planet, having its orbit very close to that of Ceres, but distinguished by a considerable excentricity, and a great inclination to the ecliptic. In this latter respect Pallas is very remarkable, as the inclination of the orbit amounts to more than 34°; while for no other planet does it exceed 6°, except for the three new ones Ceres, Juno, and Vesta, and even for these does not surpass 14°.

Struck with the approximation to equality between the major axes of the orbits of Ceres and Pallas, Olbers was led to the ingenious and fortunate conjecture, that both these bodies were only fragments of a larger planet, which had been shattered into pieces by some internal explosion, or the shock of a comet: and he remarked that if this were the case, all the fragments that revolved round the sun would pass through the descending node of the orbit of Pallas upon that of Ceres. This was verified by the discovery of a third new planet by Harding at Lilienthal, on the 2d September, 1804, very near the place where the paths of Ceres and Pallas intersect, which has received the name of Juno*. The elements of this planet agree very fairly with the hypothesis of Olbers, allowing for the different effects in the perturbations of Jupiter, arising from the difference in the inclination of the orbits.

This discovery was not the only fruit of the ingenious conjecture of Olbers; following with constancy his idea regarding the position of the nodes, he reviewed carefully, every month during the space of three years, a part of the constellations of the Virgin and the Whale. On the 29th of March, 1807, his perseverance was rewarded by the discovery of a fourth new planet in the former of these constellations, similar in all essential particulars to the three previously discovered. To this he gave the name of Vesta†.

These four small planets all resemble each other not only in the magnitude of the semi-axes of their orbits, but also in the extreme smallness of their diameter. Schröter of Lilienthal affirms, that in the case of Vesta, the apparent semi-diameter does not amount to half a second‡. Herschel has made repeated attempts to measure the diameters of all, but their extreme smallness seems to render all observation of this sort very uncertain§. The same observer

* For an account of the discovery of Ceres, see v. Zach. Monatliche Correspondenz. vol. iii., p. 592.

† Monat. Corresp. vol. v. pp. 481 and 591.

* Monat. Corresp. vol. x. p. 271.

† Monat. Corresp. vol. xv. p. 502.

‡ Philosoph. Transact. for 1807, p. 245 et seq.

§ Philosoph. Transact. for 1802, p. 213; for 1805, p. 31; for 1807, p. 260.

considers them to form a class of bodies intermediate between planets and comets; and he has proposed to give them the name of asteroids. But there seems to be no reason why a certain size should be considered as an essential character of planets; it would seem more reasonable to make the difference to consist in the magnitude of the excentricity of the orbit. Of late the discovery of comets revolving within the solar system, has made it more difficult than ever to distinguish between planets and comets; unless we take for criterion, the degree of ellipticity. This will furnish us with a distinction, at once obvious and easy of application: in the case of the orbit of Juno (the most excentric of the planets) the ratio of the excentricity to the semi-axis major does not exceed 0.26, while for the comet of Encke it is 0.85, and for that of von Biela 0.75.

It was a natural and very ancient conjecture, that each of the fixed stars is itself a sun, the centre of a system analogous to that which we inhabit. The illustrious Herschel first discovered that there exist in the sidereal spaces systems of another description, formed by the revolution of two or more fixed stars about their common centre of gravity. The telescope had shown previous observers that many stars which appear single to the naked eye, are in reality composed of two, and in some instances of three or four others. The great excellence of his telescopes and methods of micrometrical observation, enabled Herschel to go farther. On comparing the positions of certain double stars at an interval of twenty years, he found such changes as could only be explained, by the supposition of the revolution of one round the other, or of both round some intermediate point. Thus in the well-known double star Castor, the smaller star was found in 1803 to be at the same distance from the larger that it had been in 1779; but the angle of position had diminished in this time from 32° 47′ to 10° 53′, and this diminution was found to have been pretty uniform at the rate of about 56′ annually*. By a fortunate accident, the line joining the two stars of Castor had been noticed by Bradley in the year 1759, as being parallel to that joining Castor and Pollux in the heavens, which observation gives for the angle of position in 1759, 56° 32′. A comparison of this angle with that of 1803 gives an annual diminution of 1° 3′ nearly, which, considering the great difficulty of the measurements, was a very striking and satisfactory coincidence. Thus also the angle of position between the two stars forming ξ Bootis, had changed 30° 41′ in about twenty-two years*; and many similar variations were noticed in others. By an accurate analysis of the phenomena, Herschel showed that they were not generally susceptible of being explained by the effects of annual parallax, a proper motion of translation of the double star itself, or of the sun. Succeeding astronomers who have investigated this interesting subject have fully confirmed his results. We are now acquainted with a great many of these binary and multiple systems, and the accurate determination of the revolutions performed in them occupies several of the most distinguished observers of the present day.

A motion of the sun itself, and of the whole solar system in space, has been alluded to as one possible cause of some of those small motions of the fixed stars, which, not varying according to any known law, have been called their proper motions. In the year 1783, Herschel published a paper in the Philosophical Transactions, in which he endeavoured to show that the solar system is in motion towards a point in the heavens, not far from the star λ Herculis. That the sun's centre of gravity, or rather, perhaps, that of the whole system, is not a fixed point in space, appears from mechanical considerations extremely probable. This had been stated by Lalande in 1776, and even as early as 1760 the illustrious Mayer had expressed an opinion that our sun being a body analogous to the fixed stars, might have like them its proper motion. But an examination of the positions of the most remarkable stars at distant intervals, renders it very doubtful whether any such motion can be traced. This point has been examined by M. Biot†, who finds on examining the proper motions of Aldebaran, Capella, Sirius, Procyon, Pollux, Arcturus, and Lyræ, and α Aquilæ, that they cannot be explained by supposing our system in motion towards a point situated

* Philos. Trans. for 1804, p. 367.

* Philos. Trans. for 1804, p. 367.

† Astronomie Physique, vol. iii., Additions, p. 114.

near that indicated by Herschel; and the considerations which he subjoins show that the same would be the result were any other point taken. For be the point in question (which we shall call the pole of translation) where it may, it is easy to prove that the directions of the apparent motions of the fixed stars resulting from such a cause as we now suppose, will converge towards this pole. It is also evident, that there will be no motion in right ascension on the circle of declination which passes through this pole, which circle cuts the equator at two diametrically opposite points. Now, on examining the list of proper motions given by de Zach, it seems that in going from 0° to 180° of right ascension, the point where the proper motions in right ascension change their sign is between 85° and 98°; the same sign then prevails from this point to another intermediate between 219° and 230°. Yet to satisfy the condition that the two points of the equator of which we have spoken should be opposite to each other it would have been necessary to find the second change of sign between 265° and 278°, that is to say, 35° from the place where it really is.

Again, if there were a general motion towards a pole situated anywhere, stars situated very near each other would have motions nearly parallel. Yet this is found to be by no means the case: thus α Herculis and α Ophiuchi differ only 5° in right ascension, and 2° in declination; yet the variation of right ascension in forty-two years is 2″.26 for the first, and 8″.54 for the second; while in declination it is 10″.8 and +2″ respectively. The three stars α, β, and γ, of the constellation of the Eagle present discrepancies still more considerable.

The anomalies of which we have been speaking are generally most remarkable in those stars whose proper motions are the most considerable and the best ascertained. The conclusion then seems inevitable, that if the supposed motion of the solar system exist, it is as yet imperceptible, being masked by the real proper motions in various directions of the fixed stars.

A very cursory inspection shows us that the fixed stars are not equally scattered over the heavens, but rather collected in groups, some of which from their immense distance have a cloudy indistinct appearance, and are called nebulæ. Many of these nebulæ when examined with a good telescope are resolved clearly into distinct stars clustered closely together in immense numbers. The most remarkable group of this description is that which appears to surround the heavens, and bears the name of the milky way. Of this group it is supposed that our sun and the brightest of the fixed stars form a part. The immense number of stars comprised in the milky way surpasses any conception we can form upon the subject. We may, perhaps, form some idea with regard to it from the circumstance that its *depth* is supposed to exceed a thousand times the distance from Sirius to the earth. This last distance we must recollect is so great, that an observer placed in Sirius would not see the diameter of the earth's orbit (about 190,000,000 of miles) under an angle of more than 1″. Herschel estimates that in a zone of 15° in length by 2° in breadth there are comprised 50,000 stars visible in his telescope.

It must not be supposed, however, that all nebulæ are to be resolved into clusters of stars, closely packed together. Many of them resemble rather planetary bodies than the light produced by a multitude of stars. Others again appear to be composed of a single bright star placed in the centre of a less luminous mass. Herschel, observing all the varieties of appearance presented by these nebulæ, was led to the conclusion that they were all parts of a luminous substance scattered through the immensity of the heavens, which accumulates in certain points, either from mutual attraction, or that of a neighbouring star. He thought he could distinguish by the greater or less degree of sphericity, and the brilliancy of the central nucleus, as compared with the surrounding nebulosity, the progress in condensation and the relative ages of different nebulæ, just as we may follow in a forest the growth of a tree by observing individuals of all the different ages contained in it. The first stage is that of an uniformly nebulous mass; the second, that of a similar mass slightly condensed round two or three faintly luminous nuclei; these nuclei gradually become brighter; then the nebulous atmospheres of each separating by the effects of a farther condensation, there result compound nebulæ, formed of brilliant centres very near each other, and surrounded respectively by their separate atmospheres. Sometimes the luminous matter, by a more uniform condensation, forms the planetary nebulæ of which we have spoken. Lastly, a higher degree of condensation transforms the nebulæ into groups of stars thickly set together.

These observations tend to corroborate a very elegant and ingenious hypothesis of La Place, respecting the origin of the solar system. It is impossible not to be struck with the extraordinary circumstance that the motions of all the planets and satellites take place from west to east, and very nearly in the same plane; and that the motions of rotation on their respective axes are all in the direction and nearly in the plane of their motions of projection. These coincidences, to which we may add the small excentricity of all the planetary orbits, are too extraordinary to be the result of chance. They evidently depend upon some common cause, which determined the motions of the respective bodies at their formation.

La Place supposes that at some very remote period the solar atmosphere extended beyond the limits of the farthest planet. In this primitive state, the sun resembled those nebulæ described by Herschel, as composed of a bright nucleus, surrounded by nebulosity, which, as it gradually condenses at the surface of the nucleus, ultimately converts it into a star. Let us suppose such a condensation, which is evidently very gradual, to take place in the imaginary primitive solar atmosphere. The laws of dynamics show that as the condensation goes on, the sun's rotation will be accelerated, the centrifugal force at the limits of the atmosphere increased, and these limits, which depend upon the magnitude of the centrifugal force, contracted.

In this manner as the condensation proceeds, zones of vapour will be successively abandoned, which by their condensation and the mutual attraction of their particles will form so many concentric rings of vapour circulating round the sun. But the regularity that this formation requires in the arrangement of the particles of the zone, and in their cooling, must have rendered this phenomenon very rare. Accordingly we see but one instance of it in the solar system—that of the rings which circulate round Saturn. In most cases each ring of vapours would divide into several masses, which with nearly equal velocities would continue to circulate round the sun. Mechanical considerations show that these masses would assume a spheroidal form, with a motion of rotation in the same direction as the motion of revolution; and if one of them were sufficiently considerable to unite successively by its attraction all the others round its centre, we see how the ring of vapours may have been transformed into a single spheroidal mass of vapours, circulating round the sun, with a motion of rotation, in the same direction as that of revolution. The formation of the planets being conceived to take place in this manner, we may easily imagine that an ulterior condensation has produced in a similar way the satellites which revolve round the planets. If the whole system had formed in this manner with complete regularity, the orbits of the different bodies that compose it, would be circles whose planes as well as those of their respective equators would coincide with that of the solar equator. But it is easy to understand that the innumerable variations of temperature and density in the different parts of these great masses have produced the excentricities of their orbits, and the deviations of these orbits from the plane of this equator.

Chapter XXIII.

Conclusion.

If we recur for a moment to the order of the principal facts, of which it has been attempted to give an outline in this treatise, it will be seen that they may be classed in three periods sufficiently distinct. The first of these, which refers simply to the observation of celestial phenomena, and the hypotheses by which they were reduced to calculation, comprises the labours of all the astronomers antecedent to Copernicus. The second period, that of Copernicus and Kepler, embraces the discovery of the real motions of the earth and the general laws of the planetary revolutions. The third period is characterized by Newton's discovery of the causes of these laws, and the application of analysis to the theory of universal gravitation, which we owe to the great mathematicians of the last century. To the progress made within that time by the science of physical astronomy, as well as to the great improvements in the construction of instruments, is due the surprising perfection of our tables. On the accuracy with which the places of the sun, moon, and planets can be predicted for any future time, depend the most important applications of astronomy to the purposes of navigation and geography. This accuracy would have been unattainable without a knowledge of all the complicated disturbances in the motions of these bodies, produced by their mutual action: and it is the determination *à priori* of the mathematical laws followed by these disturbances, that has rendered the improvement of the tables so great and rapid during the last half century. While on the one hand the practical advantages of this perfection are so numerous and so extensively felt; on the other the solution of the great mechanical problem, to which astronomy has now been reduced, bids fair to lead to results of the highest order among philosophical and speculative truths.

Indeed the advantages derived from the perfection of astronomy are still greater in a moral than in a practical and economical view. This has been well expressed by La Place, with whose words we shall conclude. 'Astronomy, by the dignity of its object, and the perfection of its theories, is the finest monument of the human understanding, the most noble proof of its intelligence. Led away by the illusions of his senses and by his self-love, man long looked upon himself as the centre of the heavenly motions, and his vain pride has been punished by the fears that they have caused him. At last, centuries of investigation have removed the veil that hid from his eyes the system of the world. He then found himself on a planet almost imperceptible in the solar system, of which the vast extent is itself only an insensible point in the immensity of space. The sublime results to which this discovery has led him, are well calculated to console him for the rank that it assigns to the Earth; it shows him his own greatness in the smallness of the base which he has employed to measure the heavens. Let us preserve with care, let us augment the treasures of this lofty knowledge, the delight of thinking beings. They have rendered important services to navigation and geography; but their greatest benefit consists in having dissipated the fears produced by the celestial phenomena, and destroyed the errors arising from the ignorance of our true relation to nature: errors and fears which would spring up again promptly, were the torch of science extinguished.'

APPENDIX.

View of the principal Elements of the Indian Astronomy, as contained in the Sourya Siddhanta.

Obliquity of the ecliptic . . .	24° $0'$ $0''$
Greatest equation of the centre . .	2 10 32
Length of the sidereal year . .	365^d 6^h 12^m 36^s
Annual precession . . .	0° $0'$ $54''$
Sidereal revolution of the moon .	27^d 7^h 39^m 12^s
,, ,, of the apsides . .	$3232^d.833^h$
,, ,, of the nodes . .	6794.383
Greatest equation of the centre .	5° $2'$ $48''$
Lunar parallax . . .	$51'$ $40''$
Sidereal revolution of Mercury . .	87.94
,, ,, ,, Venus . . .	281.99
,, ,, ,, Mars . .	686.99
,, ,, ,, Jupiter . . .	4332.32
,, ,, ,, Saturn . .	10765.87

View of the principal Elements of the Tables of Ptolemy.

Obliquity of the ecliptic	23°	51′	0″
Greatest equation of the centre	2	23	0
Length of the tropical year	365ᵈ. 5ʰ.	55ᵐ.	12ˢ.
Annual precession	0°	0′	36″
Mean distance of the sun [the earth's radius = 1]			1210
Diurnal tropical revolution of the moon	13°	10′	34″
,, ,, ,, of the apsides	0	6	41
,, ,, ,, of the nodes	0	3	10
Greatest equation of the centre	4	59	14
Apparent diameter of the moon (constant)		32	20
Mean distance of the moon (earth's radius = 1)			= 59
Inclination of the orbit	5°	0′	0″
Mean motion in 365 days of Mercury	1493	43	13
,, ,, ,, Venus	584	46	57
,, ,, ,, Mars	191	16	54
,, ,, ,, Jupiter	30	20	22.8
,, ,, ,, Saturn	12	13	23.9

Principal Elements of the Arabian Astronomy, from El-Batani.

Obliquity of the ecliptic	23° 35′ 0″
Length of the tropical year	365ᵈ 5ʰ 46ᵐ 24ˢ
Greatest equation of the centre	1° 58′ 0″
Annual precession	0 0 54.5

The Lunar Tables of El-Batani, and his parallaxes of the sun and moon, are those of Ptolemy; however, instead of assuming the moon's diameter as constant and = 32′ 20″

he makes it = 29 30 in the apogee.
and = 38 30 in the perigee.

The following mean motions of the planets are taken from Ibn Jounis.

Motion in 30 days of Mercury	93° 12 4″1
,, ,, ,, Venus	48 29 47.2
,, ,, ,, Mars	15 43 19.7
,, ,, ,, Jupiter	2 29 38.0
,, ,, ,, Saturn	1 0 17.7

Elements of the Tables of Tycho Brahé.

Length of the tropical year	365ᵈ 5ʰ	49ᵐ	0ˢ
Greatest equation of the centre	2°	3′	15″
Obliquity of the ecliptic	23	31	30
Annual motion of the apogee	0	0	45
Mean solar parallax	0	3	0
Annual precession	0	0	51
Mean parallax of the moon	0	57	24
Greatest ditto	0	66	6
Least ditto	0	48	43
First inequality—equation of the centre	4	58	30
Annual motion in longitude	129	37	32
,, ,, anomaly	88	43	7.7
,, argument of the latitude	148	42	45.5
Greatest semidiameter	0	18	0
Least ditto	0	14	24
Horizontal refraction (for the sun)	0	34	0
Refraction at 45° of altitude	0	0	5

The following mean motions of the Planets are taken from the Rudolphine Tables of Kepler.

Mean motion of Saturn in 365 days		12°	13′	36″
,, ,,	Jupiter	.	30	20	32
,, ,,	Mars	.	131	17	8
,, ,,	Venus	.	224	47	36
,, ,,	Mercury	.	53	43	15

Additional Observations on the Indian Tables.

It has been thought advisable to reserve for the Appendix a more extended account of the Indian Tables and methods of calculation: a subject of great interest, but which it is difficult to explain without entering into details too obscure and technical to suit the text of this Treatise. It is unnecessary to recapitulate the reasons which have been assigned for refusing to the Indian Tables that antiquity which has been attributed to them by some authors; and for supposing the date of the Surya Siddhanta in particular to be several centuries posterior to the Christian æra. Nor can we enter into the discussions raised between Messrs. Bentley and Colebrooke about the age of various early Hindoo astronomers. Were the ancient Sanscrit treatises of astronomy accessible to the European public, an opinion might be formed as to the relative probability on either side of the question. But as long as these treatises (the various Siddhantas of Brahma, Surya, &c.,) remain not only unpublished, but in a language known to very few if any scientific men in this country; and as long as our knowledge of them is confined to a few scanty extracts, elicited by scientific controversy, it is next to impossible to appreciate the value of the arguments and opinions offered on the subject. In this uncertainty, we shall merely content ourselves with observing that Mr. Bentley, in his Historical View of the Hindu Astronomy*, refers the earliest Indian observations to the year 1425 B.C., at which time, according to him, the lunar mansions or constellations of the lunar zodiac were formed†. The date in question is deduced from the position of the colures: a certain mansion having received its name from being bisected by the equinoxial colure at the time of its formation‡. The next observation may be referred to the year 1181 B.C., when the sun and moon were in conjunction in the winter solstice. At this time it is said that the twelve solar months were formed, and observations made on some of the planets§. About the year 945, it was ascertained that the Colures had fallen back 3° 20′ from their positions at the former observations in 1181, thus giving an annual precession of nearly 48″.57 ‖. It is unnecessary to recapitulate various observations on the quantity of precession, the position of the planets, the heliacal rising of certain stars, made in the course of the seven subsequent centuries. About the year 204, according to Mr. Bentley ¶, considerable improvements were made in astronomy, new and more accurate tables formed, and new equations introduced. At this time, too, the Hindoo history was divided into periods for chronological purposes; and, in order to recover them by astronomical computation, should they be lost, the following system was adopted:—For the year with which each period began and ended they computed the day of the month, and the moon's age, on the day in which the sun was in conjunction with Jupiter. These chronological periods were called Yugas. The end of the first period, or Kali Yug, was fixed by a conjunction of the Sun, Moon, and Jupiter, in the beginning of Cancer, on the 26th of June, 299 B.C. At a later time, similar chronological periods, called Manwantaras, were formed, and similarly determined, with the difference that the planet Saturn was employed, instead of Jupiter. Of the Yugas there were four; and the earliest of them, the Krita Yug, began in the year 2352 B.C. Of the Manwantaras there were ten, and the earliest began 4235 B.C. The object of the introduction of the latter seems to have been to increase the supposed antiquity of the nation.

* London, 1825, 8vo.

† It appears that at first there were 28 of these mansions, each comprising an arc of 12°.51′3-7ths; but these were subsequently reduced to 27 of 13°.20′ each. Bentley, op. cit. p. 5.

‡ Bentley, p. 2. § Bentley, p. 10.

‖ This, however, was not the first observation of precession: the solstice of 1181 had shown them, that since the formation of the lunar mansions the colures had retrograded through 3°.20′.

¶ Page 75.

With the year 538 A. D., begins, according to Mr. Bentley,* the modern astronomy in Hindoostan; an astronomy characterised by the introduction of new and enormous periods. The Bramins at this time chose to select a period of 4320000000 years, which they called a Kalpa. The object of going so far back into antiquity was to obtain a mean conjunction (or at least an approximation to such) of all the planets in the beginning of the sidereal sphere, commencing with the lunar asterism Aswini. The Kalpa was subdivided into 14 Manwantaras, each of 308448000 years with the addition of 1728000 years to make up the Kalpa. The Manwantara again they divided into 71 Maha Yugas or great ages of 4320000 years each, with the addition of 1728000 to make up the Manwantara. The Maha Yuga, or great age, they divided into four others, *viz.* the Kali of 432000 years, the Dwapar of 864000 years, the Treta of 1296000, and the Krita of 1728000 years: assigning to these periods the same names that they had used in the former divisions of their history. Knowing the positions of the planets for a certain day of the year 538 A. D., the respective mean motions may be concluded by a comparison with the places at the beginning of the Kalpa; and hence their places calculated for any intermediate or subsequent time.

The most ancient astronomical system framed upon these principles is, as we have just said, referred by Mr. Bentley to the year 538 A. D. The author is supposed by some to be Brahma Acharya, by others Brahma Gupta: the system itself is given in a work called the Siddhanta Siromani, generally referred to the latter of these two authors. The next in chronological order are,—the Vasishtha Siddhanta, pretending to have been composed 1299101 years B. C., the Surya Siddhanta, claiming a pretended antiquity of 3027101 years B. C., and the Soma Siddhanta, to which the date of 3101 years B. C. is fictitiously ascribed. From astronomical data, in particular the longitude of Canopus, Mr. Bentley refers the Vasishtha Siddhanta to the year 928 A. D. In later times occur the Arya Siddhanta, (A. D. 1322) Parasara Siddhanta and others.

Such are the views of Mr. Bentley with regard to the origin and progress of astronomy among the Hindoos. In themselves they are sufficiently reasonable and probable; and the date thus assigned to the Surya Siddhanta is strongly confirmed by an examination of the errors of the tables contained in it*. But it is much to be regretted that Mr. Bentley has altogether omitted to quote his authorities for the different facts stated by him; so that it is impossible, except in a few instances, to ascertain whether they rest upon astronomical deductions or positive historical testimony. Indeed it would appear that many works now existing in Sanscrit contain passages altogether contradictory to the truth of this system: these passages Mr. Bentley declares to be forgeries and interpolations of the modern Brahmins. Indeed he is obliged to suppose that the whole of the ancient Hindoo works have been remodelled by the modern Indians to increase the pretended antiquity of their history and astronomy; that all passages unfavourable for this purpose have been struck out; and a great number of a contrary description interpolated. This hypothesis of Mr. Bentley seems rather forced: how far it is really borne out by facts must be left to the decision of the few who have opportunities of examining critically the Sanscrit literature.

We shall now proceed to consider the nature of the Indian astronomical systems and tables, without entering into any further discussions on their antiquity. The most celebrated and best known of these systems is the one known by the name of the Surya Siddhanta. We should form a false notion of this book, were we to suppose it at all similar to any European tables or treatises upon astronomy. To convey some idea of the nature of the work, we shall give an extract from it published by Mr. Davis in the second volume of the Asiatic Researches. 'Time of the denomination Murta† is estimated by respirations: six respirations make a Vicala, sixty Vicalas a Danda, sixty Dandas a Nacshatra day, and thirty Nacshatra days a Nacshatra month. The Savan month is that contained between thirty successive risings of Surya, and varies in its length according to the Lagna Bhija. Thirty Tit'his compose the Chandra month. The Saura month is that in which the Sun describes one sign of the Zodiac, and his passage through the twelve signs is one year, and one of those years is a Deva day, or day of the Gods. When it is

* See page 11.

† This is mean sidereal time. A Nacshatra, or sidereal day, is the time in which the stars appear to make one complete revolution. The Savan is the solar day: the Lagna Bhija means the right ascension. Asiat. Research. vol. ii. p. 230.

day at Asura* it is night with the Gods; and when it is day with the Gods it is night at Asura. Sixty of the Deva days multiplied by six give the Deva year; and 1200 of the Deva years form the aggregate of the four Yugas. To determine the Saura years contained in this aggregate, write down the following numbers 4, 3, 2, which multiply by 10000: the product 4320000 is the aggregate, or Maha Yuga, including the Sandhi and Sandhiansa†. This is divided into four Yugas, by reason of the different proportions of virtue prevailing on the earth in the following manner. Divide the aggregate 4320000 by 10 and multiply the quotient by 4 for the Satya Yug, by three for the Treta, by two for the Dwapar, and by one for the Cali Yug. Divide either of the Yugs by six for its Sandhi and Sandhiansa. Seventy one Yugs make a Manwantara, and at the close of each Manwantara there is a Sandhi equal to the Satya Yug, during which there is an universal deluge. Fourteen Manwantaras, including the Sandhi, compose a Calpa, and at the commencement of each Calpa there is a Sandhi equal to the Satya Yug, or 1728000 Saura years. A Calpa is therefore equal to 1000 Maha Yugs. One Calpa is a day with Brahma, and his night is of the same length; and the period of his life is 100 of his years. One half of the term of Brahma's life is expired, and of the remainder the first Calpa is begun; and six Manwantaras, including the Sandhi, are expired. The seventh Manwantara, into which we are now advanced, is named Vaivas-wata: of this Manwantara 27 Maha Yugs are elapsed, and we are now in the Satya Yug of the 28th, which Satya Yug consists of 1728000 Saura years. The whole amount of years from the beginning of the Calpa to the present time may hence be computed; but from the number of years so found must be made a deduction of one hundred times four hundred and seventy-four divine years, or of that product multiplied by three hundred and sixty for human years, that being the term of Brahma's employment in the creation; after which the planetary motions commenced.'

Making the calculations indicated in the text of the Surya Siddhanta we get

$$\text{Kali} = \frac{4320000}{10} \times 1 = 432000$$

$$\text{Dwapar} = \frac{4320000}{10} \times 2 = 864000$$

$$\text{Treta} = 3 \text{ Kalis} = 1296000$$

Satya = 4 Kalis	= 1728000
Maha Yug	= 4320000
Multiply by 71	= 302400000
71 Maha Yugs	= 306720000
Sandhiansa = Satya Yug	= 1728000
	308448000
Multiply by 14	= 1233792000
	4318272000
Sandi	= 1728000
The whole duration of the Calpa then	= 4320000000
The Sandhi at the beginning of the Calpa	= 1728000
6 Manwantaras	= 1850688000
20 Maha Yugs	= 86400000
7 Maha Yugs	= 30240000
Satya age of the 28th Maha Yug	= 1728000
	1970784000
Subtract 474 × 360	= 170640
Period elapsed from the beginning of the Calpa . . .	= 1970613360

* The South Pole.

† Morning and evening twilight.

The Surya Siddhanta then proceeds:—

'In one Yug, Surya, Buddha and Sucra perform 4320000 Madhyama revolutions through the zodiac; Mangala, Vrihaspati and Sani make the same number of Sigra revolutions through it; Chandra makes 57753336 Madhyama revolutions; Mangala 2296832 Madhyama revolutions; Buddha's Sigras are 17937060; Vrihaspati's Madhyamas 364220; Sucra's Sigras 7022376; Sani's Madhyamas are 146568. The Chandrochcha revolutions are 488203; the retrograde revolutions of the Chandrapata are 232238.'

To understand this passage it is necessary to observe that Surya means the Sun; Buddha, Mercury; Sucra, Venus; Mangala, Mars; Vrihaspati, Jupiter; Sani, Saturn; Chandra, the Moon; Chandrochcha, the Moon's apogee; Chandra Pata, the Moon's ascending node. The Madhyama revolutions of Mars, Jupiter and Saturn, and the Sigra revolutions of Venus and Mercury correspond to their revolutions about the Sun. The mean revolutions of the two latter planets are then equal to that of the Sun, as in Ptolemy. The length of the lunar month ma be found from this passage as follows:—

Chandra revolutions in a Yug	=	57753336
Madhyama revolutions of Surya	=	4320000
Lunar months in a Yug	=	53433336

And 1577917828, being the number of days in a Yug, (see below),

$$\text{Length of the lunar month} = \frac{1577917828}{53433336} = 29^d\ 12^h\ 44^m\ 2^s\ 47''' \text{ in English time.}$$

The Surya Siddhanta proceeds,—

'The time contained between sunrise and sunrise is the Bhumi Savan day: the number of those days contained in a Yug is 15777917828. The number of Nacshatra days 1582237828; of Chandra days 1603000080; of Adhi months 1593336; of Cshaya Tithis 25082252; of Saura months 51840000. From either of the planet's Nacshatra days deduct the number of its revolutions, the remainder will be the number of its Savan days contained in a Yug. The difference between the number of the revolutions of Surya and Chandra gives the number of Chandra months; and the difference between the Saura months and Chandra months gives the number of Adhi months. Deduct the Savan days from the Chandra days, the remainder will be the number of Tithi Cshayas. The number of Adhi months, Tithi Cshayas, Nacshatra, Chandra, and Savan days, multiplied severally by 1000, gives the number of each contained in a Calpa.'

To get the length of the Hindoo year, divide the number of Bhumi Savan days contained in a Yug by the solar revolutions in the same, and we obtain

$$\text{Length of the Hindoo year} = \frac{1577917828}{4320000} = 365^d\ 6^h\ 12^m\ 36^s\ 33''' \text{ in English time.}$$

To find the Moon's periodical month, divide the same number of days by the number of Chandra revolutions in a Yug, and we get

$$\frac{1577917828}{57753336} = 27^d\ 7^h\ 39^m\ 12^s\ 38'''.$$

The Chandra days, called also Tithis, are each one-thirtieth part of the Moon's synodical month, and vary in length according to the inequality of her motion. The Adhis, or intercalary lunar months, are thus found,—

Lunations in a Yug	=	53433336
Saura, or solar months	=	51840000
Adhis in a Yug	=	1593336

The Surya Siddhanta then adds,

'The number of Mandochcha revolutions, which revolutions are direct, or according to the order of the signs contained in a Calpa, is, of Surya 387; of Mangala 204; of Budha 368; of Vrihaspati 900; of Sucra 535; of Sani 39.'

'The number of revolutions of the Patas, which revolutions are retrograde, or contrary to the order of signs contained in a Calpa, is, of Mangala 214; of Budha 488; of Vrihaspati 174; of Sucra 903; of Sani 662. The Pata and Occha of Chandra, are already mentioned.'

It may be as well to repeat here that Mandochcha means the apogee; Pata the node. The Brahmins have perceived the necessity, in the course of time, of applying some corrections to these numbers. Thus the motions of the Moon's apogee and node are now increased in computations of their places by the addition of four revolutions each in a Yug, to the respective numbers above given.

If we now proceed to the consideration of the equations of the centre, we find that the Hindoos, like Ptolemy and the earlier European astronomers, employ excentric circles, and epicycles. In the excentric they suppose the excentricity equal to the sine of the greatest equation of the centre; and, generally, the sine of the equation of the centre equal to the sine of the greatest equation × sine of the anomaly. This fundamental principle is common to the Greeks and Indians. The latter, as well as the former, substitute an epicycle for the excentric, the radius of the epicycle being equal to the excentricity. But the Hindoos modify their calculation in a peculiar way, the necessity or advantage of which it is not easy to see. The equation of the centre of the Sun being 130′.32″, the Indians say,

Radius 3438′ : excentricity = 130′.32″ :: 360° : 13°.40′ = circumference of the epicycle. This circumference they call the paridhi-ansa; they take on it an arc equal to the mean anomaly. The sine and cosine of this anomaly in the little circle are in a given ratio with the sine and cosine of the same arc in the greater circle

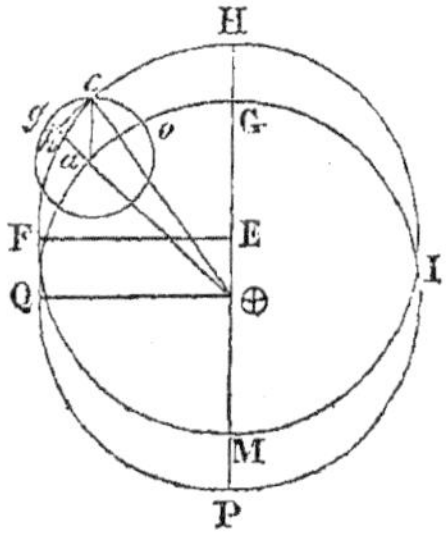

Let E be the centre of the excentric circle F M I H, which is the orbit of the Sun; ⊕, the Earth, and Q G I P a circle which has ⊕ for its centre; E ⊕ is the excentricity, H the place of the apogee. Take the Sun's mean place at *a*, and with a radius = excentricity describe the epicycle *c g o*, cutting H F M I in *c*; a line joining *c a* will clearly be parallel to G ⊕, (for the radius ⊕ *a* = the radius E H or E *c*): join ⊕*a* and prolong it to *g*, the angle *c a g* = H ⊕ *a* = the mean anomaly; call this angle ϕ, then C *h* = *ac* . sin. ϕ

$$= e \,.\, \sin.\ \phi\ [e = \text{excentricity}],$$

$$\oplus h = \oplus a + a h$$

$$= 1 + e \,.\, \cos.\ \phi.$$

Knowing, then, *c h* and ⊕*h*, we might find* C ⊕, and thence deduce the angle *c* ⊕ *a*, which angle = *a* ⊕ H − *c* ⊕ H

= mean anomaly − true anomaly

= equation of the centre.

The square of the hypothenuse $= e^2 \,.\, \sin.^2 \phi + 1 + 2e \,.\, \cos.\ \phi + e^2 \,.\, \cos.^2 \phi^2$

$$= 1 + e^2 + 2e \,.\, \cos.\ \phi,$$

$$\text{and } \sin.\ c \oplus a = \frac{e \,.\, \sin.\ \phi}{\sqrt{1 + e^2 + 2 \,.\, e \,.\, \cos.\ \phi}} = e \,.\, \sin.\ \phi - \frac{1}{2} \,.\, e^2 \sin.\ 2\phi.$$

This would be very nearly the rigorous formula. But in the case of the Sun and Moon they adopt a peculiar artifice; they adopt, in fact, a variable excentricity; they augment the circumference of the epicycle by 20′ sin. ϕ, which is the same as augmenting the radius by 3′.11″ sin. ϕ; the excentricity becomes then generally

$$= 133'.43'' - 3'.11'' \,.\, \sin.\ \phi;$$

hence the equation of the centre $= (133'.43'' - 3'.11'' \,.\, \sin.\ \phi) \sin.\phi,$

$$= 133'.43'' \,.\, \sin.\ \phi - 3'.11'' \,.\, \sin.^2\ \phi.$$

* The shortest way would be to find the arc from the tangent, but this the Indians, who possess tables of sines only, could not do.

† In the Asiat. Research., vol. ii. p. 245, it is put, by mistake, 225′.

The lunar equation is calculated by the same method; and, though the equation is double that of the Sun, with the same empirical correction $3'.11'' \sin.^2 \varphi$, which seems very extraordinary. But it is remarkable that these formulæ represent the equations in question with great accuracy; in fact, with much greater accuracy than the Indians could use, their tables of sines being only calculated for every $3^\circ\frac{1}{4}$.

Mention has already been made of these tables in the first part of this Treatise: as they are not the least curious part of the subject, it may be worth while, in this place, to give a more detailed account of them.

The precept in the Surya Siddhanta is the following. 'Divide by 8 the number of minutes contained in a sign, that is to say 1800: the quotient 225′ is the first Jyapinda, or the first of the twenty-fourth portions of half the string of the bow, (half the chord of the double arc). Divide the first Jyapinda by 233.506, subtract from the dividend the quotient 1′, and add the remainder to the first Jyapinda to form the second 449. Divide the second Jyapinda by 233.506, the quotient being $1\frac{1}{2}$ and more; subtract 2′ from the preceding 224, and add the remainder thus found to the second Jyapinda, to form the third 671. Divide this number by 233.506; subtract the quotient 3′ from the last remainder 222′; add the remainder 219 to the third Jyapinda to form the fourth 890; and thus continue till you have completed the 24 cramadjyas, (right sines).' This method is founded upon the following principle:—the second difference of sin. A,

$$\text{or } \Delta^2 . \sin. A = -4 \sin.^2 \left(\frac{\Delta A.}{2}\right) \sin. A$$

$$= -(\text{chord } A)^2 . \sin. A.$$

In this case $\Delta A = 3^\circ.45$, $\frac{1}{2} \Delta A = 1^\circ.52'.30''$

$$4 \sin.^2 \frac{\Delta A}{2} = 0.00428255$$

$$= \frac{1}{233.506}.$$

The Surya Siddhanta, as translated by Mr. Davis, gives $\frac{1}{225}$; but as this is not the right value, and yet the Indian table is accurate, there seems to be some mistake in the translation. To enable the reader to judge of the accuracy of the table, we subjoin it here, comparing it at the same time with the modern European tables*.

Arcs.	Indian sines.	Modern sines.	Arcs.	Indian sines.	Modern sines.
0. 0	000	000			
3.45	225	224.85	48.45	2585	2584.64
7.30	449	448.75	52.30	2728	2727.35
11.15	671	670.71	56.15	2859	2858.38
15.	890	889.81	60.0	2978	2977.18
18.45	1105	1105.02	63.45	3084	3083.28
22.30	1315	1315.57	67.30	3177	3176.06
26.15	1520	1520.48	71.15	3256	3255.31
30. 0	1719	1718.87	75. 0	3321	3320.68
33.45	1910	1909.91	78.45	3372	3371.69
37.30	2093	2092.77	82.30	3409	3408.44
41.15	2267	2266.66	86.15	3431	3430.38
45. 0	2431	2431.08	90.0	3438	3437.75

A Hindoo commentator on the Surya Siddhanta, after showing how to divide the quandrant of a circle into twenty-four equal parts, adds, 'The square of the bhoujajya (sine), subtracted from the square of the trijya (radius), gives the square of the cotijya (cosine).... Take the trijya = 3438′ and containing 24 jyapindas: its half is the jyapinda of a sign or 30° = 1719′, which is the eighth jyapinda or sixteenth cotijyapinda. Multiply by 3 the square of the trijya, and divide the product by 4, the square root of the quotient is the jya of two signs, or 2977′; the

* V. Delambre, Astron. Anc., vol. i. p. 459.

square root of half the square of the trijya is the jya of 45° or 2431′, which number subtracted from the trijya leaves the outcramajya (versed sine), 1007. Multiply the trijya by this outcramajya, the square root of half the product is the jya of 22°.30′. Subtract the square of this number from the square of the trijya, the square root of the difference is the jya of 67°.30′, or 3177′, which is the cotijya of 22°.30′, whose jya = 1315. This bhoujajya and cotijya, separately subtracted from the radius, leave the outcramajya of each, that is, 261 for 22°.30′ and 2123 for 67°.30′.' From this passage it results clearly, as has already been stated in Part. I. Chap. II., that the Indians are acquainted with the formula

$$R^2 = \sin.^2 A + \cos.^2 A,$$

$$\sin. 30 = \frac{R}{2}$$

$$\sin. \frac{A}{2} = \left(\frac{1 - \cos. A}{2}\right)^{\frac{1}{2}}$$

$$\sin. 60^\circ = \frac{\sqrt{3}}{2} . R.$$

The Brahmins suppose the Earth to be spherical: they suppose the diameter divided into 1600 equal parts called yojanas. To get the circumference they used anciently to multiply the diameter by three. However the rule given in the Surya Siddhanta is to multiply by the square root of ten, which supposes the ratio 1 : 3.1627. The Indian Table of sines, on the other hand, supposes the ratio to be 1 : 3.14136.

To find the latitude of a place on the Earth's surface, they observe the shadow of a vertical gnomon when the Sun is in the equator. Knowing the length of the gnomon, they calculate the hypothenuse, and, taking this for radius, they obtain the sine of the latitude = the gnomon itself. Terrestrial longitudes are found by the observation of lunar eclipses. To determine the lunar parallax, they observe the moon's rising, which they compare with the time of the calculated rising. During the difference of these times she describes a space equal to the semidiameter of the earth. This difference of times is to the moon's periodic month as 800 yojanas, or the earth's semidiameter, to the circumference of her orbit 324000. In this computation the difference between the sine and the arc through which the moon passes is neglected. In this way they made the Moon's mean horizontal parallax 53′.20″. European astronomers estimate it about 57′. As the Hindoos suppose that all planets move in their orbits with the same velocity, the orbit of the moon being thus known, they conclude by a simple proportion those of the other planets.

For the diameters of the sun and moon, it is directed to observe the time between the appearance of the limb upon the horizon, and the instant of the whole disc being risen, at the time when their apparent motion is at a mean rate, or near 90° of anomaly: then by proportion, as that time is to a natural day, so are their orbits to their diameters respectively, which of the sun is 6500 yojanas; of the moon 480 yojanas. These dimensions are increased or diminished, as they approach the lower or higher apsis, in proportion as their apparent motion exceeds or falls short of the mean, for the purpose of computing the diameter of the earth's shadow at the moon. The rest of the calculation of the eclipse, though long and tedious, does not differ essentially from that followed by Europeans. An example of it has been given at length by Mr. Davis, (Asiatic Researches, vol. ii. p. 273,) to which the reader is referred.

Having thus endeavoured to give an outline of the Indian methods of calculation, we shall conclude by adding their Tables of the Sun, Moon, and Planets, taken from Mr. Bentley's Historical View of the Hindoo Astronomy. These Tables require no explanation, as the precepts and examples subjoined to each page elucidate them sufficiently.

SUN.

Equation of the Orbit.
Argument—Sun's mean anomaly.

Degrees	$+6^s$ -0^s	$+7^s$ -1^s	$+7^s$ -1^s	Degrees
0	0° 0′ 0″	1° 6′ 02″	1° 53′ 25″	30
1	0 2 20	1 8 00	1 54 30	29
2	0 4 40	1 9 57	1 55 34	28
3	0 7 00	1 11 57	1 56 35	27
4	0 9 19	1 13 47	1 57 34	26
5	0 11 37	1 15 40	1 58 34	25
6	0 13 56	1 17 32	1 59 30	24
7	0 16 15	1 19 23	2 0 23	23
8	0 18 33	1 21 11	2 1 14	22
9	0 20 51	1 22 57	2 2 04	21
10	0 23 7	1 24 42	2 2 51	20
11	0 25 23	1 26 26	2 3 35	19
12	0 27 39	1 28 07	2 4 17	18
13	0 29 55	1 29 46	2 4 57	17
14	0 32 10	1 31 23	2 5 35	16
15	0 34 27	1 32 58	2 6 12	15
16	0 36 37	1 34 32	2 6 45	14
17	0 38 39	1 36 04	2 7 17	13
18	0 41 01	1 37 35	2 7 45	12
19	0 43 12	1 39 06	2 8 12	11
20	0 45 22	1 40 36	2 8 35	10
21	0 47 31	1 42 03	2 8 58	9
22	0 49 39	1 43 26	2 9 18	8
23	0 51 47	1 44 45	2 9 36	7
24	0 53 53	1 46 02	2 9 51	6
25	0 55 57	1 47 17	2 10 03	5
26	0 58 01	1 48 33	2 10 13	4
27	1 00 02	1 49 47	2 10 20	3
28	1 2 53	1 51 00	2 10 27	2
29	1 4 03	1 52 22	2 10 31	1
30	1 6 02	1 53 25	2 10 32	0
Degrees	-5^s $+11^s$	-4^s $+10^s$	-3^s $+9^s$	Degrees

Use of the Table.

Supp. Sun's mean long.	11^s	2°	9′	6″
Subtract the apogee .	2	17	56	38
Sun's mean anomaly .	8	14	12	28
The equation for which is +	0	2	5	42
Sun's mean longitude .	11	2	9	6
Sun's true longitude .	11	4	14	48

MOON.

Equation of the Orbit.
Argument—Moon's mean anomaly.

Degrees	$+6^s$ -0^s	$+7^s$ -1^s	$+8^s$ -2^s	Degrees
0	0° 0′ 00″	2° 32′ 2″	4° 22′ 29″	30
1	0 5 20	2 36 37	4 25 26	29
2	0 10 40	2 41 11	4 27 36	28
3	0 16 00	2 45 36	4 29 59	27
4	0 21 19	2 49 58	4 32 19	26
5	0 26 36	2 54 20	4 34 37	25
6	0 31 54	2 58 39	4 36 47	24
7	0 37 12	3 2 54	4 38 54	23
8	0 42 29	3 7 5	4 40 54	22
9	0 47 44	3 11 12	4 42 50	21
10	0 52 18	3 15 16	4 44 40	20
11	0 58 11	3 19 18	4 46 24	19
12	1 03 23	3 23 24	4 48 5	18
13	1 08 40	3 27 26	4 49 38	17
14	1 13 45	3 30 54	4 51 9	16
15	1 18 53	3 34 39	4 52 53	15
16	1 24 00	3 38 21	4 53 54	14
17	1 29 05	3 41 58	4 55 6	13
18	1 34 9	3 45 32	4 56 15	12
19	1 39 10	3 48 59	4 57 17	11
20	1 44 9	3 52 24	4 58 13	10
21	1 49 17	3 55 46	4 59 6	9
22	1 54 3	3 59 2	4 59 53	8
23	1 58 3	4 2 13	5 0 27	7
24	2 3 47	4 5 18	5 1 8	6
25	2 8 35	4 8 18	5 1 40	5
26	2 13 22	4 11 16	5 2 3	4
27	2 18 6	4 14 11	5 2 20	3
28	2 22 47	4 17 0	5 2 36	2
29	2 27 35	4 19 46	5 2 44	1
30	2 32 2	4 22 29	5 2 48	0
Degrees	-5^s $+11^s$	-4^s $+10^s$	-3^s $+9^s$	Degrees

Use of the Table.

Sup. Moon's mean long. =	11^s	8°	56′	17″
The Moon's apogee . .	2	9	38	0
Moon's mean anomaly .	8	29	18	17
The equation for which is +	0	5	2	45
Moon's mean longitude	11	8	56	17
Moon's true longitude .	11	13	59	2

MERCURY.

Equation of the Sun's Longitude.

Arg. Sun's mean long.—Aphelion of Mer.

Degrees	– 0s	– 1s	– 2s	– 3s	– 4s	– 5s	Degrees
0	′0	′134	′232	′267	′236	′139	30
1	5	138	234	268	234	135	29
2	10	142	236	268	232	131	28
3	14	145	238	267	229	126	27
4	19	149	240	267	227	122	26
5	24	153	242	267	224	117	25
6	29	157	244	267	222	113	24
7	34	160	246	266	220	108	23
8	38	164	248	266	217	104	22
9	43	167	249	265	213	99	21
10	47	171	250	264	211	95	20
11	52	174	251	264	207	90	19
12	56	178	252	263	204	85	18
13	61	181	253	262	201	80	17
14	65	185	255	261	198	75	16
15	70	191	257	260	195	70	15
16	74	194	258	259	192	65	14
17	79	197	259	258	188	60	13
18	84	200	260	257	185	55	12
19	88	203	261	256	181	51	11
20	92	206	262	254	178	46	10
21	96	209	263	252	175	41	9
22	101	212	264	251	171	36	8
23	105	214	265	249	167	31	7
24	109	217	265	248	163	26	6
25	114	220	266	246	159	21	5
26	118	222	266	244	155	15	4
27	122	225	267	242	151	10	3
28	126	227	267	240	147	5	2
29	130	230	267	238	143	1	1
30	134	232	267	236	139	0	0
Degrees	11s +	10s +	9s +	8s +	7s +	6s +	Degrees

Use of the Table.

Suppose Sun's longitude .	11s	2°	9′	16″
Mercury's aphelion subt. .	7	14	54	40
Remain	3	17	14	26
The equation for which is –	0	4	17	45
Subt. from Sun's long. =	11	2	9	6
Remain Sun's equat. long.	10	27	51	21

Note.—The true heliocentric longitude of Mercury is not used by the Hindoo astronomers.

The Elongation.

Arg. Mercury's long.—Sun's equated long.

Degrees	+ 0s	+ 1s	+ 2s	+ 3s	+ 4s	+ 5s	Degrees
0	′0	′478	′902	′1208	′1240	′906	30
1	16	494	914	1216	1235	884	29
2	32	508	926	1222	1230	860	28
3	48	522	938	1228	1235	838	27
4	64	536	950	1234	1220	808	26
5	82	552	964	1240	1218	790	25
6	98	568	976	1246	1216	762	24
7	114	582	988	1252	1214	736	23
8	130	594	1000	1256	1212	708	22
9	142	612	1010	1260	1210	684	21
10	162	628	1020	1266	1208	656	20
11	178	642	1032	1270	1204	626	19
12	194	656	1044	1274	1194	596	18
13	210	670	1054	1278	1182	566	17
14	226	686	1064	1282	1172	536	16
15	242	700	1076	1284	1160	506	15
16	258	714	1086	1286	1148	476	14
17	274	728	1096	1288	1136	444	13
18	290	742	1106	1290	1122	410	12
19	306	756	1116	1292	1106	378	11
20	324	770	1123	1290	1092	346	10
21	338	784	1134	1290	1078	302	9
22	352	796	1144	1290	1062	278	8
23	368	810	1152	1290	1044	244	7
24	384	824	1160	1290	1026	210	6
25	400	838	1170	1282	1008	174	5
26	414	848	1178	1278	990	140	4
27	430	862	1184	1262	970	104	3
28	446	876	1192	1254	948	70	2
29	462	890	1200	1246	928	34	1
30	478	902	1208	1240	906	0	0
Degrees	11s –	10s –	9s –	8s –	7s –	6s –	Degrees

Use of the Table.

Suppose Mercury's long. .	5s	20°	51′	15″
Subt. Sun's equated long.	10	27	51	21
Remains the commutation	6	22	59	54
The elong. for which is –	0	12	16	0
Sun's equated longitude .	10	27	51	21
Geocentric long. of Mercury	10	15	35	21

VENUS.

Equation of the Sun's Longitude.

Arg. Sun's mean long.—Venus's Aphelion.

Degrees	− 0^s	− 1^s	− 2^s	− 3^s	− 4^s	− 5^s	Degrees
0	′ 0	′ 56	′ 91	′ 105	′ 92	′ 54	30
1	2	57	92	105	91	52	29
2	4	59	93	105	90	51	28
3	6	60	94	105	89	49	27
4	8	62	95	105	88	47	26
5	10	63	95	105	87	46	25
6	12	65	96	105	86	44	24
7	15	66	97	105	85	42	23
8	17	67	98	105	84	39	22
9	19	68	99	104	83	38	21
10	21	70	99	104	82	37	20
11	23	71	100	104	81	35	19
12	25	72	100	103	79	33	18
13	27	73	101	103	78	31	17
14	29	74	101	103	77	29	16
15	30	75	102	102	76	28	15
16	32	76	102	101	74	26	14
17	34	78	103	101	73	24	13
18	36	79	103	101	72	22	12
19	37	80	103	100	70	20	11
20	39	81	104	99	68	18	10
21	41	82	104	99	67	16	9
22	43	83	104	98	66	14	8
23	44	84	105	97	65	12	7
24	46	85	105	97	63	10	6
25	48	86	105	95	62	8	5
26	49	87	105	95	61	6	4
27	51	88	105	94	60	4	3
28	53	89	105	93	58	2	2
29	54	90	105	93	56	1	1
30	56	91	105	92	54	0	0
Degrees	11^s +	10^s +	9^s +	8^s +	7^s +	6^s +	Degrees

Use of the Table.

Suppose Sun's long. . . .	11^s	2°	9′	6″
Venus's aphelion subtract	2	21	17	10
Remain	8	10	51	56
The equation for which is −	0	1	40	0
Sun's longitude	11	2	9	6
Sun's equated longitude .	11	3	49	6

NOTE.—The true heliocentric longitude of Venus is not used.

The Elongation.

Arg. Venus's long.—Sun's equated long.

Degrees	+ 0^s	+ 1^s	+ 2^s	+ 3^s	+ 4^s	+ 5^s	Degrees
0	′ 0	′ 754	′ 1484	′ 2152	′ 2666	′ 2656	30
1	26	778	1506	2172	2678	2636	29
2	50	802	1532	2194	2690	2608	28
3	76	828	1554	2214	2702	2578	27
4	100	852	1576	2222	2712	2544	26
5	126	876	1600	2252	2722	2510	25
6	150	900	1624	2270	2732	2470	24
7	174	928	1646	2290	2740	2430	23
8	202	954	1670	2310	2748	2384	22
9	228	976	1692	2330	2756	2336	21
10	254	1000	1714	2350	2782	2280	20
11	278	1026	1738	2368	2768	2218	19
12	302	1050	1760	2384	2772	2156	18
13	328	1074	1784	2402	2778	2090	17
14	352	1098	1804	2422	2782	2014	16
15	378	1124	1824	2440	2784	1934	15
16	404	1148	1848	2456	2784	1836	14
17	428	1172	1872	2474	2784	1746	13
18	452	1196	1892	2490	2784	1646	12
19	478	1220	1916	2510	2782	1546	11
20	504	1244	1938	2524	2778	1430	10
21	530	1268	1962	2540	2774	1316	9
22	554	1292	1982	2556	2768	1194	8
23	580	1316	2004	2570	2760	1060	7
24	604	1340	2024	2586	2750	922	6
25	630	1364	2044	2602	2740	778	5
26	654	1388	2068	2616	2726	630	4
27	678	1406	2088	2628	2702	480	3
28	704	1436	2110	2642	2694	320	2
29	728	1460	2120	2654	2676	160	1
30	754	1484	2152	2666	2656	0	0
Degrees	11^s −	10^s −	9^s −	8^s −	7^s −	6^s −	Degrees

Use of the Table.

Suppose Venus's long. .	0^s	16°	30′	11″
Subtract Sun's equat. long.	11	3	49	6
Remains the commutation	1	12	41	5
The elongation for which is −	0	17	39	34
Sun's equated longitude .	11	3	49	6
Venus's geocentric long. .	11	21	28	40

MARS.

Equation of the Orbit.
Arg. Mars's mean long.—the Aphelion.

Degrees	− 0^s	− 1^s	− 2^s	− 3^s	− 4^s	− 5^s	Degrees
0	0	320	570	689	629	381	30
1	11	330	576	690	624	370	29
2	23	340	583	691	618	359	28
3	33	350	589	691	613	348	27
4	45	359	595	692	607	336	26
5	56	369	600	692	601	325	25
6	67	378	606	692	594	313	24
7	78	387	611	692	587	301	23
8	89	396	617	692	580	289	22
9	100	405	622	691	573	277	21
10	111	415	627	690	566	265	20
11	122	423	632	690	559	253	19
12	133	432	636	689	551	240	18
13	145	440	641	687	543	228	17
14	155	449	645	685	537	215	16
15	165	458	649	683	527	202	15
16	176	466	653	680	519	189	14
17	187	475	657	679	510	176	13
18	197	483	660	676	501	162	12
19	208	492	663	673	492	149	11
20	219	498	667	670	483	136	10
21	232	506	670	667	474	123	9
22	239	514	673	667	464	109	8
23	250	521	676	661	454	96	7
24	261	528	678	657	444	82	6
25	270	536	680	652	434	69	5
26	280	543	682	648	423	55	4
27	291	550	684	644	413	42	3
28	301	557	686	639	403	27	2
29	311	563	687	634	392	14	1
30	320	570	689	629	381	0	0
Degrees	11^s +	10^s +	9^s +	8^s +	7^s +	6^s +	Degrees

Use of the Tables.

Suppose Mars's mean long.	2^s	7°	35′	18″
The Aphelion	4	8	24	57
Anomaly	9	29	10	21
The equation for which is +	0	9	35	0
Mars's mean longitude .	2	7	35	18
Mars's true helioc. long. .	2	17	10	18

Semiparallax of the Orb.
Arg. Sun's longitude—Mar's longitude.

Degrees	+ 0^s	+ 1^s	+ 2^s	+ 3^s	+ 4^s	+ 5^s	Degrees
0	0	352	687	984	1186	1096	30
1	12	363	698	994	1190	1081	29
3	24	375	709	1002	1193	1065	28
3	36	387	720	1010	1196	1048	27
4	47	398	730	1019	1199	1030	26
5	59	410	740	1026	1202	1010	25
6	71	421	751	1035	1204	990	24
7	83	432	761	1043	1205	968	23
8	95	444	773	1051	1206	944	22
9	106	455	781	1059	1208	919	21
10	118	467	792	1067	1208	892	20
11	130	478	803	1075	1208	864	19
12	142	489	813	1082	1208	833	18
13	154	500	823	1089	1207	801	17
14	164	512	833	1096	1206	767	16
15	177	523	842	1103	1204	731	15
16	189	534	852	1110	1202	694	14
17	201	545	862	1117	1199	655	13
18	213	556	872	1124	1195	613	12
19	224	568	882	1130	1191	559	11
20	237	579	892	1136	1187	525	10
21	248	590	901	1141	1183	474	9
22	259	601	610	1147	1175	430	8
23	271	612	920	1153	1168	380	7
24	283	623	930	1158	1160	329	6
25	295	634	939	1164	1152	277	5
26	306	644	948	1169	1141	223	4
27	318	655	957	1173	1131	168	3
28	329	666	966	1178	1121	112	2
29	341	677	975	1182	1109	56	1
30	352	687	984	1186	1096	0	0
Degrees	11^s −	10^s −	9^s −	8^s −	7^s −	6^s −	Degrees

Use of the Tables.

Sun's mean longitude	.	11^s	2°	9′	6″
Mars's true long. subt.	.	2	17	10	18
The commutation . .	.	8	14	58	48
The semiparallax . .	−	0	18	23	0
Its double	−	1	6	46	0
Mars's true helioc. long.	.	2	17	10	18
Mars's geocentric long.	.	1	10	24	18

JUPITER.

Equation of the Orbit.
Arg. Jupiter's mean long.—the Aphelion.

Degrees	− 0s	− 1s	− 2s	− 3s	− 4s	− 5s	Degrees
0	0′	149′	260′	305′	271′	161′	30
1	5	153	262	306	269	156	29
2	10	158	265	306	266	151	28
3	15	162	267	306	263	147	27
4	21	167	270	306	261	142	26
5	26	171	272	305	259	137	25
6	31	175	275	305	256	132	24
7	38	179	277	305	252	127	23
8	44	184	279	304	249	123	22
9	48	188	281	304	246	116	21
10	52	192	283	303	242	111	20
11	57	196	285	303	239	105	19
12	62	200	287	302	235	100	18
13	67	203	289	301	232	94	17
14	72	207	291	300	229	89	16
15	77	211	292	299	225	83	15
16	82	215	294	298	221	78	14
17	87	218	295	297	217	72	13
18	92	221	296	295	213	66	12
19	97	225	297	294	209	61	11
20	102	229	299	292	205	56	10
21	107	232	300	290	201	50	9
22	112	236	301	289	197	44	8
23	117	239	301	287	193	39	7
24	121	242	302	285	188	34	6
25	126	245	303	283	184	29	5
26	131	248	304	281	180	24	4
27	135	251	304	279	175	19	3
28	140	254	305	276	170	12	2
29	144	257	305	274	166	6	1
30	149	260	305	271	161	0	0
Degrees	11s +	10s +	9s +	8s +	7s +	6s +	Degrees

Use of the Table.

Supp. Jupiter's mean long.	10s	25°	47′	6″
Subtr. long. of the Aphelion	5	22	35	16
Remains anomaly . . .	5	3	11	50
The equation for which is	−0	2	26	0
Mean longitude of Jupiter	10	25	47	6
True heliocentric longitude	10	23	21	6

Semiparallax of the Orb.
Arg. Sun's longitude—Jupiter's longitude.

Degrees	+ 0s	+ 1s	+ 2s	+ 3s	+ 4s	+ 5s	Degrees
0	0′	145′	268′	339′	326′	203′	30
1	5	149	271	340	324	198	29
2	10	154	274	342	321	191	28
3	15	158	278	343	318	186	27
4	19	163	280	344	316	180	26
5	24	167	284	344	313	173	25
6	29	172	287	344	309	167	24
7	34	176	290	345	306	161	23
8	39	180	293	345	304	154	22
9	44	184	296	346	301	148	21
10	48	189	298	346	297	141	20
11	53	193	301	345	294	135	19
12	59	197	304	345	290	127	18
13	64	202	307	345	285	121	17
14	68	206	310	345	282	114	16
15	73	210	312	345	278	107	15
16	78	214	314	344	274	100	14
17	83	219	316	344	269	96	13
18	87	223	318	343	265	86	12
19	92	228	320	342	261	76	11
20	97	231	323	342	256	67	10
21	102	234	325	341	251	58	9
22	107	238	327	339	246	50	8
23	111	242	329	338	241	43	7
24	116	246	331	336	236	35	6
25	121	249	332	335	231	28	5
26	125	253	334	333	225	21	4
27	130	257	335	331	220	14	3
28	135	260	336	330	215	7	2
29	140	264	338	328	209	1	1
30	145	268	339	326	203	0	0
Degrees	11s −	10s −	9s −	8s −	7s −	6s −	Degrees

Use of the Table.

Sun's longitude . . .	11s	2°	9′	6″
Jupiter's true longitude .	10	23	21	6
The commutation . .	0	8	48	0
The semipar. for which is	+ 0	0	42	0
Its double . . . =	+ 0	1	24	0
True longitude of Jupiter	10	23	21	6
Geocentric longitude . .	10	24	45	6

SATURN.

Equation of the Orbit.
Arg. Saturn's mean long.—the Aphelion.

Degrees	− 0ˢ	− 1ˢ	− 2ˢ	− 3ˢ	− 4ˢ	− 5ˢ	Degrees
0	′ 0	′ 219	′ 385	′ 459	′ 411	′ 245	30
1	8	225	390	459	407	238	29
2	15	232	393	459	403	231	28
3	23	238	396	459	399	223	27
4	30	245	400	460	395	215	26
5	38	251	404	460	391	208	25
6	46	257	408	459	387	201	24
7	53	263	411	459	383	193	23
8	61	270	414	459	379	185	22
9	68	276	418	458	374	177	21
10	76	282	422	457	369	169	20
11	83	288	425	457	364	161	19
12	91	294	428	456	358	153	18
13	98	300	431	454	353	146	17
14	106	305	433	453	348	137	16
15	113	311	438	451	342	129	15
16	120	316	441	449	336	120	14
17	128	322	443	448	330	112	13
18	135	327	444	446	324	104	12
19	143	333	445	444	318	96	11
20	150	338	447	441	312	87	10
21	157	343	448	438	306	78	9
22	164	348	450	435	300	69	8
23	171	353	452	432	294	61	7
24	178	358	453	429	288	52	6
25	185	362	454	425	282	44	5
26	193	367	455	423	274	35	4
27	199	371	456	420	267	27	3
28	206	376	457	417	260	9	2
29	213	380	458	414	253	1	1
30	219	385	459	411	245	0	0
Degrees	11ˢ +	10ˢ +	9ˢ +	8ˢ +	7ˢ +	6ˢ +	Degrees

Use of the Table.

Supp. Saturn's mean long.	1ˢ 4° 12′ 56″
Saturn's aphelion . .	8 20 54 27
Mean anomaly . . .	4 13 18 29
The equation for which is	− 0 5 51 27
Saturn's mean longitude .	1 4 12 56
Saturn's true helioc. long.	0 28 21 29

Semiparallax of the Orb.
Arg. Sun's long.—Saturn's true long.

Degrees	+ 0ˢ	+ 1ˢ	+ 2ˢ	+ 3ˢ	+ 4ˢ	+ 5ˢ	Degrees
0	′ 0	′ 86	′ 156	′ 190	′ 174	′ 105	30
1	3	88	158	190	172	101	29
2	6	91	159	191	171	98	28
3	9	94	161	191	170	95	27
4	12	96	162	191	168	91	26
5	14	98	164	191	166	88	25
6	17	101	166	191	164	85	24
7	21	103	168	191	162	81	23
8	24	106	169	191	160	78	22
9	27	109	170	191	158	75	21
10	29	111	172	190	156	73	20
11	32	113	173	190	154	71	19
12	35	116	174	190	152	68	18
13	38	118	176	190	150	65	17
14	40	121	177	189	147	61	16
15	44	123	178	189	145	58	15
16	47	126	180	188	142	55	14
17	51	128	181	187	140	51	13
18	54	130	182	187	138	47	12
19	56	133	183	187	135	43	11
20	59	135	184	186	132	40	10
21	62	137	185	185	130	36	9
22	65	140	186	183	127	32	8
23	68	143	187	182	124	29	7
24	71	144	187	181	121	25	6
25	73	145	188	180	119	22	5
26	76	149	188	179	116	18	4
27	79	152	189	178	113	15	3
28	81	153	189	176	110	11	2
29	83	155	189	175	108	5	1
30	86	156	190	174	105	0	0
Degrees	11ˢ −	10ˢ −	9ˢ −	8ˢ −	7ˢ −	6ˢ −	Degrees

Use of the Table.

Saturn's true longitude .	0ˢ 28° 21′ 56″
Subtract from Sun's long.	11 2 9 6
Commutation	10 3 47 37
The semipar. of which is	− 0 2 29 39
Its double is . . .	− 0 4 59 18
Saturn's true longitude .	0 28 21 29
Saturn's geocentric long. .	0 23 22 11

MATHEMATICAL GEOGRAPHY.

CHAPTER I.

Universal Geography—Mathematical—Spherical figure of the Earth.

UNIVERSAL GEOGRAPHY is the science that conveys to us a knowledge of the earth, both as a distinct and independent body in the universe, and as connected with a system of heavenly bodies. The figure, structure and dimensions of the earth,—the properties and mutual relations of its parts—the features of its surface—its productions and inhabitants—and the laws which govern, or partially affect it as a heavenly body—are all included within the comprehensive term of universal geography. This definition, or rather description of the objects of geography, serves as the basis of M. Malte-Brun's elaborate work;* but it manifestly embraces a great variety of subjects, commonly classed and treated of under distinct heads of natural philosophy. To avoid, therefore, the confusion of ideas which the extensiveness of this definition may give rise to, it will be convenient to reduce its terms within the limits usually assigned to geography. And we are the rather induced to do this, because the interests of science have been promoted in no slight degree, by a judicious and well-defined arrangement of its parts, which at once excludes a great number of fanciful resemblances, and like a division of labour in mechanical employments, renders every branch more easy to be acquired, and more likely to be extended and improved.

In its proper and more confined sense, geography comprises a knowledge of the figure and dimensions of the earth, and the situation of places upon it—of the natural and political features and divisions of its surface—and of its various productions and inhabitants. These particulars may be arranged under three heads, namely, mathematical, physical, and general geography.

* See Malte-Brun's Universal Geography.

MATHEMATICAL GEOGRAPHY is that branch of the general science which is derived from the application of mathematical truths to the figure of the earth. Of this we shall treat first, because the other branches of geography owe to it much of their accuracy and perfection.

The figure of the earth is manifestly the first subject for inquiry;—for the principles by which we may ascertain the various truths that lie within the scope of mathematical geography, are altogether different, on the different suppositions of the earth being a flat circular plain, a cylinder, or a sphere.

A great variety of appearances, both on the surface of the earth and in the heavens, (which will be described presently) prove conclusively, that the earth is a spherical or round body. The possession of this important truth enables the geographer, by the application of the known mathematical properties of the sphere, to solve many interesting problems, the most useful of which is to determine the relative situation of places upon the earth's surface. For this and some other practical purposes, the earth is taken to be a perfect sphere; and although this supposition be not strictly true, it is sufficiently near the truth to be adopted without sensible error in the investigations into which it is commonly introduced. The nature and quantity of its deviation from a perfectly spherical shape will be for future inquiry.

At what particular period of the world the spherical figure of the earth was first discovered, cannot now be ascertained. It is natural to suppose, that the curiosity of mankind would early be directed to the shape of the earth they lived upon. But when first it engaged their attention, it fared with this as with all other parts of what is called natural philosophy. Men were led to entertain the most erroneous notions of it, by trusting too much to single appearances. Deceived by the plain-like appearance of the earth, and disregarding all other circumstances indicative of its figure, they conceived it to be an extensive plain meeting the heavens on every side. Such was, for ages, the general opinion. But there were exceptions to the prevailing ignorance, which are honorable testimonies to the value of a more enlarged and extended observation of nature. The Egyptians and Chaldeans are especially entitled to this praise. The philosophers of these nations were, in all probability, led to form a correct opinion of the figure of the earth from their great practical familiarity with the appearances of the heavenly bodies. But whatever may have been the source from whence their knowledge was derived, it is manifest that they were not ignorant of its true shape, as it must have formed an element in the calculations by which they were enabled to predict eclipses of the moon. From the Egyptians and Chaldeans, who were the fathers as well of geographical as of astronomical science, the Greek philosophers, with all their most correct notions in natural philosophy, derived also their knowledge of the earth's true shape. But (as *Sir Isaac Newton* remarks) the Greeks were of themselves more addicted to the study of philology (or language) than of nature: when therefore their communications with Egypt became less frequent, the ancient philosophy gradually declined among them; and no longer retaining the just ideas they once possessed, they put forth their own visionary speculations concerning the figure of the earth. Aristotle, the most celebrated of the Greek philosophers, did not escape the error of those who allow the suggestions of fancy to occupy the place of a severe investigation into facts; and we find him alleging the earth to be of a cylindrical shape, like a common drum. The remarkable ingeniousness of the Greeks was ever impatient of the restraint which scientific inquiry in order to a successful issue imposes upon the mind; and it is to be lamented that by reason of the admiration in which their writings were held, their errors should for so long have retained possession of the human mind, and by keeping down the spirit of inquiry retarded the full establishment of what is properly called experimental philosophy.

During the greater part of that portion of the history of Europe, called the middle or dark ages, the earth was conceived to be a flat surface extending on every side till it met the heavens. The overthrow of this popular opinion was rendered the more difficult by the Roman church admitting it into the number of articles of faith: the tenet thus became guarded with the sanction of religious belief, and by the apprehension of incurring the serious charge of heretical opinions. It is however remarkable, that the appearance of objects at sea, which are wholly inconsistent with the notion of the earth being a plain, and which lead most directly to the conclusion of its spherical shape, should not have redeemed the Venetians and Genoese, who had long been in the habit of making adventurous sea voyages, from the general ignorance. But notwithstanding the peculiar advantages enjoyed by navigators; it is evident that the best of those of the age of Columbus were not better informed of the earth's real figure. It is related as a matter of history, that the Portuguese who had arrived at the Moluccas (situated in the Pacific and to the West of America), by sailing continually in an *easterly* direction, were astonished by the appearance of Magellan's party, who reached the same point by sailing continually *west.* We may not, however, involve Columbus in this general censure; to him is properly due the glory of establishing the fact that the earth is a sphere. He was indeed eminently qualified to give a new direction to the current of opinion. In advance of the age he lived in by the extent and correctness of his information, and being at once bold in enterprise, enthusiastic in pursuit, and fertile in expedients, he possessed all the characteristics of one who is destined to overthrow a great and prevailing practical error. His persuasion that the earth was a sphere, furnished him with the happy idea of arriving at the *East* Indies by a shorter course than round the Cape of Good Hope, by sailing due *West.* He failed in his undertaking, having been misled by the error of the ancient geographers. Ptolemy's map was then in use, and the East Indies are there laid down considerably to the west of their true position. The western coast of India is by Ptolemy placed in longitude 165° east from the isle of Ferro, (one of the Canaries through which the first meridian passed,) whereas the true longitude is about 96°, thus making a difference of no less than 67°. The reasoning of Columbus was therefore right; and although he was disappointed of the immediate object of his voyage, he became the discoverer of a new world, and eventually established his own opinion of the earth's spherical shape. Magellan was the first navigator who practically demonstrated the roundness of the earth; following up the opinions which Columbus among the moderns had the merit of originating, he sailed upon the project of reaching the Moluccas by a westerly passage; but being killed in the Philippine Islands by the natives, he did not complete the entire voyage round the world. Our own countryman, Sir Francis Drake, was the first person who in one voyage circumnavigated the globe; he accomplished the voyage (undertaken however solely for purposes of plunder, and marked by rapine and bloodshed) in the space of three years; and returned to England in 1560. After these voyages, the spherical figure of the earth was generally admitted by the philosophers of Europe. A spirit of investigation soon after arose, and furnished an abundance of satisfactory proofs, which, though of daily or frequent occurrence, had hitherto been unobserved or unheeded. These proofs consist in certain remarkable appearances, either of objects upon the surface of the earth itself, or of the heavenly bodies. They are of the following description:—

If a person were situated upon an open and extensive plain, he would find, that as he departed from objects, the view of which was not hindered by any unevenness in the plain, they would *gradually* disappear from their *base* upwards; in like manner, the hull of a ship proceeding out to sea becomes invisible *first*, and afterwards the masts and rigging. The *order* in which the parts of these objects successively disappear, cannot be explained by the mere supposition that the distance between the object and the spectator gradually increasing, the object becomes first indistinct, and at last invisible; because with respect to bodies whose bulk is the same from the top to the bottom, this reason is applicable to all the parts alike, and would not account for the highest part of them being always the last visible; and with respect to bodies, the bottom part of which is the largest (as in the case of a ship), it would not only be insufficient to explain the fact, but would be directly contrary to experience, by which we are taught, that where distance alone is the cause of a body becoming first indistinct and then invisible, the larger and more bulky parts of it are seen the longest. The only supposition which can account for the *order* in which the parts of an object disappear is, that the surface of the earth is *continually and gradually* bending or curving downwards—in other words, that it is a convex surface; and the circumstance that these appearances are the same both in kind and degree all over the earth, and in whatever direction the spectator moves from the object, or the object from the spectator, proves that this convex surface is every where and in all directions precisely or very nearly the same, and, consequently, that the earth is a sphere.

The voyages of Magellan and Drake, of Anson, Cook, and Vancouver, all tend to establish the same fact; for by holding a course due west or due east, these navigators have at last arrived at the point of their departure—thus they have sailed upon a line which in one revolution returns into itself, ending where it began; and, therefore, the surface on which it was described must be a sphere, or resembling a sphere: this was further confirmed by the voyages of Captain Cook towards the South pole, from which it appeared that the course round the earth gradually diminished as it approached the pole.

The proofs derived from the appearances of heavenly bodies are even more conclusive than the foregoing. By travelling on the earth's surface from the north towards the south, a certain star in the heavens, called the pole star (which is itself almost stationary,) is observed to change its place in the heavens relatively to the spectator's horizon, and gradually *to descend;* by a movement of the spectator in the opposite direction (from south to north), the height of the same star above the horizon is observed gradually to *increase;* and in both cases this apparent change of place in the star is in proportion to the distance travelled over. This change being also observed from whatever place the movement is made (supposing it to be in a direction perpendicular to the equator or on a meridian line), cannot be otherwise accounted for than by the supposition that the earth is a sphere; and that the arc or circular space in the heavens through which the star appears to have moved, corresponds with a similar arc traced upon the surface of the earth.

Another most convincing proof is furnished by the eclipses of the moon. These eclipses are known to be caused by the earth coming between the sun and moon, and intercepting or cutting off the supply of light from the sun which illuminates the moon's surface or disk; the dark part of the moon's disk is, therefore, nothing more than a representation of the earth's shadow at the distance of the moon. In whatever position the earth happens to be at the time of an eclipse, its shadow upon the moon's disk is always in the form of a circle or of part of a circle: the earth must therefore be a sphere, since no other than a spherical body, in every position in which it can placed with respect to another body giving light, can cast a circular shadow upon a third body. If, however, the earth were shaped like a circular flat plain, its shadow upon the moon's disk would be circular only when either of its sides directly faced the moon: if turned edgeways towards the moon, the shadow would be in the form of a streak, and in all other positions it would be more or less elliptical, as the earth happened to be turned more or less obliquely towards the moon when she is eclipsed.

The supposition that the earth is a sphere, accounts for all the appearances we have described; while, on the other hand, the various suppositions which have from time to time been advanced, and which differ from this, are totally inconsistent with one or other of them. *Sir Isaac Newton*, in his "Principles of Natural Philosophy," has laid it down as a rule (and it is a rule as indisputably just as it is important), that "in experimental philosophy we are to look upon propositions collected by general induction from phenomena, as accurately or very nearly true, notwithstanding any contrary hypothesis may be imagined, till such time as other phenomena occur by which they may either be made more accurate or liable to exceptions;" he adds, "this rule we must follow, that the argument of induction may not be evaded by hypotheses." Applying this rule to the present subject, we may observe that the objections urged against the conclusion that the earth is spherical, however plausible, are entitled to no weight whatever, unless they are grounded upon some certain facts and natural appearances arising from the figure of the earth, either inconsistent with the present-received theory, or which that theory is insufficient to account for.

It is hardly necessary to remark, that the expressions occasionally to be met with in the Bible with regard to the figure of the earth, and which may appear to contradict the foregoing conclusion, have been improperly and very ignorantly applied to this subject. The object of the inspired writers who used them, was not to advance a true system of natural philosophy, or to correct the popular errors of the day in matters of mere science, but to illustrate or enforce some precept or doctrine, or to record the occurrence of some remarkable event, which could not be done intelligibly, but by adopting expressions in agreement with the opinions of the age.

The re-establishment of the old and long neglected opinion of the earth's spherical shape, may justly be regarded as furnishing an epoch in the history of modern Europe. When admitted into the number of those truths which are assumed and acted upon without proof, it had an immediate and practical effect upon the common concerns of life. To traverse boundless seas was no longer matter for apprehension: the seaman was now provided with a method of discovering his relative position upon the globe, the course he had already described, and the distance and bearing of his destined port. Navigation thence assumed a bolder and more systematic character; an extensive commerce added to the wealth, and stimulated the efforts of European nations; and the more general and frequent intercourse inseparable from commerce softened the prejudices of men, and opened to them in distant climates and countries the richest and most varied stores of knowledge. We should not perhaps be justified in placing this discovery in the same rank with the other great events which happened about this æra: the invention and general introduction of the art of printing—the reformation—and the establishment of experimental philosophy, must stand alone; but it forms together with them a class of great and brilliant events, which exhibit the human mind as once more in a state of activity, and putting forth all its energies in the attainment of whatever might most conduce to the social and moral improvement of mankind.

CHAPTER II.

The Circles of the Sphere—Extent of the Visible Horizon—Method of Drawing a Meridian Line—Circles of Position—Æquator—Latitude—First Meridian—Longitude.

THE modes of precisely fixing the situation of places upon the earth are founded upon the circumstance just now proved of its spherical form, and upon the supposition which, for the purposes to which it is applied, is not a false one, that it is enclosed in a concave or hollow sphere of the heavens, of which it occupies the middle spot or centre.

Upon the surface of the earth considered as a globe, various lines are supposed to be drawn for the purposes of geographers, and in order to determine or explain the truths of their science; and as the heavens present to us a concave sphere, having the same centre as the earth, there are also imaginary lines supposed to be traced upon the inner surface of the heavens, which exactly correspond with those traced upon the earth. By this device geography has become allied with astronomy, and has thence derived its most important improvements. We now proceed to the description of the above-mentioned lines which are supposed to divide the earth, and which are seen drawn upon the common geographical globe.

The earth has a daily motion from *west* to *east*, about one of its diameters (called the earth's *axis*), which causes all the heavenly bodies to *appear* to move daily round the earth in an opposite direction from *east* to *west*. The two extremities of this axis are called the *poles* of the earth, from a Greek word signifying a pivot; one is called the *North* pole, being that which is opposite or nearly opposite to the star in the heavens called the pole star; the other extremity of the axis is called the *South* pole; the north pole is also known by the name of the *arctic* pole, from a Greek word signifying a *bear*, the 'great bear' being the name of a constellation or collection of stars in the immediate neighbourhood of the pole star, and commonly known as Charles's wain:—the South pole has the corresponding term of the *antarctic* pole, or the pole *opposite* to the *arctic*.

Let P E p Q P (*fig.* 1) be a sphere representing the globe of the earth, O the centre, P O p the axis, P the north and p the south pole; now suppose a plane to cut this sphere into two equal

Fig. 1.

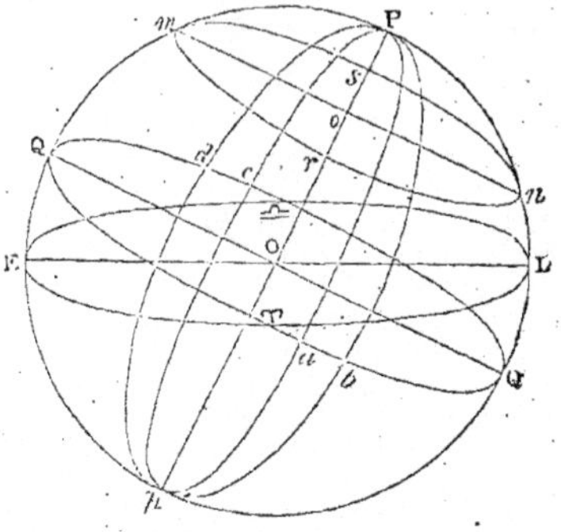

parts in a direction perpendicular to the axis, this plane will pass through the centre O, and the circle Q a b Q c d, which is the boundary of the cutting plane upon the surface of the globe, will represent the *equator*, and is every where at an equal distance from both poles. This circle and all other circles, the planes of which pass through the centre of the sphere, are called great circles. All circles such as P a p c P P b p d P, which pass through both poles P, p, of the earth, and which have the axis of the earth P o p for a common diameter, are called *meridians*, because when the centre of the sun is over or upon that one of these circles which passes through any place, it is *mid-day* or noon at that place. The plane of every meridian cuts the plane of the equator at right angles, so that the equator divides every meridian (as for instance P a p c P) into four equal parts; thus P a and a p, and p c and c P are equal to one another, and are called *quadrants* or quarters of a circle. Meridians are also called circles of latitude, because upon them the latitudes of places are measured. The Ecliptic found traced upon common globes (although it is properly an imaginary circle in the concave sphere of the heavens representing the apparent path of the sun in the course of a year) is a great circle upon the globe of the earth, the plane of which is inclined at a certain angle to the plane of the equator, and is represented in the figure by the circle E ♈ L ♎. All circles upon the sphere which do not pass through its centre are called small circles; those which are parallel to the equator, as m r n s, are called circles of longitude or parallels of latitude; and as all meridians cut the equator at right angles, they also cut all circles of longitude at right angles, which is evident from these latter circles being parallel to the equator. Every circle traced upon the earth is supposed to have a corresponding circle traced upon the concave or hollow spherical surface of the heavens. All circles, whether great or small, are divided into 360 equal parts called *degrees;* every degree is again divided into 60 equal parts called *minutes,* and every minute into 60 *seconds:* these various parts are distinguished by certain signs, thus 15 degrees is written 15°, 32 minutes is written 32′, and 5 seconds 5″; so that 15° 32′ 5″, signifies 15 degrees together with 32 minutes and 5 seconds. The magnitude of degrees is of course different in great and small circles; the amount and variation of this difference in the circles of the globe, will be explained afterwards.

The *zenith* of any place on the earth is that point in the concave surface of the heavens which is immediately opposite to the extremity of a line drawn from that place to the centre of the earth, or in the direction of a plumb line; it is the point in the heavens directly over our heads. The *nadir* is the corresponding point in the opposite hemisphere of the heavens. Of all the meridian circles, that which passes through the zenith of a place in the heavens, or through the place itself upon the earth, is the meridian of that place. The horizon of a place is the boundary of view at that place: with respect to the earth it is called the visible, sensible, or apparent horizon; with respect to the heavens it is called the rational, true, or astronomical horizon. The visible horizon is most accurately observed upon the sea where it is distinct and unbroken, and is, therefore, sometimes called the horizon of the sea. The extent of the visible horizon may easily be found if the height of the spectator's eye above the surface of the earth be known, and also the length of the earth's radius or semi-diameter. For if (*fig.* 2.) B E D F be a great circle, C the centre of the earth, A E the height of the eye, E C the semi-diameter of the earth, and A B be drawn from A, just touching the earth's surface at B, E B will be the extent in one direction of the visible horizon. If B and C be joined, B C will be perpendicular to A B. The

Fig. 2.

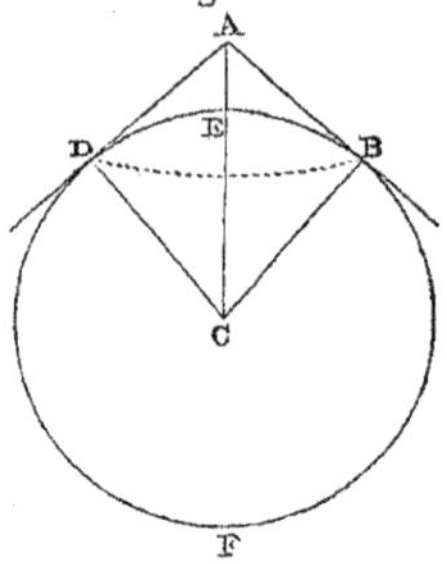

length of A E and E C or A C and also of B C, which is the semi-diameter, being known, the angle A C B may be found by a very simple mathematical process; and this angle being measured by the arc E B, the required distance is found. A B, the direct distance of the horizon at B from the spectator's eye at A, may also be found in a somewhat similar manner:—if A E be equal to 5 feet, and E C, the semi-diameter of the earth, be taken at 29,949,655 feet, the angle at C, or the arc B E, will be found to be equal to 2′ or 12,188 feet, which is nearly equal to 2 miles and 532 yards; D B is of course equal to twice B E, as the spectator sees as far one way as another, therefore D B is equal to 4 miles 1064 yards. This, however, is not quite true in practice, as by the refracting power of air and vapour, the apparent horizon is a little more extensive. The rational horizon is in every part of it 90°, or a quadrant distant from the zenith. When a heavenly body first appears above the horizon of a place, it is said to rise, and to set when it disappears or sinks below the horizon. When a heavenly body is upon the meridian of any place, it has obtained its greatest height or altitude above the plane of the horizon of that place.

The north point of the horizon is that which is nearest to the north pole of the heavens or the pole star, the point 180° distant from it is the south. The meridian line of a place passes through the north and south points. The east point is 90° distant from the north or south in that portion of the heavens where the sun, stars, &c. appear to rise; the west is 180° distant from the east point. Thus all the cardinal or principal points of the compass are determined.

By means of the observed altitudes of heavenly bodies, when at their highest or on the meridian of a place, many geographical problems are solved; it is, therefore, of great importance to ascertain the direction of the meridian line at the place of the observer. The operation in its more scientific and correct shape is one of very considerable nicety; but the following method will determine it, if much accuracy be not required. On the 15th of June, or the 24th of December, plant a stick A B in a position perpendicular to the horizon (*fig*. 3.) at an hour or two before the

Fig. 3.

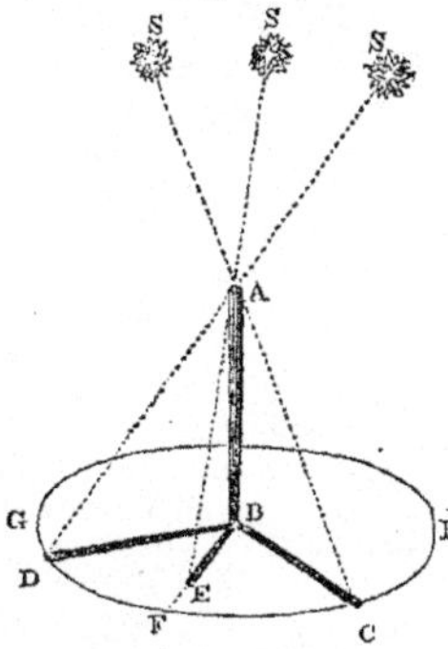

sun has arrived at its greatest altitude in the heavens, that is, at ten or eleven o'clock in the morning; mark accurately the extremity C of the shadow B C cast by the stick; then from the base B of the stick as a centre, and with the length of the shadow B C as a radius, trace a circle G H upon the ground; as the sun gradually arrives at its greatest altitude, the shadow of the stick will become gradually shorter, and will fall within the circumference of the circle which has been traced. The shadow will be at its shortest B E, when the sun is at its greatest altitude, or when it is on the meridian of the place which is the moment of noon; after this the sun will gradually decline, the shadow of the stick will become longer and longer, until at last it again reaches to the circumference of the circle in the point D, at which time in the afternoon the sun is at the same height in the heavens as it was when the shadow of the stick was of the same length B C before noon. Now it so happens, that, on the above-mentioned days, the altitude of the sun above the horizon, at one hour or two hours before noon, is equal to its altitude at the same time after noon; and as the sun has in these equal times before and after noon, described equal spaces in the heavens (supposing those spaces to be measured from the meridian), the middle point of the whole space described by the sun in the sum of those times will be that point in the heavens which the sun occupies when it is noon; at this time the sun is on the meridian. Hence, if the arc C D of the traced circle be divided into two equal parts C F and F D, and the point F of division and the base of the stick be joined, the line B F, joining these two points, and which will be the direction of the shadow of the stick at noon, will be the meridian line. The longest and most accurate meridian line in the world, is that drawn by Cassini (a celebrated astronomer and mathematician) upon the pavement of the church of St. Petronis, at Bologna, in Italy: it is 120 feet in length.

One of the principal objects in mathematical geography is to ascertain the position of any particular spot upon the earth's surface. This term *position* is strictly a relative one—applied to a body, it has no meaning unless there be some other body or mark, which is fixed, and to which the first body may be referred. If, in the midst of infinite space, there existed but a single body, it could hardly be said to have position, or at any rate the meaning of the term as applied to such a body would pass our comprehension. This may be illustrated by what is said in some books on mechanics, that motion could not exist if there were but a single body in the universe; by which is meant that in such a condition of things motion could be neither measured nor perceived: it is not intended by such expressions to assert that motion cannot exist independently of other bodies, because the existence of a foreign body cannot really affect the state or condition of motion in any moving body; it only enables us to ascertain the fact of motion, and its measure. In the idea of position, therefore, is contained a reference to something which is fixed and which is independent of the body, the position of which is required. The distance of any body from this something which is fixed being known, and also the direction given in which that distance is to be measured, its position may be determined.

In order to ascertain the situation of any spot upon the surface of the globe, it is sufficient to fix upon two great circles, the planes of which are perpendicular to each other, and from each of which the nearest distance of the spot is to be measured; these two great circles are called circles of position. Thus if (see *fig.* 4) the position of the point A upon the surface of the sphere or

Fig. 4.

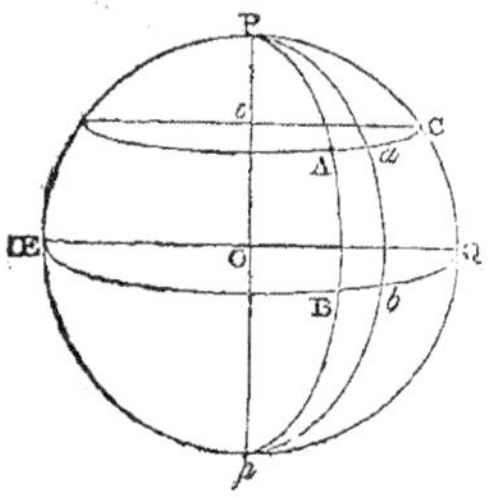

globe P Œ p P be required—it may be determined if we have given in position the great circle Œ B b Q, and the great circle P a b p, the planes of which are perpendicular to each other. For we need only make a great circle P A B p perpendicular to the circle Œ B b Q pass through A, and then a small circle pass through A parallel to Œ B b Q, and the distances of their intersection from the given great circles, viz. the arc A B being the distance from Œ B b Q, and the arc A a, or the corresponding arc B b, being the distance from P a b p, will determine the exact position of the point A.—In applying this to the practical purposes and wants of geography, it is evident that the first object is to fix the position of the two great circles Œ B b Q and P a b p, and then to devise some mode for ascertaining the distances A B, A a from each of them.

The astronomers and geographers of all countries have concurred in fixing upon the *equator* or equinoctial line (as it is sometimes called) for the position of the circle Œ B b Q. The equator has been already defined as a great circle dividing the globe into two equal parts or hemispheres, and the plane of it as perpendicular to the axis of the earth. The distance A B measured upon the meridian of A, which is a great circle perpendicular to the equator, is called the *latitude* of A. The latitude of a place is north or south latitude, as it is situated towards the north or south of the equator. It is very evident that astronomers were led to fix upon the equator for one of the great circles of position, by the circumstance of the apparent daily motion of heavenly bodies, being performed either in a circle in the heavens corresponding with the equator itself, or in circles which are parallel to it. But as there was nothing in the apparent courses of heavenly bodies, or in any particular spot upon the earth to regulate the choice of astronomers in fixing upon a first meridian or the other great circle of position perpendicular to the equator, and which is represented in the preceding figure by the circle P a b p, the consequence has been, that astronomers and geographers of different ages and countries have assumed different circles for their first meridian, from which they have measured the arc A a or B b.

The ancient geographers took for their first meridian the meridian of the Fortunate Isles, a line passing, as they conceived, through the western extremity of the habitable earth. Many of the moderns have employed the same meridian, or rather that of the island of Ferro, one of the most westerly of the Canaries. In general, however, nations adopt as their first meridian the meridian of their own metropolis, or of their principal observatory, as the English do that either of London or Greenwich, the French that of Paris. The angular distance on the arc A a or B b of any place from the first meridian is called its *longitude*, and is either east or west longitude as the place is to the east or west of the first meridian. The English map-makers frequently adopt the meridian of London instead of that of Greenwich for the first meridian, but as London (taking St. Paul's as the point referred to) is 5′ 47″ west of Greenwich, longitudes given from London may be easily reduced to longitudes reckoned from Greenwich, by adding to them 5′ 47″ if they are west longitudes, and subtracting the same quantity if they are east longitudes.

Chapter III.

General Description of the Method of finding the Latitude of a Place.

Having fixed upon the two circles of position by a reference to which the position of a place is to be determined, it will now be necessary to explain how distances from each of these circles (being the latitudes and longitudes of places) may be ascertained. This depends entirely upon the supposition that the earth is a spherical body situated in the middle or centre of the concave or hollow sphere of the heavens, and that every circle traced upon the globe of the earth has a corresponding circle in the heavens.

We shall begin with the latitude of places.

Let A be any place upon the earth's surface, and suppose in the annexed figure (*fig.* 5) that A *p r q h e* is the me-

Fig. 5.

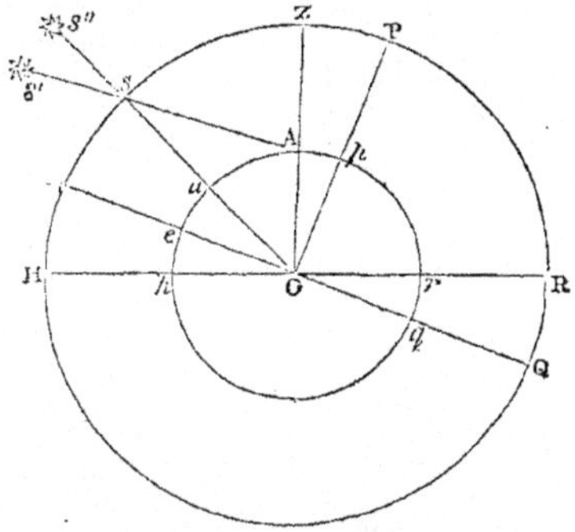

ridian of A passing through *p*, the pole of the earth, and at right angles to the equator (*e q*), the plane of the paper is the plane of the meridian, O the centre of the earth, *p* O its semi-diameter, to which (*e q*), the equator, is at right angles, H R the rational horizon of A. Then A *e*, the arc or angular distance of A from the equator measured upon the meridian of A, is the latitude of A. But *e p*, or the distance of the pole from the equator, is 90°, or a quadrant, and A *r* the distance of A from the point *r*, where the rational horizon meets the surface of the earth, is also 90°, or a quadrant. Hence *e p* is equal to A *r*; if therefore A *p*, which is common to both *ep* and A *r*, be taken away from each, the remaining quantities A *e* and *p r* will be equal; and as A *e* is the latitude of A, it follows that (*p r*) *or the height of the pole above the horizon is equal to the latitude of the place.*

Again, A *p* is the distance of the pole from A: and as A *e* is the latitude of A, and *p e* is 90°, A *p* is the difference between 90° and the latitude, so that if A *p* be known, A *e*, or the latitude, is found, by subtracting A *p* from 90°. A *p* is called the complement of the latitude, or the *co-latitude.*

Again, A *h* is 90°, and, therefore, A *e* being the latitude, *h e* is the co-latitude,—*h e, being the height of the equator above the horizon;* so that if *h e* be known, the latitude is found by subtracting *h e* from 90°.

It appears, then, that if we can find any one of the above four arcs, viz., A *e*, *p r*, A *p*, *h e*, the latitude of A will be known: and the mode of determining these arcs, *is by measuring similar arcs of corresponding circles in the heavens.* Let Z P R Q H Æ be the circle in the heavens which corresponds with the meridian circle passing through A, and Z, P, R, Q, H, Æ, points in the heavens corresponding with A, *p*, *r*, *q*, *h*, *e*. *The attention of the geographer is then transferred from the consideration of the several arcs A e, pr, A p, h e, to the corresponding arcs in the circle in the heavens, Z Æ, P R, Z P, H Æ:* for if any of these be determined in their number of degrees and parts of degrees, the latitude is found directly. Thus it is, that the geographer depends so much upon the science of astronomy for the solution of the most important geographical problems. Persons who are in the slightest degree acquainted with geometry, or with the most simple properties of the circle, will not object to the above-mentioned mode of determining the latitude of places on the earth by means of corresponding arcs in the heavens, that these corresponding arcs are of different magnitudes; for in computing the latitude, we do not so much want the actual admeasurement and linear quantity of the arc of the meridian intercepted between the given place and the equator, as the number of degrees and parts of a degree which it contains, or, in other words, the proportion which this intercepted arc bears to the whole circumference of the meridian circle. And as arcs are the measures of angles, the arcs, Z Æ and A *e*, are both measures of the same angle at *o*; and, therefore, although they are unequal in magnitude, yet they mutually bear the same proportion to the circumference of the circles of which they are parts; that is, Z Æ contains the same number of degrees as A *e*: and as the latitude of a place is always expressed in degrees and parts of degrees, the number of degrees contained in the arc in the heavens, Z Æ, which corresponds with the arc of the meridian A *e*, will be the latitude of A. If, after having ascertained the latitude in this manner, that is, in degrees and parts of degrees, the actual linear magnitude contained in the latitude, or the geographical distance between A and the equator measured upon the meridian, be required, it may be obtained thus. Let an observer at A travel upon the same meridian in a direction due north or due south, (*i. e.* from or towards the equator,) until the pole star has, with respect to the observer's horizon, been raised or sunk one degree (which may be known from observation): then as the star is itself stationary, this gain or loss of one degree in its station with respect to the horizon, has been caused by the observer having travelled exactly one degree, measured upon a meridian of the earth, nearer or farther from the north pole. If this space be actually measured, the result, expressed in linear measure, will give the magnitude of a degree of latitude in geographical miles and parts of a mile; the quantity thus found, being multiplied into the number of degrees and parts of a degree, will give the actual linear distance between A and the equator. The process thus conducted is on the supposition that the earth is perfectly spherical. A degree of latitude measured in this manner contains about sixty-nine miles.

How the spaces or arcs Z Æ, P R, Z P, H Æ in the heavens are to be measured by a spectator at the spot A on the surface of the earth, is now to be explained. This is done by means of observations made by the spectator at A, upon some heavenly body, with an instrument adapted for the purpose of measuring circular arcs: by these observations, which are made when the heavenly body is either upon the meridian of the place or not, the angular distance of the body from the zenith or from the horizon is ascertained. Thus if *s* be the sun (see *fig.* 5.) on the meridian, its angular distance *s* Z from the zenith, (called its zenith distance,) or its angular distance *s* H from the horizon, (called its altitude,) is measured, and ascertained in degrees and parts of degrees.

As, however, A is the place at which these observations are made, the angle Z A *s* is all that can be determined from observation; but this angle is not the measure of the arc Z *s*, because A is not the centre of the sphere of the heavens; but the angle ZO*s* is the proper measure of this arc, since, by the supposition, the meridian circle and the corresponding one in the heavens have the same centre, O; and it is a well known truth in geometry, that the angle Z A *s* is greater than the angle Z O *s* by the angle A S *o*. This conclusion, expressed in common language, may therefore be stated thus: that a spectator at A, looking upon a heavenly body at *s*, will see it lower down in the heavens, namely, at s', or farther removed from the zenith point Z, than a spectator situated at the centre of the earth, who would, at the same instant of time, see the same body at s''; the difference of the *apparent* places which the body, *s*, will thus occupy in the heavens, as seen from the surface of the earth, and as seen from the centre, is the angle $s'\,s\,s''$, which is equal to the angle A *s* O. This angle is formed by two lines drawn from the extremities of the earth's radius, or, in geometrical language, is the angle subtended by the earth's radius, at the distance of the body *s*. This angle is called the *parallax** of a heavenly body, and increases the zenith distance of *s*. It is obvious that parallax produces a contrary effect upon H *s*, the altitude of *s*, and that as the zenith distance is *increased* by the angle A *s* O, so the horizontal distance or altitude is *diminished* by the same angle. *The general effect, therefore, of parallax is to depress a heavenly body.* If, however, the distance of the body upon which an observation is made, be so great, that it would be seen in exactly the same position in the heavens by a spectator at the surface of the earth, and one at the centre, it is evident that the angle $s'\,s\,s''$ or A *s* O (the parallax) would be so small as to escape observation, and would, to our senses, vanish. This is the case with the fixed stars; but with respect to the sun and moon and planets, whose distances are not so great, the parallax has an observable effect upon their apparent positions, as they are seen from different parts of the earth's surface, or from the earth's surface and its centre. And this circumstance raises a necessity for correcting the *observed* distances of these heavenly bodies from the zenith or horizon of a place, in order to arrive at the *true* distance, as they would be seen from the earth's centre, and that the respec-

* From a Greek word, and thus applied, signifies simply a *change* of place.

tive arcs Z *s* and H *s* may be accurately measured. The parallax is computed and given in astronomical tables, for the purpose of making this requisite correction.

Chapter IV.

On the Methods of Determining the Latitude.

The following methods are those which are in use for finding the latitude of places on land: —

1st. By the altitudes of those stars (called circumpolar stars), which never go below the horizon of the place the latitude of which is required.

2ndly. By the greatest and least altitudes of the sun above the horizon of the place in the course of the year.

3rdly. By the observed altitude or zenith distance of a star or other heavenly body when on the meridian.

4thly. By the zenith distances of stars, which pass the meridian near to the zenith of the place.

5thly. By various altitudes of a star, observed when it is near to the meridian, and then reduced to the meridian by computation.

1st Method.—By the altitudes of circumpolar stars.

Suppose (*fig.* 6.) that Z P R Q H Æ, is that imaginary circle in the concave surface of the heavens which corresponds with the meridian of the place the latitude of which we want to find.

Fig. 6.

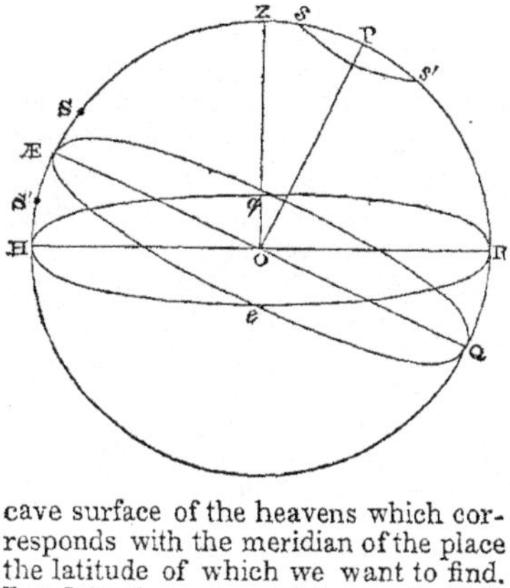

Let O be the centre of the earth, H R the rational horizon, Æ Q the circle of the equator extended to the heavens, P the north pole, and *s* the circumpolar star on which the observations are to be made. The little semicircle drawn through *s* and *s'*, parallel to the equator, will represent the apparent path of the star in its motion caused by half a daily revolution of the earth. It is evident from a mere inspection of *fig.* 6, that the star's greatest and least altitudes above the horizon will be when the star is on the meridian; its greatest when it is above, its least when below the pole P. Let *s* be its position in the first case, and *s'* in the other; then R *s* is the star's greatest altitude, R *s'* its least altitude. By means of either of the instruments called an astronomical quadrant, or an astronomical circle, R *s* and R *s'* may be observed and measured, and the number of degrees and parts of degrees contained in it be ascertained. Then as the half-circle *s' s*, which the star has described in its apparent motion from *s'* to *s*, is parallel to the equator, (for the motion of the earth, which is the cause of this apparent motion of the star, is perpendicular to the axis of the earth, so that the path of the star is also perpendicular to the axis, and therefore parallel with the equator;) and as the equator is every where at the same distance, *viz.*, 90° from the pole P, the half-circle *s' s* is also every where at the same distance from P; therefore P *s'* is equal to P *s*.

Now R *s*, which is known from observation, is equal to PR + P*s*; and R*s'*, which is also known from observation, is equal to PR − P*s'*, or PR − P*s*. Adding these two quantities, R *s* and R*s'* together, we have R*s* + R*s'*, equal to 2 P R; therefore P R, or *the height of the pole above the horizon*, (which has already been proved to be equal to the latitude of the place Z Æ,) is equal to $\frac{1}{2}$ of R *s* + R *s'*, or *one half the sum of the greatest and least altitudes of a circumpolar star*, which altitudes being known from observation, the latitude of the place is found. This mode of finding the latitude does not require any correction to be applied to the observed altitudes on account of parallax, as the body observed is a fixed star; but a correction of these altitudes is required, in consequence of the refracting power of the air and vapours which surround the earth and have effect upon the apparent places of heavenly bodies, contrary to the effect of parallax,—parallax making bodies appear lower in the heavens; whereas, a ray of light passing through the atmosphere becomes refracted and bent downwards, and the body from which the ray proceeds, appears above its true place in the heavens. The space through which a body is raised by refraction (and which is different for different altitudes), is given in tables computed for various altitudes; this correction must, of course, be *subtracted* from the *apparent observed* altitudes.

2dly. By the greatest and least altitudes of the sun above the horizon in the course of a year.

The path in which the sun's apparent yearly motion in the heavens takes place (called the ecliptic) is, at one point of it, about 23° 28′ on the north side of the equator; and, at the exactly opposite point, it is the same number of degrees and minutes on the south side of the equator. These two points of the ecliptic are the farthest off from the equator, and are exactly 90° distant from the two points where the ecliptic and equator cut each other, which are called the equinoctial points. The sun is in the former point on or about the 24th of June, and in the latter on or about the 24th of December. To all persons, therefore, living between the north pole and latitude 23° 28′, it will, on the 24th day of June, when it comes on the meridian, be the highest above the horizon, or have its greatest altitude, compared with its altitude on every other day in the year; and, in like manner, it will, on the 24th of December, have its least meridional altitude. Let S (*fig.* 6.) be its position in the former, and S′ in the latter, of these two cases. Then, as Æ is the point in the equator from which S and S′ are both distant 23° 28′, Æ S and Æ S′ are equal. The altitudes of the sun's centre in both positions are to be observed with an instrument, which observation, when corrected for parallax and refraction, will give H S and H S′, the greatest and least meridional altitudes of the sun in the course of the year. Now, H Æ = H S − Æ S, and H Æ is also = H S′ + Æ S′ or Æ S, and therefore 2 H Æ = H S + H S′; and H Æ, the height of the equator above the horizon, or *the co-latitude of the place*, is equal to ½ the sum of the greatest and least meridional altitudes of the sun in the course of the year. As, however, it seldom happens in practice that the sun, when it comes upon the meridian of the observer, is exactly at that point of its path where it is farthest from the equator, but has either already passed that point, or has not yet quite reached it, certain corrections upon the observed altitudes become necessary, in order to allow for this circumstance.

3dly. By the observed altitude, or the observed zenith distance of a star or other heavenly body, when on the meridian..

This method of finding the latitude is that which is generally employed for common geographical purposes. It is the most simple in practice, as depending only upon one observation, and is also, on this account, the most immediate in its result. It is also adapted for nautical purposes, the only difference between the modes of conducting the operation on land and at sea being in the instruments employed for making the observations, and also that, at sea, the heavenly body selected for observation is either the sun or moon, because, from the motion of the vessel, it is difficult to obtain a correct observation of the meridian altitude of any body having so small an apparent magnitude as a star. A few remarks will be made in a subsequent page, explanatory of some of the peculiarities of the modes of finding the latitude at sea: we shall, therefore, in the present instance, confine ourselves to the supposition, that the observer who is about to adopt this method of ascertaining his latitude, is on land.

Suppose (*fig.* 6.) S or S′ to be the star or other heavenly body which is selected, S being a heavenly body above the equator, S′ being a heavenly body below it; the observation is to be made when the body is on the meridian. Let Z S H represent a portion of the meridian in the heavens, and Æ *e* Q *q* represent the equator: S H or S′ H is then ascertained from observation, if the altitude be taken; or S Z or S′ Z, if the zenith distance be taken; which it is more usual to take, as, from the inequalities of the earth's surface, it is difficult to obtain on land a true horizontal boundary. These observations must be corrected for parallax and refraction, if the body be either the sun or moon; and for refraction only, if it be a fixed star. Now, the object being to ascertain either Æ H, the height of the equator above the horizon (which has been already shown to be equal to the co-latitude), or Æ Z, the zenith distance of the equator, (which is the latitude,) it is evident that if the distance of the observed heavenly body from the equator—that is, S Æ or or S′Æ, be known, the *co-latitude* will be found by subtracting S Æ, or adding S′ Æ to the observed *altitude;* and the *latitude* will be found by adding S Æ in the one case, and subtracting S′ Æ in the other, according as the body observed is above or below the equator. Now, S Æ or S′ Æ, which is the distance of a heavenly body from the equator, measured upon a meridian in the heavens, is called its declination, and is either *north* or *south* declination, according as the body is nearer and farther off the *north* pole than the equator. This declination is either computed by the observer by certain astronomical calculations, or it is taken out of astronomical tables. The Nautical Almanack gives the declination of the sun and moon for every day in the year. From the foregoing explanation of this method, the following general rule is derived for finding the latitude by means of meridian altitudes, or zenith distances of heavenly bodies. If the heavenly body have a north declination, add the declination to its observed zenith distance (corrected), or subtract it from its observed altitude (corrected), and the latitude in the first case, and the co-latitude in the other, will be obtained. If the body have a south declination, the same result will be obtained by subtracting the declination from the zenith distance, and adding it to the altitude..

4th. By the zenith distances of stars which pass the meridian near to the observer's zenith.

When this method is adopted, the observations are generally made at two places having different latitudes; and the latitude of one of the places is supposed to be previously known. It is immaterial whether both places are or not situated upon the same meridian; the star must be one which passes near the zenith of *both* places. The observations are generally made at both places on the same day; if they happen to be made on different days, various corrections become requisite, which it is as well, if possible, to avoid.

The instrument employed on this occasion is one called a zenith sector, by which small zenith distances can be measured with great exactness. Let Z, Z′, (*fig.* 7.) be the zeniths of any two places, Æ Q the equator; and suppose that the latitude of the place whose

Fig. 7.

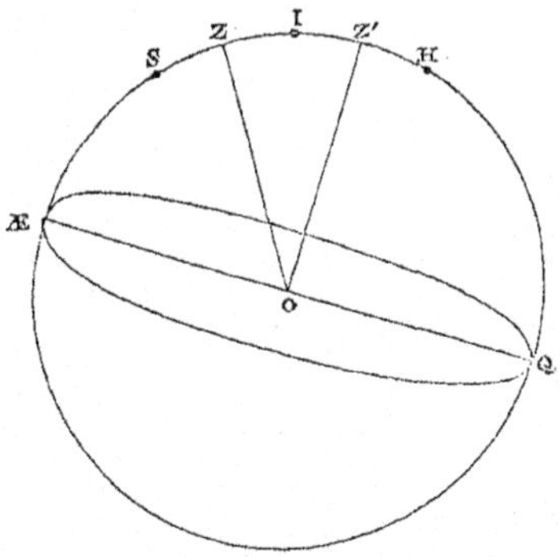

zenith is Z (that is, Z Æ,) is known, the object is to find the latitude of the place of which Z′ is the zenith (that is, Z′ Æ). The zenith distances of the star S, when it comes on the meridians of both places, must be observed. These observations will give us Z S and Z′ S. Then if S is to the south of both zeniths, as in the figure, or to the north of both, Z′ S − Z S, or the difference between the observed zenith distances, will give Z Z′; if S be to the north of one zenith, and to the south of the other, then Z S + Z′ S, or the sum of the zenith distances, will give Z Z′. Now Z Z′ is the difference of the latitudes of the two places, as is evident by an inspection of the figure; and therefore Z Æ being known, we get the latitude Z′ Æ = Z Æ + Z Z′. This method was used in the trigonometrical survey of England, and gives the latitude with great accuracy.

5th. The remaining method is by making several successive observations upon the same star at several and successive altitudes above the horizon, when it is near the observer's meridian. The various altitudes thus obtained are made the basis of a computation by which *the star's actual meridional altitude* is obtained. This is called reducing the observed altitudes to the meridian. It is a process too intricate to be introduced in this place. The star's meridional altitude is thus obtained with great exactness. The latitude is then very easily ascertained by the application of the third method. This mode of computing the latitude, by which it may be obtained to within the fraction of a second, is that which was employed by the French astronomers in their last operation of measuring an arc of the meridian.

At sea none of the preceding methods, except the third, are ever employed; *the first and fourth* are founded upon observations made with instruments requiring some nice adjustments by means of the plumb line and the spirit level, in order that the instrument may be placed exactly in the plane of the meridian and in a horizontal position—these adjustments cannot be made at sea, owing to the unsteady motion of the vessel: the fourth method is moreover not applicable to all places; the *second* method would clearly be useless, since at sea an immediate result is required; and the *fifth* is too complex in its calculations to be fitted for nautical purposes. The instrument used at sea is Hadley's sextant, by which any angles whatever may be measured, and it does not require any of the above-mentioned previous adjustments, being held in the hand of the observer. The horizon being well defined at sea, the *altitudes* of heavenly bodies are taken. The observations are taken when the sun or other body is *near* the meridian, and are continued until it is found that its altitude has attained its greatest quantity and begins to decrease: at its greatest altitude the body is on the meridian of the observer at S (*fig.* 6.); the complement of the latitude is, therefore, obtained, as in the third method, by adding or subtracting the distance of the observed body from the equator, according as it is below or above it, or has a south or north declination. The sun or moon is commonly the object observed, *and the Nautical Almanack gives the declination of these bodies for every day in the year.* The corrections for parallax and refraction must be made upon the observed altitudes. Besides these corrections, another is rendered necessary, in consequence of the observer being elevated above the surface of the sea. This elevation causes a correspondent depression or sinking of the horizon, and gives a greater apparent altitude to the observed body than it really has. This correction is called the dip.

But as it frequently happens that, at the time when the sun or moon is on the meridian, clouds prevent the observing of its *meridional* altitude, the latitude may then be obtained by observing two altitudes *out* of the meridian at different times, and noting the interval of time which elapses between the times of observation. Z P (*fig.* 6.), or the co-latitude, is then computed by the resolution of three spherical triangles, a mere mathematical process, which we need not stop to investigate.

Chapter V.

Longitude—Mode of Measuring Time—Sidereal Time—Apparent Solar Time—Mean Solar Time—Equation of Time.

Having by one of the foregoing methods ascertained the parallel of latitude in which any particular place is situated, the next inquiry is directed to the finding of the longitude, or the position which a place occupies in the parallel with respect to what is called the first meridian. In this country the meridian of the observatory at Greenwich is generally taken for the first meridian. Various are the methods which have been proposed for finding the longitudes of places; in every point of view the subject is one of very considerable interest, not only on account of its great importance in commerce and science, but also because the attempts which have for so many years been made, in order to determine the longitude with the same accuracy with which the latitude of places is found, have hitherto been unsuccessful. Since the time of Queen Anne it has been regarded as an object of great national importance; and a board, called the Board of Longitude, consisting of various official and scientific persons, was then established for the purpose of encouraging and directing attempts to determine it.

All the methods for finding the longitudes depend upon the manner in which time is measured; and in order to attain a clear notion of them, it will be proper to explain at some length how a measure of time is obtained.

Properly considered, time is, in itself, without parts, and indivisible; the flow or lapse of time is, however, capable of being measured by means of events happening in time, and which, when compared one with another, are of different continuance, taking up more or less time in their completion. Time and space are in one respect similar; space is in its nature indivisible, it does not contain within itself any marks or circumstances of division; but by means of bodies which are situated within it, we are able to consider space as though it were divided into parts. What bodies are, in this respect, to space, that precisely are events to time; they afford us the means of measurement. This is done by the comparison of those events one with another in respect of their duration; but in order to do this with accuracy, it is necessary to possess some *standard event* which always takes up exactly the same time, and to which we may refer as affording, by comparison with it, a measure of the duration of all other events: without this we should be at a loss to ascertain exactly how much time is taken up by any other event, and be left to the uncertainty of only probable conjecture. In this particular there is also a similarity between time and space; for, in measuring space, the object has been, even in remote periods of history, to fix upon a certain standard. Thus our king Henry the First commanded that the standard of measure of length should be of the exact length of his own arm, which is our present yard measure. But with him perished the standard by which the measure called a yard might afterwards be compared, corrected, and ascertained afresh. It is clear that something which was liable neither to decay nor variation was requisite to form a proper standard of measure; and accordingly by the recent Act of Parliament for weights and measures, and which proceeds upon more scientific principles, such a standard has been established in the length of a pendulum beating seconds of mean time in the latitude of London. On observing the various occurrences or events in nature, with a view to fix upon some one event, as a standard for the measure of time, it was discovered that the motion of the earth round its axis possessed all the qualifications requisite for such a purpose. This event is invariably of exactly the same continuance, and it is the only one in nature with which we are acquainted, that is so. The time spent in one revolution of the earth round its axis forms, therefore, an exact and perfect standard measure, by reference to which the time taken up by all other events may be ascertained. The beginning and the end of the revolution, and consequently the duration of it, is determined by means of the fixed stars: these stars have no motion of their own; so that their apparent daily motion is caused by the daily motion of the earth on its axis. Hence, if a fixed star be upon the meridian of a place, this motion of the earth, which is in a direction from *West to East*, gives to the star an *apparent* motion towards the *West;* and when the star next appears upon the same meridian, having moved through 360°, an entire revolution of the earth has been accomplished. The time spent in performing this revolution is the standard measure of time, and it is called a *sidereal or star day*, because it is by the appearance and re-appearance of the same star in the same place in the heavens that the completion of the revolution is ascertained. This standard being once established, it may be divided into smaller portions of time at pleasure. Portions of time measured by a reference to this standard are called *sidereal time.* Astronomical clocks are made to show sidereal time.

But it was requisite, for the sake of convenience, to obtain some other standard of measure of time, having reference to the sun, by which the common affairs of life are regulated. Now the same motion of the earth about its axis, which has already been noticed with respect to the fixed stars, gives to the sun also an apparent daily motion from the east towards the west. When the sun is upon the meridian of a place it is *apparent noon* at that place, or, in popular language, the hour of twelve in the day. After this hour, the sun, leaving the meridian, appears gradually to travel towards the west. This westerly motion continues below the horizon until it has brought the sun to a point where it rises again, and proceeding in its daily course, again reaches the same meridian on which it appeared at the hour of apparent noon on the former day. The time which has passed between these two successive appearances of the sun on the meridian of any place is called a solar day. A solar day is longer than a sidereal day; for if upon any day the sun and a fixed star be observed to be upon the meridian of a place together, the star will, on the following day, return to the meridian a few minutes before the sun. This difference in the times of the sun and a fixed star leaving and returning to a particular meridian, is caused by the sun's apparent yearly motion in the ecliptic, which being in a direction from *west to east,* and opposite to that daily motion which brings it to the meridian, makes the star, which has only the daily motion, from *east to west,* to appear on the meridian before the sun. The daily *average* amount of this yearly motion of the sun in an *easterly* direction *away from a meridian*, is 59′ 8″ (nearly one degree)

We have called this 59′ 8″ the daily *average* amount of the sun's yearly motion, because, during some parts of the year, it is more than 59′ 8″, and at others less. Hence it follows, that the intervals of time, which in the course of a year elapse between the sun's successively leaving and returning to the same meridian, are of different lengths. An *apparent* solar day, therefore, or the time between two actual successive passages of the sun over the same meridian, could not be adopted as a standard measure of time, because it is a varying, fluctuating quantity; and it is essential to a standard measure of time, that it should be a fixed quantity. But with a view to obviate this difficulty, an *artificial* solar day has been constructed, called a *mean* solar day, *the length of which is always the same*, and is the *mean* or *average* length of all the various *apparent* solar days in the course of a year; the difference in length between a mean solar day and the apparent solar day for the time being, is called the *equation of time*.

When time is reckoned with reference to the apparent solar day, it is called *apparent* time; when with reference to the mean solar day, it is called *mean time*. A common sun-dial shows the hour of apparent time. Time-keepers or chronometers, common watches and clocks, are made to show the hour of mean time. Both the apparent solar day, and the mean solar day, are divided into 24 hours; and are, for astronomical and scientific purposes, reckoned from noon to noon. The mean day is always of the same length, and although it is longer than the sidereal day, yet the quantity by which it is greater (viz. the time required for the earth by its motion on its axis to move through 59′ 8″ of space) is always the same.

Hence, the uniformity and equal length of mean days, and of seconds of mean time, really depend upon, and must at last be referred to the uniform and equal motion of the earth upon its axis, which consequently is the standard, by reference to which, the measure of time afforded by the pendulum beating seconds of mean time is ascertained, and may be corrected. It is not uninteresting to observe, that to the equable and invariable motion of the earth about its axis, we are indebted, not only for a standard measure of time, but also for all our standard measures of length, capacity, and weight; since, by the recent Act of Parliament, before referred to, all of them are referred to the length of a pendulum beating seconds of mean time in the latitude of London.

At four times in the year, and only four,—that is, on or about the 15th day of April, and the 1st of September; and on or about the 15th of June, and the 24th of December,—mean time and apparent time agree; or, which is the same thing, on these four days the sun is actually upon the meridian of some particular place, and the shadow of the style of a dial at that place is upon the hour of twelve, at the very moment that a correct time-keeper, or watch measuring mean time, and adjusted for this particular place, shows the hour of 12. Throughout the rest of the year, apparent time and mean time are different. The exact amount of this difference is easily calculated for every day: it is called the *equation of time*; because, by either adding it to, or subtracting it from, the time of the apparent solar day, the result will be, the time of the mean solar day. The equation of time is given for every day in the year in the Nautical Almanack, with directions, showing whether it is to be added to or subtracted from the *apparent* time, in order to get at the mean time.

Chapter VI.

Various Methods of Finding the Longitude.

In the application of the above-mentioned principles for reckoning the time of the day, consists the simplest method of finding the longitude of a place, or its situation in a given parallel of latitude with respect to the first meridian, the meridian of Greenwich.

As in the 24 hours into which an *apparent* solar day is divided, the sun returns to a meridian which it has left, it may be said to describe, in that time, 360 degrees of longitude; which, dividing the whole 360° by 24, and supposing the motion to be uniform, is at the rate of 15° of longitude for every hour of apparent time; so that if we find the sun to be upon the meridian of Greenwich, or it is 12 o'clock apparent time at Greenwich, it will, in one hour after of apparent time, be 15° to the west of Greenwich, in 2 hours 30° west, in 6 hours 90°, in 12 hours 180°, and so on, at which several times in succession the sun will be upon the meridians of places, or it will be apparent noon, or 12 o'clock at places situated 15°, 30°, 90°, and 180° west longitude from the meridian of Greenwich; while the several corresponding hours of apparent time at Greenwich will be 1 o'clock in the afternoon, 2 o'clock, 6 o'clock, and 12 o'clock at night, or midnight.

Beyond the 180° west longitude, east longitude commences. The only difference in the two cases is, that places to the west of Greenwich are said to have their noon later, and their reckoned time earlier: those to the east have their noon earlier and consequently their reckoned time later than at Greenwich. Hence, if, when it is the hour of apparent noon at any place situated either to the east or west of Greenwich, the corresponding hour of apparent time at Greenwich could be ascertained, the longitude of that place might be directly found by turning the difference of their times into degrees and parts of degrees, reckoning 15° for every hour of apparent time, and for proportionate parts of an hour taking proportionate parts of 15°.

But, as it has been already explained, the variation of the apparent solar day makes apparent time ill adapted as a standard to refer to for the purpose of ascertaining the difference of longitudes by the difference of the apparent times at two different meridians: it is necessary, therefore, to show how the difference of the mean time at two different meridians may be substituted in its stead. It has been stated that, at four times in the year, the equation of time is nothing, or that at some particular moment of four days in the year the hour of mean time exactly corresponds with the hour of apparent time. Thus, it appears, from the Nautical Almanack, that on the 24th of December of the present year, at the hour of apparent noon, when the sun will be on the meridian of Greenwich, the apparent time will be in advance of the mean time at Greenwich by 20″.3, that is, when it is 12 o'clock in the day, by the sun, it will want 20″.3 to 12 o'clock by the watch; so that it will be then necessary to subtract 20″.3 from the apparent time deduced from observation in order to ascertain the corresponding mean time at Greenwich for that day. But on the 25th of December, or at the hour of apparent noon at Greenwich on the following day, the apparent time will be *behind* the mean time by 9″.8, which quantity therefore must then be *added* to the apparent time to get at the mean; and the watch will be 9″.8 past 12, when it is noon by the sun. Hence, as in the space between these two successive passages of the sun over the meridian of Greenwich, the equation of time, or the difference between apparent and mean time, has, from being subtractive, become additive, it has, at some moment of that interval, been 0, or has passed through 0; or, in other words, the mean time at Greenwich having overtaken the apparent time at that place, the hour of apparent time and that of mean time will, for some one moment, between the two successive noons, be the same. Now, as the difference between mean and apparent time, or the equation of time, depends upon the variable velocity of the sun in his apparent annual motion in the ecliptic, and upon the obliquity of the ecliptic or the angle it makes with the equator; these circumstances being independent of place, the equation of time is for all parts of the earth the same that it is at Greenwich at any given moment. Hence, as at some particular moment between the noons at Greenwich of the 24th and 25th of December, the equation of time is nothing, at that moment it is also nothing at every other place upon the globe, or the apparent and mean times are then every where exactly the same. But we have already proved that the longitude might always be determined by turning the *difference of the apparent times* at Greenwich and any other place into degrees at the rate of 15° to every hour of apparent time. At the particular moment, however, when the equation of time is 0, the difference of the apparent times is the same with the difference of the mean times at Greenwich and every other place upon a different meridian. Hence, at this moment the longitudes of all places may in like manner be determined by turning the *difference of mean times* at Greenwich and at all other places into degrees at the same rate of 15° for every hour of mean time. But what is true of mean time and of the difference of mean times at one particular moment, is true always, because mean time is not variable; so that the difference of mean times at Greenwich and all other places will always give the longitudes of places; and therefore by knowing on any day in the year the mean time at Greenwich, and also the corresponding mean time at any other place, the longitude of that place will be found by converting the difference of their reckoned mean times into degrees, at the rate of 15° for every hour of mean time; it will be east longitude if the time at the place in question be later than the time at Greenwich, and west longitude if it be earlier.

All, therefore, that is required is to ascertain, 1st, the hour of mean time at the place, the longitude of which we wish to know; 2dly, the corresponding hour of mean time at Greenwich. Now the hour of mean time at any place may always be obtained by means of the corresponding *apparent* time, by adding to or subtracting from it the equation of time for the moment, which is given (or may be computed from what is given) in the Nautical Almanack. The hour of *apparent* time may always be found by means of an observed altitude of the sun, or, if the place be on land, by means of a sun-dial. The corresponding mean time at Greenwich may then be ascertained by a chronometer or time-keeper, adjusted and regulated so as to show Greenwich mean time. If, therefore, a time-piece could be made so perfect as always to show the mean time at Greenwich without error; or if its error in going were always the same, that is, if it gained or lost the same quantity every day, the longitude of places might be correctly found by such a chronometer.

This desirable object has not hitherto been attained: the most ingenious and accomplished mechanics, although prompted by the liberal rewards held out by the legislature to encourage their exertions, have failed of complete success. Time-pieces have, however, been made, which from their near approach to an equable rate of going, might appear to justify even sanguine hopes that at some period or other a *perfect* machine may be constructed; but it is highly improbable that these hopes will ever be realized. The imperfection of the human mind seems to oppose even a moral obstacle to the attainment of absolute perfection in any of its productions. In other works of art an apparent perfection may be obtained, because their defects are not visible to our senses, and we have no other means of ascertaining their existence; but in a machine which is to measure time, the smallest errors accumulate so as to become at last apparent, and in the daily equable motion of the earth on her axis, nature herself affords a perfect measure of time, by a comparison with which the errors and defects of the measure constructed by human art cannot in the long run escape detection.

At sea, where other methods cannot be resorted to with facility, chronometers are generally used for finding the longitude; but the mere circumstance that the best chronometer is liable to error, and to error which may escape notice, makes it dangerous to trust to the chronometer alone; nor ought it to be relied on but under circumstances excluding the adoption of some of the other methods of finding the longitudes.

These methods, therefore, form the next subject of consideration.

There are various appearances from time to time taking place among the heavenly bodies, that afford the means of finding the longitude nearly. These appearances are the following: 1st, Eclipses of the moon; 2d, Eclipses of Jupiter's satellites or moons; 3d, Occultations or concealments of fixed stars, by the moon's passing over them; 4th, Eclipses of the sun; 5th, The passage of the moon over the meridian of the place the longitude of which is required; 6th, The same passage compared with that of one or more stars immediately preceding or following the moon, and having nearly the same declination; 7th, The distance of the moon from particular fixed stars or from the sun. There is also another method, of limited application, by means of artificial appearances upon the earth, as explosions of gunpowder made at one place and seen at another, the longitude of which is required.

The first and second and the last of these appearances are observed at all places where they happen to be visible at the same instant of *absolute* time. The difference, therefore, in the reckoned times, either mean or apparent, at two places where they are visible, is owing to the difference of their longitude. The time at Greenwich of eclipses of the moon and of Jupiter's satellites is previously computed and set down in the Nautical Almanack, and the corresponding time at the place whose longitude is wanted, being obtained at the moment of these appearances happening, the difference turned into degrees in the usual way is the longitude. By means of explosions of gunpowder or other signals made on the earth, the difference of the longitudes of any two places not far distant from each other may be determined with very great exactness; the mean time for each place may be known by separate chronometers previously adjusted and regulated for the purpose; the difference of the times at the moment of the explosion or other signal, which is made at one place and seen instantaneously at the other, converted into degrees, will give the difference of longitudes. This method has of late become the more interesting from its having been adopted, in the course of the operations now in progress on the continent for measuring an arc of a parallel of latitude, as the best means of determining the longitude of the extremities of the arc. The space between the two extremities of this arc was divided into a great number of smaller arcs, all of such a length, that one of the extremities of each smaller arc might be made visible to an observer at the other extremity. At each point of division of the principal arc, were fixed stations, at which the requisite instantaneous signals were made and observed. The difference of times when these signals were made at one station and observed at another, gave the difference of longitudes of the extreme points of every smaller arc; and the sum of all the differences gave the difference of longitudes of the extremities of the principal arc. It is scarcely necessary to remark, that any thing answering the purpose of an instantaneous signal, may be used instead of explosions of gunpowder—such as the discharge of a rocket, or the sudden display or extinction of a lamp: a contrivance called a Heliostat (which is from two *Greek* words, and signifies any thing the position of which has some reference to the sun) has been employed on the continent: it has a strongly reflecting surface, and is placed in such a manner that the rays of the sun are reflected by it towards the desired point of observation; the reflection is then made to disappear suddenly by interposing a screen between the Heliostat and the distant spectator, and thus conveys an instantaneous signal.

The third and fourth methods, by occultations of fixed stars by the moon, and by eclipses of the sun, likewise depend upon the difference of the times at which these appearances take place at Greenwich (and which times are computed by means of tables); and of the times at which they are actually observed to take place at the spot the longitude of which is required; but with this qualification, that as these appearances are not observed at all places at the same point of *absolute* time, the difference in the absolute times of their happening must be allowed for: thus, if at Greenwich the occultation of a certain fixed star by the moon, happen at six o'clock in the morning; and at some other place to the west of Greenwich it be observed to happen at midnight, thus making a difference of six hours in the reckoned times of the appearance, it will not follow that this is all due to the longitude, and that the place in question is 90° west longitude, for the occultation does not happen at both places at the same moment of absolute time; but the star is seen at the place in question for some time after it is hidden at Greenwich. This time, which being caused by parallax may be computed, must be added to the Greenwich time, computed from the tables; and then the difference between the resulting time at Greenwich, and the time at the place at the moment of the occultation *there*, will give the true difference of corresponding reckoned times between that place and Greenwich; and from this difference the longitude may be deduced. The difference in the absolute time of these appearances occurring at different places, is owing to the sun and fixed stars shining by a light of their own, and to the moon's parallax.

The fifth method is by means of the moon's passage over the meridian.

If the sun and moon be upon the meridian of Greenwich together, on any particular day, on the following day when the sun is again on that meridian, the moon will be considerably to the *east* of it; and some time will consequently elapse before the moon reaches the meridian of Greenwich after the sun has left it. This *easterly* separation of the moon from the sun after they have been together, is caused by the moon's quicker motion in her orbit or course round the earth; and the time which elapses between the passage of the sun over the meridian of Greenwich, and that of the moon, is called the moon's retardation. The moon's motion in her orbit continuing, the distance between the sun and moon continually and gradually increases; so that if the moon's retardation be of a certain amount at the time of its passing the meridian of Greenwich, the retardation at a place to the *west* of Greenwich will be of a greater amount, in proportion to the time that is required to bring the moon from the meridian of Greenwich, to the meridian of the place to the west of Greenwich. Hence, as the increase of the moon's retardation is for 24 hours proportional to the times in which it is produced, by knowing the retardation at two different meridians, and the time during which the retardation at one of the meridians has been produced, the time during which the greater retardation at the other meridian has been produced, may be found by the rule of three. Thus, suppose that the sun and moon having been upon the meridian of Greenwich together on one day, the retardation of the moon at Greenwich on the following day, or in 24 hours, is 52′; that at a place to the west of Greenwich the retardation of the moon is observed to be 57′, or 5′ more than it was at Greenwich; then we shall have this proportion; as 52′ : 57′ :: 24 hours : 24 hours + the additional time necessary to produce the additional retardation of 5′. This additional time is due to and expresses the difference of the longitudes, and 24 hours correspond with 360° of longitude. Hence, 52′ of time : 57′ of time : : 360° : 360° + difference of the longitudes; and as in this case we have taken the meridian of Greenwich, the longitude of which is 0, we shall have 52′ of time : 57′ of time : : 360° : 360 + longitude of the place; or the longitude of the place is equal to $360 \times \frac{5'}{52'}$ of time, and expressing the time in parts of degrees at the rate of 15° to an hour, the longitude is obtained. Hence generally the longitude of a place is equal to 360°, multiplied by the difference between the retardation of Greenwich and the retardation of the place the longitude of which is required, divided by the increase of the retardation at Greenwich in the 24 hours preceding the time of observation. The increase of retardation at the place the longitude of which is required, is known from observation. The increase of retardation at Greenwich, for the 24 hours preceding, may be found by means of the Nautical Almanack.

The principle of this method is applicable to the fixed stars as well as to the sun; the only difference being, that the moon's retardation is greater with respect to the fixed stars, as they have none of the daily easterly motion which the sun has in its apparent yearly path in the heavens. The application of this principle to the fixed stars for finding the difference of the longitudes of two places, was first successfully made by M. Nicolai, a distinguished astronomer, at Manheim, and is now very generally practised on the continent. Mr. Francis Baily, in his valuable paper on this subject, lately published in the Memoirs of the Astronomical Society of London (vol. ii.), observes, " That already at several observatories, the observers have been enabled to determine their difference of meridians in a few months with as much accuracy as they formerly could in as many years." The improvement introduced by M. Nicolai consists in the choice of those stars which have very nearly the same declination or distance from the equator as the moon, and which pass the meridian very soon after, or a little before the moon. The advantages of the method are to be found in avoiding a great number of errors and troublesome calculations, which in practice were found to detract from the value of other methods, and in the frequency with which observations may be made, being every night that the moon is visible. It was employed with very great success by Lieutenant Foster on Captain Parry's last voyage but one, in determining the longitude of Port Bowen in Prince Regent's Inlet. His observations have been calculated and compared with those made at the observatories of Greenwich and Dublin, and by the late Colonel Beaufoy at Bushey Heath; and the results, which will appear in a volume of the Astronomical Society's Memoirs, show, as far as one example can do so, the great value of this method of determining the longitude on land.

None of the previous methods, however, (except that which consists in the use of chonometers,) are adapted to the situation of a person on board a ship. The late Astronomer Royal, *Dr. Maskelyne*, in his Preface to the *Nautical Almanack*, observes, " It was hoped that some means might be found of using proper telescopes on shipboard to observe these eclipses [the eclipses of Jupiter's Satellites]: and could this be effected, it would be of great service in ascertaining the longitude of a ship from time to time. In my voyage to Barbadoes, under the directions of the *Commissioners of Longitude*, in 1763,

I made a full trial of the late *Mr. Irwin's* marine chair proposed for the purpose, but could not derive any advantage from the use of it; and considering the great power requisite in a telescope for making these observations well, and the violence as well as irregularities of the motion of a ship, I am afraid the complete management of a telescope on shipboard will always remain among the desiderata."

The longitude may, however, be found at sea, when the moon is visible, by the observed distance of the moon either from the sun or from nine of the principal fixed stars mentioned in the *Nautical Almanack*. This distance is observed by means of a Hadley's Sextant. In consequence of the moon's quick motion in her orbit she is every moment changing her situation in the heavens with respect to the sun and stars. Her distance, therefore, from the sun, or a particular star, is at one moment of time different from what it was at the previous moment, and what it will be at the next; so that a particular or given distance is proper or due to a given moment, which moment will be expressed or reckoned differently at different meridians, according to the apparent time of day. This difference in the apparent times, being therefore due to the difference of meridians will, converted into degrees, give the longitude. The distance of the moon from the sun, and from nine principal fixed stars, is given in the *Nautical Almanack*, for every three hours of Greenwich time. This distance is such as it would appear at the centre of the earth; allowance having been made in computing the distance given in the Almanack as well for parallax as for refraction. The *observed* distance at the place the longitude of which is required, is in a similar manner to be reduced to the centre of the earth by correcting for the moon's and sun's parallax, and for refraction. The apparent time, at the place and moment of observation, is obtained in the usual manner, by taking the contemporary altitude of the sun or star. The difference between this apparent time and the apparent time at Greenwich, given in the tables as corresponding to the *same distance*, converted into degrees, will be the longitude of the ship. This method of finding the longitude is called the lunar method; it will generally give the longitude to within twenty miles, frequently much nearer; it was brought into general use by the exertions of Dr. Maskelyne, who, on his voyage to St. Helena, in the year 1781, employed it with great success.

The latitude and longitude of a place having thus been ascertained, the exact position of that place on the surface of the globe is determined.

Chapter VII.

To find the Direct Distance between any two Places—Decrease of Degrees of Longitude.

Knowing the latitude and longitude of two different places, the shortest distance between them, measured on the surface of the globe, may be found. Let A and B (*fig.* 8.) be two places upon

Fig. 8.

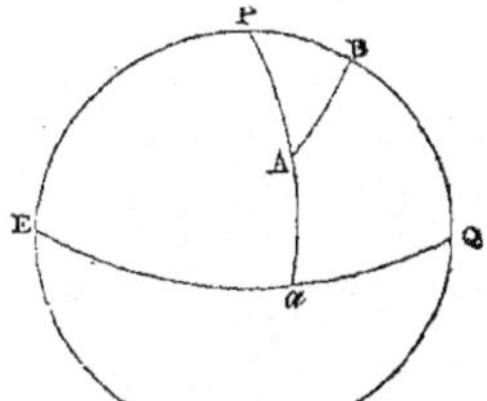

the earth's surface, E *a* Q the equator, P the pole, A *a* the latitude of A, B Q that of B; *a* Q, which is the difference of their longitudes, is known, as the longitudes of both places are supposed to be known; then A B, being the arc of a great circle passing through A and B, is the shortest distance, and may be found as a side of the spherical triangle A B P by spherical trigonometry. With a trifling inaccuracy, the distance, A B, may also be determined mechanically, by means of a common terrestrial globe and a pair of compasses. The opening of the compasses given by applying the extremity of either leg to each place on the globe, will be the measure of that arc of a great circle which lies between the two places. The number of degrees contained in this arc may then be ascertained by applying the compasses, thus open, to any graduated great circle on the globe, or one which has the degrees marked, such as the equator or ecliptic. The number of degrees thus found, being turned into geographical miles, at the rate of 69.044 miles to a degree, will give the actual distance in miles. Thus the direct distance between Paris and Buenos Aires is 99° 24′ 35″, which is equal to 99.41° nearly. Multiplying this by 69.044, will give the direct distance in miles, which is equal to 6863.66404 miles, or 6863 miles 5 furlongs nearly.

It has just been taken for granted that the shortest distance between two places on the globe is the arc of a great circle. This may be made evident by a few simple considerations. The plane of a great circle passes through the centre of the globe; that of a small circle does not: the radius of a great circle, therefore, is greater than that of a small one, and consequently the curvature or bending of the former is less than that of the latter. And as a *straight* line is the *shortest* distance between two points, so of two *curved* lines joining two points, that which is most like a straight line, *i. e.* the *less curved line* of the two, is *the shorter*; and therefore the arc of a great circle, lying between two places on the globe, is the shortest distance.

If we suppose, (as hitherto we have done,) that the earth is perfectly spherical, it is evident that a degree of latitude, being the 360th part of a meridian, which is always a great circle, must be everywhere of the same length. It is otherwise with degrees of longitude: except for places upon the equator, a degree of longitude is an arc of a small circle, and is less than a degree of longitude measured upon the equator, which is a great circle. And the magnitude of a degree of longitude becomes gradually less in proportion as the distance from the equator increases, or as the latitude increases. The radii of the circles on which the longitude is measured (called the parallels of latitude) decrease from the equator to either pole; so that the circumferences of parallels of latitude decrease in like manner: but the circumference is always equal to the number of degrees into which it is divided, multiplied by the length of each degree; and as the number of degrees is the same in all circles, the length of each degree varies with the circumference, and must, therefore, decrease from the equator to the pole. The precise measure or law of this decrease may be proved in the following manner. Let (*fig.* 9.) the arc A E represent a degree of longitude upon the equator E Q; *e a* a corresponding arc of a degree of longitude, measured upon the parallel of latitude *e a q*, in latitude A *a*.

Fig. 9.

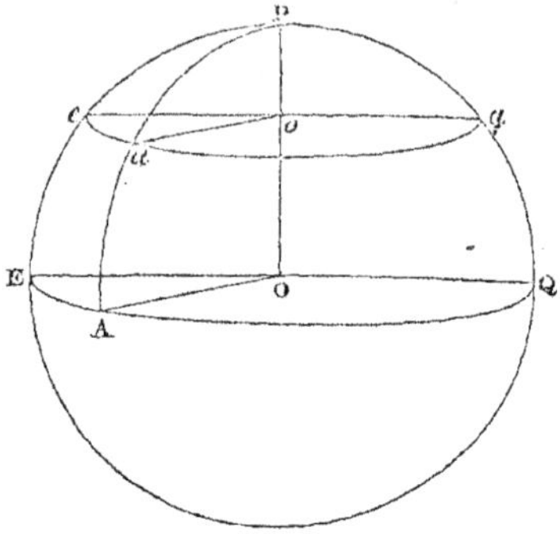

Then, as the length of a degree varies as the circumference, and the circumference as the radius, we shall have the length of A E to that of *e a* at latitude A *a*, as radius A O is to radius *a o*, where A O is the radius of the earth, and *a o* is the radius of the small circle *e a q*, the plane of which is parallel to the plane of the equator. In plane trigonometry, the radius *a o* is called the cosine of the arc A *a*, or the cosine of the latitude A *a*; hence the length of a degree of longitude at the equator is to the length of a degree at a given latitude A *a*, in the proportion of the radius of the earth to the cosine of the latitude; and as the degree at the equator and the radius of the earth are invariable, the length of a degree of longitude varies as the cosine of the latitude.

CHAPTER VIII.

Oblate-Spheroidal Figure of the Earth —Cause of this Figure—Centrifugal Force.

THE various phenomena which indicate the nature of the earth's shape have been already described. They are sufficient to establish in a general way the roundness of the earth, but they are at the same time of that vague and indefinite character, as to be incapable of solving the more difficult problem of the earth's specific and exact shape. In order to determine this, it has been necessary to resort to experiments of an extremely delicate and tedious description, and to call in the aid of complex mathematical calculations founded upon the facts which the experiments have brought to light. The attention of some of the most eminent philosophers of Europe has for many years been given to this subject; and although the true figure of the earth cannot be considered as even yet determined with all the precision that is desirable, it is now conclusively proved that the earth is not a perfect sphere, but of an oblate-spheroidal * form, bulging out at and about the equator, and flattened at the poles; and that the equatorial diameter is longer than the axis or polar diameter. The excess of the equatorial above the polar diameter represents, when compared with the whole diameter, the quantity by which the figure of the earth deviates from a perfect sphere; it is called the earth's ellipticity or compression.

The discovery and proof of the earth's elliptical shape, and the laborious undertakings engaged in for determining the quantity of it, occupy some of the most interesting pages in the history of science. It is not perhaps to be much regretted that the person who first started the idea of the earth's spheroidal shape should be unknown. The first notion of it was in all probability nothing better than one of those happy conjectures which have been verified by subsequent proofs; but the name of Newton is as intimately allied with the discovery of the earth's true figure as with that of universal gravitation. In both cases the idea had been already entertained by several philosophers, but it was Newton who redeemed the truth from conjecture, and established it upon the basis of demonstration: with respect to the figure of the earth, he proved, from admitted principles and facts, that it must of necessity be an oblate-spheroid, and he assigned a ratio between the equatorial and polar diameters.

The true figure of the earth is that which the particles composing it must assume, in order to be in a state of equilibrium or rest: the figure, therefore, depends essentially upon the forces which act upon these particles. The principal of these forces is the mutual attraction which subsists between the particles themselves. Any other force which acts in a different direction to this principal force, or with unequal intensity upon different particles, is a disturbing force; it disturbs and deranges that state or figure which the whole mass of the earth would assume if affected only by the mutual attraction of the component particles.

The method of conducting the investigation of the true figure of the earth, is one which is very usual in mechanical philosophy. The most simple and striking characteristics of the problem are singled out and considered alone, and the result obtained from them is afterwards varied and modified by the introduction of such minor and more complex conditions as are suggested by the problem in its true and practical form. In this manner, in the science of mechanics, the first principles and ground-work truths are ascertained, upon the supposition that the parts of machines are without weight, and that there is no such thing as friction, and the effect due in the practical result to these and other circumstances is estimated afterwards.

In attempting to determine the true figure of the earth, it has accordingly been assumed in the first instance, that the earth is a body consisting only of fluid particles which move easily among themselves, and that the density of the particles, or the quantity of matter contained in the same space is the same throughout, and that the earth is in a state of rest. If this were the case, the only form in which equilibrium could take place among the particles which compose the earth, and which exert a mutual and equal attraction upon each other, is a perfect sphere. This may be exemplified by adverting to the familiar circumstance of drops or globules of water, or of quicksilver, upon a perfectly smooth and dry plain; by the force of mutual and equal attraction, the particles of which they consist, dispose themselves in a globular form. It is true, that the form of these drops is not perfectly spherical, but is rather flattened at the top; because the force of gravity which acts without an equal resistance upon the upper particles, makes them press down upon the lower, and thus deranges the effect which would be produced by their mutual attraction alone: but the fact is sufficient as a popular illustration of this conclusion,—that fluid particles of an equal density, not affected by any external and partial force, would assume the form of a perfect sphere as that of a state of equilibrium.

Now in this statement it is evident, that three conditions have been introduced which are not verified in the

* An oblate spheroid is a solid body which may be conceived to be formed by the revolution of an ellipse or oval about its lesser axis.

actual state and circumstances of the earth.

These are, 1st. That the earth is a fluid body. 2d. That it is of the same density throughout. 3d. That it is in a state of rest. These conditions, therefore, not being fulfilled, it is necessary to inquire what alteration ensues in the form of equilibrium.

Let us first take into consideration the fact of the earth not being in a state of rest, but having a double motion, one in its orbit, the other about its axis. The existence of motion in inanimate matter always indicates the impression of an external force. The force represented by this motion in the earth, is therefore in addition to that of the mutual attraction of its component particles. In order, however, that it should alter the form of equilibrium, it is clearly necessary that the additional force should be not merely external but partial. The earth's motion in her orbit has therefore no effect of this kind; because, acting *equally* upon all the particles, it can have no influence upon their mutual relation. The other motion of the earth about its axis gives rise to very different considerations. This motion is performed in a plane perpendicular to the axis; so that the particles of which the earth is composed move in circles, the planes of which being perpendicular to the axis, are either coincident with, or parallel to the plane of the equator.

Thus (in *fig.* 10) the particle Q, at the

Fig. 10.

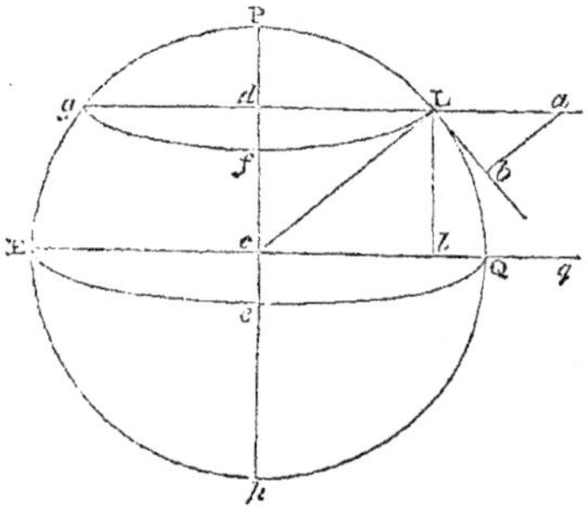

equator, moves in the equatorial circle Q *e* E; and the particle L, lying between the equator and the pole, moves in the circle L *f g*, which is parallel with the equator. And as the circles which are thus described by various particles, are all described in the same time (namely the time of an entire revolution of the earth about its axis or a sidereal day), it is evident that different parts of the earth will be differently affected by the motion of rotation:—immediately under the poles there exists (so far as this motion is concerned) a state of perfect rest; but in passing from the poles to the equator, the motion of rotation begins and becomes quicker and quicker (because in the same time a greater space or circle is described); till arriving at the equator, which is the largest circle, the plane of which is perpendicular to the axis, it has obtained its greatest velocity. This motion of rotation gives to all the particles affected by it, a tendency to fly off from the centre of rotation, and communicates to them what is therefore called a *centrifugal* force. The mutual attraction of the earth's particles, or their gravitating force, is directed *towards* * the centre of the earth, this is called a *centripetal* or centre-seeking force. The combination of these two forces, the centrifugal and centripetal, will, as they are in different directions, and as the former acts in the partial manner we have described, produce a material alteration in the form, which, under the action of the latter alone, the particles of the earth would have assumed as a state of equilibrium. For it is evident that as the centrifugal force caused by the motion of rotation is in diminution of the centripetal force of gravity, the particles affected by this centrifugal force must give way to those which are not affected by it, and thus be pushed away from the centre of the earth; those also which are more affected by it, will give way to those which are less so, and will recede still farther from the centre of the earth; so that in order now to preserve equilibrium among themselves, there must be a decrease in the *number* of particles which have the greatest *individual* gravitating force, and a proportionate increase in the number of particles which have a diminished gravitating force.

There are two reasons for the centrifugal force caused by the motion of the earth about its axis, being greatest at the equator, and gradually diminishing towards the poles. The motion of the earth on its axis being every where in a plane to which the axis is perpendicular,

* This is not strictly true: the direction of gravity is always perpendicular to the surface, and therefore, (the earth not being a sphere) it does not pass through the centre, but only near the centre, but this distinction may be disregarded in the present explanation.

the particles lying under the equator move in the circle of the equator; all other particles between the equator and the poles move in circles parallel with the equator. Hence, particles moving in the equator have the greatest velocity and the greatest centrifugal force; those nearer the equator a greater velocity and a greater centrifugal force than those particles which are more distant from the equator. Moreover, the direction of the centrifugal force communicated by this motion is always opposite to the direction of the radius of the different circles in which the particles move; thus if L *d* (*fig.*10), be the radius of the circle in which a particle at L moves, the direction of the centrifugal force at L will be L *a*, and the direction of the centrifugal force at Q, in the equator will be Q *q*, Q C being the radius of the equator and also of the earth. Now gravity or the centripetal force is always in a direction perpendicular to the surface, and therefore it is only at the equator, that this is exactly in an opposite direction to the centrifugal force. At L for instance, if L *a* be taken as in the direction, and as also representing the quantity of centrifugal force there, it may be resolved in two directions *a b*, L *b*, one of which only, *a b*, is directly opposed to the force of gravity, while the other is perpendicular to the direction of gravity, and is a tangent to the surface, and does not diminish the force of gravity. Hence it follows, that in the equatorial regions, where the directions of gravity and of the centrifugal force are both perpendicular to the surface, the two forces become altogether opposed to each other, and the whole amount of the centrifugal force of rotation takes effect in diminution of the force of gravity. But in parts distant from the equator (as at L), a portion only of the centrifugal force (viz. *a b*), and that in its diminished state, acts in opposition to the force of gravity; the rest of it (viz. L *b*) is in the direction of a tangent, and tends towards the equator.

In order, therefore, that equilibrium may now be preserved among the component particles of the earth, a great accumulation of particles takes place in the equatorial regions, which by their number compensate their deficiency in gravitating force. And this effect is increased by that part of the centrifugal force acting between the poles and the equator, which is in a tangent direction to the earth's surface, and which tends to thrust down the particles on which it acts towards the equator. Hence the equatorial regions are elevated above the polar, and the height of all other intermediate parts is in some proportion of the distance of those parts from the equator. This is the alteration produced in the figure of equilibrium, by the diurnal rotation of the earth upon its axis.

But the earth being neither altogether fluid, nor of the same density throughout, we must introduce some qualifications into the result we have just arrived at. The earth being partly solid, the particles of which it is so far composed do not move easily among themselves, but have an attraction of cohesion which opposes a certain resistance to the operation of the centrifugal force caused by its motion of rotation. This neutralizes and destroys part of the centrifugal force, and makes the earth's ellipticity to be less than it would be if the earth were altogether fluid. But the centrifugal force is not *altogether* destroyed by the attraction of cohesion; for it must have elevated even the solid parts of the equatorial regions; were it not so, the waters of the ocean, not being restrained in their motion by the same attraction of cohesion, would all have set towards the equator in order to restore the equilibrium which, by the diminution of the centripetal force of gravity there, had been disturbed, and would thus have overflowed the land at the equator and left the polar regions dry. It may here be remarked, that the two constant currents in the sea, which are observed to set from both poles towards the equator, may perhaps be accounted for, by the action of that part of the centrifugal force which is in the direction of a tangent to the earth's surface and towards the equator. The waters of the sea, having no attraction of cohesion, would obey the impulse of this force freely, as it is not opposed to the force of gravity. The *westerly* set of the same currents may be ascribed to their continually advancing into regions which have a greater *easterly* motion of rotation.

The effect produced by the varying density of the earth, which increases towards the centre, has been proved by Clairaut, a celebrated French mathematician, to be a diminution of the oblateness of the earth, so that from this cause the height of the equatorial regions is somewhat less than on the supposition of an equal density; which is contrary to what Newton supposed would be the effect of an increase of density towards the centre.

The result of Newton's inquiry into the figure of the earth was, that the equatorial diameter of the earth is to the polar as 230 : 229; from this ratio the ellipticity of the earth would be expressed by the fraction $\frac{1}{229}$; or the polar diameter would be less than the equatorial by the 229th part of the whole, and the equatorial regions would be about 17 miles higher than the polar.

A very simple mathematical process will enable us to exhibit the value of the centrifugal force at any point of the earth's surface.

Let C (*fig.* 10), be the centre of the earth, P *p* the polar axis, E Q the equator, L a particle acted upon by the centrifugal force at any latitude L Q or λ; from L let fall L *d* perpendicular to the polar axis, L *d* will be the radius of the circle of rotation at latitude λ, and C Q will be that of the equatorial circle of rotation, and is the same with the radius of the earth.

Now the whole amount of centrifugal force varies as the velocity; for motion or velocity is the producing cause of it, and the velocity of rotation of different parts of the earth's surface varies as the circle of rotation, and therefore varies as the radius of that circle:—hence the centrifugal force at any point L, varies as L *d*, which is the sine of P L, or the complement of the latitude, or the *entire* centrifugal force varies as the cosine of the latitude.

Take L *a* to represent the *whole* centrifugal force at L, resolve this in directions L *b*, *a b*, perpendicular and parallel to that of gravity: then *a b* is the only part of the centrifugal force which directly opposes the force of gravity. From L let fall L *l*, perpendicular to C Q, then L *a b* and L C *l*, are similar triangles, and *a b* : L *a* : : *c l* : C L, so that *a b*, or the centrifugal force opposed to gravity, is equal to $La \times \frac{cl}{CL} = La \times$ cos. λ = whole centrifugal force at L × cos. λ. Now the whole centrifugal force we have shown to vary as the cos. of the latitude; therefore whole centrifugal force at L : centrifugal force at the equator : : cos. λ : 1, for at the equator the latitude = *o* and cos. *o* = radius; but the centrifugal force at the equator is found to be $\frac{1}{289}$ part of the force of gravity there, or is equal to $\frac{g}{289}$, if *g* represent the force of gravity at the equator. Hence the whole centrifugal force at $L = \frac{g}{289} \cos. \lambda$, and substituting this in the equation for the centrifugal force at L, opposed to gravity, which we may call F, $F = \frac{g}{289} \cos.^2 \lambda$. and varies as $\cos.^2 \lambda$.

The value of the other part of the force L *b*, which acts in a tangent to the surface, may now be easily found, for $Lb = \sqrt{La^2 - ab^2} = \sqrt{(\frac{g}{289} \cos. \lambda)^2 - (\frac{g}{289})^2 \cos.^4 \lambda.} = \frac{g}{289} \cos. \lambda. \sqrt{1 - \cos.^2 \lambda.} = \frac{g}{289} \cos. \lambda. \sin. \lambda.$

Chapter IX.

Oblate Figure of Jupiter, Saturn and Mars—Pendulum Experiments.

In the foregoing chapter we have explained the causes which have produced the earth's ellipticity. We now proceed to the various evidences of this fact which have been derived from observation and experiment.

As the same causes must, under similar circumstances, produce similar effects, it was just to suppose that if the reasoning by which the earth's ellipticity is established be correct, the other planets of the solar system would exhibit the like appearance of a flattening at their poles and a bulging out of their equatorial parts; for their component particles are under similar circumstances of mutual attraction, of equilibrium, and of rotation about an axis*. This was the rather to be looked for, because of the analogy or resemblance which is to be traced in the principal features of the solar system. Thus, among other analogies, the orbits of all the planets revolving round the sun are elliptical; the squares of their periodic times, or the times of their revolutions, are in proportion to the cubes of their mean distances from the sun, and the forces by which they gravitate towards the sun are inversely as the squares of those distances. The supposition of an ellipticity in the planets similar to that which is observed in the earth, was first verified in the planet Jupiter. This planet completes his

* We must except from the generality of this remark the planets Uranus, or Georgium Sidus, which is too distant, and Juno, Vesta, Ceres, and Pallas, which are too small to admit of any observations by which to ascertain whether they revolve about an axis or not.

daily rotation about his axis in 9 hours and 56 minutes; the centrifugal force exerted upon him is therefore considerably greater than that which affects the earth; and the density of Jupiter is less than that of the earth. Hence the figure of Jupiter ought to be much more oblate or compressed at the poles than the earth is. This was found to be the case by an astronomical admeasurement of his diameters. The equatorial diameter of Jupiter is to the polar as 13 : 12, and is longer than the polar by about 6230 miles. Saturn and Mars exhibit the same oblate-spheroidal appearance. The other planets do not offer facilities for ascertaining the effect of a centrifugal force upon them. With respect to the moon, her motion about her axis is too slow (being performed in 29½ days) to produce enough centrifugal force to make a difference in her diameters, arising from this cause, observable. She has, it is true, a spheroidal form, but this is owing to the attraction of the earth, which is four times greater than what is caused by her motion of rotation.

The horizontal parallaxes of the moon furnish another though subordinate proof of the earth's spheroidal form. If the earth were a perfect sphere, these parallaxes would be the same for all places upon the earth's surface—if a spheroid, they would be different at different places; and this is actually the case; so that the same heavenly body which by her eclipses indicates the earth to be round, by her parallaxes shows it to be not truly spherical.

But the more direct evidence of the existence of a centrifugal force and of the earth's ellipticity, and the means of determining the amount of it, are to be looked for in the experiments made with the pendulum, and in the measurement and comparison of the lengths of degrees of the meridian in different latitudes.

The nature of the evidence afforded by pendulum experiments may be explained by a reference to a few very obvious principles.

The centrifugal force caused by the earth's rotation is (we have seen) greatest at the equator and decreases towards the poles; this centrifugal force, either in its whole quantity or in part, acts in a direction opposite to that of gravity, and therefore, being greatest at the equator, it diminishes the force of gravity at the equator by the greatest quantity. Hence, bodies are lightest at the equator, and their weight gradually increases as we proceed towards the poles, where it is greatest. The pendulum of a clock performs its vibrations, or swinging motion to and fro, by being acted upon by the force of gravity.

Fig. 11.

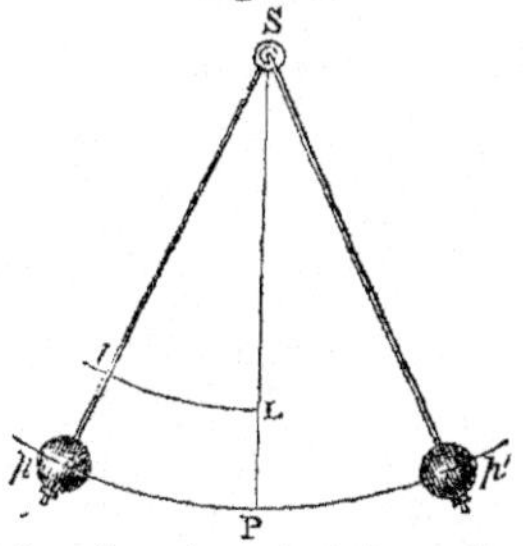

If, therefore, two pendulums S p, S p' (*fig.* 11), be taken of the same length, and the same substance or density, and be hung from the same point S, and S P be the vertical position of both, and they be made to fall from equal distances from P, it is evident they will move through the equal spaces P p and P p' in exactly the same time, because the force which causes their descent (namely that of gravity) is equal in both cases. But if one of these pendulums (S p) be made to swing at the equator (being let fall from the same height as before), and the other (S p') at some place between the equator and the poles (say at Paris), the time of S p arriving at P will be longer than the time in which S p' will move through the same space; because the force of gravity, which causes and accelerates the motions of both, is less at the equator than at Paris. But if we shorten the equator pendulum by some determinate quantity, it may be made to perform its vibrations in the same time with the Paris pendulum; for it will then have to describe a similar arc of a smaller circle or a less space, and a less accelerating force will enable it to describe this space in the same time. Such is the effect of the centrifugal force upon the vibrations of the pendulum in different latitudes. The existence of it was first detected by Richter, a French astronomer, who, in the year 1672, was sent to make astronomical observations in the island of Cayenne, which is not quite 5° north of the equator. Sir Isaac Newton has, in his Principia, described the particulars of the discovery. He says, that when Richter was, in the month of August, observing the transits of the fixed stars over the meridian, he found his clock to go slower than it ought, in respect of the mean motion of the sun, at the rate of 2′ 28″ a day. Therefore, setting up a simple pendulum to vibrate in seconds, which were measured by an excellent clock, he observed the length of that simple pendulum; and this he did over and over every week for ten months together; and upon his return to France, comparing the length of that pendulum with the length of the pendulum beating seconds at Paris, he found it shorter by $1\frac{1}{4}$ line.* In accounting for this difference in the length of the two pendulums, Newton allowed $\frac{1}{8}$th of a line as the utmost that could be attributed to the extension of the pendulum by the heat of the climate; the difference, or $1\frac{1}{18}$ line by which this pendulum was shorter than the Paris one, was made necessary by the less gravity of bodies at and near the equator. From this fact he obtained the same conclusions he had before deduced from theory alone, namely, that the equatorial diameter of the earth was greater than the polar by the 229th part of the whole diameter. Since that time observations upon the lengths of pendulums beating seconds in different latitudes, have been made with great assiduity by scientific men of all countries; but recent experiments tend to show that the earth's ellipticity is not so great as $\frac{1}{229}$; the fraction $\frac{1}{288}$ is the value which results from the latest investigations.

CHAPTER X.

Length and Measurement of Degrees upon the Earth's Surface.

THE remaining evidence of the earth's ellipticity is the different lengths of degrees of the meridian arc in different latitudes.

A degree of a meridian is that portion of it which must be travelled over, in order to change the altitude of any particular star, by the 360th part of the imaginary meridian circle in the heavens: if the spaces travelled over in different parts of the same terrestrial meridian, in order to produce this change in the altitude of a star, be not equal to one another, the terrestrial meridian cannot have the same curvature in every part, and is therefore not a circle; and consequently, the figure of the earth on the surface of which the meridian is traced cannot be a perfect sphere. Now it has been found by trial, that to raise the pole star by a quantity equal to a celestial degree, an observer must travel over a greater and increasing space as he proceeds from the equator to the pole. Hence it follows, that the degrees of a meridian line on the earth, or degrees of latitude, gradually increase from the equator to the pole; the meridian has, therefore, less curvature at the poles than at the equator, and the earth upon which it is traced is not a perfect sphere, but is flattened at the poles.

It is not to be immediately concluded from this that the earth is a regular oblate-spheroid; but it has been justly remarked, that, though it is only by experiment that the true figure of the meridian can be discovered, it has been found necessary to assume hypothetically (or by way of supposition), that its figure is the curve next in simplicity to the circle, viz. the ellipse, and also to suppose that the earth is a spheroid generated by the revolution of this ellipse about its shorter axis; for, in many complex cases, this mode of getting near the truth by probable suppositions, has been found the simplest and most convenient to be pursued; the only caution to be observed, is to submit the supposition first made to the test and correction of actual experiment. This caution has been carefully attended to in the matter we are discussing, by the measurement and comparison of degrees at various parts of the earth's surface.

In the measurement of a degree or of an arc of a meridian, many difficulties present themselves in the way of an *actual* and *mechanical* measurement. The general features of a country are such as to make any attempt of this kind unadvisable; a great number of almost conjectural allowances must be admitted into such a plan of operations, which forbid our placing much confidence in the result. The first modern measurement, having any just claim to accuracy, was, however, made in this manner. This was the measurement by Norwood, in 1635. The arc measured was that part of the meridian which lies between London and York. The difference of the latitudes of these cities was first ascertained; this gave the

* A line is a small French measure equal to the twelfth part of an inch.

number of degrees in the arc to be measured; the distance between the two cities was then actually measured; and the turnings and windings of the road, and the ascents and descents, were allowed for afterwards. The length of a degree thus determined was 122,399 English yards; which, notwithstanding the extreme liability of this method to error, is not very far from the truth; according to the latest determinations, the length of a degree between these latitudes is 121,660 yards. The only other instance of the *actual* measurement of an arc of the meridian is that of Messrs. Mason and Dixon. They measured an arc of the meridian of 179,359.313 English yards in length, in the state of Pennsylvania. An account of this measurement is given in the Philosophical Transactions for the year 1768. The other and more accurate mode of finding the length of a degree, is a combination of actual measurement and of trigonometrical calculations founded upon it. All geodesical* operations (as they are called) are now conducted according to this method. Two places are selected which lie under the same meridian, or nearly so; the difference of their latitudes, which gives the number of degrees in the arc to be measured, is then ascertained with the utmost precision. A base line of a few miles in extent, and at some little distance from the meridian arc, is then very carefully measured; this is the only *actual* measurement which need be made. The extremities of this base line are connected with the extremities of the meridian arc, by imaginary triangles; the sides of which are not measured, but, by the aid of the base line, and by means of the angles of the triangles, which are all ascertained by an instrument for measuring angles, are determined by trigonometrical computation. This mode of ascertaining the length of the meridian will, however, be set in a clearer light by following the steps of the process in the subjoined figure.

Let (*fig.* 12) M *m* represent a meridian arc; the difference of latitudes of the two extremities, M and *m*, being found, the length of a degree in the latitude of M and *m* will be the length of the whole arc divided by the number of degrees contained in it.

A level plain is then to be selected, on which a base line A B is measured; the two extremities of this line are to

Fig. 12.

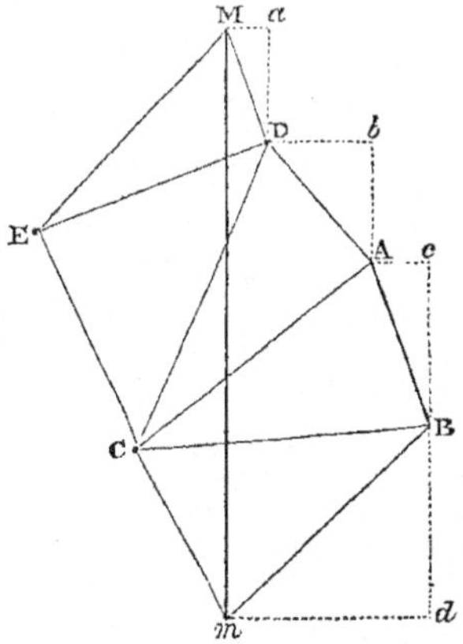

be connected with the two extremities of the arc M *m*, by a series of triangles. For this purpose convenient stations are fixed upon, such that the three stations situate in the three angles of every triangle may be visible to each other. Let C, D, E, be the stations fixed upon, these are supposed to be connected together, and with the points M, *m*, by the imaginary lines which form the various triangles ABC, BC *m*, ACD, CDE, and DEM. Then the angles by which the two stations C and B, appear to be separated from each other when viewed from the station A, is observed. This observation gives the angle C A B of the triangle A B C; the other angles of this triangle are observed and determined in the same manner, and the side A B, which is the base line, being known from measurement, the other two sides A C, B C, may be computed by plane trigonometry. By this means we obtain a side of each of the triangles B C *m*, A C D, and are enabled to continue the process without measuring any more sides. The angles of these triangles are measured as in the case of the first, and their sides are ascertained in like manner by trigonometry; and by proceeding in a similar way in the resolution of the whole series of triangles, the sides and angles of all are determined. The remaining step in the field proceedings is to ascertain the inclination of the lines M D and B *m*, to the meridian arc; astronomy affords the means of doing this. From the data

* From two Greek words, which combined, signify a dividing or apportioning of the earth.

furnished by these operations, the length of the arc M *m* is determined in the following manner:—from M and D, draw the lines M *a* perpendicular to D *a*, parallel with the meridian line, meeting each other in *a*; D *b*, A *b*, A *c*, B *c*, *m d*, B *d*, are also drawn so as to be respectively perpendicular to and parallel with the meridian. Then it is evident that the length of M *m* is equal to the sum of the lengths of *a* D, *b* A, *c* B, B *d*, which are found thus:—the inclination of M D to the meridian having been already determined by an astronomical observation, the angle D M *a* in the right-angled triangle D M *a* is known from it, and the side M D is also known, so that D *a* (which is equal to M D × sin. D M *a*) may at once be computed by trigonometrical tables. In a similar manner the sides *b* A, *c* B, *d* B are computed, and the sum of the whole gives the length of the meridian arc M *m*, and the length of a degree is the length of the whole arc divided by the number of degrees contained in it.

Picard was the first person who measured an arc of the meridian by this method. The operation was performed in the year 1670; the arc commenced near Paris, and extended northward; the result of the measurement gave, as the length of a degree in latitude 49½°, 121,627 yards, which differs only 35 yards from what is now considered as the most exact length; an accuracy which is justly supposed to be quite accidental.

Since this period arcs of meridian lines have been measured in various countries, as well in intermediate latitudes between the equator and the north pole, as near both the equator and the pole. The following table represents the length of a degree in different latitudes as determined by the five most approved measurements:—

	Latitude.	Degrees in Toises.	Degrees in Fathoms.	Country.
1	0° 0′ 0″	56749	60480.2	Peru.
2	11 0 0	56755	60486.6	India.
3	45 0 0	57011	60759.1	France.
4	52 2 2	57074	60826.6	England.
5	66 20 10	57192	60952.4	Lapland.

The following particulars will show at once the accuracy which now distinguishes geodesical operations, and some of the means taken to ensure it:—

The first base in the English measurement, of which the result is given in the above table of degrees, was about five miles in length, and was measured upon Hounslow-heath with a steel chain of exquisite workmanship. The same base had been measured three years before by General Roy, with glass rods, and the two measurements (in a length of five miles) differed only 2¾ inches. The French base was measured with rods of platina, that in Lapland with rods of iron, and an allowance was made for the changes of temperature affecting the length of the rods in the course of the operation. In a previous measurement in Lapland, the French astronomers, in order to guard against the extreme contracting effect of cold upon metals, employed rods of deal; this was the more necessary in that measurement, as it was performed in the depth of winter, and the frozen surface of a river was selected for the base line, with a view to obtain as level a plain as possible. It is usual also, in order to prove the correctness of the geodesical process, to measure, towards the conclusion, what is called a base of verification. We have already stated that all the sides of the series of triangles (with the exception of the base line A B, which is a side in the first triangle) are not measured but computed: to verify all the previous steps in the process, the length of one of the sides of the triangles, as it has been deduced from computation, is compared with its length determined by actual measurement. The side of the triangle thus measured is called a base of verification, and is taken as far distant from the first base as circumstances will admit. In the French operations the base of verification was distant between four and 500 miles from the first base, and was 7 miles in length, and yet the difference between its computed length and that obtained from its actual measurement did not amount to 12 inches.

From an inspection of the table before given, it appears that the length of a degree from the equator to the pole increases—the curvature therefore diminishes, and the earth is not a sphere but is flattened at the poles, and the polar diameter is less than the equatorial; and although the various modern measurements may not, on a comparison one with another, agree in giving to the difference of the two diameters precisely the same value, yet they all ascertain the fact of the polar diameter being less than the equatorial, and that a degree increases towards the poles; and this establishes the oblate-spheroidal figure of the earth.

The value of the compression or the fraction expressing the difference between the two diameters, as deduced from a comparison of the lengths of a meridional degree in different latitudes, determined by the most approved measurements, has been lately shown by Professor Airy, in a paper in the last volume of the *Philosophical Transactions* to be $\frac{1}{278.6}$, that is, the polar diameter is less than the equatorial by the 278.6th part of the whole diameter.

Operations are now being carried on, on the continent, which have for their object the more precise determination of the fraction of ellipticity, and of the compression of the earth. The measurement of an arc of the parallel of latitude 45°, of 15° or 16° in extent has been already accomplished. One extremity of this arc is at Marennes, on the west coast of France, and a little to the north of the Garonne, and traversing France, Piedmont, and the northern parts of Italy, its other extremity is at Fiume, in the Austrian dominions, and on the eastern shores of the Adriatic. The value of the ellipticity as deduced from these operations is $\frac{1}{282.8}$. We have already stated that the pendulum experiments give $\frac{1}{288}$. This similarity in the results afforded by such very different kinds of investigation is a strong argument in favour of the general correctness of both.

The mean degree of a meridian or the degree the length of which is as much greater than that of a degree at the equator, as it is less than that of a degree at the poles, is in latitude 45°, which is the mean latitude between the equator and the poles. Its length, according to the French measurement, is 60759.4 fathoms, or 12158.8 yards. The circumference of the elliptic meridian is found by multiplying the mean degree by 360, and is equal to 24855.84 miles. The circumference of the equator is 24896.16 miles, and is not quite 41 miles longer than the elliptic meridian.

The French measurement, in 1792, was undertaken with a view to obtain a standard measure of length, to serve as the basis of a new system of weights and measures. According to this new system, the unit, or first element of linear measure, is called a metre; and the metre was declared to be equal to ten millionth part of the quadrant of the meridian—which is a fixed and unalterable quantity in nature. The quadrant of the meridian was by this measurement found to be 5,130,740 toises, or 10,936,578 English yards: the French metre, or the ten millionth part of this quantity, would accordingly be 1.093578 yards, or 39.37 inches, nearly. This method of obtaining a standard of measure is not, perhaps, so good as that which consists in observing the length of the pendulum, which, in a certain latitude, beats seconds of mean time. For the length of this pendulum is ultimately ascertained by a reference to the equable motion of the earth upon her axis, and is, therefore, ascertainable without the aid and use of any linear measure whatever; whereas, in the very act of determining the French standard, or the quadrant of the meridian, some linear measure already in use must be employed; and thus the very basis of their new system is expressed in terms of that in the place of which it is substituted.

The importance of possessing the true length of a degree of the meridian, is not confined to investigations having for their object the determination of the figure of the earth. Upon the simple fact of the length of a degree, seemed to depend the overthrow or establishment of the theory of Universal Gravitation. The particulars connected with the discovery of a principle productive of such various effects in nature, is not the less interesting in that it illustrates the secret dependency of parts of science apparently the most distinct, and the assistance which each in its place is calculated to afford to the rest.

The corner-stone of the whole system of Universal Gravitation is, that the force which causes a heavy body to descend to the surface of the earth, is the same that retains the moon in her orbit, and makes her deflect from a straight line, or bend towards the earth. All that was requisite to establish the identity of the forces by which these two effects were produced, was to prove, that the quantity of effect produced in a certain time upon the moon in thus deflecting from a straight line, (taking into consideration the law by which the force varied, and *the distance of the moon*,) was in due proportion to the effect produced by the *force of gravity*, in the same time, upon a falling body at the surface of the earth. It is evident, therefore, that the determination of this question depended upon, and would in its solution be affected by, the distance of the moon from the earth. This distance being expressed only in a number of radii of the earth (about 60), it was necessary to ascertain the length of the earth's radius. This could only be done by means of the proportion which the radius of a circle always bears to the circumference; and the length of the circumference being 360 times that of a degree, the whole matter at last resolved itself into the geodesical operation of accurately measuring a degree upon the earth's surface. The only measure which in 1666, the time of Newton's first taking up the subject, was in existence, was that of Norwood's: this exceeded the true length of a degree by little less than 1000 yards; and as this error would be greatly multiplied in each step of the process, it is not surprising that Newton, whether he used this measure, or the still more incorrect one of 60 miles to a degree, could not reconcile the two phenomena of the falling stone and the revolving moon, so as to refer both to the same cause—namely, the attractive force of the earth. The consequence of this error in the then received length of a degree was, that for many years Newton laid aside his theory of universal gravitation. But in 1670, the measurement of an arc of the meridian, by Picard, took place; by mere accident the length of a degree, in latitude 49½, was then ascertained to within 35 yards of what is now considered the true length. This new measure brought Newton back to his favourite hypothesis. He then satisfactorily proved, that the force of gravity, and the force by which the moon is retained in her orbit, are one and the same. It is related, that towards the end of the calculation, and when he perceived its probable successful issue, he became so much agitated, as to be obliged to request a friend to assist him in completing it. Thus, by the aid of the true length of a degree, was finally established the grand theory of Universal Gravitation.

Chapter XI.

Books.

The subjects embraced in the foregoing treatise, are dispersed throughout a great number of different books, and are to be met with only in detached parts.

The proofs of the spherical figure of the earth, and the methods of finding the latitudes and longitudes of places, will be found in every Treatise of Astronomy; we shall, therefore, only refer to that part of Malte-Brun's work, which is devoted to Mathematical Geography; to the Nautical Almanack; Woodhouse's Astronomy, vol. i.. chapters 1, 5, 42 and 43; Brinkley's Elements of Astronomy, chapters 1, 3, 16 and 17; Playfair's Outlines of Natural Philosophy, vol. ii. part 1, chapters 1 and 4; and, as a popular work, to Bonnycastle's Astronomy, letters 2, 9 and 10.

For fuller information, with respect to the true figure of the earth, and the lengths of pendulums vibrating seconds in different latitudes, and measurement and lengths of degrees, we may refer to Malte-Brun; Brinkley, chap. 17; Playfair, chap. 3, of part 1, and chap. 6, of part 2; Bonnycastle, letters 15 and 16; Newton's Principia, book 3, props. 18, 19 and 20; Maclaurin's Account of Sir I. Newton's Discoveries, book 4, chap. 6; Pemberton's View of Sir I. Newton's Philosophy, book 2, chap. 6; Rees' Cyclopædia, articles, 'Earth,' and 'Degree;' various Papers in the Philosophical Transactions on the Measurement of Degrees, and on Experiments upon the Pendulum; Clairaut Figure de la Terre; Quarterly Journal of Science, for March 1827, p. 177.

PHYSICAL GEOGRAPHY.

Physical or *natural* geography might, if we regarded merely the strict meaning of the words, be limited to signify no more than a description of the principal features of the earth's surface; but it is usual, in treatises upon this branch of geography, to touch also upon the subject of climate and temperature,—to show how these, together with other natural causes, affect the condition of the human race—and to advert, in a general manner, to the animals and productions of the globe.

Geographical terms explained.

In looking over a map of the world, it is seen at once that the surface consists of various spaces of *land*, surrounded by an extensive field of water called the *sea* or *ocean*. Of these spaces of land, two are of vast extent, and on this account are termed *continents* *, (derived from a Latin word signifying, *holding together* or *connexion*). The larger of these continents includes the three divisions of Europe, Asia, and Africa, and is distinguished by the title of the *old continent*, from its having, till the discovery of America, by Columbus in the year 1492, been the only one with the existence of which Europeans were acquainted. The other, which includes North and South America, is named the *new continent*. The smaller portions of land which are scattered over the ocean are denominated *islands*. A great many islands lying together are called an *archipelago*.

In many places the land and the ocean run one into the other. When the ocean penetrates into a continent by a narrow passage, and then spreads again into a large expanse, this inland portion of the ocean is usually termed a *sea*. If the extent of such an inland sea be less, or the passage by which it communicates with the main ocean larger, it is called a *gulf* or *bay*. An inland body of water not connected with the ocean or any of its branches, is called a *lake*. A narrow passage of water leading from one sea to another is called a *strait*; a narrow neck of land lying between two seas, and connecting two masses of land greater than itself, is called an *isthmus*. When, on the other hand, a part of a continent runs out into the sea, and is joined to the main land by only a small portion of its circumference, it is named a *peninsula*, (that is, an *almost island*). If the projections of land reach but a little way into the sea, they are called *capes*, *headlands*, or *promontories*.

General View of the Globe as consisting of Land and Sea.

There is, in fact, only one continuous fluid surrounding the land, all the gulfs and inland seas being branches † of this universal ocean; but for the sake of convenience different parts of it have distinct names given to them. The following table, exhibiting the principal seas into which the ocean has been divided, will be clearly understood upon referring to the map of the world on Mercator's projection:—

* New Holland, by some geographers, is regarded as a third continent; but if we consider how much smaller it is than either of the two vast tracts above-mentioned, it will appear correct rather to assign it the first station among the *islands* of the globe. New Holland and the islands around it are, however, not unworthy of being classed as the fifth grand division of the world. English geographers have named them *Australia* (that is, *Southern lands*.)

† The Caspian *Sea*, as it is generally termed, forms no exception to this remark, because it is in fact only an immense lake.

I. The great *South Eastern* basin, the waters of which cover nearly half the globe. It includes

1. The *Antarctic Ocean*, which is comprised within the Antarctic circle, that is, between the parallel of 66° 32′ of southern latitude and the South Pole.

2. The *Southern Ocean*, the boundary of which on one side is the Antarctic circle, on the other a line drawn from Cape Horn to the Cape of Good Hope, thence to Van Diemen's Land, and again by the south of New Zealand to Cape Horn. This line forms the southern boundary of Nos. 3 and 4.

3. The *Indian Ocean*, lying between Africa on the west, and the peninsula of Malaya with the islands of Sumatra, Java, &c., and New Holland, on the east, and bounded by Persia, and Hindustan on the north. The Red Sea, or Arabian Gulf, the Persian Gulf, and the Bay of Bengal are all parts of this ocean.

4. The *Pacific Ocean*, divided by the equator into *North* and *South*, and inclosed between America on the east, and New Holland, the islands of Java and Sumatra, and the continent of Asia, on the west. On the north it terminates at Behring's strait. The seas of China, Japan, Okhotsk, &c. form parts of this ocean.

II. The *Western* basin, forming a channel between the old and new continents.

1. The *Atlantic Ocean*, commencing in the south from a line drawn from Cape Horn to the Cape of Good Hope, and terminated on the north by the Arctic circle. It is divided into *North* and *South* by the equator, and its branches are the Mediterranean, the North Sea or German Ocean, the Baltic, Baffin's Bay, Hudson's Bay, the Gulf of Mexico and the Caribbean Sea.

2. The *Arctic Ocean*, surrounding the North Pole, and bounded by the Arctic circle and the northern shores of the two continents. The White Sea, the sea of Kara, and the Gulf of Obe are parts of it.

The Ocean is spread over nearly seven-tenths of the globe; but it is remarkable how unequally the land and water are distributed. If we look at a map of the world projected upon the horizon of London, in which map, consequently, London forms the centre of the one hemisphere and the antipodes * to London, the centre of the other; the first hemisphere, it will be seen, contains a very large proportion of the whole of the land, while the second, if we except New Holland and the extremity of South America, from the twenty-ninth degree of south latitude, consists almost entirely of water. The distribution of water and land is still very unequal, if we compare only the northern and southern hemispheres, that is, the two equal parts into which the globe is divided by the equator. The following calculation will plainly exhibit this fact:

Considering the whole space included in the northern part of the torrid zone, as equal to 1, the proportion of land is	0.297
On the same supposition, the proportion of land in the northern temperate zone is	0.559
And in the northern icy zone	0.400
In the southern part of the torrid zone, the portion of land is	0.312
In the southern temperate zone	0.075
In the southern icy zone (supposed)	none.

In other words, if the quantity of land in the northern hemisphere be represented by 16, the quantity in the southern will be scarcely equal to 5.

About the middle of the last century it was asserted that a great continent must exist towards the south pole, in order to counterbalance the mass of land in the northern hemisphere; but by the voyages of Cook and others, it has been proved that the high southern latitudes contain only a few islands.—The absence of a continent near the south pole does not of itself prove that there is less land there than in the north, since it is possible that the land in general may be only rather more depressed in the south, the necessary result of which would be, that the ocean would spread itself more extensively over the surface of the earth in that quarter.

* A small island lying to the south-east of New Zealand, and called Antipodes island, is very nearly the antipodes to London.

On the Figure, &c., of the Continents.

The general direction of the land in the two continents is entirely different. In America, it is from pole to pole; in the old world, it is from south-west to north-east, and, if we keep Africa out of view, it is almost parallel to the equator. The longest straight line that can be drawn on the old continent commences on the western coast of Africa, from about Cape Verd, and extends to Behring's strait in the north-east of Asia. It is about 11,000 miles in length. A similar line, traced along the new continent, passes from the strait of Terra del Fuego, to the northern shore of North America, and is nearly 9000 miles long. In both continents the direction of the large peninsulas is similar, almost all of them running towards the south. This is the case with South America, California, Florida, Alaska, and Greenland in the New World, and in the Old with Scandinavia, Spain, Italy, Greece, Africa, Arabia, Hindustan, Malaya, Cambodia, Corea, and Kamtchatka. The only exceptions to this remark, are the peninsula of Yucatan in Mexico, and that of Jutland in the north-west of Europe. Both of these are directed towards the north; but they consist of plains and alluvial land, whereas the other peninsulas are more or less of a mountainous character. There is a further resemblance between the two continents, from each being divided into two parts by an isthmus *; but in the character of their outlines they differ very much: for while the coast of the Old World (independent of Africa) is broken equally on all sides by gulfs, bays, and inland seas, the New World has a series of openings on its eastern shore only. Of its western side, the only inlet of any magnitude is the gulf of California.

On Mountains.

Mountains are the most considerable elevations of the surface of the earth. They may be divided into two classes: those of which the chains are the most lofty, rugged, and extensive, such as the Andes, the mountains of central Asia, the Alps, &c., and those of a less majestic character, which frequently form branches as it were of the first class. The Apennines, which traverse the whole length of Italy, and the Carpathian range, which in a great measure surrounds Hungary, are of this second description. All elevations which are either of such an inferior kind, or take place in such a gradual manner as not to come within these two classes, are termed *hills* or *slopes*.

Mountains most commonly are so near to each other, and are disposed in such a manner, as to give the idea of *chains*. A chain may be defined as a series of mountains, the bases of which are continuous; but it is well to observe that the name is sometimes applied to collections of hills without much regard to its strict meaning. Sometimes chains run out from a common centre: those which proceed from the high table-land of central Asia, may be considered as an example; at others the centre mass itself is a lofty chain to which secondary chains are attached: such are the Alps. In some instances there are irregular groups of several chains, among which none can be ranked as the principal: of this the mountains in Asia Minor and Persia are examples. The most remarkable, however, are long connected chains, which, like the Andes †, continue for several thousand miles, nearly in one direction, having on both sides inferior ranges, but sending off hardly any secondary chains. These appear to be of the highest antiquity.

Some mountains are completely insulated, that is, are quite remote from any chain or group. Volcanoes are more particularly of this kind.

The character of mountains would seem to depend upon the sort of rock of which they are composed. Granite, when exposed, forms lofty and rugged elevations; gneiss is much less precipitous, and slate commonly not at all so. In this respect there is a remarkable difference, which Humboldt has noticed, between the Old and New Continents. In the former the highest points of the Alps consist of granite; but in America, granite is not found higher than 11,000 or 12,000 feet above the level of the sea, and the newest flœtz trap or whinstone, which in Europe appears only in low mountains, or at the foot of those of great magnitude, covers the tops of the Andes. Chimborazo and Antisana are crowned with vast walls of porphyry; and basalt, which in our

* The isthmus of Suez is composed of sand; that of Panama or Darien consists of stupendous rocks.

† The name of *Andes* is given only to the chain on the west coast of *South* America; the continuation of that chain in North America has other titles; but they evidently form one grand whole.

continent has not been observed higher than 4300 feet, is on the very summit of Pichincha. Other secondary formations, among which may be mentioned limestone, are also found at greater heights in the New than in the Old world.

With respect to their declivities, it is observed that most of the principal mountains have one of their sides very steep, and the other of a gradual slope. The Alps, for instance, have a much more abrupt descent on the side of Italy than on that of Switzerland; the Pyrenees are steeper towards the south than the north, while the chain of Asturias, which branches westward from the Pyrenees, is just the reverse. Mount Taurus, in the part where it approaches the Mediterranean and the Dardanelles, is abrupt on the south, but in Armenia it has a rapid descent northward. The mountains of Scandinavia are steeper towards the west and north-west, than the south and east; and the Ghauts in Hindustan are in like manner precipitous on the west, and sloping in the opposite direction. With all these chains, therefore, and indeed with most of the chains of the globe, their steepest side is found to be that which approaches most nearly to the sea, and consequently their inclination is most gradual towards the interior of the country in which they are situated.

Mountains in their course commonly make numerous curves and angles; but in most cases the *general direction* of the principal chains appears almost, if not entirely, to correspond with the greatest length of the continents to which they belong. The Andes in South and the Stony Mountains in North America exemplify this remark; as do also the chains which, with little interruption, pass from the north-eastern point of Asia to the south-west coast of Portugal, and to the western side of Africa. The secondary chains, in the same manner, frequently follow the greatest length of the large peninsulas. This is the case with the Apennines in Italy, the Dovrefield in Scandinavia, the Ghauts in Hindustan, &c.

In order to obtain a connected view of the loftiest and most extensive system of mountains upon the globe, we must suppose ourselves placed in New Holland with our face turned towards the north; America will then be on the right, Asia and Africa on the left. From Cape Horn to Behring's strait along the western coast of America, there is an almost uninterrupted range of the highest mountains: from Behring's strait again succeeds an enormous line passing in a south-westerly direction through Asia, leaving China and Hindustan to the south, somewhat interrupted as it approaches Africa, but still to be looked upon as continuing its course in the mountains of Persia and Arabia Felix. From Cape Gardafui in Africa to the Cape of Good Hope, there appears to be a chain which completes the view. The series of mountains which we have thus followed, is in the form of an immense irregular curve, which comprises within it, the Pacific and Indian Oceans, with their innumerable islands, besides a portion of Asia, including China, the Birman dominions, and the Indian peninsula. It presents a steep face towards these oceans, while, on the other side, the land very generally slopes towards the Atlantic and Arctic Oceans.

The following is a table of the height of some of the principal mountains of the globe, reckoning from the level of the sea. The elevations of those in Asia and Africa are far from having been ascertained with accuracy. Some recent measurements make the highest summits of the Himalayan range as much as 28,000 feet; but, though these calculations seem very doubtful, it is not unlikely they are at least 25,000 feet high.

Europe.

		Feet.
Mont Blanc ✝	The Alps	15,668
Mont Rosa		15,527
Ortler Spitze in the Tyrol		15,430
Aiguille d'Argenture		13,389
Eiger		13,170
Jungfrauhorn		13,730
Shreckhorn		13,310
Wetterhorn		12,500
Pass of Great St. Bernard		7,968
Ditto of Mont Cenis		6,778
Ditto of the Simplon		6,580
Nethou ✝	Pyrenees	11,427
Perdu		11,275
Canigou		9,145
Monte Corno ✝ (Apennines)		9,523
Lomnitz ✝ (Carpathian)		7,962
Sneehaetta ✝ (Dovrefield in Norway)		8,122
Mulhacen † (Sierra Nevada of Grenada in Spain)		11,678
Mont Mezin ✝ (Cevennes in the south-east of France)		6,567
Puy de Sancy ✝ (Mountains of Auvergne in France)		6,215
Etna * in Sicily		10,870
Vesuvius * near Naples		3,932

Asia.

		Feet.
Dhawalagiri	Himalaya	28,077
Jewahir	Himalaya	25,747
Mowna Roa (Sandwich Is.)		15,988
Ophir (Sumatra)		13,840
Egmont (New Zealand)		11,430
Italitzkoi (Altaian chain)		10,735
Ararat (Armenia)		9,600
Olympus (Anatolia)		6,500
Awatsha * (Kamtchatka)		9,600
Lebanon (Palestine)		9,600

Africa.

Geesh	Abyssinia	15,000
Amid Amid	Abyssinia	13,000
Lamalmon	Abyssinia	11,200
Chain of Atlas †		11,980
Peak of Teneriffe *		12,180

America.

Chimborazo †	Andes	21,425
Cayambe	Andes	19,633
Antisana *	Andes	19,136
Cotopaxi *	Andes	18,867
Yliniza	Andes	17,376
Tolima	Andes	18,324
Cotocache	Andes	16,436
Pichincha *	Andes	15,931
Farm of Antisana (inhabited)	Andes	13,437
City of Quito	Andes	9,542
Popocatepetl *	Mexico	17,720
Iztaccihuatl	Mexico	15,705
Coffer of Perote	Mexico	13,275
Lake of Toluca	Mexico	12,195
Silla de Caraccas (Venezuela)		8,633
Duida (Mountains of Parime)		8,314
Itacolumi (Brazil)		5,756
Mount Washington † (Alleghanies)		6,650
Blue Mountains (Jamaica)		7,278
Mount St. Elias		17,863
Mount Fairweather		14,736

(NOTE) Those marked * are volcanoes. And † are the highest points of the range to which they belong.

On Volcanoes.

The term *Volcano* (derived from VULCANUS, the name which the Romans gave to their imaginary god of fire) is applied to those mountains which send forth from their summits or sides, flame, smoke, ashes, and streams of melted matter called lava. Upon ascending to the top of a mountain of this kind, there is found to be an immense and deep hollow, which is denominated the *crater* or *cup*. From most of the volcanoes which are not extinct, there is a smoke more or less frequently arising; but the *eruptions*, which are discharges of stones, ashes, lava, &c., accompanied with lofty columns of fire, violent explosions and concussions of the earth, happen at irregular and sometimes very long intervals. It seems to be a very general rule that the greater the mass and the elevation of the mountain, the less frequent and more tremendous are the eruptions. Stromboli, the small volcano on one of the Lipari islands, is almost always burning; Vesuvius has more frequent eruptions than Etna; while the immense summits of the Andes, Cotopaxi, and Tunguraliua have an eruption hardly once in a century. The volcanoes of America, besides the common lava and rocks, &c., cast out scorified clay, carbon, sulphur, and water *, accompanied, in some instances, by fishes.

The eruption the most astonishing, perhaps, of any upon record is that which, in April 1815, issued from the Tomboro mountain in Sumbawa, one of the islands of the Indian Archipelago (see Notes to chap. i. of *Raffles' History of Java*); and we mention it, in order to give an idea of the violence which sometimes characterises volcanic agency. The tremulous motions which the ground underwent, during this eruption, were felt throughout a circular space of nearly 2000 miles in diameter, and the report of the explosions was heard over an equally extensive area. Within its more immediate range, embracing a space of 300 miles around it, it excited the greatest alarm. On Java, at the distance of 300 miles, the sky was overcast at midday with clouds of ashes, the showers of which covered every thing to the depth of several inches; and explosions were heard, at intervals, like the report of artillery or the noise of distant thunder. On Sumbawa itself thousands of individuals were destroyed by the fury of the eruption.

What may be termed *mud volcanoes*, from their having eruptions of mud only, are another curious phenomenon. One of these, which is situated towards the middle of the island of Java, in a plain abounding with salt springs, is thus described in the 9th volume of the *Batavian Transactions:* "On approaching it from a distance, it is first discovered by a large volume of smoke, rising and disappearing at intervals of a few seconds, resembling the vapours arising from a violent surf: a dull noise is heard, like that of distant thunder. Having advanced so near, that the vision was no longer impeded by the smoke, a large hemispherical mass was observed, consisting of black earth mixed with water, about 16 feet in diameter, rising to the height of

* Etna has been known to eject Water.

20 or 30 feet in a perfectly regular manner, and, as it were, pushed up by a force beneath—which suddenly exploded with a dull noise, and scattered about a volume of black mud in every direction. After an interval of two or three, or sometimes four or five, seconds, the hemispherical body of mud or earth rose and exploded again. In the same manner this volcanic ebullition goes on without interruption. The spot where it occurs is nearly circular and perfectly level; it is covered only with the earthy particles, impregnated with salt water, which are thrown up from below: the circumference may be estimated at about half a mile. A strong, pungent, sulphureous smell is perceived on standing near the explosion, and the mud recently thrown up is warmer than the surrounding atmosphere. During the rainy season these explosions are more violent."

The mountain of *Maccaluba* in Sicily, and some hills at the town of Taman in the Crimea, are also distinguished by eruptions of mud.

It is remarkable, that in the Old Continent, the principal chains of mountains contain no volcanoes, and that islands, and the extremities of peninsulas, are alone the seats of these convulsions; while in the New World, the immense range which runs along the shore of the Pacific Ocean possesses more volcanoes than are to be met with in the whole of the Old Continent and its adjacent islands. Professor Jameson has given the following estimate of the number of volcanoes:—

Continent of Europe	1*
Islands of ditto	12
Continent of Asia	8†
Islands of ditto	58
Continent of America	97
Islands of ditto	19

No volcano has yet been discovered on the continent of Africa, but most of its groups of islands are distinguished by them.

A line drawn round the Great Pacific Ocean, so as to include the long range of mountains on the west of America, the Asiatic peninsula of Kamtchatka, and the islands of Sumatra and Java, will have within it, by far the grandest and most extensive *volcanic system* on the globe. From Terra del Fuego, *(the land of fire,)* to the peninsula of Alaska, a complete series of volcanoes may be traced. The Aleutian islands, which stretch from that peninsula to the opposite peninsula of Kamtchatka, possess several. On Kamtchatka, there are some of great violence. The islands of Japan and Formosa have several, and beginning with Sumatra and Java, they are scattered all over that immense Archipelago which forms so remarkable a feature of the Pacific Ocean.

* Vesuvius.

† On the peninsula of Kamtchatka.

In the Indian Ocean, the Islands of St. Paul, Amsterdam, and Bourbon, have volcanoes in action. The most formidable volcanoes of the Mediterranean are Etna in Sicily, and Vesuvius upon the coast of Naples. Between these two mountains are the Lipari islands, all of volcanic character. The Atlantic Ocean contains several groups of this kind; Iceland has suffered frequently from the terrific eruptions of its volcanoes; the Azores and the Canaries, and some of the West India islands, also experience the effects of subterranean fire.

In some places, parts of the land which are covered by the waters of the ocean are the seats of volcanoes; and it has sometimes happened, that new islands have been formed during *submarine* eruptions. A recent instance of this kind occurred in 1811, in the neighbourhood of St. Michael, one of the Azores, the small group which lies about 800 miles to the west of Portugal. This new island has since disappeared. It is probable, that some clusters of islands (among which are the Azores, just mentioned, and the Lipari islands north of Sicily) owe their origin to the breaking out of submarine volcanoes.

Several mountains bear evident marks of having at some very distant period been the outlets of fires; and on this account, they are called *extinct* volcanoes.

On Valleys and Plains.

Valleys are the spaces lying between opposite ridges of mountains or of hills, and their lowest part is commonly the bed of some torrent or river, which originates in the higher grounds. Those between high mountains are in general narrow and long, resembling large clefts or fissures. In some of these valleys, among the Alps and Pyrenees, it has been observed, that the nooks or angles on each side correspond with such exactness to projections on the opposite side, that if it were possible to exert a force sufficient to bring their sides together, they would fit into each other so closely, that no trace of the opening would remain. Such valleys as these would seem to have been formed by some convulsion of nature. The narrow openings which are the entrance to the high valleys, are called *passes* or *defiles*, and these are often of the most gloomy and terrific aspect. Valleys, which are upon a lower level than the class just mentioned, are wider and more soft in their features, and gradually lose themselves in the plains. *Plains* are likewise of two kinds. Those which are extensive, but very elevated, come under the denomination of *table land*. There are several plains of this sort; but the most remarkable are those among the Andes*, those of Mexico, and the immense plains in central Asia, to the north and north-east of Hindustan. The great Himalayan and Altaian chains form the ramparts, as it were, of this extensive and desolate table-land, a large proportion of which is the desert of Gobi † or Shamo. The *low plains*, from the nature of their soil, seem formerly to have been covered by the sea. The large plain, to the south of the Baltic, is one out of several instances of this character.

On Islands.

Large islands exhibit, on a smaller scale, the same appearances as the continents: upon them, therefore, it is unnecessary to make any observations, but with respect to smaller islands, the circumstance of their commonly being in groups or chains, deserves attention. Some are banks of sand, just raised above the surface of the water. Many islands, especially those in the South Sea, owe their origin to the marine insects which produce the coral. Some groups, as has already been observed, appear to have been raised up by the action of submarine volcanoes. Since the bed of the ocean possesses as much variety of surface as the land, there is no doubt that groups or chains of islands very near to each other, and which have not been raised up by such processes as those just alluded to, are only the different summits of an extensive submarine system of mountains; and when these collections of islands lie close to mountains on shore, they may be considered as a continuation of the latter. The Aleutian isles, which run in a curve south of Behring's strait, connect in this manner the mountains of the New with those of the Old World. In some cases, where a chain of islands extends from one part of the shore of a main land to another, it would appear as if, at a remote period, the sea had overwhelmed a portion of the main land, leaving those spaces uncovered which now form the islands. This seems still more probable if the water on the land side of the chain be not very deep, or if the islands are of a lofty and mountainous character. It has been supposed that, among others, the West India islands and the Archipelago between New Holland and the opposite coast of Asia, were rendered insular by an incursion of the ocean having detached them from the continents to which (if this supposition be just) they formerly belonged.

On Springs and Rivers.

The origin of the numerous springs that break forth from beneath the earth's surface cannot be referred to one exclusive cause. The internal reservoirs by which they are supplied are, in many cases, derived from the water which the earth absorbs from rains and melted snow; from these reservoirs, wherever there is uneven or mountainous ground, the water flows out by minute fissures in the sides of the hills. But when we see springs rising up in plains, it is evident that they must have *ascended*, that is, travelled in a direction *contrary* to that produced by the force of gravity, in order to reach the surface. This, no doubt, is sometimes to be attributed to water flowing under ground from distant elevations, and to the natural tendency of a liquid to find its level. But the rising up of springs, in plains, cannot always be accounted for in this manner; and it has, therefore, been supposed, that the earth contains capillary tubes, the effect of which, in attracting liquids upwards, is explained in Chapter VIII., of the treatise upon Hydrostatics. It is also evident that such springs as suffer no diminution even from the longest continued dry weather, must be derived from a source quite independent of rains, and other external means of supply. They must, therefore, proceed from some vast body of water within the earth; and it has,

* The plains of Quito are 12,000 feet above the level of the sea.

† This word, which is usually but incorrectly written *Kobi* or *Cobi*, signifies a *naked desert*.

with apparent reason, been concluded, that many springs arise from the ocean filtering through the pores of the earth, the salt particles being lost in the passage.

Springs, which have their waters combined with mineral substances, and are, from that circumstance, called *mineral*, are very numerous, and of various kinds. *Warm* and *hot* springs are also common, especially in volcanic countries, where they are sometimes distinguished by violent ebullitions. Iceland is noted for these curious phenomena: its celebrated boiling fountain, the great *Geyser*, frequently throws out its contents to the height of more than a hundred feet, sometimes to twice that elevation.

Rivers are to be traced to springs, or to the gradual meltings of the ice and snow which perpetually cover the summits of all the most elevated ranges of mountains upon the globe. The union of various springs, or of these meltings, forms rivulets: these last follow the declivity of the ground, and commonly fall, at different stages, into one great channel, called a river, which, at last, discharges its waters into the sea, or some great inland lake. The declivities along which descend the various streams that flow into one particular river are called its *basin*, a term, therefore, which includes the whole extent of country from which the waters of the river are drawn. As mountainous regions abound in springs, we find that most rivers, more especially those of the first class, commence from a chain of mountains: each side of a chain also has its springs, and the rivers which originate on one side flow in the opposite direction to those which rise on the other. As it is the property of water to follow the most rapid descent that comes in its way, the courses of streams naturally point out the various declivities of the earth's surface, and the line from which large rivers flow in contrary directions will generally mark out the most elevated parts of the earth. In European Russia, where the rivers are very extensive, there is, however, a singular exception to this rule, the line which separates the sources of those rivers being very little above the level of the Baltic or of the Black sea. It has been observed by some writers, that the extent of a river is in proportion to the height of the range of mountains from which it descends. This is, in a certain degree, true, because the greater the bulk of the mountains, the more numerous the springs and torrents which they furnish; but the relation between the extent of a river and the surface of its basin is much closer and more invariable. Even this is not sufficiently comprehensive, for it is evident that the size of a river depends upon three circumstances—the surface of its basin—the abundance or otherwise of that surface in springs—and the degree of humidity possessed by the climate of the region from which it draws its supplies. As many springs, however, are formed by the rains, the second of these circumstances will, in some measure, vary with the last. By an attention to these remarks, the causes of the great size of the South American rivers will be apparent. The peculiar position of the Andes with respect to the plain of that continent,—the fact that by very far the largest proportion of its running waters are drained off in one general direction (towards the Atlantic)—the multiplicity of streams that intersect the country,—and the humidity of the climate, all contribute to that result. The Andes being placed so near the coast of the Pacific, the rivers which flow from them into that ocean are small; while those which flow on the other side, having such an immense space to traverse, are swelled into a most majestic volume before they reach the Atlantic. The physical circumstances of the old continent are unfavourable to the accumulation of such vast bodies of water as the rivers of South America. Europe is not of sufficient extent; Africa is oppressed by a scorching climate, and abounds in sandy deserts; in Asia the atmosphere generally is not so moist, while the more central position for the most part of the great mountainous range of that continent, and the existence of capacious inland lakes*, which are the final receptacles of the streams that fall into them, are the causes why the waters are more equally drained off in different directions than in the New World.

When water, by following a descent, has once received an impulse, the pressure of the particles behind upon those before will be sufficient to keep the stream in motion, even when there is no longer a declivity in the ground. The only effect is, that in passing along a

* The Caspian Sea and Lake Aral.

level, the course of the stream becomes gradually slower, an effect which may be perceived, more or less, in all running waters that originate in mountainous or hilly tracts, and afterwards traverse the plains. The declivity of many great rivers is much less than might at first be supposed. The Maranon or Amazons has a descent of only 10½ feet in 200 leagues of its course, that is, $\frac{1}{27}$th part of an inch for every 1000 feet of that distance. The Loire, in France, between Pouilly and Briare, falls one foot in 7,500, but between Briare and Orleans, only one foot in 13,596. Even the rapid Rhine has not a descent of more than four feet in a mile, between Schaffhausen and Strasburg, and of two feet between the latter place and Schenckenschantz. When rivers proceed through mountainous and rugged country, they frequently fall over precipices and form *cataracts*, in some cases several hundred feet in depth. The most celebrated falls in the world are those of the Niagara, in North America.

In the tropical regions most of the rivers are subject to periodical overflowings of their banks, in consequence of the rains which annually fall in such abundance in those countries during the wet season. The overflow of the Nile was considered by the ancients, who were ignorant of its cause, as one of the greatest mysteries of nature, because in Egypt, where the overflow takes place, no rain ever falls. The apparent mystery is easily explained, by the circumstance of the rains descending upon the mountains in the interior of Africa where the Nile rises. The consequent accumulation of the waters among the high grounds gradually swells the river along its whole extent, and in about two months from the commencement of the rains, occasions those yearly inundations, without which, Egypt would be no better than a desert.

The disappearance of some rivers for a certain distance under ground is accounted for with equal facility. When a river is impeded in its course by a bank of solid rock, and finds beneath it a bed of a softer soil, the waters wear away the latter, and thus make for themselves a subterraneous passage. In this way are explained the *sinking of the Rhone*, between Seyssel, and l'Ecluse, and the formation, in Virginia, of the magnificent rock bridge which overhangs the course of the Cedar creek. In Spain the phenomenon exhibited by the Guadiana, which has its waters dispersed in sandy and marshy grounds, whence they afterwards emerge in greater abundance, is to be referred to the absorbing power of the soil.

Rivers, in their junction with the sea, present several appearances worthy of notice. The opposition which takes place between the tide and their own currents occasions, in many instances, the collection at their mouths of banks of sand or mud, called *bars*, on account of the obstruction which they offer to navigation. Some streams rush with such force into the sea, that it is possible, for some distance, to distinguish their waters from those of the sea. The shock arising from the collision of the current of the majestic Amazons with the tide of the Atlantic is of the most tremendous description. Many of the largest rivers mingle with the sea by means of a single outlet, while others, for instance the Nile, the Ganges, the Volga, the Rhine, and the Orinoco, before their termination, divide into several branches*. This circumstance will depend upon the nature of the soil of the country through which a river runs; but it also frequently results from the velocity of the stream being so much diminished in its latter stage, that even a slight obstacle in the ground has power to change its course, and a number of channels are thus produced. Another cause may be assigned for the division into branches of those rivers, which, in tropical countries, periodically inundate the plains; the superfluous waters which, at those periods, spread over the country, find various outlets, which are afterwards rendered permanent by the deepening of the channels by each successive flood. In some of the sandy plains of the torrid zone the rivers divide into branches, and, from the nature of the soil and the heat of the climate, they are absorbed and evaporated, and thus never reach the sea.

On Lakes.

Lakes may be classed into *four* distinct kinds. The *first class* includes those which have no outlet, and which do not receive any running water. They are usually very small; some appear to be the craters of extinct volcanoes filled

* The triangular space formed by a river pouring itself into the sea by various mouths, is called a *Delta*, from its resemblance to the shape of the fourth letter (Δ) of the Greek alphabet.

with water. The *second class* are those which have an outlet, but which receive no running water. They have been formed by springs flowing into some large hollow: upon the water rising up to the top of the hollow, it would, of course, run over the lowest part of the edge, and thus find an outlet, and these outlets are, in some cases, the beginnings of very large rivers. As these lakes receive no stream, they must necessarily, in most cases, be in elevated situations. There is one of this kind on Monte Rotondo in Corsica, which is 9000 feet above the level of the sea.

The *third class*, which embraces all those which both receive and discharge streams of water, is much more numerous than any. Though they are the receptacles of many streams from the neighbouring country, they usually have each but one outlet, which often takes its name from the principal river that runs into the lake. The largest lakes of this class are the immense bodies of water in North America, between Canada and the United States. There are five, (Superior, Michigan, Huron, Erie, and Ontario,) almost like seas in extent, connected together, and their purity is maintained by means of the continual flow of water which is kept up from one to another. Their final outlet to the Atlantic Ocean is the great river St. Laurence. Lake Baikal, in Asiatic Russia, is also remarkable for its size; it sends forth a large stream which joins the Yenisei.

The *fourth class* of lakes comprises a very small number, but they are the most singular in their character of all. They are those which receive streams of water and often great rivers, but have no visible outlet whatever. The most celebrated are the Caspian Sea and Lake Aral, both situated in the west of Asia. The Caspian is between 600 and 700 miles long, and, in one part, between 300 and 400 miles in width; it receives some very large rivers, the chief of which are the Volga, the Ural or Yaik, and the Kur; yet, notwithstanding the plentiful supply of water which is constantly being poured into it, it not only exhibits no increase, but there is strong reason to believe, from the appearances round its shores, that its surface is now much lower than it was at a former period *. Many conjectures have been offered to account for this apparent anomaly; but, after all, the process of evaporation seems quite sufficient to explain it, especially when we consider the extensiveness of the surface which this inland sea (as it is termed) presents, to be acted upon by the atmosphere. Lake Aral is much smaller than the Caspian, but possesses the same peculiarities, and, from the character of the isthmus which separates them, it is supposed that they formerly composed one body of water. They are both *salt* lakes, and are distinguished by marine productions; from these circumstances it has been conjectured that they must, at a very remote period, have been connected with the Black Sea. If such a connection ever existed, the separation may have been occasioned by an accumulation of alluvial soil, brought down by the rivers Don and Volga.

The phenomena presented by some lakes are of a very curious kind. Several of these bodies of water are *periodical* in their appearance. In tropical countries, owing to the violent rains and the overflowing of the rivers, spaces of several hundred miles are often covered with water. South America has large lakes which are annually formed in this manner, and are again dried up by the powerful evaporation of an equatorial climate. Some lakes there are which periodically appear and disappear, owing (it is thought) to their invisible connexion with some subterranean reservoir, by the alternate increase and diminution of which they are necessarily influenced. Lake Cirknitz in Illyria is of this description. The motions and agitations which certain of these bodies of water experience are more difficult to explain. Some of them appear agitated by the escape of subterraneous gases, or by winds that blow in some cavern with which they communicate. Loch Lomond, in Scotland, and Lake Wetter, in Sweden, are often violently agitated during the calmest weather. The floating islands which exist in several lakes, seem to have been formed by the water first undermining and then detaching from the bank very light earth of the nature of peat; sometimes they are merely reeds and roots of trees woven together. Those of the lake of

* By means of observations made with the barometer upon the coasts of the several seas, the surface of the Caspian was ascertained to be 306 feet below that of the Baltic, and nearly 345 below that of the Black Sea.

Gerdau, in Prussia, are said to yield pasturage for one hundred head of cattle; and in the lake of Kolk, in Osnabruck, there is one which is covered with elm trees.

On the Changes which take place in the Earth's Surface.—Action of Running Waters.—Breaking down of Coasts.—Encroachment of Sands.—Formation of New Islands.—Volcanoes.—Earthquakes.

FROM the quiet and regular succession of natural events to which we are accustomed, and the repugnance we feel to the idea that it is possible for the course of nature to suffer interruption, we might, without due investigation, almost persuade ourselves that the physical features and condition of the globe possess an unchangeable character. So far, however, is this from being the case, that there is no country wherein traces are not discoverable of the great changes and violent revolutions of which the earth has formerly been the theatre. The confusion often exhibited in the position of the different strata or layers of which the crust of the earth is composed, the frequent discovery of the remains of animals and vegetables deeply buried in the soil, and many other appearances, testify that the surface of the globe has undergone convulsions, to the production of which none of the natural agents with which we are acquainted can be regarded as adequate; unless they once acted in a method, and with an extent of violence, of which it is impossible for us, by reference to what now exists, to form a conception. "The lowest and most level parts of the earth exhibit nothing, even when penetrated to a very great depth, but horizontal strata, composed of substances more or less varied, and containing almost all of them innumerable *marine productions*. Similar strata, with the same kind of productions, compose the lesser hills to a considerable height. Sometimes the shells are so numerous as to constitute of themselves the entire mass of the rock; they rise to elevations superior to the level of every part of the ocean, and are found in places where no sea could have carried them at the present day, under any circumstances;* they are not only enveloped in loose sand, but are often inclosed in the hardest rocks. Every part of the earth, every continent, every island of any extent, exhibits the same phenomenon." *(Cuvier's Essay on the Theory of the Earth.)* The perfect state in which these shells are generally found, and the regularity, thickness, and extent of the beds that contain them, prove that they could not have been deposited in their places by any temporary invasion of the sea, but that the water must have remained there long enough in a state of tranquillity, to have allowed them gradually to deposit themselves. Some of the strata of marine formation are much more *recent* than others; while in the midst of even the oldest strata of this kind, other strata appear full of animal or vegetable remains of *land* or *freshwater* productions. On these accounts, it would seem as if the land, now inhabited by man, had experienced various successive irruptions and retreats of the sea. There are also appearances which lead to the conclusion that the catastrophes which have occasioned these changes have been sudden and violent. To numberless living beings, they were the messengers of destruction, and of many, the very races have been utterly extinguished. Cuvier, the celebrated French geologist and natural historian, from an observation of the *fossil bones* of more than one hundred and fifty quadrupeds, has determined that upwards of ninety of these animals were of kinds *unknown* to naturalists. There can be no doubt that the revolutions in which these animals were destroyed, occasioned great changes of climate in many parts of the earth, and that in some instances, at any rate, the change took place very rapidly. Fossil plants, and animals of similar kinds to some which still exist in warm regions, have been found in countries where the cold is very much beyond what such kinds are capable of sustaining; and in the arctic zone, the carcasses of large quadrupeds have been discovered enveloped in the ice with their skin, hair, and flesh, still remaining, so that the alteration in the climate must have occurred with such suddenness, as to prevent their bodies from being decomposed by putrefaction.

Such are some of the traces that bear witness to the revolutions which the surface of the globe has undergone. These wonderful and destructive events,

* *Fossil shells* have been found on the summits of the Pyrenees; and among the Andes as much as 13 and 14,000 feet above the level of the sea.

of the immediate causes of which nothing can be declared with certainty, must have long ceased; but the earth has since experienced, and is still experiencing, changes of a very perceptible kind, which we shall now proceed to notice.

Of the several agents which contribute to these changes, water has the widest sphere of activity. In all abrupt and precipitous mountains, fragments of earth and rock are continually falling down from the higher parts, owing to the slow, but effectual action of rains, storms, &c.; and these become rounded by rolling upon each other. These fragments collect upon the sides and at the foot of the mountains, and, on some occasions, when undermined by rivulets *, have been known to slip down in immense masses, and by stopping up the course of rivers, create great devastation. But, without any such extraordinary occurrences as these, the streams that descend along the flanks of elevated grounds carry along with them some portion of the materials of their respective slopes, especially when swelled into violence by rains or the melting of snows; and such as come from mountains sweep down with them even some of the fragments of rock that have been collected in the high valleys. In proportion, however, as these streams reach the more level country, and their channels become more expanded, they deposit the fragments and stones, till at last their waters convey along only particles of mud of the minutest kind. If, therefore, these waters do not run too rapidly into the sea, or the particles in question do not previously settle in some lake through which the rivers pass, the mud is deposited at the sides of their mouths, forming low grounds, by which the shores are prolonged and encroach upon the sea; and when the waves, by casting up sand upon them, assist in their increase, whole provinces are created, capable, from their rich soil, of yielding, in the highest degree, to the support of man, and of being made the seats of wealth and civilization.

It has been concluded, with reason, that the greater part of Lower Egypt owes its formation to the alluvial matter brought down by the Nile, aided by the sand cast up by the sea. M. Dolomieu has endeavoured to show that the tongue of land on which Alexandria was built, (331 years before Christ,) did not exist in the days of Homer; (about 900 B. C.,) and that the lake Mareotis was, at the latter period, a large gulf of the sea. In the time of Strabo, the geographer, who lived about the commencement of the Christian æra, this gulf had been inclosed by land, and is described as a *lake* of six leagues in length. More certainty exists as to the changes that have occurred since that period. The sand thrown up by the sea and wind has formed, near the site of the ancient town, a narrow tongue, on which the modern Alexandria stands. It has blocked up the nearest mouth of the Nile, and reduced the lake Mareotis almost to nothing; while the rest of the shore has been very much extended by the continual deposition of alluvial matter. In the time of the ancients, the Canopian and Pelusian were the principal mouths of the Nile,† and the coast ran in a straight line from the one to the other. The water now passes out chiefly through the Bolbitian and Phatnitic mouths; and round them the greatest depositions have taken place, to which the coast is indebted for its swelling outline. The cities of Rosetta and Damietta, which were built upon these mouths close to the sea, less than 1000 years back, are now six miles distant from it. At the same time that the sediment of the Nile occasions an extension of the land, both the bed of the river and the country, which is periodically covered by the overflow of the waters, are, from the same cause, gradually being raised to a greater elevation. As a proof of this elevation of the soil, it is stated that at Cairo, before the rise of the river is

* Large masses of rock have, however, been known to detach themselves, and roll down from mountains without any apparent cause.

In Mr. Bakewell's Travels, vol. I., p. 195—202, there is an interesting description of the fall of a part of Mont Grenier, in Savoy, which took place in 1248. The ruins spread over an extent of nine square miles, and entirely buried five parishes, and the town and church of St. André. Some of the small hills or rocks, of which the ruins consist, are at the distance of three and four miles from the mountains from which they were separated. This catastrophe, Mr. Bakewell remarks, must have been caused by the gradual decay of the soft strata, of which the lower part of the mountain consists; whereby the mass of limestone above was undermined, and becoming detached, fell with destructive violence into the plain. There are appearances about Mont Grenier which threaten a renewal of the catastrophe of 1248.

† The Canopian or westernmost mouth, which used to discharge itself into the sea not far from the site of Aboukir, is now lost in the lake of Etkoo; the Pelusian, the easternmost mouth of the Nile still exists, but in a very insignificant state, flowing through the lake Menzaleh.

deemed sufficient for the purpose of irrigation, its height must exceed by 3½ feet that which was requisite ten centuries ago. According to this statement, the ground must have been raised at the rate of nearly 4¼ inches in a century. The ancient monuments of the land all have their bases, more or less, covered by the mud which has been, for ages, accumulating around them.

The delta of the Rhone undergoes a similar augmentation, and it would appear that the arms of that river have, in the course of 1800 years, become longer by three leagues; and that many places which were once situated on the brink of the sea, or of large pools, are now several miles distant from the water. In Holland and Italy, the Rhine and the Po, since they have been banked up by dikes, raise their beds and push forward their mouths into the sea with great rapidity. Many cities which, at periods within the range of history, were flourishing sea-ports, have, by the encroachments on the water, been deprived of their importance. It is with extreme difficulty that the Venetians are able to preserve the *lagunes* * by which their city is separated from the main land; and in all probability Venice is destined to experience the fate of Ravenna, which, according to Strabo, stood among lagunes in the time of the Roman emperor Augustus, but is now a league from the shore. M. Cuvier records some curious information which he obtained from M. de Prony, inspector-general of bridges and roads, who was appointed to investigate the remedies that might be applied to the devastations committed by the floods of the Po. This clearly displays some of the surprising changes which the coast of the Adriatic has undergone. At the beginning of the twelfth century, the whole waters of the Po flowed to the south of Ferrara, in the two channels called Po di Volano, and Po di Primaro; an irruption of the river to the north of that city happened not long after, and owing to this new direction of the stream, the two old channels in question had, in less than 100 years, been reduced to the comparative insignificance in which they still remain. Since the construction of the grand embankments of the Po, the formation of New Land has proceeded very rapidly, especially within the last two centuries. Such indeed has been the increase, that the city of Adria, which there is no doubt was, at a very remote date, seated on the coast of the *Adriatic*, is now more than fifteen miles distant from the nearest part of it. The distance from the same city to the extreme point of the promontory of the alluvial land, deposited round what is now the principal mouth of the Po, is upwards of twenty miles. At the same time that river has so much raised the level of its bottom, that the surface of its waters is now higher than the roofs of the houses in Ferrara; and the Adige and the Po are higher than the whole tract of country lying between them. The high level above the surrounding plain, attained by the Rhine and the Meuse in Holland, since they have been banked up; the additions of land that have been made along the shores of the North Sea, in Holstein, Friesland, Groningen, &c.; and the diminution of the sea of Azof, by the entrance of alluvial matter from the Don, are further instances of the changes which nature is able to produce by the most simple means. The *Yellow Sea* (so named from its waters being coloured by an intermixture of particles of yellow mud) affords a similar example. This sea, which lies between the peninsula of Corea and the eastern coast of China, is exceedingly shallow, as may be seen from the account of Capt. Hall, who navigated it in the year 1816—(*Voyage to Loo-Choo*, &c.) That officer states, that no land could be perceived from the mast-head at the time when his ship was in less than five fathoms water; and, before a sight of land was obtained, even this depth was considerably reduced. The bottom consisted of mud, formed of an impalpable powder, without the least sand or gravel. The fine particles, from which this mud is deposited, are brought down by innumerable streams from China and Tartary.

The alterations perceived to be taking place in many of those lakes which are traversed by rivers, proceed from the same cause as the extensions of alluvial land into the sea which we have just been considering. The matter brought down by rivers easily settles in the still water of lakes, and the necessary result is, that the basins of the latter are gradually undergoing a diminution. This process, carried on for a sufficient length of time, would end in the filling up of the lake, and in its place there would

* These are very extensive sheets of water, but so shallow that they, in no part, exceed six or seven feet in depth.

be a valley intersected, of course, by the same rivers which formerly flowed into the lake. Owing to the very long time required for the purpose, there is no instance known of a lake, of any size, ever having been filled up in this manner; but there are well authenticated cases of their being very sensibly diminished. Lake Erie, one of the vast bodies of water in North America, is rapidly decreasing; in the late survey of the boundaries between the United States and Canada, it was ascertained that Long Point, opposite Big Creek river, on the north side of the lake, had, in the space of three years, increased more than that number of miles in length by the accession of alluvial matter; and this immense basin, the average depth of which is estimated at between thirteen and seventeen fathoms, is every year becoming shallower from the influx of pebbles and earth, and the constant accumulation of reeds and shells. The diminution of the beautiful lake of Geneva is also said to have been considerable within the memory of man.

There are several instances in mountainous and marshy countries of small lakes having been dried up from different causes—such as the crystallization, or deposit of substances which the waters had previously held in solution; the gradual union of floating islands, and the collection of matter arising from the lake being the seat of animal and vegetable life; but it is evident, from their very slow progress, that the effects produced in these ways cannot be upon a very large scale.

The changes which we have hitherto traced to the action of waters have been of a beneficial kind; but others of a destructive nature are brought about by the same agency. These are the breaking down of steep coasts by the waves, and the throwing up of sand-hills, which the winds afterwards assist in pushing forward and dispersing over the adjacent land. The first is a very common occurrence; the sea detaches fragments from the foot of the cliffs, or else wastes it away, and then the upper parts, deprived of support, fall down. The broken portions that collect at the base, in consequence of these fallings down, serve, more or less, and for a shorter or longer period, according to their position and hardness of material, to protect the cliff from further ravages. The circumstances also which cause the slipping down or breaking away of masses of rock and earth, among mountains, operate in a similar manner where there are shores of a steep character. Springs filter through and displace the soft strata, and thus the more solid formations are left without support; the consequence is that, at times, large spaces of land slide or fall down from above. It is by such means as these that the 'land-slips,' on the southern shore of the Isle of Wight, have been produced. The same thing happens, but on a far grander scale, upon the coast of the Crimea; whole tracts are there carried down, sometimes bearing upon them the houses of the natives, which have, notwithstanding, been known to escape without injury.

The action of the sea, when the coast is low, and the bottom sandy, leads to very different results. The waves then push the sand forward upon the shore, where, at every ebb of the tide, it becomes partially dried; and the wind, frequently blowing from the sea, drifts it upon the beach. By little and little, hillocks or downs of sand are created, the higher parts of which are continually carried inland; so that unless the inhabitants of the country succeed in fixing them by causing suitable plants to take root in their soil, they move slowly on and overwhelm fields and dwellings with inevitable ruin. It sometimes happens that the sand cast up by the water becomes mixed with marine and other substances, which, being enveloped therein, make what have been denominated *indurated*, that is, *hardened* downs, such as are seen upon the coast of New Holland. Perhaps the most remarkable instance of the mischief occasioned by the moving downs is to be found on the French coast of the Bay of Biscay, south of the river Gironde, where they have already overwhelmed a great number of villages mentioned in the records of the middle age, and not long ago, in the single department of the *Landes*, were threatening ten with unavoidable destruction. One of these villages, named Mimisan, had been struggling against them for twenty years, with the prospect of a sand-hill of more than sixty feet in height visibly approaching it. In 1802, the pools formed by the collection of waters which these downs prevent from flowing into the sea, covered five farming establishments at the village of St. Julien. They have long been over an ancient Roman road leading from Bayonne to Bourdeaux, and which could still be seen forty years back, when the waters were low. The river Adour is now turned to the distance of more than 2000 yards out of its former course. The progress of these downs has been estimated at 60 feet yearly, and in some places, at 72 feet; at this rate, it is calculated that it will require 2000 years to enable them to reach Bourdeaux.

The coast of Elgin or Morayshire, in Scotland, also affords a striking example of the sand-flood; an account of which is given among the notes affixed to Professor Jameson's edition of Cuvier's Theory of the Earth. West of the mouth of the river Findhorn, a district of more than ten square miles in area, (chiefly included in the barony of Coubine,) which, on account of its extreme fertility, was once termed the granary of Moray, has been depopulated and rendered unproductive by the shifting of the sand-hills. It appears that the irruption of the sand commenced about the year 1677; that in 1697, not a vestige was to be seen of the manor-place, orchards, and offices of Coubine, and that two thirds of the barony were already ruined. This irruption came from Mavieston, situated on the shore, about seven miles west from the mouth of the Findhorn, where, from time immemorial, there had been large heaps of sand. These sands, which had formerly been covered with vegetation, were set at liberty by the inhabitants inconsiderately pulling up the bent and juniper for various uses, and they then drifted towards the north-east. When the wind is high, the fine particles are carried even across the bay of Findhorn. In the winter of 1816, a large portion of the only remaining farm, on the west side of the Findhorn, situated in the line of the sand's progress, was overwhelmed. The effects produced by the sand upon the river have been, and still are, of a very obvious kind. Many years ago, its mouth having become blocked up, the water cut out its present more direct channel. By this change, the old town of Findhorn, which originally stood on the east side of the river, was left upon its western bank; and the inhabitants, in consequence, removed the stones of their houses across the new channel, and erected the present village on the eastern side. The site of the old town is now covered by the sea. Even now, when the tide retires, the river almost disappears, being absorbed by the sand; and owing to the bar formed across its entrance, it is unable, at spring tides, to force its way into the sea, so that it is made to flow back and inundate a considerable extent of land at the head of the bay. Of late, however, the great accumulations of sand have disappeared from Coubine, and the ancient rich soil has, in some places, been left bare, from which it is hoped that the barony will again become serviceable land. Such a result would be rendered much more certain, if, by putting in proper kinds of plants, they were to fix the surface of the Mavieston hills, and so prevent fresh inroads of the sand from that quarter; yet, notwithstanding the destruction that has already happened, the inhabitants are still in the habit of gathering what little bent yet remains.

The same drifting of sand occurs upon several parts of the west coast of the Outer Hebrides, and the prevention of it has been attempted in two ways. Mr. A. Macleod, surgeon of North Uist, has invented the most efficacious plan, which is that of cutting thin square turfs from the neighbouring pasture grounds, and laying them down at intervals of some inches. In the course of a few years the turfs grow together, while the ground, from which they are taken, is little injured, for as the roots remain in it, a new vegetation soon rises up. The other method was introduced by Mr. Macleod, of Harris, and tried extensively upon his estate: it is to plant small bundles of *arundo arenaria*, (sand-reed) about a foot and a half apart; these take root and prevent the drifting, in a certain degree.

Another process, similar in its effects to those which have been already described, but much more extensively destructive, and depending solely upon the action of the wind, is the encroachment of the sands of the Libyan desert upon the cultivated lands of Egypt. These sands, driven by the west winds, have left no soil capable of tillage on any parts of the western banks of the Nile, which are destitute of the shelter of mountains. It would appear that, but for the ridge called the Libyan chain, which borders the left bank of that river, forming to the parts along which it runs a barrier against the sands, the western shores of the Nile would, long ago, have been made uninhabitable. Travellers have given a melancholy picture of the traces which bear witness to the ravages committed by the sand—the ruins of numerous cities and villages destroyed, and the summits of the minarets of mosques, being still visible above the surface. It is partly to these resistless invasions of the desert that the decline of Egypt from her ancient splendour is to be attributed.

The formation of new islands (to which allusion has been made in a preceding part of this treatise) constitutes a distinct and interesting class among the changes to which the surface of the globe is subject. Those which have been raised up by volcanic agency are, comparatively, few; but those of coral, which owe their origin to marine insects, (of the class of zoöphytes or *plant-animals*) are innumerable. Of the different coral tribes, the most abundant is that named the madrepore. It is most common in the tropical seas, and decreases in number and variety towards the poles; it surrounds, in vast rocks and reefs, many of the rocky islands of the South Sea and Indian Ocean, and increases their size by its daily growth. The coasts of the islands in the West Indies, of those to the east of Africa, and the shores and shoals of the Red Sea, are encircled with rocks of coral. Several navigators have furnished us with accounts of the curious manner in which these formations take place: the following is extracted from Captain Basil Hall's narrative of his voyage to the Loo Choo islands:—

"The examination of a coral reef, during the different stages of one tide, is particularly interesting. When the tide has left it for some time, it becomes dry, and appears to be a compact rock exceedingly hard and rugged; but as the tide rises, and the waves begin to wash over it, the coral worms protrude themselves from holes which were before invisible. These animals are of a great variety of shapes and sizes, and in such prodigious numbers, that, in a short time, the whole surface of the rock appears to be alive and in motion. The most common of the worms at Loo-Choo is in the form of a star, with arms from four to six inches long, which are moved about with a rapid motion, in all directions, probably to catch food. Others are so sluggish, that they may be mistaken for pieces of the rock, and are generally of a dark colour, and from four to five inches long, and two to three round. When the coral is broken, about high water mark, it is a solid hard stone; but if any part of it be detached at a spot which the tide reaches every day, it is found to be full of worms of different lengths and colours: some being as fine as a thread, and several feet long, of a bright yellow, and sometimes of a blue colour; others resemble snails, and some are not unlike lobsters in shape, but soft, and not above two inches long.

"The growth of coral appears to cease when the worm is no longer exposed to the washing of the sea. Thus, a reef rises in the form of a cauliflower, till its top has gained the level of the highest tides, above which the worm has no power to advance, and the reef, of course, no longer extends itself upwards. The other parts, in succession, reach the surface, and there stop, forming, in time, a level field with steep sides all round. The reef, however, continually increases, and being prevented from going higher, extends itself laterally in all directions. But this growth being as rapid at the upper edge as it is lower down, the steepness of the face of the reef is still preserved. These are the circumstances which render coral reefs so dangerous in navigation; for, in the first place, they are seldom seen above the water; and in the next, their sides are so steep, that a ship's bows may strike against the rock, before any change of soundings has given warning of the danger."

Captain Flinders, who, in 1801, made a survey of the coasts of New Holland, has some observations upon the formation of coral islands, particularly of Half-Way island, on the north coast of that region, which show how, after being raised up, they gradually acquire a soil and vegetation:—

"This little island, or rather the surrounding reef, which is three or four miles long, affords shelter from the south-east winds; and being at a moderate day's run from Murray's Isles, it forms a convenient anchorage for the night to a ship passing through Torres' Strait—I named it *Half-way Island*. It is scarcely more than a mile in circumference, but appears to be increasing both in elevation and extent. At no very distant period of time, it was one of those banks produced by the washing up of sand and broken coral, of which most reefs afford instances, and those of Torres' Strait a great many. These banks are in different stages of progress; some, like this, are become islands, but not yet habitable; some are above high-water mark, but destitute of vegetation; whilst others are overflowed with every returning tide.

"It seems to me, that when the animalcules which form the corals at the bottom of the ocean cease to live, their structures adhere to each other, by virtue either of the glutinous remains within, or of some property in salt water; and the interstices being gradually filled up with sand and broken pieces of coral washed by the sea, which also adhere, a mass of rock is at length formed. Future races of these animalcules erect their habitations upon the rising bank, and die in their turn, to increase, but principally to elevate this monument of their wonderful labours. The care taken to work perpendicularly in the early stages would mark a surprising instinct in these diminutive creatures. Their wall of coral, for the most part in situations where the winds are constant, being arrived at the surface, affords a shelter, to leeward of which their infant colonies may be safely sent forth; and to this, their instinctive foresight, it seems to be owing, that the windward side of a reef exposed to the open sea is generally, if not always, the highest part, and rises almost perpendicular from the depth of many fathoms. To be constantly covered with water seems necessary to the existence of the animalcules, for they do not work, except in holes upon the reef, beyond low-water mark; but the coral, sand, and other broken remnants thrown up by the sea, adhere to the rock, and form a solid mass with it, as high as the common tides reach. That elevation surpassed, the future remnants, being rarely covered, lose their adhesive property; and remaining in a loose state, form what is usually called a *Key*, upon the top of the reef. The new bank is not long in being visited by sea-birds: salt plants take root upon it, and a soil begins to be formed; a cocoa-nut, or the drupe of a pandanus, is thrown on shore; land birds visit it, and deposit the seeds of shrubs and trees; every high tide, and still more every gale, adds something to the bank; the form of an island is gradually assumed; and last of all, comes man to take possession.

"Half-way Island is well advanced in the above progressive state; having been many years, probably some ages, above the reach of the highest spring tides, or the wash of the surf in the heaviest gales. I distinguished, however, in the rock which forms its basis, the sand, coral, and shells, formerly thrown up, in a more or less perfect state of cohesion. Small pieces of wood, pumice stone, and other extraneous bodies which chance had mixed with the calcareous substances when the cohesion began, were inclosed in the rock, and, in some cases, were still separable from it without much force. The upper part of the island is a mixture of the same substances in a loose state, with a little vegetable soil; and is covered with the *casuarina* and a variety of other trees and shrubs, which give food to parroquets, pigeons, and some other birds; to whose ancestors, it is probable, the island was originally indebted for this vegetation."

It has been generally believed that the deep perpendicular reefs, very near to which the sounding line finds no bottom, consist wholly of coral; but MM. Quoy and Gaimard have adduced very satisfactory reasons, to prove that the zoöphytes, far from raising from the depths of the ocean perpendicular walls, form only layers or crusts of a few fathoms thickness. They remark that the species which always construct the most considerable banks require the influence of light to perfect them; and it is well known, that all those steep walls common in the equatorial seas, are intersected with narrow and deep openings, through which the sea enters and retires with violence; whereas, if they were entirely composed of madrepores, they would have no such openings between them, since it is the property of zoöphytes to build in masses that have no interruption. It is besides, difficult to suppose that these animals can support such different degrees of pressure and temperature as they necessarily must, if they exist at such different depths in the ocean. It is, therefore, most reasonable to conclude that the summits of submarine hills and mountains are the bases upon which the zoöphytes form layers and raise up their fabrics; a supposition which perfectly accounts for the great depth of the sea close to the reefs and islands which they have elevated to the surface of the water.

The changes occasioned by the eruptions of volcanoes are very considerable near the seat of action, but they operate over a less extensive field than any of those which have yet been mentioned.

The principal effect of the issue of subterranean fires is the elevation of the surface of the surrounding country; and the size of the mountains themselves must have been prodigiously increased by the matter thrown up during successive eruptions. Some, indeed, have gone so far as to assert that volcanoes are *entirely formed* of this matter; but even if we could admit such a theory, with respect to the isolated volcanoes of Europe, Asia, and the African islands, it would be impossible to extend it to those immense masses in America (Cotopaxi, Pichincha, &c.) which are parts of the great western chain of that continent. There are, nevertheless, very good grounds for concluding that many volcanoes owe their formation to the effects of the fires to which they give vent, and that, previous to the first appearance of those fires, there were no mountains in the places which the volcanoes in question now occupy. It would seem that the breaking out of a volcano, where none before existed, is preceded by a swelling or heaving up of the crust of the earth, owing to the expansive force of the heat, and this explains the cause of new volcanic islands sometimes emerging suddenly from the sea. The phenomenon of the swelling up of the ground was strikingly exemplified when the volcano of Jorullo, in Mexico, arose out of the plain des Playas, in the month of September, 1759. According to the representation of this mountain, annexed to the third volume of M. Humboldt's Personal Narrative, the raised up portion of the plain is about two miles in length, and in one part as much as 850 feet above its ancient level; from this part the cone of the volcano shoots up covered with ashes to a further elevation of 830 feet. The raised up ground is covered with thousands of small volcanic cones, from 6 to 10 feet high. We have thus an instance of a mountain 1680 feet in elevation above the plain upon which it stands, having been entirely formed by the heaving up of the earth when the fire originally broke out, and by the accumulation of matter which has since gathered round the principal vent of the eruptions. But the knowledge of such an instance as this does not of necessity lead to the conclusion that mountains so large as Etna, and many other volcanoes, have been entirely created by the process which gave birth to Jorullo: yet, though it might be rather hazardous to adopt such a conclusion, it is certain that those mountains must have received a vast increase, both in size and elevation, at and since the period, whenever that may have been, at which the subterranean fire first opened a passage through them.

Owing to the repeated discharges of matter upon their surface, and the violent concussions to which they are subject, the external appearance of volcanic mountains is continually varying. This is particularly the case with Etna. Among the accounts of those who have visited it, it would be difficult to find two that at all agree as to the aspect of the great crater and the different eminences about the summit of the mountain. New chasms on the sides also have, at various periods, been opened, and new cones raised up: Monte Russo, thrown up from a plain during the famous eruption of 1669, was estimated by Spallanzani, in 1788, at two miles in circumference, and (somewhat vaguely) at 150 paces in height.

The lava, stones, and ashes spread over the neighbouring country by volcanoes materially elevate its surface. The complete burial of the ancient cities of Herculaneum and Pompeii by the matter ejected from Vesuvius, in the eruption of the year 79, is a circumstance familiar to almost every one. The excavation of these cities in modern days has brought some interesting facts to light; the ruins of Herculaneum were found at a depth of 70 feet below the surface of the ground; and from the number of distinct strata of lava, one above the other, each covered with a layer of rich mould, it appeared that streams of that substance from, at least, six different eruptions had passed over it since the one which occasioned its destruction.

Vast masses of earth sometimes sink down during volcanic eruptions. In the island of Timor, a volcano of considerable elevation is said to have sunk into the ground, leaving a muddy marsh in its place. The *Papandáyang*, situated towards the western extremity of Java, was formerly one of the largest volcanic mountains of the island, but the greatest part of it was swallowed up in the earth in the year 1772. It is asserted, that, on the night of this event, an uncommonly luminous cloud was seen to envelop the mountain. Alarmed by this appearance, the inhabitants on the declivities and about the foot of the mountain took to flight, but before they could all save themselves, it began to give way, and the greatest part of it actually *fell in* and disappeared in the earth. A tremendous noise, and the discharge of showers of volcanic substances, accompanied this commotion. It is estimated, that an extent of ground, of the mountain itself and its immediate environs, 15 miles long and 6 broad, was swallowed up in the bowels of the earth.

Of the origin of the volcanic fire, nothing, of course, can be affirmed with certainty; but several explanations have been offered, more or less satisfactory. An attentive observance of the phenomena connected with eruptions, and a close examination of the substances ejected, are necessary to the attainment of any correct views upon the subject. It has been a very generally received opinion that volcanic eruptions are caused by the spontaneous combustion of pyrites; and in support of this solution are cited the experiments which have been made with sulphur and iron filings. A mixture of these substances, after being moistened, has been buried in the ground, where it became gradually heated, and at length took fire, with a loud explosion. The theory, however, which ascribes volcanic action to the inflammation of beds of coal, sulphur, or other matters, near the surface of the earth, has latterly met with less support, and powerful arguments may be urged against it. The great masses of inflammable materials are confined to the secondary and superficial strata; while, on the contrary, there can be no doubt, that the seat of volcanic fire lies far below the surface of the earth, both from the nature of the substances cast out by volcanoes, and the circumstance of the immense quantity of matter that has proceeded from many of them; such a quantity as, had it been withdrawn from the parts near the surface, would, long since, have occasioned those mountains to sink down and disappear. A different theory has been lately brought forward to account for the volcanic fire, and is well set forth and supported in a recent treatise by Dr. Daubeny, Professor of Chemistry in the University of Oxford. It is founded upon the metallic nature of the bases of the earths and alkalies, and the avidity with which these combine with oxygen, producing in that combination a high temperature and strong inflammation. It is supposed, that if these materials exist in sufficient quantity in the interior of the earth, and a sufficient body of water be admitted to them, judging from the violent effects on a small scale which we are able to produce by experiments, a heat would be engendered quite adequate to occasion all that takes place in volcanic eruptions. Now, under this hypothesis, it is requisite that water should have access to the metallic bases, and it is a curious fact that nearly all active volcanic groups are within a short distance of the sea; while even those in South America, which must be excepted from this remark, are in a range of mountains approaching in parts close to the sea. *Extinct* volcanoes, it is true, are found in situations quite beyond the access of the present ocean; but in the remote and unknown periods of their activity, it may reasonably be inferred that they were near the ocean, if not altogether beneath its surface. A further argument in favour of the present view is, that all the products which chemists know to be the result of the admission of sea-water to the metallic bases, appear under some form or other in every volcanic eruption.

Earthquakes appear to be brought about by the same causes as volcanic eruptions, but their action is much more tremendous than that of the latter. They are frequently accompanied by loud subterraneous noises, and are sometimes so violent that the ground heaves up, and undulates like an agitated sea. They are felt almost at the same instant over a most astonishing extent; though happily, compared with this extent, their destructive ravages are confined within a small range. In those parts which appear to be near the centre of their action, the most calamitous effects sometimes occur; whole cities are destroyed, and their inhabitants buried beneath the ruins; the surface of the ground undergoes violent changes; springs are stopped, and others gush out in new places; fissures are made in the earth; and enormous masses of rock and other materials sink down, or are detached from the mountains. By the earthquake experienced in Chili in 1822, a great line of coast is stated to have been lifted permanently up, to the height of several feet above its former level; while, in the interior of the country, fissures were made in the granite transversely to the direction of the earthquake.

It is generally supposed that earthquakes are produced by the disengagement of elastic vapours, which, endeavouring to escape from their confinement, heave up and agitate the crust of the earth. No doubt can exist of their connexion with volcanic eruptions; their frequency in countries where the latter take place, and the fact of the one often occurring at the same period as the other, sometimes at great distances apart, tend to establish such a connection. This is further shown by the circumstance of the shocks of earthquakes being most severe in places distant from volcanoes; as if the latter were the means of giving vent to that elastic force, which, when pent up, causes such dreadful ravages. It is also worthy of notice, that though earthquakes are sometimes felt towards the interior of continents, their terrible effects occur chiefly along the coasts, as exemplified in the earthquakes of Lima, of Lisbon, of Caraccas, and many others.

Such are the principal changes which the surface of the globe is now undergoing. It is evident, notwithstanding what some have been inclined to assert, that they could not have brought about those grand revolutions which formerly visited the earth, and in which such multitudes of the animal race were consigned to destruction. The whole of them are insufficient to alter, in any perceptible degree, the level of the sea, still less to have occasioned an overwhelming of the land by that element. Some philosophers have endeavoured to prove that a gradual and general lowering of the level of the sea takes place, and have appealed to certain observations which, if correct, tend to establish the fact of a diminution of the waters along the northern shores of the Baltic. But it must not be forgotten, that though in some places the ocean has retired, or sunk in level, in others it has encroached upon the land; while it is known that many harbours of the Mediterranean have preserved exactly the same level since the time of the ancients. It is plain, therefore, that all variations upon the coasts of the ocean are merely of a local kind, and that if the different accounts are balanced, we must arrive at the conclusion, that the general volume of the ocean, and perhaps even its superficial extent, suffer neither increase nor diminution.

On the Ocean—its Saltness, and Temperature, Tides, and Currents.

THAT vast body of water which surrounds the continents, and is the common receptacle of their running waters, is indispensably necessary to the support of animal and vegetable existence upon the earth. Its perpetual agitations purify the air, and the vapours which the atmosphere draws up from its surface, being condensed and dispersed through the upper regions, form clouds, which are the source of a constant supply of rain and moisture to the land. The ocean also, by the facilities for communication which it offers, is the means of uniting the most distant nations, while it enables them to interchange with mutual advantage the productions of their several climates.

The bottom of the sea appears to have similar inequalities to the surface of the continents; the depth of the water is, therefore, extremely various. There are vast spaces where no bottom has been found, but this, of course, does not prove that the sea is bottomless, because the line is able to reach but a comparatively small depth. Lord Mulgrave, in the Northern Ocean, let down a very heavy sounding lead, and gave out with it nearly 4700 feet of rope without finding the bottom; and Mr. Scoresby mentions having sounded in the Greenland sea as much as 7200 feet. Such experiments, however, must be of very doubtful character; it is well known how much more easily bodies may be moved along in the water than in the atmosphere, and, consequently, any current would be sufficient to carry the lead with it, and so draw the rope out of a perpendicular direction. If we were to found our opinion upon analogy, we might conclude that the greatest depth of the ocean is, at least, equal to the height of the loftiest mountains, that is, between 20,000 and 30,000 feet.

The level of the sea, if it were not for the action of external disturbing causes, would be the same every where at the same instant, owing to the equal pressure exerted by the particles of a fluid upon each other in every direction. The figure assumed by the ocean would, therefore, exhibit the true surface of our planet, that of an oblate spheroid. But it is evident that no general level of this kind can ever exist, because the tide at any given moment is at very different heights in different parts of the ocean. The level is also continually being disturbed by the operation of the wind in particular regions. Independent, however, of these circumstances, it would appear that in gulfs and inland seas which have only a slight communication with the ocean, the level of the water is usually more elevated than in the latter. This seems to be more especially the case if the only openings of these gulfs are towards the east; and it is attributed with reason to the accumulation which arises from the water being driven into these confined inlets by the general movement of the sea from east to west, a movement to which allusion will be made presently. When the French engineers were in Egypt, they made observations, according to which the waters of the Red Sea, on the east side of the isthmus of Suez, were 32½ feet higher than those of the Mediterranean on the opposite shore of the same isthmus. M. Humboldt made observations of a similar kind upon the isthmus of Panama, and his conclusion is that the waters of the Gulf of Mexico are from 20 to 23 feet higher than those of the Pacific on the other side. Of certain inland seas the level varies with the seasons; the Baltic and the Black Sea, which are in fact almost lakes, swell in the spring, from the abundance of water brought down to them at that period by the rivers.

The general colour of the sea is a deep bluish green, which becomes clearer towards the coasts. This colour is thought to arise entirely from the same cause as the azure tint of the sky; the rays of blue light, being the most refrangible, pass in the greatest quantity through the water, which, on account of its density and depth, makes them undergo a strong refraction. The other colours exhibited in parts of the sea depend on causes which are local, and sometimes deceptive. The Mediterranean in its upper part is said to have at times a purple tint. In the Gulf of Guinea the sea is white; around the Maldive Islands it is black; and in some places it has been observed to be red. These appearances are probably occasioned by vast numbers of minute marine insects, by the nature of the soil, or by the infusion of certain earthy substances in the water. The green and yellow shades of the sea proceed frequently from the existence of marine vegetables at or near the surface.

The water of the sea contains several extraneous substances, in proportions varying in different places. The component parts, in addition to pure water, are commonly muriatic or marine acid, sulphuric acid (vitriol), fixed mineral alkali, magnesia, and sulphated lime. By boiling or evaporation in the air, common salt (muriate of soda) is obtained, which for salting meat is preferred to the salt of springs. The saltness of the sea appears, with some local exceptions, to be less towards the poles than near the tropics; but the difference is very slight, and perhaps the observations made are not sufficiently numerous to justify any positive general conclusions. It is probably occasioned by the melting of the ice in the polar seas. Dr. Thomson (*Work on Chemistry*) states the proportion of salt in water taken up by Lord Mulgrave at the back of Yarmouth sands, at 3.125 per cent. of the weight of the water. He also gives, from the accounts of different observers, the following proportions:—

				Quantity of saline matter.						
N. Lat.	80°	(60 fathoms	under ice)	3.54	per cent.	S. Lat.	49°	50′	4·16	per cent.
”	74	”	”	3·60	”	”	46		4·50	”
”	60	”	”	3·40	”	”	40	30	4·00	”
”	45	”	”	4·00	”	”	25	54	do.	”
”	39	”	”	do.	”	”	20		3·90	”
”	34	”	”	do.	”	”	1	16	3·50	”
”	14	”	”	do.	”					

Mr. Scoresby, in north latitude 77° 40′, and east longitude 2° 30′, found, in a quantity of water taken from the surface, the proportion of saline matter 3·56 per cent. A statement in the Edinburgh Philosophical Journal gives the proportion in north lat. 64° 26′, east long. 0° 38′, 3·54 per cent.; in north lat. 78½°, east long. 6½°, 3·88; and in north lat. 78° 35′, east long. 6°, 3·27. Some observations which have been made tend to prove that the sea is less salt at the surface than towards the bottom.

The following table of the proportional specific gravities of different kinds of water, explains the reason of bodies being so much more buoyant in the sea than in other water: —

Distilled water	. .	1·000
Purest spring water	.	1·001 to 1·005
River water	. . . .	1·010
Sea water	. . . .	1·028

The degree of saltness in particular parts of the sea frequently varies from temporary causes. The violent tropical rains have an effect in diminishing it, especially near coasts, where an increased volume of fresh water is brought down by the rivers. The Baltic is at all times less salt than the ocean, and when a strong east wind keeps out the North sea, its waters are said tobecome almost fit for domestic uses. The most curious phenomenon of all is that of springs of fresh water rising up in the midst of the sea; Humboldt mentions that in the Bay of Xagua, on the southern coast of Cuba, springs of this kind gush up with great force at the distance of two or three miles from the land.

The bitterness which exists in sea-water, but apparently not beyond a certain depth, is with much probability considered to be owing to the vegetable and animal matter held there in a state of decomposition.

Water being a bad conductor of heat, the temperature of the sea changes much less suddenly than that of the atmosphere, and is by no means subject to such extremes as the latter. It may safely be affirmed that the temperature never in any season, or under any latitude, exceeds eighty-five or eighty-six degrees of Fahrenheit's thermometer. The existence of banks or shallows has a local effect in diminishing the temperature of the ocean, but the great agents in modifying it are currents, which mingle together the waters of different depths and regions. Thus the Gulf stream, as it is termed, which sets into the Gulf of Mexico, is much warmer than the neighbouring parts of the sea; the current of Chili is just the reverse. The following tables are extracted from M. Humboldt's Personal Narrative, (Vol. II.;) the experiments from which No. I. was drawn up, were made during the passage from Spain to the New Continent, between the 9th of June and 15th of July 1799:—

(No. I.)

North lat.	West long.	Temp. of the Atl. ocean at its surface.
39° 10′	16° 18′	59° 00′ Fahr.
34 30	16 55	61 34
32 16	17 4	63 86
30 36	16 54	65 48
29 18	16 40	66 74
26 51	19 13	68 00
20 8	28 51	70 16
17 57	33 14	72 32
14 57	44 40	74 66
13 51	49 43	76 46
10 46	60 54	78 44

(No. II.)

Table of the Temperature of the Atlantic Ocean in different degrees of Longitude.

Latitude.	Longitude.	Temp. of the Ocean.	Period of the Observation.	Observers.	Mean Temperature of the Air in the Basin of the Sea
0° 58′ S.	27° 34′ W.	80.96 Fh.	Nov. 1788	Churruca	80°·6 (Cook.)
0 57	30 11	81.86	April 1803	Quevedo	
0 33	21 20	81.86	March 1800	Perrins	
0 11 N.	84 15	82.40	Feb. 1803	Humboldt	
0 13	51 42 E.	80.78	May 1800	Perrins	
25 15 N.	20 36 W.	68.00	June 1799	Humboldt	69·8 (La Pérouse and Dalrymple.)
25 29	39 54	70.88	April 1803	Quevedo	
25 49	26 20	69.26	March 1800	Perrins	
27 40	17 4	70.88	Jan. 1768	Chappe	
28 47	18 17	74.30	Oct. 1788	Churruca	
42 34 N.	15 45 W.	51.98	Feb. 1800	Perrins	54·86 (Cook and d'Entrecasteaux.)
43 17	31 27	59.90	May 1803	Quevedo	
43 58	13 7	60.62	June 1799	Humboldt	
44 58	34 47	54.86	Dec. 1789	Williams	
45 13	4 40	59.90	Nov. 1776	Franklin	
48 11	14 18	57.74	June 1790	Williams	

These tables refer only to the Atlantic ocean; but the experiments which have been made in the South sea, and in the Indian ocean, show that within a certain distance of the equator, the general temperature of the sea follows nearly the same rule in corresponding latitudes. Within the tropics there is no sensible difference in north and in south latitudes; there is very little even as far as the thirty-fifth and fortieth degrees; but, when we advance into high latitudes, there can be no doubt that the sea is colder in the southern than in the northern hemisphere. Ice extends from five to eight degrees of latitude farther from the south than from the north pole, owing, it is probable, to the almost entire absence of land near the Antarctic circle; while the north pole is so nearly surrounded by land, that the ice of the Arctic ocean is shut up, and cannot be carried forward to such a distance by the current which sets towards the equator.

Bays, inland seas, and the spaces among clusters of islands, where the action of the waves is more confined, and the water usually of less depth, are the most favourable places for the production and accumulation of the marine ice. It is on this account that the navigation of the Baltic is annually stopped by the ice in a latitude not more northerly than that of tracts which in the main ocean are always open to the passage of ships. In severe winters, people may travel in sledges across the entrance of the Gulf of Bothnia (lat. 60°) which, including the numerous small intervening islands, is a distance of a hundred and fifteen miles. The body of ice accumulated in Sir James Lancaster's sound has defied all the attempts that have been made to accomplish the north-west passage from the Atlantic to the Pacific.

The ice of the polar seas assumes a great variety of shapes and appearances. The vast and thick sheets which are met with in high latitudes are called *fields* by navigators; they are so extensive, that their boundaries cannot be seen from a ship's mast-head; and Captain Cook found a chain of them joining Eastern Asia to North America. Sheets of less extent than fields are called *floes*. *Bergs* are islands of ice, considerably elevated above the water; and though of the most various forms, commonly perpendicular on one side, and sloping gradually down on the other, in height they are sometimes as much as two hundred feet*. There are two ways of explaining the formation of these bodies. The large masses of ice in the Polar seas, when crowded together by winds and currents, exert such an enormous pressure upon each other, that they are frequently broken, and the fragments are piled up so as to form mounds and ridges of considerable elevation; it is thus that many of the small ice-bergs originate. There is no doubt, however, that the most bulky of these bodies are detached portions of vast glaciers, such as abound on the precipitous coasts of Greenland and Spitzbergen, broken off in consequence either of their own weight, or the undermining action of the waves, and then carried by winds and currents to other parts of the ocean.

When the summer has well advanced, the masses of ice which have been frozen together during the winter gradually separate, and clear spaces of water are left. As soon as the end of September, these open spaces again begin to freeze over; but before this effect commences, the temperature of the air must be very much lowered, owing partly to the freezing point of sea being three and a half degrees (Fahrenheit) below that of common water, but more especially to the surface which the water presents to the atmosphere being repeatedly changed before its temperature is sufficiently reduced for it to freeze. This change in the surface, which is greatly assisted by the agitation of the sea, takes place in consequence of the particles of a liquid body becoming specifically heavier as they get cooled, so that they descend and are succeeded by warmer particles.

There are three kinds of movements constantly going on in the waters of the sea:—1. The agitations which its surface undergoes by the action of winds —2. Tides, which are the result of the attraction exercised on the water by the sun and moon—3. Currents, which arise from different causes, some of them existing within the element itself.

1. As the particles of a fluid press equally in every direction, it follows that when a portion of the surface of the water is displaced by a wind, the adjoining water instantly rushes in to restore the equilibrium or balance which has been destroyed. This accounts for the formation of *waves*. When a violent impulse has thus been communicated, the waves continue in motion for some hours after the gale has entirely subsided, on the same principle as a pendulum continues to swing for some time after it has been set in action. Yet the agitation occasioned by winds extends to comparatively but a little way below

* Experiments made in the Arctic Seas show that of a piece of floating ice, the proportion above water is generally about one-seventh of the thickness of the whole mass. It must not, however, be supposed from this, that an iceberg two hundred feet out of the water exhibits only one-seventh of its height, because these bodies are frequently aground.

the surface of the water: divers say that, in the roughest weather, it is calm at the depth of ninety feet.

2. The *tides* proceed from the attractive forces of the sun and moon, which diminish the gravity of the waters of the ocean, or, which is the same thing, draw or lift up the waters toward themselves. Let E (fig. 1.) represent the earth and M the moon; since the power of gravity increases as the square of the distance diminishes, it is evident

Fig. 1.

that the waters at A, the part of the earth nearest the moon, will most feel the effect of her attraction, and will be raised up to B; while those on each side of A, being farther off, will be less raised. But besides the difference occasioned by a greater or less distance, the more oblique or slanting the line of attraction, the less will be the elevation of the water acted upon, till at last the water towards and under the circle CEF will not only not be elevated but will be lowered. The reason of this is, that the force of attraction acts in straight lines; and, therefore, if we draw two straight lines from the moon's centre, MC, MF, to represent this force acting on the parts C and F, it is obvious that the water at C and at F will not be raised, but depressed by being drawn away from C to D and from F to H; and so of every part of the circle CEF. In the same manner, the water at D and H will pass to G and I, and thus the ocean will be disposed in a spheroidal form CDGBIHF. But the water will, at the same time, rise on the side of the earth away from the moon, because the earth's centre being more strongly drawn towards the moon than the point N, recedes from N, which is the same in effect as if the water at N receded or rose up from the earth's centre. The ocean, therefore, will assume a spheroidal form, CKF, on the side away from the moon, as well as on the one facing her. Thus, if we draw a line MAK from the moon's centre through the centre of the earth, the two points B, K, where it touches the earth's surface, will be those of *high water*; and if we take two more points C, F, equally distant from each of the first two, they will be points of *low water*. By the earth's rotation on its axis the part F will be carried to A and the part C to N; matters will then be just reversed, and it will be high water at F and C, and low water at A and N. According to th is it should be high water at any place *in the open sea*, when the moon is upon the meridian of that place, and low water when the moon is upon a circle cutting the meridian in question at right angles; but, in fact, the greatest and least heights of the water at such a place do not occur till about three hours *after* the periods fixed in this supposition. The delay is thus explained: the elevated parts of the sea have received such an impulse towards ascent, that they continue to rise after the earth's rotation has carried them from under the line of the direct attraction of the moon; this impulse being also aided for a time by the moon continuing to attract the water upwards, though in a less degree.

As the moon crosses the meridian of a place about every twenty-four hours fifty minutes and a half, the sea in that space of time ebbs twice and flows twice all over the world, although much less towards the poles than within the tropics, where the waters are under the direct line of the lunar attraction.

In the above remarks we have spoken only of the moon, because, though the sun is so very much larger than the moon, yet the latter, on account of her nearness to our planet, has the most powerful effect upon the tides; it is calculated that her influence is nearly triple that of the sun. The sun, however, acts upon the ocean in the same manner, though in a less degree. When these two bodies unite their influence, which they do at the seasons of new and full moon, the tides naturally rise the highest, and are then called *spring tides;* but when the moon is in her quadratures, or quarters, the action of each of the two luminaries is directly opposed to that of the other; the tides are then of course the lowest, and are called *neap*

Fig. 2.

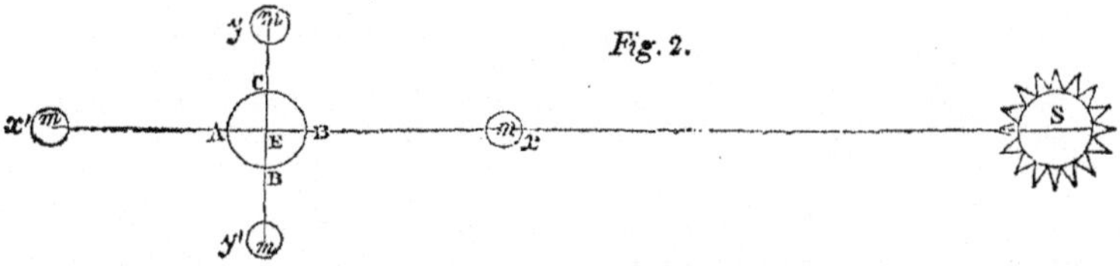

tides. To explain this more clearly, let E (fig. 2) be the earth, S the sun, and *m* the moon; when new, the moon is situated at *x*, when full at *x'*, and when in her quadratures at *y* and *y'*. It results, from what has been already said, that, both at new and full moon, the sun and moon will each assist the other in raising the ocean round A and D, and depressing it at C and B; but when the moon is in her quadratures, her tendency is to raise the waters at C and B and depress them at A and D; while that of the sun is exactly the reverse. During the moon's circuit round the earth, the spring and the neap tides each occur twice, and one after the other.

If the earth were entirely covered by a sea of uniform depth, and the sun and moon moved always in the plane of the equator, the region of the highest tides would always be directly under the equator, while at the poles there would never be any tide whatever. But the changes that occur in the positions of the sun and moon, and several other circumstances, prevent the tides from taking place in so uniform a manner.

In figure 3, let us suppose A E H to be the equator, and C and F the poles

Fig. 3.

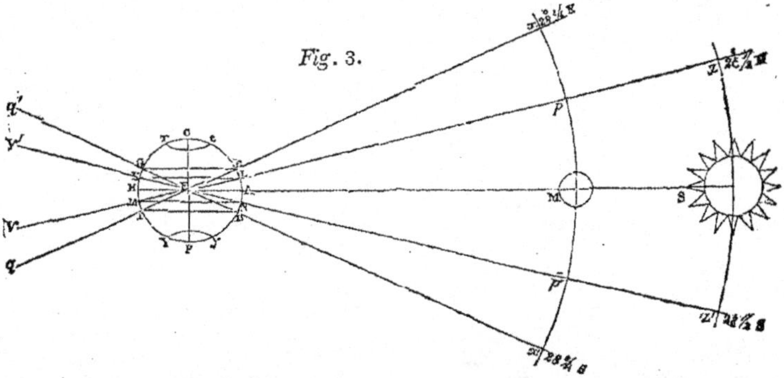

of the earth, and let M, the moon, and S, the sun, both be in the plane of the equator; in this case the water will be higher at H and A, than in any other places, and most depressed under the circle C E F. By the earth's rotation, E will come to A, and it will be high water at E, but C and F, the poles, suffer no change of position from the earth's rotation, and the waters there will, consequently, remain just as before. If the sun and moon, therefore, were to remain constantly in the plane of the equator, the highest tideswould always be at A, E, H, &c., that is, under the equator; while, as Cand F would always be situated the same with respect to the sun and moon, it would always be low water at the poles, that is, they would never have any tides at all. The sun and moon, however, are continually changing their positions, with respect both to the equator and to each other, and corresponding variations are produced in the tides. The moon, for instance, is sometimes as much as 28¾ degrees on each side of it. Suppose her to be situated at *x*, 28¾ degrees north of the equator, draw a line from *x* through the centre of the earth, and let it come out at the opposite side of the earth, the points B and I will have the highest tides, and as the earth turns round, all the parts of the circles G B and I D will successively come to B and I. The waters, under those two circles, will, therefore, have the highest lunar tides, (when the moon is at x or x') and the waters, under the circles $T\,t$ and $Y\,y$, will, in turn, be the most depressed. Results of a similar kind are obtained if we notice the changes of the sun's position. In the course of a year, that luminary ranges nearly 23½ degrees on each side of the equator. Suppose him to be situated at z, 23½ degrees north of it: a line drawn from him, through the earth's centre to the other side, will pass through L and M, and the parts under the circles L K and M N will, while the sun is at z or z', experience the highest solar tides. If, therefore, at the time when the sun is situated at z, or z', the moon happens to be at x, x', q, or q', the sea, under the four circles G B, I D, L K, and M N, will, as the earth moves round, have the highest tides on the globe; the tides of G B and I D being, however, higher than those of L K and M N; since the moon's attraction is more powerful than that of the sun. If again, while the sun is at z, or z', the moon happens to be at p, p', v, or v', the forces of both will combine to raise the tides highest under the two circles L K, M N; and when both luminaries are on the equator together, the circle A E H (that is, the equator) will, alone, be that of the highest tides. It must at the same time be kept in view, that whenever the sun and moon are not situated at the same distances from the equator, so that the circles of their highest tides do not coincide or fall together, allowance must be made for their attractive forces *counteracting*, in some degree, each other's effects upon the ocean; and as the moon completes her range on each side of the equator in about 29½ days, while the sun, to complete his, takes nearly 365¼ days, their combined motions must produce continual irregularities in the tides. Taking one year with another, the mean monthly range of the moon on each side of the equator is the same as the annual range of the sun (23° 28′); the *highest* tides are, consequently, within the tropics, and the *least* within the arctic and antarctic circles.* Within the tropics, the flood tide passes from east to west, (following the apparent course of the sun and moon,) but as the torrid zone is the seat of the highest tides, the flood in the northern temperate zone comes from the south, and in the southern temperate zone from the north. To this rule there are, nevertheless, local exceptions, caused by those derangements of the tide which we are now going to mention.

Of all irregularities in the tides, those are the greatest which are occasioned by the obstacles offered by the land to the ebb and flow of the waters. The impediments created by shallows in the ocean, and by the shores, bays, gulfs, and promontories of islands and continents are such, that the tides are greatly delayed, altered both in degree and in direction, and in many places so accumulated, that they rise to heights far exceeding what is witnessed in the open ocean. On the coasts of the islands of the South Sea, there are regular tides of only one or two feet in elevation; but on the western shores of Europe, and on the eastern shores of Asia, the tides are very strong, and have many variations. On the northern coasts of France, the flow being confined in a channel, and repelled also by the opposite coasts of England, rises to a surprising height; at St. Maloes, in Bretagne, it is said, even to 50 feet. The tide of the German Ocean is twelve hours in travelling from the mouth of the Thames to London Bridge, where it arrives about the time that there is a new tide in the German ocean. This is one instance out of many, of the effect produced upon the tide when it has to pass along a narrow channel, and to overcome an opposing current.

The explanation that has been given

---

* We have already shown, that it is high water at any place twice in every 24 hours 50½ minutes. When a place is on the *same side* of the equator as the moon, the tide, which is produced while the moon is *above* the horizon of the place, will *exceed* the tide which is produced while the moon is *under* the horizon of the place; but when a place is on the opposite side of the equator to the moon, the effect is exactly the reverse. This is explained in the following way:—LK (fig. 3) is a parallel of north, and MN a parallel of south, latitude; CHFA is a meridian; and when the moon is at K, LK and MN are also circles of the highest tides. Each place under those circles has high water twice in 24 hours 50½ minutes—once, when it is under L or M—and again, when it is under K or N; but, under M and N, the *tides are not so high* as at L and M, because we have before shown that L and M are the points of the highest tides when the moon is at p. Now, to places under L and N, the moon is *above* the horizon—and, to places under M and K, she is *below* the horizon; and, therefore, when the moon is at p, *north* of the equator, a place under LK, a parallel of *north* latitude, will have its *greatest high water* when the moon is *above* the horizon; but a place under MN, a parallel of *south* latitude, will have its *greatest high water* when the moon is *below* the horizon. When the moon is at p′, *south* of the equator, these effects will be just reversed. In summer, when the sun's declination is considerably north, the afternoon tides, *north* of the equator, are higher than the morning tides; in winter, the morning tides exceed those of the afternoon.

of the manner in which tides are created in the ocean, will enable us to perceive why it is that, in some gulfs and inland seas, there are either no tides, or such trifling ones as to be scarcely discernible. In figure 1, the waters at A are brought to B, not only by the moon raising up the parts immediately under her, but also by her drawing obliquely towards B the parts distant from B, and by the lateral flowing of the neighbouring waters (*p n*, for instance) towards B, which results from their being less attracted by the moon, and, therefore, heavier than those at B. Being heavier than B, they press upon and flow towards that part. In small collections of water, the moon acts with the same line of attraction, or nearly so, upon every portion of the surface at once, and, therefore, the whole of the waters being equally elevated at the same period, no part of them is ever higher than the other. This is one reason why the Baltic has no perceptible tides, and why even those of the Mediterranean are hardly visible.* But in addition to this, the two seas in question are so circumstanced that they cannot receive tides from the Atlantic: 1st, because their entrances are not turned towards the main direction of the Atlantic tide; 2ndly, because their entrances are so narrow, that the quantity of tide which that ocean can, in a few hours, impel into them, is insufficient, after being spread over the extensive surfaces of the two seas, to raise their level at all perceptibly. The Greeks, who accompanied Alexander the Great in his expedition to the east, having never been on any other coasts than those of the Mediterranean, were seized with complete consternation on first beholding the retreat of the strong tide which the Indian ocean sends into the river Indus. In gulfs which are differently circumstanced with respect to the direction of their entrances, and which have openings wider, as compared with their extent, the tides propagated from the ocean are sensibly felt. Hudson's and Baffin's Bays, and the Red Sea, are examples which prove the correctness of this observation.

Currents and winds (especially the latter) have, according to their direction, an influence either in quickening or retarding the tide; indeed a powerful wind will sometimes keep a tide out of very narrow channels. On the contrary, a strong wind coming from the same quarter as the tide, will raise it several feet above its usual level.

The causes which render the movements of the tides complex and irregular, may thus be summed up under four heads—1. The variations in the positions of the sun and moon, with respect to the equator and to each other. 2. The obstacles presented by the land; 3. by winds; and 4. by currents. The existence of these causes renders it impossible to lay down any general rule for calculating the level, either of high or of low water, in different latitudes.

3. *Currents* in the ocean may be occasioned in various ways: they may arise from an external impulse, (a gale of wind for instance); from a difference in temperature or saltness between two parts of the sea; from the periodical melting of the polar ice, or from the inequality of the evaporation which the surface of the sea undergoes in different latitudes. These causes may produce either constant or occasional currents, and, according as they act in concert or in opposition, will their effects be various.

The most remarkable currents are those which continually follow the same direction. There is one which sets regularly from each of the poles towards the equator; and when we get within twenty-eight or thirty degrees of the line on either side, a general movement is observed in the ocean, in a direction nearly from east to west. The existence of the two polar currents is proved by the floating of masses of ice from the frigid into the temperate regions: these masses are, at times, seen as low as the forty-fifth, or even the fortieth degree of latitude. It was the opposition of the polar current which principally occasioned the failure of the attempt made last year under Captain Parry to reach the north pole; before they desisted from their efforts, the expedition found that, as they advanced over the ice, they were being drifted *southward*, at a rate faster than that at which they were travelling northward. It is equally certain that a tropical current exists, judging not only from the direction of bodies floating on the water, but also from the circumstance that vessels, in crossing from

* The little tide which there is in the Mediterranean, seems to be formed chiefly in the part extending to the east of Malta, and to proceed northward into the Gulf of Venice. M. D'Angos observed that at Toulon, on the coast of France, the sea rose a foot about three hours and a half after the moon passed the meridian.

Europe to America, descend to the latitude of the Canary Islands, where they fall into a current and are carried rapidly to the west. In going from America to Asia across the Pacific, a similar effect is observed. It might be supposed that this was due solely to the trade-winds, but such is not the case: for it is quite possible to distinguish their effect from that of the currents, since the progress of the vessel is quicker than it could be with the aid of the wind alone.

The origin of the polar currents is, no doubt, in a great measure, to be referred to the centrifugal force which is the result of the earth's rotation. (See *Mathematical Geography*, chap. 8.) It may be further explained, when we reflect that the water towards the poles, both on account of its lower temperature and its being less attracted by the heavenly bodies, is *heavier* than the water in the tropical regions, and, moreover, that the heat of the torrid zone occasions a much more powerful evaporation of the sea than is elsewhere experienced: the consequence is, that the waters nearer the poles will move towards the equator, in order to restore the equilibrium which has, in these several ways, been destroyed. The tropical current may also, though in another manner, be explained as proceeding from the earth's rotation. The waters, as they advance from the polar seas, pass from regions where the rotatory motion of the earth's surface is very slight, to those where it is exceedingly rapid; they cannot immediately acquire the rapid motion with which the solid parts of the earth revolve in the tropical regions, and they are, accordingly, left rather behind, that is, to the *westward* (the earth turning round from west to east). The ocean, consequently, appears to retreat from the western, and advance upon the eastern coasts of the continents, or, in other words, to have a general movement from east to west; and the effect is very much assisted by the constant blowing of the trade winds.

We will now explain the modifications or changes which this grand movement in the ocean undergoes, in consequence of the obstacles presented by the land to its free progress. When it meets with shores or narrow straits to impede or turn aside its course, it forms strong and even dangerous currents. The eastern coast of America, and the West India Islands, constitute a sort of dyke to the general westward motion of the Atlantic; and it will be seen, if we refer to a map, that from Cape St. Roche, which has about five degrees of south latitude, the coast of South America stretches away in a continued line to the north-west, as far as the isle of Trinidad. Owing to this shape of the coast, the waters, as far as the tenth degree of south latitude, are, when they approach America, carried away in a current to the north-west. This current afterwards enters the gulf of Mexico, through the strait formed by the western end of Cuba, and the opposite peninsula, (from this part it is called, by navigators, the *Gulf-stream*,) and follows the bendings of the Mexican coast, from Vera Cruz to the mouth of the Rio del Norte, and thence to the mouths of the Mississippi, and the shoals west of the southern extremity of Florida. It next takes a new direction to the north, and rushes impetuously into the gulf of Florida. M. Humboldt observed in the month of May 1804, in the 26th and 27th degrees of latitude, that its velocity was eighty miles in twenty-four hours, although, at the time, there was a violent wind against it. At the end of the gulf of Florida, (north lat. 28°) it runs to the north-east, at the rate, sometimes, of five miles an hour. It may always be distinguished by the high temperature * and the saltness of its waters, their indigo-blue colour, and the quantity of sea-weed floating on the surface, and also by the heat of the surrounding atmosphere. The rapidity and temperature of the *Gulf-stream*, diminish towards the north, while, at the same time, its breadth increases.† Its further progress northward is at last checked by the southern extremity of the great bank of Newfoundland, in the 42d degree of latitude, where it turns suddenly to the east. It afterwards continues moving towards the east, and the east-south-east, as far as the Azores islands; and

---

* Humboldt observes that "the waters of the Mexican Gulf, forcibly drawn to the north-east, preserve their warm temperature to such a point, that at forty and forty-one degrees of latitude, he found them at seventy-two degrees and a half (Fahrenheit); when, out of the current, the heat of the ocean at its surface was scarcely sixty-three degrees and a half. In the parallel of New York, (forty-one degrees north) the temperature of the *Gulf stream* is, consequently, equal to that of the seas of the tropics in the eighteenth degree of latitude."

† Its breadth in latitude twenty-eight degrees and a half is seventeen leagues; (3.46 miles to a league) in the parallel of Charles town, (thirty-three degrees, nearly) from forty to fifty leagues; and on the meridian of Corvo and Flores, the westernmost of the Azores islands, it is one hundred and sixty leagues.

thence it turns towards the straits of Gibraltar, the Isle of Madeira, and the group of the Canaries, till, on reaching the parallel of Cape Blanco, it completes the round by mixing with the grand westerly current of the tropics. It is probable, however, that a branch still keeps on its course to the south and south-east, along the coast of Africa; for it is well known that ships, if they approach too near the shore, are drawn into the gulf of Guinea, and with difficulty get out again. We thus see that between the parallels of 11 and 43 degrees, the waters of the Atlantic are carried on in a continual whirlpool. Humboldt remarks that, supposing a particle of water to return to the same place from which it departed, "we can estimate, from our present knowledge of the swiftness of currents, that this circuit of three thousand eight hundred leagues is not terminated in less than two years and ten months. A boat, which may be supposed to receive no impulse from the winds, would require thirteen months, from the Canary Islands, to reach the coast of Caraccas; ten months to make the tour of the Gulf of Mexico and reach Tortoise Shoals, opposite the port of the Havannah; while forty or fifty days might be sufficient to carry it from the straits of Florida to the bank of Newfoundland. It would be difficult to fix the rapidity of the retrograde current from this bank to the coasts of Africa: estimating the mean velocity of the waters at seven or eight miles in twenty-four hours, we find ten or eleven months for this last distance." It is a curious fact, that towards the close of the 15th century, before Europeans were acquainted with the existence of America, two bodies belonging to an unknown race of men were cast by the Gulf-stream on the coasts of the Azores, and pieces of bamboo were brought by the same current to the shore of the small island of Porto Santo; by these circumstances, Columbus is said to have been strengthened in his conjectures with respect to the existence of a western continent.

An arm of the Gulf-stream in the 45th and 50th degrees of latitude, runs to the north-east, towards the coasts of Europe, and becomes very strong when the wind has blown long from the west. The fruit of trees which belong to the American torrid zone is every year deposited on the western coasts of Ireland and Norway; and on the shores of the Hebrides are collected seeds of several plants, the growth of Jamaica, Cuba, and the neighbouring continent. The most striking circumstance, perhaps, is that of the wreck of an English vessel, burnt near Jamaica, having been found on the coast of Scotland.

There are various currents in the Pacific and Indian oceans. The general westward motion of the former is impeded by a numerous archipelago, and hence it receives different directions. A strong currrent sets to the west, through each of the two straits which respectively separate New Holland from New Guinea and from Van Diemen's Land. It then gets diverted, and flows northward along the coast of Sumatra, till it reaches the bottom of the Bay of Bengal. The following appears to be the reason of its taking this course:—the general impetus of the Pacific towards the west, being encountered by New Holland and the numerous East India Isles, is broken and dispersed; while the westerly motion of the Indian sea has not, in so early a stage, acquired much strength; the polar current from the south, at the same time, presses upon the wide opening which the Indian sea presents to that quarter, and the waters on the eastern verge of that sea are, therefore, pushed into the Bay of Bengal. In the neighbourhood of Ceylon and the Maldive islands, however, the tropical motion has become powerful enough to resist the polar current. The westerly current then recommences, but is again turned out of its line and made to flow to the south-west, by the chain of islands and shallows, which reaches from the extremity of the Indian peninsula to Madagascar. After passing Madagascar, it dashes against Africa, and at the termination of that continent, mingles with the general motion of the waters.

A current afterwards sweeps from the Atlantic into the Pacific ocean, through the straits of Magellan. There can be little doubt that this is a branch of the general current from the south pole; though, at the same time, it may be partly the result of the westerly movement of the Atlantic, which, being checked by the shores of Brazil, flows to the south-west, along the South American coast.

There is a question connected with the currents of the Arctic ocean, which has engaged a good deal of attention, and been considered difficult to explain:

it is from what quarters the timber can come which is found floating on the polar seas in such large quantities, and so much of which is thrown ashore on the northern side of Iceland*. The few specimens seen of the growth of Mexico and Brazil, must have travelled to the north by means of the Gulf-stream of which we have spoken; the rest (principally pines and firs) most likely comes from Siberia and North America, along the shores of which it is drifted till it arrives at the opening into the Atlantic, in the midst of which Iceland is placed.

The existence of *under currents* different from, and even opposite in their direction to those on the surface is, by no means, improbable in some cases, though it is a matter not admitting of proof. It has been thought that the Mediterranean, which has a strong flow always setting into it through the straits of Gibraltar, sends back a portion of its waters into the Atlantic, by a concealed current. Contrary currents, passing along side by side, are not uncommon. In the Kattegat, a northern current flows out of the Baltic, along the coast of Sweden, while a southern one enters the Baltic along the coast of Denmark. When two opposite currents of about equal force meet one another, they sometimes, especially in narrow channels, turn upon a centre and assume a spiral form, giving rise to *eddies* or *whirlpools*. The most celebrated of these are the *Euripus*, near the island of Eubœa, in the Grecian Archipelago; *Charybdis*, in the strait between Italy and Sicily; and the *Maelstrom*, off the coast of Norway. The most violent of them, when agitated by tides or winds, become very dangerous to navigation.

On Currents in the Atmosphere.

It does not belong to our present subject to investigate the properties and component parts of the atmosphere, but it will be proper to notice the agitations or movements which are constantly taking place in that fluid. A change in the temperature of a portion of air; an increase or a diminution of the quantity of water which it holds in a state of vapour; in short, any circumstance which causes it either to contract or to expand, destroys the equilibrium subsisting among the different parts of the atmosphere, and occasions a rush of air, that is, a *wind*, towards the spot where the balance has been destroyed.

Every one knows that the velocity and force of winds are exceedingly various; accordingly, several methods have been suggested, and instruments invented, in order to determine their amount with some degree of exactness. In the fifty-first volume of the "Philosophical Transactions," there is a table of the different velocities and forces of winds, drawn up from a considerable number of facts and experiments; the following particulars are extracted from it:—

Velocity of the Wind.	Perpendicular force on one square foot, in Avoirdupois pounds.	Common Appellations of the forces of such Winds.
Miles in an hour.	dec. parts.	
1	.005	Hardly perceptible.
4	.079	Gentle, pleasant wind.
5	.123	
10	.492	Brisk gale.
15	1.107	
30	4.429	High wind.
35	6.027	
50	12.300	A storm.
80	31.490	A hurricane.
100	49.200	

Winds may be divided into three classes—those which flow *constantly* in the same direction, those which are *periodical*, and those which are variable. It must be observed that the terms which express the direction of winds are employed in a sense quite contrary to that in which they are used when we speak of the direction of currents in the ocean; a westerly current, for example, signifies a current flowing *towards* the west, but a westerly wind signifies a wind coming *from* that quarter.

The *permanent* winds are those which blow constantly between, and a few degrees beyond, the tropics, and are called *trade-winds*. On the north of the

* From the account which Captain Parry has given of his last voyage, it appears that there is also a great quantity of timber which has been cast by the sea upon the northern coast of Spitzbergen.

equator, their direction is from the north-east (varying at times a point or two of the compass each way); on the south of the equator, they proceed from the south-east. The origin of them is this:—the powerful heat of the torrid zone rarefies or makes lighter the air of that region; the air, in consequence of this rarefaction, rises, and to supply its place a colder atmosphere from each of the temperate zones moves towards the equator. But (as in the case of the polar currents in the ocean) these north and south winds pass from regions, where the rotatory motion of the earth's surface is less, to those where it is greater. Unable at once to acquire this new velocity, they are left behind, and instead of being north and south winds, as they would if the earth's surface did not turn round, they become *north-east* and *south-east* winds. The space included between the second and fifth degrees of north latitude, is the *internal boundary* of the two winds; and this space experiences calms, frequently interrupted, however, by violent storms. The reason why it is situated to the north of, instead of exactly at, the equator, seems to be that the northern hemisphere is warmer than the southern; for since the trade-winds are the result of the continual ascent of heated air in the equatorial parts, their internal boundary will be where the principal ascent is going on,—that is, where the annual temperature is the highest; which, on account of the above-mentioned inequality of temperature in the two hemispheres, will not be at the equator, but somewhat to the north of it. The *external limits* of the trade-winds are at a medium in about the 30th degrees of north and south latitude respectively; but each limit, as the sun approaches the neighbouring tropic, declines further from the equator. The position of the sun has an influence also on their strength and direction; for when that luminary is near the tropic of Cancer, the south-east wind becomes gradually more southerly, and stronger, and the north-east weaker, and more easterly; the effect is reversed when he gets towards the tropic of Capricorn.

The trade-winds would blow regularly round the whole globe within the distance of about 30 or 40 degrees from the equator each way, if the space within those limits were all covered with water; but the uneven surface and unequal temperature of the land divert and derange them: it is on this account that the trade-winds are constantly experienced only over the open ocean. The larger the expanse of ocean over which they range, the more steadily they blow: thus in the Pacific they are commonly more steady than in the Atlantic ocean, and in the south than in the north Atlantic. In sailing from the Canaries to Cumana, on the north coast of South America, it is hardly ever requisite to touch the sails of the vessel. The voyage across the Pacific, from Acapulco on the west coast of Mexico, to the Philippine islands, is performed with equal facility; and if there were a channel through the isthmus of Panama, a westward passage from Europe to China would be more speedy and safe than the usual navigation thither round the Cape of Good Hope; the only interruption to the evenness of this voyage would be in the Caribbean Sea and the Gulf of Mexico, where the trade wind blows impetuously, and is sometimes interrupted by westerly winds. It would not be possible, however, to return by the same route, because in sailing east way must be made to the northward, in order to get beyond the region of the trade into that of the variable winds. Both in the Atlantic and in the Pacific Ocean, the current of the trade-winds becomes broader and more directly east in its course, as it advances from one side to the other of those extensive basins. On the west coast of Africa, owing to the rarefaction which the air undergoes over that continent, the wind is mostly turned towards the shore; from Cape Bojador to Cape Verd it is generally north-west, and thence to the island of St. Thomas, under the equator, it bends gradually first to the west and then to the south-west. Along the coasts of Chili and Peru a south wind prevails. These are two instances of the interruption which the trade-winds experience in the neighbourhood of large masses of land.

In the Indian Ocean the south-east trade-wind prevails between 28° and 10° of south latitude, from within a few degrees of the east side of Madagascar, nearly to the coast of New Holland; but from the 10th degree of south latitude to the northern shores of that ocean, the uniformity of the tropical movements of the atmosphere is destroyed by the *monsoons**, which belong to the class of *periodical* winds. These blow half the year from one quarter, and the other half from the opposite direction: when they shift, variable winds and violent storms prevail for a time, which render it dangerous to put to sea. They of course suffer partial changes in particular places, owing to the form and position of the lands, and to other circumstances, but it will be sufficient to give their general limits and directions. Northward from the third degree of south latitude, a *south-west* wind blows from April to October—from October to April a *north-east;* these monsoons extend over the China sea, but here they incline more to the direction of north and south. Between the 3d and 10th degrees of south latitude, a *north-west* wind blows from October to April, and a *south-east* during the other six months of the year: the former is seldom steady in the open sea, but in December and January it sometimes extends northwards a degree or two beyond the equator. These two monsoons have the greatest strength and regularity in the Java Sea, and thence eastward towards New Guinea. The facts above exhibited may be thus summed up:—from April to October a *south-west* wind prevails north of the equator, southward of this a *south-east* wind—from October to April, a *north-east* wind north of the equator, and a *north-west* between the equator and 10° of south latitude; south of this the usual trade wind, which is in motion through the whole year.

In attempting to account for these movements of the atmosphere over the Indian Ocean, the first thing which strikes us is, that the north-east and south-east monsoons, which are found the one on the north and the other on the south side of the equator, are nothing more than the trade-winds blowing for six months, and then succeeded for the remainder of the year by winds directly opposite. It is also to be noticed that the south-west monsoon in the northern, and the north-west monsoon in the southern hemisphere, each prevails while the sun is perpendicular to their respective regions: they are, therefore, connected with the immediate presence of that luminary. If the Indian Ocean were not bounded as it is by land on the north, the trade-winds would blow over it (at least in the central parts) as they do in the Atlantic and Pacific Oceans. But it is well known that water, owing to its transparency, is very little warmed by the sun's rays, whereas the land is powerfully heated by them; consequently, when the sun is between the equator and the tropic of Cancer, India, Siam, and the adjacent countries, become much hotter than the ocean; the air over them gets rarefied and ascends; colder air then rushes in from the Indian ocean, and a *south-west* wind is produced. When the sun, however, has crossed to the south of the equator, these countries become gradually cool, and the north-east trade-wind resumes its course. At the same time the *north-west* monsoon commences in the southern hemisphere, in consequence of the air over New Holland being rarefied by the presence of the sun.

The monsoons in the Red Sea blow in the direction of the shores; and a similar effect is observed in the Mozambique channel, between Africa and Madagascar, where these winds follow the line of the channel. On the coast of Brazil, between Cape St. Augustine and the island of St. Catharine, and in the bay of Panama, on the west of the isthmus of that name, periodical winds occur somewhat similar to the monsoons of Asia.

The *land* and *sea-breezes*, which are common on coasts and islands situated between the tropics, are another kind of periodical winds. During the day, the air, over the land, is strongly heated by the sun, and a cool breeze sets in from the sea; but in the night the atmosphere over the land gets cooled, while the sea, and consequently the air over it, retains a temperature nearly even at all times: accordingly, after sunset, a land-breeze blows off the shore. The sea-breeze generally sets in about ten in the forenoon, and lasts till six in the evening; at seven the land breeze begins, and continues till eight in the morning, when it dies away. These alternate breezes are, perhaps, felt more powerfully on the coast of Malabar than anywhere—their effect there extends to a distance of twenty leagues from the land. During summer, the *sea-breeze* is very perceptible on the coasts of the Mediterranean, and sometimes even as far north as Norway.

* From the Malay word *moossin*, which signifies a *season*.

We thus perceive that within the limits of from 28 to 30 degrees on each side of the equator, the movements of the atmosphere are carried on with great regularity; but beyond these limits, the winds are extremely variable and uncertain, and the observations made have not yet led to any satisfactory theory by which to explain them. It appears, however, that beyond the region of the trade-winds, the most frequent movements of the atmosphere are from the *south-west*, in the north temperate zone, and from the *north-west*, in the south temperate zone. This remark must be limited to winds blowing over the ocean and in maritime countries; because those in the interior of continents are influenced by a variety of circumstances, among which, the height and position of chains of mountains are not the least important. These south-west and north-west winds of the temperate zones are most likely occasioned in the following manner:—In the torrid zone there is a continual ascent of air, which, after rising, must spread itself to the north and south in an opposite direction to the trade-winds below: these upper currents, becoming cooled above, at last descend and mix themselves with the lower air; part of them may perhaps fall again into the trade-winds, and the remainder, pursuing its course towards the poles, occasion the north-west and south-west winds of which we have been speaking. It has also been conjectured that these winds may frequently be caused by a decomposition of the atmosphere towards the poles, from part of the air being at times converted into water.

Hurricanes have been supposed to be of electric origin. A large vacuum is suddenly created in the atmosphere, into which vacuum the surrounding air rushes with immense rapidity, sometimes from opposite points of the compass, spreading the most frightful devastation along its track, rooting up trees, and levelling houses with the ground. They are seldom experienced beyond the tropics, or nearer the equator than the 9th or 10th parallels of latitude; and they rage with the greatest fury, near the tropics, in the vicinity of land or islands, while far out in the open ocean they rarely occur. They are most common among the West India islands, near the east coast of Madagascar, the islands of Mauritius and Bourbon, in the Bay of Bengal at the changing of the monsoons, and on the coasts of China.

Whirlwinds sometimes arise from winds blowing among lofty and precipitous mountains, the form of which influences their direction, and occasions gusts to descend with a spiral or whirling motion. They are frequently, however, caused by two winds meeting each other at an angle, and then turning upon a centre. When two winds thus encounter one another, any cloud which happens to be between them is of course condensed and turned rapidly round; and all substances sufficiently light are carried up into the air by the whirling motion which ensues. The action of a whirlwind at sea occasions the curious phenomenon called a *water-spout*, which is thus described by those who have witnessed it. From a dense cloud a cone descends in the form of a trumpet with the small end downwards; at the same time, the surface of the sea under it is agitated and whirled round, the waters are separated into vapour, and ascend with a spiral motion till they unite with the cone proceeding from the cloud; frequently, however, they disperse before the junction is effected. Both columns diminish towards their point of contact, where they are not above three or four feet in diameter. In the middle of the cone forming the water-spout, there is a white transparent tube, which becomes less distinct on approaching it, and it is then discovered to be a vacant space in which none of the small particles of water ascend; and in this, as well as around the outer edges of the water-spout, large drops of rain precipitate themselves. In calm weather, water-spouts generally preserve the perpendicular in their motion; but when acted on by winds they move on obliquely—sometimes they disperse suddenly, at others they pass rapidly along the surface of the sea, and continue a quarter of an hour or more before they disappear. A notion has been entertained that they are very dangerous to shipping, owing to the descent, at the instant of their breaking, of a large body of water sufficient to sink a ship; but this does not appear to be the case, for the water descends only in the form of heavy rain. It is true, that small vessels incur a risk of being overset if they carry much sail; because sudden gusts of wind, from all points of the compass, are very common in the vicinity of water-spouts.

On Physical Climate.

Circumstances which determine its character—Mean annual Temperature—Extremes of Heat and Cold—Isothermal Lines—Temperatures of the Southern and Northern Hemispheres compared—Quantity of Evaporation and of Rain in various Latitudes—Character of the Seasons in the different Zones.

The term *climate* is applied to the state of the air, in order to express that particular combination of temperature and moisture which exists in the atmosphere of any greater or less extent of country. The climates of different regions of the globe, and the causes which occasion their great diversity, are interesting matters of inquiry. If an uniform climate had been communicated to the whole globe, we should not have seen such wonderful variety among the animal and vegetable tribes; and many things that now raise the delight, or administer to the necessities of the human race, would have been entirely unknown. It might at first be imagined that the climate of any particular place depended solely upon the action of the sun; but, upon further consideration, we shall find that there are other circumstances to be taken into account: were it not so, any two places having the same latitude, and consequently receiving the sun's rays at the same angle, would enjoy similar climates, which is by no means the case. It is a wise ordination of Providence that the sun's action is modified in such various ways, as to produce a more equal distribution of heat over the surface of the globe than would otherwise have existed; by means of which, large regions are adapted to the residence and support of man, that would else, from extreme heat or cold, have been quite uninhabitable.

There are eight circumstances which determine physical climate:—1. The power of the sun's immediate action, which increases in proportion as we approach the equator; 2. elevation of the ground above the level of the ocean; 3. position with respect to the great seas; 4. quarter towards which the surface of the country slopes; 5. position and direction of chains of mountains; 6. nature of the soil; 7. degree of cultivation and improvement to which the country has arrived; 8. prevalent winds.

1. The amount of the immediate solar heat depends upon the position of the sun in the ecliptic, because to all places (whatever their distance from the Equator) this position determines the length of the day, and the direction in which the sun's rays strike the earth. When the sun remains a long time above the horizon, his continued action causes a powerful accumulation of heat; the nights also being short, but little of this heat escapes during his absence. On this account it is, that even within the arctic circle the summer temperature is sometimes quite oppressive*. The direction in which the rays fall upon the earth is another important consideration; their greatest force being experienced when they are perpendicular to the surface. On the contrary, when the sun is near the horizon, his rays merely glance along the ground, and many of them, before they reach it, are absorbed and dispersed, owing to the density of the lowest stratum of the atmosphere along which they have to pass. Bouguer calculated that, out of 10,000 rays falling upon the earth's atmosphere, 8123 arrive at a given point if they come perpendicularly; 7024, if the angle of direction is 50 degrees; 2831, if it is 7 degrees, and only 5, if the direction is horizontal.

2. It is well known how the temperature of a place is influenced by the elevation of the land. In proceeding from the equator towards either of the poles, without altering our height above the level of the sea, we must travel a great distance before we find the mean annual temperature reduced even a few degrees; but, by increasing our elevation, a rapid change of temperature will be experienced, till we arrive at the point where constant frost prevails. The extreme cold which exists in the upper region of the atmosphere seems to be owing to the expansion of the air (see chap. vii. of the *Treatise on Heat*); partly, also, to the circumstance of that region being beyond the reach of the heat reflected from the surface of the earth. The decreases of heat, at equal ascents, are not altogether uniform, as they take place more rapidly in the higher parts of the atmosphere. The annexed table, abridged from one drawn up by Professor Leslie, shows that even under the equator, where the sun's direct influence is most powerful, an ascent of rather more than 15,000 feet (about $2\frac{7}{8}$ miles) above the level of the sea, will bring us within the region of perpetual frost. This provision of nature of course increases considerably the number of habitable countries within the torrid zone.

* In Norway, as high as latitude 70 degrees, the thermometer has been seen to rise above 80 degrees.

Latitude.	Mean Temperature at the level of the sea.	Height of Curve of Congelation.
0	84.2 Fahr.	15,207 feet.
10	82.6	14,764
20	78.1	13,478
30	71.1	11,484
40	62.6	9001
50	53.6	6334
60	45	3818
70	38.1	1778
80	33.6	457
90	32	0

3. The effect of the sea is to *equalise* temperature, so that a maritime country is not liable to such extremes, either of heat or cold, as an inland one. The sea itself being of a very equable temperature, the winds which pass over an extent of it partake somewhat of the same character. When a *cold* wind passes over sea it receives part of the warmth of the water, the upper particles of which being thus rendered cooler, and consequently heavier than those below, descend and are succeeded by warmer particles; so that there is a *continual* tendency in the sea to temper a cold wind passing over its surface. A cold wind, blowing overland, is at first rendered warmer by the earth's surface; but this surface quickly becoming cooled, ceases to have any effect upon the wind, which, therefore, travels on with undiminished rigour. Again, a *warm* wind, in passing over sea, is cooled by the agitation which it produces bringing up cooler water from below, as well as by the constant evaporation which it occasions; the surface of the water also cannot, as that of land, be powerfully heated by the sun's rays, because it affords them a free passage, and therefore it cannot communicate heat to the atmosphere in the degree which the land does. From these circumstances it results, that, though a place situated inland, and another upon a coast may have the same mean annual temperature, the range of the thermometer at each will be very different, the summers of the latter will be cooler, and the winters milder than those of the former. It is from this cause that islands are so much more temperate than continents. It follows, too, that countries in our hemisphere will be rendered warmer by having large tracts of land to the south and sea to the north, and cooler when the relative position of these two is reversed. This fact is exemplified by a comparison of the climate of India with that of Africa north of the equator, the heats of the former country being much more supportable than those of the latter. Not only the temperature of a wind, but also its degree of moisture, depends upon the nature of the surface over which it passes. A wind coming up from the ocean is loaded with vapours, but one sweeping over an extent of land is rendered dry and parching. This explains to us why, in our own island, a south-west and an easterly wind are so opposite in character.

4. The aspect of a country has an influence upon its climate, for this reason, that the angle at which the sun's rays strike the ground, and consequently the power of those rays in heating it, varies with the exposure of the soil relatively to that luminary. When the sun is elevated on the meridian 45 degrees above the horizon, his rays fall *perpendicularly* on the side of a hill facing the south at an equal angle, while the plain below receives them at an *angle of 45 degrees.* Supposing the north side of the hill to have a similar slope, the rays would run *parallel* to its surface; and their effect be very trifling, but if the declivity were still greater, the whole surface would be in the shade. This, though an extreme case, serves to show why temperature varies with the inclination of the earth's surface. Since the warmest part of the day is not when the sun is on the meridian, but, owing to the accumulation of the heat, two or three hours afterwards, it follows that, in our hemisphere, a south-south-west or south-western aspect is the warmest, and a north-north-east, or north-eastern, the coldest, if no local circumstances exist to make it otherwise. The effect of aspect is, of course, most strikingly seen in regions covered with high mountains. In the Vallais in Switzerland, the Alps are on one side covered with ice, while vineyards and orchards flourish on the other.

5. Mountains affect a climate in more ways than one. They attract the vapours in the atmosphere, and causing them to condense, give rise to those violent rains which are often experienced in the neighbourhood of lofty ranges. They also afford shelter from winds. In narrow valleys, the sides of which in summer strongly reflect the sun's rays, this shelter sometimes renders the heat very injurious. One reason why the central and southern parts of European Russia are exposed to greater cold than their latitude and inclination southward would lead us to expect, is the absence of any chain of mountains to protect them from the full influence of the winds blowing from the White Sea and the Ural Mountains. The inhospitable climate of Siberia arises from its descent towards the north exposing it to the winds of the

Frozen Ocean, while at the same time the vast mountainous chains that cross central Asia, intercept the southern winds, whose access would tend to mitigate the rigour of the atmosphere.

6. It is evident that the nature of the soil must very materially operate upon climate. One soil acquires heat, keeps its acquired heat much longer, or reflects it more readily, than another. One, which from its porous character allows the rain descending upon it to pass freely into the earth, will emit much fewer exhalations than one which retains the waters near the surface. Thus clayey or marshy grounds lower the temperature, and especially in hot and humid climates, affect the atmosphere in a manner pernicious to health; on the other hand, those which are light, stony, or calcareous, tend to make the atmosphere salubrious. The great cold, and the unwholesome air that prevail in the Russian governments of Astracan and Orenburg, lying to the north of the Caspian Sea, are attributed partly to the saline nature of the soil; and it is well known that the arid tracts of sand in Africa and Arabia, conduce not a little to the excess of heat under which those countries labour.

7. Without cultivation, few climates would be healthy or agreeable. In countries to which the labours of civilized man have never been extended, the rivers, spreading themselves over the low grounds, form pestilential marshes, and forests, thickets, and weeds are so numerous and impenetrable, as to prevent the earth from receiving the beneficial influence of the sun's rays. The air, from these causes, is constantly filled with noxious exhalations. But the efforts of the human race, conducted with skill and perseverance, produce a surprising change: marshes are drained; rivers embanked; the soil broken up by the plough is exposed to the sun and wind, and the clearing away of the forests raises the temperature, and allows a freer circulation to the atmosphere. There is little doubt that many parts of Europe enjoy a milder climate now than they did in the time of the Romans, or even at periods much more recent. Several districts in North America have experienced, as the country has become more widely settled, a similar improvement of climate. The destruction of forests may, however, be carried to a pernicious extent, either by depriving a country of shelter from particular winds, or (especially in hot climates) by lessening too much the quantity of moisture; it being well known that there is a great evaporation from the leaves of vegetables. The sultry atmosphere and dreadful droughts of the Cape de Verde islands are owing to the destruction of the forests; and Greece, Italy, and other countries are said to have been deteriorated in climate from the same cause. It is attributed to this also that the southern part of Iceland is more accessible than formerly to the cold which proceeds from the Arctic Ocean.

8. The combined influence of the several causes of physical climate which we have been considering will be variously modified by the *prevalent winds* of a country. This is obvious enough, because we know that the character of a wind depends upon the quarter whence it comes, and the surface over which it passes. Great Britain, for example, would in a great measure lose its insular climate, if its prevailing winds came across the continent, instead of from the Atlantic Ocean.

Notwithstanding the several circumstances which we have thus pointed out as influencing climate, and which occasion numerous local irregularities, the temperature, with these exceptions, becomes gradually lower as we pass from the equator towards either of the poles. By this is not to be understood the temperature of any particular day, or even season, but the *mean annual temperature*, which is obtained by adding together the temperatures of all the months*,

* The temperature of each month is the average of all the daily temperatures in the month, and the daily temperature is the average of several observations made at stated periods, every hour or half hour, for instance, each day (24 hours). It is evident that such frequent observations would be very troublesome, and shorter methods of discovering the mean annual temperature of a place have therefore been sought after. Rules have been laid down for calculating what this is under different parallels of latitude, and the results no doubt approach very near to the truth; but it would obviously be incorrect to apply these rules to any particular place, because we should be uncertain how the climate of that place was affected by local circumstances. The best method is to ascertain at what period in each day (taking one day with another) the thermometer stands at its mean height for the day; and when this has been ascertained, one observation each day, at that period, will be sufficient. In this country it would appear that the time at which the thermometer shows the mean heat for the day, is about a quarter or half past eight in the morning. Another method of discovering the mean annual temperature at any place, is to observe the height of the thermometer in cavities at some depth below the earth's surface, it being found that this height nearly corresponds with the mean annual height in the air above. M. Lacroix, in his work on Physical Geography, states that in the caves below the Observatory at Paris, (lat. 49°) about 85 feet below the surface, Fahrenheit's thermometer constantly stands between 52° and 54°, scarcely ever varying two degrees, while above, the difference of temperature between summer and winter sometimes exceeds 90°. In the salt mines at

and dividing the sum by the number of months in the year; so that the mean annual temperature expresses that height at which the thermometer would stand at any place, if we could suppose it perfectly stationary throughout the whole year. It is not sufficient, however, to take one year only, but a series of at least ten or fifteen years, from the *mean* result of which series a conclusion nearly accurate may be drawn. Though the temperature of a place is continually varying, and though the changes occur frequently in the most sudden manner, it never differs more than a certain number of degrees either way from its mean state; and when it has reached either extreme, a reaction may shortly be expected. In the torrid zone any excessive accumulation of heat is prevented by the constant blowing of the trade-winds from cooler regions; and in the frigid zones the tendency to great extremes which arises from the continued presence of the sun in summer, and his long absence in winter, is counteracted by the circulation of the atmosphere, and by the circumstance that the fields of ice, in melting, absorb large quantities of heat, while on the other hand warmth is given out when the surface of the ocean is being frozen over.—(See chap. ix. of the *Treatise on Heat.*)

The extremes of temperature which have been witnessed in different parts of the globe are, nevertheless, very considerable. In New South Wales, Fahrenheit's thermometer sometimes rises to 100 degrees and upwards; at Pekin, in China, it has been seen at 110°, and at different places in India at 110°, and even 115°. Major Denham, in his late travels in Africa, observed it more than once at 113°; and at Belbeis, in Egypt, it is said to have risen, under the influence of the hot wind from the desert, to 125°.* These heights are intended to express the degree of heat in the *shade*. The accuracy of the observations depends upon the circumstances under which they were made, since it is requisite that the thermometer should be in a situation freely exposed to the outer air, and also where there are no surfaces immediately near to reflect the sun's rays. The most extreme cold experienced is in the northern parts of Asia and America. In Siberia, as far south as the 58th degree of latitude, M. Pallas observed the freezing of mercury*. The same phenomenon is by no means unusual at Quebec. At Hudson's Bay the spirit thermometer has sunk to — 50°, and at Melville island, (N. lat. 74¾°,) where Captain Parry wintered in his first north-western expedition, it fell, on the 15th February, 1820, to 55 degrees below *zero*.

A treatise upon *isothermal*† *lines*, published some years ago by M. Humboldt, gives several curious results drawn from various observations upon temperature made by himself and others. A few of these it will be proper to notice here, because they illustrate, in a striking manner, the fact upon which we have already remarked, that the climates of places do not depend solely upon the direct action of the sun. If it were so, all places having the same latitude would experience the same mean annual temperature. It had long been known that this was not the case, especially on comparing Europe with America; but M. Humboldt's statements will enable us to form some idea of the amount of the difference. According to that philosopher, the isothermal line which indicates the temperature of 32 degrees (the freezing point of water) passes between Ulea, in Lapland (lat. 66°) and Table Bay, on the coast of Labrador, in North America, lat. 54°. The isothermal line of 41 degrees passes near Stockholm, lat. 59½°, and St. George's Bay, Newfoundland, lat. 48°. The line of 50 degrees runs through the Netherlands, lat. 51°, and near Boston, in the United States, lat. 42½°; that of 59 between Rome and Florence, lat. 43°, and near Raleigh, in North Carolina, lat. 36°. Taking similar latitudes, the following are the differences of temperature between the west of Europe and the east of North America:—

Latitude.	Mean temperature in the West of Europe.	Mean temperature in the East of America.	Difference.
°	°	°	°
30	70.1	66.8	3.3
40	63.1	54.5	8.6
50	50.8	37.9	12.9
60	40	24	16

Wielickza, in Poland (lat. 50°), from the depth of 320 to that of 745 feet, the thermometer stands at about 50°. At Cairo, in Egypt, (lat. 30°), at the bottom of Joseph's well (210 feet deep), it stands at 70°; in the mines of Mexico (lat. 20°), 1650 feet below the surface, it stands at 74¾°. In these heights we discern how the temperature increases on approaching the equator.

* When the thermometer is raised to such an extraordinary height as this, it is probably the effect produced by very fine particles of sand which are carried along by the atmosphere. Humboldt, in the arid plains of South America, has, during a *wind of sand*, seen it at 114½ degrees nearly; while in Fezzan, in the North of Africa, it has risen, doubtless from the cause just noticed, to 125.6 degrees.

* This takes place when the mercury has sunk to 39 or 40 degrees below *zero*.

† This is derived from two Greek words, and signifies *equal heat* or *temperature*. An isothermal line, therefore, is a line drawn over places which have the *same temperature* (*annual*, unless otherwise expressed.)

From the annexed table, there appears to be nearly as much difference between the mean temperatures of the eastern and western parts of the old continent as between those of the opposite shores of Europe and America :—

Places.	Latitude.	Longitude.	Mean temperature.
St. Maloes .	48° 39′ N.	1° 57′ W.	54.5°
Amsterdam .	52 22	4 40 E.	53.4
Copenhagen .	55 41	12 30	45.7
Upsala . . .	59 51	17 48	41.9
Naples . . .	40 50	14 10	63.5
Vienna . .	48 11	16 22	50.5
Warsaw . .	52 14	21 10	48.6
Moscow . .	55 45	37 31	40.1
St.Petersburgh	59 56	30 25	38.8
Pekin (China)	39 54	116 28	54.9

M. Humboldt, after tracing the isothermal lines across America, concludes that, in California and thence northward along the western side of that continent, the temperature is nearly the same as in similar latitudes on the western side of Europe. "Europe," he then observes, "may be considered altogether as the western part of a great continent, and, therefore, as being subject to all the influence which causes the western sides of all continents to be warmer than the eastern. The same difference which has been observed between the two shores of the Atlantic, exists between the two opposite coasts of the Pacific. In the north of China, the extremes of the seasons are much more felt than in the same latitudes in New California, and at the mouth of the Columba. On the eastern side of North America, the same extremes occur as in China. New York has the summer of Rome and the winter of Copenhagen. Quebec has the summer of Paris and the winter of Petersburgh. In the same manner, at Pekin, which has the mean temperature of Britain, the heat of summer is greater than at Cairo, and the cold of winter as severe as at Upsal. This analogy between the eastern coasts of Asia and America sufficiently proves that the inequality of the seasons depends upon the prolongation and enlargement of the continents towards the pole, and upon the frequency of the north-west winds, and not upon the proximity of any elevated tracts of country." The following table illustrates the preceding remarks :—

Places.	Latitude, North.	Mean annual temperature.	Mean temperature. Winter.	Spring.	Summer.	Autumn.	Coldest M.	Hottest M.	Difference in the heat of these months.
Philadelphia	39°.56′	54°.86	33°.98	53°.06	75°.20	56°.32	32°.70	77°.00	44°.30
Pekin .	39.54	54.86	26.42	56.30	82.58	54.32	24.62	84.38	59.76
Nantes . .	47.13	54.68	40.28	54.50	68.72	55.58	39.02	70.52	31.50
Rome .	41.53	60.44	45.86	57.74	75.20	62.78	42.08	77.00	34.92
Paris . .	48.50	51.44	37.92	49.64	64.40	51.26	35.96	67.46	31.50
Quebec .	46.47	41.72	14.18	38.84	68.00	46.04	12.74	73.40	60.66
Upsala . .	59.51	41.9	24.98	39.38	60.20	42.80	24.26	61.88	37.62

The fact that places which have the same annual temperature experience very different seasons, is clearly exhibited in this comparison. From Humboldt's inquiries, it appears that the lines which mark the winter temperature deviate much more from the parallels of latitude than those which express the mean annual temperature. In Europe the latitudes of two places which have the same annual heat, never differ more than 8° or 9° : but the difference of latitude in those having the same winter temperature is sometimes no less than 18° or 19°. The winter of Scotland is as mild as that of Milan. With respect to summer, the same heat takes place at Moscow and at the mouth of the Loire, though the former is nearly 9 degrees north of the latter. Ireland is remarkable for mild winters and cold summers ; the mean temperature in Hungary for the month of August is 71°.6, while in Dublin it is no more than 60°.8.

It is generally believed that, beyond a certain distance from the equator, the temperature of the southern is lower than that of the northern hemisphere. In speaking of the temperature of the ocean, we have already observed that ice is fallen in with much sooner in sailing towards the south, than it is in approaching the north pole. Humboldt says, that near the equator, and indeed through the whole of the torrid zone, the temperature of the two hemispheres appears to be the same ; but that the difference begins to be felt in the Atlantic about 22° of latitude ; the mean temperatures of Rio Janeiro and Havannah, places at about an equal distance from the equator (23 degrees) being in the latter instance 76°.4, and in the former only 74°.5.—The southern climates generally differ from the northern with respect to the distribution of temperature through the different parts of the year. In the southern hemisphere, under the isothermal lines of 45° and 50°, there are summers which, in our hemisphere, belong to the lines $35\frac{1}{2}°$ and 41°. There is no accurate information as to the mean temperature of any place beyond 50° of south latitude; but there is every reason to suppose that it differs considerably from that of places in the same degree of north latitude.

The same writer, in the second volume of his Personal Narrative, presents the following comparison of the temperature of the air in both hemispheres. The observations employed in drawing it up were all made at sea, except those from which the mean temperature for S. lat. 34° was deduced, which were made at the Cape of Good Hope.

Latitude.	Corresponding Months.		Mean Temperature of the Months. Southern Hemisphere.	Northern Do.
0°—15°	December / June	Summer	82.4°	83.3
18	October / April	Autumn	81.5	79.7
2—26	January / July	Winter	72.5	66.74
"	September / March	Autumn	69.44	68.9
34	December / June	Winter	56.84	59.72
"	February / August	Winter	62.24	62.6
43	July / January	Summer	59.36	64.76
48	June / December	Summer	44.6	63.86
58	July / January	Summer	43.16	56.3

The coldness of the southern hemisphere has frequently been attributed to a circumstance quite inadequate to explain it, namely, that of the sun being a shorter time (by $7\frac{3}{4}$ days), on the south, than on the north side of the equator. A much greater influence than we can assign to this cause, must be ascribed to the very large proportion which the ocean bears to the land of the southern hemisphere, in consequence of which its climate differs from that of the northern, in the same way as an insular climate differs from a continental one. But even this is not altogether a sufficient explanation, and there still remains a circumstance that deserves attention. The absence from the south polar regions of any great extent of land, and the manner in which the South American continent terminates, permit the grand current of the antarctic ocean to flow freely all round that part of the globe, towards the equator. This current, being unchecked till it is lost in the westerly movement of the ocean, carries along with it the circumpolar ice into very low latitudes; and the continual absorption of heat by the melting of the ice, as it gradually advances into warmer parts, keeps the air at a lower temperature than in the northern hemisphere, where circumstances are not favourable to the passage of the polar ice out of the regions in which it is formed. Beyond the limit, however, at which the ice disappears, but little effect will be produced on the temperature by its melting, and we accordingly find that within the torrid zone, the warmth of one hemisphere is the same as that of the other, and that as far as the 35th, or even 40th degrees of latitude, there is no important difference.

The question has sometimes been agitated, whether the general temperature of the globe suffers any change. Some have gone so far as to imagine that it gradually diminishes, others have been of opinion that it receives an augmentation. Neither of these theories has very solid foundation; it is scarcely more than a century since the thermometer was rendered a correct measure of heat, and the number of observations made with it in different parts of the world is by no means sufficient to form a basis for such sweeping conclusions. If we possessed a regular series of observations taken in various countries, and extending through three or four centuries, we should most likely be enabled to discover a mean state both of temperature and moisture to which the atmosphere continually returns; and there is no doubt that if we could obtain a clear insight into the complex machinery which regulates the seasons, we should behold the same beautiful harmony, and the same system of compensation for temporary and apparent irregularities, which we are able to discern in the movements of the heavenly bodies. Independent, however, of any question as to the general temperature of the globe, a notion has been entertained that throughout Europe, a more mild and genial climate formerly prevailed: but such historical evidence as can be collected tends to prove exactly the reverse; and that the climate, as might be supposed, has, generally speaking, improved with the advance of cultivation. A discussion of this subject will be found in the first article of the Edinburgh Review, No. LIX., published in June, 1818. That article contains a list of the remarkable seasons which have taken place in Europe for several centuries past, and from the view there given we may venture to conclude, that severe cold is of much rarer occurrence than it was in former ages.

Having thus noticed the subject of temperature, it will be proper to advert to the amount of moisture which the atmosphere contains in different parts of the globe. In the course of this inquiry we shall not make use of the results given by the hygrometer*, because that instrument is neither so well known, nor brought to such a correct standard as the thermometer, but merely give the quantity of evaporation, and the depth of rain that has been observed to fall at several places upon the earth's surface.

Other things being equal, evaporation is the more abundant, the greater the warmth of the air *above* that of the evaporating body, and least of all when their temperature is the same. Neither does much take place whenever the atmosphere is more than fifteen degrees colder than the surface upon which it acts. Winds powerfully promote evaporation, because they bring the air into continual, as well as into closer and more violent contact with the surface acted upon, and also, in the case of liquids, increase, by the agitation which they occasion, the number of points of contact between the atmosphere and the liquid. It must be familiar to every person that the same quantity of water spread over a larger space, is dried up in a less period.

In the temperate zone, with a mean temperature of 52½ degrees, the annual evaporation has been found to be between 36 and 37 inches. At Cumana, on the coast of South America (N. lat. 10½), with a mean temperature of 81.86 degrees, it was ascertained to be more than 100 inches in the course of the year; at Guadaloupe, in the West Indies, it has been observed to amount to 97 inches. The degree of evaporation very much depends upon the difference (greater or less) between the quantity of vapour which the surrounding air is able to contain *when saturated*, and the quantity which it actually contains. M. Humboldt, from observations made in the passage across the Atlantic, found that in the torrid zone the quantity of vapour contained in the air, is much nearer to the point of saturation than in the temperate zone. The evaporation within the tropics is, on this account, less than might have been supposed from the increase of the temperature.

The quantity of rain falling upon the earth at any place is determined by observing the height of the water collected in a pluviometer or rain-gauge. When an inch is said to have fallen, it implies that the rain which has descended on any given surface would have acquired that depth, supposing none of it to have been absorbed by the ground, and that it received no addition by means of water flowing from the parts adjacent to that surface. The average yearly quantity of rain is greatest within the tropics; and it seems, in general, to diminish, the farther we recede from the equator. In the torrid zone it amounts, at a medium, to 100 or 110 inches, while in the north temperate zone it cannot be stated at more than 30 or 35 inches. These quantities are very differently distributed throughout the year in the two zones: the number of rainy days towards the equator is, in the majority of places, *less* than in the higher latitudes, and the rain consequently descends there in the most violent torrents: at Bombay, 16 inches have been collected in a gauge in the space of twenty-four hours. In general, much more rain falls in mountainous countries than in plains, and in countries covered with extensive forests than in those where wood is less abundant. Annexed, is a table of the annual quantities which have been observed at several places.

Places.	Latitude.	Mean annual quant. of rain.
Island of St. Domingo	19° N.	120 in.
Ditto Grenada . . .	12	112
Calcutta	22½	70 to 75
Rome	42	36
Paris	49	21
London	51½	23 or 24
Liverpool	53½	34
Kendal, Westmoreland	54¼	60
St. Petersburgh . .	60	16
Upsal	60	16

The average annual fall of rain at Bombay in the ten years 1817 to 1826, was 78.1 inches; of those years the most rainy was in 1822, in the course of which nearly 113 inches fell: whereas in 1824, a season of extreme drought and famine, the supply did not much exceed 34 inches. At Arracan, in 1825, nearly 60 inches were registered in the month of July, and above 43 in August; from which we may conclude, that the whole quantity within the year was at least 150 inches. It would seem, however, that at some places within the tropics the fall is much more copious even than this. Humboldt, on the authority of others, mentions two instances of such excessive rain as almost to induce a suspicion of the correctness of the observations. He informs us that a M. Pereira Lago, by means of a pluviometer, found the quantity of rain, in the year 1821, at San Luis do Maranhao, in Brazil (S. lat. 2½), to be 280¾

* Derived from the Greek, and signifies *measure of moisture*.

inches; and also, that Captain Roussin relates the fact of more than 160 inches having fallen at Cayenne in the single month of February.—(*See vol.* vi. *of the Translation of Humboldt's Personal Narrative, Note to p.* 276.)—At the same time, these accounts appear less surprising when we reflect, that over some of the immense forests of Guyana there is wet weather almost the whole year, and that incessant rains of four or five months are no uncommon occurrence.

It must not, however, be imagined that the climate of all hot countries is characterised by such abundant rains; for there are many which, from one year to another, are either almost or entirely destitute of rain. This is the case along an extent of several hundred miles of the coast of Peru, in Egypt and many other parts of Africa, and also in the desert tracts of Arabia. At Cumana, on the North coast of South America, the annual quantity of rain is scarcely 8 inches; and there are other places on the shores of that continent where none falls for several years, but where, nevertheless, vegetation is exceedingly strong, owing to the humidity of the atmosphere.

It is well known that the air becomes drier and less loaded with vapours, the higher we ascend. On looking from the top of the Andes towards the Pacific Ocean, a haziness is often seen, spread uniformly over the surface of the waters to the height of 9500 or 11,500 feet; and this, too, in a season when the atmosphere, beheld from the coast and at sea, *appears* quite pure and transparent. This decrease of vapour in the upper regions of the atmosphere, combined with the rarefaction of the air, is the cause of the beautiful deep tint which the sky assumes when viewed from the summits of lofty mountains. Small white fleecy clouds are sometimes, however, seen floating above the Andes at the height of 25,000 feet; from which we may judge, that even on the tops of that range the colour of the sky is not so pure as it would appear, if it were possible for an observer to attain a further elevation. In passing also from the temperate to the torrid zone, the azure hue of the sky is found to augment progressively: the transparency of climate which is so much admired in Italy and Greece is far surpassed by that which invests the plains of Quito and Peru, or the fertile islands of the Pacific Ocean.

In the torrid zone, the temperature ranges within comparatively small limits; and the various phenomena of the atmosphere occur, from one year to another, with a regular and uniform succession unknown in this part of the world. Two seasons, the *dry* and the *rainy*, divide the year. The latter depends upon the presence of the sun; countries north of the line have their wet season when that luminary is in the northern half of the ecliptic, that is, from April to October; while with southern countries it is exactly the reverse. We cannot fail to be struck with this admirable arrangement for affording shelter from the perpendicular rays of the sun, the unrestrained influence of which would be quite insupportable. Humboldt has given us an account of the atmospheric appearances which succeed each other in that part of South America lying between 4° and 10° of north latitude, and to the east of that branch of the Andes which terminates on the Atlantic side of the lake of Maracaybo. Nothing can surpass the clearness of the atmosphere from the month of December to that of January. The sky is then constantly without clouds; and if one should appear, it is sufficient to excite the whole attention of the inhabitants. The breeze from the east and the east-north-east blows with violence. The immense plains (called *Llanos*), which in the rainy season display a beautiful verdure, gradually assume the aspect of a desert; the grass is reduced to powder, the earth cracks; and the alligator and the large serpents remain buried in the dried mud till the first showers of the year awaken them from their lethargy. About the end of February, and the beginning of March, the blue of the sky becomes less intense, the hygrometer indicates greater humidity, and the stars, veiled at times by a slight vapour, lose the steady and planetary light which before distinguished them. The breeze at this period becomes less strong and regular, and is often interrupted by dead calms. The clouds accumulate toward the south-south-east, appearing like distant mountains, with strongly-marked outlines; and from time to time they detach themselves from the horizon, and traverse the vault of the sky with a rapidity that little corresponds with the feebleness of the wind below. At the end of March, the southern region of the atmosphere is illuminated by gleams of lightning; and the breeze then passes frequently, and for several hours together, to the west and south-west. This is a certain sign of the approach of the rainy season, which begins at the Oroonoko about the end of April. The sky becomes obscured, the azure disappears, and a grey tint is spread uniformly over it;—at the same time the heat progressively increases; and soon, dense vapours cover the heavens from one end to the other. The plaintive cry of the howling monkies begins to be heard before the rising of the sun. The atmosphere is at length convulsed by frequent thunder-storms, the rains descend in torrents, and the rivers, rising rapidly above their banks, overspread the plains with extensive inundations.

The occurrence of these periodical rains is capable of being explained in a very simple manner. We have remarked that they always take place in that half of the torrid zone to which the sun is vertical at the time; and that in the northern half they are preceded by the gradual subsidence of the north-easterly breezes, which are followed by calms, interrupted frequently by stormy winds from the south-east and south-west. While the north-east breeze blows with all its strength, it prevents the atmosphere over the equinoctial lands and seas north of the equator from being saturated with moisture. The hot and moist air rises above, and the north-east current continually supplies its place with colder and drier strata. In this way, the humidity of the northern torrid zone, instead of being accumulated and forming condensed vapours, ascends, and flows towards the temperate regions; and, accordingly, while the north-east breeze retains its force, which is when the sun is present in the southern signs, the sky is constantly serene. In proportion, however, as the sun passes over the equator towards the tropic of Cancer, the north-east breeze softens, and by degrees entirely ceases, because the difference in temperature between the northern temperate and the torrid zone is then at its least. The breeze having ceased, the humid air is no longer replaced by drier air from the north; and, under the powerful action of a vertical sun, the vapours rapidly accumulate, till they at length descend in violent rains. This state of things continues till the sun re-enters the southern signs; then is the commencement of cold in the temperate zone, and the current from the north sets in again,—because the difference between the warmth of the equinoctial and that of the temperate regions daily increases. By this current the air of the northern torrid zone is renewed; the rains cease, the vapours disappear, and the sky resumes its clearness and serenity of aspect.

These remarks are principally intended to refer to the seasons in the northern part of South America; but, with certain exceptions, they may very nearly be applied to those of the whole torrid zone —of course bearing in mind that, south of the equator, the rainy season is from October to April, and that the *south-east* corresponds to the *north-east* breeze of northern countries. The period of commencement of the rains is not exactly the same everywhere; and there are places where great anomalies are occasioned by the existence of chains of mountains which attract the vapours and alter the direction of the winds. In the West Indies, and also on some parts of the American continent, *two* wet seasons are distinguished; one of these, however, is of much shorter duration, and has much lighter rains than the other. In India, the rains are brought on by the south-west monsoon.

The four seasons which we distinguish in this country are known only in the temperate zones. Their succession is the most regular and perceptible from the 40th to the 60th degree of latitude; but in this we speak of Europe only, for both in America and Asia a much shorter interval separates the heat of summer from the cold of winter. That part of the northern temperate zone, which lies between the tropic of Cancer, and latitude 35°, has, in many places, a climate resembling that of regions within the tropics. In Europe, even as high as the 40th degree, the frost in the plains is neither intense nor long-continued; the trees are not stripped of their foliage above two months in the year, and although snow sometimes falls at the level of the sea, even in the 37th degree (at Malaga, for example), it is an occurrence very unusual.

From the 60th degree of latitude to the pole, only two seasons take place. A severe and protracted winter is succeeded immediately by the warmth of summer. The rays of the sun, notwithstanding the obliquity of their direction, produce powerful effects, because the great length of the days is favourable to the accumulation of heat. Even in very high latitudes, the tar on the ship's sides is sometimes melted and made to run down by the sun's action. In the north of Europe the snow is generally dissolved in three or four days, and the flowers almost immediately begin to blow. The breaking up of the thick field of ice which is annually spread over the surface of the arctic ocean, commences in the month of June, and at this season dense fogs are very common, owing to the surface of the water being colder than the air lying over it. These at length disperse, and a short interval of fine weather ensues; but, before the close of August, the approaches of winter are perceived; snow falls; and, as the temperature of the atmosphere declines more rapidly than that of the sea, fogs, called the *frost-smoke*, again arise, which disappear only when the ice has begun to extend itself over the clear spaces of the ocean. It is worthy of remark that, even in the circumpolar regions, the west of Europe still maintains its superiority of temperature over the east of North America; for the sea off North Cape in Norway, though in the 72d degree of latitude, is always open, whereas several degrees further south, off the shores of America, it is annually frozen over.

On the Distribution of Vegetables—Vegetation of the different Zones—Primitive centres of Vegetation.

THE subject of physical climate is in itself highly interesting; but it becomes still more so when we extend our view, and consider its effects upon the numerous animal and vegetable tribes which are dispersed over the earth. This dispersion has not been the result of a blind and unmeaning chance; the same wisdom which called them into such beautiful and various existence, has fixed laws for their distribution over the surface of the globe. To these laws (without entering into details which belong to botany and zoology) we shall now direct our attention.

The wide extension of vegetable life furnishes one of the most striking examples of the productive power of nature. Every climate, as we pass from the equator to the pole, or from the plains just raised above the level of the ocean to the summits which are covered with eternal snow, has its peculiar vegetation. Countries, the most inhospitable and locked up in frost nearly all the year, are not entirely destitute of it. On Melville Island (N. lat. 75°), where the duration of winter is nine or ten months, and the mean annual temperature only two degrees above zero, there are places which produce, in abundance, moss, lichen, grass, saxifrage, poppy, the dwarf willow, and the sorrel which is so valuable for its antiscorbutic qualities: the expedition under Captain Parry observed, in a sheltered spot of this island, a ranunculus in full flower in the second week of June. It is thought that even perpetual snow may be the abode of a species of vegetation; for Saussure discovered in it a reddish dust, and a red colouring matter has frequently been observed in snow by navigators in the arctic regions*.

The absence of light does not altogether prevent vegetable existence; caverns and mines produce certain plants, principally those of the cryptogamous class. In the cave of Caripe, situated to the south-east of Cumana in South America, the seeds, which are carried in by the nocturnal birds called Guacharoes, spring up at the distance of several hundred yards from the mouth of the grotto, wherever they can find mould to fix in. Blanched stalks, with some half-formed leaves, rise to the height of more than two feet; but M. Humboldt, who observed them, could not ascertain the species of these plants, their form and colour being so much changed by the absence of light. Vast fields of marine plants spring from the depths of the ocean, especially towards and within the tropics; the vine-leaved fucus vegetates at the depth of 200 feet, and, notwithstanding, has leaves as green as those of grass. In the Atlantic, between the 23d and the 35th degrees of latitude, and in the 29th and 30th of longitude, the fuci float on the surface in such numbers as to give the appearance of an immense inundated meadow. It is supposed, by many botanists, that they grow at the bottom of the sea, and float only in their ripened state, when torn off by the motion of the waves or otherwise.

Extreme heat is not destructive of vegetation, provided that it be accompanied by humidity. Plants grow, not only on the borders of hot springs, but even in the midst of waters which we should have supposed to be quite unsuited to their

* Captain Parry, in his Narrative of the attempt made in the year 1827 to reach the North Pole, mentions some striking examples of this appearance.—"In the course of this day's journey, we met with a quantity of snow, tinged, to the depth of several inches, with some red colouring matter. This circumstance recalled to our recollection our having frequently before, in the course of this journey, remarked that the loaded sledges, in passing over hard snow, left upon it a light rose-coloured tint, which, at the time, we attributed to the colouring matter being pressed out of the birch of which they were made. To-day, however, we observed that the runners of the boats, and even our own footsteps, exhibited the same appearance; and on watching it more narrowly afterwards, we found the same effect to be produced, in a greater or less degree, by heavy pressure, on almost all the ice over which we passed, though a magnifying glass could detect nothing to give it this tinge. The colour of the red snow, which occurred only in two or three spots, appeared somewhat different from this, being rather of a salmon than a rose colour, but both were so striking as to be the subject of constant remark." This colouring substance has generally been thought to belong to the order *Algæ*.

existence. Examples of this sort occur in Iceland and many other countries. Even sulphureous exhalations are not fatal to vegetation: it is reported that the interior of the crater of Vesuvius, after a long period of repose, was in 1611 covered with shrubs. The greatest obstacle to it is the absence of moisture; those sandy tracts where rain seldom or never falls, and where the soil is constantly being shifted by the winds, exhibit a hopeless sterility. The verdure of the *oases*, or islands of vegetation, scattered over some parts of the African desert, is maintained by springs which rise up to the surface of the ground. The chemical nature of the soil influences the size and vigour of plants rather than sets limits to their cultivation. Common salt, however, dissolved, and scattered over the earth in large quantities, almost entirely prevents their growth. The fusion which lava undergoes is probably the reason why the progress of vegetation on its surface is so long retarded; whereas, from the ashes thrown out by volcanoes, the most abundant crops are raised.

The scale of atmospherical heat is that which ordinarily determines the character and progress of vegetation. Hence, under the fierce climate of the torrid zone, we need only ascend lofty mountains, to a certain height, in order to behold the trees, fruits, and flowers of the temperate zone; while still higher are found those of the frigid zone. The low vallies of the Andes, towards the equator, are adorned with bananas and palm-trees, while the elevated parts of the chain produce oaks, firs, and several other tribes common to the north of Europe. Near the equator, the oak grows at an elevation of 9200 feet above the sea, and never descends lower than one of 5500 feet; but, in the latitude of Mexico, it is seen as low as 2600 feet. From the height of about 15,000 feet, to the boundary of perpetual congelation, lichens are the only plants visible. Similar gradations, on a smaller scale, are observed among the Alps; on ascending which, chesnuts, beeches, oaks, and pines occur in succession, the last gradually becoming stunted till they disappear not far from the border of perpetual snow. The vegetation which covers the sides of mountains may be divided into distinct zones or bands, each zone containing its peculiar tribes. On the volcano of Teneriffe, one of the Canary Islands (N. lat. 28½°), as many as five of these zones are distinguished*:—(1) the region of *vines;* (2) of *laurels;* (3) of *pines;* (4) of the *retama* (an alpine broom); and (5) the region of *grasses.* These zones are arranged in stages, one above another, and occupy, on the declivity of the Peak, a perpendicular height of 11,200 feet.

* Humboldt's Personal Narrative, vol. i.

In the equinoctial region where, in respect of warmth, the seasons differ little from each other, the geographical distribution of plants is regulated almost entirely by the mean temperature of the whole year; but in the temperate zone this distribution depends not so much upon the mean temperature of the year as upon that of the summer season. In Lapland, there are fine forests on the continent at Enontekies, where the mean annual temperature is only 27 degrees, while on the island of Mageroe, where it is more than 32 degrees, only a few scanty shrubs are to be seen. The more vigorous vegetation of Enontekies is the effect of a warmer summer; the mean temperature of July being there 59½°; whereas, at the isle of Mageroe, it is only 46¾°. Some plants in summer require a certain degree of warmth only for a short period; for others, a more moderate warmth is sufficient, if it be of longer duration. The birch and the pine afford an example of this difference. The former tree does not put forth its leaves till the temperature has risen to about 53 or 54 degrees; and in all places where the mean summer heat falls short of this, the birch cannot flourish, however great may be the mildness of the winters. Such is the case on the island just mentioned, and in other parts of Lapland. The pine, on the contrary, requires a long rather than a warm summer. In the interior of Lapland, where the summer, though short, is warm, the birch rises much nearer the line of perpetual congelation than the pine; but in the Alps and other high chains in lower latitudes, where the summer is of longer continuance but colder, the pine is seen after the birch has entirely disappeared.

The frigid zone contains but few species of plants, yet of these the vegetation in summer is extremely rapid. The verdure of those countries, which lie within the polar circle, is confined chiefly to the hills having a southern aspect, and the trees are of very diminutive growth. Besides mosses and lichens, there exist ferns, creeping plants, and some shrubs yielding berries of an agreeable flavour. The arctic regions of Europe are peculiarly favoured; for, in certain parts of Lapland, there are fine forests, and even rye and leguminous plants are produced.

In the high latitudes of the northern temperate zone are the pine and the fir, which show their adaptation to a cold climate by retaining their verdure in the midst of the rigours of winter. To these, on advancing southward, succeed the oak, the elm, the beech, the lime, and other forest trees. Several fruit trees, among which are the apple, the pear, the cherry, and the plum, grow better in the northern half of this zone; while to its most southern part especially belong the more delicate fruits, such as the olive, the lemon, the orange, and the fig; and, amongst trees, the cedar, the cypress, and the cork.

The space, comprised between the 30th and the 50th parallels of latitude, may be considered as the country of the vine and the mulberry. Wheat extends as far north as the 60th degree; oats and barley a few degrees further. In the southern part of this zone, maize and rice are more commonly cultivated.

The vegetation of the torrid zone is characterised by a wealth, a variety, and a magnificence, which are nowhere to be found in the other regions of the globe. Under the beams of a tropical sun, the most juicy fruits and the most powerful aromatics arrive at perfection; and innumerable productions supply the wants and administer to the luxuries of man. There the grounds yield the sugar-cane, the coffee-tree, the palm, the bread-tree, the pisang, the immense baobab, the date, the cocoa, the vanilla, the cinnamon, the nutmeg, the pepper, the camphor tree, &c. &c. In South America, is the remarkable tree called the *cow-tree*, which, when incisions are made in its trunk, yields abundance of a glutinous and nourishing milk*. There are also various sorts of dyewood, and several species of corn peculiar to hot climates; while this zone is not destitute (in its elevated tracts) of every kind which grows in the plains of temperate countries.

Under the equator, the climate best suited for the culture of all kinds of European grain lies between the altitudes of 6000 and 9000 feet above the level of the ocean. Wheat will seldom form an ear below the elevation of 4500 feet, or ripen above that of 10,800. With respect, however, to the lowest height at which corn can be raised between the tropics, there are great irregularities, which tend to prove that the augmentation of heat is not prejudicial to its cultivation, unless attended with an excess either of drought or of moisture. In the environs of La Victoria, a town of Venezuela (lat. 10¼° N.), fields of corn are seen mingled with plantations of sugar-canes, coffee, and plantains, at the height of not more than from 1700 to 1900 feet above the sea. The district of Quatro Villas, in the interior of the island of Cuba, furnishes a still more remarkable example; there, fine harvests are raised almost at the level of the ocean. In nearly the same latitudes, on the other side of the Mexican Gulf, the fine fields of wheat are generally between 3800 and 7700 feet of elevation; while on the slope of the mountains of Mexico and Xalapa, vegetation, even at the height of 4320 feet, is so luxuriant, that wheat does not form ears. It is erroneous to suppose that grain degenerates in advancing towards the equator, or that the harvests are more abundant in northern climates. On the contrary, it has been found that nowhere to the north of the 45th parallel of latitude is the produce of wheat so considerable, as it is on the northern coasts of Africa, and in America, on the table-lands of New Grenada, Peru, and Mexico*. Near the town of La Victoria above-mentioned, the average produce is three or four times as great as that of northern countries. It is equally large at Buenos Ayres in the 35th degree of south latitude.

The vegetable forms near the equator are in general more majestic and imposing, and the varnish of the leaves is more brilliant. The largest trees are adorned with flowers larger, more beautiful, and more odoriferous than those of herbaceous plants in our zone. It is scarcely possible for an inhabitant of temperate regions to picture to himself the beauty and the grandeur of the vast forests of equinoctial America. Trees which attain a stupendous height and size are covered with

* Humboldt, in the 4th vol. of his Personal Narrative, has given an account of this tree. The fecundity of nature in the torrid zone strikingly appears, when we consider the circumstances under which this vegetable milk is produced. "On the barren flank of a rock grows a tree with coriaceous and dry leaves. Its large woody roots can scarcely penetrate into the stone. For several months of the year, not a single shower moistens its foliage. Its branches appear dead and dried; but when the trunk is pierced, there flows from it a sweet and nourishing milk. It is at the rising of the sun that this vegetable fountain is most abundant. The Blacks and natives are then seen hastening from all quarters, furnished with large bowls to receive the milk, which grows yellow, and thickens at its surface."—*Personal Narrative*, vol. iv. pp. 216 and 217.

* The mean temperature for three months of summer is, in the north of Europe, from 59° to 66°; in Barbary and in Egypt, from 80½° to 84°; and within the tropics, at between 8950 and 3840 feet of height, from 57° to 78°.

profusion of climbing plants, and the same lianas as creep on the surface of the earth reach the tops of the trees, and pass from one to another at the height of more than a hundred feet. By this continual interlacing of parasite plants, the botanist is often led to confound the flowers, the fruits, and leaves which belong to different species. M. Humboldt gives the following striking description of the woods on the banks of the Cassiquiare, on approaching the point where that river branches off from the Oroonoko. "The luxuriousness of the vegetation increases in a manner of which it is difficult, even for those who are accustomed to the aspect of the forests between the tropics, to form an idea. There is no longer a beach: a palisade of tufted trees forms the bank of the river. You see a canal 200 toises (426 yards) broad, bordered by two enormous walls, clothed with lianas and foliage. We often tried to land, but without being able to step out of the boat. Toward sunset we sailed along the bank for an hour, to discover, not an opening (since none exists), but a spot less wooded, where our Indians, by means of the hatchet and manual labour, could gain space enough for a resting-place for 12 or 13 persons."

A large proportion of the trees of these majestic forests are more than 100 feet in height; while some, especially of the palms, have an elevation of 150 to 200 feet*. Various instances are recorded of the enormous growth of trees in tropical climates. Humboldt measured on the banks of the Atabapo, a bombax ceiba more than 120 feet high, and 15 feet in diameter. Near the village of Turmero, which lies to the south-west of the city of Caraccas, is the famous *Zamang del Guayre*, a species of mimosa, known throughout the province for the great extent of its branches, which form a hemispherical head more than 600 feet in circumference. The height of its trunk is about 63 feet, and its thickness between nine and ten. The branches extend like an immense umbrella, and bend towards the ground, from which they remain at an uniform distance of 12 or 15 feet. The circumference of this head is so regular, that M. Humboldt traced different diameters, and found them 204 and 198 feet in length. The dragon-tree (*dracæna*) at Oratava, in the island of Teneriffe, is another specimen of enormous growth. Its trunk is about 50 or 60 feet high, and its girth near the roots almost 48 feet. Its average girth is stated by M. de Borda to be 35 feet, 9 or 10 inches. The baobabs are of still greater dimensions than the above. At Senegal, and in the islands of Cape Verd, some were remarked which had a circumference of from 56 to 60 feet, and in another part of Africa one was seen whose diameter was 34 feet (more than 100 in circumference.)

* Such is the force of vegetation exhibited by American plants in every zone, that in the latitude of 57° N., on the same isothermal line with St. Petersburgh and the Orkney Islands, the Canadian Pine (*Pinus Canadensis*) displays trunks above 150 feet high, and more than six feet in diameter.

The distribution of plants cannot be explained solely by the influence of climate or by the distribution of temperature; for it frequently happens that similar climates are found in different parts of the globe without identity of productions. The climate of the high mountains of the torrid zone is analogous to that of our temperate zone; yet Humboldt did not discover one indigenous rose-tree in all South America, and it also appears that this shrub is entirely wanting in the southern hemisphere. The genus *erica* (heath) is quite peculiar to the Old World; of the 137 species known, not one is to be met with in the new continent. They seem to be very rare even in Asia. On the other hand, the *cactus* (Indian fig) is confined to the New World. It is true that a similarity exists in respect to their vegetation between very distant countries, where the physical circumstances are alike; but in some instances it is only a general resemblance of the vegetable forms. In many cases the same *genera* recur; but there are comparatively few examples in which identical *species* have been recognized in countries far remote from each other*. Species of pine, beech, elm, &c., are found in America, differing, however, from the Asiatic and European species. "The lofty mountains of equinoctial America (says M. Humboldt) have certainly plantains, valerians, arenarias, ranunculuses, medlars, oaks, and pines, which from their physiognomy we might confound with those of Europe; but they are all *specifically different*." The Antarctic birch (*betula Antarctica*) of Tierra del Fuego corresponds to, but does not exactly resemble the dwarf birch (*betula nana*) of Northern Europe.

According to Humboldt, the species of plants at present known amount to 44,000. Of these, 6000 are *cryptoga-*

* For an explanation of the terms *genus* and *species* see page 59.

*mous**. The remaining 38,000 *phanerogamous*† plants are thus distributed :—

In Europe	7000
Temperate regions of Asia .	1500
Asia, within the tropics and islands	4500
In Africa	3000
Both the temperate regions of America	4000
In America, between the tropics	13,000
New Holland, and the islands of the Pacific	5000

He also states the proportions of plants which grow in latitudes 0°, 45°, and 68°, to be as the numbers 12, 4, and 1. The mean annual temperatures in these latitudes are respectively 81½°, 55½°, and 32½°, and the mean summer temperatures 82½°, 70°, and 53½°. Within the tropics, the *monocotyledonous*‡ plants are to the *dicotyledonous*§, as 1 to 6 ; between the latitudes 36° and 52°, as 1 to 4 ; and at the polar circle as 1 to 2. The *annual* monocotyledonous and dicotyledonous plants in the temperate zone amount to one-sixth of the whole phanerogamous class ; in the torrid zone they scarcely form one-twentieth, and in Lapland one-thirtieth part. A circumstance worthy of remark is the extreme rarity of the *social* plants between the tropics, that is, of those plants which, like the heath of Europe, live together and cover large tracts of land‖. In the torrid zone, they are found only on the sea shore and upon elevated plains.

Among the vegetable forms there are some which become more common from the equator towards the poles, as the ferns, the heaths, and the rhododendrons; others, on the contrary, increase from the poles to the equator, among these are the *rubiaceæ*¶, the *euphorbiæ*, and the leguminous plants ; while others, such as the *cruciferæ***, the *umbelliferæ*††, &c., attain their maximum in the temperate zone, and diminish towards the equator and the poles. The vegetable forms present (under the same *isothermal lines*) such constant relations, that when upon any point of the globe we know the number of species belonging to one of the great families, both the whole number of phanerogamous plants, and the number of species composing the other vegetable families, may be estimated with considerable accuracy.

It has been a question discussed among philosophers, in what way the various vegetable tribes were originally diffused over the surface of the earth. *Three* different hypotheses have been maintained upon this subject. The *first* supposes that there was only one primitive centre of vegetation; all species of plants having had their existence originally confined to one tract of the earth, whence they were gradually dispersed over all countries.

This hypothesis was adopted by the celebrated Linnæus. He imagined the habitable world to have been at the commencement limited to one spot, in which were collected the originals of all the species of plants, together with the first parents of all animals and of the human race. As such various natures would require a diversity of climates for their support, he supposed this tract to have been situated in a warm region, and to have contained a lofty mountain range, between the base and the summit of which were to be found all temperatures and climates, from the temperature and climate of the torrid to those of the frigid zone. Linnæus endeavoured to support this hypothesis by referring to the means which are provided for the multiplication and dispersion of plants. Winds, rivers, and marine currents, are all, more or less, instrumental in the conveyance of seeds from one country to another. The former carry the lighter kinds of seeds to immense distances, and the two latter sometimes transport others from the most remote parts. The naturalist just mentioned remarks that the *Erigeron Canadense* was first introduced into the gardens near Paris from Canada ; the seeds being scattered by the wind, this plant was, in the course of a century, spread over all France, Italy, Sicily, Belgium, and Germany. The migration of plants by means of currents is also well ascertained, and many instances of it are recorded. On the shores of the Hebrides are collected at times the seeds of the *mimosa scandens*, *dolichos urens*, and several other plants of Jamaica, the isle of Cuba, and the neighbouring continent. These are not the only methods in which the dispersion of species is effected: it is also known that some seeds (as the

* Having neither blossoms nor visible fructification.
† Having visible organs of fructification.
‡ Having only one cotyledon or *seed-lobe*.
§ Having two *seed-lobes*.
‖ Through Jutland, Holstein, Hanover, Westphalia, and Holland, a long chain of hills may be traced, entirely covered with common heath, and the *erica tetralix*. Very little success has attended the efforts made by the farmers to oppose the inroads of these plants.
¶ One of Jussieu's natural orders of plants, so named from *rubia* (madder).
** An order of plants, so named from their petals, four in number, being disposed in the form of a *cross*.
†† Having an *umbella* or *umbel*. This term is used to designate a particular mode of flowering, which consists of several flower-stalks, or rays, nearly equal in length, spreading from a common point or centre. The flowers of the hemlock and parsley are *umbellate*.

misseltoe and juniper) are capable of preserving their vitality in the stomachs of birds, and are thus propagated. Lastly, man has introduced various plants into countries where they previously had no existence.

The *second* hypothesis is, that each species of plants originated in and was diffused from a single primitive centre; but that there were several of these centres situated in different parts of the globe, each centre the seat of a particular number of species. The *third* and last hypothesis is, that wherever a suitable soil and climate existed, there the vegetable tribes sprang up; and that plants of the same species were, from the first, spread over different regions.

We proceed to relate some facts which have been observed, and which will enable us to form some opinion as to which of the three preceding hypotheses has the best foundation. The greater number of these facts are taken from the opening part of Dr. James Prichard's work, entitled "Researches into the Physical History of Mankind," where they are brought forward in a similar discussion to the present.

Those plants whose structure is the most simple are found to be very generally diffused. Among the cryptogamous tribes, (such as mosses, lichens, &c.,) which form the lowest order of the vegetable creation, the same species are often met with in the most distant regions*. Two-thirds of the lichens observed in Australia, are also natives of Europe; and of the ferns of New Holland, which constitute rather more than 100 species, twenty-eight have been discovered in other countries. Many of the monocotyledonous tribes are also widely spread. Several grasses are common to Europe and Australia. In South America too, not only the mosses, but likewise several grasses, are the same as European species. It is not so, however, when we view the distribution of the more perfect, or of the dicotyledonous plants, there being a very small number of such, which are common to countries distant from each other. With respect to the dicotyledonous tribes, Humboldt has maintained that all the indigenous kinds in those parts of America visited by him are peculiar to that continent, and that the only exceptions to this rule are plants of the sea-coasts, the migration of which is easily to be explained. The observations of Mr. Brown on the botany of Terra Australis (*Southern Land*) tend nearly to the same point. Of the plants already known in that country, 400 species are cryptogamous, 860 monocotyledonous, and 2900 dicotyledonous. Of the 400 cryptogamous, more than 120, that is nearly *one-third* part, are also indigenous in Europe. Of the 860 monocotyledonous, only 30, or about *one twenty-ninth* part, have been found in Europe, and more than half of these are grasses and *cyperoids*. But of the 2900 dicotyledonous species, only 15, or about the *one hundred and ninety-third* part, are the same in Australia as in Europe. Results no less striking have been obtained on comparing the vegetation of other southern countries with that of Europe and the northern regions. Though the proportion of European plants in Australia is so small, it appears to be greater than that which is observed in the south of Africa. The proportion of European species in South America is probably still less than it is in Southern Africa.

From the preceding remarks, it is to be gathered, that the most simply organized tribes of plants are very widely dispersed; that plants of the more perfect or more complex forms are, on the contrary, limited to particular countries; and that the monocotyledonous, which may be considered as tribes of an intermediate class, are neither so extensively spread as the former, nor confined within such narrow limits as the latter.

Some exceptions to this general rule have lately been brought to light by the botanical discoveries made during the expedition to the river Zaire on the west coast of Africa (about 6° S. lat.) From Mr. Brown's observations upon upwards of 600 plants collected in the neighbourhood of that river, it appears that about one-twelfth of the collection consists of species, which are also met with either in India, or on the opposite shores of Guyana and Brazil; and it is a curious fact, that in this number the *more perfect plants are in the greatest proportion*. This apparent anomaly is probably to be explained by the transportation of seeds from one shore to another by means of currents in the inter-tropical seas. Mr. Brown remarked that most of

* To explain the extensive diffusion of these species, Linnæus supposed that their seeds, being invisible particles, might be carried to incalculable distances by the winds. It may, however, be remarked, that in the less perfect tribes of plants, the specific distinctions, not being so strongly marked as in the more complex forms, may escape detection; and thus two plants found in distant places may be set down as of the same species, when there is really some minute difference between them.

those plants in the African collection which are also natives of other countries, were seen only on the lower parts of the river Zaire, where they bear but a small proportion to the whole vegetation; and that most of the dicotyledonous species are such as produce seeds capable of retaining the germ of life during a long immersion in the waters of the ocean.

It will be proper in this place to mention the phenomena belonging to the vegetation of islands. In small islands the most remote from continents the species of plants are very few, and sometimes quite peculiar. Thus in Kerguelen's land, or the island of Desolation, when visited by Captain Cook, the whole flora was found to contain only 16 or 18 plants, all of which were considered to be peculiar to the island. Not a shrub was seen in the whole country. The flora of islands, as far as it is not peculiar to them, generally consists of the same species which grow on the nearest main lands. The different groups seated in the great Southern Ocean which lies between America and Eastern Asia serve as an example; the easternmost islands contain more plants of American families or species, and the western, of those tribes peculiar to India. Islands placed in the neighbourhood of two continents comprise the vegetation of both. Malta and Sicily have plants which belong to Europe, and others of an African stock. The vegetation of the Cape de Verd islands is intermediate between the flora of the Canary isles and that of the African coast.

The facts which have been introduced in the course of this inquiry forbid us to adopt the hypothesis of Linnæus, which considers all plants to have originated from one common centre. The propagation of the several tribes of plants has certainly taken place from a number of different points; since, of various parts of the world, separated by vast distances, each possesses a vegetable kingdom in a great measure peculiar to itself. The third hypothesis to which we alluded is equally untenable; since it is seen that plants are confined to particular tracts, till their seeds are conveyed elsewhere. Numerous instances have occurred in which plants by transportation have acquired a new country, and there become abundant; a striking example of this kind is the dispersion of the *Erigeron Canadense* over Europe, which we have already related. This shows that in the first instance plants of the same species were not produced in all regions possessing a soil and climate suitable for their growth. We have also seen that, when the same species are observed to exist in countries widely separated, the circumstances generally are such as authorise us to infer that they were dispersed from one point. The only hypothesis which remains, and which is reconcileable with all the phenomena observed, is, that the vegetable creation was originally divided into different provinces, and that each country (probably each principal range of mountains) had its peculiar tribes which, at first, existed nowhere else. This conclusion is strengthened by the circumstance of particular plants having an entirely local and insulated existence, growing naturally on some particular mountain, and nowhere else. The cedar of Lebanon is one among several examples of plants of this description. Such instances alone might be deemed conclusive in favour of the hypothesis for which we have been arguing.

On the distribution of Animals—their original dispersion from distinct centres.

In the ascent from the vegetable to the animal world, and from one rank of animal existence to another, the most admirable order is manifest. We are not surprised by sudden steps or encountered by violent contrasts; an evident connection pervades the whole; and though there is a vast diversity when we compare the meanest specimen of organic life with its most perfect and majestic forms, yet between the two an harmonious chain may be traced, and we pass from one extreme to the other by a regular and scarcely perceptible gradation.

The lowest class is that of the *zoophytes* (plant animals), which raise up the coral islands spoken of in a former part of this treatise. They may be regarded as confused masses of beings, none of them endued with a separate life. Nevertheless, there is reason to believe, from the observations of MM. Péron and Lesueur, that each description of zoophyte has its place of residence determined by the temperature requisite for its support. The *molluscæ*, whether naked or covered with shells (*testaceous*)*, possess each an individual existence, and of these it is unquestionable that different species belong to different countries. The pearl oyster arrives at perfection only in the equatorial seas.

* The term *testaceous* is in strictness applied only to such fish as have strong, thick, and entire shells; those which have shells, soft, thin, and consisting of several pieces joined together, as the lobster, &c., being called *crustaceous*.

Insects are the next in the scale of animal existence. In the midst of the exuberant vegetation of the torrid zone, the largest and the most splendid of these tribes are to be seen; the butterflies of Africa, of the East Indies, and of America, are adorned with the most brilliant colours; and in the tropical forests, especially those of South America, millions of shining flies present at night almost the appearance of an extensive conflagration. In these countries some races of insects exist in such multitudes, and are armed with such destructive or venomous qualities, as to enable them to lay waste the fruits of the earth through large tracts of country, or to become a source of the most serious personal annoyance and discomfort to man. The white ants (*termites*) raise lofty hillocks; and where they much abound, have been known to excavate the soil to such a degree, as to endanger the safety of houses which happen to stand above the seat of their operations. They devour paper and parchment so rapidly, that whole provinces of Spanish America (Humboldt informs us) do not afford one written document that dates a hundred years back. The *mosquitoes* and other species of the family of *tipulæ* are also formidable enemies to the human race in these climates. Amidst the forests of South America, especially along the banks of particular rivers, there are large tracts which are almost uninhabitable, owing to the thick swarms of these insects and the unceasing torment which they occasion. The lower strata of air to the height of nearly 20 feet from the ground are sometimes so filled with them as to give the appearance of a condensed vapour. During the day, the atmosphere teems with the *mosquitoes*, which are small venomous flies; these are succeeded, at night, by a species of gnats called *zancudoes*. The distribution of these insects is very remarkable, and frequently depends on local circumstances which cannot be explained. They are, in general, found to shun those rivers which have what the Spaniards call black waters (*aguas negras*), and also dry and unwooded spots. They swarm most upon the banks of rivers, and they nearly disappear where the elevation of the ground exceeds two or three thousand feet above the sea*. The annoyance occasioned by insects of this description is not confined to the torrid zone; for even in the Arctic regions of Greenland and Lapland, the short heats of summer give birth to swarms of gnats of another species.

With respect to *fishes*, it is probable that every basin of the ocean has its particular tribes; while, indeed, the regions which some inhabit are well known. Thus the cod, which are distributed over all the northern seas between Europe and América, congregate chiefly upon the great sand-banks to the south-east of Newfoundland. The most remarkable species of fish are met with in the torrid zone and its vicinity. The flying fish hardly extends to any part of the ocean so high as the 40th parallel; but in the voyage to America, hundreds of them are seen by navigators after passing the tropic of Cancer. The largest and most powerful of those fish which possess electrical properties, also live within the torrid zone. The Mediterranean contains four species of electrical torpedoes; but the shocks which they communicate cannot be compared in violence to those of the *gymnoti* (electric eels), which inhabit several of the rivers, and also the stagnant pools, in the llanos of South America. All the inhabitants of the waters dread the society of these animals. It is related that, some years ago, it became necessary to change the direction of a road near Uritucu, in consequence of the number of mules of burden annually lost in fording a river in which the eels were very numerous. The temperature of the waters in which the gymnoti habitually live, is from 78 to 80 degrees: their electric force is said to diminish in colder waters.

The seas of the warm regions contain the shark, which is noted for its extreme ferocity; but the most enormous in size are the whale tribes, which belong more particularly to the high latitudes.

The migration of fishes seems to be occasioned by their seeking for shallow water, in order to deposit their spawn. The herrings, which are supposed to come from the bottom of the Arctic Ocean, proceed every year to the coasts of the British islands, Norway, Sweden, Denmark, Holland, and the United States; and also to those of Kamtchatka and the neighbouring islands. The opinion is, that their innumerable shoals follow the direction of the chains of submarine banks and rocks which they meet with in their progress. Tunnies also migrate regularly every year from the Atlantic Ocean to the Mediterranean.

The hot regions of the globe, and those of America in particular, contain

* Humboldt's Personal Narrative, vol. v., p. 86—116.

the largest and the most venomous of the order of *reptiles*, including the rattle-snake, and several other kinds armed with deadly poisons; and the boa constrictor, which destroys even the great quadrupeds by the force with which it coils round their bodies. Here, too, the lizard tribe, under the various names of crocodiles, gavials, alligators, and caymans, attain to an immense growth. The largest is the crocodile of the river Nile, which measures, when full grown, even thirty feet in length. It is worthy of remark, that the dry season near the equator has the same effect upon several of the reptile race as the cold of northern countries: in South America, when the swelling of the rivers subsides, and the surface of the llanos becomes parched by the heat, boas and crocodiles bury themselves beneath the mud, and await, in a state of lethargy, the periodical rains. As we proceed into the higher latitudes, reptiles diminish both in number and magnitude, and are, even the worst of them, comparatively harmless.

With respect to *birds*, we might at first be inclined to infer, from the powers of locomotion with which they are gifted, that the existence of each species is not limited to a certain region; and it is true that some of them, including several of the vulture tribe, spread themselves almost over the whole world. But it appears to be generally the case, that particular kinds are confined to a very small range, especially such as have heavy bodies and weak powers of flight. Even the *condor*, which frequently soars at an elevation of four miles, never forsakes the chain of the Cordilleras of Peru and of Mexico; and the great eagle does not quit the ridges of the Alps. The torrid zone possesses a variety of the most beautiful birds, including the humming-bird of America, the cockatoos, the bird of Paradise, the Lories, and several others of the parrot genus. The bird of Paradise is never met with beyond New Guinea and the neighbouring islands. Parrots, in the New World, are seen as high north as the 35th degree; but on the old continent, they do not appear to reach farther than the 28th parallel. Of the birds which cannot fly, each equatorial region insulated by the ocean has its particular kinds. The ostrich of Africa and Arabia, the cassowary of Java and of New Holland, and the Brazilian ostrich, are distinct species, possessing a general similarity of organisation.

The frozen zone has its own kinds of birds, among which are the *strix Lapponicus* (Lapland owl), and the *anas mollissima* (eider duck), which frequents the shores of the Arctic seas, and from whose nests the eider-down is obtained. The several species of *sea-birds* do not wander beyond certain limits assigned to each. The albatross is seen flitting along the surface of the waves, as we approach the 40th parallel of latitude. The sea-swallows and the tropical birds keep within the torrid zone. The penguin of the Northern differs from the manchot of the South Seas.

The migration of birds from one country to another, in consequence of the changes of the seasons, is a remarkable phenomenon. The direction and extent of these migrations are still, in most cases, but imperfectly known. On the approach of winter, swallows, storks, and cranes abandon the northern countries of Europe for the warmer climates of the south. In the equinoctial zone, which is nearly of the same warmth during the whole year, the variations of drought and humidity appear to influence the habits of animals in the same manner as the great changes of temperature in our climates. In South America, when the Oroonoko begins to swell with the rains, an innumerable quantity of ducks remove from eight and three degrees of north, to one and four degrees of south latitude, toward the south-south-east. These birds quit the valley of the Oroonoko at this period, doubtless because the increasing depth of the waters and the inundations of the shores prevent them from catching fish and insects. In the month of September, when the Oroonoko decreases and retreats within its bed, they return from the Amazon and the Rio Branco towards the north. The southern coasts of the West India islands also receive every year, at the season of the inundations of the great rivers of terra firma (the continent), numerous flights of the fishing birds of the Oroonoko, and of its tributary streams. It is in obedience to a similar instinct, that, during the heats of summer, the humming-birds advance in pursuit of insects into the northern parts of the United States, and even into Canada.

Quadrupeds are an order of animals more perfectly organised than any of those which have been under consideration; while, as they are in many instances immediately connected with man, and altogether come more under his observance, their distribution presents a more ample subject for investigation. The hot regions towards the equator furnish this order in the utmost number and variety; and many of its tribes are there distinguished for their size, their amazing strength, or the ferocity of their dispositions. The lion, the tiger, the elephant, the rhinoceros, the hippopotamus, the panther, the leopard, the hyæna, and the camelopard, are all inhabitants of the torrid zone and its vicinity. In the temperate regions, the animals are of much smaller dimensions; and the only beasts of prey are the wolf, the bear, the lynx, and the wild boar; but there the domestic tribes are reared in all their perfection. The white or Polar bear, which is quite different from the common bear and much more formidable, inhabits the coasts of the Arctic Ocean; so that, under both extremes of temperature, the animal creation assumes a character of excessive ferocity.

The *domestic* animals have been conveyed by man to various parts of the world, and are therefore very widely dispersed. Under this title are included the dog, the cow, the sheep, the goat, the horse, the ass, the pig, and the cat. Humboldt states (contrary to what has been supposed by some), that the cows in the equinoctial parts of South America will yield as rich a milk as in temperate countries. The ass is not capable of enduring cold so well as others of the domestic races: when beheld in the northern regions of Europe, he is quite a degenerate animal; south of the 40th parallel of latitude, under the influence of a more genial climate and better treatment, he is large, lively, and docile. The horse, originally a native of the central parts of the old continent, is now spread from the confines of the Arctic Circle to beyond the 50th degree of south latitude. It exists as high as Norway and Iceland, where it is small and of a peculiar variety, and extends even into the desolate regions of Patagonia. This animal was introduced into South America by the Spaniards, in their early visits to that continent; it has since greatly multiplied, and immense herds now rove wild over the *llanos.* The existence of these creatures is exposed to the most severe sufferings. "In the rainy season," says M. Humboldt, "the horses that wander in the savannah, and have not time to reach the rising grounds of the llanos, perish by hundreds amidst the overflowings of the rivers. The mares are seen, followed by their colts, swimming, during a part of the day, to feed upon the grass, the tops of which alone wave above the waters. In this state they are pursued by the crocodiles; and it is by no means uncommon to find the prints of the teeth of these carnivorous reptiles on their thighs.

"We cannot reflect," he proceeds, "on the effects of these inundations, without admiring the prodigious pliability of the organisation of the animals that man has subjected to his sway. In Greenland, the dog eats the refuse of the fisheries; and, when fish are wanting, feeds on sea-weed. The ass and the horse, originally natives of the cold and barren plains of Upper Asia, follow man to the New World, return to the savage state, and lead a restless and painful life in the burning climate of the tropics. Pressed, alternately, by excess of drought and of humidity, they sometimes seek a pool in the midst of a bare and dusty soil, to quench their thirst; and at other times flee from water, and the overflowing rivers, as menaced by an enemy that encounters them on every side. Harassed during the day by gadflies and mosquitoes, the horses, mules, and cows find themselves attacked at night by enormous bats*, that fasten on their backs, and cause wounds which become dangerous because they are filled with acaridæ, and other hurtful insects. In the time of great drought, the mules gnaw even the thorny melocactus (melon-thistle), in order to drink its cooling juice; and draw it forth as from a vegetable fountain. During the great inundations, these same animals lead an amphibious life, surrounded by crocodiles, water-serpents, and manatees. Yet, such are the immutable laws of Nature, their races are preserved in the struggle with the elements, and amid so many sufferings and dangers. When the waters retire, and the rivers return into their beds, the savannah is spread over with a fine odoriferous grass; and the animals of old Europe and Upper Asia seem to enjoy, as in their native climate, the renewed vegetation of spring."

It has been observed by Buffon, that the largest quadrupeds, such as the elephant, the rhinoceros, the hippopotamus, the camelopard, the camel, and most of the ox kind, are possessed exclusively by the old world. In America, the *fossil* remains of some

* In Brazil, in the province of Ciara, the bats cause such destruction among the cows, that rich farmers are said from this cause to be sometimes reduced to indigence.

large animals have been discovered; but of living species, there are very few of considerable bulk. It has also been remarked, that the tribes which are the most powerful and perfect in their structure belong chiefly to the old world; those of the new having, in general, a character of organisation which assigns them a lower rank in the scale of animated beings. Such carnivorous animals, for example, as have the greatest vigour and courage (among which are the lion, the tiger, and the hyæna) are confined to Asia and Africa. The American tribes which approach most nearly to these, are, in general, much more gentle and feeble than the African and Asiatic species. It must be admitted, however, that Buffon's assertion respecting the cowardice of the feline race of America must be taken with some limitation: for it appears that the jaguar will sometimes attack men; and, when assailed by armed numbers, he has been known to offer an obstinate resistance. The swiftest, as well as the most graceful and beautiful quadrupeds (the antelopes, for example) also chiefly belong to the old continent; while those kinds which are the most useful to man, including the goat, the horse, the ox, and the ass, were unknown in America till their introduction into that country by the Spaniards.

Confining our view to wild animals, we may divide the earth into a number of zoological regions or provinces, each of which is the residence of a distinct set of quadrupeds*.

The first of these provinces, if we commence from the north, is the Arctic region, which contains the white bear, the rein-deer, the Arctic fox, and other tribes common to both of the great continents. The circumstance of their being common to both continents, is accounted for by the communication which, during winter, is established between the shores of Asia and America by means of the ice, over which a passage from one to the other becomes practicable to such animals as are fitted to endure the intense cold of the circumpolar countries.

The northern temperate zone is divided by the ocean into two great districts. The same tribes are found to be spread from the western to the eastern parts of the old continent; but the quadrupeds which inhabit the temperate climate of America are peculiar races.

The equatorial region contains three extensive tracts, widely separated from each other by the sea. These are the intertropical parts of Africa—of America—and of continental India. Each of the three tracts in question has a distinct nation of quadrupeds. The Indian isles, particularly the Sunda and Molucca islands, may also be considered as a separate region.

Beyond the Indian Archipelago is Papua, under which name it is usual to include New Guinea, New Britain, and New Ireland. These countries, with the islands which are formed by a continuation of their mountain chains, namely, the archipelago of Solomon's Islands, Louisiade, and the New Hebrides, together with the more remote groups in the great Southern ocean, may be regarded as one zoological province. It is remarked, that all this extensive region seems almost wholly destitute of native warm-blooded quadrupeds, except a few species of bats, and some small domestic animals in the possession of the natives.

The large region of Australia forms another zoological province, in which are contained many indigenous tribes of a very singular description; and, lastly, the southern extremities of America and Africa are each distinguished by the possession of peculiar races.

Of these several provinces, into which the animal world admits of division, none is peopled with so remarkable a stock of animals as Australia, including, under that designation, New Holland, and the adjacent islands to the southward. It possesses several entire genera of quadrupeds which have been discovered in no other part of the world; and it further deserves notice, that most of the tribes peculiar to New Holland, though on the whole very different from each other, have some striking characters of organisation common to all. It was assumed by Linnæus, that the great class of warm-blooded quadrupeds was, without exception, *viviparous* and *mammiferous*—two terms, the first of which denotes their production of their offspring in a living and perfect state; and the second, their being supplied with organs for suckling their young. On this latter account, it received the name of *mammalia*, by which it is known among naturalists. It appears, however, that, in New Holland, a tribe of warm-blooded animals has been discovered, to which that name is not applicable, because it is *oviparous* (that is, produces *eggs*), and is therefore unprovided with organs of the description above-mentioned. This curious tribe is, as far as our present knowledge extends,

* See Dr. Prichard's work, before mentioned.

quite confined to New Holland. Another remarkable tribe is the *marsupial*, which term comprises such as produce their young in an immature state, and keep them for a time attached to their bodies, chiefly in abdominal *bags* or *pouches*, given them by nature for that purpose. This tribe also is met with principally in New Holland. One genus of it, indeed, the opossum or didelphis, is peculiar to the warm parts of America, and some species of phalangers are seen in the Moluccas; but over the Australian regions there are distributed several genera of the marsupial order, comprehending more than 40 species. Among these are the wombat, the kangaroo-rat, the kangaroos, and the dasyuri or Australian opossums.

The didelphis, or American opossum, differs from the Australian opossum in several respects; one of which is the having a long prehensile or muscular tail, which serves as a fifth limb, and is of great use to animals which inhabit forests so extensive and lofty as those of Guyana*. In the same part of America there are other animals which resemble the opossum in this respect; these are the sapajous, a numerous tribe of monkeys, the ant-eaters, the kinkajou, and the hystrix prehensilis (prehensile porcupine). Herein we behold striking instances of the structure of animals being fitted to the nature of the country in which they reside. The monkeys of Africa and of India are distinguished by no such peculiarity, for in those parts of the world it is not requisite.

We have already mentioned that the new continent is, compared with the old, nearly destitute of the most powerful and perfect tribes of quadrupeds. In their place are found most of those singular races, in the formation of which the ordinary rules of nature seem most widely to have been relinquished. Such are the tribes which Cuvier has termed Edentes, or quadrupeds defective with respect to teeth, some of them being entirely destitute of those organs. Thus America contains the whole family of sloths, the ant-eaters, which are quite unprovided with teeth, and the armadilloes, which have grinding teeth, but no tusks or cutting teeth.

That part of Southern Africa which extends beyond the tropic of Capricorn, forms quite a distinct zoological province, separated as it is by the intervention of the torrid zone, from the milder climates north of the equator. Accordingly the animal creation of this region assumes a character almost as peculiar as that which is displayed by its vegetation. Of the order termed Mammalia, Southern Africa contains several peculiar genera, which occupy various distances towards the north, according to the degree in which they are capable of enduring a hot climate. In many instances the same *genera* are found in this region as in temperate countries north of the equator; but it is particularly to be observed, that the southern *species* differ from the northern. Thus the quagga, the zebra, and some others of the horse kind, answer to the ass and the jiggetai of Asia. From the southern tropic to the Cape of Good Hope, the continent of Africa stretches into fine level plains, over which roam a great diversity of hoofed quadrupeds. Besides five of the horse genus, there are also peculiar species of rhinoceros, of the hog and the hyrax; and among ruminating animals, the giraffe or camelopard, the Cape buffalo, and several remarkable antelopes, as the spring-bock, the gnou, the leucophoe, and others.

The animals of the Indian Archipelago have in some respects a different character from those of continental India, and approach towards those of Africa. The Sunda isles are said to contain a hippopotamus —an animal which does not exist in the rivers of Asia. The rhinoceros of Sumatra resembles the African more than the Indian species, but is specifically different from both. It is the same with the crocodile tribe, which is divided into three sub-genera, the crocodile proper, the alligator or cayman, and the gavial. The alligator belongs to America—the gavial inhabits the Ganges, and probably other rivers in continental India—of the crocodile proper there are six species, of which some belong to Africa, others to the isles of the Indian ocean, and one is said to have been discovered in the West Indies. Among flying quadrupeds, the flying macauco or lemur, is seen only in the islands of the Indian ocean; two species of flying squirrel, and two remarkable genera of the family of bats, also reside there; besides which some of the flying phalangers are supposed to belong to the Moluccas.

It will be proper, in the next place, to inquire more particularly into the manner in which the most numerous families of quadrupeds are distributed over different parts of the world. In this enumeration, it is of course not intended to include those animals which man has been the means of conveying from one country to another, whence the same species have

* The phalangers of Australia and the Moluccas have the prehensile tail.

in some instances become scattered over the most distant parts of the earth.

Most of those animals which Linnæus comprised in his order *Primates*, inhabit warm or temperate climates. The two grand divisions of this order are the bat and the monkey tribes. The latter has been subdivided by Cuvier into two chief branches, very different from each other—the simiæ proper, which are confined to the old continent—and the sapajous, which are peculiar to America. The African simiæ are distinguished from the Indian; and most of their species are limited each to a comparatively small tract. In the island of Madagascar, no true simiæ exist, but in their place are the makis, a tribe of lemurs. Of the bat tribe, the rousettes or frugivorous (*fruit-eating*) bats, inhabit the Indian archipelago and Australia, and the vampyres, or blood-sucking bats, nine species of which have been mentioned, are all peculiar to the hot parts of America. The most numerous genus of bats are the vespertiliones; some of these are very extensively dispersed, but this is not the case with the majority, and not one species is common to the old and new Continents.

The *feræ*, or carnivorous quadrupeds properly so termed, are extensively spread, although most of them belong to hot climates. Of the twenty-eight species of the cat kind, which have been enumerated, not one is the same in America and in the Old Continent; even the lynx of Canada is now believed to be a distinct kind from the European. The African species are generally confined to Africa, and the Indian to the eastern side of the Indus. The tiger is found only in Asia, extending as high as Chinese Tartary; but it is by far most common in India, living in ravines and jungles. Africa, although destitute of tigers, possesses panthers and leopards. The lion is most formidable in Africa, where there are two species, the Barbary and the Senegal; it is also said to belong to India, and it inhabits those parts of Arabia and Persia, which border on the Tigris and Euphrates from the Persian Gulf as far as Bagdad. The Arabian species is smaller than the others, and the males have no mane. The *couguar* or *puma*, a native of South and North America, which is sometimes called the American lion, is a very different animal from the real lion. Of the dog kind, several species endure an arctic climate, and are common to the high latitudes of both continents. The *lagopus* or *isatis* (arctic fox) is found at Spitzbergen, and may be traced through the north of Asia to Kamtchatka, and thence to the shores of America, Hudson's Bay, and Greenland. The wolf and the *lycaon* (black fox) are also common to all the arctic countries. Others of the dog-kind require a warm or a temperate climate, and these occupy a limited space either in Asia, in Africa, or in America. The dog of the Falkland islands (*canis antarcticus*), which was the only quadruped discovered there, has by some been considered a separate species, while others have held it to be the same as a species which inhabits Chili.

The *pachydermatous* (thick-skinned) tribes, live only in warm or temperate regions. There exist two species of elephant, the Indian and the African. Of the rhinoceros there are several species, but none of them are possessed in common by Asia and Africa. Those with two horns inhabit southern Africa, those with one horn belong to India and China, and to some of the islands of the Indian Archipelago. In America, the only representative of these large pachydermatous animals is the tapir. The hyrax and the hog tribes do not extend into cold climates; the wild boar, which ranges further towards the north than any of his tribe, is spread over various parts of Europe, but is never seen to the north of the Baltic. The domestic hog, since its introduction into America, has run wild and formed large herds in that Continent.

Among ruminating animals, the goat, the antelope *, and the giraffe, or camelopard, are limited to the Old Continent; but of sheep, some peculiar species are possessed by America, as, for instance, the paco, which in its domestic state is called bicunua or vigonia, and is an inhabitant of Peru. The camelopard, which is so remarkable for its height, its swan-like neck, and its gentle disposition, is a native of southern Africa. The antelope, of which the species are very numerous, is almost confined to Asia and Africa, none of them being found in Europe, except the chamois and the saiga. They inhabit as well the torrid zone as those parts of the temperate zones which are not very remote from the tropics. The dromedary, or camel with one hump, is a native both of Africa and Asia; the two-humped or Bactrian camel, so called because it is supposed

* Several American species have been described which are considered to be nearly related to the antelope.

to have originally come from Bactriana, belongs to much more northern climates than the other: it lives in the Crimea and in central Asia, in countries where the winter is very severe. The llama, or guanaco, which has been named by some the camel of the New World, is a widely different kind from either of the preceding. The musk resides in the mountains of Asia, from Cashmere, along the Altaï, to the mouth of the river Ameer. This animal is not seen in the New Continent, although there are tribes which bear a relation to it. Some species of the deer and ox inhabit very cold climates, and these have passed along the arctic regions from one continent to the other. Those which are unable to support such severe seasons, are confined to certain local tracts in either continent.

We shall conclude our inquiry into the distribution of quadrupeds, with the mention of some facts respecting the description of animals discovered in islands. Small islands lying at a great distance from continents are in general quite destitute of land quadrupeds, except such as appear to have been conveyed to them by men. Kerguelen's Land, the islands of Juan Fernandez, the Gallapagos, &c., are instances of this fact. Among all the fertile groups in the Pacific Ocean, dogs, rats, hogs, and a few bats, are the only quadrupeds which have been seen. The Indian isles near New Guinea, abound in oxen, buffaloes, goats, deer, cats, rats, hogs, and dogs; but, according to all accounts, none of these have reached New Guinea, the two latter excepted. In Easter island, the most remotely seated in the Pacific, there are no domestic animals, except fowls and rats, which are eaten by the natives.

The quadrupeds of islands situated in the neighbourhood of continents, are generally the same as those of the adjacent main-land. This remark may be made respecting the animals of the British and of the Mediterranean isles, as well as of those in Madagascar, and in the islands near New Holland. When in such islands any quadrupeds are met with, which do not exist in the neighbouring continents, they are usually distinct species which occur no where else, and either have always had a local existence, or have been entirely destroyed from the main-land.

There is thus reason to conclude, that islands in general derived their quadrupeds from the continents in the neighbourhood of which they are placed.

Having now stated the facts which relate to the distribution of the principal races of animals, it remains only to inquire what inference we are entitled to draw concerning the manner of their original dispersion over the earth: whether that dispersion took place from a single spot, or whether, as in the case of the vegetable tribes, it commenced from a variety of distinct centres.

The local existence of insects is so closely connected with that of the plants which not only yield them sustenance, but also, in many instances, furnish their only place of abode, that we might at once expect to find the same laws prevailing in the dispersion of this part of the animal creation, as in that of the vegetable tribes. This conjecture has been confirmed by positive researches. M. Latreille, by whom the subject has undergone a full investigation, states, that the whole or the greater part of the arachnides and insects which inhabit countries of similar soil and temperature, but widely distant from each other, consist in general of different kinds. All the insects and arachnides which have been brought from the eastern parts of Asia, under whatever latitude, are distinct from those of Europe and of Africa. It further appears, that with insects, as with plants, where the species are different, the *genera* are nevertheless often the same. In this manner, the entomology of America approximates to that of the Austral countries, and the east of Asia. The insects of New Holland are often of the same genera with those of the Moluccas, and the south-eastern parts of India; they show much affinity to those of New Zealand and New Caledonia, and, as just observed, include similar genera to those of America: yet the entomology of New Holland is, notwithstanding, marked by a peculiar character. A large number of insects are found near the Cape of Good Hope, which are unknown in other countries: M. Lichtenstein collected there between 600 and 700 species, of which 340 were ascertained to be entirely new.

In adverting to the dispersion of the various tribes which inhabit the waters of the ocean, including the marine mammalia, as well as fishes and molluscæ, it is to be remarked that, in the descriptions of these tribes, great vagueness and inaccuracy has long prevailed. Therefore it is that this department of the animal kingdom contains so many species which are said to inhabit indiscriminately all parts of the ocean. The common whale (*balæna mysticetus*) has been supposed to belong equally to the frozen seas of Spitzbergen and to those of the antarctic circle. The sea-calf (*phoca vitulina*) is reported by several writers to be a native not only of both circumpolar regions, but also of the seas of the torrid zone; while some have gone so far as to assert that it exists in the Caspian, and even in the fresh-water lakes of Baikal, Onega, and Ladoga. But the fact of such extensive dispersion of marine animals rests entirely on the authority of incompetent persons. The celebrated naturalists Lesueur and Péron, who personally collected and examined a vast number of marine species in the southern hemisphere, have come to the conclusion, that the arctic ocean does not contain one tribe, well known and described, which is not *specifically distinct* from those animals most analogous to it in the antarctic seas. This remark applies not only to the cetaceous and phocaceous tribes, but also to the lower departments of marine animals; and, descending through a variety of worms and molluscæ, even to the shapeless sponges of the antarctic waters, these naturalists assert that, among all this immense assemblage, not one species will be found which exists in the seas of the northern hemisphere. It further appears that those maritime animals, which possess little power of self-extension, prevail within very narrow limits. Each species of the family of *medusæ* is seen in abundance in particular districts, and occurs in no other place. It is the same with the numerous *testacea* which adorn the shores of the southern seas. The shores of Timor present a great multitude and variety of beautiful testacea; but not one of these extends so far as the southern coast of New Holland.

With respect to reptiles, birds, and quadrupeds, the facts already stated concerning their distribution are sufficient to show that different regions of the world are each in possession of peculiar kinds. Many entire genera are wholly confined to certain districts; but when, as it frequently happens, the same genus is discovered among the wild and native animals of two distant regions, to a communication between which natural obstacles are opposed, it is not the same *species* that inhabit both countries, but corresponding species of the same *genus*. Thus the American species of the cat kind differ from the African and the Asiatic; and the species of horse, ox, antelope, rhinoceros, and elephant, of Africa, are distinct from those of the same genera in Asia.

Under these circumstances, we can arrive at no other conclusion than that the first dispersion of animals, like that of vegetables, took place from divers points. It is probable that, at least, each of the great mountainous chains and table lands was originally furnished by the Creator with a stock of animals. The offspring of each species have since spread themselves to as remote a distance from the first spot of their existence as their locomotive powers, their capability of bearing changes of climate, or the absence of physical obstacles to their further progress, may have allowed them to wander.

On Man in his Physical Character—His universal Dispersion over the Earth—Unity of his Species—Terms Genus and Species explained—Varieties of the Human Race, and manner in which they may be accounted for—Influence of Climate.

THE physical character of man, although it be not such as to exempt its possessor from those laws of generation, of growth, and of dissolution, which prevail among the inferior tribes of animals, is nevertheless of a peculiar and pre-eminent kind. His organisation, more perfect and complex than theirs; his erect and noble aspect; his form, better suited for rendering obedience to the impulses of a rational and intelligent mind;—all essentially distinguish him from the brutes over whom he exercises dominion. Under such circumstances as these, it is not a little surprising that there should ever have existed naturalists who pretended to confound the human species with tribes of the lower animal creation.

In some respects it may appear that the organisation of man subjects him to great disadvantages: the extreme feebleness of the human frame at the first period of its existence; the slowness of its growth; the multiplicity of its wants; the variety of ills and infirmities to which through life it is exposed, have no parallel among the beasts of the field. Yet who that considers the present moral imperfection of man, can deny the good which results from these physical disadvantages inseparable from his condition? Endued with the strength of the lion or the elephant, or clothed with a skin impervious to cold and moisture, he would probably have remained sunk in selfish indolence, and ignorant of all the arts which embellish life. But the feeling of his wants and weakness has aroused faculties which would else have lain dormant in his mind—has, by uniting him to his fellows, given rise to the most endearing ties and the most useful forms of society; and has so called forth his inventive resources, that he has, to a considerable extent, acquired the command and the direction of the powers of nature.

The researches of modern navigators have shown that the human race is spread nearly over the whole earth. It has been found in the midst of the most sultry regions, in the vicinity of the pole, and upon islands which a boundless ocean would have seemed to cut off from communication with the rest of the world. The islands of Spitzbergen and of Nova Zembla to the *north*, and Sandwich Isle, the Isles of Falkland, and Kerguelen's Land to the *south*, are the only countries of considerable extent which have been found entirely destitute of human inhabitants. In the north, the habitations of man stretch nearly to the 75th degree of latitude; while in the south a miserable race (that of the Petcheres) exists on the bleak and barren shores of Terra del Fuego. The oases, or islands of verdure, scattered over the sands of Africa, are also the seats of population. In one part of the world the human body supports a heat higher than that which makes ether boil; and, in another, a cold which occasions the congelation of mercury.

Notwithstanding the dissimilarities of structure and complexion which are observed upon comparing the natives of different countries, there are the strongest reasons to believe that the human race forms not only a single *genus*, but also a single *species*; or, in other words, that all the several varieties of men sprung originally from one pair of individuals. Though there exist independent grounds for this opinion (as will presently be shown), it is proper, before we proceed to a statement of these, to remark that the whole tenor of Revelation is against any other supposition.

We may discern in the difference between the means adopted for peopling the earth with the human race, and those provided for covering it with the inferior creatures, the traces of that wisdom which uniformly pervades the arrangements of Providence. Had there been, in the first instance, no more than one pair of each genus of animals, and one individual of each tribe of plants, and had these been called into being upon only one spot of the earth, large regions separated by wide seas and lofty chains of mountains from the country containing that single spot, would for ever have remained almost, if not entirely, destitute of plants and animals, unless at the same time means had been provided for their dispersion far more effectual than any which we behold in operation. To prevent a result so little in harmony with what appears to be the general system of the universe, each separate region of the globe was supplied with a distinct stock of plants and animals. But in the instance of the human race, such a plan of proceeding was not requisite. Man was endued with a constitution capable of accommodating itself to the greatest changes of climate, and with the power of inventing methods for protecting himself against atmospheric influence: he was also enabled, by the aid of the same power of invention, to transport himself over the most extensive seas and across the most formidable ranges of mountains. Furnished with these capabilities, his race was originally placed in only one spot of the world.

In order that the question may be fully understood, it is requisite to explain what is signified by the terms *genus* and *species*, of which frequent use has been made in the course of the present treatise. A race of animals, or a tribe of plants, marked by any peculiarities of structure, which from one generation to another have always been constant and undeviating, form a *species;* and two races are held to be *specifically distinct*, if they are distinguished from each other by some peculiarities, which, in the lapse of generations, the one cannot be supposed to have acquired, or the other to have lost, through any known operation of physical causes: so that, under the word species, are comprised all those animals which are concluded to have sprung, in the first instance, from a single pair*. The term *genus* has a more extensive application. There are several species which so resemble each other as immediately to suggest the idea of some near relation between them. The horse, the ass, the zebra, and others of the horse kind, are one instance of this remark; the different species of elephants are another; and a third is furnished by the several kinds of oxen, buffaloes, bisons, &c., all belonging to the ox genus, and bearing a striking resemblance one to the other. As no physical causes are known which could have operated so as to produce those differences of structure, which exist between the several species of one genus, it is concluded that they origin-

* It is certainly possible that what we term one species may have sprung from two or more original pairs *exactly alike;* but, judging from all that has been observed, it is by no means probable.

ally sprung from different individuals. A *genus*, consequently, is a collection of several species on a principle of resemblance; and it may, therefore, comprise a greater or less number of species, according to the particular views of the naturalist. It is not, however, always easy to determine what races of animals are of the same, and what of separate species. It is a well known fact, that considerable varieties arise within the limits of one species; and such varieties are often transmitted to the offspring, and become, in a great measure, permanent or fixed in the race. Hence the difficulty, in some cases, of deciding whether two races of animals, of the same genus, and similar in many particulars, but differing in others, are merely what are termed *varieties* of one species, their diversities having proceeded from the action of external, or other causes, on a stock originally the same, or tribes of an entirely distinct origin from the beginning.

Dr. Prichard, in an able work entitled "Researches into the Physical History of Mankind," to which we have before had occasion to refer, mentions the criteria which may be used in order to determine whether all the races of men belong to the same species. The first of these is furnished by a reference to the general laws of their animal economy: since, should it appear that, in two races of animals, the duration of life is the same; that their natural functions observe the same laws; that they are subject to the same diseases, and susceptible of the same contagions—there is a very strong presumption that they are of the same species. Another way of examining the subject, is, to inquire whether the diversities in mankind are strictly analogous to those varieties in form, colour, &c., which occur in the lower departments of the animal creation, within the limits of the same species. The method, however, which bears most directly upon the question, is to ascertain what is positively known respecting the springing up of varieties among the human race.

First, with respect to the laws of the animal economy, it may be inquired whether there are any peculiarities which distinguish one race of men from another, of such a description as to render it probable that they constitute distinct species. Certain writers have supposed that there is a difference between Europeans and some other nations, in the duration of life. It is not to be denied that savage nations are, in general, shorter lived than the inhabitants of civilized countries; but this circumstance is sufficiently explained by their addiction to intemperate habits, and their constant exposure to fatigue and hardships. When they have not such disadvantages to contend with, their term of life seems to be as long as that of any other race of men. Several instances are cited, both of negroes in the West Indies, and of native Africans, having attained to a very advanced age; and among the natives of America, cases of longevity are far from uncommon. Humboldt mentions the name of a Peruvian who lived to the age of one hundred and forty-three, and who, only thirteen years before his death, was accustomed to travel on foot from three to four leagues daily. The Laplanders, again, are said to be rather remarkable for long life.

As to diseases, it seems an undoubted fact that all human contagions, and all epidemic complaints, exert their pernicious influence on every tribe of men, though the natives of particular climates suffer more than others. And as to constitutional complaints, the difference is only one of predisposition, and analogous to what is witnessed in different families in the same nation, some being constitutionally more liable to certain disorders than others. The constitution of the American is observed to be the most torpid, that of the European in general the most irritable.

The conclusion which results from the first method of inquiry, as will be more particularly seen by referring to the work above-mentioned, is, that the grand laws of the animal economy are the same in their operation upon all the races of mankind. The deviations which occur are not greater than the common varieties in constitution which exist within the limits of the same family. Here, then, is one strong presumption in favour of the inference that all men belong to one species; for it appears that, among animals, neighbouring species, so closely allied as to have often been taken for mere varieties of the same stock (as the wolf and the dog), differ materially in the laws of their animal functions.

The next mode of inquiry suggested, is, how far the diversities of complexion, figure, and stature, seen among the several races of men, are analogous to those varieties which, in the inferior animals, often occur without marking any specific difference, and, indeed, originate before our eyes within the limits of the same species. In this place we shall accordingly note the most remarkable instances of variety which appear in mankind.

One, and that the most obvious to common observers, is the *diversity of colour*. It is well known that there is a correspondence between the colours of the skin, the hair, and the eyes of individuals. With few exceptions, light-coloured eyes are joined to a fair complexion and light hair, but a relation of the colour of the skin to that of the hair is perhaps universal. The women of Barbary and Syria are said to be often very white, although they have black hair; but this is to be attributed to art, and careful protection from the sun. Dr. Prichard, taking the hue of the hair as the leading character, divides mankind into *three* principal varieties of colour, which he calls the *melanic*, the *xanthous*, and the *albino*. The first includes all individuals or races who have *black* hair; the second, those who have either brown, auburn, *yellow*, flaxen, or red hair; and the third, those who have *white* hair, and who are also distinguished by red eyes. "The *melanic* variety forms by far the most numerous class of mankind. It is the complexion generally prevalent; except in some particular countries, chiefly in the northern regions of Europe and Asia, where races of the xanthous variety have multiplied; and it may be looked upon as the natural and original complexion of the human species. The hair of the head, in the melanic races, is of various texture and growth, from the long and lank hair of the native Americans, to the fine crisp hair of the African negroes. The hue of the skin varies from a deep black, which is that of some African nations, to a much lighter or more dilute shade. The dusky hue is combined, in some nations, with a mixture of red, in others with a tinge of yellow. The former are the copper-coloured nations of America and Africa; the latter, the olive-coloured races of Asia. In the deepness or intensity of colour we find every shade or gradation, from the black of the Senegal negro, or the deep olive and almost jet black of the Malabars and some other nations of India, to the light olive of the northern Hindoos. From that we still trace every variety of shade among the Persians and other Asiatics, to the complexion of the swarthy Spaniards, or of European brunettes in general." Examples of the *albino* variety have been noticed in almost all countries. In Europe they are by no means uncommon; sixteen instances have been seen in Germany by Professor Blumenbach. The fact of their occurrence in other parts of the world is also fully established. It appears that they are frequent among the copper-coloured native Americans on the isthmus of Darien; and they have been observed in many islands both of the Indian and of the great Southern Ocean. Among the Hindoos they are regarded with peculiar horror. The most remarkable circumstance, yet not an unusual one, is the birth of white negroes among the black races of Africa: they are looked upon as curiosities, and are often collected by the native monarchs. Their hair is of a woolly character, and many of them are true *albinos*. The *xanthous* variety may be considered as intermediate between the other two. It chiefly prevails in the temperately cold regions of Europe and Asia, where it sometimes runs through whole tribes; but it also springs up out of every melanic race. The Jews, like the Arabs, are generally black haired, but many may be seen with light hair and eyes; and the same remark will apply to the Russians. The xanthous variety also appears among the South Sea islanders and the natives of America, and even among the negro races of Africa, both in their own climate and in the places to which they have been transported.

Varieties of form, more especially of the *shape of the skull*, furnish another grand instance of diversity among the races of men. The ingenious Professor Blumenbach has made the varieties in the construction of the skull the basis of a division of mankind into five principal races or departments, which he denominates, 1. the Caucasian; 2. the Mongolian; 3. the Ethiopian or Negro; 4. the American; and 5. the Malay and South Sea Island *. The *first* is that variety to which the nations of Europe and some of the western Asiatics belong; in this class the head is almost round, and of the most symmetrical shape, the cheek bones without any projection, the face oval, and the features moderately prominent. In the *second* class, the head is almost square, the cheek bones projecting *outwards*, the nose flat, the face broad and flattened, with the parts imperfectly distinguished, the internal angle of the eye depressed towards the nose. In the *third* class, the head is narrow and compressed at the sides, the forehead very convex, the cheek bones projecting *forwards*, the nostrils wide, the jaws lengthened, the skull in general

* The first three are the most strongly marked varieties in the form of the skull; the remaining two being only approximations to the preceding.

thick and heavy, the face narrow, projecting towards the lower part, the nose spread and almost confounded with the cheeks, the lips, particularly the upper one, very thick. The *fourth* variety approaches to the Mongolian: the cheekbones are prominent, but more arched and rounded than in the skull of the Mongole, the form of the forehead and of the top of the head is often altered by means of artificial pressure during infancy, the face is broad without being flat, the forehead low, and the eyes deeply seated. In the *fifth* variety, the summit of the head is slightly narrowed, the forehead a little arched, the upper jaw somewhat projecting, the face less narrow, and the features more prominent and better marked than in the negro.

These descriptions give a clear idea of five principal varieties in the form of the human head; nevertheless, the attempt to assign them as the distinctive characters of so many races of men is open to strong objections; since, whether we take as a standard the figure of the skull, or perhaps any other peculiarity of structure, it is impossible, with reference to that standard, to divide the human species into departments, such as can be regarded with probability as so many separate races or families. The third, or Ethiopian variety of skull, is found in the greatest degree among the tribes inhabiting the coast of Guinea and other western countries of Africa; but it must not be set down as common among all the Negro nations; for there are many black and woolly headed races in Africa which come under the designation of Negro, who display a very different shape of the head and features from that described under the *third* variety. Again, the skulls of the New Hollanders are almost as much compressed as those of any Negroes, and they would, under the preceding classification, be brought within the Ethiopic race; yet no one could assert that the New Hollanders and the woolly-headed Africans, differing so much in other physical characters, and separated by so vast a distance, should be included in the same department of the human species. On this account, Dr. Prichard has proposed a division of the varieties of the skull into three classes, distinguished by names derived from their forms, and not from any supposed origin of the nations to which they respectively belong.

In the *shape of the body*, as well as in the *size and proportion of the limbs*, and consequently in the degree of strength and agility which they possess, there are some remarkable varieties among nations. Some Negro tribes, the Australian or New Holland savages, and the Kalmucks, seem to be those which differ most in figure from Europeans. According to numerous measurements, the arm below the elbow is somewhat longer in the Negro, in proportion to the upper arm, and to the height of the stature, than it is in the European. It has therefore been remarked, that in this respect the generality of Negroes approach more to the structure of the ape; but if we descend to individuals, many Europeans will be found in whom the fore arm is as long as in the majority of Negroes, and, on the other hand, Negroes in whom it is as short as in the majority of Europeans. A clumsy form of the legs, broad and flat feet, and large hands, are also described as peculiarities of the Negro.

With respect to *stature*, the difference between one nation and another is generally not very considerable. From all accounts, the tallest race of men existing are the Patagonians. They are usually more than six, and, in some instances, as much as seven feet in height. On the contrary, the natives of Tierra del Fuego are described as miserable and puny savages. The Esquimaux, in the north of America, are likewise diminutive, being generally under five feet. Africa also contains some small races. Of the Bosjesmans or Bushmen, who are said to be the most deformed of mankind, Lichtenstein saw two individuals who were scarcely four feet high. This unhappy race, who were plundered of their property and hunted down like wild beasts by the early settlers in the Cape colony, have since lived among the rocks and woods on the northern frontier of the settlement, where they support themselves in a great measure by depredation. These instances tend to show that when whole races are deformed and stunted, it is to be attributed to exposure to constant hardships or an inclement climate, and to the wretched and precarious nature of their subsistence. Both extremes of stature which have been observed among nations, are frequently surpassed by individual examples in the inhabitants of different countries. Many natives of Europe, from eight to nine feet high, have been exhibited as objects of curiosity, and there have been dwarfs of less than four, and even three feet.

The only other varieties in the human

race which require notice, are those in the *texture of the skin*, and *the character of the hair*. The skin of Negroes is said to be always cooler than that of Europeans in the same climate, and to be distinguished for its sleekness and velvet-like softness. A similar observation has been made of several African tribes, and also of the Otaheitans. These qualities of the skin seem to be connected with the existence of the dark matter by which, among these nations, it is coloured; for in *albinos*, both African and Otaheitan, the skin becomes rough and inflamed, and cracks upon being exposed to the sun. The contrasts between the hair of different races are exceedingly striking. In the Negro, the Hottentot, and some other races, the hair is short and crisp, somewhat approaching to the nature of wool; in other nations it is long and lank; and between these two kinds there are numerous gradations. In Africa, the Kaffers have hair like that of the Negroes; some tribes have it longer; some again, who are black and otherwise similar to Negroes, have hair curled, but not crisp. The Papuas (the name by which the inhabitants of New Guinea and the neighbouring islands are distinguished) have crisp hair; but, unlike that of the Africans, it grows very long, and admits of being spread out into an immense bush. The hair of the natives of Van Diemen's Land is as crisp as that of the Africans, but the New Hollanders have straight hair; while in the New Hebrides it is of an intermediate character, and varies considerably in the men of the same island. The hair of the natives of America is generally lank; in a few instances curled; but in none crisp or woolly.

In the next place, we shall compare the diversities, of which a sketch has been given, as existing in the appearance of the human race, with the variations in form, colour, and structure, which are seen in the lower animals, and especially in the domesticated kinds. The differences in colour which quadrupeds of the same species exhibit, are so familiar to the eye, that it is unnecessary to do more than allude to them. Among horses, oxen, dogs, cats, rabbits, &c., we continually behold hues which are analogous either to the melanic, the xanthous, or the albino varieties in mankind. Among several kinds of wild animals, a white not unfrequently springs up. In many instances, certain colours are prevalent in particular breeds, and in some cases we might be warranted in concluding that the colours depend upon the local circumstances of the countries in which the breed is placed. Blumenbach mentions several examples which will illustrate this remark. He states that all the swine of Piedmont are black; those of Normandy, white; and those of Bavaria of a reddish brown colour. In Hungary, the oxen are of a greyish white; in Franconia, they are red. The turkeys of Normandy are black; those of Hanover almost all white. In Guinea, the dogs and the gallinaceous,* fowls are as black as the human race. Even in our own country, certain colours may be seen prevailing in the cattle of particular districts. Doctor F. Buchanan says, that in Mysore the sheep exhibit three sorts of colour: red, black, and white, and these are not distinct breeds. Don Felix de Azzara relates some curious circumstances respecting the colour of the horses and oxen in Paraguay, where, as well as in other parts of South America, both these races have run wild, and become very numerous. He says that all the *wild* horses are of one colour (a chestnut or bay-brown), whereas the domestic horses are of all colours, as in other countries. He makes a similar kind of observation concerning the oxen. As to form and the structure and proportion of parts, the diversities which arise in the same race of animals, far surpass those which subsist between one nation and another among men. Alluding to the hog tribe, Professor Blumenbach remarks, that "no naturalist has carried his scepticism so far as to doubt the descent of the domestic swine from the wild boar. It is certain that, before the discovery of America by the Spaniards, swine were unknown in that quarter of the world, and that they were first carried thither from Europe. Yet, notwithstanding the comparative shortness of the interval, they have in that country degenerated into breeds wonderfully different from each other, and from the original stock. These instances of diversity, and those of the hog kind in general, may therefore be taken as clear and safe examples of the variations which may be expected to arise in the descendants of one stock." He afterwards observes that the difference between the craniums of the Negro and of the European is not greater than that between the craniums of the wild boar and of the domestic swine. In the

* From *gallus*, a cock.

breeds of oxen, sheep, and horses, we may discern additional examples of deviation from an original standard. Some breeds of sheep and oxen are destitute of horns; others, on the contrary, are distinguished by the large size of their horns. In Paraguay there are breeds of oxen without horns, descended from the common horned race. With respect to horses, Blumenbach again observes that there is less difference in the form of the skull in the most dissimilar of mankind, than between the elongated head of the Neapolitan horse and the skull of the Hungarian breed, which is remarkable for its shortness and the extent of the lower jaw. The varieties in the covering of animals are not less worthy of notice than those to which reference has already been made. In the same race of sheep, some are clothed with wool, others with hair. It is known that if a flock is neglected, the fine wool is succeeded by a much coarser growth, intermixed with strong hairs; the breed, being no longer kept up with care, seems gradually to degenerate towards the characters of the *argali*, or wild sheep of Siberia, which naturalists consider to be the stock whence all domestic sheep have proceeded. A striking specimen of the changes which occur in breeds is afforded by the sheep of the West India islands, which, although descended from the woolly sheep of Europe, are covered with coarse hair. The deterioration has usually been attributed to the heat of the climate; but it must also be referred to the circumstance of their breed having been neglected. Other animals, such as goats and dogs, display a similar variety in the nature of their covering.

The preceding facts clearly prove that, in the lower animal creation, there spring up, *in the same species*, varieties of an analogous or similar kind to those which mark the different races of men. The existence of this analogy confirms still further the opinion expressed as to the unity of the human species. It now only remains to inquire whether it is absolutely known that varieties have arisen in a family or race of men similar to those diversities which distinguish one nation from another.

It is a well-attested fact that, among negroes and other dark-coloured tribes, individuals of the albino and xanthous complexions are not unfrequently born; and with respect to form and structure, and the texture of the skin and hair, many instances are recorded wherein surprising peculiarities have made their appearance in a race or family, and some in which these have been transmitted to descendants. The description of such cases would exceed the limits of the present treatise; but an account of several may be seen by referring to the author* whom we already cited.

It appears, therefore, that if we apply to the subject under discussion the several criteria stated at the outset of this inquiry, the results, every one, lead to the inference that the various nations of the globe are descended from the same stock. This inference is drawn, *first*, from the observed uniformity in the grand laws of their animal economy, allowance, of course, being made for the effects of climate and of particular habits; *secondly*, from the existence in the same species among the inferior tribes of the creation, of varieties analogous to those which occur in mankind; and *thirdly*, from the fact of varieties being really known to have sprung up among men, more or less similar to those which distinguish different nations. There is, nevertheless, a point at which the similarity between the two cases obviously terminates: the peculiarities which arose in the human species at a remote and unknown period have become the characteristic marks of large nations, whereas those which have made their appearance in later times, have, in general, extended very little beyond the individuals in whom they first showed themselves, and certainly have never attained to anything like a prevalence throughout whole communities. But this is a circumstance which it does not seem difficult to explain, if we consider that, ever since the population of the world has been of large amount, the possessors of any peculiar organisation have borne such a very small numerical proportion to the nation to which they belonged, that it is no way surprising that they should soon have been lost in the general mass, still less that they should have failed to impress it with their own peculiar characters. In the early period of the world, when mankind, few in numbers, were beginning to disperse themselves in detached bodies over the face of the earth, the case was altogether different, and we can easily understand how, if any varieties of colour, form, or structure, then originated in the human

* Dr. Prichard's "Researches into the Physical History of Mankind."

race, they would naturally, as society multiplied, become the characteristics of a whole nation. These considerations may suggest to us the *manner* in which national diversities first obtained their ascendancy. The *causes* of those diversities are, and probably ever will remain, enveloped in mystery; and the inference as to the unity of the human kind is not weakened by our inability to assign those causes, since we are ignorant of the occasions even of the varieties which sometimes display themselves within the limits of a single family.

It will be seen, however, upon a comprehensive survey, that in the distribution of the different races of men there is a certain relation to *climates*. We may observe that the *black* races of men are principally situated within the torrid zone; and the *white* races in the regions approaching towards the pole; and that the countries bordering on the torrid zone are generally inhabited by nations of a *middle* complexion. It further appears that the natives of mountainous and elevated tracts are usually of lighter colour than the natives of the low and hot plains on the sea-coast. In Africa, most of the races between the tropics are either black or of a very deep colour; while beyond the tropics the prevailing complexion is either brown or red. The people of Fezzan, who are of a black hue, form an exception to this rule; but it has been remarked of them, that they are chiefly slave-dealers, and have been intermixed with the negro race brought from the interior of Africa. In other countries of the globe, the majority of the nations near the equator are almost black. In India there are black tribes in Malabar; and the Hindoos of the Deccan are generally very dark, as are also the inhabitants of Ceylon. In most of the islands of the Indian ocean, the *aborigines* are of a black colour; and still further eastward are the Papuas of New Guinea and the black inhabitants of Solomon's islands and the New Hebrides. In equatorial America, the natives are not so dark as in other parts of the torrid zone; but its elevated mountains, vast rivers, and extensive forests, impart a peculiar character to the climate of the New World, which may probably account for the difference. In the low countries of California it is remarked that the population are nearly as dark as negroes.

It is a very general opinion, that the origin of the diversities of *colour* in mankind is to be referred entirely to the gradual influence of climate and of the sun's rays in darkening the complexion; it being a commonly-observed fact, that the skin, even of white men, becomes embrowned by constant exposure to the heat. But there are circumstances which militate against this opinion. There is positive testimony that the offspring of individuals, darkened by the sun in hot countries, is born with the *original complexion*, and not with the acquired hue of the parents *; besides which, it is known that white races of men, who have been removed from a cold to a hot climate, and have not intermarried with the natives, have retained for ages their original colour; while, on the other hand, black families, when transplanted into more temperate countries, have remained for generations of exactly the same hue as their African progenitors. Dr. Prichard has also remarked that the above supposition is contrary to a general law of the animal economy, according to which, *acquired* varieties † are not transmitted from parents to their offspring, but terminate in the generation in which they have had their origin.

Yet although it seems that the existence of varieties cannot be attributed to the slow and gradual operation of climate upon successive generations, numerous facts lead to the conclusion that there is a natural tendency among races, both of men and animals, to the production of varieties suited in form and constitution to the local circumstances of the country where they arise. Or it may, perhaps, be better explained, in some cases, by supposing that, whatever varieties occur, the ability to establish a footing in any country belongs to those only which possess a constitution adapted to local circumstances. Thus men of the *xanthous* variety of colour are known to spring up among the negroes in Africa; but their constitution being entirely unsuited to the climate, we cannot believe that they would ever become numerous in that continent. In the temperately-cold regions of the world they would be favourably circumstanced; and we accordingly see that this variety has multiplied there to a considerable extent.

* Dr. Prichard's "Researches," vol. ii. pp. 532, &c.

† *Acquired* varieties are opposed to those which a person brings into the world at his birth.

NAVIGATION.

Introduction.

NAVIGATION is the science which teaches the mariner how to conduct his ship from any one port or place to any other. An application of its principles will enable him at any time to discover the situation of his vessel, the direction in which he ought to steer towards his intended destination or port, and his distance from such port and from the place from whence he sailed:—This science explains also the method of applying the different instruments used in its practice, and of discovering their errors and imperfections.—In short, it is that branch of the seaman's education which may be learnt on shore, and apart from the ship; and furnishes a material part of all that is requisite to be known by him, on whom devolves the anxious responsibility of directing the course of a vessel through that vast and treacherous expanse of waters, which forms the greater portion of the surface of our globe, deprived of that assistance which the traveller on land receives from beaten tracks, and known landmarks, but enabled to rely with confidence on other guides, which the bounty of his Creator has provided, and which the ingenuity of man has converted to his use under such circumstances of apparent difficulty.

The term *Navigation* is derived from the Latin word *navigo*, which signifies literally, *to work or manage a ship;* but the art itself consists chiefly in the application to practice of a branch of mathematics, and a branch of the science of astronomy, which owes its name to that application.

PART I.

CHAPTER I.

On Angles and Triangles.—On the Mode of solving Triangles by Construction.

(1.) IN the course of our present inquiries we shall find it necessary to possess the means of ascertaining the exact size of an angle;* this is effected through the medium of circles. Conceive any number of circles, three, for instance, to be drawn around the angle A″ C M″, in *fig*. 5, (page 11,) and let them be so drawn, that the point C, at which the angle is formed, may be the common centre of all the circles. Then the lines A″ C and M″ C are said to *intercept*, that is, to *take between them*, a portion of each of the circles: in the largest and outer circle they intercept the portion A″ M″, which is called an arc, from its resemblance to a bow: for as we are indebted for many of our scientific terms to the Arabians, who were a people of warlike habits, they exhibit many metaphors drawn from the art of war. Now the curved line A″ M″, and the corresponding intercepted portions A M, A′ M′, are all said to measure the angle A″ C M″, which is formed by the intercepting lines A″ C, M″ C; it matters not which of these arcs we take as a measure; for they are all the same fractional part of the respective circles to which they belong: thus, if A″ M″ be an eighth part of the greatest and outer circle, A′ M′ is an eighth part of the next greatest, and A M an eighth of the smallest and innermost circle. Mathematicians conceive circles to be divided into 360 equal parts, which they call *degrees;* each of these degrees is again supposed to be divided into sixty equal parts, called *minutes*, and each of these minutes into sixty equal parts, called *seconds;* and, if smaller parts are required, the seconds may be divided into thirds, and the thirds into fourths, &c. each denomination being always sixty times the value of that which follows it in the order of succession; these divisions for the sake of brevity are written as follows:—thus, 57 degrees, 31 minutes, 42 seconds, 37 thirds, 48 fourths, and 32 fifths, are written

$$57^{\circ}\ 31'\ 42''\ 37'''\ 48^{iv}\ 32^{v}.$$

An angle is said to contain the same number of degrees, &c. which the lines containing it intercept on the circumscribed circles; if the lines A″ C, M″ C intercept 40° on each circle, A″ C M″ is an angle of 40°.

(2.) When the lines which contain an angle intercept a quarter of the circles described around it, the angle is called a *right angle*, and the lines containing

* This word is from the Latin *angulus*, which signifies a *corner*.

it are said to be *perpendicular*, or at *right angles* to one another; the quarter of the circle is called a *quadrant*.* Thus in *fig.* 5 the lines E″ C and A″ C are represented as cutting off a quarter of each of the three circles, and the angle A″ C E″ is a right angle, and the lines A″ C and E″ C are perpendicular to one another: and as the arcs, which lines containing a right angle intercept, are a quarter of the circles to which they respectively belong, which circles contain 360°, every right angle must contain 90 of those degrees.

(3.) Every triangle has six parts, the three angles and the three sides. There are various kinds of triangles—the *right angled* triangle is that which has one of its angles a right angle. Thus the triangle A B C (*fig.* 1) is a right angled triangle, having the right angle A C B as one of its angles. In this species of triangle, the sides are distinguished by mathematicians by particular names; the side A B, opposite to the right angle, is called the *hypotenuse*,† and the two sides B C and A C are called either base or perpendicular, according to the position in which the triangle may happen to be placed: thus, in the present position of the triangle A B C, A C would be termed the base, and C B the perpendicular, C A B the angle at the base, and A B C the angle at the vertex or top.

Fig. 1.

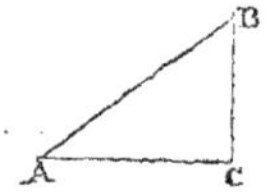

(4.) *Oblique angled* triangles are those which have not any of their three angles right angles, in short, which are not right angled.

(5.) The angles of any triangle, formed by straight lines, are together equal to 180°,‡ or two right angles; therefore in any right angled triangle A B C (*fig.* 1) as the angle A C B is equal to 90°, the angle at B will be equal to the angle at A subtracted from 90°, and the angle at A will be equal to the angle at B subtracted from 90°: these angles are called the *complements* to each other, for the one makes up what the other wants of being a right angle, or 90°.

(6.) There are mathematical instruments by which we are enabled to draw angles containing whatever number of degrees we please; but we cannot depend on their accuracy to a less quantity than about the quarter of a degree or 15′; that is, were we to attempt to draw an angle containing exactly 43° 14′ 16″, the small quantity (44″) by which that angle differs from an angle of 43° 15′ would be wholly imperceptible and lost, owing to the imperfection of the means we make use of to effect our object; probably the very thickness of the lines containing our angle would be equal to nearly a quarter of a degree—it follows, therefore, that we cannot depend on an angle, thus drawn, to less than that quantity.

(7.) There is an instrument also called a *plain diagonal decimal scale;* this is a ruler divided into a certain number of equal parts, and these parts, by a particular contrivance, are again subdivided *decimally*, or into tenths and hundredths of those parts. By this instrument we are enabled, but with no great degree of accuracy, to draw lines that shall bear to each other nearly any proportions we may please to assign them.

(8.) The use of the above mentioned instruments may be illustrated as follows: suppose we know, in a particular case, that the hypotenuse of a right angled triangle is 1100 miles, and the angle at the vertex equal to 53°: then, with our plain scale, we draw a line proportional to 1100 miles, or containing 1100 equal parts, each part being any length we choose to fix upon, and which then becomes our unit of length: let it be the line B A in *fig.* 1; we draw also another line B C making, with the line B A, an angle of 53°: we do not know the length the line B C ought to be when the triangle is right angled, or where we ought to stop in drawing it, but we may soon ascertain it by our ruler or scale in the following manner:—as the angle at B is 53°, if the triangle is right angled, the angle at A must be 37°, (Art. 5,) draw therefore the line A C, making an angle of 37° with the hypotenuse A B, it is obvious that the drawing this line, at this particular angle, determines the length that the line B C ought to be, when the angle at C is a right angle; and in order to ascertain the *proportional* lengths of the lines B C and A C, as compared with A B, we need only measure them with our scale. This is called *resolving the triangle* A B C *by construction*. We say the *proportional* lengths of the lines B C and A C, because it matters not what unit of length we make use of, in other words, what the length of each of the

---

* From the Latin word *quadrans*, which signifies the quarter of a small Roman coin, called an *as*.

† From a Greek word, signifying to *extend under*.

‡ Simson's Euclid, book i. prop. 32.

1100 parts is, or how that unit is represented on our scale, provided that we apply the same unit and measure to all the three sides of the triangle. Thus it is of no consequence whether the 1100 parts of the hypotenuse be 1100 miles, or yards, or feet, &c. or whether an inch, or half an inch, &c. on our scale represent a mile, provided that in taking all the three measures the inch or half inch represents the same quantity: if half an inch, for instance, represent 100 miles when we measure A B, it must represent the same quantity of miles when we measure B C or A C.

(9.) In order, therefore, to resolve a triangle by construction, it is only necessary to have the means of constructing the data, that is, the things given, and then we measure the lines or angles that are unknown: for if the data be sufficient, the very act itself of representing them on paper enables us also to represent those lines and angles that are not given, and when those unknown quantities are represented on paper, and on a scale proportional to the known quantities, we have but to measure them and their values are then ascertained. The examples, which it will be necessary to subjoin in this Part, of this mode of resolving triangles, will sufficiently illustrate the above explanation of the phrase *resolving a case by construction.*

Chapter II.

Preliminary Considerations.—On Traverses.

(10.) If we often interrupted the course of useful pursuits to analyze the manner in which we perform some of the complex though common operations of life, a very small part only of that which is actually accomplished by human energies and industry would then be achieved: yet there are cases in which, in order to obtain a clear conception of matters of science or art, with which we are not familiar, it is expedient to advert to those every day occupations, which have some connexion more or less remote with the object of inquiry, and to consider the different processes, which lead to their completion, and which, generally, through the influence of habit, make no impression on our minds.

Thus the very common operation of walking from any one spot to another, not in sight at the commencement of our journey, is connected with the infinitely more difficult one of directing the course of a ship through the ocean. The traveller on land has, in common with the mariner, to ascertain his course and distance: but the great point, in which journeying upon land and sailing on the sea differ in this respect, is, that on land the course and distance are usually determined by means of fixed marks on the surface of the earth, with the appearance of which the traveller is either acquainted, or is enabled to avail himself of their assistance by the aid of certain general considerations drawn from experience; which is indeed the source of the greater part of that more humble knowledge which all mankind possess in common. In traversing the ocean, on which vessels leave no permanent trace behind them of the path they have pursued, and the same difficulties present themselves to each succeeding voyager, the sky and the sea alone remain to direct the mariner in his course.

It is easy, however, to imagine a case in which the traveller on land may be wholly dependent on those guides, which a knowledge of some of the first principles of navigation alone can furnish.

(11.) Thus (*fig.* 2) we will suppose

Fig. 2.

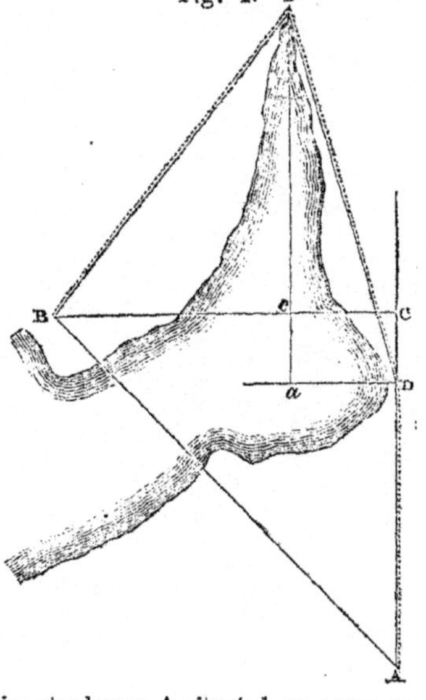

him at a house A situated on some extensive and barren plain, on which there is no object that can by possibility serve as a landmark to direct his steps, and that he wishes to traverse the plain to a village B at the distance of sixty *nautical* (Art. 44) miles from the spot on which the house is placed: let a line A C be conceived to be drawn through that spot towards the North Pole, or North point of the heavens; and let a line A B join the house and the village; and let the direction of the village from the house, or the line A B, make an angle of 45° with the line A C, that is, let the angle B A C equal 45°: draw also the line B C at right angles to A C. The angle B A C and the distance B A (=60 nautical miles) are determined, by means which will be explained hereafter, (Art. 54.)

Let us further suppose that the most eastern point of a large body of water, shown in *fig.* 2 by the irregular and crooked line, is situated due North of the house A, and that the situation and extent of this lake obliges our traveller, in his journey from A to B, to pursue the track marked out by the dotted line. Such a track is called by seamen a *traverse;* and it will be readily seen, that none of the ordinary modes which are in daily practice, of finding the way from one spot to another, will avail a person so circumstanced as in our example.

(12.) Before, however, we proceed to explain all the contrivances which are to be substituted for such ordinary modes, we shall define certain terms which it will be convenient to make use of hereafter; we shall also give a more simple example than that proposed in the last article.

Few of our readers can be ignorant, that such lines as B C and A B in *fig.* 2, supposed to be drawn on the round surface of the earth, cannot be straight lines;—they must be curved lines. But as the earth is a globe of very large dimensions, (Art. 44,) a very small curved distance, such as sixty miles, differs very slightly from a straight line; and therefore in our investigations we may assume the line A B, and of course all the other lines in the figure smaller than A B, to be straight lines. We shall find hereafter, (Art. 52,) that no error can *possibly* arise from this supposition in questions similar to those solved in examples I. and II.; and such an assumption, it is evident, must materially facilitate our inquiries. Draw the right angled triangles *a b* D, *b c* B. We shall call the line A C, drawn towards the north pole, and all lines parallel to it, as *a b*, *meridian* lines; and as those lines are drawn towards the top of our page, that top will represent the north, the right of the paper the east, the left the west, and the bottom the south. This supposition is invariably made with respect to all figures drawn to illustrate problems in navigation, so that we are enabled to discover the nature of lines by the position in which we find them, with reference to the top of the page. We perceive that the lines A B, D *b*, *b* B all make certain angles with meridian lines; these angles are called *courses;* the angle C A B is called the course from A to B; and as it equals 45°, the course from A to B is said to be N* 45° W, or 45° from the north, on the side of the west; and as this is just half way between the two, it is N W. The course from one place to another is also called the *bearing* of the latter place from the former; thus B is said to *bear* N 45° W from A, and A, S 45° E from B. For the present we shall call all lines drawn at right angles to meridian lines, *departures*. Thus B C, *a* D, and B *c* are termed *departures;* also the lines which make with the meridian lines the angles called courses are called *nautical distances;* such are the lines A B, D *b*, and *b* B. The true nature of courses, departures, and nautical distances, will be better understood hereafter, (Art. 51 and 52.)

Example I.

(13.) Let a man (*fig.* 1) be supposed to travel, or a ship be supposed to sail, eleven miles, (from B to A); let the course (A B C) make an angle of 53° with the meridian line (B C); and let it be required to find the length of the line B C, or, in other words, the distance made good in the direction B C, or from north towards the south.

It is plain, that as the ship's motion is compounded of a southern and western course, she must sail farther than the length of the line B C before she makes good that distance towards the south; we see, in fact, that she sails on the hypotenuse of the triangle instead of the perpendicular. The course also is by the last article S 53° W.

Draw the line B A proportional to eleven miles, making an angle of 53° with another line B C drawn from the

* It seems scarcely necessary to mention, that the letters N, S, E, and W stand for north, south, east, and west respectively. The letter *b* stands for *by*: thus, north by west is written N b W.

north towards the south; now the angle B A C, the complement of the course, (Art. 5) is 37°, draw therefore the line A C, making an angle of 37° with the line A B, and *measure* B C with the same scale with which A B was measured or set off; and B C will be found to be about six miles and a half, and A C the departure to be about eight miles and three quarters. The line B C is called the *difference of latitude;* and the four parts or elements of the triangle A B C, which are of importance in navigation, are the three sides A B, B C, and A C, and the angle A B C, which represent *the nautical distance, difference of latitude, departure,* and *course:* if any two of these be given, the other two may be found by construction; and, as we shall see hereafter, by *inspection* or *calculation.** The line A C, or departure, represents the amount of distance made good towards the west.

EXAMPLE II.

(14.) We proceed to the traverse drawn in *fig.* 2. This case, we shall presently perceive, is, in effect, nothing more than two solutions of the same question that was proposed in example I, and a solution of one, in which the course is the element of the triangle sought after. The bearing of the village B from the house A, (see Art. 11 and 12,) or the direct course from the latter to the former, is N 45° W; with the course N 45° W and the nautical distance A B = 60, we shall find by the mode of proceeding adopted in the last example A C = to about 42½ miles, and B C = 42½ miles, for in this case the departure equals the difference of latitude. So that our traveller on starting is enabled to discover by our method that he has 42½ miles of *northing*, or difference of latitude, to make, and the same quantity of *westing* or departure. Having ascertained this fact before he sets off, he might now proceed, had he the means of preserving any particular direction in which he might wish to travel, and of measuring the distance he walks; therefore the mode of effecting these two objects must be next explained. He can determine his course, or the angle his track makes, with every meridian he crosses, either by the heavenly bodies, or by a compass; a description of which instrument will be given in Chapter III. of the Second Part. The compass will enable him to ascertain at any time the direction in which he is walking;* and, consequently, at all times to preserve whatever direction he may find it expedient to take; thus, for instance, it will enable him the first day to travel due north, for the purpose of *making*, as the mariner terms it, the eastern point of the lake, the *bearing* of which we have supposed him to know before he sets out. With respect to the distances travelled, these may be discovered as follows: take a long string so divided into equal parts that each part shall be the 120th part of a nautical mile, and to the end of the string attach a piece of lead. Now as a minute is $\frac{1}{60}$ of an hour, half a minute is $\frac{1}{120}$ of an hour; or half a minute is the same part of an hour, that the divisions on the string are parts of a mile. Let pieces of cloth be tied round the string to mark the divisions; the first piece of cloth at the distance of about 30 feet from the lead; the first 30 feet of string is called the *stray* line. Let the traveller at the end of every hour throw down the lead on the ground and walk on, leaving the lead behind him, and allowing the string attached to it to pass through his hand as he walks; let him mark the time by a seconds watch when the first piece of cloth at the end of the stray line passes, and at the end of half a minute from that time let him stop the string, catching it up in his hand; and the number of divisions and parts of divisions contained between his hand and the piece of cloth at the end of the stray line will represent the number of miles he has walked during the last hour, on the supposition that he has always walked at the same pace, (*viz.* the pace at which he performed the experiment,) during that hour: for that number represents the pace at which he walks during the time the string is running through his hand; those divisions being the same part of a mile that half a minute is of an hour; and if $\frac{1}{120}$ part of a mile, or one division of the string, is described in $\frac{1}{120}$ of an hour, and during the preceding part of the hour the same rate of going has been preserved, 120 times that division, or a mile, has been performed in 120 times half a minute, or an hour; so that if the distance between the hand and cloth at the end of the stray line is one division, the pace is

* By calculation, B C = 6.62, and A C = 8.79.

* We make no mention in the text of the *variation* or *deviation* of the compass from the true north, this will be explained hereafter.

one mile per hour; if two divisions, two miles, and so forth. The line we have described resembles very nearly a seaman's log-line. Sailors call the pieces of cloth and divisions *knots*, so that knot and mile mean in effect the same thing: when a ship is said to be sailing eight knots an hour, the meaning is, that she runs eight divisions on the log line from the stray line in half a minute, that is, is sailing at the rate of eight miles an hour. At sea the lead is enclosed within a piece of wood, which floats on the surface of the water. On the assumption that the line is thrown down every hour, add together the distances due to each hour of the day during which the journey is prosecuted, and this will give the whole distance to be set down for each day. If our traveller were not aware of the existence of the lake, he would travel or steer NW; but we suppose him acquainted with the bearing of its eastern point; he will naturally, however, keep as near to the true direction, or course, as he can. So ships, when they are impeded by contrary winds, keep as near to their true course as circumstances will permit. Now we have seen, (Art. 11,) that the nearest course the traveller can hold at first will be due north; and the first day, or during the first 24 hours, we will suppose him to have walked 33 miles in that direction, when he arrives at D the eastern extremity of the lake; let him now alter his course, and steer for the point *b*, the most northern point of the lake; let him ascertain his second and third days' course to have been N 15° W, and the distance travelled, or D *b*, 47 miles. During the second and third days' journey our traveller has a mark to direct his steps, *viz.* the margin of the lake; but when he arrives at *b*, he will have no means of determining his direct and shortest course from thence to B, except by means of the distances travelled and courses steered during the first three days. We have seen that before B can be reached, $42\frac{1}{2}$ miles of northing and westing must be made; now on the first day 33 miles northing are made and no westing, therefore at the end of the first day $9\frac{1}{2}$ miles northing remain, and $42\frac{1}{2}$ westing. To ascertain the result of the second and third days' work, we have the distance D *b* = 47 miles, and the angle C D *b* = 15°; but the angle *a b* D = the angle *b* D C, for it is a property of parallel lines,* that when a straight line, as *b* D, meets them, it makes the alternate angles equal; and the angles *a b* D and *b* D C are called the alternate angles: those angles both represent the same course, but taken in an opposite direction, that is, *a b* D represents a southern and eastern course, whilst C D *b* represents an equal northern and western course. Therefore in the right angled triangle *a b* D we have the angle *a b* D = 15° and D *b* = 47 miles, to find *a b* and *a* D. Draw a line proportional to 47 miles, and draw another line making an angle of 15° with the former; draw a third line at the other extremity of the first line, making an angle of 75°, the complement of 15°, with that line: then we shall have constructed the data of our question, (Art. 9,) and the triangle *a b* D in the figure represents such a right angled triangle as we have been describing; measure therefore *a b* and *a* D as before, and *a b* will be found equal to about $45\frac{1}{2}$ miles, and *a* D to about $12\frac{1}{4}$. We perceive, therefore, that with respect to difference of latitude, the result of the second and third days' work is 36 miles more northing than necessary; now *a* D = *c* C,* and *c a* = C D, for the four-sided figure *a c* C D is a parallelogram,† take *c a* = C D = $9\frac{1}{2}$ from *a b*, and *c b* will remain = 36 miles; take *a* D = *c* C = $12\frac{1}{4}$ miles from B C (= $42\frac{1}{2}$), and B *c* will remain = $30\frac{1}{4}$ miles, so that of departure $30\frac{1}{4}$ miles remain, and at the end of the third day's journey there will remain to be performed 36 miles of southing, and $30\frac{1}{4}$ of westing; and in the right angled triangle B *c b* we have *b c*=36 miles, and B *c*=$30\frac{1}{4}$ miles given, to find the course B *b c* and distance *b* B. Draw therefore two straight lines perpendicular to each other, the one proportional to 36 miles and the other proportional to $30\frac{1}{4}$ miles; let B *c* and *c b* in the triangle B *c b* be those lines, measure the angle B *b c*, and it will be found = 40°, and *b* B, or the nautical distance, will be found to be 47 miles, or = D *b*; so that the traveller discovers he must now for the rest of his journey steer S 40° W, and as he has 47 miles left to complete his task, we may suppose that he takes two days more to perform it. We shall put the first three days' journey into a table, which is called a *traverse*

* Simson's Euclid, book i. prop. 29.

* Simson's Euclid, book i. prop. 28 and 34.

† From two Greek words signifying 'a *parallel drawing*. A parallelogram is *defined* to be a four-sided figure having its opposite sides parallel, and the opposite sides and angles of such a figure are *proved* to be equal.

table, and the construction of which will be sufficiently explained by that which has preceded.

Traverse Table.

Courses.	Dist.	Diff. Lat.		Departure.	
		N.	S.	E.	W.
N.	33	33			
N 15° W.	47	45½			12¼
		78½			12¼

(15.) If there had happened to have been southing during these three days, we should have subtracted it from the northing: the same remark applies to the Departure column, in case there had been easting: for our object was to ascertain the total quantity of difference of latitude and departure made good in the first three days, for the purpose of discovering how much of each remained to be accomplished, and thus we found B *c* and *c b*.*

(16.) On account of the great use of right angled triangles in solving problems in navigation, tables have been calculated called *Traverse Tables*, or *Difference of Latitude and Departure Tables*, which give the value of the perpendicular and base of a right angled triangle to every angle of an integral number of degrees, which the hypotenuse can form with them, and to every hypotenuse from 1 to 300. Solving right angled triangles by these tables is called solving by *inspection*.

(17.) The same principles will of course apply, whether we suppose the traverse to have been performed on land, or by a ship at sea. Thus, traverse sailing, or in other words, resolving cases of sailing in which a traverse is performed, consists simply in finding out by the solution of two or more right angled triangles, the total amount of difference of latitude and departure actually made good during the traverse; bearing it in mind, that opposite courses destroy each other, and therefore in these cases the less quantity must be subtracted from the greater: for instance, suppose that a ship during a traverse of eight days, which she is compelled to make by foul winds, sails 239 miles E, 240 W, 230 N, and 230 S; it is manifest, that the two last distances, being equal and in a contrary direction, destroy each other, and that the ship after sailing 939 miles, is only one mile distant from the place from whence she sailed; that is, she will be found one mile to the west of that place at the end of the traverse, for the total amount of westings exceeds the eastings by one mile. This is called *resolving a traverse.*

(18.) We shall conclude the subject of traverses by observing, that when several different courses are steered during the day, the total distance *on each course* must appear in the traverse table for that day, and in a line with it, the difference of latitude and departure due to that distance and course. In example II. it was our object as much as possible to preserve an analogy between the two cases of travelling on land and water; but we preferred for obvious reasons illustrating the subject of traverses by an example of a journey on land instead of a voyage at sea.

CHAPTER III.

The Plane Chart.—Pilotage.—Sailing in Tides and Currents.

(19.) THE plane chart is a representation of a small portion of the earth's surface on a large scale. In order to construct a chart of this description, we must ascertain the bearings of the objects there represented from each other; this may be done by surveying the space included within the chart; we then lay down the objects on the chart according to their several ascertained bearings. Figure 2, for instance, may be considered as a chart of the surface of a plain, in which the position of a village, house, and lake are laid down according to their several bearings. Such a chart it will be readily seen would have been of great service to the traveller in the example; as by it he could have ascertained the course he ought to have taken on each day, without the trouble of going through the process we have described: and if there had been drawn thereon a scale on which the length of a mile in the chart had been exhibited, he could thereby have measured all the distances. A scale is therefore generally drawn on a map of this description, and also a figure of a compass, for the purpose of enabling the mariner the more easily to determine the relative bearing of one spot on the chart from another. The meridian lines, or meridians, are drawn parallel to one another, (Art. 12.)

(20.) These plane charts are of great

* By calculation, AC = 42.43, *a b* = 45.4, *a* D = 12.17, *b c* = 35.97, B c = 30.26, B *b* = 47, and B *b c* = 40° 4′.

use to pilots and the masters of coasting vessels, who are not guided so much by the principles we have explained in Chapter II. as by buoys, soundings, and the bearings of the different headlands, lighthouses, and other objects on the land, in sight of which they steer their vessels. At night they are generally directed either by the bearings of the lights on the coast, or by the compass: for they are taught that, in order to avoid certain dangers, they must steer a particular course from one point to another. The bearing of any object, in sight from the deck of a ship, is ascertained immediately by looking at the compass. (Art. 45.)

(21.) As many rocks, shoals, and other dangers usually exist in the neighbourhood of land, and they are for the most part hidden below the surface of the water, even at low tide, it is the peculiar province of the pilot to be acquainted with their exact situation. The exact situation of a point may always be determined by the intersection of two straight lines, drawn from any two positions, and meeting in the required point: to illustrate this, let (*fig.* 3) represent an anchorage within

Fig. 3.

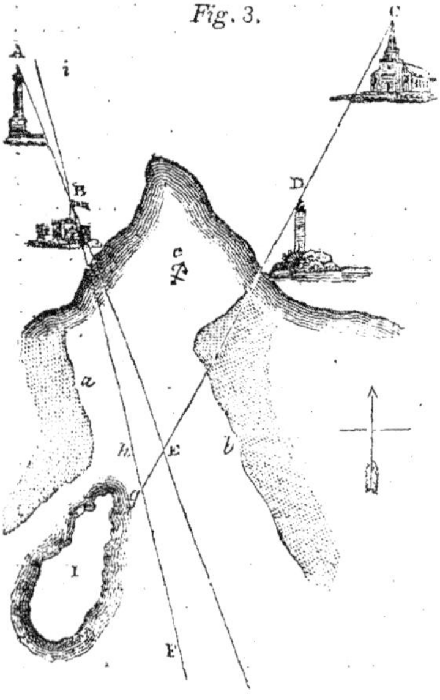

a bay: an anchor, in charts of this description, always denotes the exact spot at which the anchor should be dropped, this spot is *c*; *a* and *b* are shoals marked out by dots and the irregular lines; I, an island; A, B, C, and D are objects on shore; and E points out the situation of a hidden rock in the very middle of the channel to the harbour, which channel lies between the two shoals *a* and *b*. A, B, and E, are supposed to have such a position with respect to each other, as to lie in the same straight line. But the situation of the rock E is not determinable by the line A B E alone; for if the pilot know only, that it is situated somewhere in that line, he would still be ignorant of the exact spot which it occupied. But if the objects D and C, and the rock E, are so situated as to lie in the same straight line; then it is obvious that the point E is determined by the intersection of the two lines A B E, C D E. The pilot would then have information sufficient to enable him to avoid that rock in conducting a ship to this anchorage: for, in coming in from the south, suppose him so to place his vessel as to have the objects A and B in one line, or *in one*, as it is termed by seamen; now, he knows that whilst he continues to have these objects in one, he is in the exact line of the rock; but so long as D remains open to the left of C he will not strike upon it.* A safe mode of proceeding might be suggested in this case:—let him keep the object B at least a ship's length open to the right of A, and he will be in the line F *g* *h* B *i*; indeed he might keep it as far open to the right as he pleases, provided he avoids the island I, and the shoal *a*. When, in sailing on this line, he arrives at *g*, he brings D and C in one, and he knows the exact bearing of the rock E from his ship, but he cannot be said to be clear of E until he has passed the point *h*, which is situated due west of the rock; at that point D will be open to the right of C a certain quantity, say a ship's length; the pilot's direction may now be expressed in technical language:

"In coming from the south, keep B a ship's length open to the east of A until the most northern point of island I bears SSW, and D is open a ship's length to the east of C; you are then

* Smugglers are said to make a successful application of the principle we are here illustrating, in sinking their cargoes; they sail on a line A B E, keeping two objects A and B in that one line, until they arrive at a point in which two other conspicuous objects, as D and C, become *in one*, and there they sink their goods: by the intersection of these two lines they can find them again at any time, when they are relieved from the fear of detection.

clear of all dangers, and may steer NNE for the anchorage at *c*." The straight lines in the figure show the marks that are to be used as guides in entering the bay, and the marks themselves are called the *leading marks*.

(22.) Piloting vessels is however rendered in many cases very difficult, by the necessity which exists of taking into account, and allowing for the effects of tides and currents. These effects may be shown by supposing A B D C (*fig.* 4)

Fig. 4.

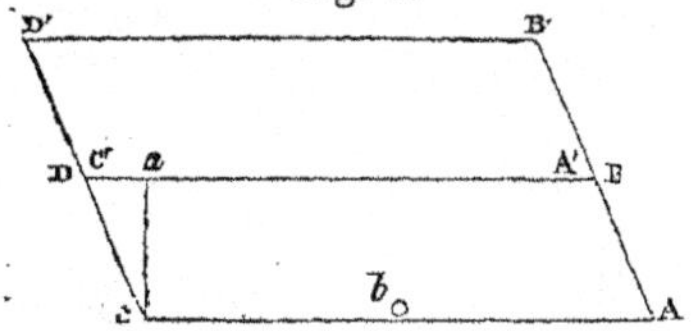

to be a trough containing water; let a light body *b* float on the surface of the water along the line A C, whilst the trough itself is being carried in the direction A B B′, or C D D′: at the same instant of time at which the trough takes up the new position A′ B′ D′ C′, let the body *b* arrive at the point C: now the point C is transferred to C′, and at that point will the body *b* be found at this time. This is exactly what happens in the case of a ship acted on by a current or tide; whilst the ship performs any particular course, due west for instance as in the figure, the current compels her to perform another course in the direction of its set,* and a distance proportional to its velocity; this new course and distance are entered in the traverse table, and taken into account, as if the ship had made them without the assistance of the current:—it amounts to the same thing in practice; for as the whole surface of water A B D C is supposed to move in the direction A B B′, the same quantity of difference of latitude and departure is made good in that manner, as if the ship *b* after its arrival at C had sailed directly from C to D, and the water had remained stationary; in both cases C *a* will represent the difference of latitude made, and *a* D the departure.

Chapter IV.

On Tides and Winds.

(23.) The alternate rise and fall of the waters of the ocean, called tides, is produced by the attractive forces of the sun and moon: now, as those attractive forces vary inversely as the square of the distances of the bodies attracting from the bodies attracted, (that is, become less in the same proportion as the squares of those distances become greater,) and as the moon, though much smaller, is nearer to us than the sun, her effect in raising the waters is greater than that of the sun, which is comparatively very small. The earth by its diurnal motion round its axis from west to east (see Art. 62) causes an apparent daily revolution of the moon round the earth in a contrary direction. The waters of the sea follow the moon in this her apparent course, so far as the irregularities of the land and shores will admit of. It might be supposed, therefore, that under these circumstances it must be always high water at any place when the moon is over it: but this is not the case, for when the waters have once begun to rise they will continue to rise after the cause which influences them has produced its maximum effect, for it does not then cease to act entirely. High water, therefore, is observed to happen about three hours after the moon has passed any place, or, as it is called, *passed the meridian of that place*.

The moon produces two tides of high water at the same time, one at places on the earth's surface nearest to her, and another at those on the opposite side of the globe, which are the most removed from her: for she attracts the centre of the earth more than the sea on that opposite side, as being nearer to her, the effect of which is to draw that centre away from the sea, and as the sea is left behind, it appears to rise. When the sun and moon are together, as at new moon, they combine their forces in causing the tides, and make *spring* tides; and the same thing happens at full moon when the moon is opposite to the sun; but when the moon is in quadratures, or half-way between the change and full and full and change, the whole action of the sun in causing tides is directly opposed to that of the moon, and this produces *neap* tides.

We have seen that there is high water at two spots on the surface of the earth, immediately opposite to each other, at the same instant of time; and there-

* The direction of a current is called its *set*; a current that flows *towards the* NNW *quarter* is said to set NNW: the velocity of a current is called its *drift*.

will be low water precisely at the same time, at all places which are 90° distant from the two places of high water, (Art. 35 and 36.) The action of the sun sometimes accelerates, and sometimes retards, the tides produced by the moon. The time of high water at any particular place is much affected by local circumstances; but, having ascertained the time of high water at that place, at the full and change of the moon,* we are enabled to find it at any other time, supposing the longitude of the place known, by means of the *Nautical Almanac*, and a table calculated for the purpose.

(24.) In a belt extending about 30° on each side of the equator, the wind is observed to blow all the year round from nearly the same quarter of the heavens: to the north of the equator it blows nearly from the NE quarter, and to the south of the equator from the SE quarter. These winds, from the great assistance which they afford to commerce, are called the NE and SE *trade* winds. When ships are bound from Europe to the West Indies, or to any part of North America, south of the parallel of about 38°, they seek the aid of these winds, but when they return, they keep away to the northward for the purpose of avoiding them. In December the NE trade is found to the south of the line, and on the other hand in June, the SE trade makes its appearance to the north of the line; but the SE trade is at other seasons of the year often found as far to the north of the line as the second degree of north latitude. The space between the second and fifth degrees of north latitude, is the inward boundary of these winds; in this space, storms, sudden squalls, and violent rains, are of frequent occurrence; but here likewise the wind often blows from the eastward, and also from the SW. There are *periodical* winds, which blow half the year in one direction, and half the year in the opposite direction, these are called *monsoons*: they are found in the Bay of Bengal, the Arabian Sea, the Mozambique Channel, on the coasts of Sumatra and Java, along the coast of China, and off the western coast of New Holland. October and April are the two months, in which the change in the direction of these winds usually takes place.

* It is on this account that, in books of voyages, the time of high water at different places at the full and change of the moon is so often to be met with.

PART II.

Chapter I.

On Plane and Spherical Trigonometry, —Logarithms.

(25.) We have seen, that the various lines, the magnitudes of which were the subjects of investigation in the first Part, constituted the sides of triangles, supposed to be drawn upon a plane surface. When we proceed to take into consideration the globular or round figure of the earth, it will not be necessary on that account to make any alteration in our hypothesis, with respect to the nature of the lines, the values of which it is necessary to ascertain; for the magnitude of the *curved* lines, which a vessel *makes* in traversing the round surface of our globe, is calculated in practice, through the medium of certain imaginary *straight* lines, which are proved to be either equal to such *curved* lines, or to have a known relation to them; such imaginary lines form the sides of plane* triangles, and present no greater difficulties to the calculator, than the lines, the values of which have been hitherto obtained by the method of construction.

(26.) In this our second Part, those imaginary lines will be invented, and the triangles resolved, which they constitute. It is clear, therefore, that the particular branch of mathematical science, which professes to teach the proportions and rules, derived from the relation which subsists between the sides and angles of triangles, must be of primary importance in our present inquiries. The name of that branch is *Trigonometry*,† and it forms the subject of the first chapter.

(27.) Every triangle having six parts, the three sides, and the three angles, (Art. 3,) one of the principal objects of the science of trigonometry is to discover rules, by which, if some of those parts be given the others can be found —*Rules have accordingly been discovered, by which, if any three of those parts be given, a side being always one, the other three can be found.* In right angled triangles, as the right angle is always given, if any *two* of the remaining parts be given, a side being one, the rest can be found.

* A *plane* triangle is one whose sides are straight lines; and the term is used in contradistinction to a *spherical* triangle, the sides of which are formed by portions of the circles of a sphere. This will be explained hereafter.

† From two Greek words, signifying *the measure of triangles*.

(28.) It appears from article 1, that whatever number of arcs may be described and intercepted between the lines containing any angle, (which are called its legs or sides,) they contain an equal number of divisions or degrees, &c. If the arc is large, the divisions are of course proportionally large, as the *number* of such divisions is the same in all circles; and it is found in fact, that the size of any division is in exact proportion to the radius* of the circle, or to the length of a string with which every circle may be conceived to be described. This truth is expressed *technically*, by saying, *the circumferences of circles, and corresponding parts of them, vary as their radii.*

(29.) Certain relations subsist between the sides and angles of all triangles; and they are determined through the medium of circles and right angled triangles. Construct *fig.* 5, by drawing the three circles having the same centre C,

Fig. 5.

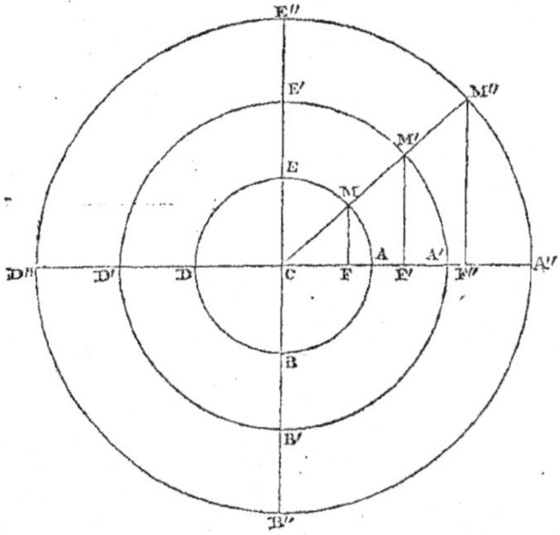

and their diameters at right angles to each other, as there represented; draw the line C M M' M'', and from the points M, M', M'' draw the lines M F, M' F', M'' F'' perpendiculars to the diameter D'' A'', then the three right angled triangles F M C, F' M' C, and F'' M'' C will be *equiangular*, that is, will have all the angles of the one equal to all the angles of the other, each to each; for the right angle in every one is of course equal to the right angles in the others, and the angle at C forms the angle at the base to every one of the three triangles, that is, it is common to all the three; and as all the angles of a plane triangle are together equal to two right angles (Art. 5) the remaining or third angle must be equal in all the triangles; for that angle is the complement (Art. 5) of the angle at C in each of the triangles. Now all plane triangles which are equiangular, have the sides which contain the *corresponding* equal angles proportional; that is, the sides which contain the right angle in each are proportional to the sides containing the right angles in the others, and the same is true of the other angles;* hence C M : M F :: C M' : M' F', C M : M F :: C M'' : M'' F'', and C M' : M' F' :: C M'' : M'' F'', and the same is true of lines similarly drawn in any number of circles; but the lines C M, C M', and C M'', are radii (see Art. 28) of the respective circles, and the lines M F, M' F', and M'' F'' are called the *sines*† of the angle at C; the line M F is called the sine of the angle at C to radius C M, M' F' the sine of the same angle to radius C M', and M'' F'' the sine of the same angle to radius C M''; also the three equiangular triangles have the sides about the angle at C proportional, or C M : C F :: C M' : C F' : C M'' : C F'';

* From the Latin word *radius*, (the plural of which is radii,) which signifies the spoke of a wheel; these radii are of course all equal to one another.

* Simson's Euclid, book 6. prop. 4. When triangles are equiangular, and have the sides containing corresponding equal angles proportional, the triangles are said to be *similar*, and the corresponding sides are called *homologous*, that is, having the same ratio. *The proposition referred to in this note is one of the most useful and important in geometry.*

† The origin of this word is doubtful, it is probably from the Latin word *sinus*, which is itself a translation of the Arabian word Jeib a *bosom*, the Arabian name for a sine. See Preface to Hutton's Logarithms.

now the lines C F, C F′, and C F″ are termed the *cosines* of the angle at C to the three radii C M, C M′, and C M″; so that the sine of any angle may be defined to be, a line drawn from one extremity of the arc on which the angle stands, perpendicular to a diameter of the circle to which the arc belongs, passing through the other extremity of the arc; and the cosine of an angle to be that part of the same diameter which is intercepted between the sine and the centre of the circle: the proportions before stated may now be more technically expressed as follows, radius (C M): the sine or cosine of any angle (M F or C F) to radius C M : : any other radius (C M′) : the sine or cosine of the same angle (M′ F′ or C F′) to that other radius C M′. So that if we could only know the value of one sine or cosine of a certain number of angles, and the radius to which these sines or cosines were calculated, we should then have two terms in a proportion, or rule of three sum, and then, having a third, we should be able, without difficulty, to find the fourth term. Now, in fact, we do know the values of one radius and one sine and cosine to every second in the circle; we have tables calculated which give the proportional value of the sine and cosine of each angle to an assumed radius; and therefore when we wish to know the value of any other sine or cosine, the only other thing requisite to be known besides the angle itself, is the value of the radius, to which this latter sine or cosine is referred; and on the other hand, if we wish to know the value of a new radius, we must know the value of a sine or cosine to that radius. To illustrate this, suppose the angle at C = 43°, and the line C M or radius = in value to the radius in the tables, or unity; now as the tables give the value of the sine of any angle to the radius C M (= 1), if we wish to know the length of the line M′ F′, and we know the length of the line C M′ (suppose C M′ = 2) we can immediately find the line M′ F′ by the following proportion, C M (1) : tabular sine of 43° (= 0,682) : : C M′ (2) : the answer = 1,364.

(30.) We are also enabled in the same manner to ascertain the values of certain other lines which we shall now proceed to define. The *tangent* of an angle or arc* is a line drawn from that extremity of the arc through which the diameter of the circle passes, (which extremity is called the beginning of the arc,) perpendicular to that diameter, touching the circle but not cutting it,* and terminated by the radius of the circle produced, and passing through the other extremity of the arc: the radius thus produced until it meets the tangent, is called the *secant* of the angle or arc, from the Latin word *secans* which signifies *cutting*; thus, in *fig.* 6, A T is the tangent of the arc A M

Fig. 6.

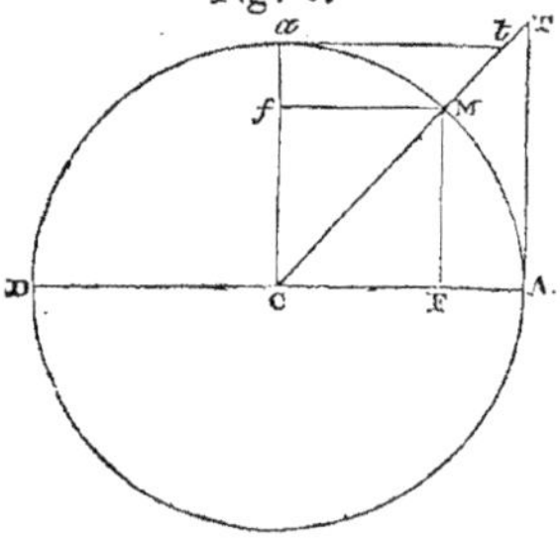

and angle A C M, C T the secant, M F the sine, and C F the cosine; also the difference between any arc and 90° is called the *complement* of that arc; and the difference between any arc or angle and 180°, is called the *supplement* of that arc or angle; thus in *fig.* 6, *a* M and the angle *a* C M are the complements of the arc M A and the angle M C A, and D M and D C M are the supplements of the same arc and angle. Now it is easily demonstrated† (see note Art. 14) that C F the cosine of the angle M C A = *f* M the sine of the angle *a* C M, and that M F the sine of the angle M C A = *f* C the cosine of the angle *a* C M; that is, the cosine of any arc or angle equals the sine of its complement, and the sine of any arc or angle equals the cosine of its complement; and the lines C F, *a t* and C *t* derive the names of *co*sine, *co*tangent, and *co*secant, from the circumstance that they respectively represent the sine, tangent, and secant, of the complement of the arc or angle named; thus the line C *t* is called the cosecant of M C A, or the secant of its complement M C *a*; and the line *a t* is the cotangent of the angle M C A, or the tangent of its complement.

* The lines M F, C F, M′ F′, C F′, &c. are called sines and cosines to the arcs, as well as to the angle which the arcs measure.

* From this circumstance the line derives its name, it being called *tangent* from the Latin word *tangens*, which signifies *touching*.

† Simson's Euclid, book 1, prop. 28 and 34.

The terms we have used are abbreviated as follows:

For sine of A M, or A C M, write sin. A M, or sin. A C M, and in the same manner, for

Cosine	write	Cos.
Tangent	. . . .	Tan.
Cotangent	. . .	Cot.
Secant	. . . . .	Sec.
Cosecant	. . . .	Cosec.
Radius	. . . . .	Rad.

(31.) Now, as the triangles F C M and A C T are *similar*, we have the proportion C F : C M :: C A : C T, or cos.: rad. :: rad. : sec.: and by similar triangles *a* C *t*, M C F, C *t* : *a* C :: C M : F M, or cosec. : rad. :: rad. : sin. Also (Art. 29 and 30) tabular rad.: tab. sin. cos. tan. or sec. :: new or given rad. : sin. cos. tan. or sec. to the new or given rad., or sin. cos. tan. or sec. to new or given

$$\text{rad.} = \frac{\text{tab. sin. \&c.} \times \text{new rad.}}{\text{tab. rad.}};*$$

and if the tab. rad. = 1, the sin. cos. &c. to the new rad. = tab. sin. &c. × new rad. So that if any three *parts* of the right angled triangles C F M and C A T are given, a side being always one, the other three may be found.

(32.) There are two kinds of trigonometrical tables, the first kind contains the sines, cosines, &c. of angles, calculated to the radius unity. The sines, cosines, tangents, and secants in these tables to radius 1, are called *natural*, to distinguish them from those contained in the second description of trigonometrical tables, which are called *artificial* or *logarithmic.*†

(33.) The artificial or logarithmic sines, &c. are far more frequently used than the natural; for, by a peculiar artifice, they are so constructed, that in all cases in which, were it not for this invention, it would be necessary to multiply natural sines, &c. by each other, we need only add the logarithmic sines, &c.; and, in cases in which we should be compelled to divide a natural sine or cosine, &c. by another, we need only subtract the logarithmic sine, &c. which represents the divisor from the logarithmic sine, &c. which represents the dividend; this contrivance, it will readily be seen, must often save much time and labour. There are also logarithms of numbers of the same construction, and presenting the same facilities to the computer. This word *logarithm* in calculations is always abridged by writing *log.*; and the logarithmic sine of the arc A M is written, log. sin. A M.

We shall hereafter have occasion to employ these logarithms, and to illustrate their use and nature by examples.

The radius of the logarithmic tables is 10,000,000,000; our equation in Art. 31, when used with these tables, becomes log. new sin., &c. = log. tab. sin., &c. + log. new rad.—log. tab rad. The logarithm of the radius in the artificial tables is 10.

(34.) There are also proportions, by means of which, when any three parts of an *oblique* angled plane triangle are given (except the three angles) the other three may be found. In the case of *spherical* triangles also, both right angled and oblique angled, there are proportions which enable us, three parts being given, to determine the other three: and with respect to spherical triangles there is this peculiarity, that even where the three angles only are given, the other parts, *viz.* the three sides, may be found, and the whole triangle determined.

(35.) A *sphere* is a solid shaped like a ball, or, to use the language of science, bounded by a curve surface of such a nature, that every point thereon is situated at an equal distance from a point within the solid called the centre. A *great* circle of a sphere is that whose plane passes through the centre of the sphere; it divides the sphere into two equal parts, and has the same centre as the sphere itself. In *fig.* 7 the circle A E B C D, the plane of which passes

Fig. 7.

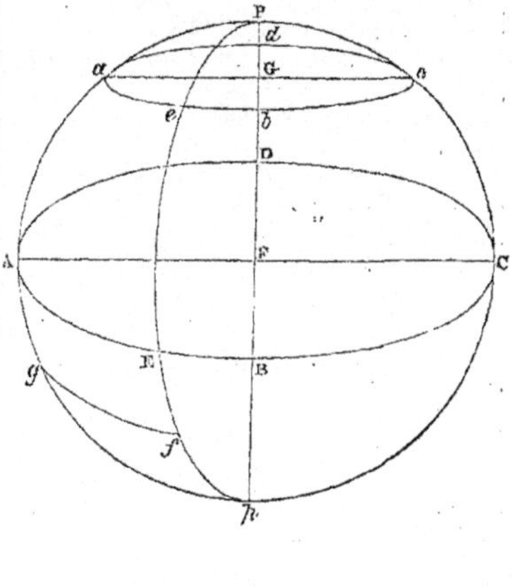

* The product of the second and third terms divided by the first. See the Treatise on Algebra, (Art. 127 and 128.)

† From two Greek words, *logos* and *arithmos*, which signify the *number of ratios or proportions*, or the number showing the proportion.

through the centre F of the sphere A P C *p*, is a great circle; P A *p* and P E *p* are respectively halves of great circles; all great circles of the same sphere are equal. The plane of a *small circle*, does not pass through the centre of the sphere. In *fig.* 7, *a b c d* represents such a circle. A spherical triangle is conceived to be formed on the surface of some sphere, three great circles of which intersect each other in such a manner that their arcs enclose on that surface a triangular space; thus A E P is a spherical triangle. If a diameter of a sphere (P F *p*) be drawn perpendicular to the plane of a great circle of that sphere (A B C D), the two extremities of that diameter P and *p* are called the *poles* of that great circle. Every spherical angle is supposed to be contained between the arcs of *great* circles; and the angle thus formed on the surface of the sphere is equal to the angle contained between the planes of the great circles which form it; thus the spherical angle A P E is equal to the plane angle contained between the semicircular planes P A *p*, P E *p*, at their intersection P *p*; and it is measured by the arc A E of the great circle A E B C D intercepted between these two planes, the poles of which great circle are the two extremities P and *p* of that diameter, which forms the common intersection of the two planes.

A great circle passing through the poles of another great circle is called a *secondary* to it, and secondaries are perpendicular to the great circle to which they are secondaries; also the arcs of a secondary intercepted between the poles and the circumference of the great circle to which it is secondary, are all quadrants, or contain 90°. Thus the great circle P A *p* C, passing through the poles P and *p* of the great circle A E B C D is called a secondary to that circle; and the arcs P A and P C, *p* A and *p* C, as also the arcs of all other secondaries, intercepted between the poles P and *p*, and the circumference of the great circle A B C D, are quadrants.*

* The above is to be considered merely as a short view of the most practical and elementary part of trigonometry. The subject cannot be fully treated until the student has made considerable progress both in algebra and geometry. We refer him to the treatise on trigonometry, which will be published hereafter. In a subsequent part of the algebra the manner of constructing logarithmic tables, and their use, will be fully explained.

Chapter II.

Definitions.—Latitude and Longitude —Magnitude of the Earth.

(36.) We shall take *fig.* 7 to represent the globe or sphere of the earth;* the small circle *a b c d* is supposed to be parallel to the great circle A E B C D, and consequently perpendicular to its secondaries; *a* G *c* and A F C are diameters of those circles, and the point *e* is a point in which the small circle meets the great circle P E *p*.

(37.) This being premised, P and *p* may represent the north and south *poles* of the earth, the great circle A E B C D will then represent the *equator*, or *equinoctial*, or, as it is called by seamen, *the line;* the poles of which are the poles of the earth. All secondaries to the equator, as P A *p* C, P E *p*, &c. are called *meridians;* the diameter P F *p* will represent the *axis*, round which the earth turns every 24 hours; and all small circles, as *a b c d*, parallel to the equator, are called *parallels of latitude.*

(38.) Let *a*, *b*, *d*, *e*, *f*, and *g* be six places on the earth's surface; then the meridian P *e* E *f p* passing through the places *e* and *f*, is called the meridian of those places; and the meridian P *a* A *p* is called the meridian of the place *a*, and of all other places situated on the semicircle P *a* A *p;* for the meridian of any place is the secondary to the equator which passes through it; the arc of that secondary intercepted between the place and the equator is called the *latitude* of that place; thus the arc *a* A represents the latitude of the place *a*, the arcs *e* E and *f* E the latitudes of the places *e* and *f*. If the place is situated on the north side of the equator, its latitude is called north latitude; if on the south side, its latitude is called south latitude; thus *e* is in north latitude, and *f* is in south latitude.

(39.) Places situated on the same parallel of latitude, as *a* and *e*, have the same latitude; for the arcs of secondaries, intercepted between their great circle and a small circle parallel to it, or between two small circles parallel to their great circle, are equal.

(40.) The *longitude* of any place is its distance from a particular meridian, called the *first* meridian, measured on

* The earth is not strictly spherical, but for the purposes of navigation it may be so considered. (See the Treatise on Mathematical Geography.)

the arc of the equator, intercepted between the first meridian and the meridian of the place; suppose P *a* A *p* to be the particular meridian which is called the first meridian, then the arc of the equator A E will represent the longitude of the places *e* and *f*; for it follows from the above definition, that all places on the same meridian must have the same longitude.

(41.) If the meridian passing through the Royal Observatory at Greenwich, be that which is called the first meridian, places will be said to have east or west longitude, according as they are situated to the east or west of that meridian; thus, *e*, *f*, and *b*, have all east longitude, but a place *d* on the opposite side of the globe has west longitude.

(42.) The arc *e f*, intercepted between the two places *e*, *f*, represents their *difference of latitude;* and it follows from that which has preceded, that the arc of any meridian or secondary to the equator, intercepted between the parallels of latitude passing through any two places, will represent their difference of latitude, (art 39;) thus, if *g f* be part of a parallel passing through *f* and *g*, *g a* will also represent the difference of latitude between *e* and *f*, and between *g* and *e*, and *f* and *a*.

(43.) The *difference of longitude* between any two places is represented by the arc of the equator, intercepted between the meridians passing through the two places; which arc also measures the spherical angle at the poles included between the two meridians, (Art. 35;) thus, the arc A E, or the spherical angle A P E, represents the difference of longitude between the places *a* and *e*, and *g* and *e*, and *f* and *a*.

(44.) The magnitude of the earth is that of a sphere of about 7916 English miles in diameter; and a sphere which has a diameter of that size, must be about 24869 English miles in circumference. Now the great circles of the earth are supposed to be divided into 360°, that is, into 21600′, (Art. 1) and a *geographical* or *nautical** mile is $\frac{1}{60}$ of a degree of a great circle of the earth, or one minute of such a circle, (Art. 1;) therefore a great circle contains only 21600 *nautical* miles, whilst it contains 24869 *common or statute* miles; or a common mile is to a nautical mile, as 21600 to 24869; to express this proportion in feet we must say,

21600 : 24869 :: 5280 feet : 6079,089 feet.

Hence a geographical or nautical mile is about 6079 English feet.

Chapter III.

On the Mariner's Compass.—The half Minute Glass.—Mariner's Scales.—Mercator's Chart.

(45.) The course of a ship, or the angle which her track makes with the meridians of the globe, is determined by the compass. A circular card, represented in *fig.* 8, has a magnetized bar of steel attached to the back of it, so as to form one of its diameters; this diameter is supported horizontally on a point, and is exactly balanced on its centre, so as to allow it to turn round freely with the card in every direction; that diameter of the circular card which is marked with the letters N and S, standing for north and south, and with a star at each extremity, shows the position of the magnet below: it is the property of such a bar, when thus balanced, to point *nearly* to the North Pole, and it carries the letters N and S on the card along with it into the same position. The horizon of every place is supposed to be divided into thirty-two parts, or points, called by the names represented by letters in *fig.* 8. When the bar of steel has carried the horizontal card into the position above described, the thirty-two lines or points marked in the figure will be opposite to thirty-two corresponding points in the horizon, which are called by the same names as those abridged in the card, so that the circular card is an artificial representation of the horizon of any place. These points are each subdivided into four parts, called *quarter points*. If there be an object opposite to one of these lines of the card, say that marked SW b S or south west by south, that object is said to *bear* (Art. 12) SW b S from the spectator at the compass. This, however, may not be the true direction or bearing of that object, it may be only the *magnetic* bearing; for the magnet rarely points truly north, it is subject to two errors from different causes, the one a cause which acts continually, and the other local; the effect produced by the former is called the *variation*, and that produced by the latter the *deviation;* these errors will be adverted to hereafter; their amount is easily ascertained; and they establish the distinction between *true* and

* From the Greek word ναῦς, *naus*, a *ship*.

Fig. 8.

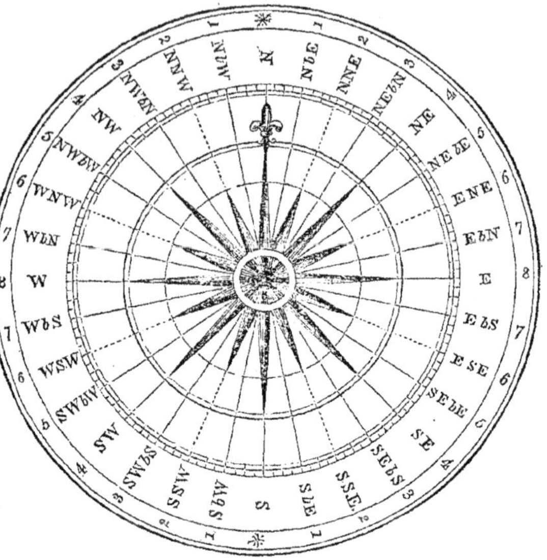

magnetic courses or bearings. The card we have described is by a particular contrivance so suspended, that it remains perfectly horizontal, notwithstanding the various irregular motions and concussions to which a ship at sea is liable.

In the inside of the box containing the card, there are two black vertical lines, which lie in an imaginary straight line drawn through the ship, from head to stern, and which coincide respectively with the N and S points of the card or compass, when the ship's head is towards the *magnetic* north, or the north which that compass shows; so that when the ship's head is directed towards any other point, the points, or quarter points, which are then opposite to these vertical lines, show the points, or quarter points, of the horizon to which the head and stern of the ship are then directed; now as the circle of the compass card is divided into 32 equal parts, each point must contain 11° 15′; so that if the points marked on the card N b W and S b E coincide with the vertical lines in the new position of the ship, (the N b W with the vertical line towards the head, and the contrary,) the straight line drawn through the ship makes an angle of 11° 15′ with the line drawn towards the magnetic north, or the magnetic meridian, as it is called: if the ship were now to set sail, and to sail with her head in this position, her apparent course would be N b W, and her track or wake would make an angle of 11° 15′ with the magnetic meridian;* in this manner are ships steered by the compass:—the card of some compasses is divided into degrees; a compass of this description would show at once the angle which the line drawn through the ship's head makes with the magnetic meridian.

(46.) In heaving the log at sea, the half minute is measured, not by a seconds watch, but by a half minute glass, which differs from a common hour-glass in nothing but the time the sand takes to run out.

(47.) Besides the plain diagonal decimal scale, of which we have spoken, (Art. 7,) there are others, the most useful of which is that called Gunter's. But as these scales only enable the mariner to solve problems, in which no great accuracy is required, we shall do no more than mention them, and not stop to explain their utility.

(48.) We have already described a plane chart, (Art. 19,) it remains to explain the principles on which a chart is constructed of much greater practical

* This would be the case if the ship were making no leeway at the time. See Art. 55.

utility, called a *Mercator's Chart.** This chart exhibits a most convenient and ingenious manner of representing the surface of the globe on a plane. The meridians of the globe, or secondaries to the equator, are drawn parallel to one another in this chart, as in the plane chart; and therefore the parts of the parallels of latitude, intercepted between any two meridians, must be equal to one another, and to the intercepted part of the equator in all latitudes; for these parallels of latitude cross the parallel meridians at right angles, and form a number of four-sided figures, which are all right angled parallelograms; (see note Art. 14.) the difference of longitude between any two places may therefore be measured on these parallels, as well as on the equator itself; and the distance between any two meridians is in all latitudes the same, and equal to their difference of longitude: the proportion, however, which subsists between the parts of meridians, and the same elementary parts of parallels, on the globe, is accurately preserved in a Mercator's Chart in all latitudes. If we look at *fig.* 7, we shall perceive, that though $a\ e$ in that figure is the same part of the small circle $a\ e\ b\ c\ d$, that A E is of the circle A E B C D, or equator; yet that it is much shorter in length; and therefore one minute, for example, or any other elementary part or division of a circle, must be smaller on that circle than on the larger circle, (see Art. 28,) which represents the equator; in fact, the length of A E : length of $a\ e$: : rad. of equator : rad. of parallel; that is, A E : $a\ e$: : A F : a G, for those lines are the radii of the equator and the parallel; but a G is the sine of the arc a P to radius A F, which arc is the complement of the latitude of the place a, therefore a G is the cosine of the latitude of a; (Art 29 and 38 ;) also the length of a minute, on a meridian of the globe, is equal to the length of a minute on the equator, for all great circles of a sphere are equal. Hence we have,

Length of 1′ on meridian : length of 1′ on parallel : : rad. : cos. lat. of parallel. Therefore in the globe the proportion between the length of an elementary part of a meridian and the length of the same elementary part of a parallel is, as rad. : cos. of the latitude in which that parallel is situated. *It is this proportion of radius to the cosine of the latitude of the parallel, which is preserved in a Mercator's Chart.* As on the globe the length of 1′, taken on a parallel, becomes very small as we approach the pole, compared with its length on a meridian; it is evident that, as all the parallels in our chart are equal to the equator, in order to preserve this proportion, our meridians must be very considerably lengthened; and the more so as we approach nearer to the pole, in the neighbourhood of which the natural parallels are so very small. It is the mode of lengthening these meridians in proportion as the parallels on the chart are lengthened beyond their natural size, we have now to investigate.

Now, cos. lat. : rad. : : rad. : sec. lat. (Art. 31,)

and our proportion becomes,

Length of 1′ on parallel : length of 1′ on meridian : : rad. : sec. lat.

We began by observing that the parts of meridians on our chart were to have the same proportion to the same elementary parts of parallels, as they have on the globe: that proportion we have just expressed; and, therefore, in our chart also,

1′ on parallel : 1′ on meridian : : rad. : sec. lat. or,

1′ on parallel : rad. : : 1′ on meridian : sec. lat. (See Algebra, Art. 127 and 128.)

Let us commence, therefore, the construction of the chart by drawing a line to represent the equator, divide it into equal parts, which we will call 1′ each; draw parallels, and through the equal divisions of the equator draw the meridians parallel to one another, and at right angles to the equator; and the parallels will likewise be divided into equal parts of 1′ each. Now if we suppose that the line which represents 1′ of longitude on all the parallels, or on the equator of our chart, represents also the length of the radius of the natural trigonometrical tables, it is evident, looking at our proportion, that 1′ on a meridian should also represent the natural secant of the latitude in which the measure is taken; or the parallel of the latitude of one minute, for example, ought to intercept between itself and the equator a line equal to the secant of 1′ to a radius equal to 1′ on a parallel; for the two first terms in our proportion are made equal, the other two therefore must be equal: in the same manner, when we draw the pa-

* Mercator never divulged the principles on which he constructed the charts which bear his name. It was to Mr. Edward Wright, that science was indebted for the first exposition of those principles.

rallel of the latitude of 2′, we must make it intercept between itself and the parallel of 1′ a line equal to the natural secant of 2′. Hence to find the distance of any parallel in the chart from the equator, or the number of lines representing elementary parts of parallels, which the part of the meridian intercepted between them contains, we have only to add the natural secants together, until we arrive at the secant of the given parallel: thus the part of the meridian intercepted between the equator and the parallel of 5′, or the *projected* meridian, as it is called,

$=$ sec. 1′ + sec. 2′ + sec. 3′ + sec. 4′ + sec. 5′.

This distance, which we have found for the parallel of 5′, is called also the *meridional parts* of 5′; the meridional parts or distances are computed to every minute of the quadrant as far as 86°, and inserted in tables.* It is frequently necessary, in the solution of nautical problems, to refer to the number of these meridional parts intercepted between the parallels of two places, in lieu of their *real* or *natural* difference of latitude. This circumstance has introduced the terms, *proper difference of latitude*, and *meridional difference of latitude;* the latter term being used to express the number of meridional parts intercepted between the parallels of any two places on the earth's surface, when they are delineated on a Mercator's Chart.

Chapter IV.

Principles of Navigation.—Invention and Construction of the Four Triangles.—Proportions derived therefrom.—Examples.

(49.) When a ship sails due north or due south she does not alter her longitude, and when she sails due east or west she does not alter her latitude. These are the two most simple cases that can be proposed, in which it is necessary to take into account the spherical figure of the earth, and, therefore, with these cases we will commence.

Let us suppose the ship (*fig.* 7) to sail on a meridian from the place *f* to the place *e*, now the longitude of these two places is the same, (Art. 40,) and, consequently, the ship has not changed her longitude; the whole of her progress during the voyage, has been effective only in changing her latitude: the *nautical distance* also which she has made, is the arc of a great circle of the earth, and the number of miles in that distance, which is ascertained by the process of heaving the log, is the number of minutes in the arc *f e*, and represents the difference of latitude; so that by that process alone, the ship's place at *e* can be determined. In this instance only, and when she sails on the equator, is a ship, steered by a compass, able to trace out the arc of a great circle of the earth; and, consequently, these are the only two cases in which a ship, sailing on any given course from point to point, takes the shortest possible road between those points:—for the arc of a great circle is the shortest line which can connect two points on a sphere. (See the Treatise on Math. Geog. chap. 7.)

(50.) Next let us suppose the ship sails due west from *e* to *a*; she evidently does not change her latitude, for she sails on a parallel of latitude all the points in which have the same latitude, (Art. 39.) This second case, however, is not quite so simple as the former, unless indeed the ship sails on the equator itself; for here the heaving the log only gives the length of the arc of the parallel *e a*, whilst the mariner, in computing the place of his ship, must know the change of longitude, which so much nautical distance sailed on that particular parallel produces; that is, he must know the length of the arc E A on which the difference of longitude is measured; but we have seen (Art. 48) that,

$$\text{cos. lat.} : \text{rad.} :: e\,a : \text{E A},$$

and therefore $\text{E A} = e\,a \times \dfrac{\text{rad.}}{\text{cos. lat.}}$;

consequently, from this equation, E A may be found, and the place of the ship determined. This is called a case of *parallel sailing*.

If we take, however, any right angled triangle B E F, (see *fig.* 11, page 22.) B E : E F :: rad. : cos. B E F, or B E

$= \text{E F} \times \dfrac{\text{rad.}}{\text{cos. B E F}}$, (Art. 29.) Make

therefore the angle B E F equal to any given latitude, and if E F represents the length of the part of the parallel which a ship has traversed in that latitude, or the arc *e a*, B E will represent the difference of longitude made, or the arc E A. Suppose that a ship sails twenty miles in the parallel of London, then if B E F=

* In computing these tables, however, the addition of the natural secants is not the method resorted to, as they can be calculated more conveniently and correctly from the expression,
log. merid. part = 3,8984895 + log. (10 − log. tan. ½ complement latitude.) See a paper by Dr. Halley, Phil. Trans. No. 219.

the latitude of London, B E the difference of longitude $= \frac{\text{rad.} \times 20}{\text{cos. lat.}} =$ about 32 miles. So that, in resolving cases of parallel sailing, we need not concern ourselves with the sphere, but may advert to this right angled triangle, by which we may see exhibited, and *may measure*, the comparative lengths of the corresponding parts of the parallel and equator. We shall now show *that all cases of sailing whatever may be resolved through the medium of right angled triangles.*

(51.) We proceed to the third and by far the most difficult case, where the ship sails neither on a meridian, nor the equator, nor on a parallel; in which case she is continually changing both her latitude and longitude.

In *fig.* 9, let P represent the north pole of the earth, P F and P E parts of meridians, and A G, M L, and F E portions of parallels of latitude; let the ship sail from A to E. It is required to investigate rules by which, if the nautical distance sailed and course be given, the

Fig. 9.

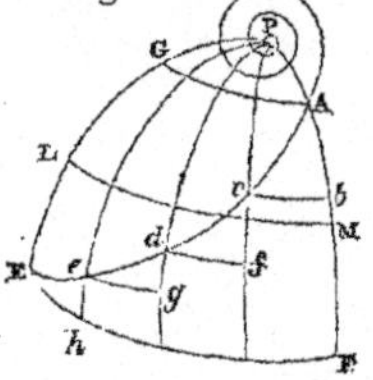

Fig. 10.

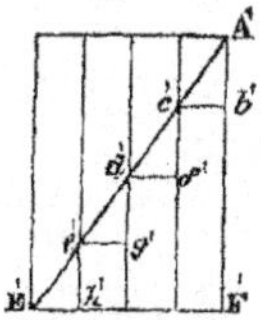

differences of latitude and longitude between any two places, as A and E, may be found; and conversely, if the differences of latitude and longitude be given, the nautical distance and course may be found. It will be necessary, however, first to define some lines of which we shall have occasion to speak.

The curve A E in the figure, which is the line the ship traces out in sailing from A to E, without altering her course by compass, is called a *rhumb** line. A rhumb line may be defined to be *the shortest line which can join two points on the globe, cutting all the meridians which it crosses at the same angle.*

From the nature of the compass it is evident, that a ship's way, so long as she steers the same course† by a compass, must make the same angle with every meridian she crosses; she therefore must trace out the line we have defined; and the angles c A b, $d c f$, &c. which her track makes with the meridians, or courses, must be all equal to one another. The angles a ship's way makes with the equator and parallels, when she sails on a meridian, or with meridians, when on the equator or parallels, are equal, since they are always right angles. The *oblique* rhumb line is called also the *Loxodromic* curve, from two Greek words signifying an *oblique course*, and is a line of a very peculiar nature; it is a spiral, and has the remarkable property of winding round and round the pole of the earth, constantly approaching, yet never reaching it: so that if a ship could sail on the same oblique course for ever, she would approach infinitely near, either to the north or south pole, but could never actually reach them.

In the triangle A F E, A E represents the nautical distance, and therefore *the portion of a rhumb line intercepted between any two places through which the rhumb line passes, is their nautical distance;* A F also represents the difference of latitude, and the angle E A F the course.

The *meridian distance* made, *is the distance between the meridian arrived at and the meridian left, measured on the equator, or the parallel on which the ship is;* thus, when a ship sails from A to E, F E is the meridian distance she has made; but when she sails from E to A, then G A is the meridian distance; and when she sails on a parallel, or the equator, the arc of the parallel or equator she describes is itself her meridian distance. In parallel sailing, therefore, the ship *actually measures* her meridian distance as she proceeds; but in sailing on an oblique rhumb line, it is the oblique rhumb line which she measures.

(52.) Construct *fig.* 9 thus, suppose the rhumb line A E divided into four equal parts in the points c, d, and e; and through those points draw three meridians; then draw the parts of parallels $b c$, $f d$, and $g e$; the result is the formation of the four small triangles A b c, $c f d$, $d g e$, and $e h$ E; but if instead of four small triangles, constructed in the manner above mentioned, we were to imagine many thousand, in short, an indefinite number so constructed, by a continual subdivision of the rhumb line A E, and a continual drawing of meridians through the points of division, and to consider the triangle

* This name is derived from the Portuguese word *rumbo*, or *rumo*, which signifies a course.

† When there is no leeway, (see Art. 55.)

A *b c* one of such infinitely small triangles; then the bases *b c*, &c. of these numerous triangles added together, would represent what is termed the *departure* made in sailing from A to E; and as an arc of a parallel of latitude, or of the equator, between any two places, may be supposed to consist of an infinite number of such small lines, it represents the departure between the two places by which it is intercepted. That this departure answers to the line which we have called the departure in the first Part, will appear from the following considerations: as the triangles A *b c*, &c, are indefinitely small, the lines composing them may be assumed to be straight lines; suppose, therefore, the ship to sail at first only from A to *c*, an indefinitely small distance; in this case the line *b c* represents the line we have hitherto called the departure, for it is a straight line perpendicular to a meridian; so when the ship continues her vogage and arrives at E, she has sailed over an indefinite number of lines, all equal to A *c*; now the very small right angled triangles A *b c*, &c. have their hypotenuses equal, and all the angles of the one equal to all the angles of the others, each to each, therefore the triangles are equal and similar;* therefore the ship has also made an indefinite number of small departures all equal to and identical with *b c*; therefore their sum *b c* + &c. must be the departure to the sum A *c* + &c. or A *c* taken as many times as *b c*, or the nautical distance A E.

As the several small elementary triangles A *b c*, &c. are equal, whatever number of times the line A E contains the line A *c*, the same number of times will the sum A *b* + &c. contain A *b*, and the sum *b c* + &c. contain *b c*; hence we have sum of all the A *c*'s : sum of all the A *b*'s : : A *c* : A *b*; and sum of all A *b*'s : sum of all *b c*'s : : A *b* : *b c*. But the sum A *b* + &c. = A F the difference of latitude; (Art. 42;) and the sum *b c* + &c. = the departure. Let there be a *plane* triangle A″ E″ F″,† having the angle at its vertex A″, or E″ A″ F″, equal to the angles *c* A *b*, &c. or the course, and having the side A″ E″ equal to the number of nautical miles in the spiral line A E, and the side A″ F″ equal to the number of nautical miles in the difference of latitude A F; in other words, let A″ E″ = A E, and A″ F″ = A F: we have already proved that, sum of all the A *c*'s : sum of all the A *b*'s : : A *c* : A *b*, that is, A *c* : A *b* : : A″ E″: A″ F″. Now it is demonstrated in treatises on geometry,* that if two triangles have one angle of the one equal to one angle of the other, and the sides about the equal angles proportionals, the triangles are equiangular, and consequently similar; we have seen that the two triangles A *b c*, A″ F″ E″ are so circumstanced, and therefore they are equiangular, but A *b c* is a right angled triangle, therefore A″ F″ E″ is also a right angled triangle; and A *b* : *b c* : : A″ F″: F″ E″; and we have shown, A *b* : *b c* : : sum of all the A *b*'s : sum of all the *b c*'s, but the sum of all the A *b*'s = A″ F″, and the sum of all the *b c*'s = the departure; therefore the last proportion becomes, A *b* : *b c* : : A″ F″ : departure,

hence departure $= \dfrac{A'' F'' \times b\,c}{A\,b}$, but F″E″

$= \dfrac{A'' F'' \times b\,c}{A\,b}$, hence F″E″ = the departure.

So that we have arrived at this conclusion, that when a ship sails on an oblique rhumb line, and the nautical distance made forms the hypotenuse, and the course forms the angle at the vertex of a right angled plane triangle; the perpendicular of that triangle will represent the difference of latitude, and the base the departure made; if any two therefore of these four parts be given, the others can be found; the triangle being a right angled triangle. It appears, therefore, that the *departure* made by a ship, may be defined to be, *the sum of all the successive elementary meridian distances, when the nautical distance is assumed to be divided into an indefinite number of equal parts.*

The departure therefore is a species of *imaginary* quantity, the result of an hypothesis made for the purpose of obtaining a straight line to represent it. We shall presently discover, however, that it is of great utility as a connecting link; for it joins and adapts the rule for the solution of cases of parallel sailing to cases of sailing on an oblique rhumb line.

Let the meridians in *fig.* 10 be the representation on a Mercator's Chart of the meridians in *fig.* 9, marked with the same letters without the dashes; then a straight line A′ E′ will represent the rhumb line A E on such a chart; for that line has been defined to be the shortest line that, in connecting two places, as A′ and E′, cuts all the meridians it

* Simson's Euclid, book 1. prop. 26. and book 6. prop. 4. and (Art. 5.)

† The triangle A″ E″ F″ is not drawn.

* Simson's Euclid, book 6. prop. 6.

crosses at the same angle; and as the meridians on a Mercator's Chart are parallel, and a straight line is the shortest line that can be drawn between two points; it follows, that the straight line A′ E′ (which is not only the shortest line that can be drawn between the two points A′ and E′, but also in crossing the meridians makes equal angles with them all)* answers completely to our definition of a rhumb line; and those angles or courses must be equal to the courses of the former figure; divide this rhumb line, as before, indefinitely into equal parts in c′, &c. and let the meridians pass through these divisions; complete the elementary triangles by drawing the parts of parallels; it is evident, that these new right angled triangles A′ b′ c′, &c. are also all equal, similar, and identical triangles; and they are all similar not only to the triangle A′ F′ E′, but also to the triangles A b c, &c. in the former figure, and, lastly, to the triangle A″ F″ E″ for every one of these triangles has the angle at its vertex the same; and they are right angled, consequently their third angles, (Art. 5,) or the angles at the base, must be equal; therefore also they must be all similar to one another, and have their sides about the equal angles proportionals, hence we have

A′ F′ : F′ E′ : : A″ F″ : F″ E″,

but A′ F′ is the meridional difference of latitude, F′ E′ the difference of longitude (Art. 48,) A″ F″ the proper difference of latitude, and F″ E″ the departure. Hence we have

mer. diff. lat. : diff. long. : : prop. diff. lat. : departure, (A.)

Also in the right angled triangle A′F′E′,

A′ F′ : F′ E′ : : rad : tan E′ A′ F′ (Art. 30,)

that is,

mer. diff. lat. : diff. long. : : rad. : tan. course (B.)

These proportions marked A and B are one mode of connecting the nautical distance and course made by a vessel, when sailing on an oblique rhumb line, with her change of longitude.

Cases of sailing on an oblique rhumb line resolved by these two proportions, are said to be resolved by *Mercator's Sailing*.†

We shall now explain another mode by which the solution of such cases may be effected through the medium of the imaginary line called the *departure.* This method is called *middle latitude sailing.*

Let M L (*fig.* 9) be an arc of the parallel of latitude half way between the parallels of F and A, or the parallel of the middle latitude. We perceive on inspecting the figure, that all the elementary parts of the departure b c, &c. with the exception of the last, are less than corresponding intercepted parts of the meridian distance F E, but greater than the corresponding intercepted parts of the meridian distance A G; that is, the one meridian distance is too great, and the other too small to represent the departure. This is not the case, however, with respect to the arc of the middle parallel; its corresponding parts are neither *all* greater, with one exception, nor *all* smaller than the parts of departure, but some are greater and some smaller; that is, on one side of the arc of the middle parallel are parts of departure less, and on the other, parts of departure greater; and inasmuch as this is caused by the *regular* divergency of the meridians as we approach the equator, we may fairly infer, that the defect of the parts of the arc of the middle parallel which are too small, is nearly supplied by the excess of the parts too great, and that the arc of the middle parallel intercepted between the meridians of any two places is *nearly* equal to the departure made in sailing from the one place to the other on an oblique rhumb line. Having obtained, therefore, an arc of a parallel in a known* latitude, which will nearly represent our departure, the case is resolved into a case of parallel sailing, (Art. 50;) for we may consider the vessel to have sailed the sum of all the elementary parts of the departure on the parallel of M and L; and in that case (on the supposition that M L is really equal to the departure) her change of longitude will be the same as if she had described A E, and may be found by the proportion

cos. lat. : rad. : : dep. : diff. long.

But we must remember that the latitude in this proportion, of which the cosine is taken, is the middle latitude between A

* A straight line crossing parallel lines makes equal angles with them all. See Simson's Euclid, book i. prop. 29.

† On a Mercator's chart, the course between any two places is immediately found by measuring the angle, which the straight line joining them forms with the meridians; the distance between those places may be found correctly by applying a scale of cosines of middle latitudes to a meridian on the chart. The modern Mercator's Charts have these scales and directions explanatory of their use.

* For the latitudes of the places A and F are assumed to be known, and the latitude of M is found by taking a latitude as much greater than that of F, as it is smaller than that of A, or the *mean* latitude.

and F; therefore, in resolving cases of middle latitude sailing, we should mark this by writing the proportion as follows,

cos. mid. lat. : rad. : : dep. : diff. long.

It appears, therefore, that if we suppose the angle B E F (*fig.* 11) in the triangle B E F to be equal to the middle latitude, and F E equal to the departure made in sailing the distance A E on the oblique course E A F, then, on the supposition that M L = that departure, the hypotenuse of that triangle, B E, will represent the difference of longitude made. (Art. 50.) It is evident, however, this method of middle latitude sailing is but an *approximation*, the departure actually made is not *exactly* equal to the arc of the middle parallel, and the principles of parallel sailing require *that the departure should be reckoned in the parallel to which it truly belongs.* The parallel to which it truly belongs is clearly that parallel, which will give the difference of longitude actually made.

Therefore the above proportion should contain the cosine of the *true* middle latitude, or middle latitude in which the departure should be reckoned, instead of the cosine of the *mean* middle latitude, as it is called. Mr. Benjamin Workman, in a small Tract published in 1805, under the patronage of Dr. Maskelyne, gave a very useful table for converting *mean* middle latitudes into *true* middle latitudes; and as his improvements are not so well known, or so duly appreciated as they deserve to be, we have inserted that table at the end of this treatise; and shall explain, before we close this chapter, the principles on which it is constructed.

Cases of sailing in low latitudes, and when the given course is within one point of due east or west, may be very properly solved by this method of middle latitude sailing, without applying the corrections from Mr. Workman's table; and whenever the given course is less than 50°, or, not being less, when the difference of latitude exceeds 20°, or 1200 miles, the method of Mercator's sailing should be used. In all other cases of sailing on an oblique rhumb line recourse should be had to middle latitude sailing as improved by Mr. Workman.

Let the triangle A F E (*fig.* 11) represent the auxiliary right angled triangle A″ F″ E″, we shall call it, for distinction, the *plane* triangle. Let B E F be another right angled triangle having the same base E F, and let the angle B E F be equal to the *true* middle latitude, or the latitude in which the departure F E should be reckoned; then will the right angled triangle B E F be a proper triangle to resort to for finding the difference of longitude made by sailing the distance A E on the oblique course E A F, when the departure F E is reckoned in the latitude B E F; and also when the difference of longitude made, in sailing a nautical distance equal to the departure F E on the parallel of the latitude B E F, is required. (Art. 50.) This right angled triangle, therefore, is a triangle fitted to resolve cases both of parallel sailing and middle latitude sailing; we shall call it the *parallel* triangle.

Fig. 11.

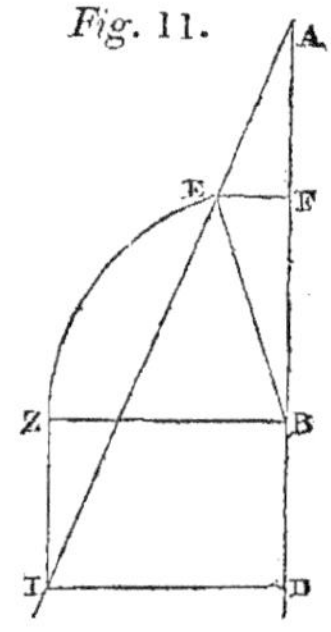

Produce A E and A B indefinitely; with B as a centre and B E as a radius describe the arc of a circle E Z; at the point B draw B Z perpendicular to A B produced, and meeting the arc E Z in Z; through the point Z draw Z I parallel to A B produced; it must meet A E produced either on the one or the other side of B Z.* Let them meet in I, and from I draw I D at right angles to A B produced; then the four-sided figure Z I D B is a parallelogram, and B Z = D I;† but B Z = B E, for they are both radii of the same circle, therefore B E = D I, and D I in the right angled triangle A D I represents the difference of longitude made on sailing the distance A E with the oblique course E A F, when the departure F E is reckoned in the latitude B E F; and as

A D : D I : : rad. : tan. course,

therefore by proportion B, A D equals the meridional or projected difference of latitude made in sailing the same course and distance, when the true middle latitude equals the angle B E F. Now the right angled triangle A D I answers in

* Simson's Euclid, book i. prop. 29.

† Ibid. book i. prop. 34.

every respect to the triangle A′ F′ E′ in *fig.* 10; it is the triangle therefore by reference to which we must resolve all cases that should, according to our rule, be resolved by Mercator's sailing, we call it therefore the *Mercator* triangle. The seven important elements of navigation are, the course E A F, nautical distance A E, difference of latitude A F, departure F E, difference of longitude B E, or D I, middle latitude B E F, and meridional difference of latitude A D; of which the plane triangle contains the four first; the parallel triangle the fourth, fifth, and sixth; and the Mercator triangle the first, fifth, and seventh. If the reader remembers this, and our rule as to the courses to which particular modes of solution are adapted, he will know to which of these three triangles he ought to turn, when any case of sailing is presented to him for solution.

There is, however, a fourth triangle, B E A, which we shall call the *oblique* triangle; it contains the course, nautical distance, and difference of longitude. From this triangle we derive one of two useful proportions which remain to be stated, and which are necessary also in explaining the principles on which Mr. Workman's table is constructed; these two proportions we shall now proceed to investigate. In the triangle B E F,

B E : F E : : rad. : sin. E B A, (Art. 29,) also

F E : A E : : sin. E A B : rad.

Hence (Algebra, Art. 127 and 128) we have

B E : A E : : sin. E A B : sin. E B A,

and

A E : sin. E B A : : B E : sin. E A B,

But the angle E B A is the complement of B E F the mid. lat., therefore,

sin E B A = cos. mid. lat.

therefore

dist. : cos. mid. lat. : : diff. long. : sin. course (C.)

In the parallel triangle B E F,

F E : B E : : sin. B* or cos. mid. lat. : rad.

In the plane triangle,

A F : F E : : rad. : tan. A or tan. course,

Hence (Algebra Art. 127 and 128)

A F : B E : : cos. mid. lat. : tan. course,

or

diff. lat. : diff. long. : : cos. mid. lat. : tan. course. (D.)

The manner of finding an algebraical expression that shall represent the cosine of the true middle latitude may now be explained.

Put l = diff. lat., L = diff. long., m = mer. diff. lat., c = cos. mid. lat., R = rad., T = tan. course, c'' = cosec. course, c' = cos. course, s = sin. course, s' = sec. of the course, d = dist., and D = dep.; then by proportion B we have

$$m : \mathrm{L} : : \mathrm{R} : \frac{\mathrm{L\,R}}{m} = \mathrm{T}.$$

By proportion D we have,

$$l : \mathrm{L} : : c : \frac{\mathrm{L}\,c}{l} = \mathrm{T}.$$

Hence, $\frac{c\,\mathrm{L}}{l} = \frac{\mathrm{L\,R}}{m}$,

multiplying both sides of the equation by $\frac{l}{\mathrm{L}}$, $c = \frac{\mathrm{R}\,l}{m}$,

and log. c = log. R + log. l − log. m.

So that to find the true middle latitude, we must add 10 to the logarithm of the proper difference of latitude, and subtract from the sum the logarithm of the meridional difference of latitude, the remainder is the logarithmic cosine of the true middle latitude, (Art. 33.) From this equation, therefore, calculating m as before mentioned, (note, Art. 48,) Table 2 might be computed. The true middle latitude always exceeds the mean.

By proportion C, $s = \frac{c\,\mathrm{L}}{d}$, but $s = \frac{\mathrm{R}^2}{c''}$ (Art. 31.) Substituting this value

$$\frac{\mathrm{R}^2}{c''} = \frac{c\,\mathrm{L}}{d},$$

multiplying both sides of this equation by $\frac{d\,c''}{\mathrm{R}^2}$, we obtain

$$d = \frac{c\,\mathrm{L}\,c''}{\mathrm{R}^2}.$$

The last equation gives Mr. Workman's rule for finding the nautical distance from the difference of longitude, course, and true middle latitude. It is this, *Add together the logarithmic* c, *the logarithmic* c″, *and the logarithm of* L, *and subtract* 20 *from the sum, the remainder is the logarithm of* d, *the nautical distance.*

Table 1 contains the artificial or logarithmic sines, tangents, and secants, to every point and quarter point of the compass.

Table 2 is so simple in its construction, that it needs but little explanation; look for the nearest degree of the mean

* Sin. B is an abbreviation for the sin. of the angle E B A.

middle latitude at the side, and the nearest degree of the difference of latitude at the top, and the correction is found under the latter in a line with the former, which correction, added to the mean middle latitude, gives the true middle latitude. If there are odd minutes, a proportional part may be allowed for them. We shall presently illustrate these tables by an example.

(53.) Before we proceed to solutions by actual computation, we shall restate the four proportions we have obtained. In addition to these, we must remember that the three right angled triangles of figure 11 furnish several others, on the principles explained in Art. 29, 30, and 31.

$m : L :: l : D,$ (A.)

$m : L :: R : T,$ (B.)

$d : c :: L : s,$ (C.)

$l : L :: c : T,$ (D.)

EXAMPLE III.

(54.) A ship has sailed from the Lizard 700 miles on a WSW course (true,) required her latitude and longitude, or, as it is called, *the latitude and longitude in.*

The Lizard is in latitude 49° 57′ 44″ N, longitude 5° 11′ 5″ W.

By the plane triangle, $R : c' :: d : l = \frac{c' d}{R}$, and therefore log. l, or diff. lat. = log. d, or dist. + log. c', or cos. course, — log. R, or radius.

2,845098	log. dist.
9,582840	log. cos. of 6 points.
12,427938	
10,000000	log. rad.
2,427938	log. diff. lat. = 267′, 88 = 4° 27′ 54″.

49° 57′ 44″ —
4 27 54
= 45 29 50 = the latitude in.

The true middle lat. is found thus,

45° 29′ 50″ = the latitude in +
2 13 57 = the half difference of the two latitudes
= 47 43 47 = the mean middle lat.

Now in Workman's Table, (Table II.) in a line with 48° and under 4° is found 3′, and under 5° is found 4′; therefore we will call the correction + 3′ 30″, but 47° 43′ 47″ + 3′ 30″ = 47° 47′ 17″ = the true middle latitude, which in this case, as the course is so nearly west, differs but little from the mean.

By proportion D we have,

$c : T :: l : L = \frac{T\, l}{c}$, and therefore log. L = log. T + log. l — log. c.

10,382776	log. tan. 6 points.
2,427938	log. diff. lat.
12,810714	
9 827328	log. cos. mid. lat.
2,983386	log. diff. long. = 962′,47 = 16° 2′ 28″.

5° 11′ 5″ W +
16 2 28
= 21 13 33 W = longitude in.

If we had taken the mean middle latitude in this case instead of the true, it would have given 961′,54 for the difference of longitude, or 16° 1′ 32″.

EXAMPLE IV.

The ship during the whole of her passage from the Lizard, which she made at the rate of 7 knots an hour, has been sailing in a current, which set to the N b E (true) with a drift of one mile per hour. An error therefore has been committed in the computations of the latitude and longitude, by not allowing for this current: it is now required to find the latitude and longitude in, when the proper allowance is made for the effect of this current.

As before, $l = \frac{c' d}{R}$. And as by the hypothesis the ship has been sailing for 100 hours in a current with a drift of one mile per hour, the distance, the surface of the water has been carried in that interval of time, (Art. 22,) = 100 miles.

2,000000	log. dist.
9,991574	log. cos. 1 point.
11,991574	
10,000000	log. rad.
1,991574	log. diff. lat. = 98′,08 = 1° 38′ 5″.

By Mercator triangle, proportion B, we have

$R : T :: m : L = \frac{T\, m}{R}.$

9,298662	log. tan. 1 point.
2,149219	log. mer. diff. lat. = 141.
11,447881	
10,000000	log. rad.
1,447881	log. diff. long. = 28′,05 = 28′ 3″.

By inspection, (Art. 16,) diff. lat. = 98′,1 and diff. long. = 28′,06, so that these quantities might have been found by this method with great exactness, without the trouble of a calculation; and indeed in practice the diff. lat. and diff. long. are often found by inspection. As in order to answer the question we must resolve a traverse, (Art. 22,) we will put the results into a Traverse table.

The distance is found by Workman's rule (Art. 52) as follows:

10,015845 log. of c''.
9,820550 log. of c.
2,970542 log. of L.

22,806937
20,000000

2,806937 log. of d = 641,12 miles.

Courses.	Dist.	Diff. Lat. N.	Diff. Lat. S.	Diff. Long. E.	Diff. Long. W.
WSW.	700		267,88		962,47
N b E.	100	98,08		28,05	
			267,88		962,47
			98,08		28,05
True course S 74° 37′ W Dist. 641,12 miles.			169,8 S.		934,42 W.

$$\begin{array}{l} 49^\circ\ 57'\ 44'' - \\ \ \ 2\ \ 49\ \ 49 \\ \hline = 47\ \ \ 7\ \ 55 \end{array}$$ = the true latitude in.

$$\begin{array}{l} \ \ 5^\circ\ 11'\ \ 5''\ \mathrm{W} + \\ 15\ \ 34\ \ 25 \\ \hline = 20\ \ 45\ \ 30\ \mathrm{W} \end{array}$$ = the true longitude in.

With the *true* diff. lat. and long. above obtained by resolving the traverse, the *true* course is found as in Example V.

The true course S 74° 37′ W is nearly WSW ¾ W; supposing the variation 2¼ points westerly, this will give a compass course W b N; but the course by compass from the Lizard to the Southernmost of the Scilly Islands is WNW, consequently the ship will pass one point clear of these islands, so far as our current affects her; but should the tides cooperate with the current, and produce a more rapid drift to the Northward, the ship might be lost on these islands, or the rocks around them; which illustrates our observation in Article 22, with respect to the judgment and skill requisite in a pilot, to whom the difficult task belongs of estimating and allowing for the probable effects of currents and tides, which are ever varying in their power and direction, and therefore embarrass, and often destroy, the most experienced mariners.

In steering across a river, for instance, from point A to point B on the opposite shore in a tideway, steer for a point on that side of B *from* which the tide flows, and as much above or below B, as in your judgment you conceive the tide would have carried the boat on the side of B *towards* which the tide flows, during the passage across, had it been constantly steered in a direction parallel to the line joining A and B.

Example V.

Required the *direct* course, and distance on a rhumb line, from London to Naples.

London 51° 30′ 49″ N latitude, and 5′ 47″ W longitude.
Naples 40 50 15 and 14 15 45 E.

10 40 34 = diff. lat. = 640,56 miles.

14 21 32 = diff. long. = 861,53 miles.

from proportion B, we have,

$$T = \frac{R\,L}{m},$$

2,9352704 log. of L.
10,0000000 log. R.

12,9352704
2,9684829 log. of m.

9,9667875 log. T 42° 49′
or the course is S 42° 49′ E.

and by the plane triangle, $d = \dfrac{s'\,l}{R}$

10,1345808 log. of s'.
2,8065598 log. of l.

12,9411406
10,0000000 log. of R.

2,9411406 log. of d.
or d = 873,25 nautical miles.

The distance, measured on the arc of a great circle, will be found to be 872,22 nautical miles, and 1003,63 statute miles, 69,04 being reckoned to a degree: this is, however, what is termed in common conversation the distance *as the crow flies.*

CHAPTER V.

On Leeway, and Plying to Windward.

(55.) When a ship is sailing near to the point from whence the wind blows, a considerable part of the force of the wind is employed in driving her away from that point, or to *leeward;* but as this action of the wind cannot turn the *head* of the ship round, or alter her apparent course, the effect produced is a continual drifting of the vessel from the wind with the head still turned in the same direction as if no drifting took place, and consequently the compass showing the same course: but if the ship drifts in this manner, her keel will make a track or wake in the water in a direction opposite to the point towards which she is really moving. Let therefore the figure of a compass be drawn on the stern of the ship, and so placed that the line joining the north and south points of the card shall be in the direction of the keel, or the fore and aft line of the ship; the angle included between this line and the wake is the difference between the ship's apparent and her true course by compass, and is called the *leeway :* this leeway therefore is always to be allowed for from the wind; that is, if a ship is steering WNW, with the wind at north, the leeway is reckoned to the left of WNW from the wind; and if in this case the angle, or leeway, is found to be two points, the ship's true compass course is due west.

The only method that ought to be relied on in practice of ascertaining the amount of this correction of the apparent course of a ship, is that of actually measuring the angle formed as before mentioned; which angle is in fact the bearing of the wake by the fixed compass: means therefore should be devised for enabling the proper officer to take such a bearing with correctness, and in this respect practical navigation seems to be deficient.

(56.) There are few large vessels that can lie within less than six points of the wind; and therefore, when the wind blows from any point within six points of the bearing of a port for which a vessel is bound, she must *tack* or *ply to windward;* that is, she must steer a course as near to the bearing of the port as the wind and other circumstances will admit of, and she must steer that course, until the bearing of her port is altered so as to become a course, which the direction of the wind will allow her to steer. Suppose, for instance, a ship is bound to a place bearing E, but the wind is ENE, that is, two points from the bearing of her port; the ship's course must be SE or N; for these courses are respectively six points from the direction of the wind, and are, therefore, nearer to the bearing of the port than any other courses which the ship can describe; and the ship, supposing the wind to remain in the same quarter, must sail SE, until the bearing of her port becomes from the alteration of her place due N, or N till it becomes SE; and then she must tack and steer N or SE, according as her first course has been SE or N; but if it should be more convenient so to do, the ship may make a great number of tacks, or SE and N courses, before she arrives at her port; and the whole distance sailed will not be greater in this case than in the other.

PART III.

On Nautical Astronomy.

(57.) The two first Parts of this treatise have shown, how the situation of a ship on the surface of the globe may be found by a reckoning kept on board of the courses steered, and the number of miles sailed on each course. We have supposed, however, the errors of the compass to be known, and it will be necessary therefore hereafter to point out how they may be found.

The object of the third Part is to explain, how the relative angular positions of the celestial bodies, with respect to each other, and the horizon and meridian of any place, enable the mariner to determine the situation of that place, and thus correct the errors of the *reckoning.* So that these two different methods serve as a check upon each other, and have together been found amply sufficient for all the purposes of the practical navigator.

That part of the heavens which is visible to us, and in which the celestial bodies appear, is in the shape of a hollow hemisphere; and it will facilitate greatly the comprehension of nautical astronomy, if we imagine the earth placed in the centre of a hollow glass sphere, which has the heavenly bodies exhibited on its surface;—that the earth, instead of revolving round its axis every twenty-four hours from West to East, remains at rest; and that the hollow glass sphere with the heavenly bodies upon it revolves uniformly round the earth in the same time from East to West: the suppositions we have made will accurately represent the *diurnal* motions of those heavenly bodies:—the axis round which the hollow sphere revolves is the axis of the earth produced from each of its extremities till it meets the sphere: let us imagine great circles traced out on the hollow sphere corresponding with and opposite to the great circles of the earth already described; corresponding to the equator, let there be a *celestial* equator; corresponding to the meridians, *celestial* meridians; corresponding to the poles, *poles of the heavens* at the extremities of the axis of the sphere; let the parallels also be represented in the same manner: if the circular plane of the equator of the earth were to be enlarged and extended so as to reach the glass sphere, then should its circumference coincide with the celestial equator traced out upon that sphere, and the same observation applies to all the other circles above described; and it is in this sense, therefore, that we use the term *corresponding*. If a straight line joining the earth's centre, and any place on the earth's surface, be produced until it meet the hollow glass sphere, the point at which it meets that sphere is called the *zenith* of that place; and as the sphere revolves, every point of it which successively touches the extremity of that line will successively become a zenith to the place; and stars on the sphere, which in succession pass that line, will be said to pass over the zenith of the place. The celestial meridians are called circles of *declination:* for the arc of a celestial meridian intercepted between any heavenly body and the celestial equator is not called its latitude, but its *declination*. As each celestial meridian is brought in succession opposite to the terrestrial meridian of any place, by the revolution of the sphere, it acquires the name of the celestial meridian of that place.

Great circles drawn on the hollow sphere passing through the zenith of any place, are called *vertical* circles; and that particular vertical circle which passes through the East and West points is called the *prime* vertical; the angle contained between a vertical circle passing through a heavenly body and the celestial meridian opposite to a place, is called its *azimuth*. The *sensible* horizon is that line in the heavens which is the intersection of a circular plane touching the earth at the point at which the spectator is placed and extended to the heavens, or hollow sphere; and the *rational* horizon is the intersection with the heavens of a circular plane parallel to the former, similarly extended, and passing through the centre of the earth. When a heavenly body, the sun for example, by the revolution of the glass sphere, is brought opposite to the meridian of any place, it is said to be *on the meridian* of that place; and when it is at a distance from that meridian, the angle contained at the pole between a celestial meridian passing through that object, and the celestial meridian then opposite to the place is called its *hour angle;* for it expresses, or rather is proportional to, the number of hours which must elapse before the sun is upon the meridian. When the sun is on the meridian of any place, it is twelve o'clock at that place: and, as the sun is supposed to revolve round the earth *uniformly* in twenty-four hours, if his hour-angle can be ascertained, the time at the place will be ascertained; for the *whole* hour-angle made in the course of his revolution is equal to 360°, which is described in twenty-four hours; hence, 15° of an hour-angle will be performed in an hour, &c.; and by the rule of proportion any number of degrees of an hour-angle may be converted into time, and this time will express the time *before*, or the time *after* apparent noon at the place of observation.

But though the sun may not have reached the meridian of a place *a*, (*fig.* 7,) it may be on the meridian of a place *e* to the east of *a*, for the glass sphere revolves *from* the east *towards* the west; it is therefore noon at the place *e*, before it is noon at the place *a:* but if a mariner placed at *e* could, when the sun was on his meridian, ascertain the hour angle of the sun from the meridian of *a*, or how much it wanted of being noon at *a* at that time, he would then know his longitude from *a*; for the longitude is measured by the arc E A, which arc measures likewise the angle contained between the meridians of *a* and *e* at the pole P, (Art. 35;) and the celestial meridians, as they correspond with those meridians, must contain the same angle at the celestial pole, which those meridians contain at the terrestrial; but those celestial meridians contain the hour angle of the sun from the meridian of *a*, as appears from the definition of an hour angle; therefore the two angles are one and the same angle: if the mariner at *e*, therefore, be possessed of a watch, which shows the hour angle at *a*, or what o'clock it is at *a*, and he have the means also of ascertaining at the same instant of time what o'clock it is at *e*, he can thus determine the arc A E, or the longitude of *e*; and it is evident also that any other means of ascertaining what o'clock it is at *a*, besides a watch, would answer the same purpose. Suppose that the mariner knew that at twelve o'clock at night at *a*, two heavenly bodies, *whose distance from each other is continually varying*, would be found at a certain distance from each other, say 30°, on a particular day; suppose that on that day at the spot *e* he watches those heavenly bodies, and ascertains them to be 30° of the arc of a great circle, drawn on the hollow sphere, distant from each other; then let him ascertain the time at *e*, if he finds that time to be three o'clock in the morning, as three hours difference in time answer to 45° of an hour angle, he may thence conclude, that he is situated at a spot 45° east of Greenwich. The moon is one of those heavenly bodies, which is continually changing its place on the surface of our sphere, moving from some fixed stars and towards others; it seems to revolve rapidly round the sphere from west to east in a great circle, which makes an angle of about 28° with the equator: the sun seems to move also, but not so rapidly, from west to east, and the great circle he seems to trace out is called the *ecliptic*, making an angle of about 23° with the equator. Now the distances of the moon from the sun and nine principal fixed stars near to her apparent path, are, in fact, computed and set down in the Nautical Almanac, as they would appear at the centre of the earth every three hours of Greenwich time, on those days when the moon is visible; and these distances therefore may be found for any other time at Greenwich by allowing a proportional difference. This mode of ascertaining the longitude by measuring these distances is called the *lunar method.*

(58.) Let the circle H E Z P H′ Q (*fig.* 12.) represent one of the great circles of our glass sphere; let it represent the celestial meridian of the place *e* in *fig.* 7, that is, it represents in succession all those

Fig. 12.

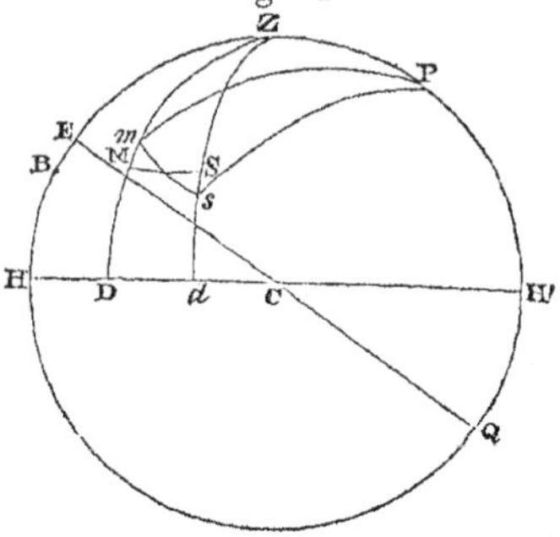

circles of declination, which by the revolution of the sphere come in succession opposite to the meridian of *e*. Now as the straight line which determines the zenith of a place passes through its meridian, and lies also in the circular plane of its meridian, the circular plane itself extended, or the celestial meridian, must pass through its zenith. Let Z be the zenith of *e*, let P be the north pole of the sphere, and E C Q the celestial equator,* then Z D and Z *d* are arcs of *vertical* circles; if *s* be the sun, then the angle H Z *s* represents his azimuth, and *s* P E his hour angle from the meridian of *e*, or the time before noon at *e*; if *m* be the moon, *m s* may represent the distance between the moon and the sun measured on the arc of a great circle *m s*, also H D C H′ represents the *rational* horizon of *e*. Let M and S, also, be two places of the moon and sun; then S *d* and M D are called the *altitudes* of the sun and moon; they are the arcs of vertical circles intercepted between those heavenly bodies and the rational horizon: but we can have no means of *measuring* how high a celestial object is above that horizon; we can ascertain it, however, by measuring its height above the sensible horizon, and adding a correction to that altitude for the difference between its heights above the two horizons; this correction is called *parallax*.

Besides parallax, there are two more corrections to be applied to the observed altitudes of heavenly bodies; these are first *refraction*, which is to be subtracted from those altitudes; for it is the quantity which heavenly bodies appear raised above the sensible horizon, in consequence of their light having to pass

* When the eye is in the plane of a circle it appears a straight line.

through the atmosphere of the earth before it reaches us; and 2ndly the *dip*:—the dip is a correction rendered necessary by the peculiar nature of the instruments with which arcs of the great circles of the concave sphere are measured at sea; these instruments are called *quadrants or sextants*: the dip is also to be *subtracted* from the observed altitudes, for it is occasioned by the circumstance of the eye of the observer being elevated above the plane of the sensible horizon, from which altitudes are measured by the sextant. In the case of the moon the parallax is greater than the refraction and dip together; in the case of the other heavenly bodies, it is less. Thus the moon appears lower than its true place, while the others, on the contrary, appear higher; therefore *m* and *s* may represent the true places of the moon and sun respectively, and M and S the apparent places; then *m* P and *s* P will be the complements of the declinations of those two bodies: the distance *m s*, or true distance at the centre of the earth, is not equal to M S the apparent distance. P *m s*, M Z S, &c. represent spherical triangles, (Art. 35,) drawn on the concave surface of the heavenly sphere, and if three parts of such triangles be given, the other three can be found. The fixed stars have no parallax, they are found to be at so great a distance from the earth, that the radius of the earth is a mere point compared with that distance.

When we speak of the altitudes of such heavenly bodies as the sun and moon, or their distances, we mean the altitudes or distances of their *centres*; but altitudes or distances of their limbs can always be converted into altitudes or distances of their centres, by means of the values of their semidiameters given in the Nautical Almanac.

(59.) The latitude of a place *e*, or its distance from the equator, measured on its meridian, must be equal to the distance of its zenith from the celestial equator, or Z E, for they are corresponding arcs of circles, (Art. 1.) Let B be a heavenly body on the meridian of *e*, whose declination is given in the Nautical Almanac; let the mariner at *e* obtain its true altitude B H, by observing its altitude above the sensible horizon, and making the proper allowances for parallax, &c.; let him look for the value of B E, or the declination, in that almanac, and add it to B H, this will give the arc E H, which is the complement of E Z, the latitude. If B had been to the north of the equator, or had had *north declination*, the declination must have been subtracted; this is the most usual mode of finding the latitude by observation at sea. There is another, however, which remains to be explained; let *s* and *m* be two places of the sun before noon; obtain from observation the two altitudes *s d*, *m* D, and consequently the complements of those altitudes, called *zenith distances*, Z *s* and Z *m*; note also the time which elapses between the two observations, which will be the angle *m* P *s*, then in the spherical triangle *m* P *s*, *m* P, and *s* P, the complements of the sun's declination are given by the Nautical Almanac, and the included angle *m* P *s* is given, making together three parts, therefore the other parts may be found; that is, the angle P *m s*, and the side *m s* may be found; then in the triangle Z *m s*, all the three sides are given to find the angle Z *m s*; take away the known angle P *m s*, from the known angle Z *m s*, and the angle Z *m* P remains, which is therefore known; then in the triangle Z *m* P, *m* Z and *m* P are known, and the included angle Z *m* P, to find the side Z P, which also equals the complement of the latitude. If the ship changes her latitude, as well as place, in the interval between the two observations, a correction is applied, for the difference of latitude made during that interval.

(60.) Suppose the latitude known, and that we wish to find the hour angle Z P *s*, for the purpose of obtaining the time at *e*, and also the azimuth *s* Z H; obtain *s* Z from observation, as before, and *s* P by means of the Nautical Almanac; then in the triangle Z *s* P, all the three sides are given to find the angle Z P *s*, which is the hour angle, and the angle *s* Z P, which is the azimuth from H′, and the supplement of the azimuth from H; but the azimuth of the sun, or any other object, is its true bearing from the true meridian of a place; for the true bearing of an object, *s*, is evidently measured by the arc H *d*, which measures the angle *s* Z H. If this true bearing be compared with the bearing by compass, the difference will be the *variation* of the compass.

(61.) This variation of the compass is the result of a slow progressive alteration of the position of the needle, with respect to the true meridian. It is observed to move towards the west, until it arrives at its maximum on that side;

it then returns, passes over the true meridian, and moves easterly, until it arrives at its maximum, towards the east, it then returns as before. In the year 1660, in London, the needle pointed to the true north, since which time it has travelled about $24\frac{1}{2}°$ to the westward, and has lately begun to return. The variation, however, is very different in different parts of the globe, and must, therefore, be determined at sea, by comparing the true bearing of a celestial object with its bearing by compass. For the purpose of taking these magnetic bearings, a particular kind of compass is employed, called an *azimuth compass:* sometimes the angular distance of the sun from the prime vertical is observed with these compasses when he rises or sets, which is called his *amplitude;* and then his true distance is computed, and compared with the observed: this is another mode of finding the variation.

(62.) Suppose the latitude of *e* to be known, and that the longitude is required; then if we suppose M and S to represent the apparent places of the moon and sun, and *m* and *s* the true; we can obtain M Z and S Z, by observation as before, and also M S the apparent distance; therefore in the triangle M Z S, the three sides will be given to find the angle M Z S; then in the triangle *m* Z *s*, the angle *m* Z *s* will be known, and also *m* Z and *s* Z, (for they are the zenith distances corrected for parallax, refraction, &c.) therefore *m s*, which is the true distance, may be found.

Having found the true distance, and the time at *e* by Article 60, find out by the Nautical Almanac, at what time at Greenwich the moon and sun are at the true distance *m s* from each other that day; this time will be the time at Greenwich; and having the corresponding time at *e*, the longitude is known, (Art. 57.) The reader must not, however, imagine that in finding his latitude, variation, time, and longitude, by observation, the mariner is actually compelled in practice to perform all the troublesome calculations we have adverted to, and to resolve all these different triangles by the rules of spherical trigonometry: for various ingenious devices and contrivances have from time to time been invented by scientific men, for the purpose of shortening his labour, and comprising within a small volume all the requisite tables:—A more useful application of great learning and talents cannot well be imagined, and too much praise can scarcely be awarded to the exertions of those, who are willing to devote their time to the laudable object of devising means of imparting their knowledge to others, comparatively unlearned; and the rather as the task of instruction has never yet been enumerated among the pleasures of science.

We have supposed the circles, on which the altitudes, distances, &c. of heavenly bodies are measured, to be drawn on the concave surface of a hollow glass sphere revolving round the earth: it is evident, however, that the same phenomena will take place, and the same investigations will apply, if the earth be supposed to revolve from west to east every twenty-four hours, and all the heavenly bodies, except the sun and planets, to be at rest. We may likewise suppose the circles that have been defined, to be traced out in the heavens themselves.

(63.) The *deviation* of the compass was first observed by Mr. Wales, the astronomer of Capt. Cook: it is occasioned by the iron on board a ship attracting the compass. A simple and ingenious method of discovering the amount of this deviation, has been invented by Mr. Barlow of the Royal Military Academy.

Fig. 13 represents a vessel, her compass is represented at C. Now it is found that there is a point in which the whole action of the iron of a ship is concentrated, it is called the centre of attraction: let *p* be this point. When the head of the ship is directed towards the magnetic north, then, if the iron is equally distributed on each side of the compass, it

Fig. 13.

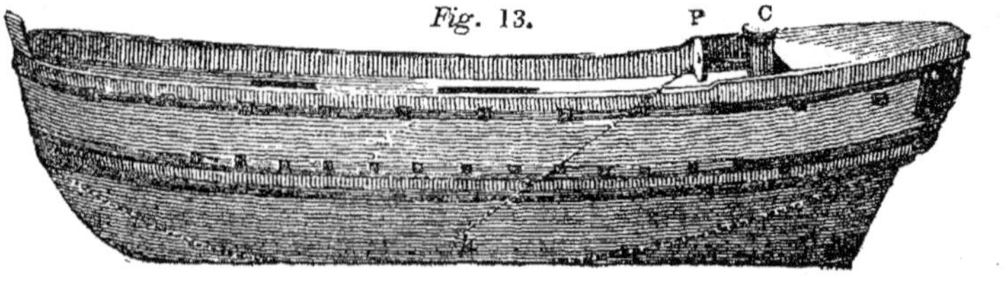

cannot be affected by the deviation; for the attractions of the point *p* and the earth in this case coincide: but when the head of the vessel is directed towards the magnetic east or west, then nearly the maximum of deviation will be produced, for the iron of the ship will draw the needle towards the point *p*, and in a direction perpendicular to the line of the attraction of the earth.

When the ship is in harbour, place her head in the magnetic north, and take the bearing of some object; this bearing will be free from deviation, as the two attractions coincide: now warp the ship's head round till it be directed towards the magnetic east, and again take the bearing of the object; the difference between this bearing and the former one will be the deviation, when the ship's head is east: repeat this experiment in different positions of the ship's head with respect to the magnetic north, and note the deviations produced in each position; these are the deviations that will be produced at sea when the ship steers those particular courses: now take her compass on shore and place it on the top of a log of wood, in which a number of holes are bored at the distance of eight, nine, ten, &c. inches from the top; take also a circular plate of iron P, of about a foot in diameter, screwed on a brass leg, and fix it in one of these holes; turn the log round towards different points of the compass; and if the plate when fixed in that position with respect to the ship's compass, produces the same effect upon it at every point, that is produced on board the ship by her iron, the experiment is finished; if not, try different positions of the plate, until the precise situation is ascertained in which the same deviations are produced; then measure the distance of the centre of the plate below the card of the compass, and also its distance from the vertical line drawn through the centre of the card perpendicular to the earth, and take the compass on board again with the plate. It is found by experiment, that when the plate is fixed on board the ship in a position with respect to the compass, similar to that it occupied on shore, when it produced *corresponding deviations*, the deviation on shipboard, in the different positions of the ship's head, is very nearly *doubled;* for very nearly the same deviations are produced by the plate, that were produced by it on shore; and those deviations were equal to the deviations produced by the ship's iron, the attraction of which is now added to the attraction of the plate.

The proper position of the plate, therefore, having been ascertained as above mentioned, or by the tables accompanying it,* take a bearing of any object, then fix the plate in its position, and observe its effect on the ship's compass; allow a deviation equal in amount to the effect produced, but so that the bearing without the plate may be intermediate between the bearing with the plate, and the bearing taken as the true one. For example, suppose a headland to bear due west, affix the plate to the compass; suppose that the same headland now bears W b N, that is, the plate produces a deviation of a point westerly—but whatever deviation the plate produces, the iron of the ship must have produced nearly the same deviation before the plate was attached—therefore the iron of the ship must have drawn the compass one point to the west of its true position; so that the true bearing of the headland, independent of the effect of both iron and plate, must be W b S.

Another method of fixing the plate has been tried with success.—This method consists in fixing it behind, or *abaft*, the compass, and in a position in which its attraction will produce opposite effects, and neutralize that of the ship's iron: indeed some late investigations by the distinguished mathematician, M. Poisson, would seem to show, that this mode of *correcting* the deviation is to be preferred to the other plan of discovering its amount, and allowing for it;† especially in all cases where the deviation is likely to be considerable. In fact, this mode of fixing the plate was adopted in the trials made on board H. M. S. Griper, and in Captain Parry's two last expeditions, in high northern latitudes, where the deviation was very considerable, and the success of the experiments was notwithstanding very great.

We close our treatise with a description of this simple and beautiful contrivance; it is the last improvement which genius has furnished to the practical navigator, and an invention of our own times. With respect to the subject of this our third Part, we may observe that the navigator is indebted to the science of astronomy for the most important branch of his art; and that the

* Plates are sold by W. and J. Gilbert, 148, Leadenhall-street, accompanied by tables, which show the attractions of the plates for several positions.

† See Quarterly Journal of Science, vol. xix. p. 128.

astronomer owes to navigation the best practical illustration of the utility of his pursuits, and the gratifying conviction that the labours of his predecessors, in the career of fame, have conferred a real benefit on mankind.

The principal English Treatises on the subject of Navigation itself are, Robertson's Elements of Navigation; Riddle's Treatise on Navigation and Nautical Astronomy; Professor Inman's Treatise on Navigation and Nautical Astronomy; Kerrigan's Young Navigator's Guide to the Sidereal and Planetary Parts of Nautical Astronomy; Norie's Complete Epitome of Practical Navigation; Bowditch's Improved Practical Navigator, edited by Kirby; and Moore's Practical Navigator.

The works on subjects connected with Navigation are far too numerous for insertion here: we may mention, however, Dr. Maskelyne's Preface to the Nautical Almanac, and the Preface to Taylor's Logarithms by the same author. Some additions to the Nautical Almanacs of different years by Maskelyne, Wales, Campbell, Lyons, Witchell, Blair, and Brinkley; Lax's Tables to be used with the Nautical Almanac, &c.; the Nautical Almanac; Connoissance des Tems; Professor Schumacher's Ephemeris, &c.; Thomson's Lunar and Horary Tables; Workman's Navigation Improved; Investigations relative to the problem for clearing the Apparent Distance, &c. by Dr. Brinkley, in the Gentleman's Mathematical Companion for the Year 1815; Mendoza's Paper in the Philosophical Transactions for 1797; Kelly's Practical Introduction to Spherics and Nautical Astronomy; Barlow's Essay on Magnetic Attractions; Bain, on the Compass; Several valuable Papers in the Philosophical Transactions by Halley, Maclaurin, Kater, Sabine, Foster, &c. &c. The foreign authors on Navigation are also very numerous, the modern treatises are, Bezout, Traité de Navigation, and Dubourguet, Traité de Navigation: the more ancient are those of Bartolomew Crescenti, Rome, 1607; John Baptist Riccioli, Bologna, 1661, &c. &c.; and that of M. Peter Bouguer, 1760, abridged and improved by M. de la Caille.

TABLE I.

Logarithmic Sines, Tangents, and Secants, to every Point and Quarter Point of the Compass.

Points.	Sine.	Co-sine.	Tangent.	Co-tang.	Secant.	Co-sec.	Points.
0	0.000000	10.000000	0.000000	Infinite.	10.000000	Infinite.	8
0¼	8.690796	9.999477	8 691319	11.308681	10.000523	11.309204	7¾
0½	8.991302	9.997904	8.993398	11.006602	10.002096	11.008698	7½
0¾	9.166520	9.995274	9.171247	10.828754	10.004726	10.833480	7¼
1	9.290236	9.991574	9.298662	10.701338	10.008426	10.709764	7
1¼	9.385571	9.986786	9.398785	10.601215	10.013214	10.614429	6¾
1½	9.462824	9.980885	9.481939	10.518061	10.019115	10.537176	6½
1¾	9.527488	9.973841	9.553647	10.446353	10.026159	10.472512	6¼
2	9.582840	9.965615	9.617224	10.382776	10.034385	10.417160	6
2¼	9.630992	9.956163	9 674829	10.325171	10.043837	10.369008	5¾
2½	9.673387	9.945430	9.727957	10.272043	10.054570	10.326613	5½
2¾	9.711050	9.933350	9.777700	10.222300	10.066650	10.288950	5¼
3	9.744739	9.919846	9.824893	10.175107	10.080154	10.255261	5
3¼	9.775027	9.904828	9.870199	10.129801	10.095172	10.224973	4¾
3½	9.802359	9.888185	9.914173	10.085827	10.111815	10.197641	4½
3¾	9.827084	9.869790	9.957295	10.042705	10.130210	10.172916	4¼
4	9.849485	9.849485	10.000000	10.000000	10.150515	10.150515	4
	Co-sine.	Sine.	Co-tang.	Tangent.	Co-sec.	Secant.	

TABLE II.

A Table of Corrections to be added to the Mean Middle Latitude to find the True Middle Latitude.

Mid. Lat. Deg.	DIFFERENCE OF LATITUDE.* 3°	4°	5°	6°	7°	8°	9°	10°	11°	12°	13°	14°	15°	16°	17°	18°	19°	20°
	° ′	° ′	° ′	° ′	° ′	° ′	° ′	° ′	° ′	° ′	° ′	° ′	° ′	° ′	° ′	° ′	° ′	° ′
15	0 02	0 03	0 04	0 06	0 09	0 12	0 15	0 19	0 23	0 27	0 31	0 35	0 40	0 45	0 51	0 58	1 06	1 14
16	0 02	0 03	0 04	0 06	0 09	0 12	0 15	0 18	0 22	0 26	0 30	0 34	0 38	0 43	0 49	0 56	1 03	1 11
17	0 02	0 03	0 04	0 06	0 08	0 11	0 14	0 17	0 21	0 25	0 28	0 32	0 37	0 42	0 48	0 54	1 01	1 08
18	0 02	0 03	0 04	0 06	0 08	0 11	0 14	0 17	0 20	0 24	0 27	0 31	0 36	0 41	0 46	0 52	0 58	1 06
19	0 02	0 03	0 04	0 06	0 07	0 10	0 13	0 16	0 19	0 23	0 26	0 30	0 34	0 40	0 45	0 50	0 56	1 03
20	0 02	0 03	0 04	0 06	0 07	0 09	0 12	0 15	0 18	0 22	0 25	0 29	0 33	0 38	0 43	0 48	0 54	1 00
21	0 02	0 03	0 04	0 06	0 07	0 09	0 12	0 15	0 18	0 21	0 25	0 29	0 33	0 37	0 42	0 47	0 53	0 58
22	0 02	0 03	0 04	0 06	0 07	0 09	0 12	0 15	0 17	0 20	0 24	0 28	0 32	0 36	0 41	0 46	0 51	0 56
23	0 02	0 03	0 04	0 06	0 07	0 09	0 12	0 15	0 17	0 20	0 24	0 28	0 32	0 36	0 40	0 45	0 50	0 55
24	0 02	0 03	0 04	0 06	0 07	0 09	0 11	0 14	0 16	0 19	0 23	0 27	0 31	0 35	0 39	0 44	0 48	0 53
25	0 02	0 03	0 04	0 05	0 07	0 09	0 11	0 14	0 16	0 19	0 23	0 27	0 31	0 35	0 39	0 43	0 47	0 52
26	0 02	0 03	0 04	0 05	0 07	0 09	0 11	0 14	0 16	0 19	0 22	0 26	0 30	0 34	0 38	0 42	0 47	0 52
27	0 02	0 03	0 04	0 05	0 07	0 08	0 11	0 14	0 16	0 19	0 22	0 26	0 30	0 33	0 38	0 42	0 46	0 51
28	0 02	0 03	0 04	0 05	0 06	0 08	0 10	0 13	0 15	0 18	0 21	0 25	0 29	0 33	0 37	0 41	0 46	0 51
29	0 02	0 03	0 04	0 05	0 06	0 08	0 10	0 13	0 15	0 18	0 21	0 25	0 29	0 32	0 36	0 41	0 45	0 50
30	0 02	0 03	0 04	0 05	0 06	0 08	0 10	0 13	0 15	0 18	0 21	0 25	0 28	0 32	0 36	0 41	0 45	0 50
31	0 02	0 03	0 04	0 05	0 06	0 08	0 10	0 13	0 15	0 18	0 21	0 25	0 28	0 32	0 36	0 41	0 45	0 50
32	0 02	0 03	0 04	0 05	0 06	0 08	0 10	0 13	0 15	0 18	0 21	0 25	0 28	0 31	0 36	0 41	0 45	0 50
33	0 02	0 03	0 04	0 05	0 06	0 08	0 10	0 13	0 15	0 18	0 21	0 24	0 27	0 31	0 35	0 40	0 44	0 49
34	0 02	0 03	0 04	0 05	0 06	0 08	0 10	0 13	0 15	0 18	0 21	0 24	0 27	0 31	0 35	0 40	0 44	0 49
35	0 02	0 03	0 04	0 05	0 06	0 08	0 10	0 13	0 15	0 18	0 21	0 24	0 27	0 31	0 35	0 40	0 44	0 49
36	0 02	0 03	0 04	0 05	0 06	0 08	0 10	0 13	0 15	0 18	0 21	0 24	0 27	0 31	0 35	0 40	0 44	0 49
37	0 02	0 03	0 04	0 05	0 06	0 08	0 10	0 13	0 15	0 18	0 21	0 24	0 27	0 31	0 35	0 40	0 44	0 49
38	0 02	0 03	0 04	0 05	0 06	0 08	0 10	0 13	0 15	0 18	0 21	0 24	0 27	0 31	0 36	0 40	0 45	0 50
39	0 02	0 03	0 04	0 05	0 06	0 08	0 10	0 13	0 15	0 18	0 21	0 25	0 28	0 32	0 36	0 41	0 45	0 50
40	0 02	0 03	0 04	0 05	0 06	0 08	0 10	0 13	0 15	0 18	0 22	0 25	0 28	0 32	0 36	0 41	0 45	0 50
41	0 02	0 03	0 04	0 05	0 06	0 08	0 10	0 13	0 15	0 18	0 22	0 25	0 28	0 32	0 37	0 41	0 45	0 50
42	0 02	0 03	0 04	0 05	0 06	0 08	0 10	0 13	0 15	0 18	0 22	0 26	0 29	0 33	0 37	0 42	0 46	0 51
43	0 02	0 03	0 04	0 05	0 07	0 09	0 11	0 14	0 16	0 19	0 23	0 26	0 30	0 34	0 38	0 42	0 46	0 51
44	0 02	0 03	0 04	0 05	0 07	0 09	0 11	0 14	0 16	0 19	0 23	0 27	0 30	0 34	0 38	0 43	0 47	0 52
45	0 02	0 03	0 04	0 05	0 07	0 09	0 11	0 14	0 16	0 19	0 23	0 27	0 31	0 35	0 39	0 43	0 47	0 52
46	0 02	0 03	0 04	0 05	0 07	0 09	0 11	0 14	0 16	0 19	0 23	0 27	0 31	0 35	0 39	0 44	0 48	0 53
47	0 02	0 03	0 04	0 05	0 07	0 09	0 11	0 14	0 16	0 20	0 23	0 27	0 31	0 35	0 40	0 44	0 49	0 54
48	0 02	0 03	0 04	0 05	0 07	0 09	0 11	0 14	0 16	0 20	0 23	0 27	0 31	0 35	0 40	0 45	0 50	0 55
49	0 02	0 03	0 04	0 05	0 07	0 09	0 11	0 14	0 17	0 21	0 24	0 28	0 32	0 36	0 41	0 45	0 51	0 57
50	0 02	0 03	0 04	0 05	0 07	0 09	0 11	0 14	0 17	0 21	0 24	0 28	0 32	0 36	0 41	0 46	0 52	0 58
51	0 02	0 03	0 04	0 05	0 07	0 09	0 11	0 14	0 17	0 21	0 24	0 28	0 32	0 37	0 42	0 47	0 53	0 59
52	0 02	0 03	0 04	0 05	0 07	0 09	0 12	0 15	0 18	0 22	0 25	0 29	0 33	0 37	0 42	0 48	0 54	1 00
53	0 02	0 03	0 04	0 06	0 07	0 09	0 12	0 15	0 18	0 22	0 25	0 29	0 33	0 38	0 43	0 49	0 55	1 01
54	0 02	0 03	0 04	0 06	0 08	0 10	0 13	0 16	0 19	0 23	0 26	0 30	0 34	0 39	0 44	0 50	0 56	1 02
55	0 02	0 03	0 04	0 06	0 08	0 10	0 13	0 16	0 19	0 23	0 26	0 30	0 35	0 40	0 45	0 51	0 57	1 03
56	0 02	0 03	0 04	0 06	0 08	0 10	0 13	0 16	0 20	0 24	0 27	0 31	0 36	0 41	0 46	0 52	0 58	1 04
57	0 02	0 03	0 04	0 06	0 08	0 11	0 14	0 17	0 20	0 24	0 28	0 32	0 37	0 42	0 48	0 54	1 00	1 06
58	0 02	0 03	0 04	0 06	0 09	0 11	0 14	0 17	0 21	0 25	0 29	0 33	0 38	0 44	0 50	0 55	1 02	1 08
59	0 02	0 03	0 04	0 06	0 09	0 12	0 15	0 18	0 22	0 26	0 30	0 34	0 39	0 45	0 51	0 57	1 04	1 10
60	0 02	0 03	0 04	0 06	0 09	0 12	0 15	0 19	0 23	0 27	0 31	0 35	0 40	0 46	0 52	0 59	1 06	1 13
61	0 02	0 03	0 05	0 07	0 09	0 12	0 15	0 19	0 23	0 27	0 31	0 36	0 41	0 47	0 54	1 01	1 08	1 15
62	0 02	0 03	0 05	0 07	0 09	0 12	0 16	0 20	0 24	0 28	0 32	0 37	0 42	0 49	0 56	1 03	1 10	1 18
63	0 02	0 04	0 05	0 07	0 09	0 13	0 16	0 20	0 24	0 29	0 33	0 39	0 44	0 51	0 58	1 05	1 12	1 21
64	0 02	0 04	0 06	0 08	0 09	0 13	0 17	0 21	0 25	0 29	0 34	0 40	0 46	0 53	1 00	1 07	1 14	1 24
65	0 02	0 04	0 06	0 08	0 10	0 13	0 17	0 21	0 25	0 30	0 35	0 41	0 48	0 55	1 02	1 09	1 17	1 27
66	0 02	0 04	0 06	0 08	0 10	0 14	0 18	0 22	0 26	0 31	0 37	0 43	0 50	0 58	1 05	1 12	1 21	1 31
67	0 02	0 04	0 06	0 08	0 11	0 15	0 18	0 23	0 27	0 33	0 38	0 45	0 53	1 00	1 07	1 16	1 25	1 35
68	0 02	0 04	0 06	0 08	0 11	0 15	0 19	0 24	0 28	0 34	0 40	0 48	0 55	1 02	1 10	1 19	1 30	1 39
69	0 02	0 05	0 06	0 09	0 12	0 16	0 20	0 25	0 30	0 36	0 44	0 50	0 58	1 05	1 13	1 23	1 34	1 44
70	0 03	0 05	0 06	0 09	0 13	0 17	0 21	0 26	0 31	0 38	0 44	0 52	1 00	1 08	1 17	1 28	1 39	1 50
71	0 04	0 06	0 07	0 09	0 13	0 18	0 22	0 27	0 33	0 40	0 46	0 55	1 03	1 12	1 22	1 32	1 44	1 56
72	0 04	0 06	0 08	0 10	0 14	0 19	0 23	0 29	0 35	0 42	0 49	0 58	1 06	1 16	1 27	1 38	1 50	2 04

* When the difference of latitude is under 2°, the correction may be neglected; when it is 2° and under 3°, add a correction of 0° 01′.

EXPLANATION OF SCIENTIFIC TERMS

MADE USE OF IN THIS VOLUME.

N.B. Many of the terms are common to this and the two preceding volumes; but those explanations that were given in the former Glossaries will be merely referred to in the present, except in a few cases, where the definitions were not supposed to be sufficiently explicit. Several omissions may be supplied by consulting the Indexes.

ABERRATION. The name given to an apparent change of place in the fixed stars, consequent upon the time taken up during the passage of a ray of light to the eye compared with that of the earth's annual motion. This is familiarly explained in the preceding glossaries, and more scientifically in the present volume:—*Astronomy*, pages 64 and 146, and *History of Astronomy*, page 89.

ACCELERATION OF MOTION.—See *Glossary* I. Acceleration may be either regular or variable, that is, by equal or by unequal accessions in equal times. The motions of the planets are continually accelerated, according to a known law, from their aphelion to their perihelion; while there is a continued retardation in a like ratio in the opposite half of their orbits: the average of these motions through the whole circuit (the space divided by the time) is called the *mean motion.*

ACOTYLEDONOUS.—See *Cotyledon.*

ACRONYCAL (a Greek compound denoting the *point of night*) in its literal acceptation denotes the moment of the sun's setting or of his rising,—the beginning or the end of night. In modern astronomy it is confined to the former; and a star is said to rise or to set *acronically* when it rises or sets at the instant of sunset. On the contrary, when a star rises or sets in the morning at the moment of sunrise, it is said to rise or to set *cosmically;* but how this Greek derivative has been so applied we have not been able to determine. In the case of its *cosmical* rising or setting, the star is never visible; but when it appears in the morning a little before the sun, or sets in the evening a little after him, it is said to rise or to set *heliacally* (Greek *helios*, the sun); and it is only in his *heliacal* rising or setting, that the planet Mercury is ever visible to the naked eye.

ALMANAC.—See *Calendar*. The word is Arabic, signifying the *reckoning*.

ALTITUDE of the sun, or of a star.—See *Glossary* II., article *Horizon;* and *Astronomy*, page 16.

AMPLITUDE of the sun, or of a star.—See *Glossary* II., article *Horizon; Astronomy*, page 256; and *Navigation*, page 30.

ANGLE.—See *Glossary* I.

———OF ELONGATION.—See *Elongation.*

ANNUAL PARALLAX.—See *Parallax.*

ANNULAR, an adjective, from the Latin *annulus*, a ring. Those eclipses of the sun, in which a ring of light is visible around the dark body of the moon, are termed *Annular Eclipses;* they are also *Central Eclipses,* because the centres of the sun and of the moon appear to coincide.

ANTARCTIC.—See *Arctic.*

ANTIPODES (Greek *anti*, against, and *podos*, a foot). If a straight line be supposed drawn from any point of the earth's surface, through its centre, so as to terminate in a point on the opposite surface, two persons standing on those points would be *Antipodes* to one another.

APHELION.—See *Glossary* II. and *Astronomy*, page 127.

APOGEE.—See *Glossary* II. and *Astronomy*, page 19.

APPARENT TIME.—See *Time.*

APSIDES.—See *Orbit.*

ARC OF A CIRCLE.—See *Glossary* I. *Angle.*

ARCTIC. The Greek *arctos*, a bear, is the origin of the two scientific adjectives *Arctic* and *Antarctic* (anti-arctic) which are equivalent to northern and southern. An imaginary line passing through the centre of the earth, and on which it turns in its diurnal rotation, is called its *axis;* and the two ends of that line, where they are supposed to terminate at the surface, are the *Poles* or pivots (Latin *poli*), and are, respectively, the north and south (the *Arctic and Antarctic*) poles of our globe. If we suppose the line to be extended in both directions, it will become an *axis* to the spherical concavity of the sky; and will mark two points in the heavens that are also called the north and the south poles, around which, or rather their line of junction, all the fixed stars appear to revolve. The north pole of the heavens (the only one visible in our latitude) is a point, situated in the constellation called *Ursa Minor*, or the little bear; and a bright star, in the tip of the tail of this imaginary animal, is called the *Pole-star*, or *Polar-star*, because near to the real pole. Circles supposed to be drawn round the Arctic, and the Antarctic pole, at the distance of about 23½ degrees, are termed, respectively, the *Arctic* and the *Antarctic Circle*. On the terrestrial globe, those circles surround what are called the two *Frozen Zones*, or girdles of the earth.

AREA.—See *Glossary* II.

ASCENSION, RIGHT.—See *Right Ascension*.

ASTRONOMICAL HORIZON.–See *Glossary* II., *Horizon*.

ATMOSPHERE.—See *Glossary* I.

———— REFRACTION OF.—See *Glossary* II.

ATTRACTION.—See *Glossary* I.

AXIS.—See *Arctic*, and *Astronomy*, page 6.

—— OF AN ELLIPSIS, PARABOLA, CONE, &c.—See *Glossary* I., *Cone* and *Conic Sections*.

—— MAJOR AND AXIS MINOR OF AN ELLIPSIS, are the same as the *Transverse* and *Conjugate Diameters*, which see in *Glossary* I.

AZIMUTH.—See *Glossary* II., *Astronomy*, pp. 36 and 256, and *Navigation*, page 27.

———— COMPASS.—See *Glossary* II., and *Navigation*, page 30.

————, MAGNETIC.—See *Glossary* II., article *Horizon*.

BEARING, in *Navigation*, is the situation of one place from another with respect to the points of the compass. Thus if *A* lies in the direction of south-west from *B*, then to an observer at *B*, *A* is said to *bear* south-west, or to have a south-west *Bearing*; while to an observer at *A*, the point *B* will *bear* north-east, or have a north-east *Bearing*.—See *Navigation*, page 15.

BERGS. The Swedish *Berg* is a hill or mountain; and hence the name of *Bergs*, or more generally *Icebergs*, is given to the mountains of ice which are met with in the Polar Seas. Flat sheets of wide-spread ice are termed *Fields*; and small portions (because found floating) are *Floes*.

BISSEXTILE.—See *Calendar*.

CALENDAR OR KALENDAR. This term is understood to have been derived from the Greek *kaleo*, I call or proclaim, because the first appearance of the new moon was watched for and proclaimed; and hence the first days of the several months were marked *Calendæ* (the *Calends*) in the *Calendar*, or *Almanac* of the Romans.

Time, as measured by the revolutions of the sun and moon, is divided into days, months, and years. The *Months* (*Mooneths*) were at first meant to denote the period of a lunar revolution, or the time from one new moon to another; and the years were counted by the revolutions of the sun:—from the shortest day (*daylight*) to the shortest day again, or from the point of time of the vernal equinox to its return. The solar year, however, does not contain an even number of lunations; and calendars were made to shew the connexion between them. The earliest known Roman year consisted of 304 days, divided into ten months, or moons; and hence the names September, October, November, and December, from the Latin words for 7, 8, 9, and 10. Two more moons were prefixed by Numa, making up 354 days, or about 12 lunations. This period, however, being still less than a revolution of the sun, Julius Cæsar reformed the calendar, by lengthening the several months (to the same extent as they now are), so as the year should consist of 365 days; and as the sun is nearly six hours longer in performing his apparent annual circuit, every fourth year was made 366 days, by adding a day to the month of February. This addition of a day to every fourth year is known at present by the name of *Leap-year*; but the Romans did not add it to the end of February, as we do, but *intercalated* it after the twenty-third, so that the twenty-fourth day was counted twice: this day, usually termed *Sextilis*, was therefore *Bis-sextilis*; and consequently the *Bissextile* is another name for *Leap-year*.

The subsequent reformation of the calendar by Pope Gregory XIII., now called the *Gregorian Calendar*, and the reasons for that reformation, are amply, though concisely explained, in the *History of Astronomy*, at pp. 37, 47, and 48, to which we refer.

CALENDAR MONTHS contain the number of days marked to each in the calendar, as distinguished from *Lunar Months*, which are *legally* of four weeks or twenty-eight days: of course, neither of these periods corresponds to a lunation.

CANICULAR PERIOD. The *Dog-star* is otherwise called *Sirius* and *Canicula* (Latin, *Canis*, a dog); and what is connected with that star is termed *Canicular*. The *Dog-days* (or *Canicular days*) in the ancient calendars were forty days; reckoning twenty before and twenty after, the heliacal rising of Sirius. That period, being then the hottest of the year, was, by the Greeks, accounted the season of "fevers, plagues, and death." The time of the heliacal rising of the dog-star, however, (what the astrologers of old never dreamt of,) varies in consequence of the precession of the equinoxes; so that, instead of happening in the warmest season, it has gradually advanced towards the autumn. The modern almanac makers have, therefore, regardless of the star, marked the dog-days as commencing on the third of July and ending on the eleventh of August.

The Egyptian year consisted of 365 days; and supposing it to have begun at the heliacal rising of the dog-star (which they called *Soth*), its next heliacal rising would be about six hours later every year; so that a period of 1461 years would elapse before that star would again rise heliacally at the beginning of their year. This they knew, and it has been termed the *Sothic*, or *Canicular period*.—See *Astronomy*, pp. 15 and 43.

CAPE.—See *Headland.*

CENTRAL ECLIPSE.—See *Annular.*

CENTRIFUGAL FORCE.—See *Glossary* I., and *History of Astronomy*, pp. 76, 77.

CENTRIPETAL FORCE.—See *Glossary* II., and *Mathematical Geography*, page 24.

CHARTS, in *Geography* and *Navigation*, are representations of portions of the earth's surface on paper, according to scales which regulate the relative proportions of the parts. Geographical charts are more usually called *Maps*, being general; but *Nautical*, or *Marine Charts* are particularly appropriated to delineations of a coast and part of the adjacent sea. *Marine Charts* are constructed by two different methods. In a *Plane Chart*, the meridians as well as the latitudes are represented by equidistant and straight parallel lines; and, consequently, the longitudes and latitudes appear equal, differing from the fact. In a *Mercator's Chart*, the meridians as well as the circles of latitude are also represented by straight parallel lines; but the distance of the latter increase in a determinate ratio from the equator to the poles.—See *Navigation*, pages 7 and 17.

CHORD OF AN ARC.—See *Glossary* I., *Angle.*

CIRCLE, DIVISION AND PARTS OF. —See *Glossary* I., *Angle*, and *Astronomy*, page 8, note.

———GREAT, OF A SPHERE. A great circle of a sphere is any circle which is formed by a plane supposed to pass through its centre.

CIRCLES, CONCENTRIC AND ECCENTRIC.—See *Concentric Circles.*

——— OF DECLINATION.—See *Declination* and *Navigation*, page 27.

———, SECONDARY.—See *Astronomy*, page 16.

———, VERTICAL.—See *Glossary* I. *Horizon* and *Astronomy*, page 16.

CIRCUMPOLAR STARS. On referring to the explanation of *Arctic*, we see why all the fixed stars appear to revolve round the *poles*. At every place of the earth, except on the equinoxial, one or other of the poles is always elevated above the horizon to a degree equal to the latitude of that place. Certain stars, according to their vicinity to the pole, notwithstanding they also apparently move round it, will never fall so low as the horizon; those never set, and are termed *Circumpolar Stars.*

CLEPSYDRA, an instrument used by the ancients for measuring time. It received various improvements, but the general principle was that of the gradual dropping of water from one vessel into another, in the manner of a sand-glass; to which latter, the name of *Clepsydra* was sometimes incongruously given.

CLIMATE. In its geographical and technical application, *Climate*, or *Clime*, denotes an imaginary narrow belt of the globe, parallel to the equator; and is so called from the Greek *clima*, inclination, because the difference of climates depends on the inclination, or obliquity of the sphere. The belts constituting the several climates are small, depending on the average length of the longest day: that of each increasing by half an hour, from the equator to the polar circles, when the climates are counted by months, till they reach the poles. In a more popular and accurate sense, a *Climate* is designated (as hot or cold, dry or moist) from the comparative degree of heat and moisture which generally exists in its atmosphere. —See *Physical Geography*, page 34.

COLURES.—See *Ecliptic* and *Astronomy*, page 62.

COMPASS.—See *Glossary* II.

COMPLEMENT OF AN ANGLE OR ARC is what it wants to complete the quadrant, or right angle.

CONCENTRIC CIRCLES are such as have the same centre, the one surrounding the other, as with a ring. Circles that are wholly, or partially, included in another, but have different centres, are termed *Eccentric.*

CONCENTRIC THEORY.—See *Epicycle.*

CONE, RIGHT AND OBLIQUE.—See *Glossary* I., *Cone.*

CONIC SECTIONS.—See *Glossary* I.

CONJUGATE DIAMETERS.—See *Glossary* I., *Conic Sections.*

CONJUNCTION OF THE SUN AND PLANETS. The planets, relatively to the earth, are separated into two divisions, *Inferior* and *Superior:* the former having their orbits *within*, and the latter *without*, that of the earth. When a planet, as seen from the earth, is in the same direction as the sun, it is said to be in *conjunction with the sun.* This, however, in the case of an inferior planet, may be either when it passes between the sun and the earth, or when it is on the further side of the sun; the former is the *Inferior*, and the latter the *Superior Conjunction*. A superior planet, never passing between the sun and the earth, is only once in conjunction with the sun during its revolution. In the point of its orbit, when the earth is between it and the sun, the planet is said to be in *Opposition* to the sun. The *Conjunctions* and *Oppositions* of the moon have the general name of *Syzygies*.—See *Quadratures*. The angle under which we see the distance of a planet from the sun (reduced to the ecliptic) is called its *elongation*.—See *Astronomy*, page 75.

CONSTELLATIONS. The whole of the fixed stars are apportioned into separate groups termed *Constellations*, each of which is included within the outline of a certain figure (chiefly of an animal) which is imagined to be drawn on the concave of the sky. Forty-eight of these constellations are of unknown antiquity. The twelve that occupy the zodiac are termed the *Twelve Signs*.—See *Zodiac*.

CO-SECANTS, CO-SINES, CO-TANGENTS. These are the *Secants*, *Sines*, and *Tangents*, of arcs that are the complements of those in question.

COSMICAL, a term applied to the rising or setting of a star, and opposed to *Acronycal*, which *see*.

COTYLEDON. The embryo, which constitutes the vital principle of a vegetable seed, is termed the *Corculum*, being a Latin diminutive, signifying *a little heart*, which in many cases its shape resembles. This *Corculum* consists of two parts: the *Rostellum*, or radicle, which, descending, becomes the root; and the *Plumula*, or feather, which, ascending, becomes the stem and leaves. A pair (for some eminent botanists assert that they are never single) of roundish or compressed bodies constituting the chief bulk of most seeds, and immediately attached to the *Corculum*, have been named *Cotyledons* or *Seed-lobes*, from a Greek word signifying a cavity. These generally rise out of the ground with the *Plumula*, and assume the appearance of leaves (though unlike those of the future plant); and when the real foliage comes forth, they droop and die. In the natural classification of Jussieu, all plants are divided, in the first instance, into three divisions: those of which the seeds have only *one Cotyledon*, termed *Monocotyledonous;* those which have *two Cotyledons*, termed *Dicotyledonous;* and those in which *Cotyledons* are altogether wanting, or *Acotyledonous*.

CRUSTACEOUS. For the explanation of this term, and its distinction from *Testaceous*, see *Physical Geography*, page 49, note.

CRYPTOGAMOUS. In the sexual system of Linnæus the fructification of plants is ascribed to certain essential parts of the flower; and such plants as, having no flowers, wanted those parts, were classed under one head by the name of *Cryptogamia*, or *hidden marriages*. *Cryptogamous* plants are therefore those that bear no flowers, and flowering plants have sometimes been termed *Phanerogamous*.

CULMINATION is the transit of a star over the meridian, or the point of its highest altitude.

CURVATURE, RADIUS OF.—See *Radius of Curvature*.

CURVE.—See *Glossary* I.

CUSPS.—See *Phases*.

CYCLE is a period of time during which certain natural phenomena complete their round; so as to begin anew, and continually occupy another cycle equal to the past. The Greek *cyclos* is literally a *circle*.

——— METONIC.—See *Epact*.

CYCLOID.—See *Glossary* I., and *Trochoid* in the present.

DAY. In common language, *day* is opposed to *night*, as light to darkness; but, in this usage, its lengths are very unequal, varying with the latitude of the place, and the time of the year; for, within the polar circles, there are weeks and months in which the sun never sets, and equal periods in which he never rises. In all other climates, however, the day and night together, that is, from sunrise to sunrise, always make up a period, nearly equal, which is called the *Solar Day*. This day is divided into twenty-four equal hours; and these hours, counted from noon and enumerated from one to twenty-four, make the solar day of the astronomers. The *Civil Day* begins at midnight, and is counted in two portions of twelve hours each: from midnight to noon, and from noon to the succeeding midnight.

Though in the course of the year the earth must have revolved $365\frac{1}{4}$ times on its axis, making so many real solar days, those days are not of uniform length, some being longer and some shorter than the average. This average is termed *Mean Time*, and a chronometer regulated to this average agrees with the true solar day only at four points of the year. The accumulated difference of the two modes of measurement is called the *Equation of Time*.

——— SIDEREAL, is the time which elapses between that of a star being in the meridian of a place to the moment when it arrives at the meridian again. This period is always the same, not being affected by the motion of the earth in her orbit, as the solar day is. The sidereal day is about four minutes less than the mean solar day.

DECLINATION. The declination of a celestial body is its perpendicular distance from the equator, measured on a meridian or great circle passing through the object and the poles of the heavens. Such a great circle is called a *Circle of Declination.* Declination on the celestial globe corresponds with latitude on the terrestrial; and the *Parallels of Declination* are similar to the *Parallels of Latitude.*

DEFERENT CIRCLE.—See *Epicycle.*

DENSITY.—See *Glossary* I.

DELTA. The mouths of certain rivers so called.—See *Physical Geography,* page 9, note.

DEPARTURES.—See *Navigation,* page 4.

DIAMETERS, TRANSVERSE AND CONJUGATE.—See *Glossary* I. *Conic Sections.*

DICHOTOMISE, a Greek compound signifying to cut in two equal and similar parts.—See *Quadratures.*

DICOTYLEDONOUS.—See *Cotyledon.*

DIGIT (Latin *digitus,* a finger) is an old measure of a finger's breadth. In Astronomy it is accounted the twelfth part of the diameter of the sun or moon; thus we say that the moon is six digits eclipsed, when half of her face is covered by the earth's shadow.

DIRECT AND INVERSE PROPORTION, or *Ratio.* See *Glossary* II. *Ratio.*

DISC. The *Discus* of the ancients was a circular piece of wood, stone, or metal, which they used in their games. In Astronomy it is a name for the apparent face of the sun or moon. The faces of the planets may also be termed *Discs* when viewed through the telescope.

DRIFT IN NAVIGATION denotes the angle which the line of a ship's motion makes with the nearest meridian, when she *drives* with her side to the wind, and is not governed by the power of the helm; and also the distance which the ship drives on that line only in a storm. The *Drift* of a current is its velocity and the direction of its motion.

EBB-TIDE.—See *Tides.*

ECCENTRIC CIRCLES.—See *Concentric Circles.*

ECCENTRICITY. Referring to this word in *Glossary* II. we observe an error: for "the distance between," read "half the distance between."

ECLIPTIC. The position of this great circle in the heavens is given under the head of *Equator.* The Ecliptic, like all other great circles, is supposed to be divided into degrees and minutes, and has its poles, through which other imaginary circles are drawn. These latter may be compared to the circles of longitude, or meridians, on the terrestrial globe; the longitude of a star being counted on the Ecliptic, in degrees and minutes, from the first degree of Aries (which is the vernal equinoctial point) in the same way as the terrestrial longitude is counted upon the equinoctial line from the standard meridian.

The lowest point of the Ecliptic, that in which the sun seems to pause in his orbit and begins again to ascend, is called the *Winter Solstice;* and the highest point in his career, when he begins again to descend, is the *Summer Solstice,* which two *Solstices* (Latin *Sol,* the sun, and *sto,* to stand) make, respectively, in our latitude, the shortest and the longest day. Circles drawn parallel to the Equator through these points are termed *Tropical Circles,* from the Greek *trope,* a turning; and the included corresponding portion of the earth, forming the *Torrid Zone,* is said to lie between the *Tropics.* In the celestial sphere, the Tropical circle at the Summer Solstice is the *Tropic of Cancer;* and that of the Winter Solstice is the *Tropic of Capricorn:* because (at one time) terminating respectively at those signs of the Zodiac. We say, *at one time,* for these points, though they retain the same designation, continually, though slowly, change their situation in the heavens, on account of the *Precession of the Equinoxes.* A great circle, crossing the Equator at right angles, and passing through the two *Solstitial points,* is called the *Solstitial Colure;* while another, passing through the *Equinoctial points,* is the *Equinoctial Colure.*—See *Astronomy,* pages 17 and 62.

ELLIPSIS.—See *Orbit,* and *Glossary* I.

ELONGATION.—See *Conjunction.*

EPACT. The year of 365¼ days contains twelve lunations and nearly eleven days more; so that, were it to begin with the new moon, she would be eleven days old on the first day of the succeeding year; the next year she would be twenty-two days; and on the third new year's day she would have passed a whole lunation and about three days more. The age of the moon (thus varying) on the first day of any year is termed the *Epact,* from a Greek word signifying *adscititious.* Those Epacts will form a varying series for nineteen years, when the new moon will again nearly coincide with the close of the year. This period of nineteen years is called the *Metonic Cycle,* from *Meton* its inventor; and the number of the years that have passed since the last coincidence (when the Epact was nothing) is called the *Golden Number.* —See *Calendar.*

EPICYCLE, a Greek derivative signifying a *Circle* upon a *Circle.* It was a prejudiced opinion among the ancient astronomers, that the motions of the heavenly bodies must necessarily be in circles: and, in order to make that doctrine tally with observation, they invented, in succession, the two theories of *Epicycles* and *Eccentrics.* In the former, called also the *Concentric Theory,* the earth was supposed to be placed in the centre of a circle on the circumference of which the centre of another circle revolved; and on the circumference of this second circle (called an

Epicycle) the planet was imagined to move: a supposition which accounted, in some degree, for the apparent irregularities of its motion. The primary circle was called the *Deferent*. In the *Eccentric theory*, the earth was also placed stationary in the centre of the starry sphere; but the sun was carried round in a circle, the centre of which was eccentric from that of the earth. Thus, says Milton:—

"They gird the sphere
With *centric* and *eccentric* scribbled o'er,
Cycle and epicycle, orb in orb."

See on this subject, *History of Astronomy*, pages 26—31.—See also *Concentric Circles*.

EPICYCLOID.—See *Cycloid*. If a circle roll upon the circumference of another circle instead of a straight line, points either on, within, or without its circumference, if on the same plane, will form varieties of *Epicycloids*.

EQUATION OF THE CENTRE.—See *Orbit*, and *Radius Vector*.

——— OF TIME.—See *Day*.

EQUATOR. An imaginary great circle of a sphere, equally distant from the poles of its rotatory motion, is termed, in Astronomy, the *Equator*, whatever that sphere may be: thus we speak of the *Equator* and the *Equatorial* portion of Jupiter. On terrestrial globes and maps of the earth it is usually called the *Equinoctial Line*, or simply the *Line*.

On celestial globes that figure the concavity of the heavens, the Equator is crossed at an angle of about twenty-three and a half degrees, by another great circle called the *Ecliptic*, which represents the apparent path of the sun, through the twelve signs, in his annual course. The Ecliptic crosses the Equator in two opposite points called the *Equinoxes* (Latin *equi* and *noctes*), because it is only when the sun is in one or other of those points in the heavens that the length of the day is exactly equal to that of the night. The two Equinoxes are denominated, one the *Vernal* and the other the *Autumnal*, because they are crossed by the sun respectively in the spring and autumn. The *Equinoctial Points*, where the ecliptic thus intersects the equator, are not stationary with respect to the fixed stars, but are regularly, though slowly, moving backwards; and this retrograde motion is called the *Precession of the Equinoxes*. For a particular explanation of this latter subject, see *Astronomy*, pages 36—40.

EQUINOX, AND EQUINOCTIAL POINTS.—See *Equator*.

EVECTION is one of the most considerable of the lunar irregularities, and was discovered by Ptolemy. It is periodical, running through all its changes in about twenty-seven days.—See *Astronomy*, page 200.

EXTREME AND MEAN RATIO.—See *History of Astronomy*, page 63. A line is so divided, when the rectangle under the whole line and the lesser segment is equal to the square of the greater segment; and hence the whole line is to the greater segment as that greater segment is to the lesser. The segments of such a division, being incommensurable with the whole line, cannot be exactly given in numbers; but the geometrical construction is easy. —See *Euclid's Elements*, Book II. prop. xi.

FIELDS OF ICE.—See *Bergs*.

FLOES.—See *Bergs*.

FLOOD-TIDE.—See *Tides*.

FOCUS.—See *Glossary* I.

FORCE, CENTRIFUGAL.—See *Glossary* I.

——— CENTRIPETAL.—See *Glossary* II.

FORMULA.—See *Glossary* II.

GENUS. In Natural History, a number of objects, such as animals and plants, are found to possess certain characteristics in common; and such are classed together, by the makers of systems, under one head, or kind, termed a *Genus*. The other *permanent* differences between the individuals of the same genus constitute *Species*; and the *accidental* differences found among the species are termed *Varieties*. *Genus* (kind) is Latin; and, in transferring the word into English, we have also adopted its plural, *genera*.

GEOCENTRIC. The Greek *ge*, the earth, is the root of a numerous class of well-known scientific terms, such as *Geography*, *Geology*, *Geometry*, &c. *Geocentric* is having the same centre as the earth, or having the earth for its centre. Thus the moon's orbit is Geocentric; but the orbits of the other planets, and of the earth itself, are *Heliocentric* (Greek *helios*, the sun), having the sun as their centre of motion. The *Geocentric place* of a planet is the place in which it would appear to an eye in the centre of the earth. The *Geocentric Latitude* of a planet is its latitude as seen from the earth; or it is the inclination of a line connecting the planet and the earth to the plane of the ecliptic. The *Geocentric Longitude of a planet* is the distance, measured on the Ecliptic, in the order of the signs, between the *Geocentric place* and the first point of Aries.

GEODESICAL, OR GEODETICAL, denotes something belonging to, or connected with, the mensuration of the earth's surface.

GIBBOUS. The Latin *gibbus* is protuberant, in the manner of a *Hunchback*. In English, the term is applied to designate that appearance of the moon (some days before and after the full) in which more than half her disc is enlightened; the line between light and dark being curved, or bulged outwards.—See *Phases*.

GNOMON, the name of an upright pillar, from the shadow of which the ancient astronomers determined the altitude of the sun. By its means they also calculated the altitudes of the other heavenly bodies. Gnomons of great height, with meridian lines attached to them, are still common in France and Italy. The style of a dial is likewise termed a gnomon.

GOLDEN NUMBER.—See *Epact.*

GRAVITY.—See *Glossary* I.

GREAT CIRCLE OF A SPHERE.—See *Circle.*

GREGORIAN CALENDAR.—See *Calendar.*

GULF. A *Gulf* or *Gulph*, is a portion of the ocean running up into the land between two promontories, and spreading out into a capacious *Bay*. *A Bay* (*Bow*) is a projection of the ocean into the land, but is not necessarily a *Gulph*, which includes the idea of a sort of abyss where the waters are *engulphed*, or swallowed up.

HEADLAND. Any projection of the land into the sea, may be termed a *Headland*; but a *Cape* is a Headland which has been distinguished from others by a particular designation: thus the *Cape of Good Hope* (called by way of eminence the *Cape*) and *Cape Horn* are in fact *Headlands*. A *Promontory* is also a Headland; but agreeing with its Latin derivation (*promontorium*) it has the name of Promontory only when the projecting head of land is a high point or a rock.

HELIACAL and HELIACALLY (see *Acronycal*) are formed from the Greek *helios*, the sun. *Helix*, a spiral, has the adjective and adverb, *Helical* and *Helically*.

HELIOCENTRIC expresses the same relation to the sun, that *Geocentric* (which *see*) does to the earth. Thus the *Heliocentric place* of a planet is the point of the ecliptic in which that planet, viewed from the sun, would appear to be, and therefore coincides with its *Longitude* as seen from the sun's centre. The *Heliocentric Latitude* of a planet is the inclination of a line, drawn between the centres of the sun and planet, to the plane of the Ecliptic.

HOMOCENTRIC (Greek *homos*, alike,) is equivalent to *Concentric*, which *see*.

HORIZON, *rational and sensible*. See *Glossary* II.

HOROSCOPE, a Greek compound, denoting a *view* of *the hour*, was the name, given by the astrologers, to a scheme, or figure, of the twelve signs of the zodiac at any particular hour; generally that in which a man was born, by which it was pretended to predict his fortune through life. The signs were called *Houses*, as being the monthly abodes of the sun, and, besides, every house was appropriated to some planet, every planet having two. In a more particular application the *Horoscope* denoted the point and sign of the Ecliptic which rose above the horizon at the hour in question; that point was the *Ascendant*; and the planet to which the sign was appropriated was termed the *Lord of the Ascendant*, and had its influence over the fate of the new-born child.—See *History of Astronomy*, page 36.

HYGROMETER, a Greek compound, signifying a *meter* or measurer of moisture.—See *Glossary* II.

HYPOTENUSE.—See *Triangle*

HYPOTHESIS.—See *Glossary* II. *Induction.*

INCIDENCE AND REFRACTION OF LIGHT, Laws of.—See *Glossary* I. *Refractive Power.*

INTERCALATION.—See *Calendar.*

INVERSE RATIO.—See *Glossary* II. *Ratio.*

ISOTHERMAL (Greek *isos*, equal, and *thermos*, warm) having the same heat or temperature. For the *Isothermal lines*, or lines on the globe that pass through places of equal temperature, see *Physical Geography*, page 37.

JULIAN CALENDAR.—See *Calendar.*

KNOTS.—See *Log-line.*

LAGUNES, extensive sheets of shallow water, particularly described at page 13 of *Physical Geography.*

LAKES. For a clear account of the four classes of Lakes, see *Physical Geography*, page 9.

LARBOARD. When standing at the stern of a ship and looking towards the prow, the left-hand side is termed the *larboard* and the right hand side the *starboard.*

LATITUDE, on the earth, is the expression of the distance of a place from the Equator, measured in degrees and minutes of a great circle. See *Declination.*

LATITUDE, GEOCENTRIC.—See *Geocentric.*

———— HELIOCENTRIC. See *Heliocentric.*

LAW OF THE SINES.—See *Glossary* I. *Refractive Power.*

LEAP-YEAR.—See *Calendar.*

LEEWARD. IN NAVIGATION, the *lee-side* of a ship is that half of a ship (divided lengthways) which is opposite to that on which the wind blows when it crosses her course, and which is called the *weather-side*. All objects on the *lee-side* are said to be to the *leeward*, and those on the *weather-side* to the *windward* of the vessel.

LEEWAY is the angle made by the line on which the ship should run, according to the point of the compass steered upon, and the real line of the ship's way occasioned by contrary winds, rough sea, or the set of a current.

LEGUMINOUS, a general denomination

for those plants that bear *Legumes*, or pods, such as peas and beans; the common English term is *pulse*. They constitute an *order* in the botanical system of Jussieu.

LEVEL.—See *Glossary* I.

LIBRATION OF THE MOON. Though the moon always presents nearly the same face to the earth, yet we sometimes see more of the eastern hemisphere, and at other times, more of the western. The same variation is occasionally observable in the northern and southern hemispheres, and this oscillation is called her Libration. See *Astronomy*, pages 83 and 84.

LIGHT.—See *Glossary* I.

LIMB, with Astronomers, designates the curved edge of a circle, such as the divided *limb* of a quadrant, and the outermost border of the sun or moon. In their observations, for example, they speak of the *moon's* lower or upper *limb*, and even of her eastern or western *limb*; and especially in the case of an eclipse, when a portion of her disc is obscured.

LINE OF THE NODES. See *Orbit*.

LOG-LINE. The *Log*, in sea-language, is the name of a piece of wood in the form of the sector (usually a quadrant) of a circle of five or six inches radius. It is about a quarter of an inch thick, and so balanced by means of a plate of lead nailed upon the circular part, as to swim perpendicularly in the water with about two-thirds immersed under the surface. The *Log-line* is a small cord of about one hundred fathoms in length, one end of which is fastened (by means of two legs) to the centre and to the arched part of the *Log*, while the other is wound round a reel in the gallery of the ship. The Log thus poised keeps its place in the water while the line is unwound from the reel by the ship's sailing; and the length of line unwound in a given time gives the rate of the ship's course. This is calculated by *knots* made on the line at between forty and fifty feet distance, while the time is measured by a sand-glass of a certain number of seconds. The length between the knots is so proportioned to the time of the glass, that the number of knots unwound shows the number of miles which the ship is sailing in the hour.

LONGITUDE. The Longitude of any heavenly body is measured on an arc of the ecliptic, intercepted between the vernal equinoctial point and a great circle passing through the body, and perpendicular to the ecliptic. *Longitude on the earth* is measured in a similar manner upon an arc of the equator, and counted to the east or west, from a certain meridian.—See *Equinox*.

LONGITUDE, GEOCENTRIC.—See *Geocentric*.

LOXODROMIC CURVE.—See *Rhumb-line*.

MAGNETIC AZIMUTH.—See *Glossary* II. *Horizon* and *Mariner's Compass*.

——————— MERIDIAN.—See *Glossary* II. *Horizon* and *Mariner's Compass*.

MAMMALIA, in the Linnæan System, the denomination of that class of animals which suckle their young. See *Physical Geography*, page 53.

MAMMIFEROUS ANIMALS. Such as have teats (Latin *mammæ*) for nourishing their young.—See *Physical Geography*, page 53.

MARINER'S COMPASS.—See *Glossary* II.; also *Navigation*, pages 15 and 16.

MARSUPIAL, (Greek *marsupos*, a purse) the designation given by naturalists to a tribe of the class *mammalia*, of which the *mammæ* and young of most of the species are, for a time, inclosed within an external pouch, or second womb, the pouch being supported by two *marsupial bones*.—See *Physical Geography*, page 54.

MEAN DAY.—See *Day*.

MEAN MOTION.—See *Motion*.

MEAN TIME.—See *Day*.

MERCATOR'S CHART.—See *Chart*.

MERIDIAN LINE.—See *Glossary* II. *Horizon* and *Astronomy*, pages 4 and 16.

METONIC CYCLE.—See *Epact*.

MOLLUSCA, the name of one of the orders of the Linnæan Class of *Vermes*, or worms. They are simple animals, furnished with limbs; some are naked and others *testaceous*, that is, covered with shells. They are chiefly inhabitants of the sea.

MOMENTUM.—See *Glossary* II. A different view of the *Momentum*, or impetus, of a moving body is taken when the motion is supposed to be accelerated. See *Astronomy*, page 12.

MONOCOTYLEDONOUS.—See *Cotyledon*.

MONSOONS are periodical winds in the East Indies, blowing constantly the same way during six months of the year, and the contrary way during the remaining six. They are a species of *Trade-winds*, which, in some quarters, blow constantly in one direction throughout the whole year. For the causes of this difference, see *Physical Geography*, page 31.

MONTH.—See *Calendar*.

MOTION. For a general definition, see *Glossary* II. *Mean motion* is understood to be a calculated average of a series of known variable motions.—See *Acceleration* and *Astronomy*, page 31.

NADIR. The *Zenith* and the *Nadir* (two Arabic words) are scientific names for two opposite points of direction in space. They are the poles of an interminably extended straight line, passing upright or downright, from the feet to the head, or from the head to the feet, of the person whose zenith and nadir they are: the former ending in a point of the sky above, and the latter (through the centre of the earth) in a point of the sky below. The zenith of one place is the nadir of its antipodes.

NAUTICAL DISTANCE.—See *Navigation*, page 4.

NEAP-TIDES.—See *Tides*.

NODES.—See *Orbit*.

NONIUS.—See *Glossary* I.

NORMAL, from the Latin *Norma*, a square, or rule, signifies literally a perpendicular; but it is generally used to denote the perpendicular to a curve at some particular point, at which point the normal is also perpendicular to a tangent.

NUCLEUS is literally the kernel of a nut (Latin *Nux*); but is used by astronomers to designate the apparently solid part or *Body* of a comet, as seen through the hazy atmosphere which surrounds it. It is sometimes also called the *Head*, in contradistinction to its train or *Tail*.

NUTATION is a sort of tremulous motion of the axis of the earth, whereby its inclination to the plane of the ecliptic is not always the same, but varies backwards and forwards some seconds. The period, or cycle, in which all these variations are completed, is nine years.—See *Astronomy*, pages 62, 63, and 158.

OBLATE AND OBLONG SPHEROIDS. —See *Glossary* I. *Oblong Spheroids* are also termed *Prolate Spheroids*.

OBLIQUE CONE.—See *Glossary* I. *Cone*.

OCCULTATION is when a fixed star or planet is hid from our sight, by the interposition of the moon, or some other planet. —See *Transit*.

OCTANT.—See *Quadrant*.

OPPOSITION OF THE PLANETS.—See *Conjunction*.

ORBIT. The Latin *Orbis* is a circle, as also a globe; and hence the paths of the planets round the sun are termed *Orbits*, and the planets themselves *Orbs*, though the former are now understood to be elliptical, and the latter spheroids.

The planets, though subjected to many disturbances, move round the sun in tracks that are calculated as ellipses, having the sun in one of their foci. The path of the sun apparently, but of the earth in reality, when traced in space, is the *Ecliptic*. It is the earth's orbit; and a plane, supposed to pass through this course, and to be extended indefinitely, is the *Plane of the Ecliptic*. In a similar way we may suppose planes to pass through the orbits of the other planets. All these planes will pass through the sun's centre; but all of them will cut the plane of the ecliptic, though at different angles, which are respectively called the *Inclination of the Orbits*. The two points in which the orbit of a planet cuts the plane of the ecliptic are the *Nodes* of that planet. In its revolution, the point in which the planet rises to the north of the ecliptic is the *Ascending Node*, and the other is the *Descending Node*. A straight line, uniting the two, is the *Line of the Nodes*, which passes through the centre of the sun.

In the same manner the secondary planets, or *Satellites*, move around their *Primaries*, which are also planets placed in one of the foci of those secondary elliptic orbits. The average inclination of the moon's orbit (for it is variable) is about five degrees; it is only when she is in one of her nodes that an eclipse can take place, and it is hence that the ecliptic has its name.

Either of the two points of a planetary orbit, which is at the greatest or at the least distance from the centre of motion, is called an *Apsis*, a Greek word, signifying the curved link of a chain. The two points, when spoken of together, are termed the *Apsides*; and the diameter which joins them is the *Line of the Apsides*. These names occasionally coincide with other terms. In the orbit of the earth (or of any primary planet) which has the sun as its centre of motion, its *Aphelion* is the same as its *Higher Apsis*, and its perihelion is the *Lower Apsis*; while in the moon's orbit, the *Higher Apsis* is equivalent to the *Apogee*, and the *Lower Apsis* to the perigee. —See those several Articles.

OVALS.—See *Glossary* II. for a general definition; but we may add, that ovals similar to the Cartesian kind may be formed in unlimited variety; for example, that which Cassini imagined for the planetary orbits supposes two foci, as in the ellipse, but that the two lines drawn from them to any point in the curve, instead of their sum, shall have their products always equal.—See *Ellipsis*.

OVIPAROUS (Latin *ovum*, an egg, is a term applied to such animals as produce their young from eggs, in opposition to *Viviparous Animals* which bring forth their young alive.—See *Physical Geography*, page 53, bottom of column 2.

PACHYDERMATOUS (from the Greek *pachos*, thick, and *derma*, skin).—See *Physical Geography*, page 55. Cuvier has formed a separate order, containing nine genera of *Pachydermata*, or thick-skinned quadrupeds.

PARABOLA.—See *Glossary* I. *Conic Sections*.

PARALLAX is an arc of the heavens intercepted between the true place of a star and its apparent place. Thus, suppose the true place of a star to be that point in which it would appear to an eye placed in the centre of the earth, an eye placed at the surface, which is a semi-diameter distant from the centre, would see it in a different point; and the arc between these is the measure of the parallax.

This parallax is greatest in the horizon, and diminishes as the altitude increases; for in the zenith a star has no parallax at all, the lines of observation from the centre and from the surface of the earth coinciding. The fixed stars have no parallax. This may be easily accounted for from their immense distance compared with the semi-diameter of the earth; but even when seen at opposite points of the earth's orbit, where what is called the *Annual Parallax* might have been expected, none has been satisfactorily observed. For further particulars on this subject, see *Astronomy*, pages 54—61, and 146.

PARALLELOGRAM.—See *Glossary* II.

PARALLELS OF DECLINATION AND OF LATITUDE.—See *Declination* and *Latitude*.

PENUMBRA.—See *Glossary* II.

PERIGEE.—See *Glossary* II. and *Astronomy*, page 19.

PERIHELION.—See *Glossary* II. and *Astronomy*, page 127.

PERPENDICULAR.—See *Normal*, and *Glossary* I. *Angle*.

PHANEROGAMOUS.—See *Cryptogamous*.

PHASES (Greek *phaino*, to shine). A term denoting the several appearances, or shapes, of the illumination of certain heavenly bodies, such as the moon, Venus, Mercury, &c. Of these the lunar phases (except perhaps Saturn, on account of his ring) are the most varied. She appears *circular* when full, *gibbous* a few days before and after the full, a *semicircle* at the quadratures, and *horned* at a few days before and after the new moon. In the latter situation, when visible, the tips or *horns* of her *Crescent* (which are at the extremities of her diameter) are termed *Cusps*, from the Latin *Cuspis*, a point. *Crescent* (Latin *crescare*) is growing, and the *Crescent Moon* is, literally, the *Growing Moon*, when she appears like a *bow*; but the same appearance is exhibited before her change, when she is said to be *waning*. For a representation of those several phases, see *Astronomy*, page 72.

PHYSICS.—See *Glossary* II.

PLANE CHART.—See *Chart*.

PLANET (Greek *planetes*, wandering) is a name given to those heavenly bodies which change their position with respect to the fixed stars, and are found to revolve round the sun as a centre. These are properly termed *Primary Planets*; for other wandering stars circulate about these primaries, and are therefore called *Secondary Planets*, or *Satellites*.

PLANETS, INFERIOR and SUPERIOR.—See *Conjunction of the Sun and Planets*.

PLANETARY MOTIONS.—For an illustration of the law by which they are said to describe equal areas in equal times, see *Radius Vector*.

PLUVIOMETER.—See *Rain-gage*.

POLES, NORTH AND SOUTH.—See *Arctic*.

PRECESSION OF THE EQUINOXES.—See *Equator*.

PREHENSILE. (Latin *prehendo*, to seize) an epithet given by naturalists to certain animals which are capable of grasping with their tails as with a claw.

PRIME VERTICAL.—See *Vertical Circles*.

PROLATE SPHEROID. The same as *Oblong Spheroid*, which see in *Glossary* I.

PROMONTORY.—See *Headland*.

PROPORTION.—See *Glossary* I.

QUADRANT. A quadrant in Geometry is merely the quarter of a circle, and as such is noticed in *Glossary* I. under the head *Angle*. The term is also applied to an instrument for measuring angles, which is a quarter circle of wood or metal, having its circular part, or limb, divided into 90 parts, or degrees, and these again subdivided into minutes, &c., by means of a *Nonius* or a *Vernier*. Hadley's quadrant is properly an *Octant*, or eighth part of a circle, in which the angles are taken by means of the reflexion of light, and when the limb is extended to 60 degrees (the sixth of a circle) the instrument is called a *Sextant*. These are severally described in the Society's treatise on *Optical Instruments*.

QUADRATURES.—See *Orbit*. When the moon is in either of the middle points of her orbit, between her conjunction and opposition (lines from the earth to the moon and to the sun, including a quadrant, or 90 degrees), she is said to be in her quadrature. Her face is then half shown;—it is bisected, or *dichotomized*. The places of her orbit, where she is either in conjunction or opposition, are her *syzygies*, a Greek compound signifying conjunction.—See *Conjunction*.

RADIUS.—See *Glossary* I. *Angle*.

RADIUS OF CURVATURE. In speaking of the concavity of other curves than the circle, the radius of curvature at a given point is the radius of a circle that has the same curvature as the curve has at that point. It is the varying length of the thread B D, when forming the involute A D E, (*Fig.* 9) *Glossary* I. See Articles *Curves, Evolutes, and Involutes*.

RADIUS VECTOR. The radius vector is a right line drawn from the centre of force (in any curve, on which a body is supposed to move by centripetal force) to that point of the curve where the body is supposed to be. It is a general radius to the curve, and has the addition of *Vector* (Latin, a carrier) because it is imagined to carry forward the body to which it is attached. The earth, for example, moves in an elliptic orbit, of which the sun (the centre of force) is in one of the foci; and, of consequence, the radius vector is continually increasing in length during her course from the perihelion to the aphelion, and decreasing in the same proportion in the progress of her return.

Let the ellipses A M N B O P represent the earth's orbit; of which the aphelion is at A, the perihelion at B, and sun's centre at S: A B being the line of the Apsides and S C the eccentricity of the orbit.

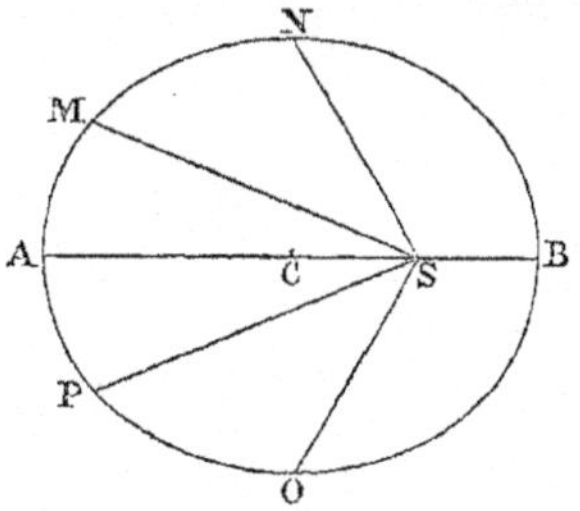

When the earth is in the aphelion, at A, the radius vector is at its *maximum*, and equal to A S; but, in moving from A to B, in the curve A M N B, it is regularly shortened until it reaches its *minimum* B S. The law of the planetary motions, as *guessed* by Kepler and *demonstrated* by Newton, is that the radius vector passes over equal areas of the orbit in equal times; and, if we suppose the revolution to be completed in twelve months, the semi-ellipse A M N B being divided into three equal triangular areas, A S M, M S N, and N S B, by the radius vector at M S and N S, will mark three portions (A M, M N, and N B) of the elliptic curve, each corresponding to two months of the time of the revolution. These curvilineal bases of the three equal triangles are, obviously, themselves of very unequal lengths, owing to the varying lengths of the other sides; but, by the law of motion just mentioned, they must each, nevertheless, be run over by the earth in the same period of time; and, consequently, the motion must be continually accelerated as she approaches her perigee at B. In the progress of her return to A, through the other half of her orbit, B O P A, the velocity of her motion will be continually diminished in a corresponding proportion.

The angle A S M formed at the sun by the line of the apsides and the radius vector, at any point of the orbit, is termed the *Anomaly*. It increases irregularly through the whole of that semi-ellipsis (decreasing in its opposite) as compared with time. A calculated medium angular increase gives the *Mean Anomaly;* and the difference between the *True Anomaly* and the *Mean Anomaly* is the *Equation of the Centre*.—See *Astronomy*, pp. 32 and 124.

RAIN-GAGE, an instrument for ascertaining the comparative quantity of rain which falls in different places, and in different seasons. Rain-gages, of a simple construction, are common throughout Europe, and are sometimes called *Pluviometers*, from the Latin *pluvia*, rain.

RATIO.—See *Glossary* II.

———, EXTREME AND MEAN.—See *Extreme and Mean Ratio.*

RAY OF LIGHT.—See *Glossary* I.

REFLEXION OF LIGHT.—See *Glossary* I.

REFRACTION.—See *Glossary* I., and *Astronomy*, pages 46—54.

RHUMBS are the thirty-two points of the horizon, as marked on the circle of the mariner's compass; and serve to calculate the angle which a ship's course makes with the magnetic meridian.—See *Navigation*, pages 15 and 16.

RHUMB-LINE is a line prolonged from any point of the compass (in a nautical chart), except the four Cardinal points. It cuts all the meridians under the same angle; and when delineated on the globe, it forms a curve termed the *Loxodromic Curve*.—See *Navigation*, page 19.

RIGHT ANGLE.—See *Glossary* I. *Angle.*

RIGHT ASCENSION OF THE SUN OR STAR, is that degree and minute of the equinoctial, counted from the vernal equinox (the first degree of Aries), which comes to the meridian with the sun, star, or other point of the heavens, whose right ascension is required.—See *Astronomy*, page 16.

SATELLITES, secondary planets which circulate round some primary one, as the moon does about the earth.—See *Planets.*

SECANT.—See *Glossary* I. *Angle.*

SECONDARY CIRCLES, are such as are in planes that are perpendicular to those circles of which they are the secondaries.

SEXTANT.—See *Quadrant.*

SIDEREAL DAY.—See *Day.*

——— YEAR.—See *Year.*

SINE AND VERSED SINE.—See *Glossary* I. *Angle.*

SINES, LAW OF THE.—See *Glossary* I. *Refractive Power.*

SOLAR DAY.—See *Day.*

——— YEAR.—See *Year.*

SOLSTICE.—See *Ecliptic.*

SOLSTITIAL COLURE.—See *Ecliptic.*

SOTHIAC PERIOD.—See *Canicular Period.*

SPECIES.—See *Genus.*

SPHERE, SPHERICAL, AND SPHEROID.—See *Glossary* I.

SPRING-TIDES.—See *Tides.*

STARBOARD.—See *Larboard.*

STYLE, NEW AND OLD.—See *Calendar* and *History of Astronomy*, pages 47 and 48.

SUBTEND.—See *Triangle.*

SUPPLEMENT OF AN ANGLE, OR ARC, is what it wants of a semicircle, or 180 degrees.—See *Glossary* I. *Angle.*

SYNODIC MONTH, a complete *Lunation*, or the period from one conjunction of the moon with the sun to another, being 29 days, 12 hours, and 44 minutes. The Greek *synodus* is a meeting or convention.

SYZYGY.—See *Quadrature.*

TABLE-LANDS, the name given to extensive plains highly elevated amid mountains.

TANGENT AND TANGENTIAL PLANE.—See *Glossary* I. *Tangent.*

TEMPERATURE.—See *Glossary* I. and *Physical Geography*, page 37.

TESTACEOUS, covered with a shell. For the distinction of this term from *Crustaceous*, see *Physical Geography*, page 49.

THERMOMETER.—See *Glossary* I.

TIDES. The rising and falling of the waters of the ocean, which occur twice in twenty-four hours, throughout all parts of the earth, are called the *Tides.* It is an alternate rise and fall, a *flux* and a *reflux*, a *flow* and an *ebb* of the water in respect to the land: the rise being called the *Flood-tide*, and the fall the *Ebb-tide.* When the *flux* is at its height, and about to recede at any particular place, it is there *High-water;* and when at its lowest, and about to rise, it is *Low-water.* The cause of the tides, as explained in *Astronomy*, from page 23 to 27, being the attraction of the sun and moon (chiefly the latter), the situations of those luminaries, with respect to one another, have an effect on the height of those swells of the waters. When the moon is in *conjunction*, or in *opposition*, the powers of both bodies being united, the tides are highest, and called *Spring-tides;* but when the moon is in her quadratures, they are lowest, and called *Neap-tides.*

TRADE-WINDS.—See *Monsoons.*

TRANSIT (Latin *transive*, to pass over) in Astronomy, is the passage of one heavenly body over the disc of a larger one. When the nearer body has a greater apparent diameter, so as to hide the other, the passage is termed an Occultation of the latter. —See *Occultation.*

TRANSLUCENT and TRANSPARENT. See *Glossary* II.

TRAVERSE. *Traverse-sailing*, or the working of a traverse, is the method of calculating a ship's place after she has made two or more short courses on different points of the compass.

TRIANGLE. A surface contained under three lines, has necessarily three corners, or angles; and is, therefore, called a *Triangle.* When these lines are straight, the figure being on a plane, is called a *Plane Triangle;* but when they are circular, lying on the surface of a globe, it is termed a *Spherical Triangle.* Each of the sides, in either case, is opposite to an angle, and is said to *subtend* that angle to which it is opposed. When one of the angles is a *Right Angle*, the side which subtends it is necessarily the longest of the three, and is called the *Hypotenuse*, from a Greek verb, signifying to *subtend.* The science of triangles is *Trigonometry.*

TROCHOID. The *Trochoid* and the *Cycloid* are similar curves, as corresponding with the derivations of their names: the one being from the Greek *trochos*, a wheel, and the other from *cyclos*, a circle. The cycloid is described in Glossary I., and the following may be considered as a supplement to that description.

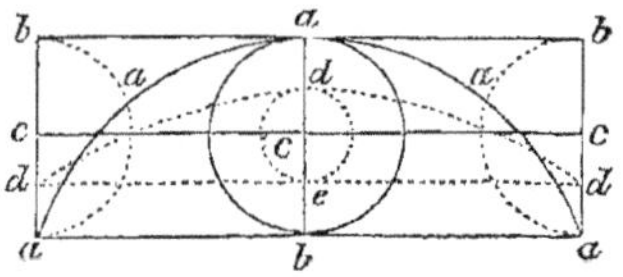

The cycloid *a a a a a* is formed by the point *a*, while the circle is rolling along the base line *a b a;* but if, instead of the tracing point being at the extremity of the diameter, it were taken within the circle, as at *d*, it would then describe the dotted curve line *d d d*, which is called a *Trochoid.* The right line *d e d* joining the ends of the curve is *the base of the Trochoid:* it is parallel to the path of the rolling circle, and equal in length to its circumference. The perpendicular *d e*, from the vertex *d* to the base, is the *Axis.* It divides the *Trochidal Space d d d e* into two equal portions; and the axis itself is bisected by the right line *c c c*, which is the path of the centre of the generating circle. The *Trochoid* is sometimes called a *Protracted Cycloid;* and a curve formed by a point *without* the circle (upon the diameter extended) is termed a *Contracted Cycloid*: the base of this latter, too, is of the same length; being always equal to the circumference of the generating circle in all the forms of the cycloid. If in the trochoid a circle be drawn round the centre *c*, having the axis *d e* as a diameter, the two circles will represent the *wheel and its nave*, in the famous Aristotelian paradox. —See *Epicycloid.*

TROPICS.—See *Ecliptic.*

TROPICAL YEAR.—See *Year.*

UMBELLIFEROUS.—See *Physical Geography*, page 47.

VAPOUR.—See *Glossary* I.

VARIATION OF THE COMPASS.—See *Glossary* II.

VARIETIES.—See *Genus.*

VELOCITY is the comparative celerity, or swiftness, of a moving body.

VERNIER.—See *Glossary* I.

VERSED SINE.—See *Glossary* I.

VERTICAL CIRCLES.—See *Glossary* II. *Horizon.* A vertical circle, passing through the east and west points of the horizon, is called the *Prime Vertical.*

VIVIPAROUS, a general designation for such animals as bring forth their young alive, in opposition to *Oviparous*, which *see.*

VORTICES.—See *Glossary* II.

WANING. Declining in power. The term is applied to the moon, as decreasing in her light, from the full to change.

WEATHERSIDE OF A SHIP.—See *Leeward.*

WINDWARD.—See *Leeward.*

YEAR. The period of time in which the earth performs her revolution round the sun, or that in which the sun apparently moves from a point in the ecliptic until he returns to the same point, is the *Solar Year.* It is also termed the *Tropical Year*, and consists of 365 days, 5 hours, and nearly 49 minutes. The *Sidereal Year*, that between the departure of the sun from any fixed star to his return to the same star, is about 17 minutes longer. The *Anomalistical Year* is the time that elapses from the sun's leaving his apogee till he returns to it, and is 365 days, 6 hours, and about 14 minutes.—See *Calendar.*

ZENITH.—See *Nadir.*

ZENITH DISTANCE. The complement of the altitude of a heavenly body is its *Zenith distance.*

ZODIAC. The ecliptic (as is stated under that head) crosses the equator at an angle of about 23½ degrees. It is the earth's orbit, and the apparent path of the sun. The orbits of the other planets cut the ecliptic at different angles; but (until lately, that Juno, Pallas, Ceres, and Vesta were discovered), the orbits of all passed within seven degrees on either side of the earth's course. A zone of sixteen degrees in width, having the ecliptic for its central line, was, therefore, conceived to include the whole of the planetary orbits. It was called the *Zodiac* (Greek *Zodiacos*, from *Zöon*, an animal), because it contained the figures of all the animals, &c. which formed the twelve signs.

ZONE (Latin *Zona*, a girdle), in *Geography*, is the denomination given to each of the five parallel belts into which the earth is imagined to be divided in respect to temperature. The *Torrid Zone* includes all the space which lies between the tropics, being nearly 47 degrees, or 23½ degrees on each side of the equinoctial line. Two *Frigid Zones* occupy those parts which lie between the poles and the polar circles; and two divisions that lie between those circles and the torrid zone, are called the *Temperate Zones.*

ZOOLOGY (Greek *zöon*, an animal, and *logos*, a discourse), is that division of Natural History which treats of animals, and

ZOOPHYTES (Greek *phyton*, a plant) are such natural productions as are supposed to participate of the qualities both of plants and animals.

GENERAL INDEX.

The references are given to the treatise or treatises in which the article is found. Of the contractions, HIST. OF ASTRO. stands for *History of Astronomy*; MATH. GEOG. for *Mathematical Geography*; and PHYS. GEOG. for *Physical Geography*.

Printed in Dunstable, United Kingdom

91218510R20318